VOLKSWAGEN NEW BEETLE
1998-05 REPAIR MANUAL

CHILTON'S

Covers U.S. and Canadian models of Volkswagen New Beetle Gasoline and 1.9L ECOdiesel engines

Does not include information specific to models with the 1.9L TDI diesel engine

by Bob Henderson

CHILTON Automotive Books

PUBLISHED BY **HAYNES NORTH AMERICA, Inc.**

Manufactured in USA

ISBN-13: 978-1-56392-711-9
ISBN-10: 1-56392-711-X
Library of Congress Control No. 2008923465

Haynes Publishing Group
Sparkford Nr Yeovil
Somerset BA22 7JJ England

Haynes North America, Inc
861 Lawrence Drive
Newbury Park
California 91320 USA

ABCDE
FGHIJ
KLMNO
P

6M4

Contents

1998 Volkswagen New Beetle

ACKNOWLEDGMENTS

Technical writers who contributed to this project include Eric Godfrey, Jeff Kibler, Jay Storer and Larry Warren. Wiring diagrams originated by Valley Forge Technical Information Services.

08-432

About this manual

ITS PURPOSE

The purpose of this manual is to help you get the best value from your vehicle. It can do so in several ways. It can help you decide what work must be done, even if you choose to have it done by a dealer service department or a repair shop; it provides information and procedures for routine maintenance and servicing; and it offers diagnostic and repair procedures to follow when trouble occurs.

We hope you use the manual to tackle the work yourself. For many simpler jobs, doing it yourself may be quicker than arranging an appointment to get the vehicle into a shop and making the trips to leave it and pick it up. More importantly, a lot of money can be saved by avoiding the expense the shop must pass on to you to cover its labor and overhead costs. An added benefit is the sense of satisfaction and accomplishment that you feel after doing the job yourself.

USING THE MANUAL

The manual is divided into Chapters. Each Chapter is divided into numbered Sections. Each Section consists of consecutively numbered paragraphs.

At the beginning of each numbered Section you will be referred to any illustrations which apply to the procedures in that Section. The reference numbers used in illustration captions pinpoint the pertinent Section and the Step within that Section. That is, illustration 3.2 means the illustration refers to Section 3 and Step (or paragraph) 2 within that Section.

Procedures, once described in the text, are not normally repeated. When it's necessary to refer to another Chapter, the reference will be given as Chapter and Section number. Cross references given without use of the word "Chapter" apply to Sections and/or paragraphs in the same Chapter. For example, "see Section 8" means in the same Chapter.

References to the left or right side of the vehicle assume you are sitting in the driver's seat, facing forward.

Even though we have prepared this manual with extreme care, neither the publisher nor the author can accept responsibility for any errors in, or omissions from, the information given.

➡NOTE

A *Note* provides information necessary to properly complete a procedure or information which will make the procedure easier to understand.

⁂ CAUTION

A *Caution* provides a special procedure or special steps which must be taken while completing the procedure where the Caution is found. Not heeding a Caution can result in damage to the assembly being worked on.

⁂ WARNING

A *Warning* provides a special procedure or special steps which must be taken while completing the procedure where the Warning is found. Not heeding a Warning can result in personal injury.

Introduction to the Volkswagen New Beetle

The New Beetle combines current Volkswagen mechanical and technical details with an all new body, styled like the famous air-cooled original Beetle, or "Bug."

Engines covered in this manual include the 1.8L turbocharged gasoline four-cylinder, the 2.0L gasoline four-cylinder, and the 1.9L turbocharged ECOdiesel engine.

The chassis layout is conventional by today's standards, with the engine mounted at the front and the wheels driven through either a 5-speed or 6-speed manual transmission, or a 4-speed or 6-speed automatic transmission, via independent driveaxles.

These models feature independent front suspension with coil springs and MacPherson struts. At the rear, all models have a torsion beam axle supported by coil springs and shock absorbers.

The power-assisted steering is rack-and-pinion, mounted behind the engine.

All models have a power assisted brake system with disc brakes at the front and either disc or drum brakes at the rear. An Anti-lock Brake System (ABS) is optional.

Vehicle identification numbers

VEHICLE IDENTIFICATION NUMBERS

Modifications are a continuing and unpublicized process in vehicle manufacturing. Since spare parts manuals and lists are compiled on a numerical basis, the individual vehicle numbers are essential to correctly identify the component required.

VEHICLE IDENTIFICATION NUMBER (VIN)

The Vehicle Identification Number (VIN), which appears on the Vehicle Certificate of Title and Registration, is also embossed on a plate located on the upper left (driver's side) corner of the dashboard, near the windshield (see illustration). The VIN tells you when and where a vehicle was manufactured, its country of origin, make, type, passenger safety system, line, series, body style, engine and assembly plant. The VIN is also located on the right side of the floor, under the rear seat's bottom cushion.

VIN ENGINE AND MODEL YEAR CODES

Two particularly important pieces of information found in the VIN are the engine code and the model year code. Counting from the left, the engine code is the 5th character; the model year code is the 10th character.

The engine codes are:

B................................ 2.0L four-cylinder (gasoline)
F................................ 1.9L four-cylinder (turbo-diesel)
D................................ 1.8L four-cylinder (gasoline turbo)

On the models covered by this manual the model year codes are:

W.............................. 1998
X.............................. 1999
Y.............................. 2000
1.............................. 2001
2.............................. 2002
3.............................. 2003
4.............................. 2004
5.............................. 2005

VEHICLE SAFETY CERTIFICATION LABEL

The Vehicle Safety Certification label is attached to the front of the driver's door post (see illustration). The label contains the name of the manufacturer, the month and year of production, the Gross Vehicle Weight Rating (GVWR), the Gross Axle Weight Rating (GAWR) and the certification statement. On most models, the label also includes the OEM tire sizes and pressures.

SERVICE PARTS IDENTIFICATION LABEL

Located on the inside of the spare tire well in the trunk, this label contains information about the options on your vehicle and the paint and trim codes (see illustration). This information is important when ordering parts or when bodywork and repainting is done.

The VIN plate is visible from the outside of the vehicle, through the driver's side of the windshield

The Vehicle Safety Certification label is affixed to the drivers side door post

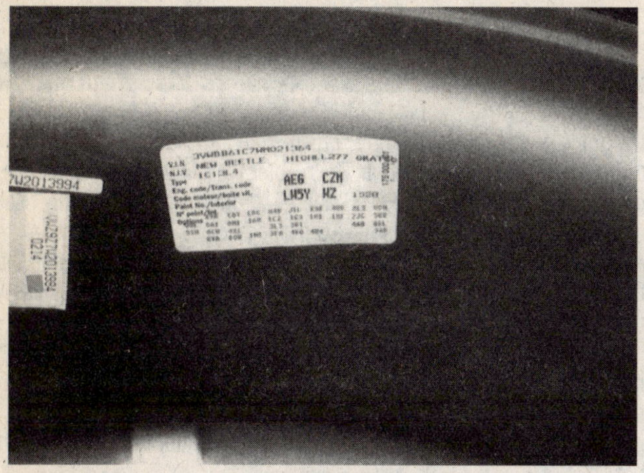

The Service Parts Identification label contains information on options and trim/paint codes - label is in the spare tire well in the trunk

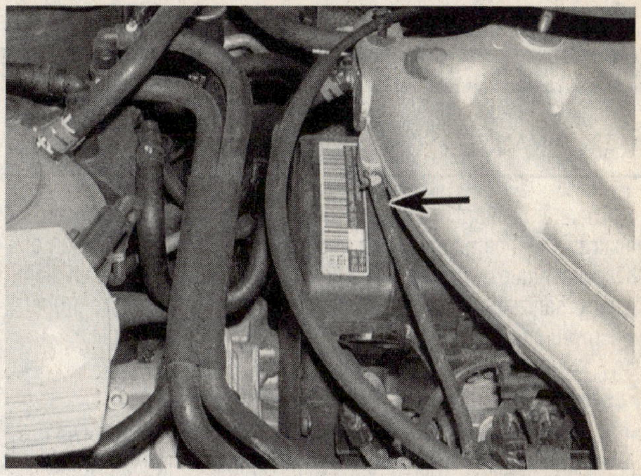

The engine identification label (arrow) is located on the front of the valve cover - a stamped number is on the back of the block at the engine/transaxle juncture

ENGINE IDENTIFICATION NUMBER (EIN)

The Engine Identification Number (EIN) for the gasoline engine is stamped into the top-rear of the engine block just where it meets the transaxle. It is also on a decal located on the front of the engine (see illustration). The number starts with three letters, such as AEG (code for the 2.0L gasoline engine), followed by a six-digit number.

TRANSAXLE IDENTIFICATION NUMBER (TIN)

The Transaxle Identification Number (TIN) is stamped into a machined pad on the front side of the transaxle (see illustration).

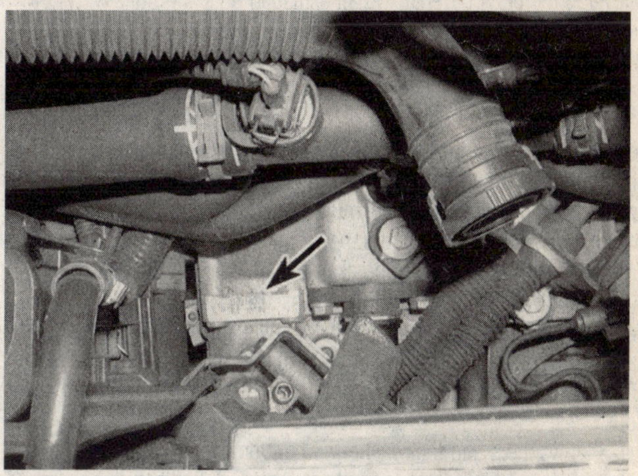

Typical transaxle identification number location (arrow) - manual transaxle shown, automatic transaxle similar

Buying parts

Replacement parts are available from many sources, which generally fall into one of two categories - authorized dealer parts departments and independent retail auto parts stores. Our advice concerning these parts is as follows:

Retail auto parts stores: Good auto parts stores will stock frequently needed components which wear out relatively fast, such as clutch components, exhaust systems, brake parts, tune-up parts, etc. These stores often supply new or reconditioned parts on an exchange basis, which can save a considerable amount of money. Discount auto parts stores are often very good places to buy materials and parts needed for general vehicle maintenance such as oil, grease, filters, spark plugs, belts, touch-up paint, bulbs, etc. They also usually sell tools and general accessories, have convenient hours, charge lower prices and can often be found not far from home.

Authorized dealer parts department: This is the best source for parts which are unique to the vehicle and not generally available elsewhere (such as major engine parts, transmission parts, trim pieces, etc.).

Warranty information: If the vehicle is still covered under warranty, be sure that any replacement parts purchased - regardless of the source - do not invalidate the warranty!

To be sure of obtaining the correct parts, have engine and chassis numbers available and, if possible, take the old parts along for positive identification.

Maintenance techniques, tools and working facilities

MAINTENANCE TECHNIQUES

There are a number of techniques involved in maintenance and repair that will be referred to throughout this manual. Application of these techniques will enable the home mechanic to be more efficient, better organized and capable of performing the various tasks properly, which will ensure that the repair job is thorough and complete.

Fasteners

Fasteners are nuts, bolts, studs and screws used to hold two or more parts together. There are a few things to keep in mind when working with fasteners. Almost all of them use a locking device of some type, either a lockwasher, locknut, locking tab or thread adhesive. All threaded fasteners should be clean and straight, with undamaged threads and undamaged corners on the hex head where the wrench fits. Develop the habit of replacing all damaged nuts and bolts with new ones. Special locknuts with nylon or fiber inserts can only be used once. If they are removed, they lose their locking ability and must be replaced with new ones.

Rusted nuts and bolts should be treated with a penetrating fluid to ease removal and prevent breakage. Some mechanics use turpentine in a spout-type oil can, which works quite well. After applying the rust penetrant, let it work for a few minutes before trying to loosen the nut or bolt. Badly rusted fasteners may have to be chiseled or sawed off or removed with a special nut breaker, available at tool stores.

If a bolt or stud breaks off in an assembly, it can be drilled and removed with a special tool commonly available for this purpose. Most automotive machine shops can perform this task, as well as other repair procedures, such as the repair of threaded holes that have been stripped out.

Flat washers and lockwashers, when removed from an assembly, should always be replaced exactly as removed. Replace any damaged washers with new ones. Never use a lockwasher on any soft metal surface (such as aluminum), thin sheet metal or plastic.

Fastener sizes

For a number of reasons, automobile manufacturers are making wider and wider use of metric fasteners. Therefore, it is important to be able to tell the difference between standard (sometimes called U.S. or SAE) and metric hardware, since they cannot be interchanged.

All bolts, whether standard or metric, are sized according to diameter, thread pitch and length. For example, a standard 1/2 - 13 x 1 bolt is 1/2 inch in diameter, has 13 threads per inch and is 1 inch long. An M12 - 1.75 x 25 metric bolt is 12 mm in diameter, has a thread pitch of 1.75 mm (the distance between threads) and is 25 mm long. The two bolts are nearly identical, and easily confused, but they are not interchangeable.

In addition to the differences in diameter, thread pitch and length, metric and standard bolts can also be distinguished by examining the bolt heads. To begin with, the distance across the flats on a standard bolt head is measured in inches, while the same dimension on a metric bolt is sized in millimeters (the same is true for nuts). As a result, a standard wrench should not be used on a metric bolt and a metric wrench should not be used on a standard bolt. Also, most standard bolts have slashes radiating out from the center of the head to denote the grade or strength of the bolt, which is an indication of the amount of torque that can be applied to it. The greater the number of slashes, the greater the strength of the bolt. Grades 0 through 5 are commonly used on automobiles. Metric bolts have a property class (grade) number, rather than a slash, molded into their heads to indicate bolt strength. In this case, the higher the number, the stronger the bolt. Property class numbers 8.8, 9.8 and 10.9 are commonly used on automobiles.

Strength markings can also be used to distinguish standard hex nuts from metric hex nuts. Many standard nuts have dots stamped into one side, while metric nuts are marked with a number. The greater the number of dots, or the higher the number, the greater the strength of the nut.

Metric studs are also marked on their ends according to property class (grade). Larger studs are numbered (the same as metric bolts), while smaller studs carry a geometric code to denote grade.

It should be noted that many fasteners, especially Grades 0 through 2, have no distinguishing marks on them. When such is the case, the only way to determine whether it is standard or metric is to measure the thread pitch or compare it to a known fastener of the same size.

Standard fasteners are often referred to as SAE, as opposed to metric. However, it should be noted that SAE technically refers to a non-metric fine thread fastener only. Coarse thread non-metric fasteners are referred to as USS sizes.

Since fasteners of the same size (both standard and metric) may have different strength ratings, be sure to reinstall any bolts, studs or nuts removed from your vehicle in their original locations. Also, when replacing a fastener with a new one, make sure that the new one has a strength rating equal to or greater than the original.

Tightening sequences and procedures

Most threaded fasteners should be tightened to a specific torque value (torque is the twisting force applied to a threaded component such as a nut or bolt). Overtightening the fastener can weaken it and cause it to break, while undertightening can cause it to eventually come loose. Bolts, screws and studs, depending on the material they are made of and their thread diameters, have specific torque values, many of which are noted in the Specifications at the end of each Chapter. Be sure to follow the torque recommendations closely. For fasteners not assigned a specific torque, a general torque value chart is presented here as a guide. These torque values are for dry (unlubricated) fasteners threaded into steel or cast iron (not aluminum). As was previously mentioned, the size and grade of a fastener determine the amount of torque that can safely be applied to it. The figures listed here are approximate for Grade 2 and Grade 3 fasteners. Higher grades can tolerate higher torque values.

Fasteners laid out in a pattern, such as cylinder head bolts, oil pan bolts, differential cover bolts, etc., must be loosened or tightened in sequence to avoid warping the component. This sequence will normally be shown in the appropriate Chapter. If a specific pattern is not given, the following procedures can be used to prevent warping.

Initially, the bolts or nuts should be assembled finger-tight only. Next, they should be tightened one full turn each, in a criss-cross or diagonal pattern. After each one has been tightened one full turn, return to the first one and tighten them all one-half turn, following the same

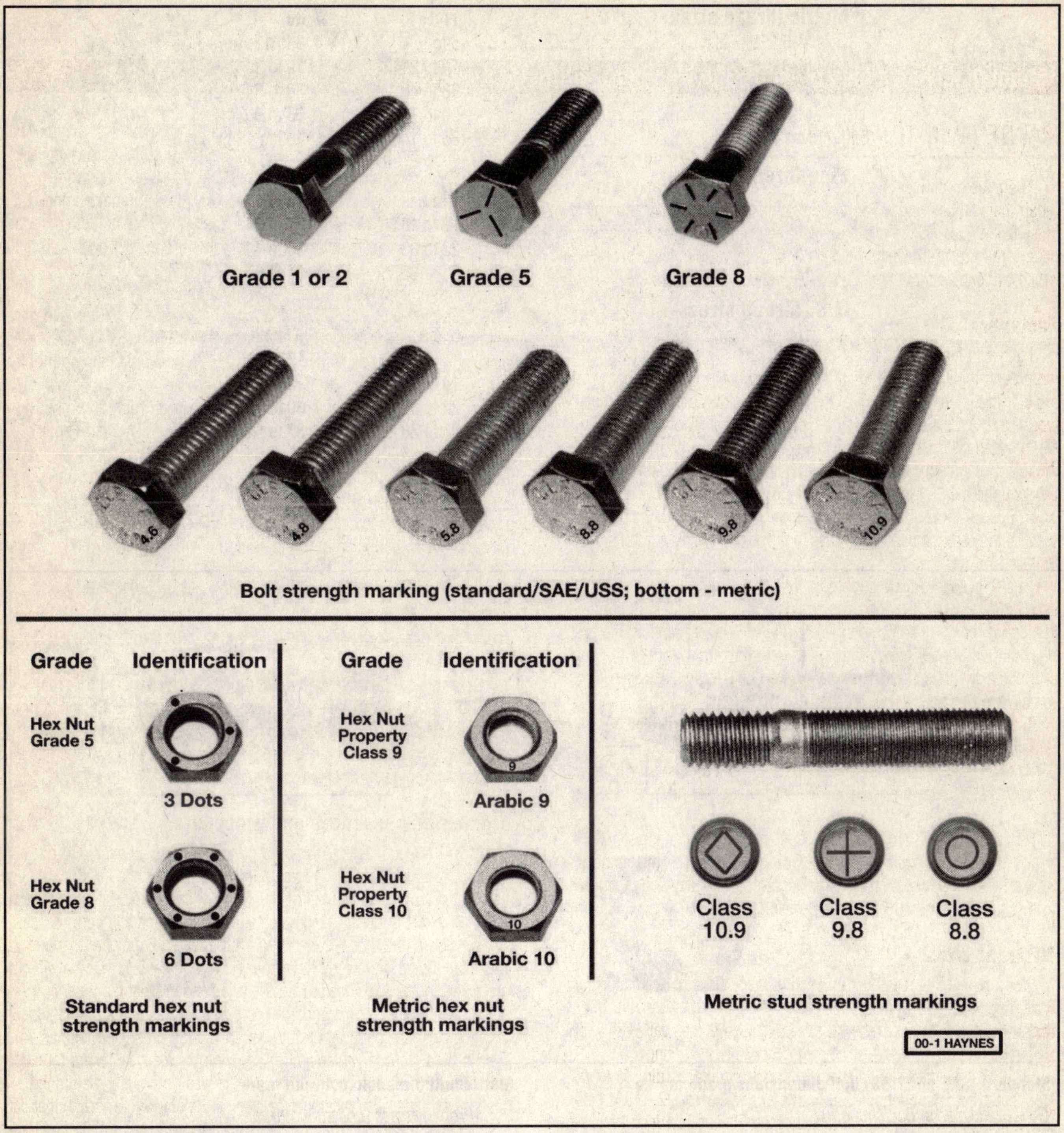

Bolt strength marking (standard/SAE/USS; bottom - metric)

Standard hex nut strength markings

Metric hex nut strength markings

Metric stud strength markings

pattern. Finally, tighten each of them one-quarter turn at a time until each fastener has been tightened to the proper torque. To loosen and remove the fasteners, the procedure would be reversed.

Component disassembly

Component disassembly should be done with care and purpose to help ensure that the parts go back together properly. Always keep track of the sequence in which parts are removed. Make note of special characteristics or marks on parts that can be installed more than one way, such as a grooved thrust washer on a shaft. It is a good idea to lay the disassembled parts out on a clean surface in the order that they were removed. It may also be helpful to make sketches or take instant photos of components before removal.

When removing fasteners from a component, keep track of their locations. Sometimes threading a bolt back in a part, or putting the washers and nut back on a stud, can prevent mix-ups later. If nuts and bolts cannot be returned to their original locations, they should be kept in a compartmented box or a series of small boxes. A cupcake or muffin tin is ideal for this purpose, since each cavity can hold the bolts and nuts from a particular area (i.e. oil pan bolts, valve cover bolts, engine

Metric thread sizes	**Ft-lbs**	**Nm**
M-6	6 to 9	9 to 12
M-8	14 to 21	19 to 28
M-10	28 to 40	38 to 54
M-12	50 to 71	68 to 96
M-14	80 to 140	109 to 154
Pipe thread sizes		
1/8	5 to 8	7 to 10
1/4	12 to 18	17 to 24
3/8	22 to 33	30 to 44
1/2	25 to 35	34 to 47
U.S. thread sizes		
1/4 - 20	6 to 9	9 to 12
5/16 - 18	12 to 18	17 to 24
5/16 - 24	14 to 20	19 to 27
3/8 - 16	22 to 32	30 to 43
3/8 - 24	27 to 38	37 to 51
7/16 - 14	40 to 55	55 to 74
7/16 - 20	40 to 60	55 to 81
1/2 - 13	55 to 80	75 to 108

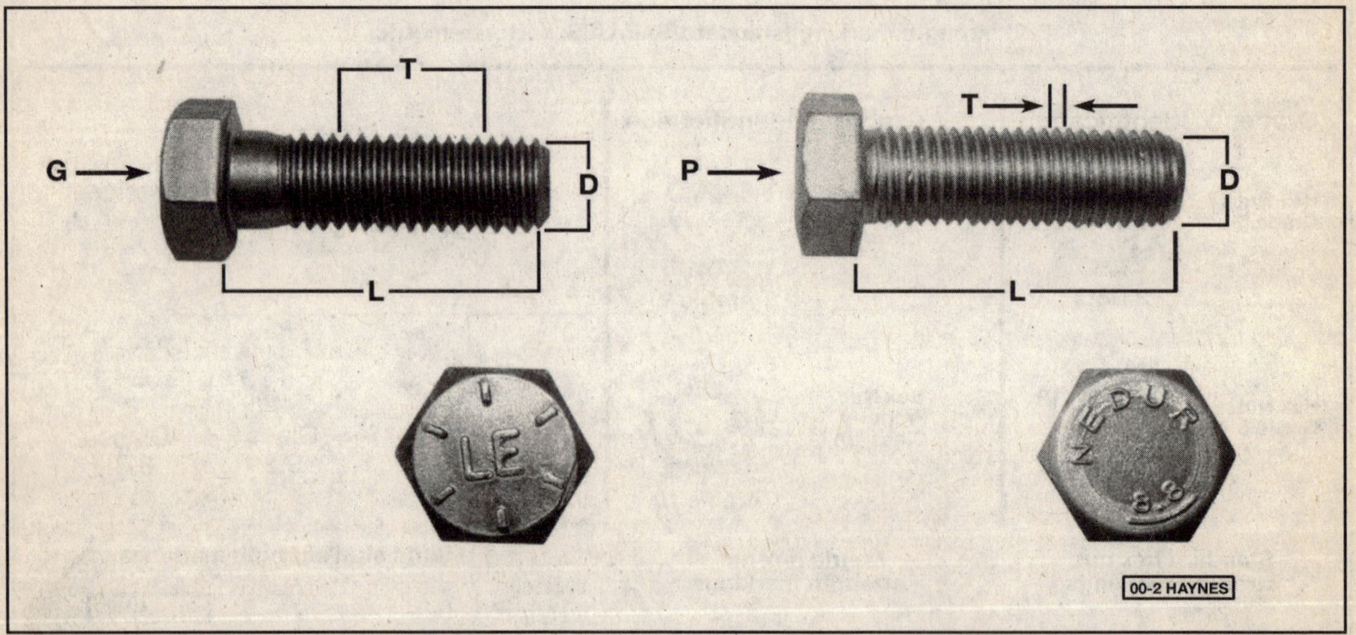

Standard (SAE and USS) bolt dimensions/grade marks

G Grade marks (bolt strength)
L Length (in inches)
T Thread pitch (number of threads per inch)
D Nominal diameter (in inches)

Metric bolt dimensions/grade marks

P Property class (bolt strength)
L Length (in millimeters)
T Thread pitch (distance between threads in millimeters)
D Diameter

mount bolts, etc.). A pan of this type is especially helpful when working on assemblies with very small parts, such as the carburetor, alternator, valve train or interior dash and trim pieces. The cavities can be marked with paint or tape to identify the contents.

Whenever wiring looms, harnesses or connectors are separated, it is a good idea to identify the two halves with numbered pieces of masking tape so they can be easily reconnected.

Gasket sealing surfaces

Throughout any vehicle, gaskets are used to seal the mating surfaces between two parts and keep lubricants, fluids, vacuum or pressure contained in an assembly.

Many times these gaskets are coated with a liquid or paste-type gasket sealing compound before assembly. Age, heat and pressure can sometimes cause the two parts to stick together so tightly that they

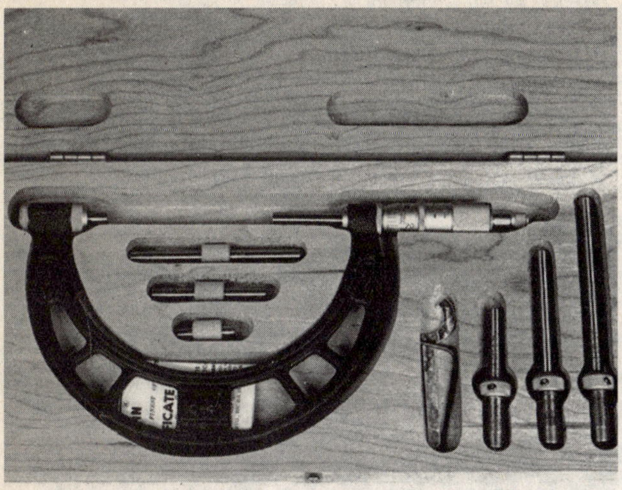

Micrometer set

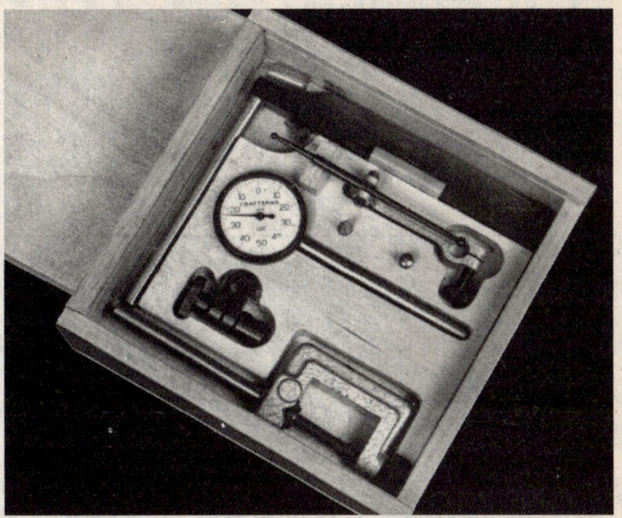

Dial indicator set

are very difficult to separate. Often, the assembly can be loosened by striking it with a soft-face hammer near the mating surfaces. A regular hammer can be used if a block of wood is placed between the hammer and the part. Do not hammer on cast parts or parts that could be easily damaged. With any particularly stubborn part, always recheck to make sure that every fastener has been removed.

Avoid using a screwdriver or bar to pry apart an assembly, as they can easily mar the gasket sealing surfaces of the parts, which must remain smooth. If prying is absolutely necessary, use an old broom handle, but keep in mind that extra clean up will be necessary if the wood splinters.

After the parts are separated, the old gasket must be carefully scraped off and the gasket surfaces cleaned. Stubborn gasket material can be soaked with rust penetrant or treated with a special chemical to soften it so it can be easily scraped off. A scraper can be fashioned from a piece of copper tubing by flattening and sharpening one end. Copper is recommended because it is usually softer than the surfaces to be scraped, which reduces the chance of gouging the part. Some gaskets can be removed with a wire brush, but regardless of the method used, the mating surfaces must be left clean and smooth. If for some reason the gasket surface is gouged, then a gasket sealer thick enough to fill scratches will have to be used during reassembly of the components. For most applications, a non-drying (or semi-drying) gasket sealer should be used.

Hose removal tips

⁂ WARNING:

If the vehicle is equipped with air conditioning, do not disconnect any of the A/C hoses without first having the system depressurized by a dealer service department or a service station.

Hose removal precautions closely parallel gasket removal precautions. Avoid scratching or gouging the surface that the hose mates against or the connection may leak. This is especially true for radiator hoses. Because of various chemical reactions, the rubber in hoses can bond itself to the metal spigot that the hose fits over. To remove a hose, first loosen the hose clamps that secure it to the spigot. Then, with slip-joint pliers, grab the hose at the clamp and rotate it around the spigot. Work it back and forth until it is completely free, then pull it off. Silicone or other lubricants will ease removal if they can be applied between the hose and the outside of the spigot. Apply the same lubricant to the inside of the hose and the outside of the spigot to simplify installation.

As a last resort (and if the hose is to be replaced with a new one anyway), the rubber can be slit with a knife and the hose peeled from the spigot. If this must be done, be careful that the metal connection is not damaged.

If a hose clamp is broken or damaged, do not reuse it. Wire-type clamps usually weaken with age, so it is a good idea to replace them with screw-type clamps whenever a hose is removed.

TOOLS

A selection of good tools is a basic requirement for anyone who plans to maintain and repair his or her own vehicle. For the owner who has few tools, the initial investment might seem high, but when compared to the spiraling costs of professional auto maintenance and repair, it is a wise one.

To help the owner decide which tools are needed to perform the tasks detailed in this manual, the following tool lists are offered: *Maintenance and minor repair, Repair/overhaul and Special.*

The newcomer to practical mechanics should start off with the *maintenance and minor repair* tool kit, which is adequate for the simpler jobs performed on a vehicle. Then, as confidence and experience grow, the owner can tackle more difficult tasks, buying additional tools as they are needed. Eventually the basic kit will be expanded into the *repair and overhaul* tool set. Over a period of time, the experienced do-it-yourselfer will assemble a tool set complete enough for most repair and overhaul procedures and will add tools from the special category when it is felt that the expense is justified by the frequency of use.

Maintenance and minor repair tool kit

The tools in this list should be considered the minimum required for performance of routine maintenance, servicing and minor repair work. We recommend the purchase of combination wrenches (box-end and open-end combined in one wrench). While more expensive than open end wrenches, they offer the advantages of both types of wrench.

Combination wrench set (1/4-inch to 1 inch or 6 mm to 19 mm)

Adjustable wrench, 8 inch
Spark plug wrench with rubber insert
Spark plug gap adjusting tool
Feeler gauge set
Brake bleeder wrench
Standard screwdriver (5/16-inch x 6 inch)
Phillips screwdriver (No. 2 x 6 inch)
Combination pliers - 6 inch
Hacksaw and assortment of blades
Tire pressure gauge
Grease gun
Oil can
Fine emery cloth
Wire brush
Battery post and cable cleaning tool
Oil filter wrench

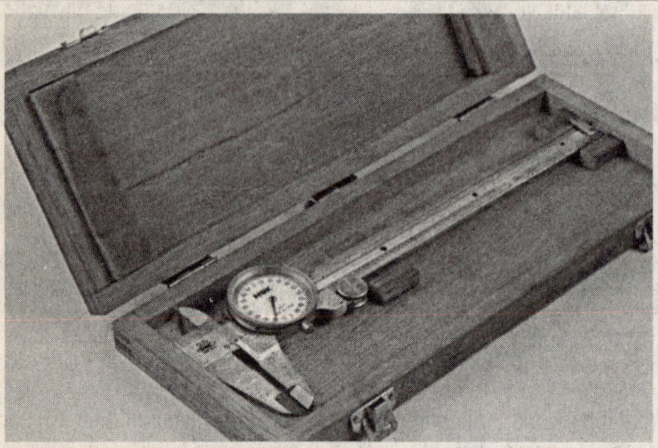

Dial caliper

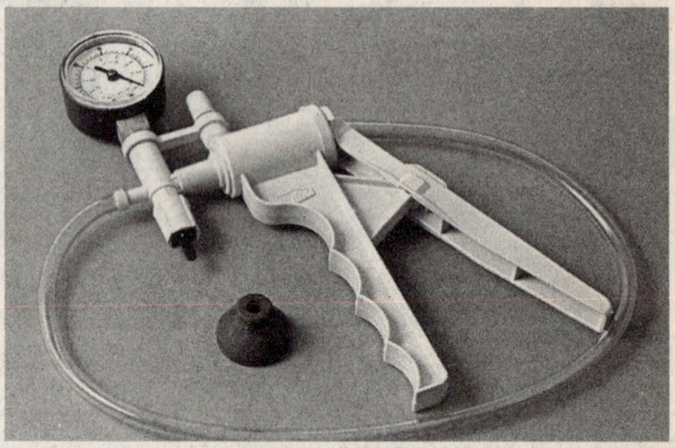

Hand-operated vacuum pump

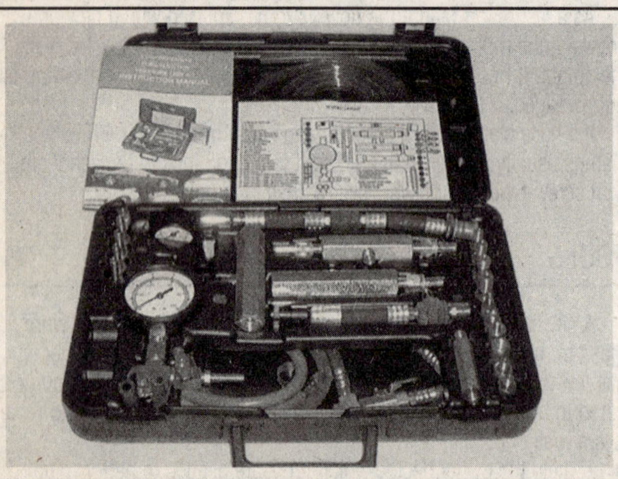

Fuel pressure gauge kit

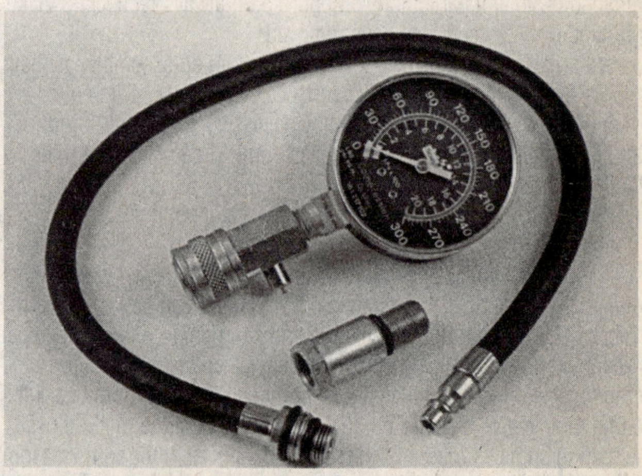

Compression gauge with spark plug hole adapter

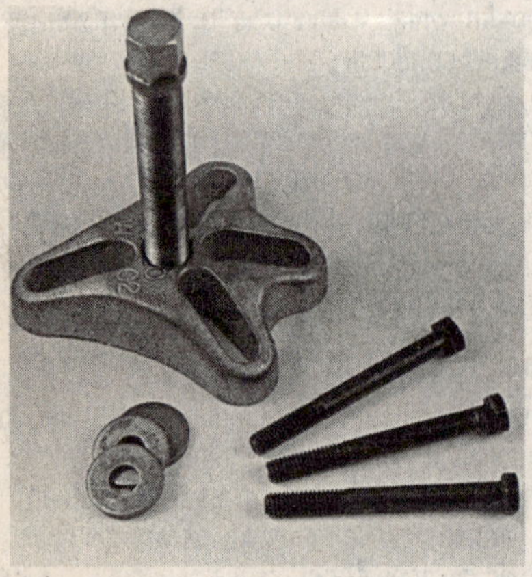

Damper/steering wheel puller

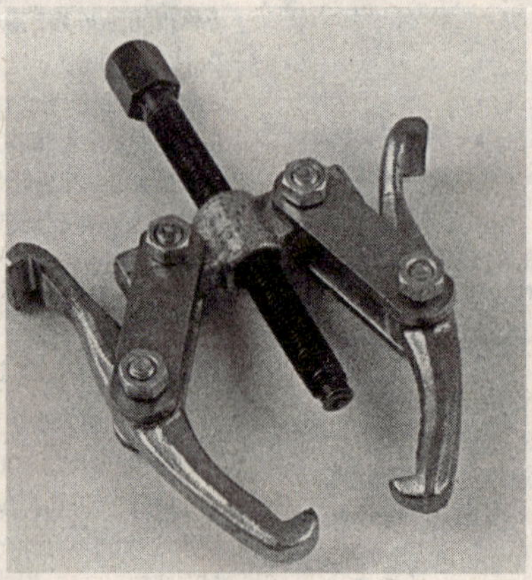

General purpose puller

Hydraulic lifter removal tool

Funnel (medium size)
Safety goggles
Jackstands (2)
Drain pan

➡Note: If basic tune-ups are going to be part of routine maintenance, it will be necessary to purchase a good quality stroboscopic timing light and combination tachometer/dwell meter. Although they are included in the list of special tools, it is mentioned here because they are absolutely necessary for tuning most vehicles properly.

Repair and overhaul tool set

These tools are essential for anyone who plans to perform major repairs and are in addition to those in the maintenance and minor repair tool kit. Included is a comprehensive set of sockets which, though expensive, are invaluable because of their versatility, especially when various extensions and drives are available. We recommend the 1/2-inch drive over the 3/8-inch drive. Although the larger drive is bulky and more expensive, it has the capacity of accepting a very wide range of large sockets. Ideally, however, the mechanic should have a 3/8-inch drive set and a 1/2-inch drive set.

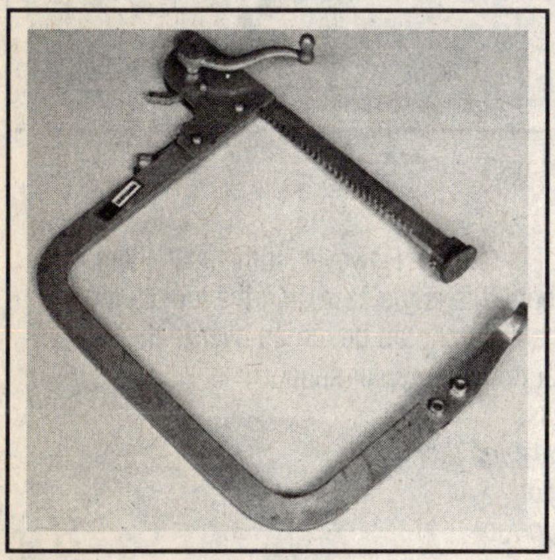
Valve spring compressor

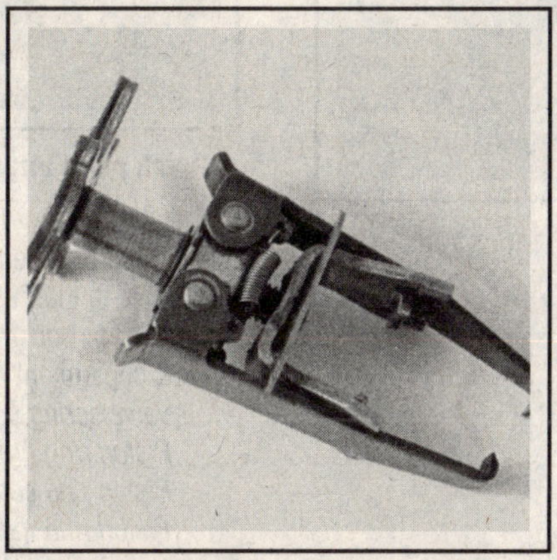
Valve spring compressor

Ridge reamer

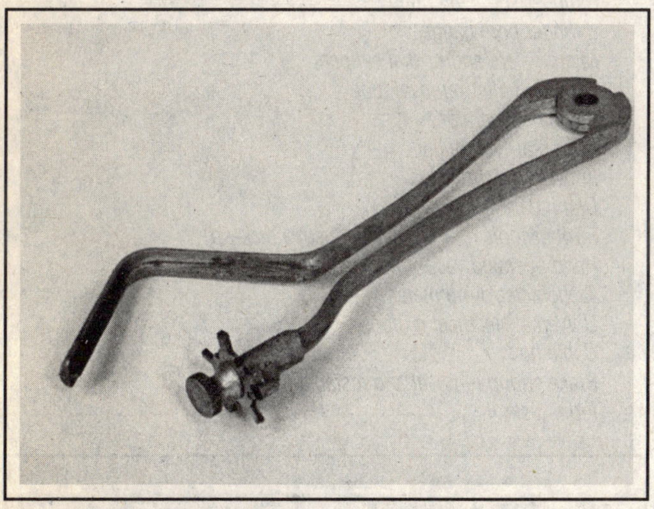
Piston ring groove cleaning tool

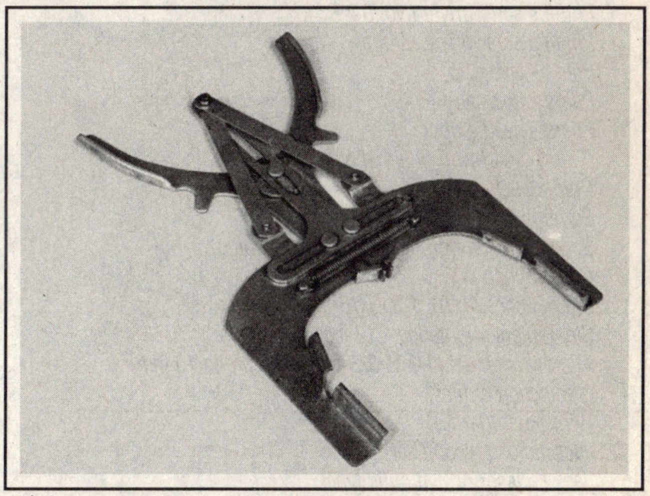
Ring removal/installation tool

Ring compressor

Cylinder hone

Brake hold-down spring tool

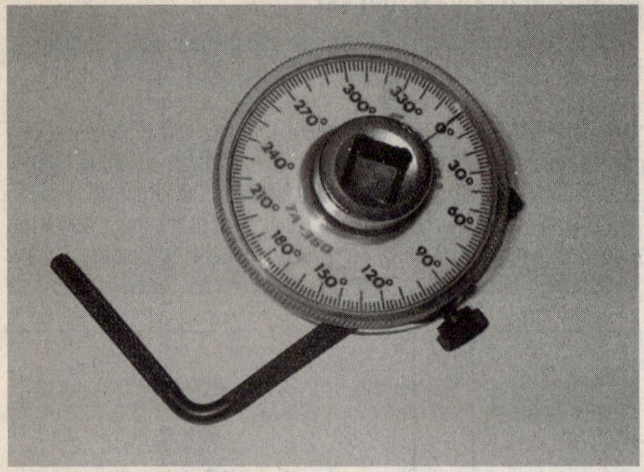

Torque angle gauge

Socket set(s)
Reversible ratchet
Extension - 10 inch
Universal joint
Torque wrench (same size drive as sockets)
Ball peen hammer - 8 ounce
Soft-face hammer (plastic/rubber)
Standard screwdriver (1/4-inch x 6 inch)
Standard screwdriver (stubby - 5/16-inch)
Phillips screwdriver (No. 3 x 8 inch)
Phillips screwdriver (stubby - No. 2)
Pliers - vise grip
Pliers - lineman's
Pliers - needle nose
Pliers - snap-ring (internal and external)
Cold chisel - 1/2-inch
Scribe
Scraper (made from flattened copper tubing)
Centerpunch
Pin punches (1/16, 1/8, 3/16-inch)
Steel rule/straightedge - 12 inch
Allen wrench set (1/8 to 3/8-inch or 4 mm to 10 mm)
A selection of files
Wire brush (large)
Jackstands (second set)
Jack (scissor or hydraulic type)

➡**Note: Another tool which is often useful is an electric drill with a chuck capacity of 3/8-inch and a set of good quality drill bits.**

Special tools

The tools in this list include those which are not used regularly, are expensive to buy, or which need to be used in accordance with their manufacturer's instructions. Unless these tools will be used frequently, it is not very economical to purchase many of them. A consideration would be to split the cost and use between yourself and a friend or friends. In addition, most of these tools can be obtained from a tool rental shop on a temporary basis.

This list primarily contains only those tools and instruments widely available to the public, and not those special tools produced by the vehicle manufacturer for distribution to dealer service departments. Occasionally, references to the manufacturer's special tools are included in the text of this manual. Generally, an alternative method of doing the job without the special tool is offered. However, sometimes there is no alternative to their use. Where this is the case, and the tool cannot be purchased or borrowed, the work should be turned over to the dealer service department or an automotive repair shop.

Valve spring compressor
Piston ring groove cleaning tool
Piston ring compressor
Piston ring installation tool
Cylinder compression gauge
Cylinder ridge reamer
Cylinder surfacing hone
Cylinder bore gauge
Micrometers and/or dial calipers
Hydraulic lifter removal tool
Balljoint separator
Universal-type puller
Impact screwdriver
Dial indicator set
Stroboscopic timing light (inductive pick-up)
Hand operated vacuum/pressure pump
Tachometer/dwell meter
Universal electrical multimeter
Cable hoist
Brake spring removal and installation tools
Floor jack

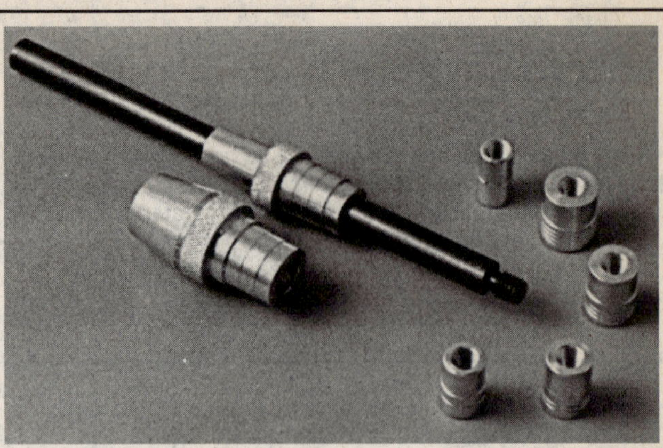

Clutch plate alignment tool

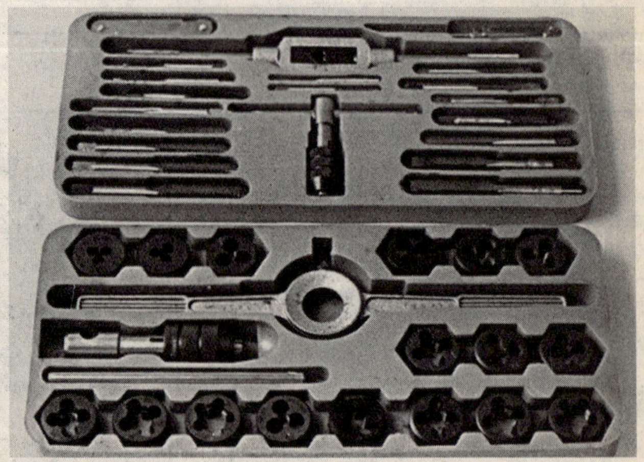

Tap and die set

Buying tools

For the do-it-yourselfer who is just starting to get involved in vehicle maintenance and repair, there are a number of options available when purchasing tools. If maintenance and minor repair is the extent of the work to be done, the purchase of individual tools is satisfactory. If, on the other hand, extensive work is planned, it would be a good idea to purchase a modest tool set from one of the large retail chain stores. A set can usually be bought at a substantial savings over the individual tool prices, and they often come with a tool box. As additional tools are needed, add-on sets, individual tools and a larger tool box can be purchased to expand the tool selection. Building a tool set gradually allows the cost of the tools to be spread over a longer period of time and gives the mechanic the freedom to choose only those tools that will actually be used.

Tool stores will often be the only source of some of the special tools that are needed, but regardless of where tools are bought, try to avoid cheap ones, especially when buying screwdrivers and sockets, because they won't last very long. The expense involved in replacing cheap tools will eventually be greater than the initial cost of quality tools.

Care and maintenance of tools

Good tools are expensive, so it makes sense to treat them with respect. Keep them clean and in usable condition and store them properly when not in use. Always wipe off any dirt, grease or metal chips before putting them away. Never leave tools lying around in the work area. Upon completion of a job, always check closely under the hood for tools that may have been left there so they won't get lost during a test drive.

Some tools, such as screwdrivers, pliers, wrenches and sockets, can be hung on a panel mounted on the garage or workshop wall, while others should be kept in a tool box or tray. Measuring instruments, gauges, meters, etc. must be carefully stored where they cannot be damaged by weather or impact from other tools.

When tools are used with care and stored properly, they will last a very long time. Even with the best of care, though, tools will wear out if used frequently. When a tool is damaged or worn out, replace it. Subsequent jobs will be safer and more enjoyable if you do.

HOW TO REPAIR DAMAGED THREADS

Sometimes, the internal threads of a nut or bolt hole can become stripped, usually from overtightening. Stripping threads is an all-too-common occurrence, especially when working with aluminum parts, because aluminum is so soft that it easily strips out.

Usually, external or internal threads are only partially stripped. After they've been cleaned up with a tap or die, they'll still work. Sometimes, however, threads are badly damaged. When this happens, you've got three choices:

1) *Drill and tap the hole to the next suitable oversize and install a larger diameter bolt, screw or stud.*
2) *Drill and tap the hole to accept a threaded plug, then drill and tap the plug to the original screw size. You can also buy a plug already threaded to the original size. Then you simply drill a hole to the specified size, then run the threaded plug into the hole with a bolt and jam nut. Once the plug is fully seated, remove the jam nut and bolt.*
3) *The third method uses a patented thread repair kit like Heli-Coil or Slimsert. These easy-to-use kits are designed to repair damaged threads in straight-through holes and blind holes. Both are available as kits which can handle a variety of sizes and thread patterns. Drill the hole, then tap it with the special included tap. Install the Heli-Coil and the hole is back to its original diameter and thread pitch.*

Regardless of which method you use, be sure to proceed calmly and carefully. A little impatience or carelessness during one of these relatively simple procedures can ruin your whole day's work and cost you a bundle if you wreck an expensive part.

WORKING FACILITIES

Not to be overlooked when discussing tools is the workshop. If anything more than routine maintenance is to be carried out, some sort of suitable work area is essential.

It is understood, and appreciated, that many home mechanics do not have a good workshop or garage available, and end up removing an engine or doing major repairs outside. It is recommended, however, that the overhaul or repair be completed under the cover of a roof.

A clean, flat workbench or table of comfortable working height is an absolute necessity. The workbench should be equipped with a vise that has a jaw opening of at least four inches.

As mentioned previously, some clean, dry storage space is also required for tools, as well as the lubricants, fluids, cleaning solvents, etc. which soon become necessary.

Sometimes waste oil and fluids, drained from the engine or cooling system during normal maintenance or repairs, present a disposal problem. To avoid pouring them on the ground or into a sewage system, pour the used fluids into large containers, seal them with caps and take them to an authorized disposal site or recycling center. Plastic jugs, such as old antifreeze containers, are ideal for this purpose.

Always keep a supply of old newspapers and clean rags available. Old towels are excellent for mopping up spills. Many mechanics use rolls of paper towels for most work because they are readily available and disposable. To help keep the area under the vehicle clean, a large cardboard box can be cut open and flattened to protect the garage or shop floor.

Whenever working over a painted surface, such as when leaning over a fender to service something under the hood, always cover it with an old blanket or bedspread to protect the finish. Vinyl covered pads, made especially for this purpose, are available at auto parts stores.

Jacking and towing

JACKING

⁂ WARNING:

The jack supplied with the vehicle should only be used for changing a tire or placing jackstands under the frame. Never work under the vehicle or start the engine while this jack is being used as the only means of support.

The vehicle should be parked on level ground with the wheels blocked, the parking brake applied and the transmission in Park (automatic) or Reverse (manual). If the vehicle is parked alongside the roadway, or in any other hazardous situation, turn on the emergency hazard flashers. If a tire is to be changed, loosen the wheel bolts one-half turn before raising the vehicle off the ground.

Place the jack under the vehicle in the indicated position (see illustrations). Operate the jack with a slow, smooth motion until the wheel is raised off the ground. Remove the wheel bolts, pull off the wheel, install the spare and thread the wheel bolts back on. Tighten the wheel bolts snugly, lower the vehicle until some weight is on the wheel, tighten the wheel bolts completely in a criss-cross pattern and remove the jack.

TOWING

As a general rule, the vehicle should be towed with the front (drive) wheels off the ground.

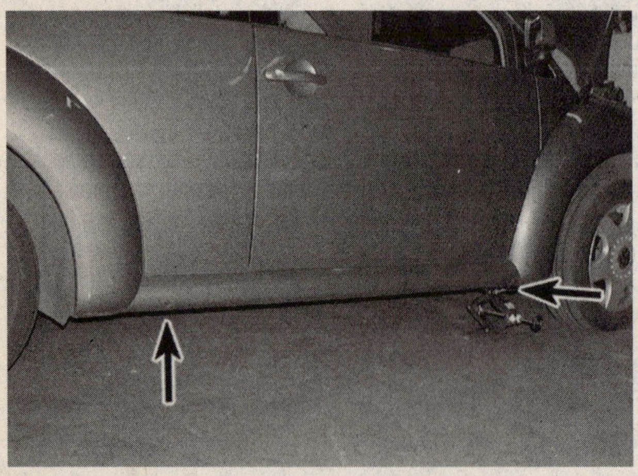

Jacking positions with factory jack (arrows)

Vehicles equipped with an automatic transaxle can be towed from the front only with all four wheels on the ground, provided that speeds don't exceed 30 mph and the distance is not over 40 miles. Otherwise, a towing dolly should be used. If towed on a dolly from the rear, the ignition key must be in the ACC position, since the steering lock mechanism isn't strong enough to hold the front wheels straight while towing. The tow truck operator will attach a purpose-built steering wheel holder suitable for towing conditions.

⁂ CAUTION:

Never tow a vehicle with an automatic transaxle from the rear with the front wheels on the ground.

When towing a vehicle equipped with a manual transaxle with all four wheels on the ground, be sure to place the shift lever in neutral and release the parking brake, and turn the ignition key to the first position to unlock the steering wheel. Also, check the transaxle lubricant to make sure it is up to the proper level.

Equipment specifically designed for towing should be used. It should be attached to the main structural members of the vehicle, not the bumpers or brackets.

Safety is a major consideration when towing and all applicable state and local laws must be obeyed. A safety chain system must be used at all times.

The jack and factory tools are located in a tray above the spare tire

Booster battery (jump) starting

Observe these precautions when using a booster battery to start a vehicle:

- *a) Before connecting the booster battery, make sure the ignition switch is in the Off position.*
- *b) Turn off the lights, heater and other electrical loads.*
- *c) Your eyes should be shielded. Safety goggles are a good idea.*
- *d) Make sure the booster battery is the same voltage as the dead one in the vehicle.*
- *e) The two vehicles MUST NOT TOUCH each other!*
- *f) Make sure the transaxle is in Neutral (manual) or Park (automatic).*
- *g) If the booster battery is not a maintenance-free type, remove the vent caps and lay a cloth over the vent holes.*

The battery on these vehicles is located at the left front corner of the engine compartment, under a cover.

Connect the red-colored jumper cable to the positive (+) terminal of the booster battery and the other end to the positive (+) terminal of the dead battery. Then connect one end of the black jumper cable to the negative (-) terminal of the booster battery, and the other end of that cable to a good ground point on the engine of the disabled vehicle, preferably not too near the battery.

Start the engine using the booster battery and let the booster vehicle run at 2000 rpm for a few minutes to put some charge into the weak battery, then, with the engine running at idle speed, disconnect the jumper cables in the reverse order of connection.

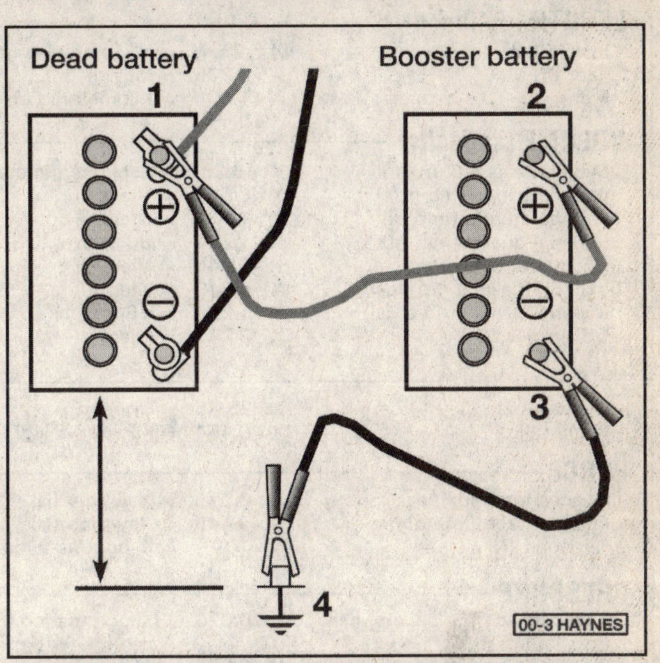

Make the booster battery cable connections in the numerical order shown (note that the negative cable of the booster battery is NOT attached to the negative terminal of the dead battery)

Anti-theft audio system

GENERAL INFORMATION

1 All models are equipped with a stereo that has an anti-theft feature that will render the stereo inoperative if stolen or if the battery is disconnected. If the power source to the stereo is cut, the stereo will be inoperative. Even if the power source is immediately re-connected, the stereo will not function.

2 Do not disconnect the battery, remove the stereo or disconnect related components unless you have the individual ID (code) number for the stereo.

UNLOCKING THE STEREO AFTER A POWER LOSS

3 When the power is restored to the stereo, the stereo won't operate. Enter your ID code to reactivate it, using the following Steps.

4 Turn the radio ON. The lower display window (where radio stations are displayed) should show "SAFE." After a few seconds, the SAFE message should go out and the numerals "1000" should display.

5 At the bottom of the radio, the numbered buttons used to preset stations are used to enter your anti-theft code.

6 Press the number 1 preset button repeatedly until the first number of your four-digit code is displayed, then enter the second number of your code on button #2, and so on until all four code numbers are displayed.

7 Press the SEEK button until you hear a sound from the radio.The display will read "LSM" if you've entered the correct code.

8 If you have entered the wrong code during the above procedure, the display will read "SAFE," but you have one more try. If the second code is incorrect, the radio will stay locked for one hour, at which time two more code attempts can be made. Leave the radio on for an hour (while it is locked) and the key in the ignition switch, then try again.

9 You should have the code written down in a secure place, for use in unlocking the anti-theft feature.

CONVERSION FACTORS

LENGTH (distance)

Inches (in)	X 25.4	= Millimeters (mm)	X 0.0394	= Inches (in)
Feet (ft)	X 0.305	= Meters (m)	X 3.281	= Feet (ft)
Miles	X 1.609	= Kilometers (km)	X 0.621	= Miles

VOLUME (capacity)

Cubic inches (cu in; in^3)	X 16.387	= Cubic centimeters (cc; cm^3)	X 0.061	= Cubic inches (cu in; in^3)
Imperial pints (Imp pt)	X 0.568	= Liters (l)	X 1.76	= Imperial pints (Imp pt)
Imperial quarts (Imp qt)	X 1.137	= Liters (l)	X 0.88	= Imperial quarts (Imp qt)
Imperial quarts (Imp qt)	X 1.201	= US quarts (US qt)	X 0.833	= Imperial quarts (Imp qt)
US quarts (US qt)	X 0.946	= Liters (l)	X 1.057	= US quarts (US qt)
Imperial gallons (Imp gal)	X 4.546	= Liters (l)	X 0.22	= Imperial gallons (Imp gal)
Imperial gallons (Imp gal)	X 1.201	= US gallons (US gal)	X 0.833	= Imperial gallons (Imp gal)
US gallons (US gal)	X 3.785	= Liters (l)	X 0.264	= US gallons (US gal)

MASS (weight)

Ounces (oz)	X 28.35	= Grams (g)	X 0.035	= Ounces (oz)
Pounds (lb)	X 0.454	= Kilograms (kg)	X 2.205	= Pounds (lb)

FORCE

Ounces-force (ozf; oz)	X 0.278	= Newtons (N)	X 3.6	= Ounces-force (ozf; oz)
Pounds-force (lbf; lb)	X 4.448	= Newtons (N)	X 0.225	= Pounds-force (lbf; lb)
Newtons (N)	X 0.1	= Kilograms-force (kgf; kg)	X 9.81	= Newtons (N)

PRESSURE

Pounds-force per square inch (psi; lbf/in^2; lb/in^2)	X 0.070	= Kilograms-force per square centimeter (kgf/cm^2; kg/cm^2)	X 14.223	= Pounds-force per square inch (psi; lbf/in^2; lb/in^2)
Pounds-force per square inch (psi; lbf/in^2; lb/in^2)	X 0.068	= Atmospheres (atm)	X 14.696	= Pounds-force per square inch (psi; lbf/in^2; lb/in^2)
Pounds-force per square inch (psi; lbf/in^2; lb/in^2)	X 0.069	= Bars	X 14.5	= Pounds-force per square inch (psi; lbf/in^2; lb/in^2)
Pounds-force per square inch (psi; lbf/in^2; lb/in^2)	X 6.895	= Kilopascals (kPa)	X 0.145	= Pounds-force per square inch (psi; lbf/in^2; lb/in^2)
Kilopascals (kPa)	X 0.01	= Kilograms-force per square centimeter (kgf/cm^2; kg/cm^2)	X 98.1	= Kilopascals (kPa)

TORQUE (moment of force)

Pounds-force inches (lbf in; lb in)	X 1.152	= Kilograms-force centimeter (kgf cm; kg cm)	X 0.868	= Pounds-force inches (lbf in; lb in)
Pounds-force inches (lbf in; lb in)	X 0.113	= Newton meters (Nm)	X 8.85	= Pounds-force inches (lbf in; lb in)
Pounds-force inches (lbf in; lb in)	X 0.083	= Pounds-force feet (lbf ft; lb ft)	X 12	= Pounds-force inches (lbf in; lb in)
Pounds-force feet (lbf ft; lb ft)	X 0.138	= Kilograms-force meters (kgf m; kg m)	X 7.233	= Pounds-force feet (lbf ft; lb ft)
Pounds-force feet (lbf ft; lb ft)	X 1.356	= Newton meters (Nm)	X 0.738	= Pounds-force feet (lbf ft; lb ft)
Newton meters (Nm)	X 0.102	= Kilograms-force meters (kgf m; kg m)	X 9.804	= Newton meters (Nm)

VACUUM

Inches mercury (in. Hg)	X 3.377	= Kilopascals (kPa)	X 0.2961	= Inches mercury
Inches mercury (in. Hg)	X 25.4	= Millimeters mercury (mm Hg)	X 0.0394	= Inches mercury

POWER

Horsepower (hp)	X 745.7	= Watts (W)	X 0.0013	= Horsepower (hp)

VELOCITY (speed)

Miles per hour (miles/hr; mph)	X 1.609	= Kilometers per hour (km/hr; kph)	X 0.621	= Miles per hour (miles/hr; mph)

FUEL CONSUMPTION *

Miles per gallon, Imperial (mpg)	X 0.354	= Kilometers per liter (km/l)	X 2.825	= Miles per gallon, Imperial (mpg)
Miles per gallon, US (mpg)	X 0.425	= Kilometers per liter (km/l)	X 2.352	= Miles per gallon, US (mpg)

TEMPERATURE

Degrees Fahrenheit = (°C x 1.8) + 32

Degrees Celsius (Degrees Centigrade; °C) = (°F - 32) x 0.56

***It is common practice to convert from miles per gallon (mpg) to liters/100 kilometers (l/100km), where mpg (Imperial) x l/100 km = 282 and mpg (US) x l/100 km = 235**

FRACTION/DECIMAL/MILLIMETER EQUIVALENTS

DECIMALS TO MILLIMETERS

Decimal	mm
0.001	0.0254
0.002	0.0508
0.003	0.0762
0.004	0.1016
0.005	0.1270
0.006	0.1524
0.007	0.1778
0.008	0.2032
0.009	0.2286
0.010	0.2540
0.020	0.5080
0.030	0.7620
0.040	1.0160
0.050	1.2700
0.060	1.5240
0.070	1.7780
0.080	2.0320
0.090	2.2860
0.100	2.5400
0.110	2.7940
0.120	3.0480
0.130	3.3020
0.140	3.5560
0.150	3.8100
0.160	4.0640
0.170	4.3180
0.180	4.5720
0.190	4.8260
0.200	5.0800
0.210	5.3340
0.220	5.5880
0.230	5.8420
0.240	6.0960
0.250	6.3500
0.260	6.6040
0.270	6.8580
0.280	7.1120
0.290	7.3660
0.300	7.6200
0.310	7.8740
0.320	8.1280
0.330	8.3820
0.340	8.6360
0.350	8.8900
0.360	9.1440
0.370	9.3980
0.380	9.6520
0.390	9.9060
0.400	10.1600
0.410	10.4140
0.420	10.6680
0.430	10.9220
0.440	11.1760
0.450	11.4300
0.460	11.6840
0.470	11.9380
0.480	12.1920
0.490	12.4460

Decimal	mm
0.500	12.7000
0.510	12.9540
0.520	13.2080
0.530	13.4620
0.540	13.7160
0.550	13.9700
0.560	14.2240
0.570	14.4780
0.580	14.7320
0.590	14.9860
0.600	15.2400
0.610	15.4940
0.620	15.7480
0.630	16.0020
0.640	16.2560
0.650	16.5100
0.660	16.7640
0.670	17.0180
0.680	17.2720
0.690	17.5260
0.700	17.7800
0.710	18.0340
0.720	18.2880
0.730	18.5420
0.740	18.7960
0.750	19.0500
0.760	19.3040
0.770	19.5580
0.780	19.8120
0.790	20.0660
0.800	20.3200
0.810	20.5740
0.820	21.8280
0.830	21.0820
0.840	21.3360
0.850	21.5900
0.860	21.8440
0.870	22.0980
0.880	22.3520
0.890	22.6060
0.900	22.8600
0.910	23.1140
0.920	23.3680
0.930	23.6220
0.940	23.8760
0.950	24.1300
0.960	24.3840
0.970	24.6380
0.980	24.8920
0.990	25.1460
1.000	25.4000

FRACTIONS TO DECIMALS TO MILLIMETERS

Fraction	Decimal	mm	Fraction	Decimal	mm
1/64	0.0156	0.3969	33/64	0.5156	13.0969
1/32	0.0312	0.7938	17/32	0.5312	13.4938
3/64	0.0469	1.1906	35/64	0.5469	13.8906
1/16	0.0625	1.5875	9/16	0.5625	14.2875
5/64	0.0781	1.9844	37/64	0.5781	14.6844
3/32	0.0938	2.3812	19/32	0.5938	15.0812
7/64	0.1094	2.7781	39/64	0.6094	15.4781
1/8	0.1250	3.1750	5/8	0.6250	15.8750
9/64	0.1406	3.5719	41/64	0.6406	16.2719
5/32	0.1562	3.9688	21/32	0.6562	16.6688
11/64	0.1719	4.3656	43/64	0.6719	17.0656
3/16	0.1875	4.7625	11/16	0.6875	17.4625
13/64	0.2031	5.1594	45/64	0.7031	17.8594
7/32	0.2188	5.5562	23/32	0.7188	18.2562
15/64	0.2344	5.9531	47/64	0.7344	18.6531
1/4	0.2500	6.3500	3/4	0.7500	19.0500
17/64	0.2656	6.7469	49/64	0.7656	19.4469
9/32	0.2812	7.1438	25/32	0.7812	19.8438
19/64	0.2969	7.5406	51/64	0.7969	20.2406
5/16	0.3125	7.9375	13/16	0.8125	20.6375
21/64	0.3281	8.3344	53/64	0.8281	21.0344
11/32	0.3438	8.7312	27/32	0.8438	21.4312
23/64	0.3594	9.1281	55/64	0.8594	21.8281
3/8	0.3750	9.5250	7/8	0.8750	22.2250
25/64	0.3906	9.9219	57/64	0.8906	22.6219
13/32	0.4062	10.3188	29/32	0.9062	23.0188
27/64	0.4219	10.7156	59/64	0.9219	23.4156
7/16	0.4375	11.1125	15/16	0.9375	23.8125
29/64	0.4531	11.5094	61/64	0.9531	24.2094
15/32	0.4688	11.9062	31/32	0.9688	24.6062
31/64	0.4844	12.3031	63/64	0.9844	25.0031
1/2	0.5000	12.7000	1	1.0000	25.4000

Automotive chemicals and lubricants

A number of automotive chemicals and lubricants are available for use during vehicle maintenance and repair. They include a wide variety of products ranging from cleaning solvents and degreasers to lubricants and protective sprays for rubber, plastic and vinyl.

CLEANERS

Carburetor cleaner and choke cleaner is a strong solvent for gum, varnish and carbon. Most carburetor cleaners leave a dry-type lubricant film which will not harden or gum up. Because of this film it is not recommended for use on electrical components.

Brake system cleaner is used to remove brake dust, grease and brake fluid from the brake system, where clean surfaces are absolutely necessary. It leaves no residue and often eliminates brake squeal caused by contaminants.

Electrical cleaner removes oxidation, corrosion and carbon deposits from electrical contacts, restoring full current flow. It can also be used to clean spark plugs, carburetor jets, voltage regulators and other parts where an oil-free surface is desired.

Demoisturants remove water and moisture from electrical components such as alternators, voltage regulators, electrical connectors and fuse blocks. They are non-conductive and non-corrosive.

Degreasers are heavy-duty solvents used to remove grease from the outside of the engine and from chassis components. They can be sprayed or brushed on and, depending on the type, are rinsed off either with water or solvent.

LUBRICANTS

Motor oil is the lubricant formulated for use in engines. It normally contains a wide variety of additives to prevent corrosion and reduce foaming and wear. Motor oil comes in various weights (viscosity ratings) from 0 to 50. The recommended weight of the oil depends on the season, temperature and the demands on the engine. Light oil is used in cold climates and under light load conditions. Heavy oil is used in hot climates and where high loads are encountered. Multi-viscosity oils are designed to have characteristics of both light and heavy oils and are available in a number of weights from 5W-20 to 20W-50.

Gear oil is designed to be used in differentials, manual transmissions and other areas where high-temperature lubrication is required.

Chassis and wheel bearing grease is a heavy grease used where increased loads and friction are encountered, such as for wheel bearings, balljoints, tie-rod ends and universal joints.

High-temperature wheel bearing grease is designed to withstand the extreme temperatures encountered by wheel bearings in disc brake equipped vehicles. It usually contains molybdenum disulfide (moly), which is a dry-type lubricant.

White grease is a heavy grease for metal-to-metal applications where water is a problem. White grease stays soft under both low and high temperatures (usually from -100 to +190-degrees F), and will not wash off or dilute in the presence of water.

Assembly lube is a special extreme pressure lubricant, usually containing moly, used to lubricate high-load parts (such as main and rod bearings and cam lobes) for initial start-up of a new engine. The assembly lube lubricates the parts without being squeezed out or washed away until the engine oiling system begins to function.

Silicone lubricants are used to protect rubber, plastic, vinyl and nylon parts.

Graphite lubricants are used where oils cannot be used due to contamination problems, such as in locks. The dry graphite will lubricate metal parts while remaining uncontaminated by dirt, water, oil or acids. It is electrically conductive and will not foul electrical contacts in locks such as the ignition switch.

Moly penetrants loosen and lubricate frozen, rusted and corroded fasteners and prevent future rusting or freezing.

Heat-sink grease is a special electrically non-conductive grease that is used for mounting electronic ignition modules where it is essential that heat is transferred away from the module.

SEALANTS

RTV sealant is one of the most widely used gasket compounds. Made from silicone, RTV is air curing, it seals, bonds, waterproofs, fills surface irregularities, remains flexible, doesn't shrink, is relatively easy to remove, and is used as a supplementary sealer with almost all low and medium temperature gaskets.

Anaerobic sealant is much like RTV in that it can be used either to seal gaskets or to form gaskets by itself. It remains flexible, is solvent resistant and fills surface imperfections. The difference between an anaerobic sealant and an RTV-type sealant is in the curing. RTV cures when exposed to air, while an anaerobic sealant cures only in the absence of air. This means that an anaerobic sealant cures only after the assembly of parts, sealing them together.

Thread and pipe sealant is used for sealing hydraulic and pneumatic fittings and vacuum lines. It is usually made from a Teflon compound, and comes in a spray, a paint-on liquid and as a wrap-around tape.

CHEMICALS

Anti-seize compound prevents seizing, galling, cold welding, rust and corrosion in fasteners. High-temperature anti-seize, usually made with copper and graphite lubricants, is used for exhaust system and exhaust manifold bolts.

Anaerobic locking compounds are used to keep fasteners from vibrating or working loose and cure only after installation, in the absence of air. Medium strength locking compound is used for small nuts, bolts and screws that may be removed later. High-strength locking compound is for large nuts, bolts and studs which aren't removed on a regular basis.

Oil additives range from viscosity index improvers to chemical treatments that claim to reduce internal engine friction. It should be noted that most oil manufacturers caution against using additives with their oils.

Gas additives perform several functions, depending on their chemical makeup. They usually contain solvents that help dissolve gum and varnish that build up on carburetor, fuel injection and intake parts. They also serve to break down carbon deposits that form on the inside surfaces of the combustion chambers. Some additives contain upper cylinder lubricants for valves and piston rings, and others contain chemicals to remove condensation from the gas tank.

MISCELLANEOUS

Brake fluid is specially formulated hydraulic fluid that can withstand the heat and pressure encountered in brake systems. Care must be taken so this fluid does not come in contact with painted surfaces or plastics. An opened container should always be resealed to prevent contamination by water or dirt.

Weatherstrip adhesive is used to bond weatherstripping around doors, windows and trunk lids. It is sometimes used to attach trim pieces.

Undercoating is a petroleum-based, tar-like substance that is designed to protect metal surfaces on the underside of the vehicle from corrosion. It also acts as a sound-deadening agent by insulating the bottom of the vehicle.

Waxes and polishes are used to help protect painted and plated surfaces from the weather. Different types of paint may require the use of different types of wax and polish. Some polishes utilize a chemical or abrasive cleaner to help remove the top layer of oxidized (dull) paint on older vehicles. In recent years many non-wax polishes that contain a wide variety of chemicals such as polymers and silicones have been introduced. These non-wax polishes are usually easier to apply and last longer than conventional waxes and polishes.

Safety first!

Regardless of how enthusiastic you may be about getting on with the job at hand, take the time to ensure that your safety is not jeopardized. A moment's lack of attention can result in an accident, as can failure to observe certain simple safety precautions. The possibility of an accident will always exist, and the following points should not be considered a comprehensive list of all dangers. Rather, they are intended to make you aware of the risks and to encourage a safety conscious approach to all work you carry out on your vehicle.

ESSENTIAL DOS AND DON'TS

DON'T rely on a jack when working under the vehicle. Always use approved jackstands to support the weight of the vehicle and place them under the recommended lift or support points.

DON'T attempt to loosen extremely tight fasteners (i.e. wheel lug nuts) while the vehicle is on a jack - it may fall.

DON'T start the engine without first making sure that the transmission is in Neutral (or Park where applicable) and the parking brake is set.

DON'T remove the radiator cap from a hot cooling system - let it cool or cover it with a cloth and release the pressure gradually.

DON'T attempt to drain the engine oil until you are sure it has cooled to the point that it will not burn you.

DON'T touch any part of the engine or exhaust system until it has cooled sufficiently to avoid burns.

DON'T siphon toxic liquids such as gasoline, antifreeze and brake fluid by mouth, or allow them to remain on your skin.

DON'T inhale brake lining dust - it is potentially hazardous (see *Asbestos* below).

DON'T allow spilled oil or grease to remain on the floor - wipe it up before someone slips on it.

DON'T use loose fitting wrenches or other tools which may slip and cause injury.

DON'T push on wrenches when loosening or tightening nuts or bolts. Always try to pull the wrench toward you. If the situation calls for pushing the wrench away, push with an open hand to avoid scraped knuckles if the wrench should slip.

DON'T attempt to lift a heavy component alone - get someone to help you.

DON'T rush or take unsafe shortcuts to finish a job.

DON'T allow children or animals in or around the vehicle while you are working on it.

DO wear eye protection when using power tools such as a drill, sander, bench grinder, etc. and when working under a vehicle.

DO keep loose clothing and long hair well out of the way of moving parts.

DO make sure that any hoist used has a safe working load rating adequate for the job.

DO get someone to check on you periodically when working alone on a vehicle.

DO carry out work in a logical sequence and make sure that everything is correctly assembled and tightened.

DO keep chemicals and fluids tightly capped and out of the reach of children and pets.

DO remember that your vehicle's safety affects that of yourself and others. If in doubt on any point, get professional advice.

ASBESTOS

Certain friction, insulating, sealing, and other products - such as brake linings, brake bands, clutch linings, torque converters, gaskets, etc. - may contain asbestos. Extreme care must be taken to avoid inhalation of dust from such products, since it is hazardous to health. If in doubt, assume that they do contain asbestos.

FIRE

Remember at all times that gasoline is highly flammable. Never smoke or have any kind of open flame around when working on a vehicle. But the risk does not end there. A spark caused by an electrical short circuit, by two metal surfaces contacting each other, or even by static electricity built up in your body under certain conditions, can ignite gasoline vapors, which in a confined space are highly explosive. Do not, under any circumstances, use gasoline for cleaning parts. Use an approved safety solvent.

Always disconnect the battery ground (-) cable at the battery before working on any part of the fuel system or electrical system. Never risk spilling fuel on a hot engine or exhaust component. It is strongly recommended that a fire extinguisher suitable for use on fuel and electrical fires be kept handy in the garage or workshop at all times. Never try to extinguish a fuel or electrical fire with water.

FUMES

Certain fumes are highly toxic and can quickly cause unconsciousness and even death if inhaled to any extent. Gasoline vapor falls into this category, as do the vapors from some cleaning solvents. Any draining or pouring of such volatile fluids should be done in a well ventilated area.

When using cleaning fluids and solvents, read the instructions on the container carefully. Never use materials from unmarked containers.

Never run the engine in an enclosed space, such as a garage. Exhaust fumes contain carbon monoxide, which is extremely poisonous. If you need to run the engine, always do so in the open air, or at least have the rear of the vehicle outside the work area.

If you are fortunate enough to have the use of an inspection pit, never drain or pour gasoline and never run the engine while the vehicle is over the pit. The fumes, being heavier than air, will concentrate in the pit with possibly lethal results.

THE BATTERY

Never create a spark or allow a bare light bulb near a battery. They normally give off a certain amount of hydrogen gas, which is highly explosive.

Always disconnect the battery ground (-) cable at the battery before working on the fuel or electrical systems.

If possible, loosen the filler caps or cover when charging the battery from an external source (this does not apply to sealed or maintenance-free batteries). Do not charge at an excessive rate or the battery may burst.

Take care when adding water to a non maintenance-free battery and when carrying a battery. The electrolyte, even when diluted, is very corrosive and should not be allowed to contact clothing or skin.

Always wear eye protection when cleaning the battery to prevent the caustic deposits from entering your eyes.

HOUSEHOLD CURRENT

When using an electric power tool, inspection light, etc., which operates on household current, always make sure that the tool is correctly connected to its plug and that, where necessary, it is properly grounded. Do not use such items in damp conditions and, again, do not create a spark or apply excessive heat in the vicinity of fuel or fuel vapor.

SECONDARY IGNITION SYSTEM VOLTAGE

A severe electric shock can result from touching certain parts of the ignition system (such as the spark plug wires) when the engine is running or being cranked, particularly if components are damp or the insulation is defective. In the case of an electronic ignition system, the secondary system voltage is much higher and could prove fatal.

HYDROFLUORIC ACID

This extremely corrosive acid is formed when certain types of synthetic rubber, found in some O-rings, oil seals, fuel hoses, etc. are exposed to temperatures above 750-degrees F (400-degrees C). The rubber changes into a charred or sticky substance containing the acid. Once formed, the acid remains dangerous for years. If it gets onto the skin, it may be necessary to amputate the limb concerned.

When dealing with a vehicle which has suffered a fire, or with components salvaged from such a vehicle, wear protective gloves and discard them after use.

Troubleshooting

CONTENTS

This section provides an easy reference guide to the more common problems that may occur during the operation of your vehicle. These problems and their possible causes are grouped under headings denoting various components or systems, such as Engine, Cooling system, etc. They also refer you to the chapter and/or section that deals with the problem.

Remember that successful troubleshooting is not a mysterious black art practiced only by professional mechanics. It is simply the result of the right knowledge combined with an intelligent, systematic approach to the problem. Always work by a process of elimination, starting with the simplest solution and working through to the most complex - and never overlook the obvious. Anyone can run the gas tank dry or leave the lights on overnight, so don't assume that you are exempt from such oversights.

Finally, always establish a clear idea of why a problem has occurred and take steps to ensure that it doesn't happen again. If the electrical system fails because of a poor connection, check the other connections in the system to make sure that they don't fail as well. If a particular fuse continues to blow, find out why - don't just replace one fuse after another. Remember, failure of a small component can often be indicative of potential failure or incorrect functioning of a more important component or system.

ENGINE

1 Engine will not rotate when attempting to start

1 Battery terminal connections loose or corroded (Chapter 1).
2 Battery discharged or faulty (Chapter 1).
3 Automatic transaxle not completely engaged in Park (Chapter 7B) or clutch pedal not completely depressed (Chapter 8).
4 Broken, loose or disconnected wiring in the starting circuit (Chapters 5 and 12).
5 Starter motor pinion jammed in flywheel ring gear (Chapter 5).
6 Starter solenoid faulty (Chapter 5).
7 Starter motor faulty (Chapter 5).
8 Ignition switch faulty (Chapter 12).
9 Starter pinion or flywheel teeth worn or broken (Chapter 5).
10 Transmission Range (TR) switch malfunctioning (Chapter 7B).

2 Engine rotates but will not start

1 Fuel tank empty.
2 Battery discharged (engine rotates slowly) (Chapter 5).
3 Battery terminal connections loose or corroded (Chapter 1).
4 Leaking fuel injector(s), faulty fuel pump, pressure regulator, etc. (Chapter 4).
5 Broken or stripped timing belt (Chapter 2).
6 Ignition components damp or damaged (Chapter 5).
7 Worn, faulty or incorrectly gapped spark plugs (Chapter 1).
8 Broken, loose or disconnected wiring in the starting circuit (Chapter 5).
9 Broken, loose or disconnected wires at the ignition coils or faulty coils (Chapter 5).
10 Defective MAF sensor (see Chapter 6).
11 Broken, loose or disconnected wires at the fuel shutdown solenoid (diesel) (Chapter 4B).
12 Defective fuel injection pump or fuel injector (diesel) (Chapter 4B).
13 Incorrect fuel injection pump timing (diesel) (Chapter 4B).
14 Contaminated fuel.

3 Engine hard to start when cold

1 Battery discharged or low (Chapter 1).
2 Malfunctioning fuel system (Chapter 4).
3 Faulty coolant temperature sensor or intake air temperature sensor (Chapter 6).
4 Injector(s) leaking (Chapter 4B).
5 Faulty ignition system (Chapter 5).
6 Defective MAF sensor (see Chapter 6).
7 Defective fuel injection pump or fuel injector (diesel) (Chapter 4B).
8 Incorrect fuel injection pump timing (diesel) (Chapter 4B).

4 Engine hard to start when hot

1 Air filter clogged (Chapter 1).
2 Fuel not reaching the fuel injection system (Chapter 4).
3 Corroded battery connections, especially ground (Chapter 1).
4 Faulty coolant temperature sensor or intake air temperature sensor (Chapter 6).
5 Low cylinder compression (Chapter 2).
6 Defective fuel injection pump or fuel injector (diesel) (Chapter 4B).
7 Incorrect fuel injection pump timing (diesel) (Chapter 4B).

5 Starter motor noisy or excessively rough in engagement

1 Pinion or flywheel gear teeth worn or broken (Chapter 5).
2 Starter motor mounting bolts loose or missing (Chapter 5).

6 Engine starts but stops immediately

1 Loose or faulty electrical connections at distributor, coil or alternator (Chapter 5).
2 Insufficient fuel reaching the fuel injector(s) (Chapters 1 and 4).
3 Vacuum leak at the gasket between the intake manifold/plenum and throttle body (Chapters 1 and 4).
4 Intake air leaks, broken vacuum lines (see Chapter 4).
5 Defective fuel injection pump or fuel injector (diesel) (Chapter 4B).
6 Contaminated fuel.

7 Oil puddle under engine

1 Oil pan gasket and/or oil pan drain bolt washer leaking (Chapter 2).
2 Oil pressure sending unit leaking (Chapter 2).
3 Valve covers leaking (Chapter 2).
4 Engine oil seals leaking (Chapter 2).
5 Oil pump housing leaking (Chapter 2).

8 Engine lopes while idling or idles erratically

1 Vacuum leakage (Chapters 2 and 4).
2 Leaking EGR valve (Chapter 6).
3 Air filter clogged (Chapter 1).
4 Fuel pump not delivering sufficient fuel to the fuel injection system (Chapter 4).
5 Leaking head gasket (Chapter 2).
6 Timing belt and/or sprockets worn (Chapter 2).
7 Camshaft lobes worn (Chapter 2).
8 Defective fuel injection pump or fuel injector (diesel) (Chapter 4B).
9 Incorrect fuel injection pump timing (diesel) (Chapter 4B).

9 Engine misses at idle speed

1 Spark plugs worn or not gapped properly (Chapter 1).
2 Faulty spark plug wires (Chapter 1).
3 Vacuum leaks (Chapters 2 and 4).
4 Uneven or low compression (Chapter 2).
5 Problem with the fuel injection system (Chapter 4).
6 Faulty ignition coils (Chapter 5).
7 Defective fuel injection pump or fuel injector (diesel) (Chapter 4B).
8 Incorrect fuel injection pump timing (diesel) (Chapter 4B).

10 Engine misses throughout driving speed range

1 Fuel filter clogged and/or impurities in the fuel system (Chapter 1).
2 Low fuel output at the fuel injector(s) (Chapter 4).
3 Faulty or incorrectly gapped spark plugs (Chapter 1).
4 Defective spark plug wires (Chapters 1 or 5).
5 Faulty emission system components (Chapter 6).
6 Low or uneven cylinder compression pressures (Chapter 2).
7 Weak or faulty ignition system (Chapter 5).
8 Vacuum leak in fuel injection system, throttle body, intake manifold or vacuum hoses (Chapter 4).
9 Defective fuel injection pump or fuel injector (diesel) (Chapter 4B).
10 Incorrect fuel injection pump timing (diesel) (Chapter 4B).

11 Engine stumbles on acceleration

1 Spark plugs fouled (Chapter 1).
2 Problem with fuel injection system (Chapter 4).
3 Fuel filter clogged (Chapters 1 and 4).
4 Intake manifold leak (Chapters 2 and 4).
5 EGR system malfunction (Chapter 6).

12 Engine surges while holding accelerator steady

1 Intake air/vacuum leak (Chapter 4).
2 Problem with fuel injection system (Chapter 4).
3 Problem with the emissions control system (Chapter 6).

13 Engine stalls

1 Idle speed incorrect (Chapter 4).
2 Fuel filter clogged and/or water and impurities in the fuel system (Chapters 1 and 4).
3 Faulty emissions system components (Chapter 6).
4 Faulty or incorrectly gapped spark plugs (Chapter 1).
5 Faulty spark plug wires (Chapter 1).
6 Vacuum leak in the fuel injection system, intake manifold or vacuum hoses (Chapters 2 and 4).
7 Defective fuel injection pump or fuel injector (diesel) (Chapter 4B).

14 Engine lacks power

1 Faulty spark plug wires or coils (Chapters 1 and 5).
2 Faulty or incorrectly gapped spark plugs (Chapter 1).
3 Problem with the fuel injection system (Chapter 4).
4 Plugged air filter (Chapter 1).
5 Brakes binding (Chapter 9).
6 Automatic transaxle fluid level incorrect (Chapter 1).
7 Clutch slipping (Chapter 8).
8 Fuel filter clogged and/or impurities in the fuel system (Chap-ters 1 and 4).
9 Emission control system not functioning properly (Chapter 6).
10 Low or uneven cylinder compression pressures (Chapter 2).
11 Obstructed exhaust system (Chapters 2 and 4).
12 Defective fuel injection pump or fuel injector (diesel) (Chapter 4B).
13 Incorrect fuel injection pump timing (diesel) (Chapter 4).
14 Defective turbocharger or wastegate (diesel or gas-turbo) (Chapter 4B).

15 Engine backfires

1 Emission control system not functioning properly (Chapter 6).
2 Faulty plug wires or coils (Chapters 1 and 5).
3 Problem with the fuel injection system (Chapter 4).
4 Vacuum leak at fuel injector(s), intake manifold, air control valve or vacuum hoses (Chapters 2 and 4).
5 Valves sticking (Chapter 2).

16 Pinging or knocking engine sounds during acceleration or uphill

1 Incorrect grade of fuel.
2 Ignition timing incorrect (Chapter 5).
3 Fuel injection system faulty (Chapter 4).
4 Improper or damaged spark plugs or wires (Chapter 1).
5 EGR valve not functioning (Chapter 6).
6 Vacuum leak (Chapters 2 and 4).
7 Knock sensor malfunctioning (Chapter 6A).

17 Engine runs with oil pressure light on

1 Low oil level (Chapter 1).
2 Idle rpm below specification (Chapter 4).
3 Short in wiring circuit (Chapter 12).
4 Faulty oil pressure sender (Chapter 2).
5 Worn engine bearings and/or oil pump (Chapter 2).

18 Engine continues to run after switching off

1 Excessive engine operating temperature (Chapter 3).
2 Excessive carbon deposits on valves and pistons (Chapter 2).
3 Fuel shut-off valve malfunctioning (diesel) (Chapter 4B).

ENGINE ELECTRICAL SYSTEM

19 Battery will not hold a charge

1 Alternator drivebelt defective or not adjusted properly (Chapter 1).
2 Battery electrolyte level low (Chapter 1).
3 Battery terminals loose or corroded (Chapter 1).
4 Alternator not charging properly (Chapter 5).
5 Loose, broken or faulty wiring in the charging circuit (Chapter 5).
6 Short in vehicle wiring (Chapter 12).
7 Internally defective battery (Chapters 1 and 5).

20 Alternator light fails to go out

1 Faulty alternator or charging circuit (Chapter 5).
2 Alternator drivebelt defective or out of adjustment (Chapter 1).
3 Alternator voltage regulator inoperative (Chapter 5).

21 Alternator light fails to come on when key is turned on

1 Warning light bulb defective (Chapter 12).
2 Fault in the printed circuit, dash wiring or bulb holder (Chapter 12).

FUEL SYSTEM

22 Excessive fuel consumption

1 Dirty or clogged air filter element (Chapter 1).
2 Engine management problem (Chapter 6).
3 Emissions system not functioning properly (Chapter 6).
4 Fuel injection system not functioning properly (Chapter 4).
5 Low tire pressure or incorrect tire size (Chapter 1).

23 Fuel leakage and/or fuel odor

1 Leaking fuel feed or return line (Chapters 1 and 4).
2 Tank overfilled.
3 Problem with the evaporative emissions control system (Chapters 6).
4 Problem with the fuel injection system (Chapter 4).

COOLING SYSTEM

24 Overheating

1 Insufficient coolant in system (Chapter 1).
2 Water pump defective (Chapter 3).
3 Radiator core blocked or grille restricted (Chapter 3).
4 Thermostat faulty (Chapter 3).
5 Electric coolant fan inoperative or blades broken (Chapter 3).
6 Expansion tank cap not maintaining proper pressure (Chapter 3).
7 Blown head gasket (Chapter 2).

25 Overcooling

1 Faulty thermostat (Chapter 3).
2 Inaccurate temperature gauge sending unit (Chapter 3)

26 External coolant leakage

1 Deteriorated/damaged hoses; loose clamps (Chapters 1 and 3).
2 Water pump defective (Chapter 3).
3 Leakage from radiator core or coolant reservoir tank (Chapter 3).
4 Cylinder head gasket leaking (Chapter 2).

27 Internal coolant leakage

1 Leaking cylinder head gasket (Chapter 2).
2 Cracked cylinder bore or cylinder head (Chapter 2).

28 Coolant loss

1 Too much coolant in system (Chapter 1).
2 Coolant boiling away because of overheating (Chapter 3).
3 Internal or external leakage (Chapter 3).
4 Faulty expansion tank cap (Chapter 3).

29 Poor coolant circulation

1 Defective water pump (Chapter 3).
2 Restriction in cooling system (Chapters 1 and 3).
3 Thermostat sticking (Chapter 3).

CLUTCH

30 Pedal travels to floor - no pressure or very little resistance

1 Hydraulic release system leaking or air in the system (Chapter 8).
2 Broken release bearing or fork (Chapter 8).

31 Unable to select gears

1 Faulty transaxle (Chapter 7).
2 Faulty clutch disc or pressure plate (Chapter 8).
3 Faulty release lever or release bearing (Chapter 8).
4 Faulty shift lever assembly or cables (Chapter 8).
5 Faulty clutch release system.

32 Clutch slips (engine speed increases with no increase in vehicle speed)

1 Clutch plate worn (Chapter 2 and 8).
2 Clutch plate is oil soaked by leaking rear main seal (Chapter 2 and 8).
3 Clutch plate not seated (Chapter 8).
4 Warped pressure plate or flywheel (Chapter 8).
5 Weak diaphragm spring in pressure plate (Chapter 8).
6 Clutch plate overheated. Allow to cool.
7 Piston stuck in bore of clutch release cylinder, preventing clutch from fully engaging (Chapter 8).

33 Grabbing (chattering) as clutch is engaged

1 Oil on clutch plate lining, burned or glazed facings (Chapter 8).
2 Worn or loose engine or transaxle mounts (Chapters 2 and 7).
3 Worn splines on clutch plate hub (Chapter 8).
4 Warped pressure plate or flywheel (Chapter 8).
5 Burned or smeared resin on flywheel or pressure plate (Chapter 8).

34 Transaxle rattling (clicking)

1 Release fork loose (Chapter 8).
2 Low engine idle speed (Chapter 1).

35 Noise in clutch area

Faulty bearing (Chapter 8).

36 Clutch pedal stays on floor

1 Broken release bearing or fork (Chapter 8).
2 Hydraulic release system leaking or air in the system (Chapter 8).
3 Over-center spring in clutch pedal assembly broken (Chapter 8).

37 High pedal effort

1 Piston binding in bore of release cylinder (Chapter 8).
2 Pressure plate faulty (Chapter 8).

MANUAL TRANSAXLE

38 Knocking noise at low speeds

Worn driveaxle constant velocity (CV) joints (Chapter 8).

39 Noise most pronounced when turning

Differential gear noise (Chapter 7A).*

40 Clunk on acceleration or deceleration

1 Loose engine or transaxle mounts (Chapters 2 and 7A).
2 Worn differential pinion shaft in case.*
3 Worn or damaged driveaxle inboard CV joints (Chapter 8).

41 Clicking noise in turns

Worn or damaged outboard CV joint (Chapter 8).

42 Vibration

1 Rough wheel bearing (Chapter 10).
2 Damaged driveaxle (Chapter 8).
3 Out-of-round tires (Chapter 1).
4 Tire out of balance (Chapters 1 and 10).
5 Worn CV joint (Chapter 8).

43 Noisy in neutral with engine running

1 Damaged input gear bearing (Chapter 7A).*
2 Damaged clutch release bearing (Chapter 8).

44 Noisy in one particular gear

1 Damaged or worn constant mesh gears (Chapter 7A).*
2 Damaged or worn synchronizers (Chapter 7A).*
3 Bent reverse fork (Chapter 7A).*
4 Damaged fourth speed gear or output gear (Chapter 7A).*
5 Worn or damaged reverse idler gear or idler bushing (Chapter 7A).*

45 Noisy in all gears

1 Insufficient lubricant (Chapter 7A).
2 Damaged or worn bearings (Chapter 7A).*
3 Worn or damaged input gear shaft and/or output gear shaft (Chapter 7A).*

46 Slips out of gear

1 Worn or improperly adjusted linkage (Chapter 7A).
2 Transaxle loose on engine (Chapter 7A).
3 Shift linkage does not work freely, binds (Chapter 7A).
4 Input gear bearing retainer broken or loose (Chapter 7A).*
5 Dirt between clutch cover and engine housing (Chapter 7A).
6 Worn shift fork (Chapter 7A).*

47 Leaks lubricant

1 Side gear shaft seals worn (Chapter 7A).
2 Excessive amount of lubricant in transaxle (Chapters 1 and 7A).
3 Loose or broken input gear shaft bearing retainer (Chapter 7A).*
4 Input gear bearing retainer O-ring and/or lip seal damaged (Chapter 7A).*
5 Striking rod seal leaking (Chapter 7A).
6 Vehicle speed sensor O-ring leaking (Chapter 7A).

48 Hard to shift

Shift cable(s) worn (Chapter 7A).

** Although the corrective action necessary to remedy the symptoms described is beyond the scope of this manual, the above information should be helpful in isolating the cause of the condition so that the owner can communicate clearly with a professional mechanic.*

AUTOMATIC TRANSAXLE

➡Note: Due to the complexity of the automatic transaxle, it is difficult for the home mechanic to properly diagnose and service this component. For problems other than the following, the vehicle should be taken to a dealer or transaxle shop.

49 Fluid leakage

1 Automatic transaxle fluid in these vehicles is transparent yellow in color. Fluid leaks should not be confused with engine oil, which can easily be blown onto the transaxle by air flow.

2 To pinpoint a leak, first remove all built-up dirt and grime from the transaxle housing with degreasing agents and/or steam cleaning. Then drive the vehicle at low speeds so air flow will not blow the leak far from its source. Raise the vehicle and determine where the leak is coming from. Common areas of leakage are:

a) Pan (Chapters 1 and 7B).
b) Dipstick tube (Chapters 1 and 7B).
c) Transaxle oil lines (Chapter 7B).
d) Speed sensor (Chapter 7B).
e) Driveaxle oil seals (Chapter 7B).

50 Transaxle fluid brown or has a burned smell

Transaxle fluid overheated (Chapter 1).

51 General shift mechanism problems

1 Chapter 7, Part B, deals with checking and adjusting the shift cable on automatic transaxles. Common problems that may be attributed to poorly adjusted cable are:

a) Engine starting in gears other than Park or Neutral.
b) Indicator on shifter pointing to a gear other than the one actually being used.
c) Vehicle moves when in Park.

2 Refer to Chapter 7B for the shift cable adjustment procedure.

52 Transaxle will not downshift with accelerator pedal pressed to the floor

The transaxle is electronically controlled. This type of problem - which is caused by a malfunction in the control unit, a sensor or solenoid, or the circuit itself - is beyond the scope of this book. Take the vehicle to a dealer service department or a competent automatic transmission shop.

53 Engine will start in gears other than Park or Neutral

Transmission Range (TR) switch malfunctioning (Chapter 7B).

54 Transaxle slips, shifts roughly, is noisy or has no drive in forward or reverse gears

There are many probable causes for the above problems, but the home mechanic should be concerned with only one possibility - fluid level. Before taking the vehicle to a repair shop, check the level and condition of the fluid as described in Chapter 1. Correct the fluid level as necessary or change the fluid and filter if needed. If the problem persists, have a professional diagnose the cause.

DRIVEAXLES

55 Clicking noise in turns

Worn or damaged outboard CV joint (Chapter 8).

56 Shudder or vibration during acceleration

1 Excessive toe-in (Chapter 10).
2 Incorrect spring heights (Chapter 10).
3 Worn or damaged inboard or outboard CV joints (Chapter 8).
4 Sticking inboard CV joint assembly (Chapter 8).

57 Vibration at highway speeds

1 Out-of-balance front wheels and/or tires (Chapters 1 and 10).
2 Out-of-round front tires (Chapters 1 and 10).
3 Worn CV joint(s) (Chapter 8).

BRAKES

➡Note: Before assuming that a brake problem exists, make sure that:

a) The tires are in good condition and properly inflated (Chapter 1).
b) The front end alignment is correct (Chapter 10).
c) The vehicle is not loaded with weight in an unequal manner.

58 Vehicle pulls to one side during braking

1 Incorrect tire pressures (Chapter 1).
2 Front end out of alignment (have the front end aligned).
3 Front, or rear, tire sizes not matched to one another.
4 Restricted brake lines or hoses (Chapter 9).
5 Malfunctioning drum brake or caliper assembly (Chapter 9).
6 Loose suspension parts (Chapter 10).
7 Loose calipers (Chapter 9).
8 Excessive wear of brake shoe or pad material or disc/drum on one side.

59 Noise (high-pitched squeal when the brakes are applied)

Front and/or rear disc brake pads worn out (Chapter 9).

60 Brake roughness or chatter (pedal pulsates)

1 Excessive lateral runout (Chapter 9).
2 Uneven pad wear (Chapter 9).
3 Defective disc (Chapter 9).

61 Excessive brake pedal effort required to stop vehicle

1 Malfunctioning power brake booster (Chapter 9).
2 Partial system failure (Chapter 9).
3 Excessively worn pads or shoes (Chapter 9).
4 Piston in caliper or wheel cylinder stuck or sluggish (Chapter 9).
5 Brake pads or shoes contaminated with oil or grease (Chapter 9).
6 Brake disc grooved and/or glazed (Chapter 1).
7 New pads or shoes installed and not yet seated. It will take a while for the new material to seat against the disc or drum.
8 Vacuum pump not operating properly (diesel) (Chapter 9).

62 Excessive brake pedal travel

1 Partial brake system failure (Chapter 9).
2 Insufficient fluid in master cylinder (Chapters 1 and 9).
3 Air trapped in system (Chapters 1 and 9).

63 Dragging brakes

1 Master cylinder pistons not returning correctly (Chapter 9).
2 Restricted brakes lines or hoses (Chapters 1 and 9).
3 Incorrect parking brake adjustment (Chapter 9).

64 Grabbing or uneven braking action

1 Malfunction of proportioning valve (Chapter 9).
2 Malfunction of power brake booster unit (Chapter 9).
3 Binding brake pedal mechanism (Chapter 9).

65 Brake pedal feels spongy when depressed

1 Air in hydraulic lines (Chapter 9).
2 Master cylinder mounting bolts loose (Chapter 9).
3 Master cylinder defective (Chapter 9).

66 Brake pedal travels to the floor with little resistance

1 Little or no fluid in the master cylinder reservoir caused by leaking caliper piston(s) (Chapter 9).
2 Malfunctioning master cylinder (Chapter 9).
3 Loose, damaged or disconnected brake lines (Chapter 9).

67 Parking brake does not hold

Parking brake linkage improperly adjusted (Chapters 1 and 9).

SUSPENSION AND STEERING SYSTEMS

➡Note: Before attempting to diagnose the suspension and steering systems, perform the following preliminary checks:

a) Tires for wrong pressure and uneven wear.
b) Steering universal joints from the column to the rack-and-pinion for loose connectors or wear.
c) Front and rear suspension and the rack-and-pinion assembly for loose or damaged parts.
d) Out-of-round or out-of-balance tires, bent rims and loose and/or rough wheel bearings.

68 Vehicle pulls to one side

1 Mismatched or uneven tires (Chapter 10).
2 Broken or sagging springs (Chapter 10).
3 Wheel alignment out of specifications (Chapter 10).
4 Front brake dragging (Chapter 9).

69 Abnormal or excessive tire wear

1 Wheel alignment out of specifications (Chapter 10).
2 Sagging or broken springs (Chapter 10).
3 Tire out-of-balance (Chapter 10).
4 Worn strut damper or shock absorber (Chapter 10).
5 Overloaded vehicle.
6 Tires not rotated regularly.

70 Wheel makes a thumping noise

1 Blister or bump on tire (Chapter 10).
2 Improper strut damper or shock absorber action (Chapter 10).

71 Shimmy, shake or vibration

1 Tire or wheel out-of-balance or out-of-round (Chapter 10).
2 Loose or worn wheel bearings (Chapters 1, 8 and 10).
3 Worn tie-rod ends (Chapter 10).
4 Worn balljoints (Chapters 1 and 10).
5 Excessive wheel runout (Chapter 10).
6 Blister or bump on tire (Chapter 10).

72 Hard steering

1 Defective balljoints, tie-rod ends or rack-and-pinion assembly (Chapter 10).
2 Front wheel alignment out of specifications (Chapter 10).
3 Low tire pressure(s) (Chapters 1 and 10).

73 Poor returnability of steering to center

1 Defective balljoints or tie-rod ends (Chapter 10).
2 Binding in steering gear or column (Chapter 10).
3 Lack of lubricant in steering gear assembly (Chapter 10).
4 Front wheel alignment out of specifications (Chapter 10).

74 Abnormal noise at the front end

1 Defective balljoints or tie-rod ends (Chapters 1 and 10).
2 Damaged strut mounting (Chapter 10).
3 Worn control arm bushings or tie-rod ends (Chapter 10).
4 Loose stabilizer bar (Chapter 10).
5 Loose wheel bolts (Chapter 1 Specifications).
6 Loose suspension bolts (Chapter 10).

75 Wander or poor steering stability

1 Mismatched or uneven tires (Chapter 10).
2 Defective balljoints or tie-rod ends (Chapters 1 and 10).
3 Worn strut assemblies (Chapter 10).
4 Loose stabilizer bar (Chapter 10).
5 Broken or sagging springs (Chapter 10).
6 Wheels out of alignment (Chapter 10).

76 Erratic steering when braking

1 Wheel bearings worn (Chapter 10).
2 Broken or sagging springs (Chapter 10).
3 Defective wheel cylinder or caliper (Chapter 9).
4 Warped discs or drums (Chapter 9).
5 Front end alignment incorrect.

77 Excessive pitching and/or rolling around corners or during braking

1 Loose stabilizer bar (Chapter 10).
2 Worn strut dampers, shock absorbers or mountings (Chapter 10).
3 Broken or sagging springs (Chapter 10).
4 Overloaded vehicle.
5 Front end alignment incorrect.

78 Suspension bottoms

1 Overloaded vehicle.
2 Worn strut dampers or shock absorbers (Chapter 10).
3 Incorrect, broken or sagging springs (Chapter 10).

79 Cupped tires

1 Worn strut dampers or shock absorbers (Chapter 10).
2 Wheel bearings worn (Chapter 10).
3 Excessive tire or wheel runout (Chapter 10).
4 Worn balljoints (Chapter 10).

80 Excessive tire wear on outside edge

1 Inflation pressures incorrect (Chapter 1).
2 Excessive speed in turns.
3 Front end alignment incorrect (excessive toe-in). Have professionally aligned.
4 Suspension arm bent or twisted (Chapter 10).

81 Excessive tire wear on inside edge

1 Inflation pressures incorrect (Chapter 1).
2 Front end alignment incorrect (toe-out). Have professionally aligned.
3 Loose or damaged steering components (Chapter 10).

82 Tire tread worn in one place

1 Tires out-of-balance.
2 Damaged wheel. Inspect and replace if necessary.
3 Defective tire (Chapter 1).

83 Excessive play or looseness in steering system

1 Wheel bearing(s) worn (Chapter 10).
2 Tie-rod end loose or worn (Chapter 10).
3 Steering gear loose or worn (Chapter 10).
4 Worn or loose steering intermediate shaft (Chapter 10).

84 Rattling or clicking noise in steering gear

1 Steering gear loose or worn (Chapter 10).
2 Steering gear defective.

1 TUNE-UP AND ROUTINE MAINTENANCE

1 Volkswagen New Beetle Maintenance schedule

The following maintenance intervals are based on the assumption that the vehicle owner will be doing the maintenance or service work, as opposed to having a dealer service department do the work. These are the minimum maintenance intervals recommended by the factory for vehicles that are driven daily. If you wish to keep your vehicle in peak condition at all times, you may wish to perform some of these procedures even more often. Because frequent maintenance enhances the efficiency, performance and resale value of your car, we encourage you to do so. If you drive in dusty areas, tow a trailer, idle or drive at low speeds for extended periods or drive for short distances (less than four miles) in below freezing temperatures, shorter intervals are also recommended.

When the vehicle is new, follow the maintenance schedule to the letter, record the maintenance performed in your owners manual and keep all receipts to protect the new vehicle warranty. In many cases the initial maintenance check is done at no cost to the owner (check with your dealer service department for more information).

⁂ CAUTION 1:

These models are equipped with an anti-theft radio. Before performing a procedure that requires disconnecting the battery, make sure you have the proper activation code.

⁂ CAUTION 2:

Disconnecting the battery can cause driveability problems that require a scan tool to remedy. See Chapter 5, Section 1 for the use of an auxiliary voltage input device before disconnecting the battery.

EVERY 250 MILES OR WEEKLY, WHICHEVER COMES FIRST

Check the engine oil level (Section 4)
Check the coolant level (Section 4)
Check the windshield washer fluid level (Section 4)
Check the brake and clutch fluid levels (Section 4)
Check the tires and tire pressures (Section 5)

EVERY 3000 MILES OR 3 MONTHS, WHICHEVER COMES FIRST

All items listed above, plus . . .
Check the power steering fluid level (Section 6)
Change the engine oil and filter (Section 7)

EVERY 6000 MILES OR 6 MONTHS, WHICHEVER COMES FIRST

All items listed above, plus . . .
Check the seat belts (Section 8)
Inspect the windshield wiper blades (Section 9)
Check and service the battery (Section 10)
Check the engine drivebelt (Section 11)
Inspect underhood hoses (Section 12)
Check the cooling system (Section 13)
Rotate the tires (Section 14)
Drain the water separator (diesel models) (Section 25)

EVERY 10,000 MILES OR 9 MONTHS, WHICHEVER COMES FIRST

Check the timing belt, diesel engines (Section 15)

EVERY 15,000 MILES OR 12 MONTHS, WHICHEVER COMES FIRST

All items listed above, plus . . .
Check the fuel system (Section 16)
Check the brake system (Section 17)*
Check the exhaust system (Section 18)
Check the manual transmission lubricant level (Section 19)
Replace the interior ventilation filter (Section 20)*

EVERY 20,000 MILES OR 18 MONTHS, WHICHEVER COMES FIRST

Replace the fuel filter (diesel models) (Section 25)
Check for trouble codes in the on-board computer (See Chapter 6)

EVERY 30,000 MILES OR 30 MONTHS, WHICHEVER COMES FIRST

All items listed above, plus . . .
Check the automatic transmission fluid level (Section 21)
Differential lubricant level check, automatic transaxles (Section 22)
Change the brake fluid (Section 23)
Replace the air filter (Section 24)*
Replace the spark plugs, 2.0L engines (Section 26)**
Service the cooling system (drain, flush and refill) (green-colored ethylene glycol antifreeze only) (Section 27)
Check the steering, suspension and driveaxle boots(Section 28)

EVERY 40,000 MILES OR 36 MONTHS, WHICHEVER COMES FIRST

Replace the timing belt and tensioner (diesel models) (Chapter 2)

EVERY 60,000 MILES OR 48 MONTHS, WHICHEVER COMES FIRST

Replace the fuel filter (gasoline models) (Section 25)
Replace the spark plugs, 1.8L engines (Section 26)**
Change the manual transaxle lubricant (Section 29)
Change the differential lubricant, automatic transaxle models (Section 30)**

EVERY 100,000 MILES OR 60 MONTHS, WHICHEVER COMES FIRST

Service the cooling system (drain, flush and refill) (red-colored, silicate and phosphate-free antifreeze only) (Section 27)
Inspect/replace the spark plug wires (Section 31)
Change the automatic transaxle fluid and filter (Section 32)**

Replace the timing belt (gasoline models) (Chapter 2)

* *This item is affected by "severe" operating conditions, as described below. If the vehicle is operated under severe conditions, perform all maintenance indicated with an asterisk (*) at half the indicated intervals. Severe conditions exist if you mainly operate the vehicle . . .*

in dusty areas
towing a trailer
idling for extended periods
driving at low speeds when outside temperatures remain below freezing and most trips are less than four miles long

** *Perform this procedure at half the recommended interval if operated under one or more of the following conditions:*

in heavy city traffic where the outside temperature regularly reaches 90-degrees F or higher
in hilly or mountainous terrain
frequent trailer towing
if the vehicle has been driven through deep water

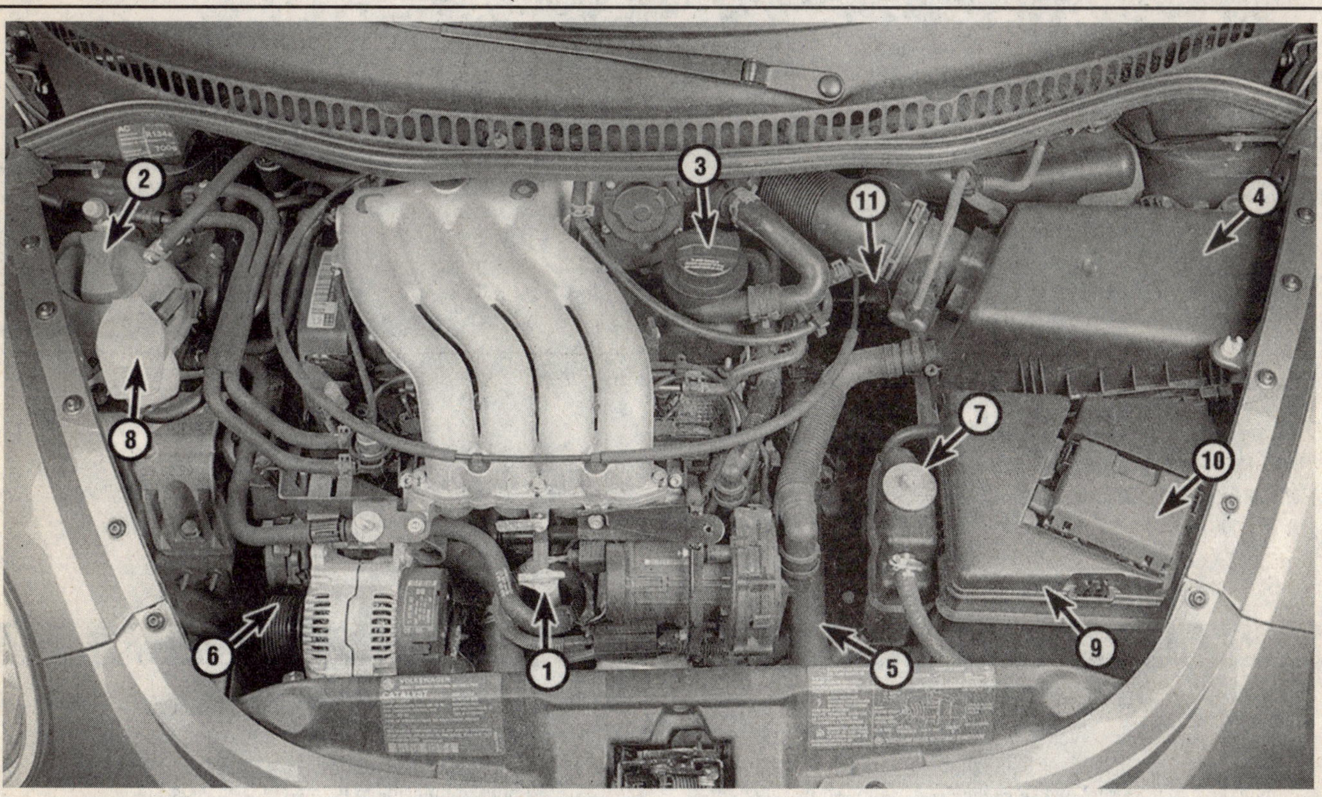

Engine compartment components (gasoline model shown, diesel similar)

1 *Engine oil dipstick*
2 *Surge tank coolant pressure cap*
3 *Oil filler cap*
4 *Air filter housing*
5 *Upper radiator hose*
6 *Drivebelt*
7 *Power steering fluid reservoir*
8 *Windshield washer fluid reservoir*
9 *Battery*
10 *Underhood fuse box*
11 *Brake fluid reservoir (below air duct)*

Typical engine compartment underside components

1 *Balljoint*
2 *Engine oil pan drain plug*
3 *Manual transaxle drain plug*
4 *Engine oil filter*
5 *Catalytic converter*
6 *Tie-rod end*
7 *Driveaxle boots*
8 *Stabilizer bar*
9 *Engine cooling fans*
10 *Manual transaxle check/fill plug*

2 Introduction

This Chapter is designed to help the home mechanic maintain the Volkswagen New Beetle vehicle with the goals of maximum performance, economy, safety and reliability in mind.

Included is a master maintenance schedule, followed by procedures dealing specifically with each item on the schedule. Visual checks, adjustments, component replacement and other helpful items are included. Refer to the accompanying illustrations of the engine compartment and the underside of the vehicle for the locations of various components.

Servicing your vehicle in accordance with the mileage/time maintenance schedule and the step-by-step procedures will result in a planned maintenance program that should produce a long and reliable service life. Keep in mind that it's a comprehensive plan, so maintaining some items but not others at the specified intervals will not produce the same results.

As you service your vehicle, you will discover that many of the procedures can - and should - be grouped together because of the nature of the particular procedure you're performing or because of the close proximity of two otherwise unrelated components to one another.

For example, if the vehicle is raised for chassis lubrication, you should inspect the exhaust, suspension, steering and fuel systems while you're under the vehicle. When you're rotating the tires, it makes good sense to check the brakes since the wheels are already removed. Finally, let's suppose you have to borrow or rent a torque wrench. Even if you only need it to tighten the spark plugs, you might as well check the torque of as many critical fasteners as time allows.

The first step in this maintenance program is to prepare yourself before the actual work begins. Read through all the procedures you're planning to do, then gather up all the parts and tools needed. If it looks like you might run into problems during a particular job, seek advice from a mechanic or an experienced do-it-yourselfer.

Typical rear underside components

1 *Shock absorber*
2 *Coil spring*
3 *Brake caliper*
4 *Rear axle assembly*
5 *Fuel tank*
6 *Fuel filter (gasoline models)*
7 *Parking brake cable*
8 *Muffler*

3 Tune-up - general information

The term tune-up is used in this manual to represent a combination of individual operations rather than one specific procedure that will maintain a gasoline engine in proper tune.

If, from the time the vehicle is new, the routine maintenance schedule is followed closely and frequent checks are made of fluid levels and high wear items, as suggested throughout this manual, the engine will be kept in relatively good running condition and the need for additional work will be minimized.

More likely than not, however, there may be times when the engine is running poorly due to lack of regular maintenance. This is even more likely if a used vehicle, which has not received regular and frequent maintenance checks, is purchased. In such cases, an engine tune-up will be needed outside of the regular routine maintenance intervals.

The first step in any tune-up or diagnostic procedure to help correct a poor-running engine is a cylinder compression check. A compression check (see Chapter 2C) will help determine the condition of internal engine components and should be used as a guide for tune-up and repair procedures. If, for instance, the compression check indicates serious internal engine wear, a conventional tune-up won't improve the performance of the engine and would be a waste of time and money. Because of its importance, the compression check should be done by someone with the right equipment and the knowledge to use it properly.

The following procedures are those most often needed to bring a generally poor-running engine back into a proper state of tune.

MINOR TUNE-UP

Check all engine related fluids (Section 4)
Clean, inspect and test the battery Section 10)
Check and adjust the drivebelt (Section 11)
Check all underhood hoses (Section 12)
Check the cooling system (Section 13)
Check the air filter (Section 24)
Replace the spark plugs (Section 26)
Inspect the spark plug wires (Section 31)

MAJOR TUNE-UP

All items listed under Minor tune-up, plus . . .
Replace the air filter (Section 24)
Replace the spark plug wires (Section 31)
Check the ignition system (Chapter 5)
Check the charging system (Chapter 5)

4 Fluid level checks (every 250 miles or weekly)

➡Note: The following are fluid level checks to be done on a 250 mile or weekly basis. Additional fluid level checks can be found in specific maintenance procedures that follow. Regardless of intervals, be alert to fluid leaks under the vehicle, which would indicate a fault to be corrected immediately.

1 Fluids are an essential part of the lubrication, cooling, brake, clutch and windshield washer systems. Because the fluids gradually become depleted and/or contaminated during normal operation of the vehicle, they must be periodically replenished. See *Recommended lubricants and fluids* at the end of this Chapter before adding fluid to any of the following components.

4.2 The engine oil dipstick (arrow) is clearly marked

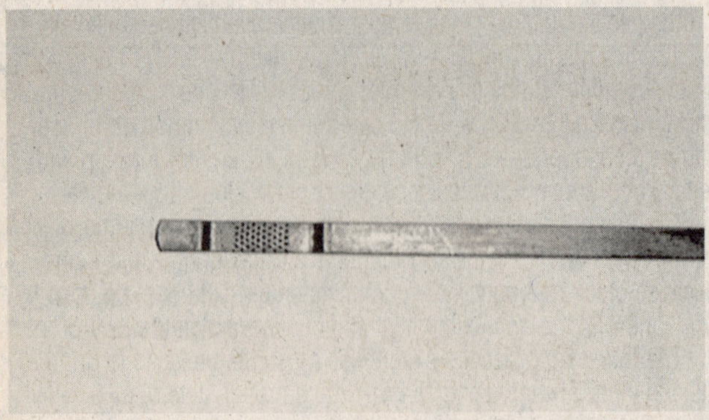

4.4 The oil level must be maintained between the marks at all times - it takes one quart of oil to raise the level from the MIN to MAX mark

➡Note: The vehicle must be on level ground when fluid levels are checked.

ENGINE OIL

➧ Refer to illustrations 4.2, 4.4 and 4.6

2 The engine oil level is checked with a dipstick that extends through a tube and into the oil pan at the bottom of the engine (see illustration).

3 The oil level should be checked before the vehicle has been driven, or about 5 minutes after the engine has been shut off. If the oil is checked immediately after driving the vehicle, some of the oil will remain in the upper engine components, resulting in an inaccurate reading on the dipstick.

4 Pull the dipstick out of the tube and wipe all the oil from the end with a clean rag or paper towel. Insert the clean dipstick all the way back into the tube, then pull it out again. Note the oil at the end of the dipstick. Add oil as necessary to keep the level in the shaded area on the dipstick (see illustration).

5 Do not overfill the engine by adding too much oil since this may result in oil-fouled spark plugs, catalytic converter failure, oil leaks or oil seal failures.

6 Oil is added to the engine after unscrewing a cap from the valve cover (see illustration). A funnel may help to reduce spills.

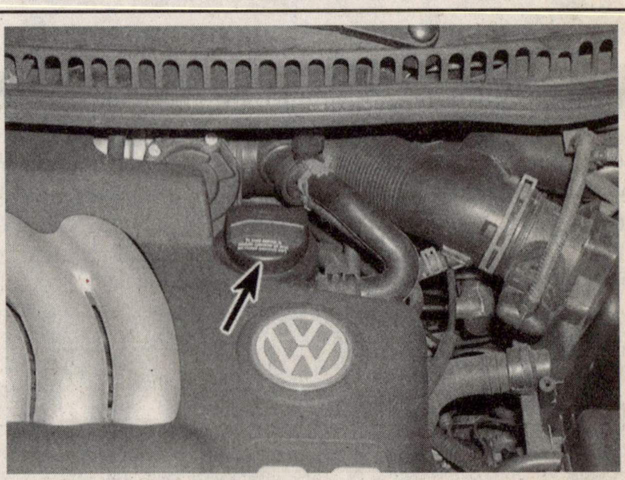

4.6 Oil is added to the engine after unscrewing the oil filler cap (arrow) - always make sure the area around the opening is clean before removing the cap to prevent dirt from contaminating the engine

7 Checking the oil level is an important preventive maintenance step. A consistently low oil level indicates oil leakage through damaged seals, defective gaskets or past worn rings or valve guides. If the oil looks milky or has water droplets in it, the cylinder head gasket may be blown or the head or block may be cracked. The engine should be checked immediately. The condition of the oil should also be checked. Whenever you check the oil level, slide your thumb and index finger up the dipstick before wiping off the oil. If you see small dirt or metal particles clinging to the dipstick, the oil should be changed (see Section 7).

ENGINE COOLANT

Refer to illustration 4.9

⁂ WARNING:

Do not allow antifreeze to come in contact with your skin or painted surfaces of the vehicle. Rinse off spills immediately with plenty of water. Antifreeze is highly toxic if ingested. Never leave antifreeze lying around in an open container or in puddles on the floor; children and pets are attracted by its sweet smell and may drink it. Check with local authorities on disposing of used antifreeze. Many communities have collection centers that will see that antifreeze is disposed of safely.

⁂ CAUTION:

Never mix green-colored ethylene glycol antifreeze and red-colored silicate and phosphate-free coolant because doing so will destroy the efficiency of the red coolant which is designed to last for 100,000 miles or five years.

➡Note: Non-toxic antifreeze is now manufactured and available at local auto parts stores, but even this type should be disposed of properly.

8 All vehicles covered by this manual are equipped with a coolant expansion tank, located at the right side of the engine compartment, and connected by hoses to the radiator and cooling system.

9 The coolant level in the expansion tank should be checked regularly.

⁂ WARNING:

Do not remove the pressure cap to check the coolant level when the engine is warm.

The level of coolant in the expansion tank varies with the temperature of the engine. When the engine is cold, the coolant level should be at or slightly above the MIN mark on the tank (see illustration). Once the engine has warmed up, the level should be at or near the MAX mark. If it isn't, add coolant to the expansion tank. To add coolant, simply twist open the cap and add a 50/50 mixture of antifreeze and water (see Caution).

10 Drive the vehicle and recheck the coolant level. If only a small amount of coolant is required to bring the system up to the proper level, water can be used. However, repeated additions of water will dilute the antifreeze and water solution. In order to maintain the proper ratio of antifreeze and water, always top up the coolant level with the correct mixture. An empty plastic milk jug or bleach bottle makes an excellent

4.9 The coolant expansion tank is located on the right (passenger's) side of the engine compartment - keep the level near the MAX mark (arrow) or MIN mark on the side of the reservoir, depending on engine temperature

container for mixing coolant. Do not use rust inhibitors or additives.

11 If the coolant level drops consistently, there may be a leak in the system. Inspect the radiator, hoses, filler cap, drain plugs and water pump (see Section 13). If no leaks are noted, have the pressure cap tested by a service station.

12 If you have to remove the pressure cap, wait until the engine has cooled completely, then wrap a thick cloth around the cap and unscrew it slowly. If coolant or steam escapes, let the engine cool down longer, then remove the cap.

13 Check the condition of the coolant as well. It should be relatively clear. If it is brown or rust colored, the system should be drained, flushed and refilled. Even if the coolant appears to be normal, the corrosion inhibitors wear out, so it must be replaced at the specified intervals. If the system is filled with standard green coolant/water, it must be flushed and replaced more frequently than if the original silicate and phosphate-free red coolant is retained.

WINDSHIELD WASHER FLUID

Refer to illustration 4.14

14 Fluid for the windshield washer system is located in a plastic reservoir in the right side of engine compartment (see illustration).

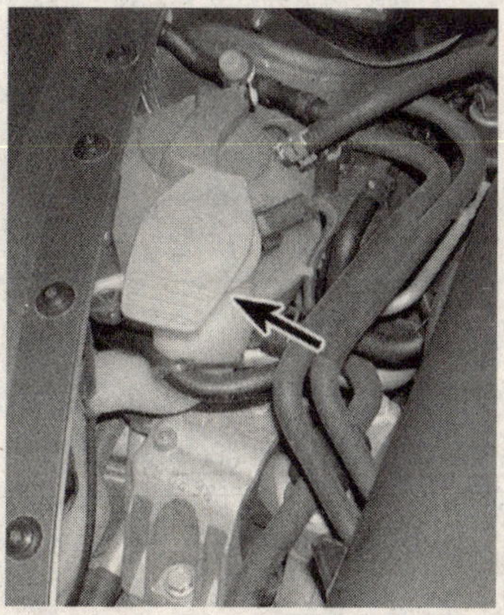

4.14 Flip open the cap (arrow) to check the fluid level in the windshield washer tank

4.17 Lift up the battery cover and remove the screw-in caps (arrows) to check the battery electrolyte level

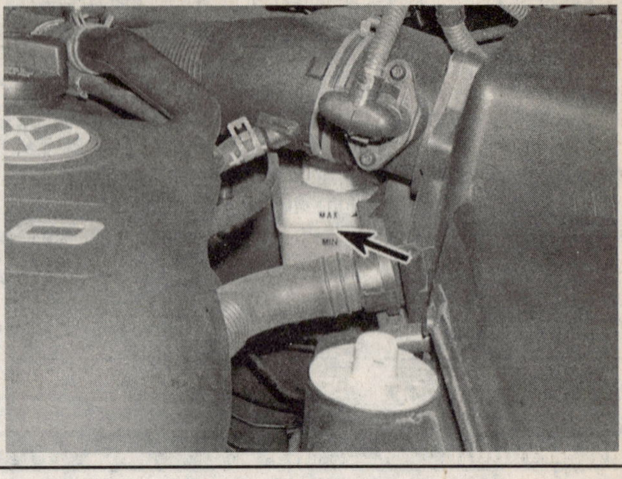

4.19 Never let the brake fluid level drop below the MIN mark (arrow)

15 In milder climates, plain water can be used in the reservoir, but it should be kept no more than 2/3 full to allow for expansion if the water freezes. In colder climates, use windshield washer system antifreeze, available at any auto parts store, to lower the freezing point of the fluid. Mix the antifreeze with water in accordance with the manufacturer's directions on the container.

**** CAUTION:**

Don't use cooling system antifreeze - it will damage the vehicle's paint.

16 To help prevent icing in cold weather, warm the windshield with the defroster before using the washer.

BATTERY ELECTROLYTE

➧ Refer to illustration 4.17

17 These vehicles are equipped with a battery which has a translucent plastic case. You should be able to read the electrolyte level by looking at the side of the case, and comparing the level to the Minimum and Maximum markings on the case, but most models have the battery mounted in a black plastic box. Instead, read the fluid level by removing the screw-in plastic caps on top of the battery (see illustration). This check is most critical during the warm summer months. Add only distilled water to any battery, and fill only to the ledge below the filler caps.

➡Note: The filler caps are equipped with O-rings. Make sure the O-rings are in place when reinstalling the caps.

BRAKE AND CLUTCH FLUID

➧ Refer to illustration 4.19

18 The brake master cylinder is mounted on the upper left of the engine compartment firewall. The clutch master cylinder on manual transaxle vehicles is mounted at the clutch pedal assembly, but receives its hydraulic fluid directly from the brake master cylinder reservoir, via a hose. Checking the fluid at the brake master cylinder reservoir effectively checks the level of brake and clutch fluid.

19 The translucent plastic reservoir allows the fluid inside to be checked without removing the cap (see illustration). Be sure to wipe the top of the reservoir cap with a clean rag to prevent contamination of the brake and/or clutch system before removing the cover.

20 When adding fluid, pour it carefully into the reservoir to avoid spilling it on surrounding painted surfaces. Be sure the specified fluid is used, since mixing different types of brake fluid can cause damage to the system. See *Recommended lubricants and fluids* at the end of this Chapter or your owner's manual.

**** WARNING:**

Brake fluid can harm your eyes and damage painted surfaces, so use extreme caution when handling or pouring it. Do not use brake fluid that has been standing open or is more than one year old. Brake fluid absorbs moisture from the air. Moisture in the system can cause a dangerous loss of brake performance.

21 At this time, the fluid and master cylinder can be inspected for contamination. The system should be drained and refilled if deposits, dirt particles or water droplets are seen in the fluid.

22 After filling the reservoir to the proper level, make sure the cap is on tight to prevent fluid leakage.

23 The brake fluid level in the master cylinder will drop slightly as the pads at the front wheels wear down during normal operation. If the master cylinder requires repeated additions to keep it at the proper level, it's an indication of leakage in the brake system or clutch release system, which should be corrected immediately. Check all brake lines and connections (see Section 17 for more information). Inspect the clutch release system, too (see Chapter 8).

24 If, upon checking the master cylinder fluid level, you discover the reservoir empty or nearly empty, the brake system should be bled and thoroughly inspected (see Chapter 9). Inspect the clutch release system, too (see Chapter 8).

5 Tire and tire pressure checks (every 250 miles or weekly)

➧ Refer to illustrations 5.2, 5.3, 5.4a, 5.4b, 5.8a and 5.8b

1 Periodic inspection of the tires may spare you the inconvenience of being stranded with a flat tire. It can also provide you with vital information regarding possible problems in the steering and suspension systems before major damage occurs.

2 The original tires on this vehicle are equipped with 1/2-inch wide wear bands that will appear when tread depth reaches 1/16-inch, at which point the tires can be considered worn out. Tread wear can be monitored with a simple, inexpensive device known as a tread depth indicator (see illustration).

3 Note any abnormal tread wear (see illustration). Tread pattern irregularities such as cupping, flat spots and more wear on one side than the other are indications of front end alignment and/or balance problems. If any of these conditions are noted, take the vehicle to a tire shop or service station to correct the problem.

4 Look closely for cuts, punctures and embedded nails or tacks. Sometimes a tire will hold air pressure for a short time or leak down very slowly after a nail has embedded itself in the tread. If a slow leak persists, check the valve stem core to make sure it's tight (see illustration). Examine the tread for an object that may have embedded itself in the tire or for a "plug" that may have begun to leak (radial tire punctures are repaired with a plug that's installed in a puncture). If a puncture is suspected, it can be easily verified by spraying a solution of soapy water onto the puncture area (see illustration). The soapy solution will bubble if there's a leak. Unless the puncture is unusually large, a tire shop or service station can usually repair the tire.

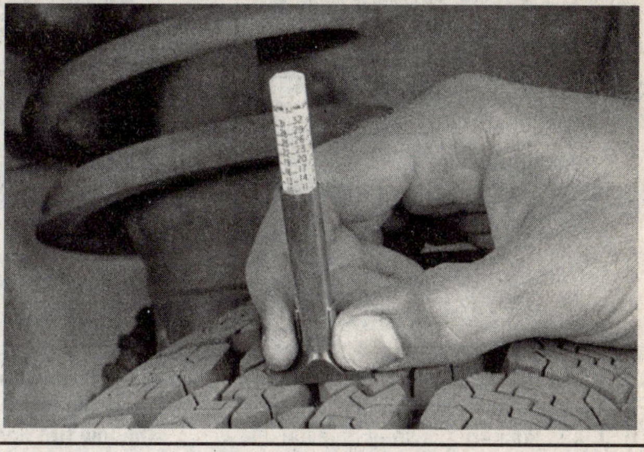

5.2 Use a tire tread depth indicator to monitor tire wear - they are available at auto parts stores and service stations and cost very little

5 Carefully inspect the inner sidewall of each tire for evidence of brake fluid leakage. If you see any, inspect the brakes immediately.

6 Correct air pressure adds miles to the lifespan of the tires, improves mileage and enhances overall ride quality. Tire pressure cannot be accurately estimated by looking at a tire, especially if it's a radial. A tire pressure gauge is essential. Keep an accurate gauge in the

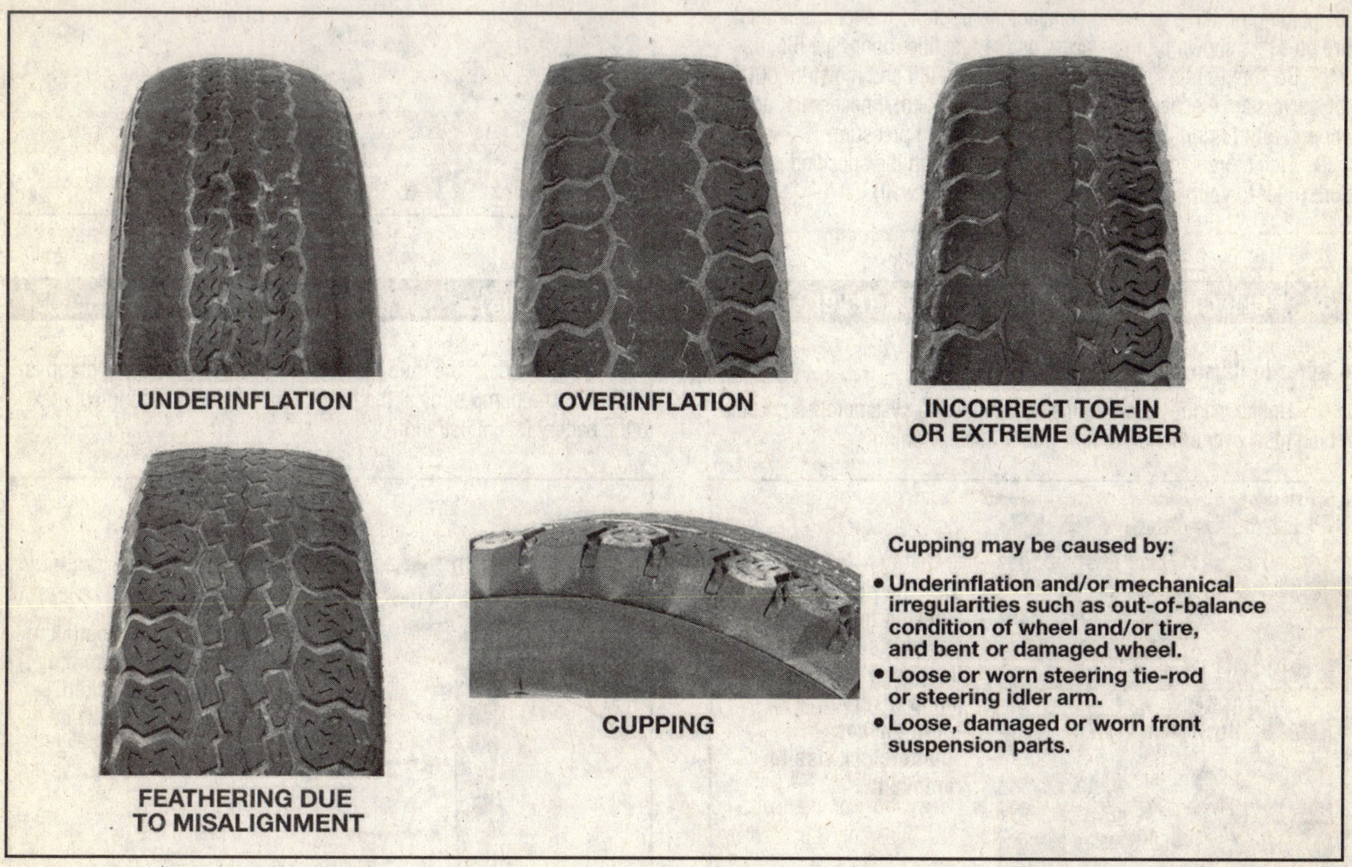

5.3 This chart will help you determine the condition of the tires and the probable cause(s) of abnormal wear

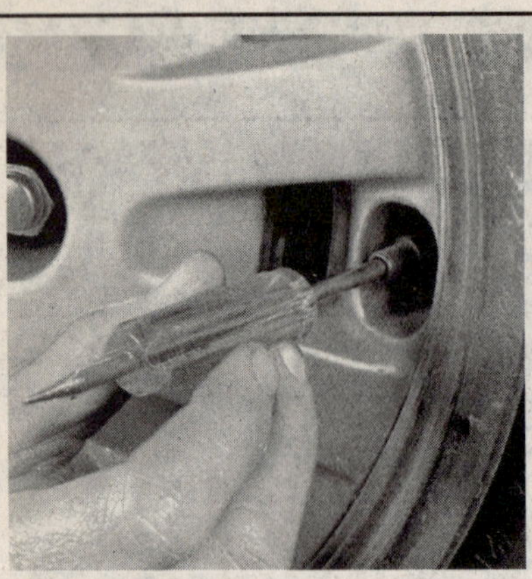

5.4a If a tire loses air on a steady basis, check the valve stem core first to make sure it's snug (special inexpensive wrenches are commonly available at auto parts stores)

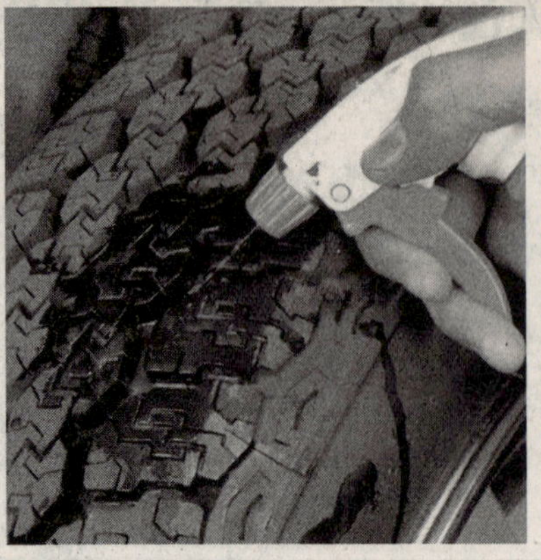

5.4b If the valve stem core is tight, raise the corner of the vehicle with the low tire and spray a soapy water solution onto the tread as the tire is turned slowly - leaks will cause small bubbles to appear

5.8a To extend the life of the tires, check the air pressure at least once a week with an accurate gauge (don't forget the spare!)

vehicle. The pressure gauges attached to the nozzles of air hoses at gas stations are often inaccurate.

7 Always check tire pressure when the tires are cold. Cold, in this case, means the vehicle has not been driven over a mile in the three hours preceding a tire pressure check. A pressure rise of four to eight pounds is not uncommon once the tires are warm.

8 Unscrew the valve cap protruding from the wheel or hubcap and push the gauge firmly onto the valve stem (see illustration). Note the reading on the gauge and compare the figure to the recommended tire pressure shown on the placard on the fuel filler door (see illustration). Be sure to reinstall the valve cap to keep dirt and moisture out of the valve stem mechanism. Check all four tires and, if necessary, add enough air to bring them up to the recommended pressure.

9 Don't forget to keep the spare tire inflated to the specified pressure (refer to your owner's manual or the tire sidewall).

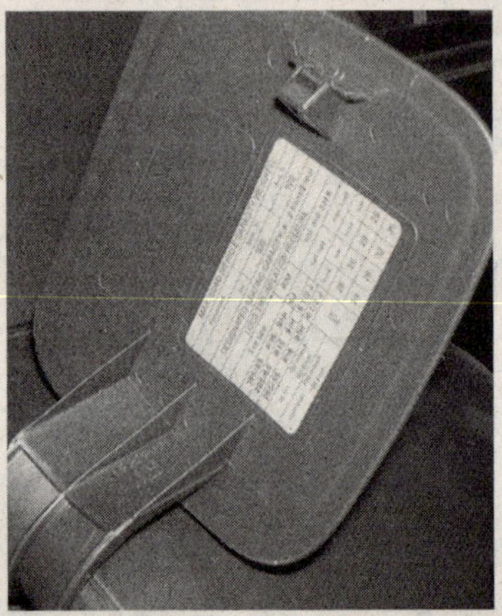

5.8b The tire pressure chart is located on the fuel filler lid

6 Power steering fluid level check (every 3000 miles or 3 months)

Refer to illustrations 6.2 and 6.6

1 Unlike manual steering, the power steering system relies on fluid which may, over a period of time, require replenishing.

2 On all models, the fluid reservoir for the power steering pump is located on the pump body at the front of the engine compartment, next to the battery (see illustration).

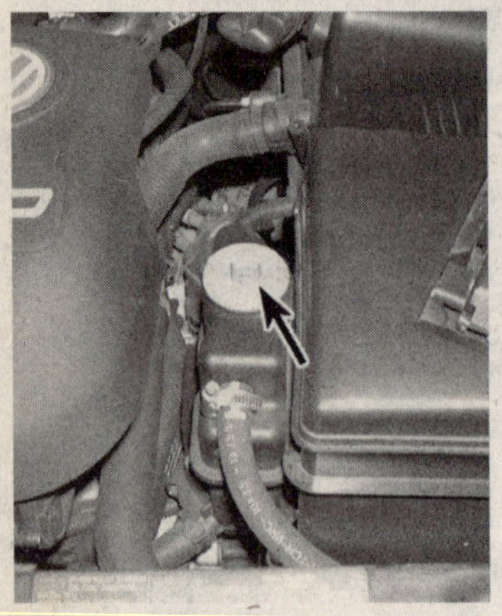

6.2 The power steering fluid dipstick (arrow) is located in the power steering pump reservoir - turn the cap counterclockwise to remove it

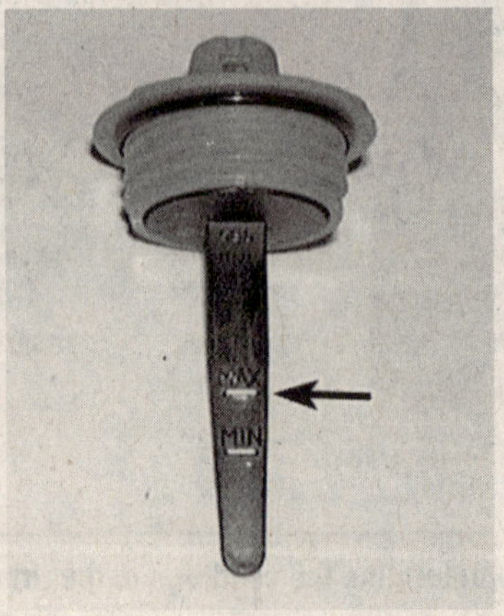

6.6 The power steering fluid dipstick has marks on it so the fluid can be checked hot (arrow at MAX) or cold (MIN)

3 For the check, the front wheels should be pointed straight ahead and the engine should be off.

4 Use a clean rag to wipe off the reservoir cap and the area around the cap. This will help prevent any foreign matter from entering the reservoir during the check.

5 Twist off the cap and check the temperature of the fluid at the end of the dipstick with your finger.

6 Wipe off the fluid with a clean rag, reinsert the dipstick, then withdraw it and read the fluid level. The fluid should be at the proper level, depending on whether it was checked hot or cold (see illustration). Never allow the fluid level to drop below the lower mark on the dipstick.

7 If additional fluid is required, pour the specified type directly into the reservoir, using a funnel to prevent spills.

8 If the reservoir requires frequent fluid additions, all power steering hoses, hose connections, steering gear and the power steering pump should be carefully checked for leaks.

7 Engine oil and filter change (every 3000 miles or 3 months)

➧ Refer to illustrations 7.3, 7.8, 7.10, 7.15a, 7.15b and 7.19

1 Frequent oil changes are the most important preventive maintenance procedures that can be done by the home mechanic. As engine oil ages, it becomes diluted and contaminated, which leads to premature engine wear.

2 Although some sources recommend oil filter changes every other oil change, we feel that the minimal cost of an oil filter and the relative ease with which it is installed dictate that a new filter be installed every time the oil is changed.

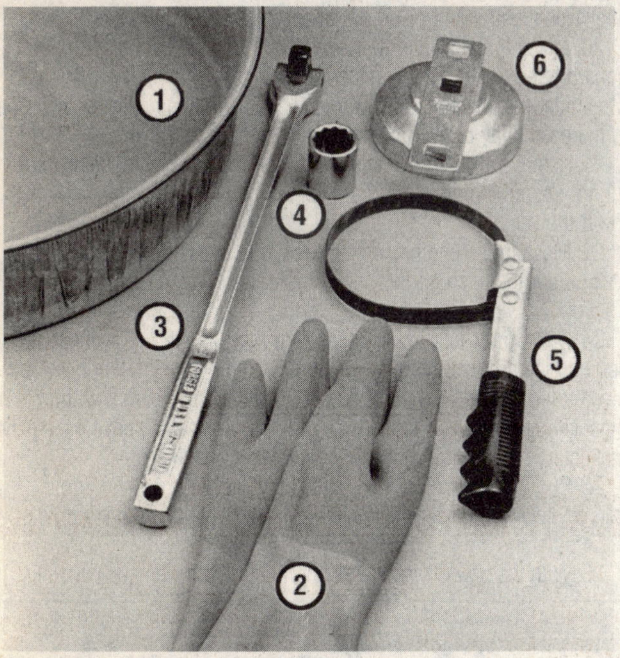

7.3 These tools are required when changing the engine oil and filter

1 ***Drain pan*** *- It should be fairly shallow in depth, but wide to prevent spills*

2 ***Rubber gloves*** *- When removing the drain plug and filter, you will get oil on your hands (the gloves will prevent burns)*

3 ***Breaker bar*** *- Sometimes the oil drain plug is tight, and a long breaker bar is needed to loosen it*

4 ***Socket*** *- To be used with the breaker bar or a ratchet (must be the correct size to fit the drain plug - six-point preferred)*

5 ***Filter wrench*** *- This is a metal band-type wrench, which requires clearance around the filter to be effective*

6 ***Filter wrench*** *- This type fits on the bottom of the filter and can be turned with a ratchet or breaker bar (different-size wrenches are available for different types of filters)*

3 Gather together all necessary tools and materials before beginning this procedure (see illustration).

4 You should have plenty of clean rags and newspapers handy to mop up any spills. Access to the under side of the vehicle may be improved if the vehicle can be lifted on a hoist, driven onto ramps or supported by jackstands.

⁂ WARNING:

Do not work under a vehicle which is supported only by a bumper, hydraulic or scissors-type jack.

5 If this is your first oil change, familiarize yourself with the locations of the oil drain plug and the oil filter.

6 Warm the engine to normal operating temperature. If the new oil or any tools are needed, use this warm-up time to gather everything necessary for the job. The correct type of oil for your application can be found in *Recommended lubricants and fluids* at the end of this Chapter.

7 With the engine oil warm (warm engine oil will drain better and more built-up sludge will be removed with it), raise and support the vehicle. Make sure it's safely supported!

8 Remove the splash shield under the engine (see illustration).

9 Move all necessary tools, rags and newspapers under the vehicle. Set the drain pan under the drain plug. Keep in mind that the oil will initially flow from the pan with some force; position the pan accordingly.

10 Being careful not to touch any of the hot exhaust components, use a wrench to remove the drain plug near the bottom of the oil pan

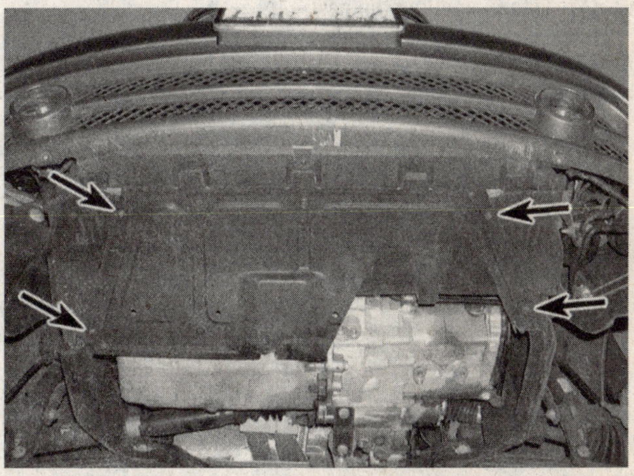

7.8 Remove the screws (arrows) and lower the engine splash shield to access the oil filter

7.10 Use the proper size box-end wrench or socket to remove the oil drain plug to avoid rounding it off

(see illustration). Depending on how hot the oil is, you may want to wear gloves while unscrewing the plug the final few turns.

11 Allow the oil to drain into the pan. It may be necessary to move the pan as the oil flow slows to a trickle.

12 After all the oil has drained, wipe off the drain plug with a clean rag. Small metal particles may cling to the plug and would immediately contaminate the new oil.

13 Clean the area around the drain plug opening and reinstall the plug. Tighten the plug securely with the wrench.

14 Move the drain pan into position under the oil filter.

15 Use the oil filter wrench to loosen the oil filter (see illustrations).

16 Completely unscrew the old filter. Be careful; it's full of oil. Empty the oil inside the filter into the drain pan, then lower the filter.

17 Compare the old filter with the new one to make sure they're the same type.

18 Use a clean rag to remove all oil, dirt and sludge from the area where the oil filter mounts to the engine. Check the old filter to make sure the rubber gasket isn't stuck to the engine. If the gasket is stuck to

7.15b On diesel engines, unscrew the filter top (1) and remove the filter element (4) - when installing the new filter, use new rubber seals (2 and 3)

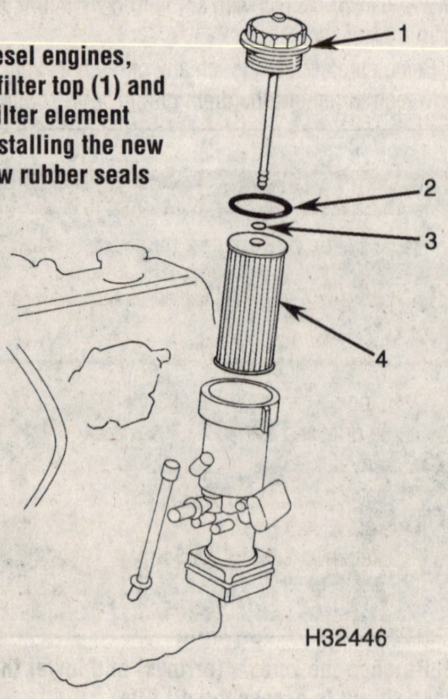

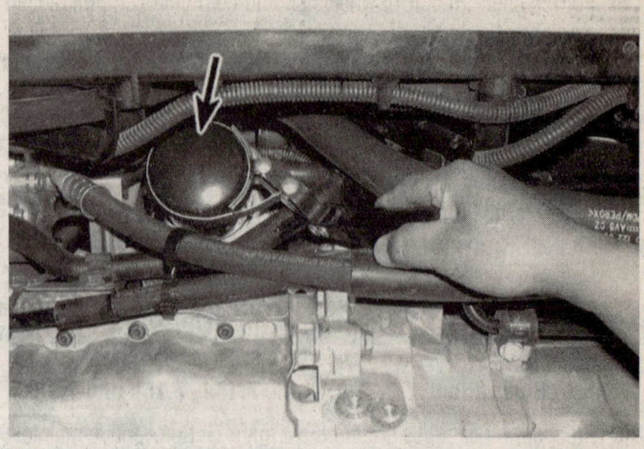

7.15a Since the oil filter (arrow) is on very tight, you'll need a special wrench for removal - DO NOT use the wrench to tighten the new filter (gasoline model shown)

the engine (use a flashlight if necessary), remove it.

19 Apply a light coat of clean oil to the rubber gasket on the new oil filter (see illustration), or on diesel engines, apply oil to the O-rings.

20 Attach the new filter to the engine, following the tightening directions printed on the filter canister or packing box (gasoline engines). Most filter manufacturers recommend against using a filter wrench due to the possibility of overtightening and damage to the seal.

21 Remove all tools, rags, etc. from under the vehicle, being careful not to spill the oil in the drain pan, reinstall the splash shield, then lower the vehicle.

22 Move to the engine compartment and locate the oil filler cap.

23 Pour the fresh oil through the filler opening. A funnel may be helpful.

24 Refer to the engine oil capacity in this Chapter's Specifications and add the proper amount of fresh oil into the engine. Wait a few minutes to allow the oil to drain into the pan, then check the level on the oil dipstick (see Section 4 if necessary). If the oil level is above the hatched area, start the engine and allow the new oil to circulate.

25 Run the engine for only about a minute and then shut it off. Immediately look under the vehicle and check for leaks at the oil pan drain plug and around the oil filter.

26 With the new oil circulated and the filter now completely full,

7.19 Lubricate the oil filter gasket with clean engine oil before installing the filter on the engine

recheck the level on the dipstick and add more oil as necessary.

27 During the first few trips after an oil change, make it a point to check frequently for leaks and proper oil level.

28 The old oil drained from the engine cannot be reused in its present state and should be disposed of. Check with your local auto parts store, disposal facility or environmental agency to see if they will accept the oil for recycling. After the oil has cooled it can be drained into a container (capped plastic jugs, topped bottles, milk cartons, etc.) for transport to one of these disposal sites. Don't dispose of the oil by pouring it on the ground or down a drain!

8 Seat belt check (every 6000 miles or 6 months)

1 Check seat belts, buckles, latch plates and guide loops for obvious damage and signs of wear.

2 Where the seat belt receptacle bolts to the floor of the vehicle, check that the bolts are secure.

3 See if the seat belt reminder light comes on when the key is turned to the Run or Start position. A chime should also sound.

WARNING:

The lap and shoulder belt adjusters incorporate a small explosive device that is triggered by the airbag system. The adjusters automatically tighten up the seat belts in case of a collision sufficient to set off the airbags. Do not use a hammer or impact driver of any kind near the seat belt or airbag module unless the airbag system has been disabled (see Chapter 12).

9 Wiper blade inspection and replacement (every 6000 miles or 6 months)

Refer to illustration 9.3

1 The windshield wiper blade elements should be checked periodically for cracks and deterioration.

2 Lift the wiper blade assembly away from the glass.

3 Press the release lever and slide the blade assembly out of the hook in the end of the wiper arm (see illustration).

4 Squeeze the two rubber prongs at the end of the blade element, then slide the element out of the frame.

Note: These elements can be replaced by hand, without pliers.

5 Compare the new element with the old for length, design, etc. Some replacement elements come in a three-piece design (two metal strips, one on either side of the rubber) that is held together by several small plastic sleeves. Keep the sleeves in place on this design until you start sliding the element into the frame. Remove each of the plastic sleeves as needed when they reach the frame.

6 Slide the new element into the frame, notched end last and secure the clips into the notches of the frame.

7 Reinstall the blade assembly on the arm, wet the windshield and test for proper operation.

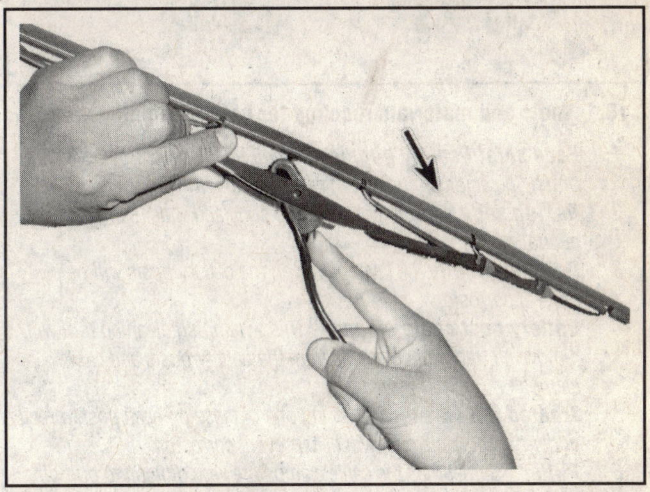

9.3 Depress the release lever (finger is on it here) and slide the wiper assembly down the wiper arm and out of the hook in the end of the arm

10 Battery check, maintenance and charging (every 6000 miles or 6 months)

Refer to illustrations 10.1, 10.5, 10.7a and 10.7b

WARNING:

Certain precautions must be followed when checking and servicing the battery. Hydrogen gas, which is highly flammable, is always present in the battery cells, so keep lighted tobacco and all other open flames and sparks away from the battery. The electrolyte inside the battery is actually dilute sulfuric acid, which will cause injury if splashed on your skin or in your eyes. It will also ruin clothes and painted surfaces. When removing the battery cables, always detach the negative cable first and hook it up last!

CAUTION 1:

These models are equipped with an anti-theft radio. Before performing a procedure that requires disconnecting the battery, make sure you have the proper activation code.

CAUTION 2:

Disconnecting the battery can cause driveability problems that require a scan tool to remedy. See Chapter 5, Section 1 for the use of an auxiliary voltage input device before disconnecting the battery.

1 A routine preventive maintenance program for the battery in your vehicle is the only way to ensure quick and reliable starts. But before performing any battery maintenance, make sure that you have the proper equipment necessary to work safely around the battery (see illustration).

2 There are also several precautions that should be taken whenever battery maintenance is performed. Before servicing the battery, always turn the engine and all accessories off and disconnect the cable from the negative terminal of the battery.

3 The battery produces hydrogen gas, which is both flammable and explosive. Never create a spark, smoke or light a match around the bat-

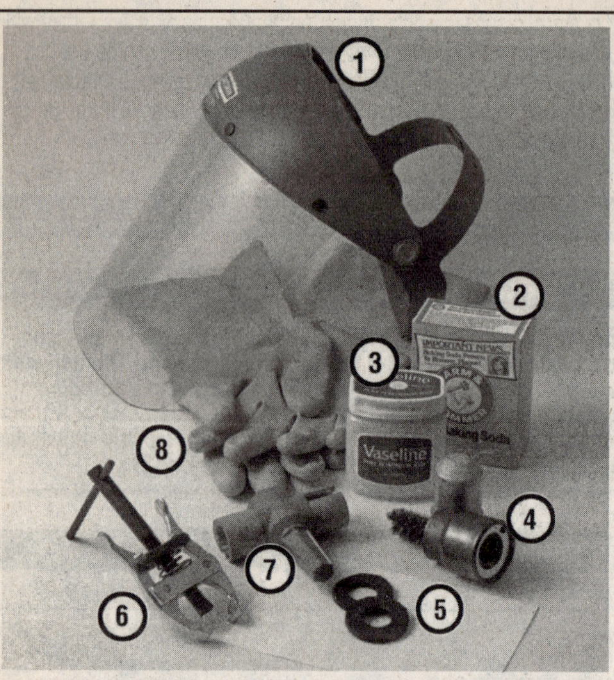

10.1 Tools and materials required for battery maintenance

1 ***Face shield/safety goggles*** *- When removing corrosion with a brush, the acidic particles can easily fly up into your eyes*
2 ***Baking soda*** *- A solution of baking soda and water can be used to neutralize corrosion*
3 ***Petroleum jelly*** *- A layer of this on the battery posts will help prevent corrosion*
4 ***Battery post/cable cleaner*** *- This wire brush cleaning tool will remove all traces of corrosion from the battery posts and cable clamps*
5 ***Treated felt washers*** *- Placing one of these on each post, directly under the cable clamps, will help prevent corrosion*
6 ***Puller*** *- Sometimes the cable clamps are very difficult to pull off the posts, even after the nut/bolt has been completely loosened. This tool pulls the clamp straight up and off the post without damage*
7 ***Battery post/cable cleaner*** *- Here is another cleaning tool that is a slightly different version of Number 4 above, but it does the same thing*
8 ***Rubber gloves*** *- Another safety item to consider when servicing the battery; remember that's acid inside the battery!*

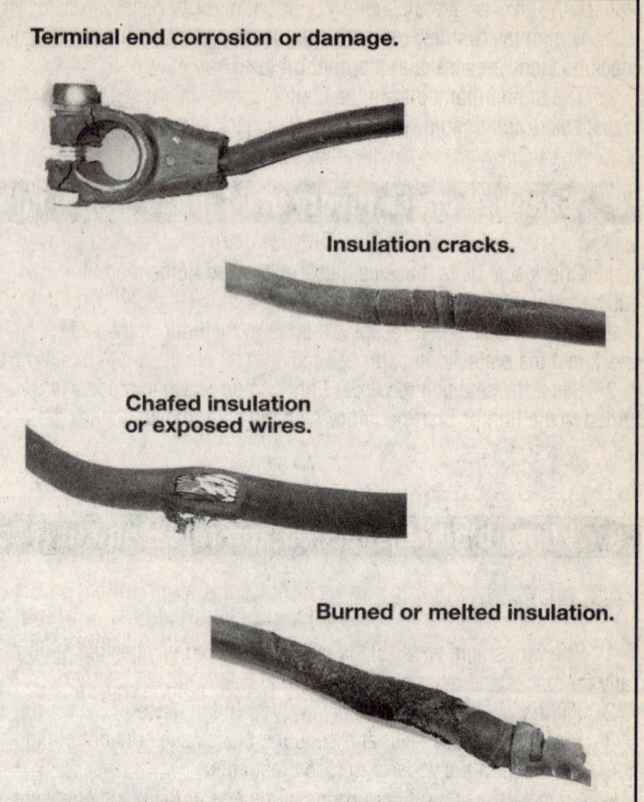

10.5 Typical battery cable problems

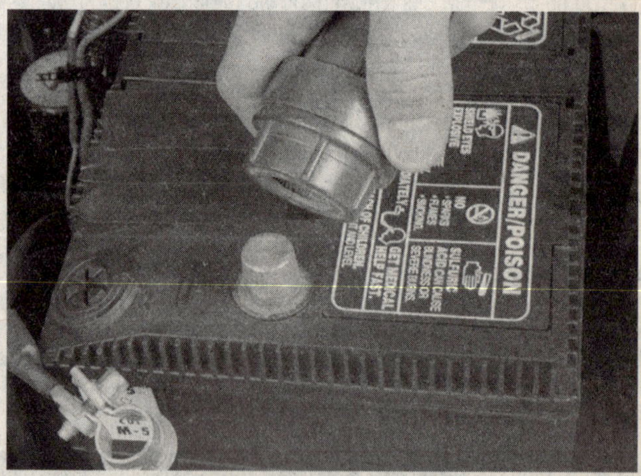

10.7a A tool like this one (available at auto parts stores) is used to clean the top-terminal type battery posts

tery. Always charge the battery in a ventilated area.

4 Electrolyte contains poisonous and corrosive sulfuric acid. Do not allow it to get in your eyes, on your skin or on your clothes. Never ingest it. Wear protective safety glasses when working near the battery. Keep children away from the battery.

5 Note the external condition of the battery. If the positive terminal and cable clamp on your vehicle's battery is equipped with a rubber protector, make sure that it's not torn or damaged. It should completely cover the terminal. Look for any corroded or loose connections, cracks in the case or cover or loose hold-down clamps. Also check the entire length of each cable for cracks and frayed conductors (see illustration).

6 If corrosion, which looks like white, fluffy deposits is evident, particularly around the terminals, the battery should be removed for cleaning. Loosen the cable bolts with a wrench, being careful to remove the ground cable first, and slide them off the terminals. Refer to Chapter 5 for specific battery removal procedures.

➡**Note: Read the Cautions in Section 1 before disconnecting the battery cables.**

7 Clean the cable ends thoroughly with a battery brush or a terminal cleaner and a solution of warm water and baking soda. Wash the terminals and the battery case with the same solution but make sure that the solution doesn't get into the battery. When cleaning the cables, terminals and battery case, wear safety goggles and rubber gloves to

prevent any solution from coming in contact with your eyes or hands. Wear old clothes too - even diluted, sulfuric acid splashed onto clothes will burn holes in them. If the terminals have been corroded, clean them up with a terminal cleaner (see illustrations). Thoroughly wash all cleaned areas with plain water.

8 Make sure that the battery tray is in good condition and the hold-down clamp bolts are tight. If the battery is removed from the tray, make sure no parts remain in the bottom of the tray when the battery is reinstalled. When reinstalling the hold-down clamp bolts, do not overtighten them.

9 Any metal parts of the vehicle damaged by corrosion should be covered with a zinc-based primer, then painted.

10 Information on removing and installing the battery can be found in Chapter 5. Information on jump starting can be found at the front of this manual.

CHARGING

⁂ WARNING:

When batteries are being charged, hydrogen gas, which is very explosive and flammable, is produced. Do not smoke or allow open flames near a charging or a recently charged battery. Wear eye protection when near the battery during charging. Also, make sure the charger is unplugged before connecting or disconnecting the battery from the charger.

➡Note: The manufacturer recommends the battery be removed from the vehicle for charging because the gas that escapes during this procedure can damage the paint. Fast charging with the battery cables connected can result in damage to the electrical system.

11 Slow-rate charging is the best way to restore a battery that's discharged to the point where it will not start the engine. It's also a good way to maintain the battery charge in a vehicle that's only driven a few miles between starts. Maintaining the battery charge is particularly important in the winter when the battery must work harder to start the engine and electrical accessories that drain the battery are in greater use.

12 It's best to use a one or two-amp battery charger (sometimes called a "trickle" charger). They are the safest and put the least strain on the battery. They are also the least expensive. For a faster charge, you can use a higher amperage charger, but don't use one rated more than 1/10th the amp/hour rating of the battery. Rapid boost charges that claim to restore the power of the battery in one to two hours are hardest on the battery and can damage batteries not in good condition. This type of charging should only be used in emergency situations.

13 The average time necessary to charge a battery should be listed in the instructions that come with the charger. As a general rule, a trickle charger will charge a battery in 12 to 16 hours.

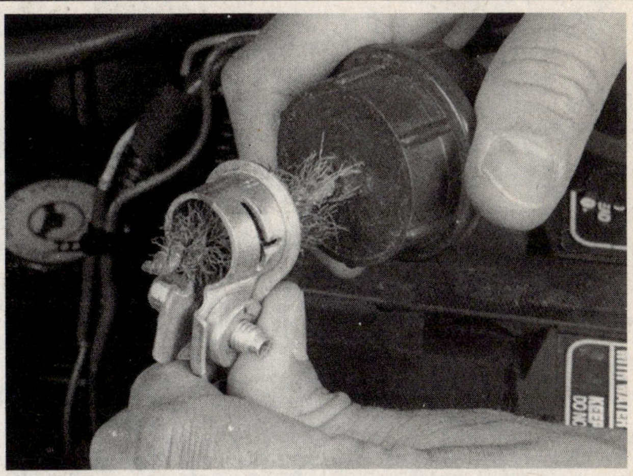

10.7b Use the brush side of the tool to clean inside the cable end

14 Remove all the cell caps (if equipped) and cover the holes with a clean cloth to prevent spattering electrolyte. Disconnect the negative battery cable and hook the battery charger cable clamps up to the battery posts (positive to positive, negative to negative), then plug in the charger. Make sure it is set at 12-volts if it has a selector switch.

15 If you're using a charger with a rate higher than two amps, check the battery regularly during charging to make sure it doesn't overheat. If you're using a trickle charger, you can safely let the battery charge overnight after you've checked it regularly for the first couple of hours.

16 If the battery has removable cell caps, measure the specific gravity with a hydrometer every hour during the last few hours of the charging cycle. Hydrometers are available inexpensively from auto parts stores - follow the instructions that come with the hydrometer. Consider the battery charged when there's no change in the specific gravity reading for two hours and the electrolyte in the cells is gassing (bubbling) freely. The specific gravity reading from each cell should be very close to the others. If not, the battery probably has a bad cell(s).

17 Some batteries with sealed tops have built-in hydrometers on the top that indicate the state of charge by the color displayed in the hydrometer window. Normally, a bright-colored hydrometer indicates a full charge and a dark hydrometer indicates the battery still needs charging.

18 If the battery has a sealed top and no built-in hydrometer, you can hook up a digital voltmeter across the battery terminals to check the charge. A fully charged battery should read 12.5 volts or higher.

19 Further information on the battery and jump-starting can be found in Chapter 5 and at the front of this manual.

11 Drivebelt check, adjustment and replacement (every 6000 miles or 6 months)

➧ Refer to illustrations 11.2, 11.4, 11.5a and 11.5b

1 A serpentine drivebelt is located at the front of the engine and plays an important role in the overall operation of the engine and its components. Due to its function and material make up, the belt is prone to wear and should be periodically inspected. The serpentine belt drives the alternator, power steering pump, and air conditioning compressor, if equipped.

2 With the engine off, open the hood and use your fingers (and a flashlight, if necessary) to move along the belt, checking for cracks and separation of the belt plies. Also check for fraying and glazing, which gives the belt a shiny appearance (see illustration). Both sides of the belt should be inspected, which means you will have to twist the belt to check the underside.

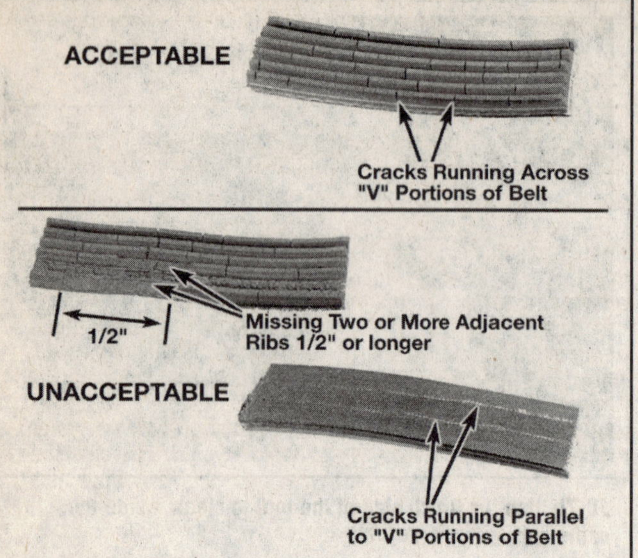

11.2 Check ribbed (serpentine) belts for signs of wear like these - if it looks worn, replace it

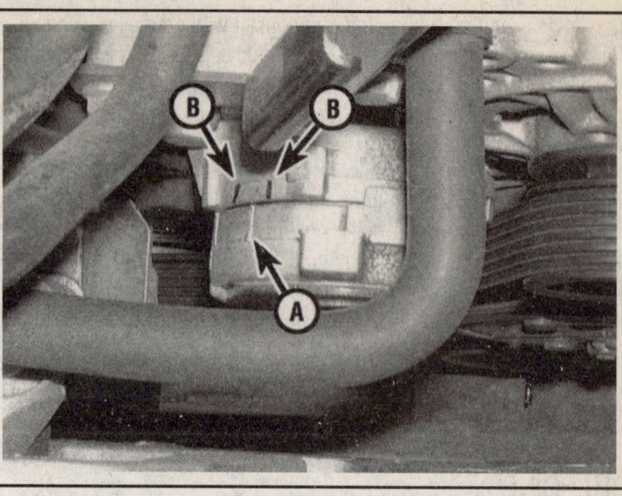

11.4 The indexing mark (A) on the belt tensioner must remain between the marks (B) on the tensioner assembly

3 Check the ribs on the underside of the belt. They should all be the same depth, with none of the surface uneven.

4 The tension of the belt is maintained by a spring-loaded tensioner assembly and isn't adjustable. The belt should be replaced when the indexing arrow is outside the range of the indexing marks on the tensioner assembly (see illustration).

5 To replace the belt, rotate the tensioner to release belt tension (see illustrations). On gasoline models, the tensioner can be locked in the released position by inserting a drill bit through the hole at the top of the tensioner.

➡**Note: On diesel engines, access to the drivebelt tensioner is from below, and you may have to remove the pipe between the intercooler and turbocharger (see Chapter 4B).**

6 Remove the belt from the tensioner and auxiliary components and slowly release the tensioner.

7 Route the new belt over the various pulleys, again rotating the tensioner to allow the belt to be installed, then release the belt tensioner.

TENSIONER REPLACEMENT

8 To replace a tensioner that doesn't fall into the proper tension range, even with a new belt, or that exhibits binding, remove the two mounting bolts.

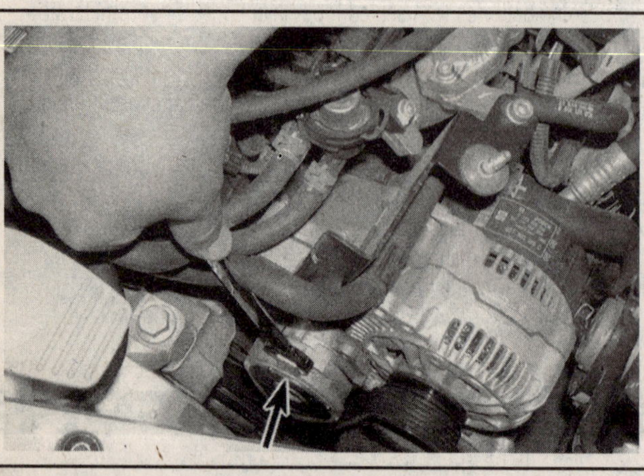

11.5a On gasoline models, the belt tensioner (arrow) is accessed from above - use a long 15mm wrench to rotate it for belt removal

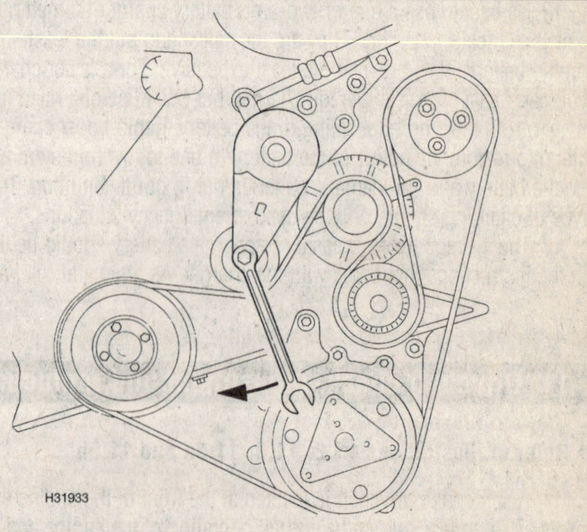

11.5b Use a box-end wrench from below to turn the tensioner for belt removal on diesel models

12 Underhood hose check and replacement (every 6000 miles or 6 months)

GENERAL

⁂ WARNING:

Replacement of air conditioning hoses must be left to a dealer service department or air conditioning shop that has the equipment to depressurize the system safely and recover the refrigerant. Never remove air conditioning components or hoses until the system has been depressurized.

1 High temperatures in the engine compartment can cause the deterioration of the rubber and plastic hoses used for engine, accessory and emission systems operation. Periodic inspection should be made for cracks, loose clamps, material hardening and leaks. Information specific to the cooling system hoses can be found in Section 14.

2 Some, but not all, hoses are secured to their fittings with clamps. Where clamps are used, check to be sure they haven't lost their tension, allowing the hose to leak. If clamps aren't used, make sure the hose has not expanded and/or hardened where it slips over the fitting, allowing it to leak.

VACUUM HOSES

3 It's quite common for vacuum hoses, especially those in the emissions system, to be color-coded or identified by colored stripes molded into them. Various systems require hoses with different wall thickness, collapse resistance and temperature resistance. When replacing hoses, be sure the new ones are made of the same material.

4 Often the only effective way to check a hose is to remove it completely from the vehicle. If more than one hose is removed, be sure to label the hoses and fittings to ensure correct installation.

5 When checking vacuum hoses, be sure to include any plastic T-fittings in the check. Inspect the fittings for cracks and the hose where it fits over the fitting for distortion, which could cause leakage.

6 A small piece of vacuum hose (1/4-inch inside diameter) can be used as a stethoscope to detect vacuum leaks. Hold one end of the hose to your ear and probe around vacuum hoses and fittings with the other end, listening for the "hissing" sound characteristic of a vacuum leak.

⁂ WARNING:

When probing with the vacuum hose stethoscope, be very careful not to come into contact with moving engine components such as the drivebelt, cooling fan, etc.

FUEL HOSE

⁂ WARNING:

Gasoline is extremely flammable, so take extra precautions when you work on any part of the fuel system. Don't smoke or allow open flames or bare light bulbs near the work area, and don't work in a garage where a gas-type appliance (such as a water heater or clothes dryer) is present. Since gasoline is carcinogenic, wear latex gloves when there's a possibility of being exposed to fuel, and, if you spill any fuel on your skin, rinse it off immediately with soap and water. Mop up any spills immediately and do not store fuel-soaked rags where they could ignite. When you perform any kind of work on the fuel system, wear safety glasses and have a Class B type fire extinguisher on hand. The fuel system is under pressure, so if any lines must be disconnected, the pressure in the system must be relieved first (see Chapter 4 for more information).

7 Check all rubber fuel lines for deterioration and chafing. Check especially for cracks in areas where the hose bends and just before fittings, such as where a hose attaches to the fuel filter and fuel injection unit.

8 High quality fuel line, specifically designed for high-pressure fuel injection applications, must be used for fuel line replacement. Never, under any circumstances, use regular fuel line, unreinforced vacuum line, clear plastic tubing or water hose for fuel lines.

9 Spring-type (pinch) clamps are commonly used on fuel lines. These clamps often lose their tension over a period of time, and can be "sprung" during removal. Replace all spring-type clamps with screw clamps whenever a hose is replaced.

METAL LINES

10 Sections of metal line may be routed along the frame, between the fuel tank and the engine. Check carefully to be sure the line has not been bent or crimped and no cracks have started in the line.

11 If a section of metal fuel line must be replaced, only seamless steel tubing should be used, since copper and aluminum tubing don't have the strength necessary to withstand normal engine vibration.

12 Check the metal brake lines where they enter the master cylinder and brake proportioning unit for cracks in the lines or loose fittings. Any sign of brake fluid leakage calls for an immediate and thorough inspection of the brake system.

13 Cooling system check (every 6000 miles or 6 months)

➧ **Refer to illustration 13.4**

⁂ CAUTION:

Never mix green-colored ethylene glycol antifreeze and red-colored silicate and phosphate-free coolant because doing so will destroy the efficiency of the red coolant, which is designed to last for 100,000 miles or five years.

1 Many major engine failures can be attributed to a faulty cooling system. If the vehicle is equipped with an automatic transmission, the cooling system also cools the transmission fluid and thus plays an important role in prolonging transmission life.

2 The cooling system should be checked with the engine cold. Do this before the vehicle is driven for the day or after it has been shut off for at least three hours.

3 Remove the pressure cap on the expansion tank by slowly turning

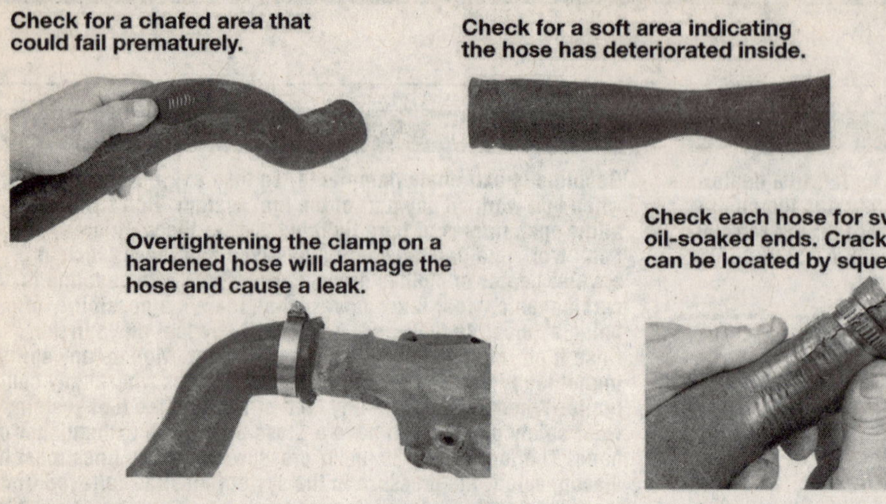

13.4 Hoses, like drivebelts, have a habit of failing at the worst possible time - to prevent the inconvenience of a blown radiator or heater hose, inspect them carefully as shown here

it counterclockwise. If you hear any hissing sounds (indicating there is still pressure in the system), wait until it stops. Thoroughly clean the cap, inside and out, with clean water. Also clean the filler neck on the tank. All traces of corrosion should be removed. The coolant inside the tank should be relatively transparent. If it is rust colored, the system should be drained, flushed and refilled (see Section 27). If the coolant level is not up to the top, add additional antifreeze/coolant mixture (see Section 4).

4 Carefully check the large upper and lower radiator hoses along with any smaller diameter heater hoses that run from the engine to the firewall. Inspect each hose along its entire length, replacing any hose that is cracked, swollen or shows signs of deterioration. Cracks may become more apparent if the hose is squeezed (see illustration).

5 Make sure all hose connections are tight. A leak in the cooling system will usually show up as white or rust-colored deposits on the areas adjoining the leak. If wire-type clamps are used at the ends of the hoses, it may be wise to replace them with more secure, screw-type clamps.

6 Use compressed air or a soft brush to remove bugs, leaves, etc. from the front of the radiator or air conditioning condenser. Be careful not to damage the delicate cooling fins or cut yourself on them.

7 Every other inspection, or at the first indication of cooling system problems, have the cap and system pressure tested. If you don't have a pressure tester, most gas stations and repair shops will do this for a minimal charge.

14 Tire rotation (every 6000 miles or 6 months)

Refer to illustrations 14.2a and 14.2b

1 The tires should be rotated at the specified intervals and whenever uneven wear is noticed.

2 Radial tires must be rotated in the recommended pattern (see illustrations).

Note: Most vehicles are sold with non-directional radial tires, but some replacement performance tires are available that are directional, and have a different rotation pattern. Directional tires have an arrow on the sidewall indicating the direction they must turn when mounted on the vehicle.

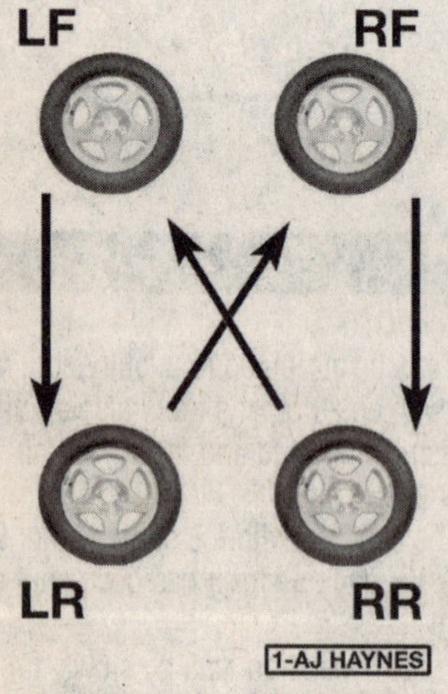

14.2a The recommended four-tire rotation pattern for non-directional radial tires

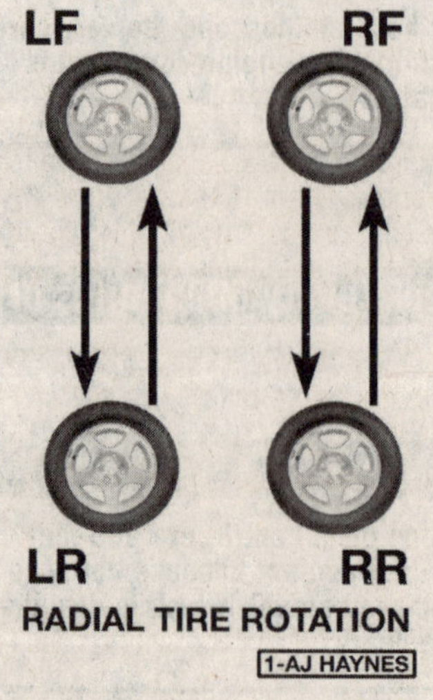

14.2b The recommended four-tire rotation pattern for directional radial tires

3 Refer to the information in *Jacking and towing* at the front of this manual for the proper procedures to follow when raising the vehicle and changing a tire. If the brakes are to be checked, don't apply the parking brake as stated. Make sure the tires are blocked to prevent the vehicle from rolling as it's raised.

4 Preferably, the entire vehicle should be raised at the same time. This can be done on a hoist or by jacking up each corner and then lowering the vehicle onto jackstands placed under the frame rails. Always use four jackstands and make sure the vehicle is safely supported.

5 After rotation, check and adjust the tire pressures as necessary. Tighten the wheel bolts to the torque listed in this Chapter's Specifications.

15 Timing belt check, diesel engines (every 10,000 miles or 9 months)

Refer to illustration 15.2

1 Refer to Chapter 2B for the procedure to remove the upper timing belt cover.

2 Check the timing belt for wear, especially on the thrust side of the teeth (see illustration).

3 Look for cracks, missing teeth, splits, fraying, or any sign that oil or diesel fuel has gotten onto the belt. If at all in doubt about the condition of the belt, replace it (see Chapter 2B).

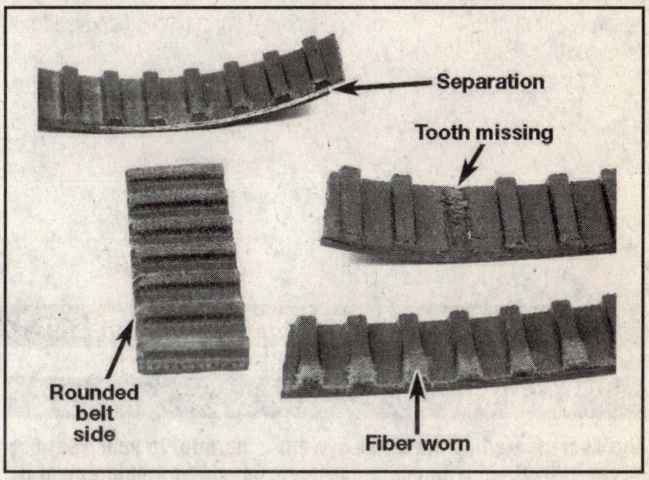

15.2 Timing belt defects

16 Fuel system check (every 15,000 miles or 12 months)

Refer to illustration 16.7

WARNING:

Gasoline is extremely flammable, so take extra precautions when you work on any part of the fuel system. Don't smoke or allow open flames or bare light bulbs near the work area, and don't work in a garage where a gas-type appliance (such as a water heater or clothes dryer) is present. Since gasoline is carcinogenic, wear latex gloves when there's a possibility of being exposed to fuel, and, if you spill any fuel on your skin, rinse it off immediately with soap and water. Mop up any spills immediately and do not store fuel-soaked rags where they could ignite. When you perform any kind of work on the fuel system, wear safety glasses and have a Class B type fire extinguisher on hand. The fuel system is under constant pressure, so, before any lines are disconnected, the fuel system pressure must be relieved (see Chapter 4). Diesel fuel is only slightly less dangerous than gasoline, so take the same precautions when dealing with diesel fuel.

1 If you smell fuel while driving or after the vehicle has been sitting in the sun, inspect the fuel system immediately.

2 Remove the fuel filler cap and inspect it for damage and corrosion. The gasket should have an unbroken sealing imprint. If the gasket is damaged or corroded, install a new cap.

3 Inspect the fuel feed and return lines for cracks. Make sure that the connections between the fuel lines and the fuel injection system are tight.

WARNING:

Your vehicle is fuel injected, so you must relieve the fuel system pressure before servicing fuel system components. The fuel system pressure-relief procedure is outlined in Chapter 4.

4 If the fuel injectors are visible, look for signs of fuel leakage (wet spots) around any of the injectors, they may need new O-rings (see Chapter 4).

5 Since some components of the fuel system - the fuel tank and part of the fuel feed and return lines, for example - are underneath the vehicle, they can be inspected more easily with the vehicle raised on a hoist. If that's not possible, raise the vehicle and support it on jackstands.

6 With the vehicle raised and safely supported, inspect the fuel tank and filler neck for punctures, cracks and other damage. The connection between the filler neck and the tank is particularly critical. Inspect all fuel tank mounting brackets and straps to be sure that the tank is securely attached to the vehicle.

WARNING:

Do not, under any circumstances, try to repair a fuel tank (except rubber components).

7 Carefully check all rubber hoses and nylon lines leading away from the fuel tank (see illustration). Check for loose connections, dete-

riorated hoses, crimped lines and other damage. Repair or replace damaged sections as necessary (see Chapter 4).

8 The evaporative emissions control system can also be a source of fuel odors. The function of the system is to store fuel vapors from the fuel tank in a charcoal canister until they can be routed to the intake manifold where they mix with incoming air before being burned in the combustion chambers.

9 The most common symptom of a faulty evaporative emissions system is a strong odor of fuel. If a fuel odor has been detected, and you have already checked the areas described above, check the charcoal canister, located in the right rear fenderwell, and the hoses connected to it (see illustration 16.7). The rear fenderwell liner must be removed for access (see Chapter 11).

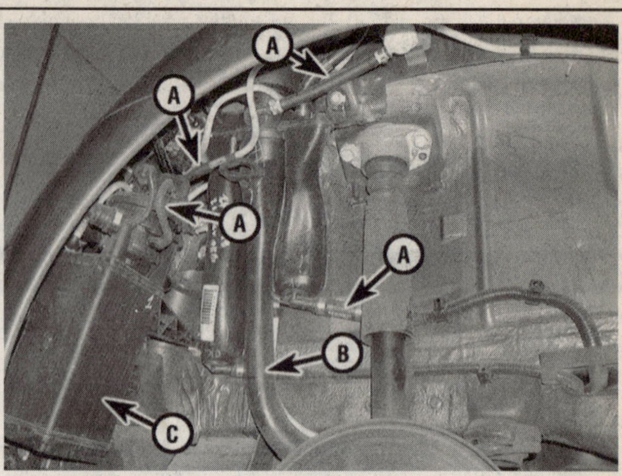

16.7 Inspect the fuel tank mounting straps, the various fuel and vapor lines (A), the plastic filler hose (B), and the evaporative emissions canister (C)

17 Brake system check (every 15,000 miles or 12 months)

WARNING:

The dust created by the brake system is harmful to your health. Never blow it out with compressed air and don't inhale any of it. An approved filtering mask should be worn when working on the brakes. Do not, under any circumstances, use petroleum-based solvents to clean brake parts. Use brake system cleaner only! Try to use non-asbestos replacement parts whenever possible.

➡Note: For detailed photographs of the brake system, refer to Chapter 9.

1 In addition to the specified intervals, the brakes should be inspected every time the wheels are removed or whenever a defect is suspected.

2 Any of the following symptoms could indicate a potential brake system defect: The vehicle pulls to one side when the brake pedal is depressed; the brakes make squealing or dragging noises when applied; brake pedal travel is excessive; the pedal pulsates; or brake fluid leaks, usually onto the inside of the tire or wheel.

DISC BRAKES

Refer to illustrations 17.7a, 17.7b, 17.9 and 17.11

3 Loosen the wheel bolts.

4 Raise the vehicle and place it securely on jackstands.

5 Remove the wheels (see *Jacking and towing* at the front of this book, or your owner's manual, if necessary).

6 There are two pads (an outer and an inner) in each caliper. The pads are visible with the wheels removed. Most vehicles covered by this manual have disc brakes front and rear.

➡Note: On 2001 and later models, there are wear indicators in the pads, and a brake pad wear warning light on the instrument panel.

7 Check the pad thickness by looking at each end of the caliper and through the inspection window in the caliper body (see illustrations). If the lining material is less than the thickness listed in this Chapter's Specifications, replace the pads.

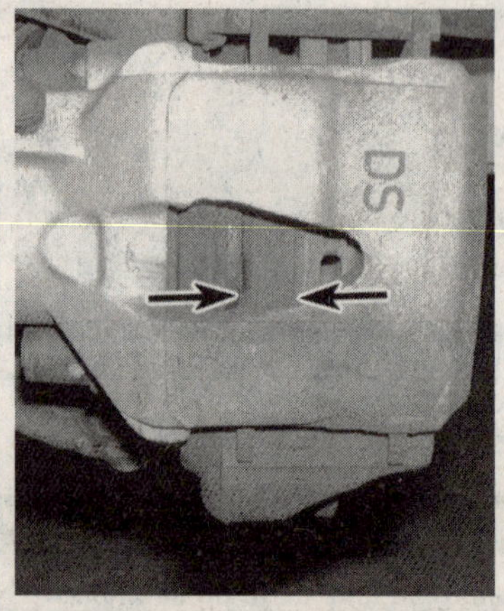

17.7a With the wheel off, check the thickness of the inner pad (arrows) through the inspection hole (front caliper shown, rear caliper similar)

➡Note: Keep in mind that the lining material is riveted or bonded to a metal backing plate and the metal portion is not included in this measurement.

8 If it is difficult to determine the exact thickness of the remaining pad material by the above method, or if you are at all concerned about

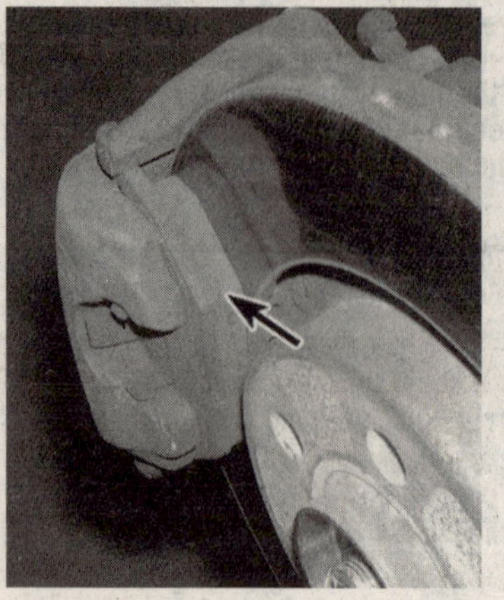

17.7b The outer pad (arrow) is more easily checked at the edge of the caliper

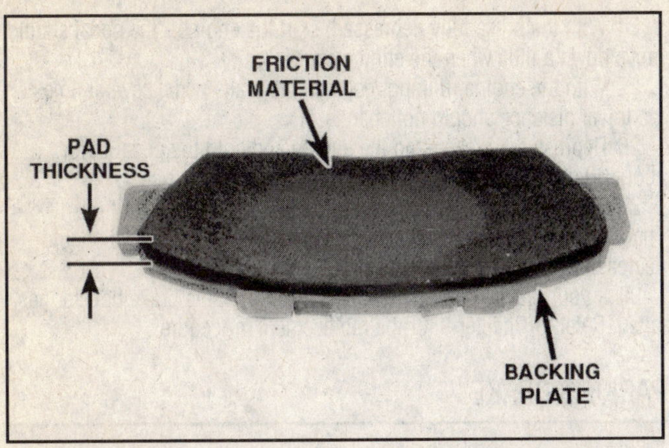

17.9 If a more precise measurement of pad thickness is necessary, remove the pads and measure the remaining friction material

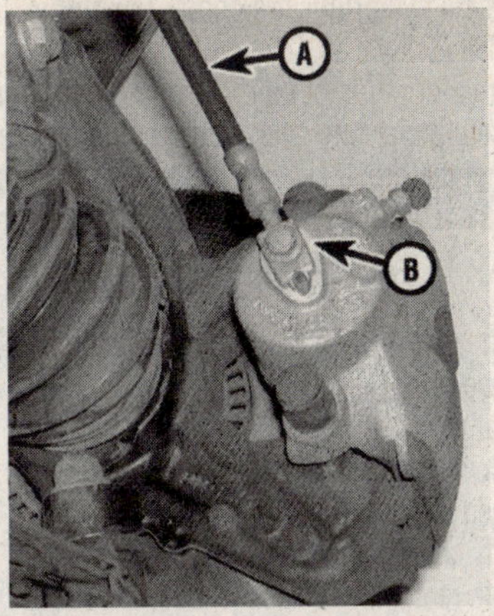

17.11 Check along the brake hoses (A) and at each fitting (B, caliper end) for deterioration, cracks and leakage

the condition of the pads, remove the caliper(s), then remove the pads from the calipers for further inspection (refer to Chapter 9).

9 Once the pads are removed from the calipers, clean them with brake cleaner and re-measure them with a ruler or a vernier caliper (see illustration).

10 Measure the disc thickness with a micrometer to make sure that it still has service life remaining. If any disc is thinner than the specified minimum thickness, replace it (refer to Chapter 9). Even if the disc has service life remaining, check its condition. Look for scoring, gouging and burned spots. If these conditions exist, remove the disc and have it resurfaced (see Chapter 9).

11 Before installing the wheels, check all brake lines and hoses for damage, wear, deformation, cracks, corrosion, leakage, bends and twists, particularly in the vicinity of the rubber hoses at the calipers (see illustration). Check the clamps for tightness and the connections for leakage. Make sure that all hoses and lines are clear of sharp edges, moving parts and the exhaust system. If any of the above conditions are noted, repair, reroute or replace the lines and/or fittings as necessary (see Chapter 9).

DRUM BRAKES

▸ Refer to illustrations 17.17a and 17.17b

12 Some models are equipped with drum rear brakes. Loosen the wheel bolts, raise the vehicle and support it securely on jackstands. Block the front tires to prevent the vehicle from rolling; however, don't apply the parking brake or it will lock the drums in place.

13 Remove the wheels, referring to *Jacking and towing* at the front of this manual if necessary.

14 Mark the hub so it can be reinstalled in the same position. Use a scribe, chalk, etc. on the drum, hub and backing plate.

15 Remove the brake drum (see Chapter 9 if necessary).

16 With the drum removed, carefully clean the brake assembly with brake system cleaner.

⁂ WARNING:

Don't blow the dust out with compressed air and don't inhale any of it (it is harmful to your health).

17 Note the thickness of the lining material on both front and rear

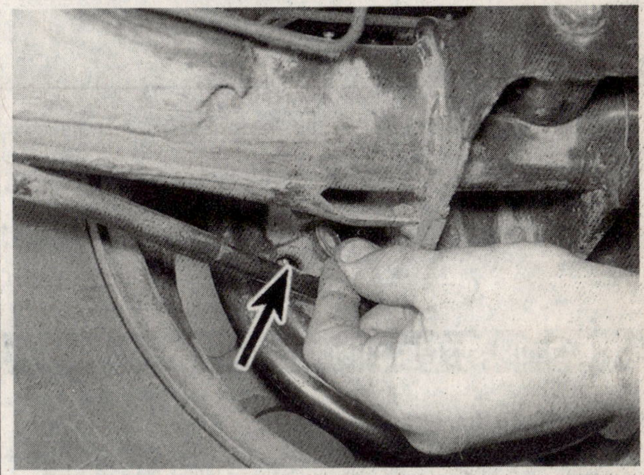

17.17a Removing the rubber plug on the back of drum brake equipped models allows for inspection of the brake shoe lining thickness through the hole (arrow)

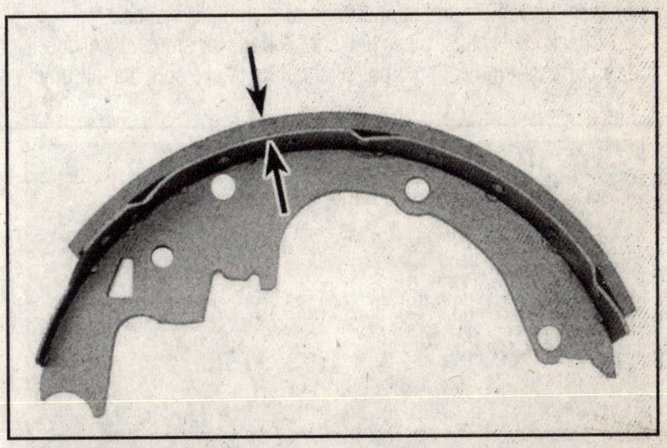

17.17b A more precise measurement of brake shoe thickness can be made by measuring the shoes directly with the brake drum removed

brake shoes. If the material has worn away to within 3/32-inch (2.3 mm) of the recessed rivets or metal backing, the shoes should be replaced (see illustrations). The shoes should also be replaced if they're cracked, glazed (shiny areas), or covered with brake fluid.

18 Make sure all the brake assembly springs are connected and in good condition.

19 Check the brake components for signs of fluid leakage. With your finger or a small screwdriver, carefully pry back the rubber boots on the wheel cylinder located at the top of the brake shoes. Any leakage here is an indication that the wheel cylinders should be overhauled immediately (see Chapter 9). Also, check all hoses and connections for signs of leakage.

20 Wipe the inside of the drum with a clean rag and brake cleaner. Again, be careful not to breathe the dust.

21 Check the inside of the drum for cracks, score marks, deep scratches and "hard spots" which will appear as small discolored areas. If imperfections cannot be removed with fine emery cloth, the drum must be taken to an automotive machine shop for resurfacing.

22 Repeat the procedure for the remaining wheel. If the inspection reveals that all parts are in good condition, reinstall the brake drums, install the wheels and lower the vehicle to the ground.

BRAKE BOOSTER CHECK

23 Sit in the driver's seat and perform the following sequence of tests.

24 With the brake fully depressed, start the engine - the pedal should move down a little when the engine starts.

25 With the engine running, depress the brake pedal several times - the travel distance should not change.

26 Depress the brake, stop the engine and hold the pedal in for about 30 seconds - the pedal should neither sink nor rise.

27 Restart the engine, run it for about a minute and turn it off. Then firmly depress the brake several times - the pedal travel should decrease with each application.

28 If your brakes do not operate as described, the brake booster has failed. Refer to Chapter 9 for the replacement procedure.

PARKING BRAKE

29 The parking brake cables operate the rear calipers or brake shoes mechanically, and there is no adjustment needed at the rear brakes unless cables or other parts are being replaced. One method of checking the parking brake is to park the vehicle on a steep hill with the parking brake set and the transmission in Neutral (stay in the vehicle for this check!). If the parking brake cannot prevent the vehicle from rolling, it's in need of adjustment (see Chapter 9).

18 Exhaust system check (every 15,000 miles or 12 months)

Refer to illustrations 18.2a and 18.2b

1 With the engine cold (at least three hours after the vehicle has been driven), check the complete exhaust system from the manifold to the end of the tailpipe. Be careful around the catalytic converter, which may be hot even after three hours. The inspection should be done with the vehicle on a hoist to permit unrestricted access. If a hoist isn't available, raise the vehicle and support it securely on jackstands.

2 Check the exhaust pipes and connections for signs of leakage and/or corrosion indicating a potential failure. Make sure that all brackets and hangers are in good condition and tight (see illustrations).

3 Inspect the underside of the body for holes, corrosion, open seams, etc. which may allow exhaust gasses to enter the passenger compartment. Seal all body openings with silicone sealant or body putty.

4 Rattles and other noises can often be traced to the exhaust system, especially the hangers, mounts and heat shields. Try to move the pipes, mufflers and catalytic converter. If the components can come in contact with the body or suspension parts, secure the exhaust system with new brackets and hangers.

18.2a Inspect the muffler (A) for signs of deterioration, and all hangers (B)

18.2b Inspect all flanged joints (arrow indicates front exhaust pipe joint) for signs of exhaust gas leakage

19 Manual transaxle lubricant level check (every 15,000 miles or 12 months)

➧ Refer to illustration 19.2

1 The manual transmission has a filler plug which must be removed to check the lubricant level. If the vehicle is raised to gain access to the plug, be sure to support it safely on jackstands - DO NOT crawl under a vehicle that is supported only by a jack! Be sure the vehicle is level or the check may be inaccurate.

2 Using the appropriate wrench, unscrew the plug from the transmission; most models require a 17mm Allen wrench (see illustration).

3 Use your little finger to reach inside the housing to feel the lubricant level. The level should be at or near the bottom of the plug hole. If it isn't, add the recommended lubricant through the plug hole with a syringe or squeeze bottle.

4 Install and tighten the plug. Check for leaks after the first few miles of driving.

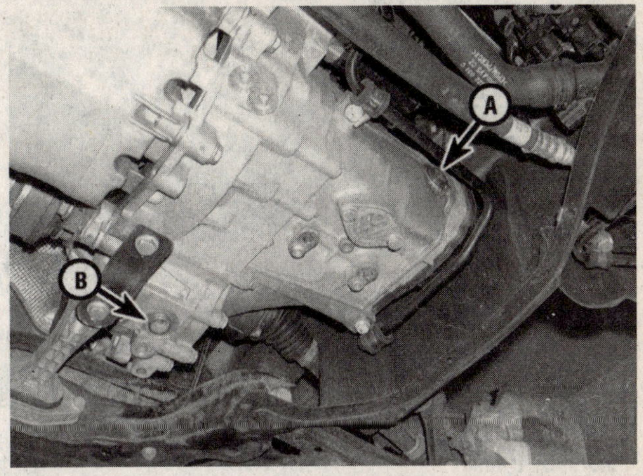

19.2 The manual transaxle check/fill plug (A) is located at the front side of the transaxle - B is the transaxle drain plug

20 Interior ventilation filter replacement (every 15,000 miles or 12 months)

➧ Refer to illustration 20.3

1 Most of the models covered by this manual are equipped with a pair of air filters under the top cover of the dash that clean the interior air from the heating/air conditioning system.

2 Refer to Chapter 11 for the procedures to remove the center instrument panel cover, the right-side plenum cover and plenum close-out panel to access the filters.

3 Release the two tabs on the filter frame and pull it up, then remove the filter (see illustration).

4 The remainder of the installation is the reverse of the removal procedure.

➡Note: Unless you drive in extra-dusty conditions, the replacement interval is not critical. The factory recommends replacing the filter at 10,000 miles on 1998 models, and at 20,000 miles on 1999 and later models.

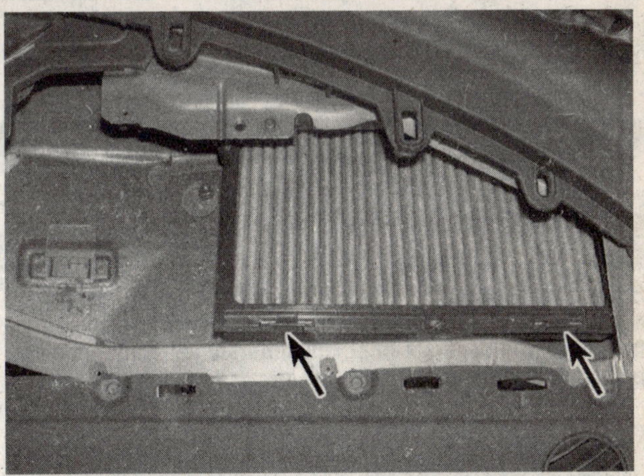

20.3 Release the tabs (arrows) to raise the filter frame, then take out the filter

21 Automatic transaxle fluid level check (every 30,000 miles or 30 months)

➧ Refer to illustrations 21.3 and 21.4

1 The automatic transmission fluid level should be carefully maintained. Low fluid level can lead to slipping or loss of drive, while overfilling can cause foaming and loss of fluid.

➡Note: Checking the fluid on the New Beetle isn't easy. The transmission is considered by the manufacturer to be a "sealed" unit, to which fluid doesn't need to be added unless a leak is evident. There is no conventional dipstick in the engine compartment, but rather a fill plug and a level-inspection plug accessible only from below the transaxle. Make sure you have a new seal for each plug before you begin.

2 Transaxle fluid expands as it warms up, and the fluid check should only be begun on a cold drivetrain. The transaxle fluid must only be checked at a transaxle temperature of 95 to 113 degrees F. With the parking brake set, start the engine, then move the shift lever through all the gear ranges, ending in Park. The fluid level must be checked with the vehicle level.

3 When the transaxle temperature reaches 95 to 113 degrees F,

21.3 Location of the level-check plug (arrow) on automatic transaxles

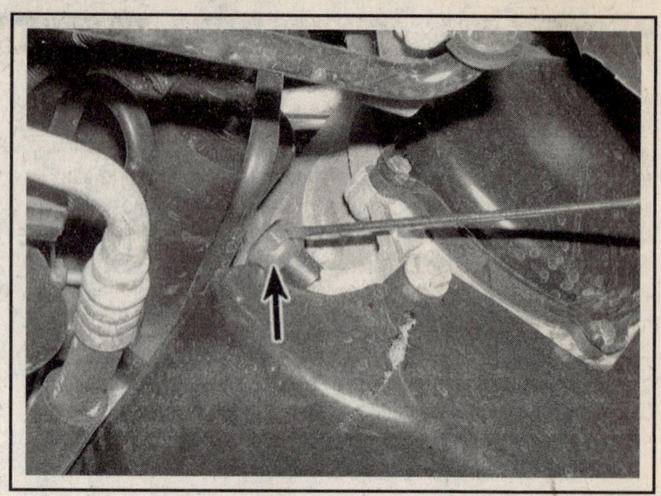

21.4 Use a large screwdriver to pry the automatic transaxle fill plug (arrow) out of the transaxle from below

remove the level-check plug (see illustration). If fluid just comes out of the hole as drips, the fluid level is OK. If more fluid comes out, the transmission may have been overfilled.

4 If no fluid comes out at 113 degrees F, remove the filler plug and add a small amount of fluid, until it just drips out of the level-check hole (see illustration).

5 When the correct level is achieved at the specified temperature, install new seals on the level-check plug and screw it in, and install the fill plug with a new seal. The cap over the fill plug is easily damaged during removal, so have a new one of these on hand as well.

6 The condition of the fluid should also be checked along with the level. If the fluid is a dark reddish-brown color, or if it smells burned, it should be changed. If you are in doubt about the condition of the fluid, purchase some new fluid and compare the two for color and smell.

➡Note: When the work is done at a dealership, the factory scan tool is used to read the transmission fluid temperature. To perform the job at home, you may need a cooking or darkroom thermometer that has a long probe, which you can insert in the level-check hole to read the fluid temperature.

22 Differential lubricant level check, automatic transaxles (every 30,000 miles or 30 months)

➧ Refer to illustrations 22.2 and 22.3

➡Note: On manual transaxle models, the differential lubricant is not separated from the transmission lubricant; filling the transmission fills the differential, too (see Section 19). The differential has a separate filler, and different lubricant, on automatic transaxle models.

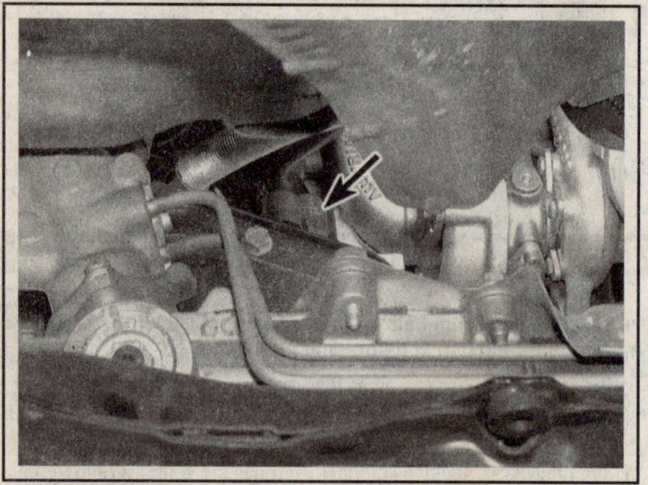

22.2 Remove the Vehicle Speed Sensor (arrow) from the top of the speedometer driveshaft on the differential, then remove the driveshaft

1 If the vehicle is raised to perform this procedure, be sure to support it safely on jackstands - DO NOT crawl under the vehicle when it's supported only by the jack. Be sure the vehicle is level or the check may not be accurate.

2 Refer to Chapter 6 and remove the Vehicle Speed Sensor (VSS) from the transaxle. Using a wrench, remove the speedometer driveshaft from the transaxle (see illustration).

3 The speedometer driveshaft is used as a dipstick for the differential lubricant (see illustration). If the level is too low, use a pump or squeeze bottle to add the recommended lubricant until it just starts to run out of the opening.

4 Install the driveshaft securely and reinstall the VSS.

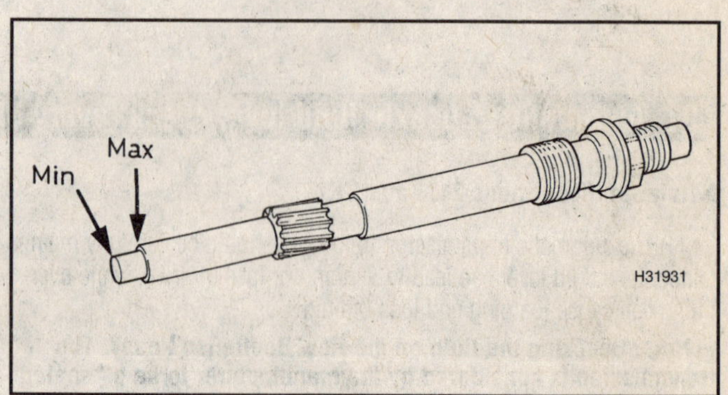

22.3 The differential lubricant level should be between these points (arrows) on the speedometer driveshaft

23 Brake fluid change (every 30,000 miles or 24 months)

⁂ WARNING:

Brake fluid can harm your eyes and damage painted surfaces, so use extreme caution when handling or pouring it. Do not use brake fluid that has been standing open or is more than one year old. Brake fluid absorbs moisture from the air. Excess moisture can cause a dangerous loss of braking effectiveness.

1 At the specified intervals, the brake fluid should be drained and replaced. Since the brake fluid may drip or splash when pouring it, place plenty of rags around the master cylinder to protect any surrounding painted surfaces.

2 Before beginning work, purchase the specified brake fluid (see *Recommended lubricants and fluids* at the end of this Chapter).

3 Remove the cap from the master cylinder reservoir.

4 Using a hand suction pump or similar device, withdraw the fluid from the master cylinder reservoir.

5 Add new fluid to the master cylinder until it rises to the line indicated on the reservoir.

6 Bleed the brake system as described in Chapter 9 at all four brakes until new and uncontaminated fluid is expelled from each bleeder screw. Be sure to maintain the fluid level in the master cylinder as you perform the bleeding process. If you allow the master cylinder to run dry, air will enter the system.

7 Refill the master cylinder with fluid and check the operation of the brakes. The pedal should feel solid when depressed, with no sponginess.

⁂ WARNING:

Do not operate the vehicle if you are in doubt about the effectiveness of the brake system.

24 Air filter replacement (every 30,000 miles or 24 months)

➧ Refer to illustrations 24.3a and 24.3b

1 At the specified intervals, the air filter element should be replaced with a new one.

2 On all models, the air filter is housed in a black plastic box mounted on the inner fenderwell on the left side of the engine compartment. On models so equipped, disconnect the hose between the air intake and the air filter housing cover.

3 Disconnect the Mass Airflow Sensor connector, remove the screws and pull the air filter housing cover up, then lift the air filter element out of the housing (see illustrations). Wipe out the inside of the air filter housing with a clean rag.

➡Note: There is some "wrestling" involved in getting the air filter cover out, since it has to come away from the air induction hose and away from the fender at almost the same time.

4 While the cover is off, be careful not to drop anything down into the air filter housing.

5 Place the new filter element in the air filter housing. Make sure it seats properly in the groove of the housing.

6 Installation is the reverse of removal.

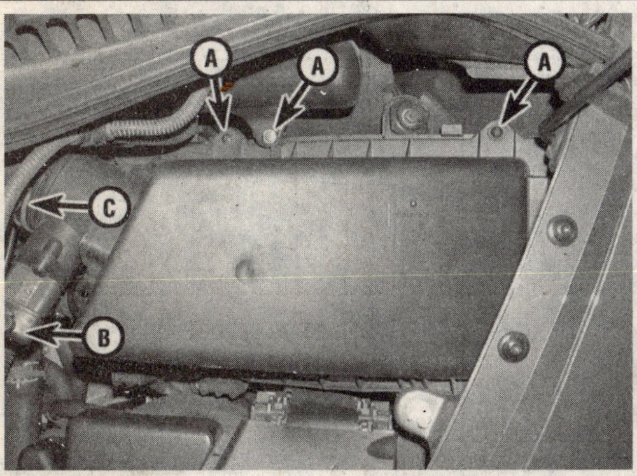

24.3a Remove the screws (A), disconnect the MAF sensor connector (B), release the clamp (C) and lift out the air filter housing cover

24.3b Set the cover aside and pull out the filter element

25 Fuel filter, draining and replacement (see Maintenance schedule for service intervals)

DIESEL ENGINES

**** WARNING:**

Diesel fuel isn't as volatile as gasoline, but it is flammable, so take extra precautions when you work on any part of the fuel system. Don't smoke or allow open flames or bare light bulbs near the work area. Don't work in a garage or other enclosed space where there is a gas-type appliance (such as a water heater or clothes dryer). Finally, when you perform any work on the fuel system, wear safety glasses, latex gloves and have a Class B type fire extinguisher on hand. If you spill any diesel fuel on your skin, rinse it off immediately with soap and water.

**** CAUTION:**

Keep a container handy below the work area, since spilled diesel fuel can ruin asphalt pavement. Diesel fuel can also deteriorate rubber hoses, so clean up spills immediately if it gets on fuel, coolant or brake hoses.

Draining

Refer to illustration 25.4a through 25.4f

1 Water collects frequently in diesel fuel tanks, but since water is heavier than diesel fuel, it will separate at the fuel filter/separator. Regular draining at the fuel filter will keep the water from getting to the engine.

2 Raise the vehicle and support it securely on jackstands.

3 The fuel filter is mounted in the engine compartment, on the right side in front of the windshield washer tank.

4 Pull the retaining clip that secures the control valve to the top of the filter assembly (see illustrations). Pull up the control valve and its hoses, away from the filter.

5 At the bottom of the fuel filter, loosen the drain plug, while holding a container underneath the filter to catch the diesel fuel and water.

➡Note: Have spare rags or a small container to catch or wipe up extra fuel that will spill from the filter assembly.

6 The water will come out first. When you have drained a few ounces of fuel/water, twist the drain plug closed. Pour a few ounces of

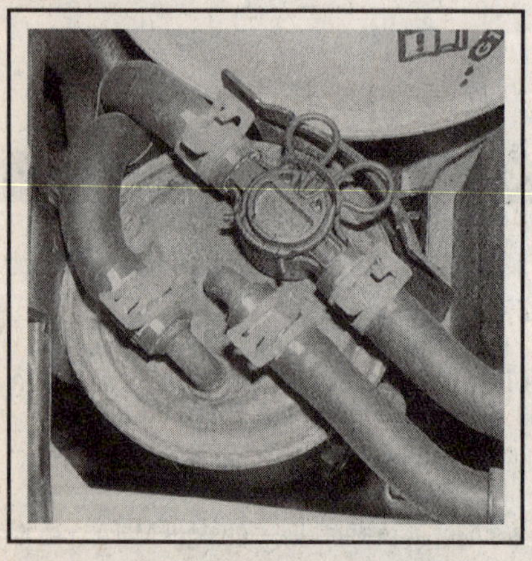

25.4a The fuel filter is mounted on the inner fender, above the righthand wheelarch

25.4b Release the clip . . .

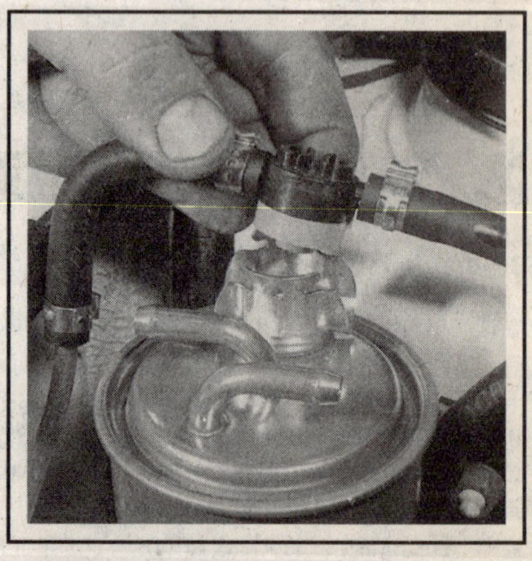

25.4c . . . and lift out the control valve, leaving the fuel hoses attached to it

25.4d Loosen the securing screw . . .

25.4e . . . and raise the filter out of its retaining bracket

25.4f Reconnect the fuel supply and delivery hoses

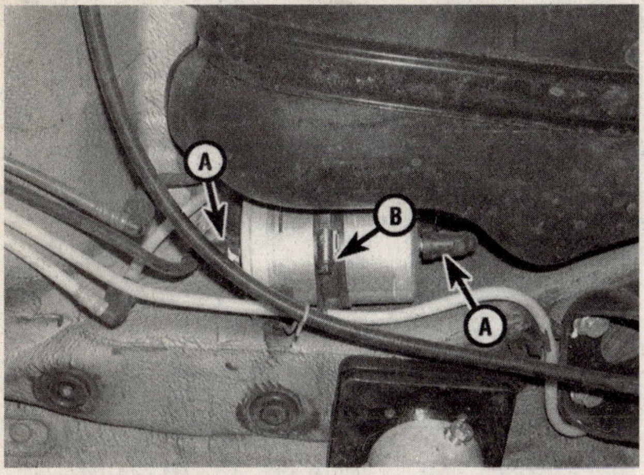

25.18 Detach the fittings (A) at the fuel filter, then loosen the clamp (B) to remove the filter

fresh diesel fuel in the top of the filter.

7 Use a new O-ring when reattaching the control valve to the filter top.

8 When the vehicle is running, inspect for any leaks around the filter assembly.

Replacement

9 Disconnect the fuel control valve from the top of the filter.

10 Mark the other two fuel lines with tape and disconnect them.

➡Note: The lines may be secured with crimp-type clamps, which may have to be cut off with wire-cutting pliers to remove the hoses.

11 Remove the fasteners and take the filter and mounting bracket from the vehicle. Loosen the bracket and remove the filter.

12 Fill a new filter with fresh diesel fuel and install it in the bracket, then reattach the bracket to the vehicle. Attach the control valve with a new O-ring and secure the hoses with screw-type clamps.

13 When the vehicle is running, inspect for any leaks around the filter assembly.

GASOLINE ENGINES

➧ Refer to illustration 25.18

⁂ WARNING:

Gasoline is extremely flammable, so take extra precautions when you work on any part of the fuel system. Don't smoke or allow open flames or bare light bulbs near the work area, and don't work in a garage where a gas-type appliance (such as a water heater or clothes dryer) is present. Since gasoline is carcinogenic, wear latex gloves when there's a possibility of being exposed to fuel, and, if you spill any fuel on your skin, rinse it off immediately with soap and water. Mop up any spills immediately and do not store fuel-soaked rags where they could ignite. When you perform any kind of work on the fuel system, wear safety glasses and have a Class B type fire extinguisher on hand. The fuel system is under pressure, so if any lines must be disconnected, the pressure in the system must be relieved first (see Chapter 4 for more information).

14 The gasoline fuel filter is located under the rear of the vehicle, at the front of the fuel tank.

15 The manufacturer does not give a replacement interval, but our suggested interval (see Section 1) is based on experience with many other vehicles. The gasoline fuel filter is not expensive, so replacing it is like inexpensive insurance.

16 Relieve the fuel system pressure (see Chapter 4), then disconnect the cable from the negative terminal of the battery (see Chapter 5). Raise the vehicle and support it securely on jackstands.

17 Use compressed air or carburetor cleaner to clean any dirt surrounding the fuel inlet and outlet line fittings.

18 There are quick-connect fittings at each end of the filter (see illustration). Depress the tab on each fitting to release them, then pull them off the filter.

➡Note: Have spare rags or a small container to catch or wipe up extra gasoline that will spill from the filter.

19 Loosen the screw-clamp and remove the filter.

20 Installation is the reverse of removal. Makes sure the arrow on side of the new filter is pointing toward the engine side of the fuel system, and check for leaks after running the vehicle.

26 Spark plug replacement (see *Maintenance schedule* for service intervals)

➧ Refer to illustrations 26.2, 26.4, 26.5a, 26.5b, 26.6a, 26.6b, 26.6c, 26.8, 26.9a, 26.9b and 26.10

1 Original equipment spark plugs are conventional-type on 2.0L engines, and long-lasting platinum-tipped type on the 1.8L turbo engine. The replacement interval for conventional plugs is about half that of platinum-tipped plugs, although any engine can be fitted with platinum-tipped replacement spark plugs to take advantage of the longer period between plug changes. The spark plugs are threaded into the front side of the cylinder head, adjacent to the intake ports on 2.0L engines, and on top of the engine on 1.8L turbo models, which have individual ignitions coils mounted over each spark plug.

2 In most cases, the tools necessary for spark plug replacement include a spark plug socket which fits onto a ratchet (spark plug sockets are padded inside to prevent damage to the porcelain insulators on the new plugs), various extensions and a gap gauge to check and adjust the gaps on the new plugs (see illustration). A special plug boot removal tool is available for separating the wire boots from the spark plugs. A torque wrench should be used to tighten the new plugs.

3 The best approach when replacing the spark plugs is to purchase the new ones in advance, adjust them to the proper gap and replace them one at a time. When buying the new spark plugs, be sure to obtain the correct plug type for your particular engine. This information can be found in your owner's manual and the Specifications at the end of this Chapter.

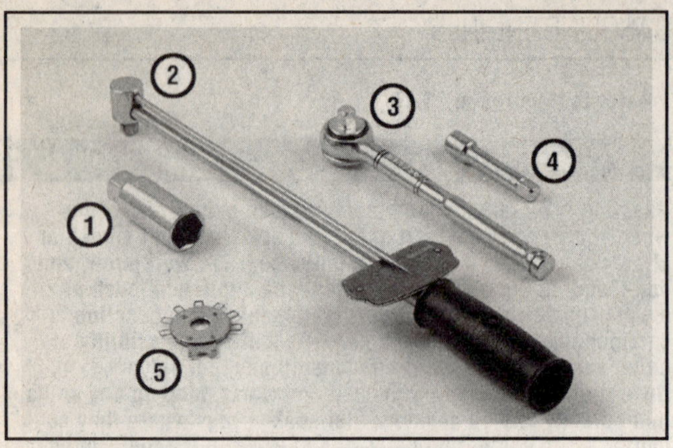

26.2 Tools required for changing spark plugs

1 ***Spark plug socket*** *- This will have special padding inside to protect the spark plug's porcelain insulator*
2 ***Torque wrench*** *- Although not mandatory, using this tool is the best way to ensure the plugs are tightened properly*
3 ***Ratchet*** *- Standard hand tool to fit the spark plug socket*
4 ***Extension*** *- Depending on model and accessories, you may need special extensions and universal joints to reach one or more of the plugs*
5 ***Spark plug gap gauge*** *- This gauge for checking the gap comes in a variety of styles. Make sure the gap for your engine is included*

4 Allow the engine to cool completely before attempting to remove any of the plugs. Remove the plastic cover at the top of the engine for access (see illustration). While you're waiting for the engine to cool, check the new plugs for defects and adjust the gaps.

5 The gap is checked by inserting the proper thickness gauge between the electrodes at the tip of the plug (see illustration). The gap between the electrodes should be the same as the one specified on the Emissions Control Information label or in this Chapter's Specifications. The gauge should just slide between the electrodes. If the gap is incorrect, use the adjuster on the gauge body to bend the curved side

26.4 Remove the plastic plugs (arrows) with a small screwdriver to access three screws retaining the engine cover

electrode slightly until the proper gap is obtained (see illustration). If the side electrode is not exactly over the center electrode, bend it with the adjuster until it is. Check for cracks in the porcelain insulator (if any are found, the plug should not be used).

➡Note: Manufacturers recommend using a tapered thickness gauge when checking platinum-type spark plugs. Other types of gauges may scrape the thin platinum coating from the electrodes, thus dramatically shortening the life of the plugs.

6 With the engine cool, remove the spark plug wire from one spark plug. Pull only on the boot at the end of the wire - do not pull on the wire. A plug wire removal tool should be used if available (see illustrations).

➡Note: You will have to reach around the intake plenum to remove the number 2 and 3 plug wires and plugs (2.0L engine).

To extract the number 2 plug, you may have to disconnect the electrical connector at the injector nearest it, and swivel the injector to provide more working room.

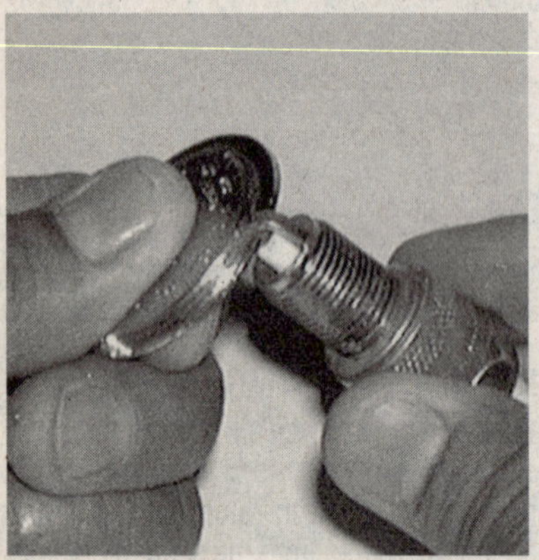

26.5a Spark plug manufacturers recommend using a tapered thickness gauge when checking the gap - slide the thin side into the gap and turn until the gauge just fills the gap, then read the thickness on the gauge - do not force the tool into the gap or use the tapered portion to widen a gap

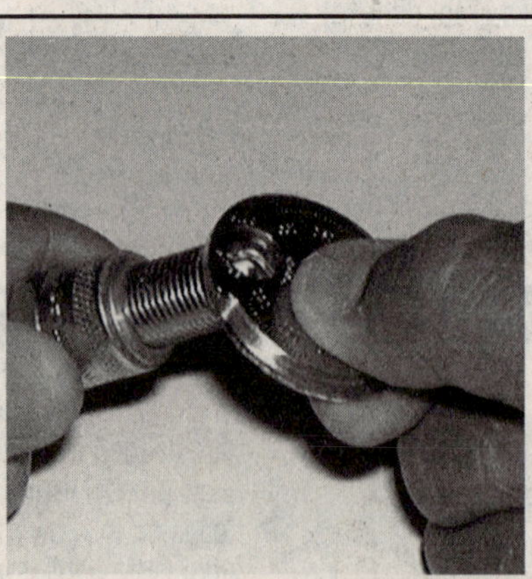

26.5b To change the gap, bend the side electrode only, using the adjuster hole in the tool, and be very careful not to crack or chip the porcelain insulator surrounding the center electrode

26.6a A tool like this one (arrow) makes the job of removing the spark plug boot easier - twist it back-and-forth and pull only on the boot

26.6b On 1.8L turbo models, remove the two screws (arrows) securing each individual ignition coil . . .

26.6c . . . then pull the coil/boot unit up and off the spark plug

26.8 Use a socket and extension to unscrew the spark plugs - various length extensions and perhaps a flex-joint may be required to reach some plugs

⁂ CAUTION:

Remember to reconnect the injector when you are finished.

7 If compressed air is available, use it to blow any dirt or foreign material away from the spark plug hole. The idea here is to eliminate the possibility of debris falling into the cylinder as the spark plug is removed.

8 Place the spark plug socket over the plug and remove it from the engine by turning it in a counterclockwise direction (see illustration).

9 Compare the spark plug with the chart (see illustration) to get an indication of the general running condition of the engine. Before installing the new plugs, it is a good idea to apply a thin coat of anti-seize compound to the threads (see illustration). Don't get any on the bottom threads or it could run (during hot engine operation) down onto the porcelain or electrodes and potentially ruin the plug.

10 Thread one of the new plugs into the hole until you can no longer turn it with your fingers, then tighten it with a torque wrench (if available) or the ratchet. Where plugs are harder to reach, it might be a good idea to slip a short length of rubber hose over the end of the plug to use as a tool to thread it into place (see illustration). The hose will grip the plug well enough to turn it, but will start to slip if the plug begins

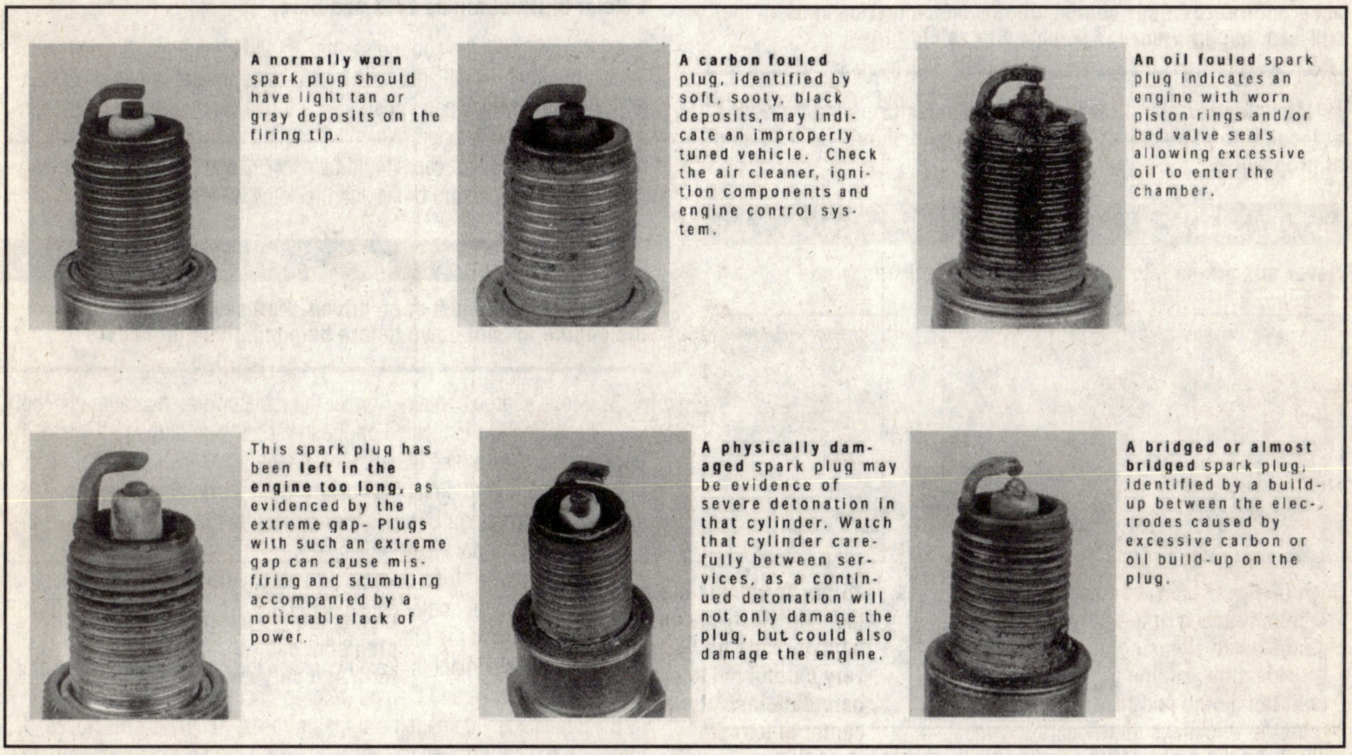

26.9a Inspect the spark plug to determine engine running conditions

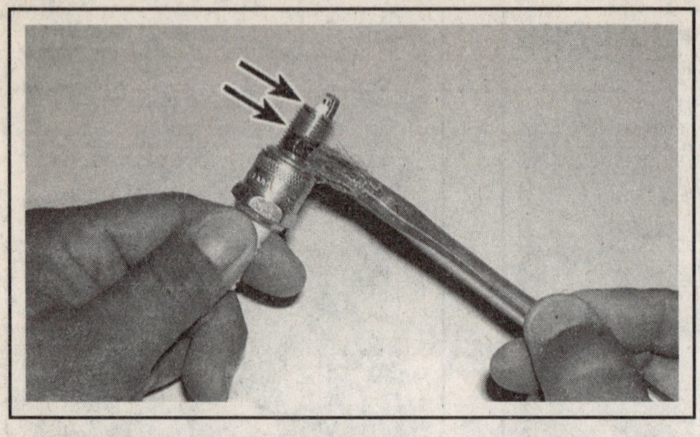

26.9b Apply a thin coat of anti-seize compound to the spark plug threads, being careful not to get any near the lower threads (arrows)

to cross-thread in the hole - this will prevent damaged threads and the accompanying repair costs.

11 Attach the plug wire to the new spark plug, again using a twisting motion on the boot until it's seated on the spark plug.

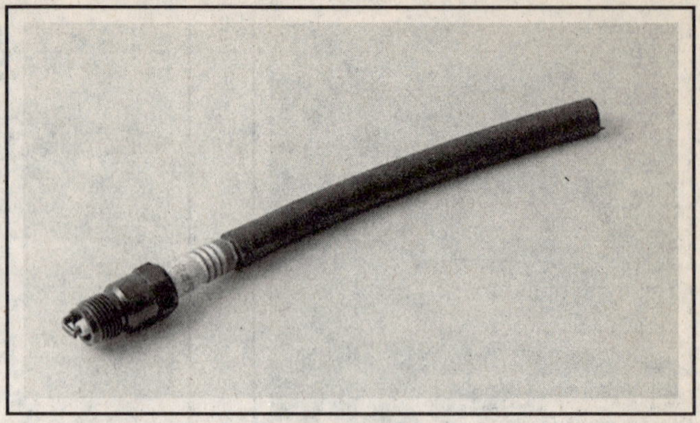

26.10 A length of snug-fitting rubber hose will save time and prevent damaged threads when installing the spark plugs

12 Repeat the procedure for the remaining spark plugs, replacing them one at a time to prevent mixing up the spark plug wires.

27 Cooling system servicing (draining, flushing and refilling) (see *Maintenance schedule* for service intervals)

⁂ WARNING:

Do not allow antifreeze to come in contact with your skin or painted surfaces of the vehicle. Rinse off spills immediately with plenty of water. Antifreeze is highly toxic if ingested. Never leave antifreeze lying around in an open container or in puddles on the floor; children and pets are attracted by its sweet smell and may drink it. Check with local authorities on disposing of used antifreeze. Many communities have collection centers that will see that antifreeze is disposed of safely.

➡Note: Non-toxic antifreeze is now manufactured and available at local auto parts stores, but even this type should be disposed of properly.

⁂ CAUTION:

Never mix green-colored ethylene glycol antifreeze and red-colored silicate and phosphate-free coolant because doing so will destroy the efficiency of the red coolant, which is designed to last for 100,000 miles or five years.

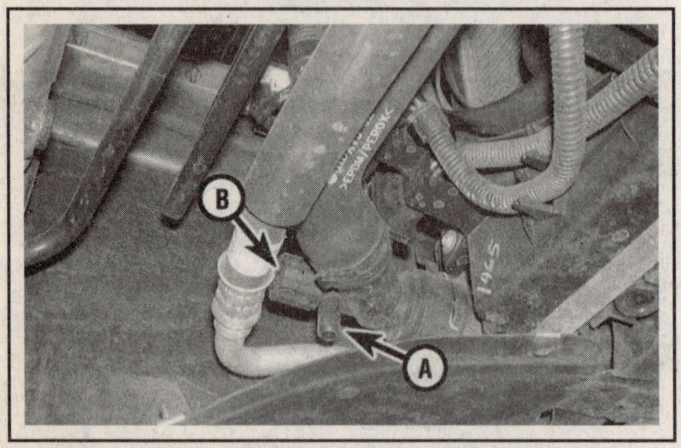

27.3 Attach a hose to the drain (A), then turn the cooling system drain knob (B), located at the lower radiator hose

DRAINING

➧ Refer to illustrations 27.3 and 27.4

1 Periodically, the cooling system should be drained, flushed and refilled to replenish the antifreeze mixture and prevent formation of rust and corrosion, which can impair the performance of the cooling system and cause engine damage. When the cooling system is serviced, all hoses and the radiator cap should be checked and replaced if necessary.

2 Apply the parking brake and block the wheels.

⁂ WARNING:

If the vehicle has just been driven, wait several hours to allow the engine to cool down before beginning this procedure.

3 Move a large container under the radiator drain to catch the coolant. The drain plug is located on the lower left side of the radiator, at the lower radiator hose (see illustration). Attach a 3/8-inch diameter hose to the drain fitting to direct the coolant into the container, then open the drain fitting (a pair of pliers may be required to turn it). Remove the coolant expansion tank cap.

4 After coolant stops flowing out of the radiator, move the container under the engine oil cooler (see illustration). Remove the hose and allow the coolant in the block to drain.

5 While the coolant is draining, check the condition of the radiator hoses, heater hoses and clamps (refer to Section 13 if necessary).

6 Replace any damaged clamps or hoses. Reconnect the hoses, replacing the spring-type clamps with screw-type ones if necessary to get a tight connection.

FLUSHING

➧ **Refer to illustration 27.9**

7 Once the system is completely drained, remove the thermostat from the engine (see Chapter 3). Then reinstall the thermostat housing without the thermostat. This will allow the system to be thoroughly flushed.

8 Reinstall the lower hose on the oil cooler and tighten the radiator drain plug. Turn your heating system controls to Hot, so that the heater core will be flushed at the same time as the rest of the cooling system.

9 Disconnect the upper radiator hose, then place a garden hose in the upper radiator inlet and flush the system until the water runs clear at the upper radiator hose (see illustration).

10 In severe cases of contamination or clogging of the radiator, remove the radiator (see Chapter 3) and have a radiator repair facility clean and repair it if necessary.

11 Many deposits can be removed by the chemical action of a cleaner available at auto parts stores. Follow the procedure outlined in the manufacturer's instructions.

➡**Note: When the coolant is regularly drained and the system refilled with the correct antifreeze/water mixture, there should be no need to use chemical cleaners or descalers.**

REFILLING

12 To refill the system, install the thermostat and reconnect any radiator hoses.

13 Place the heater temperature control in the maximum heat position.

14 Make sure to use the proper coolant listed in this Chapter's Specifications. Slowly fill the expansion tank with the recommended mixture of antifreeze and water until it reaches the MIN mark. Wait five minutes and recheck the coolant level in the tank, adding if necessary.

15 Leave the expansion tank cap off and run the engine in a well-ventilated area until the thermostat opens (coolant will begin flowing through the radiator and the upper radiator hose will become hot).

16 Turn the engine off and let it cool. Add more coolant mixture to bring the level back up to the MIN level on the expansion tank.

17 Squeeze the upper radiator hose to expel air, then add more coolant mixture if necessary. Install the expansion tank cap.

18 Place the heater temperature control and the blower motor speed control to their maximum setting.

19 Start the engine, allow it to reach normal operating temperature and check for leaks.

20 Let the engine cool, then recheck the coolant level, adding as necessary to bring the level just above the MIN mark.

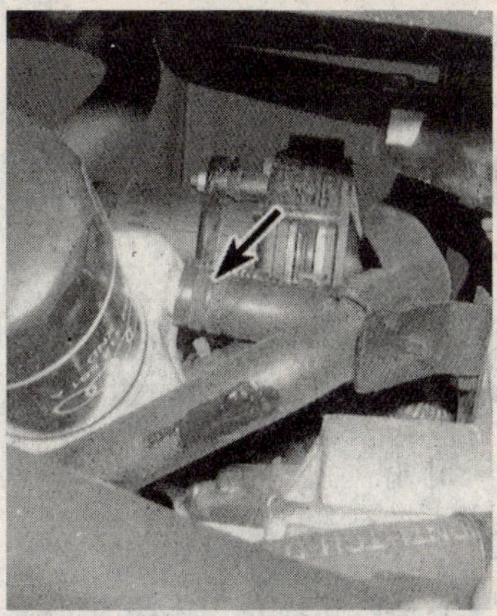

27.4 Remove the lower hose (arrow) from the engine oil cooler to drain the cylinder block (seen here from below)

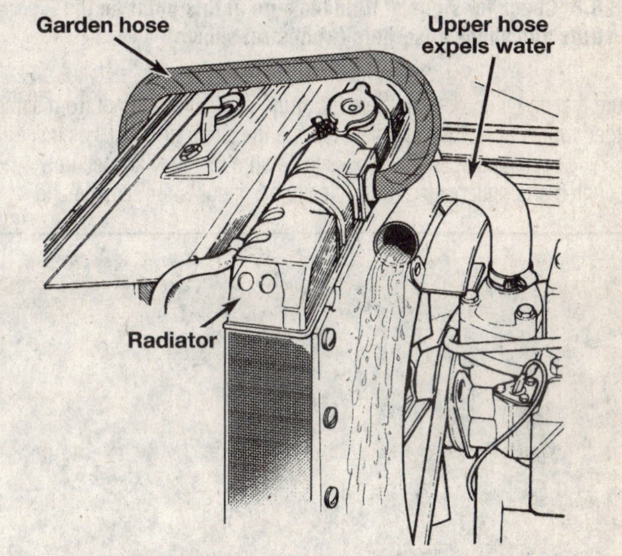

27.9 With the thermostat removed, disconnect the upper radiator hose and flush the radiator and engine block with a garden hose

28 Suspension, steering and driveaxle boot check (every 30,000 miles or 30 months)

➡**Note: The steering linkage and suspension components should be checked periodically. Worn or damaged suspension and steering components can result in excessive and abnormal tire wear, poor ride quality and vehicle handling and reduced fuel economy. For detailed illustrations of the steering and suspension components, refer to Chapter 10.**

STRUT/SHOCK ABSORBER CHECK

➧ **Refer to illustration 28.6**

1 Park the vehicle on level ground, turn the engine off and set the parking brake. Check the tire pressures.

2 Push down at one corner of the vehicle, then release it while noting the movement of the body. It should stop moving and come to rest in a level position within one or two bounces.

3 If the vehicle continues to move up-and-down or if it fails to return to its original position, a worn or weak strut or shock absorber is probably the reason.

4 Repeat the above check at each of the three remaining corners of the vehicle.

5 Raise the vehicle and support it securely on jackstands.

6 Check the struts and shock absorbers for evidence of fluid leakage (see illustration). A light film of fluid is no cause for concern. Make

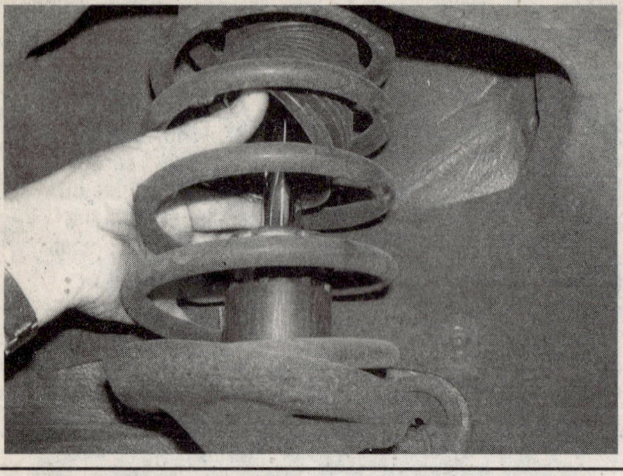

28.6 Check for signs of fluid leakage at this point on the struts and shock absorbers (front strut shown)

sure that any fluid noted is from the struts or shocks and not from some other source. If leakage is noted, replace the struts or shocks as a set.

7 Check the struts and shocks to be sure that they are securely mounted and undamaged. Check the upper mounts for damage and

28.9b Inspect the tie-rod ends (A), the lower balljoints (B), and stabilizer link bushings (C)

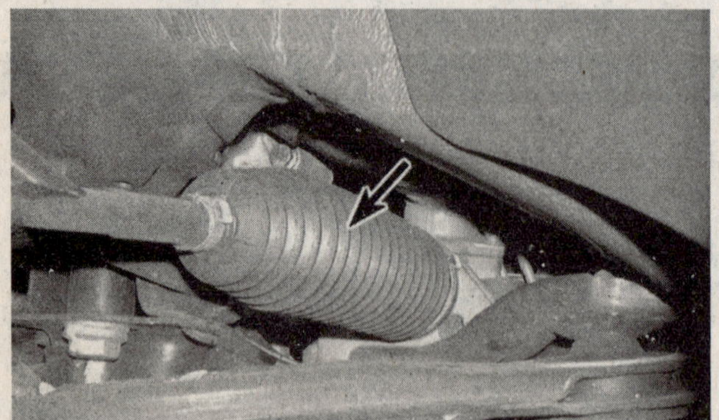

28.9c Inspect the steering gear boots (arrow) for signs of cracking or lubricant leakage

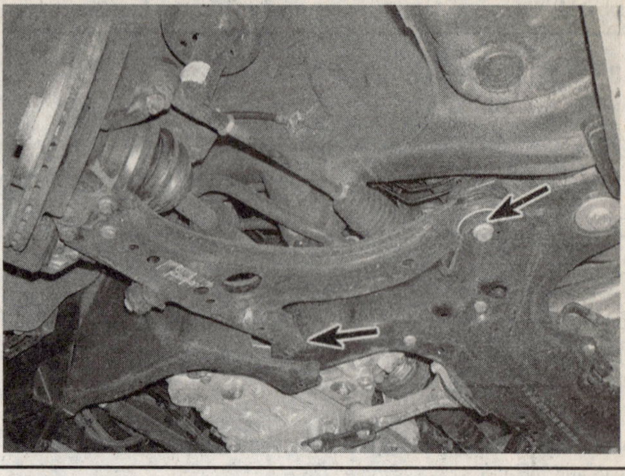

28.9a Examine the mounting points (arrows) for the lower control arms on the front suspension subframe

wear. If damage or wear is noted, replace the struts and shocks as a set (front or rear).

8 If the struts or shocks must be replaced, refer to Chapter 10 for the procedure.

STEERING AND SUSPENSION CHECK

➧ **Refer to illustrations 28.9a, 28.9b, 28.9c, 28.9d and 28.11**

9 Visually inspect the steering and suspension components (front and rear) for damage and distortion. Look for damaged seals, boots and bushings and leaks of any kind. Examine the bushings where the control arms meet the chassis (see illustrations).

10 Clean the lower end of the steering knuckle. Have an assistant grasp the lower edge of the tire and move the wheel in-and-out while you look for movement at the steering knuckle-to-control arm balljoint. If there is any movement the suspension balljoint(s) must be replaced.

11 Grasp each front tire at the front and rear edges, push in at the front, pull out at the rear and feel for play in the steering system components. If any freeplay is noted, check the rack mounts and the tie-rod ends for looseness (see illustration).

12 Additional steering and suspension system information and illustrations can be found in Chapter 10.

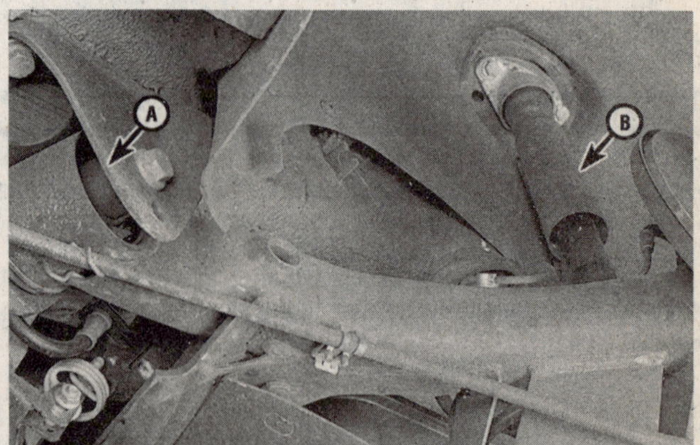

28.9d At the rear suspension, check the rear axle pivot bushings (A) and the shock absorbers (B)

28.11 With the steering wheel in the locked position and the vehicle raised, grasp the front tire as shown and try to move it back-and-forth - if any play is noted, check the steering gear mounts and tie-rod ends for looseness

28.14 Inspect the inner and outer driveaxle boots for loose clamps, cracks or signs of leaking lubricant (inner boot shown)

DRIVEAXLE BOOT CHECK

➧ **Refer to illustration 28.14**

13 The driveaxle boots are very important because they prevent dirt, water and foreign material from entering and damaging the constant velocity (CV) joints. Oil and grease can cause the boot material to deteriorate prematurely, so it's a good idea to wash the boots with soap and water. Because it constantly pivots back and forth following the steering action of the front hub, the outer CV boot wears out sooner and should be inspected regularly.

14 Inspect the boots for tears and cracks as well as loose clamps (see illustration). If there is any evidence of cracks or leaking lubricant, they must be replaced as described in Chapter 8.

29 Manual transaxle lubricant change (every 60,000 miles or 48 months)

1 This procedure should be performed after the vehicle has been driven so the lubricant will be warm and therefore will flow out of the transmission more easily. Raise the vehicle and support it securely on jackstands.

2 Move a drain pan, rags, newspapers and wrenches under the transaxle.

3 Remove the transaxle drain plug at the differential portion of the case and allow the lubricant to drain into the pan (see illustration 19.2).

4 After the lubricant has drained completely, reinstall the plug and tighten it securely.

5 Remove the fill plug from the side of the transmission case (see Section 19). Using a hand pump, syringe or squeeze bottle, fill the transmission with the specified lubricant until it just reaches the bottom edge of the hole. Reinstall the fill plug and tighten it securely.

6 Lower the vehicle.

7 Drive the vehicle for a short distance, then check the drain and fill plugs for leakage.

30 Differential lubricant change, automatic transaxles (every 60,000 miles or 48 months)

1 This procedure should be performed after the vehicle has been driven, so the lubricant will be warm and therefore will flow out of the differential more easily.

2 Raise the vehicle and support it securely on jackstands. You'll be draining the lubricant by removing the speedometer driveshaft (see Section 22), so move a drain pan, rags, newspapers and wrenches under the vehicle.

3 Remove the speedometer driveshaft and use a suction pump (available at auto parts stores) or syringe to withdraw the old differential lubricant.

4 Use a hand pump, syringe or squeeze bottle to fill the differential housing with the specified lubricant, through the speedometer driveshaft hole. Use the bottom of the speedometer driveshaft as a dipstick to check the level (see illustration 22.3).

5 When the level is correct, reinstall the speedometer driveshaft and Vehicle Speed Sensor (see Section 22 and Chapter 6).

31 Spark plug wires (2.0L engine)- replacement (every 100,000 miles or 60 months)

➧ **Refer to illustration 31.6**

1 The spark plug wires should be checked at the recommended intervals and whenever new spark plugs are installed in the engine. The 2.0L gasoline engine has an ignition coil pack (no distributor is used) and short plug wires from each coil terminal to the corresponding spark plug. The 1.8L turbo gasoline engine has individual coils mounted over each spark plug. These coils have a boot that fits directly onto the spark plug, with no conventional secondary wires.

2 Remove the engine cover (see illustration 26.4). Begin this procedure by making a visual check of the spark plug wires while the

engine is running. In a darkened garage (make sure there is adequate ventilation) start the engine and observe each plug wire. Be careful not to come into contact with any moving engine parts. If there is a break in the wire, you will see arcing or a small spark at the damaged area. If arcing is noticed, make a note to obtain new wires, then allow the engine to cool and check the distributor cap and rotor.

3 Disconnect the plug wire from one spark plug (with the engine Off). To do this, grab the rubber boot, twist slightly and pull the wire free. Do not pull on the wire itself, only on the rubber boot. A boot pulling tool is helpful (see illustration 26.6a).

4 Check inside the boot for corrosion, which will look like a white crusty powder. Push the wire and boot back onto the end of the spark plug. It should be a tight fit on the plug. If it isn't, remove the wire and use a pair of pliers to carefully crimp the metal connector inside the boot until it fits securely on the end of the spark plug.

5 Using a clean rag, wipe the entire length of the wire to remove any built-up dirt and grease. Once the wire is clean, check for holes, burned areas, cracks and other damage. Don't bend the wire excessively or the conductor inside might break.

6 Disconnect the wire from the coil pack. Pull the wire straight off the coil (see illustration). Pull only on the rubber boot during removal. Check for corrosion and a tight fit in the same manner as the spark plug end. Reattach the wire to the coil.

7 Check the remaining spark plug wires one at a time, making sure they are securely fastened at both ends when the check is complete.

8 If new spark plug wires are required, purchase a new set for your specific engine model. Wire sets are available pre-cut, with the rubber boots already installed. Remove and replace the wires one at a time to avoid mix-ups in the firing order. The wire routing is extremely important, so be sure to note exactly how each wire is situated before removing it. Compare your old ones to the new ones to insure obtaining the correct replacements. Consult the cylinder and coil numbering illustration at the beginning of this Chapter for the correct routing of the wires.

31.6 Use a spark plug boot pulling tool to remove a spark plug wire from the coil pack - never pull on the wire itself

32 Automatic transaxle fluid and filter change (every 100,000 miles or 60 months)

➧ Refer to illustration 32.9

1 At the specified intervals, the transmission fluid should be drained and replaced. Since the fluid will remain hot long after driving, perform this procedure only after the engine has cooled down completely.

2 Before beginning work, purchase the specified transmission fluid (see *Recommended lubricants and fluids* at the end of this Chapter) and a new filter and pan gasket.

3 Other tools necessary for this job include a floor jack, jackstands to support the vehicle in a raised position, a drain pan capable of holding at least eight quarts, newspapers and clean rags.

4 Raise the vehicle and support it securely on jackstands.

5 Place the drain pan underneath the transmission pan.

6 Remove the transmission pan mounting bolts, then carefully pry the transmission pan loose with a screwdriver, leaving one end loosely attached so that the fluids flows out at one end of the pan.

⁂ WARNING:

The transmission fluid may be hot.

7 Remove the pan completely and carefully clean the gasket surface of the transmission to remove all traces of the old gasket and sealant.

8 Clean the pan with solvent and wipe it dry.

➡Note: Some models are equipped with magnets in the transmission pan to catch metal debris. Clean the magnet thoroughly. A small amount of metal material is normal at the magnet. If there is considerable debris, consult a dealer or transmission specialist.

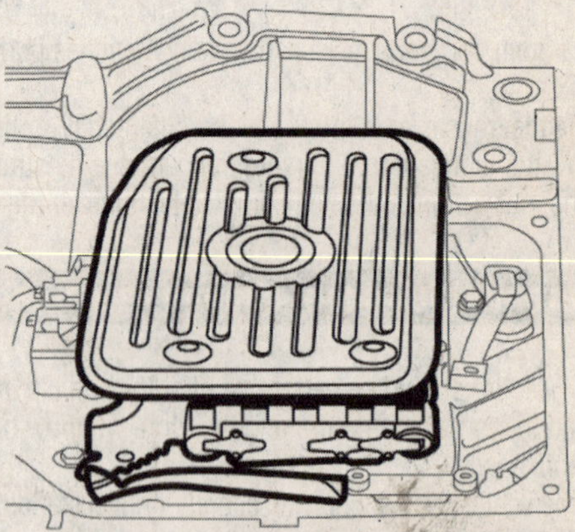

H32327

32.9 Pull the filter down from the valve body, then replace the rubber seal

9 Remove the filter from the valve body inside the transmission (see illustration).

➡Note: Be very careful not to gouge the delicate aluminum gasket surface on the valve body.

10 Install a new seal and filter. On many replacement filters, the gasket is attached to the filter to simplify installation.

11 Make sure the gasket surface on the transmission pan is clean, then install a new gasket on the pan. Put the pan in place against the transmission and, working around the pan, tighten each bolt a little at a time to the torque listed in this Chapter's Specifications.

12 Reinstall the components removed for access to the pan bolts.

13 Lower the vehicle and add approximately 3 quarts of the specified type of automatic transmission fluid through the filler tube (see Section 21).

14 With the transmission in Park and the parking brake set, run the engine at a fast idle, but don't race it.

15 Move the gear selector through each range and back to Park, then let the engine idle for a few minutes. Check the fluid level. It will probably be low. Add enough fluid to bring the level up until it just drips out of the level-check hole.

➡Note: The fluid level must be checked at a certain temperature (see Section 21).

16 Check under the vehicle for leaks during the first few trips.

Specifications

Recommended lubricants and fluids

➡Note: Listed here are manufacturer recommendations at the time this manual was written. Manufacturers occasionally upgrade their fluid and lubricant specifications, so check with your local auto parts store for current recommendations.

Engine oil	
Gasoline engines	API "certified for gasoline engines"
Diesel engine	API service CF4 or CG4
Viscosity	See accompanying chart
Fuel	
Gasoline engines	Unleaded gasoline, 87 octane minimum
Diesel engine	No. 2 diesel fuel
Automatic transaxle fluid	VW automatic transmission fluid
Manual transaxle lubricant	SAE 75W90 synthetic gear oil
Differential (automatic transaxles)	SAE 75W90 synthetic gear oil
Power steering fluid	VW hydraulic oil
Brake fluid	DOT 4 brake fluid
Engine coolant	50/50 mixture of silicate and phospate-free antifreeze and demineralized water

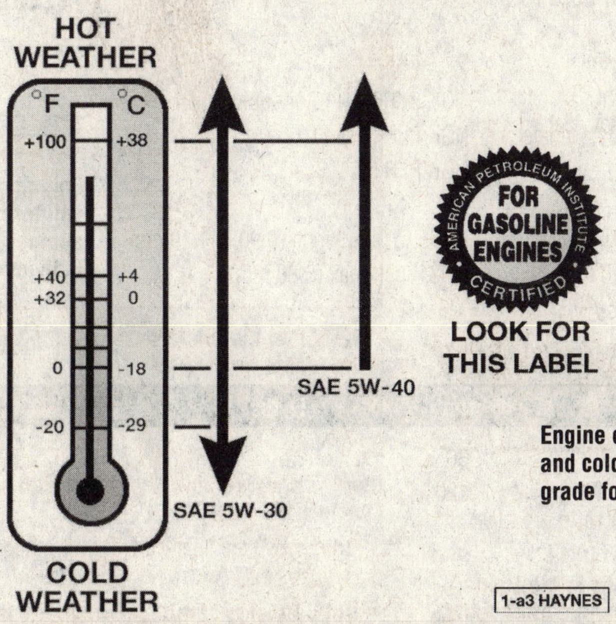

Engine oil viscosity chart - for best fuel economy and cold starting, select the lowest SAE viscosity grade for the expected temperature range

Recommended lubricants and fluids (continued)

Hood and trunk hinge lubricant	Lubriplate, lubricant aerosol spray
Door hinge and check spring grease	NLGI no. 2 multi-purpose grease
Key lock cylinder lubricant	Graphite spray
Hood latch assembly lubricant	NLGI no. 2 multi-purpose grease
Door latch lubricant	NLGI no. 2 multi-purpose grease or equivalent

Capacities*

Engine oil (including filter)	
1.8L	
Through 2000	4.4 qts (4.2 liters)
2001 through 2003	4.8 qts (4.5 liters)
2004 and 2005	4.4 qts (4.2 liters)
2.0L	4.4 qts (4.2 liters)
1.9L diesel engine	4.8 qts (4.5 liters)
Manual transaxle	
02J model	2.1 qts (2.0 liters)
02M model	2.4 qts (2.3 liters)
Automatic transaxle	
01M 4-speed	3.2 qts (3.0 liters)
09G 6-speed	Filled for life with approximately 7.5 quarts (7.0 liters)
02E 6-speed direct shift	5.5 quarts (5.2 liters)
Differential lubricant, 02M and 01M models	0.8 qts (0.75 liters)
Cooling system	
Gasoline engines	5.3 qts (5.0 liters)
Diesel engines	6.3 qts (6.0 liters)

**All capacities approximate. Add as necessary to bring to appropriate level.*

Brakes

Disc brake pad wear limit (lining only)	1/8 inch (3 mm)
Drum brake shoe wear limit (lining only)	3/32 inch (2.3 mm)

Ignition system

Spark plug type	
2.0L	
Code AEG	NGK BKR6 ET-10 or equivalent
All other engine codes, 2004 and later	NGK PZFR 5D-11
1.8L	NGK PFR 6Q
Spark plug gap	
2.0L engine	0.035 to 0.043 inch (0.90 to 1.10 mm)
1.8L engine	0.031 inch max (0.80 mm)
Firing order	1-3-4-2

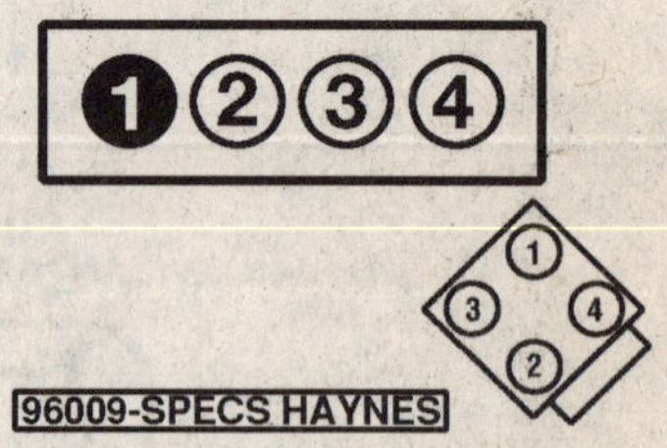

Engine cylinder location and coil terminal number (gasoline engines, except 1.8L has individual coils at each spark plug)

Torque specifications

Torque specifications	Nm	Ft-lbs
Spark plugs	30	22
Wheel bolts	120	87

2A GASOLINE ENGINES

Section

Reference to other Chapters

CHECK ENGINE light on - See Chapter 6
Cylinder compression check - See Chapter 2C
Drivebelt check, adjustment and replacement - See Chapter 1
Engine - removal and installation - See Chapter 2C
Engine oil and filter change - See Chapter 1
Engine overhaul - general information - See Chapter 2C
Spark plug replacement - See Chapter 1
Water pump - removal and installation - See Chapter 3

1 General information

➧ **Refer to illustration 1.1**

⁂ CAUTION 1:

Avoid disconnecting the battery whenever possible! Disconnecting the battery can cause severe driveability problems that require a special scan tool to remedy. See Chapter 5, Section 1 for the use of an auxiliary power source before disconnecting the battery.

⁂ CAUTION 2:

These models are equipped with an anti-theft radio. Before performing a procedure that requires disconnecting the battery, make sure you have the proper activation code.

➡**Note: The engine cover must be removed before performing many of the procedures in this Chapter (see illustration).**

This Part of Chapter 2 is devoted to in-vehicle repair procedures for the 1.8L and the 2.0L four-cylinder gasoline engines. These engines utilize cast-iron engine blocks with aluminum cylinder heads. The 1.8L engine is turbocharged and utilizes dual overhead camshafts. The 2.0L engine is normally aspirated and utilizes a single overhead camshaft. Hydraulic lifters are used to actuate the valves on both engines. The aluminum cylinder heads are equipped with pressed-in valve guides and hardened valve seats. The oil pump is mounted at the front of the engine and is driven by a chain from the crankshaft.

To positively identify these engines, locate the engine designation on the engine cover under the hood.

All information concerning engine removal and installation and engine block and cylinder head overhaul can be found in Part C of this Chapter.

1.1 The engine cover is fastened to the engine by several retaining nuts (arrows) - pry out the plastic caps, remove the nuts and detach the cover from the engine - it will also be necessary to remove the oil dipstick on 2.0L engines

The following repair procedures are based on the assumption that the engine is installed in the vehicle. If the engine has been removed from the vehicle and mounted on a stand, many of the steps outlined in this Part of Chapter 2 will not apply.

The Specifications included in this Part of Chapter 2 apply only to the procedures contained in this Part. Part C of Chapter 2 contains the Specifications necessary for cylinder head and engine block rebuilding.

2 Repair operations possible with the engine in the vehicle

Many major repair operations can be accomplished without removing the engine from the vehicle.

Clean the engine compartment and the exterior of the engine with some type of degreaser before any work is done. It will make the job easier and help keep dirt out of the internal areas of the engine.

Depending on the components involved, it may be helpful to remove the hood to improve access to the engine as repairs are performed (refer to Chapter 11 if necessary). Cover the fenders to prevent damage to the paint. Special pads are available, but an old bedspread or blanket will also work.

If vacuum, exhaust, oil or coolant leaks develop, indicating a need for gasket or seal replacement, the repairs can generally be made with the engine in the vehicle. The intake and exhaust manifold gaskets, oil pan gasket, crankshaft oil seals and cylinder head gasket are all accessible with the engine in place.

Exterior engine components, such as the intake and exhaust manifolds, the oil pan, the oil pump, the water pump, the starter motor, the alternator and the fuel system components can be removed for repair with the engine in place.

Since the cylinder head can be removed without pulling the engine, camshaft and valve component servicing can also be accomplished with the engine in the vehicle. Replacement of the timing belt and pulleys is also possible with the engine in the vehicle.

In extreme cases caused by a lack of necessary equipment, repair or replacement of piston rings, pistons, connecting rods and rod bearings is possible with the engine in the vehicle. However, this practice is not recommended because of the cleaning and preparation work that must be done to the components involved.

3 Top Dead Center (TDC) for number one piston - locating

➧ Refer to illustrations 3.5 and 3.6

1 Top Dead Center (TDC) is the highest point in the cylinder that each piston reaches as it travels up-and-down when the crankshaft turns. Each piston reaches TDC on the compression stroke and again on the exhaust stroke, but TDC generally refers to piston position on the compression stroke. The timing marks on the vibration damper/crankshaft pulley installed on the front of the crankshaft are referenced to the number one piston at TDC.

2 Positioning the piston(s) at TDC is an essential part of procedures such as timing belt and sprocket replacement.

3 In order to bring any piston to TDC, the crankshaft must be turned using one of the methods outlined below. When looking at the timing belt end of the engine, normal crankshaft rotation is clockwise.

⁂ WARNING:

Before beginning this procedure, be sure to place the transmission in Park or Neutral, set the parking brake and remove the ignition key.

- a) *The preferred method is to turn the crankshaft with a large socket and breaker bar attached to the large bolt threaded into the center of the crankshaft pulley.*
- b) *A remote starter switch, which may save some time, can also be used. Attach the switch leads to the S (switch) and B (battery) terminals on the starter motor. Once the piston is close to TDC, use a socket and breaker bar as described in the previous paragraph.*
- c) *If an assistant is available to turn the ignition switch to the Start position in short bursts, you can get the piston close to TDC without a remote starter switch. Use a socket and breaker bar as described in Paragraph a) to complete the procedure.*

4 Disable the ignition system by disconnecting the primary electrical connectors at the ignition coil pack/modules (see Chapter 5).

5 Remove the spark plugs (see Chapter 1) and install a compression gauge in the number one cylinder (see illustration). Turn the crankshaft clockwise with a socket and breaker bar as described above.

6 When the piston approaches TDC, compression will be noted on the compression gauge. Continue turning the crankshaft until the notch in the crankshaft damper is aligned with the TDC mark on the front cover (see illustration). At this point number one cylinder is at TDC on the compression stroke. If the marks aligned but there was no compression, the piston was on the exhaust stroke; continue rotating the crankshaft 360-degrees (1-turn) and line-up the marks.

➡Note: If a compression gauge is not available, TDC for the No.1 piston can be obtained by simultaneously aligning the marks on the camshaft (timing belt) sprocket with the marks on the rear timing cover and the marks on the crankshaft damper with the TDC mark on the front cover (see illustration 5.9).

7 After the number one piston has been positioned at TDC on the compression stroke, TDC for any of the remaining cylinders can be located by turning the crankshaft 180 degrees and following the firing order (refer to the *Specifications*). Rotating the engine 180 degrees past TDC #1 will put the engine at TDC compression for cylinder #3.

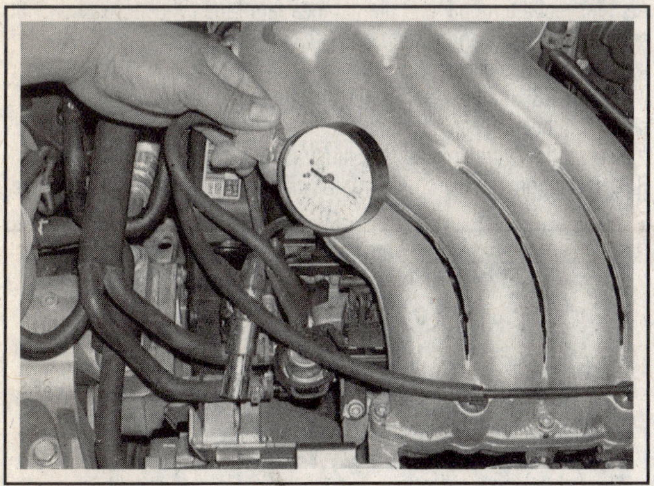

3.5 A compression gauge can be used in the number one plug hole to assist in finding TDC

3.6 Align the notch on the crankshaft drivebelt pulley with the mark on the timing belt cover (arrows)

4 Valve cover - removal and installation

REMOVAL

➧ Refer to illustrations 4.2, 4.3 and 4.6

1 Remove the engine cover (see illustration 1.1).

2 On 1.8L engines, remove the secondary air injection valve solenoid, the overrun recirculation valve solenoid, the vacuum reservoir and vacuum hoses from the top of the valve cover (see illustration). Also remove the ignition coils (see Chapter 5) and the breather hose from the top of the valve cover.

3 On 2.0L engines, remove the upper intake plenum (see Section 8) and the crankcase breather valve from the top of the valve cover (see illustration)

4 Remove the upper timing belt cover from the engine (see illustration 5.5).

5 On 1.8L engines, remove the ground strap from the valve cover.

6 Remove the retaining nuts and detach the valve cover from the cylinder head. On 2.0L engines it will be necessary to detach the upper rear timing belt cover and the valve cover sealing flange reinforcement strips before removing the valve cover (see illustration).

7 If the cover is stuck to the head, bump the end with a block of wood and a hammer to jar it loose. If that doesn't work, try to slip a flexible putty knife between the head and cover to break the seal.

CAUTION:

Don't pry at the cover-to-head joint or damage to the sealing surfaces may occur, leading to oil leaks after the cover is reinstalled.

INSTALLATION

➧ Refer to illustrations 4.9a, 4.9b and 4.9c

8 The mating surfaces of the housing or cylinder head and cover must be clean when the cover is installed. Use a gasket scraper to remove all traces of sealant and old gasket material, then clean the mating surfaces with lacquer thinner or acetone. If there's residue or oil on the mating surfaces when the cover is installed, oil leaks may develop. Also inspect the rubber end plug at the rear of the cylinder head on 1.8L engines for cracks and damage. Now would be a good time to replace it, if damage has occurred.

9 On 1.8L engines, apply RTV sealant to the corners of the camshaft front bearing cap and to the camshaft drive chain tensioner where they meet the cylinder head (see illustrations). On 2.0L engines, apply RTV sealant only to the corners of the camshaft front bearing cap where

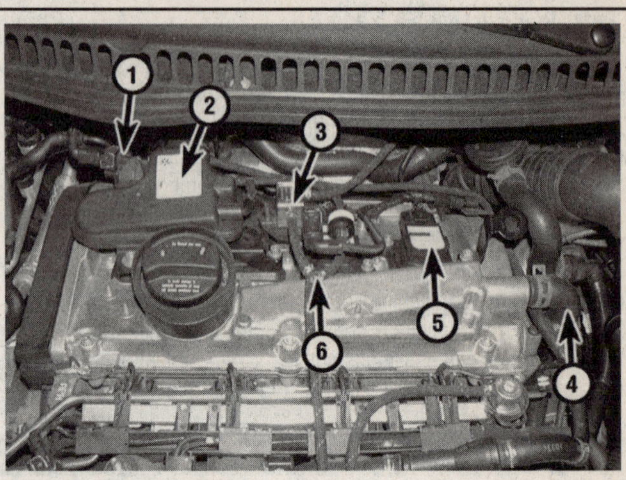

4.2 On 1.8L engines, detach the following components to access the valve cover

1 Secondary air injection solenoid
2 Vacuum reservoir
3 Overrun recirculation solenoid
4 Crankcase breather hose
5 Ignition coil(s)
6 Vacuum hoses

4.3 On 2.0L engines, detach the crankcase breather valve from the valve cover

4.6 Valve cover mounting details - 2.0L engine

A Upper rear timing belt cover
B Valve cover reinforcement strips
C Valve cover mounting bolts
D Air injection tube bracket

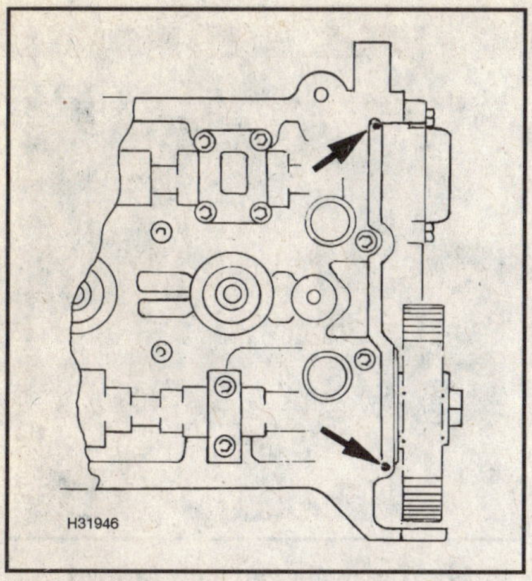

4.9a On 1.8L engines, apply sealant to the two points at the front camshaft bearing cap . . .

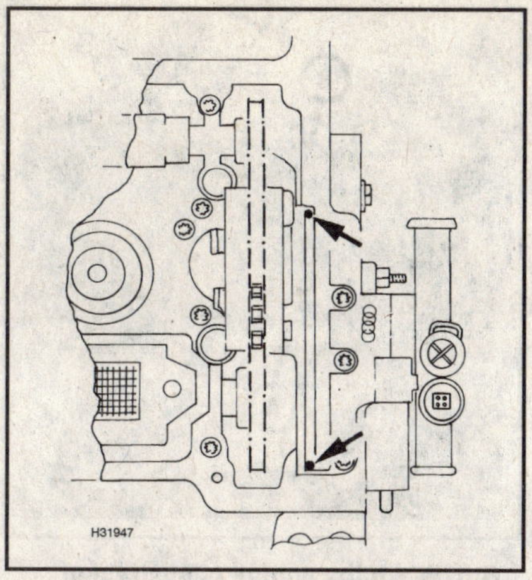

4.9b . . . and to the corners of the camshaft drive chain tensioner where they meet the cylinder head

4.9c On 2.0L engines, apply sealant to the corners of the front camshaft bearing cap only

it meets the cylinder head (see illustration).

10 Position a new valve cover gasket over the studs on the cylinder head. On 1.8L engines, install the spark plug tube grommet gasket over the studs on the cylinder head with the index marks facing the timing belt end of the engine.

11 Install the valve cover and any brackets removed, then tighten the retaining nuts to the torque listed in this Chapter's Specifications in several steps.

12 Reinstall the remaining parts, run the engine and check for oil leaks.

5 Timing belt and sprockets - removal, inspection and installation

⁂ WARNING:

Wait until the engine is completely cool before beginning this procedure.

➡Note: Do not rotate the crankshaft or the camshaft separately during this procedure with the timing belt removed as damage to valves may occur. Only rotate the camshaft a few degrees as necessary to align the camshaft sprocket marks with the marks on the rear timing cover.

REMOVAL

➧ Refer to illustrations 5.3, 5.5, 5.9, 5.11a, 5.11b, 5.12a, 5.12b, 5.13, 5.14a, 5.14b, 5.16 and 5.17

1 Relieve the fuel system pressure (see Chapter 4A). Remove the engine cover (see illustration 1.1).

2 Partially drain the engine coolant (see Chapter 1) and remove the coolant reservoir and hoses from the engine compartment (see Chapter 3).

3 Remove the fuel lines from the fuel distribution block/leak detection pump located on the passenger side of the engine compartment (see illustration).

4 Remove the spark plugs and drivebelt (see Chapter 1).

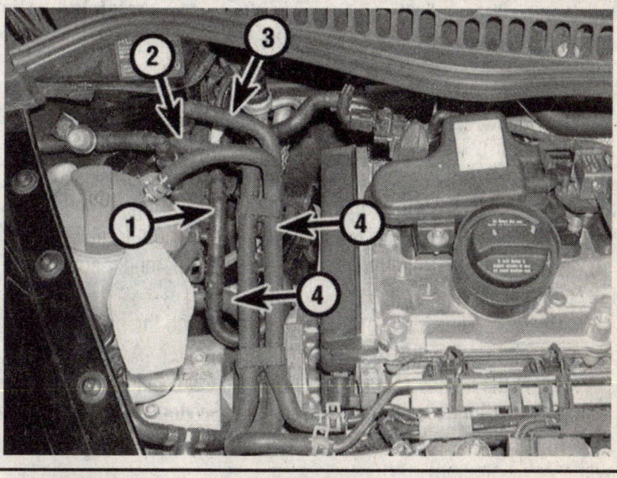

5.3 Remove the following hoses from the engine compartment to access the front of the engine - be sure to mark the hoses to ensure correct reinstallation

1 *Vacuum hose for the leak detection pump*
2 *Fuel return line*
3 *Fuel supply line*
4 *Coolant reservoir hoses*

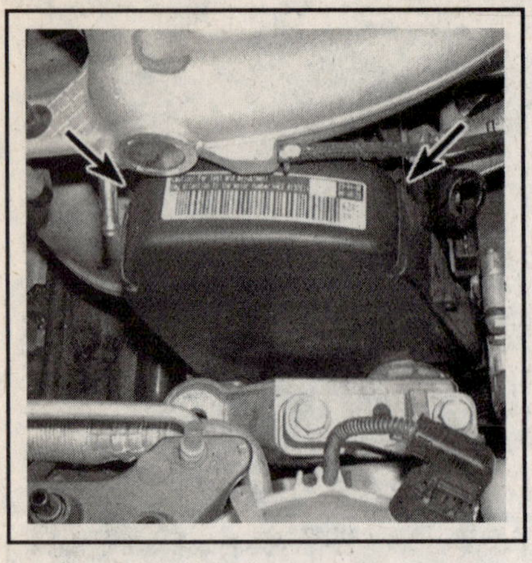

5.5 The upper timing belt cover is retained by two clips (arrows)

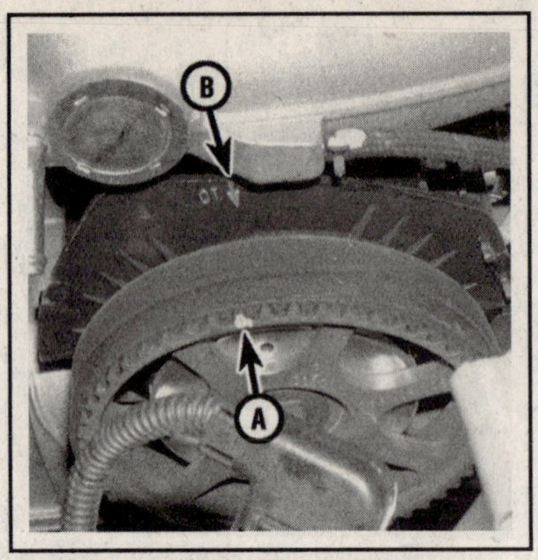

5.9 When the engine is positioned at TDC for the No.1 cylinder on the compression stroke, the camshaft sprocket mark (A) will be aligned with the rear cover mark (B)

5.11a The upper two bolts (arrows) securing the engine mount support bracket can be accessed from above . . .

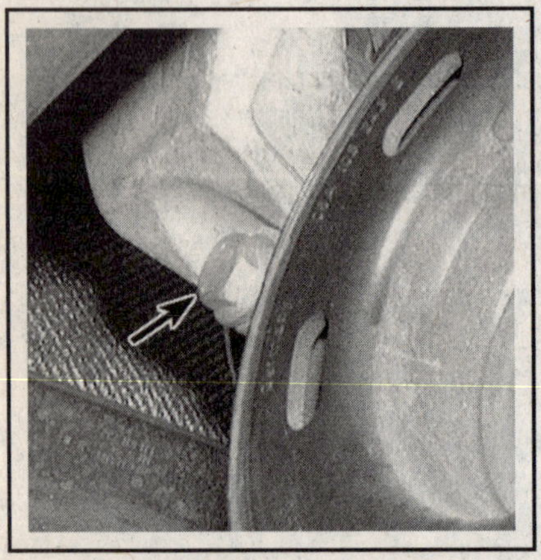

5.11b . . . and the lower bolt is accessed from below - because of the lack of clearance, it will be necessary to loosen all three bolts and remove the support bracket and the bolts together

5.12a Center timing belt cover retaining bolts (arrows)

5.12b Lower timing belt cover retaining bolts (arrows)

5 Unclip and remove the upper timing belt cover from the engine (see illustration).

6 Block the rear wheels and set the parking brake. Loosen the lug nuts on the right front wheel and raise the vehicle. Support the front of the vehicle securely on jackstands and remove the right front wheel.

7 Remove the engine under cover and the lower fender apron to allow access to the bottom of the engine.

8 On 1.8L engines, remove the lower hoses and the pipes leading from the intercooler to the turbocharger (see Chapter 4A, if necessary).

9 Rotate the engine in the normal direction of rotation (clockwise) until the No.1 cylinder is located at TDC (see Section 3). Verify that the camshaft sprocket mark is aligned with the mark on the rear timing belt cover (see illustration).

10 Use a strap wrench to hold the crankshaft pulley from rotating. Loosen the crankshaft drive sprocket retaining bolt and the crankshaft pulley bolts, then remove the crankshaft pulley (see Section 11). After the bolts are loosened, verify that the crankshaft has not moved from TDC.

➡Note: Loosening the drive sprocket bolt is only required if the crankshaft drive sprocket is expected to be removed. It is not typically necessary to remove the drive sprocket when you're simply replacing a timing belt, but it will need to be removed if you are replacing the crankshaft front oil seal or housing. If you do remove the drive sprocket, obtain a new bolt (the manufacturer doesn't recommend re-using it).

11 Support the engine from underneath with a jack and a block of wood, then remove the right (passenger side) engine mount (see Section 17) and the engine mount support bracket from the front of the engine (see illustrations).

12 Detach the retaining screws from the center and lower timing belt covers and remove the covers (see illustrations).

13 If you plan to re-use the timing belt, apply match marks on the sprocket and belt and an arrow indicating direction of travel on the belt (see illustration).

14 Release the timing belt tensioner (see illustrations).

➡Note: On 1.8L engines, it will be necessary to purchase an M5 x 55 stud, nut and washer at your local hardware store to compress the tensioner enough to lock it in place. After compressing the tensioner it can be locked into place using a small drill bit or similar tool.

15 Remove the timing belt from the engine, taking care to avoid twisting or kinking it excessively.

16 If you're removing the upper part of the belt only, for camshaft

5.13 If you intend to re-use the timing belt, apply directional marks on the belt and the rear timing belt cover

5.14a On 2.0L engines, loosen the timing belt tensioner retaining nut (arrow) and let the tensioner rotate to release the tension on the belt

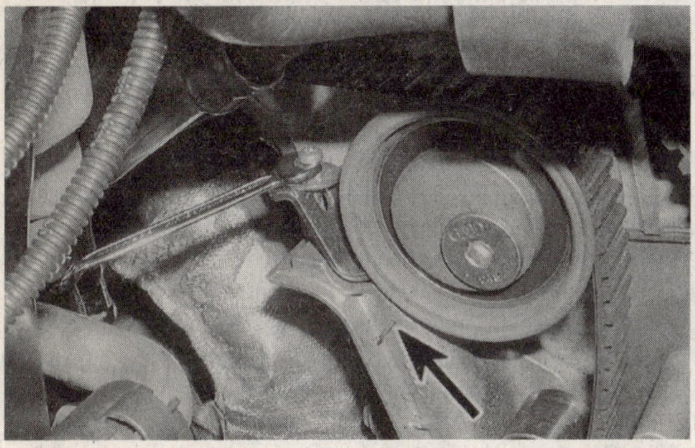

5.14b On 1.8L engines, use an M5 x 55 stud, nut and washer to compress the tensioner piston until the holes in the piston and the tensioner housing align, then lock the tensioner in place with a small drill bit or similar device (arrow)

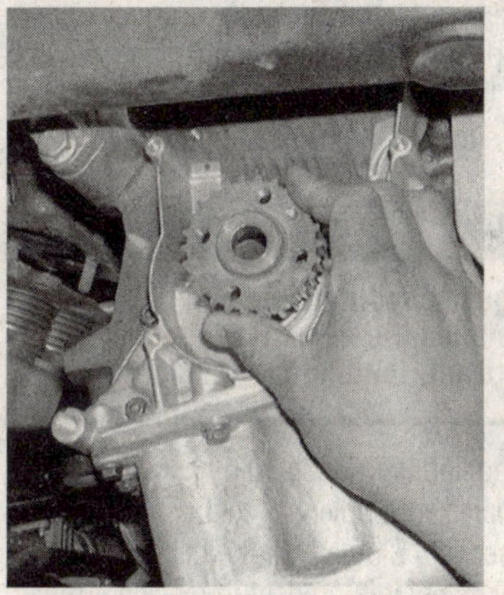

5.16 With the crankshaft drive sprocket retaining bolt removed, the crankshaft sprocket is easily removed from the engine

seal replacement or cylinder head removal, it isn't necessary to detach the belt from the crankshaft sprocket. If the sprocket is worn or damaged, or if you need to replace the crankshaft front oil seal, remove the drive sprocket retaining bolt which was loosened in Step 10 and detach the crankshaft sprocket from the crankshaft (see illustration).

17 If the camshaft sprocket is damaged or needs to be removed for other procedures such as cylinder head removal, use a spanner wrench or similar tool to hold the sprocket in place as the sprocket retaining bolt is loosened, then remove the camshaft sprocket from the end of the camshaft (see illustration).

INSPECTION

➧ Refer to illustration 5.19

CAUTION:

Do not bend, twist or turn the timing belt inside out. Do not allow it to come in contact with oil, coolant or fuel. Do not turn the crankshaft or camshaft more than a few degrees (if necessary for tooth alignment) while the timing belt is removed.

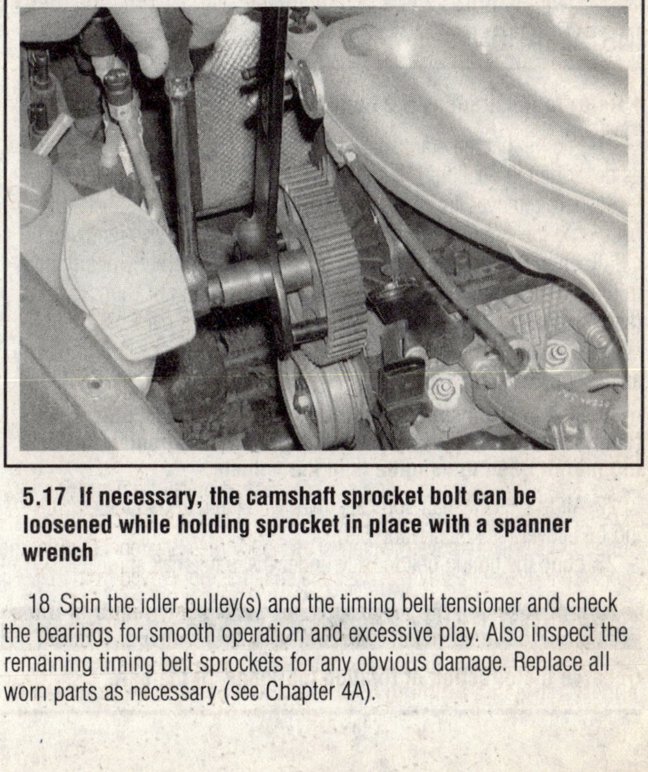

5.17 If necessary, the camshaft sprocket bolt can be loosened while holding sprocket in place with a spanner wrench

18 Spin the idler pulley(s) and the timing belt tensioner and check the bearings for smooth operation and excessive play. Also inspect the remaining timing belt sprockets for any obvious damage. Replace all worn parts as necessary (see Chapter 4A).

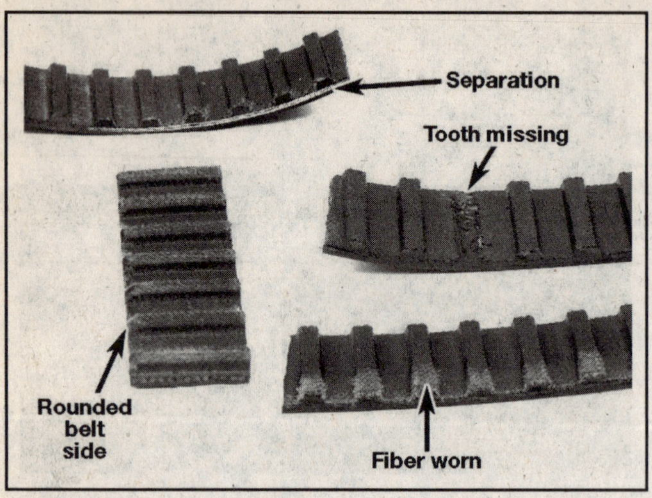

5.19 Check the timing belt for cracked and missing teeth - wear on one side of the belt indicates sprocket misalignment problems

19 Examine the belt for evidence of contamination by coolant or lubricant. If this is the case, find the source of the contamination before progressing any further. Check the belt for signs of wear or damage, particularly around the leading edges of the belt teeth (see illustration).

CAUTION:

If the belt appears to be in good condition and can be re-used, it is essential that it is reinstalled the same way around, otherwise accelerated wear will result, leading to premature failure.

20 Replace the belt if its condition is in doubt; the cost of belt replacement is negligible compared with potential cost of the engine repairs, should the belt fail in service. Similarly, if the belt is known to have covered more than 80,000 miles, it is prudent to replace it regardless of condition, as a precautionary measure.

INSTALLATION

Refer to illustrations 5.21 and 5.26

21 Ensure that the crankshaft is still set to TDC on No. 1 cylinder, as described in Section 3. If any of the timing sprockets or idler pulleys where removed for inspection or needed replacement, install them back onto the engine now. If the timing belt tensioner was removed on a 2.0L engine, be sure the rear tab on the roller is properly positioned in the rear timing belt cover and install the retaining nut hand tight (see illustration). If the timing belt tensioner was removed on a 1.8L engine, make sure it is reinstalled with the tensioner locked into place as described in Step 14.

Note: A small inspection mirror mounted on a flexible shaft (available at auto parts stores) is helpful in examining the timing marks when the engine is in the vehicle.

22 Align the camshaft sprocket mark with the mark on the rear timing belt cover (see illustration 5.9).

23 Loop the timing belt loosely under the crankshaft sprocket.

CAUTION:

Observe the direction of rotation markings on the belt.

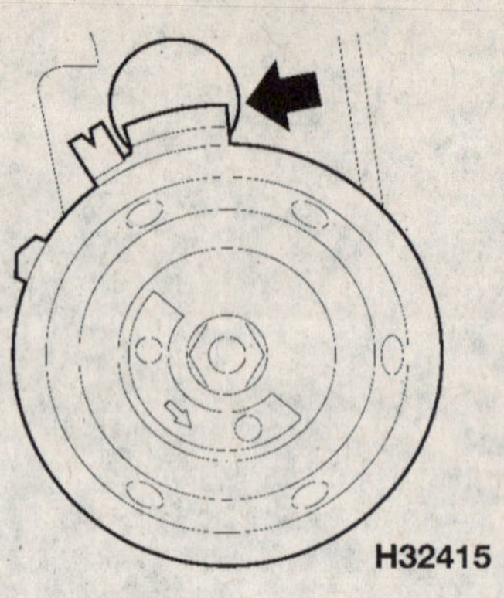

5.21 On 2.0L engines, the timing belt tensioner is properly seated when the tab on the tensioning roller is engaged in the hole on the rear timing belt cover

24 Engage the timing belt teeth with the crankshaft sprocket, then maneuver it into position over the idler pulley, water pump and the camshaft sprocket. Ensure the belt teeth seat correctly on the sprockets, then install the belt around the timing belt tensioner.

Note: Slight adjustments to the position of the camshaft sprocket may be necessary to achieve this.

25 Ensure that the 'front run' of the belt is taut and all the slack is in the section of the belt that passes over the tensioner roller.

26 On 2.0L engines, tension the belt by turning the eccentrically-mounted tensioner counterclockwise until the upper marks align; two holes are provided in the side of the tensioner hub for this purpose - a pair of sturdy right-angled snap-ring pliers is a suitable substitute for the correct VW tool (see illustration). After the tensioner marks are aligned properly, tighten the tensioner locknut to the specified torque.

27 On 1.8L engines, remove the tensioner locking pin and loosen the nut on the tensioner compressing tool fabricated earlier. Make sure to take all the slack out of the timing belt while loosening the nut (see illustration 5.14b).

28 At this point, double check to make sure that the crankshaft is still set to TDC on No. 1 cylinder (see Section 3) and the camshaft sprocket mark is aligned with the mark on the rear timing belt cover.

29 Rotate the crankshaft through two complete revolutions. Reset the engine to TDC on No. 1 cylinder, with reference to Section 3 and check the alignment marks again. Also re-check the timing belt tension and adjust it, if necessary.

5.26 Tensioning the timing belt using a pair of snap-ring pliers in the belt tensioner - the belt is tensioned correctly when the notch on the front of the tensioner is aligned with the raised mark (2.0L engines only)

30 Install the lower and center timing belt covers, then install the engine mount support bracket to the engine. Install the passenger engine mount (see Section 17) to secure the engine as the remaining components are installed onto the vehicle.

31 The remainder of the installation is the reverse of removal.

6 Camshaft and lifters - removal and installation

➡Note 1: The camshaft and lifters should always be thoroughly inspected before installation and camshaft endplay should always be checked prior to camshaft removal. Although the hydraulic lifters are self adjusting and require no periodic service, there is an in-vehicle procedure for checking excessively noisy hydraulic lifters. Refer to Chapter 2C for the camshaft and lifter inspection procedures.

➡Note 2: 2.0L engines have only one camshaft. 1.8L engines have two; the exhaust camshaft is driven by the timing belt, while the intake camshaft is driven by a chain connected to the rear of the exhaust camshaft.

REMOVAL

➧ Refer to illustrations 6.2, 6.4, 6.5, 6.6a, 6.6b, 6.7, 6.9, 6.10, 6.11a and 6.11b

1 Remove the engine cover (see illustration 1.1).

2 Remove the valve cover (see Section 4). Also remove the oil deflector(s) to expose the camshaft(s) (see illustration).

3 Remove the timing belt and camshaft sprocket (see Section 5).

6.2 The plastic oil deflector(s) are easily removed by simply lifting them off the cylinder head (2.0L engine shown, 1.8L engine similar)

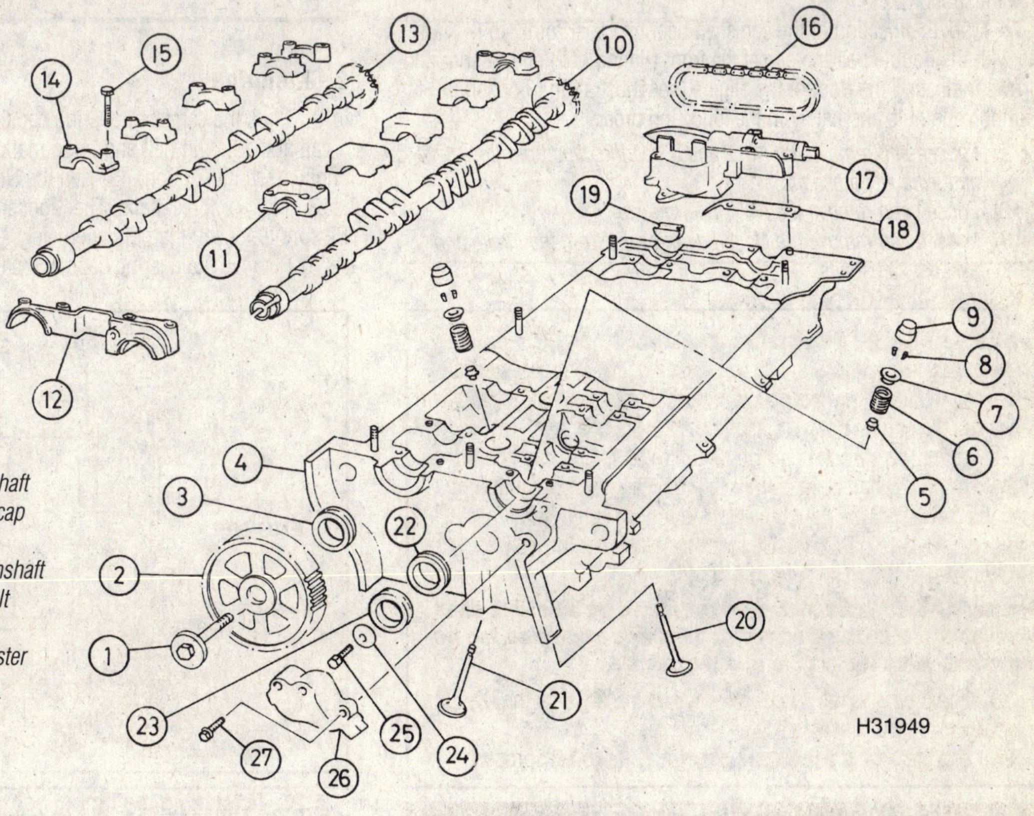

6.4 Exploded view of the camshafts and related components - 1.8L engine

1 *Camshaft sprocket bolt*
2 *Camshaft sprocket*
3 *Oil seal*
4 *Cylinder head*
5 *Valve stem oil seal*
6 *Valve spring*
7 *Valve cap*
8 *Split keepers*
9 *Hydraulic lifter*
10 *Intake camshaft*
11 *Bearing cap, intake camshaft*
12 *Front combined bearing cap*
13 *Exhaust camshaft*
14 *Bearing cap, exhaust camshaft*
15 *Camshaft bearing cap bolt*
16 *Drive chain*
17 *Automatic camshaft adjuster*
18 *Rubber seal*
19 *Rubber grommet*
20 *Exhaust valve*
21 *Intake valve*
22 *Oil seal*
23 *Camshaft position sensor rotor*
24 *Tapered washer*
25 *Rotor securing bolt*
26 *Camshaft position sensor*
27 *Camshaft position sensor securing bolt*

6.5 With the notches in the rear drive chain sprockets aligned with the arrows on the rear bearing caps, apply match marks on the chain with a permanent marker - be sure to wipe the oil from the chain and sprockets first, so the marker will adhere to the components - the number of rollers between the marks should be exactly 16

1.8L engine

4 Remove the camshaft position sensor (see Chapter 6A). Remove the camshaft sensor reluctor ring from the end of the intake camshaft (see illustration).

5 Mark the position of the camshaft drive chain in relationship to the sprockets and the marks on the rear bearing cap (see illustration). This will ensure that the drive chain is installed in exactly the same direction and position from which it was removed.

6 Using a special tool, compress the camshaft drive chain tensioner (see illustration).

7 Mark the location of the camshaft bearing caps from 1 to 6, starting with the double bearing cap at the front (timing belt) end. Also mark arrows indicating the front of the engine (see illustration). Loosen the bearing caps nuts alternately in the following order:

1) *Loosen and remove the No. 2 and 4 bearing caps from the intake and exhaust camshafts*
2) *Loosen and remove the No. 1 bearing cap*
3) *Loosen and remove the No. 6 bearing caps from the intake and exhaust camshafts*
4) *Loosen and remove the drive chain tensioner mounting bolts*
5) *Loosen and remove the No. 3 and 5 bearing caps from the exhaust camshafts*
6) *Loosen and remove the No. 3 and 5 bearing caps from the intake camshafts*

8 Remove the camshafts and the drive chain tensioner as an assembly from the cylinder head. Separate the tensioner and the drive chain from the camshafts on a workbench.

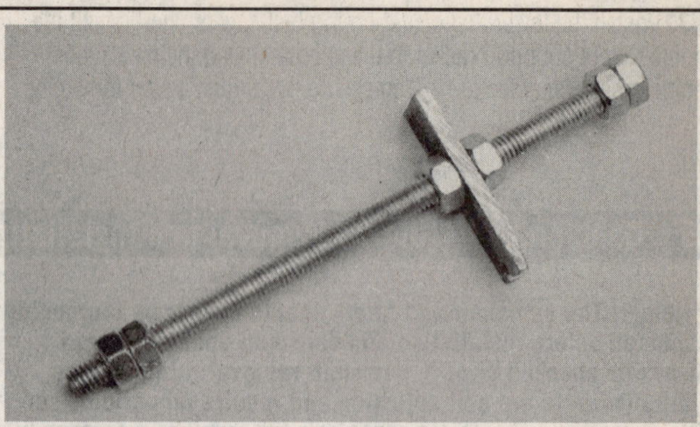

6.6a Home-made tool for locking the camshaft adjuster, or chain tensioner, in position - 1.8L engine

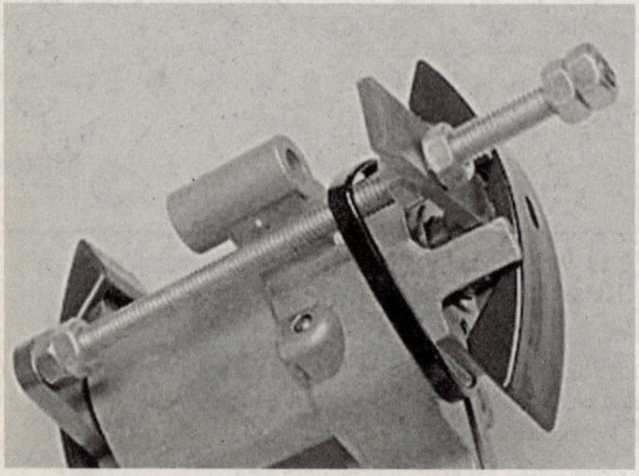

6.6b Home-made tool in position, locking the camshaft adjuster in its compressed condition (shown with camshaft adjuster removed for clarity) - 1.8L engine

CAUTION:

Keep the caps in order. They must go back in the same location and direction they were removed from.

2.0L engine

9 Mark the camshaft bearing caps from 1 to 5, starting with the No. 1 cap at the timing belt end. Also mark arrows indicating the front of the engine (see illustration). Loosen the No. 1, 3 and 5 camshaft bearing caps in two or three steps. Then loosen the No. 2 and 4 bearing caps. Be sure to loosen the nuts alternately and evenly.

10 Remove the bearing caps and camshaft.

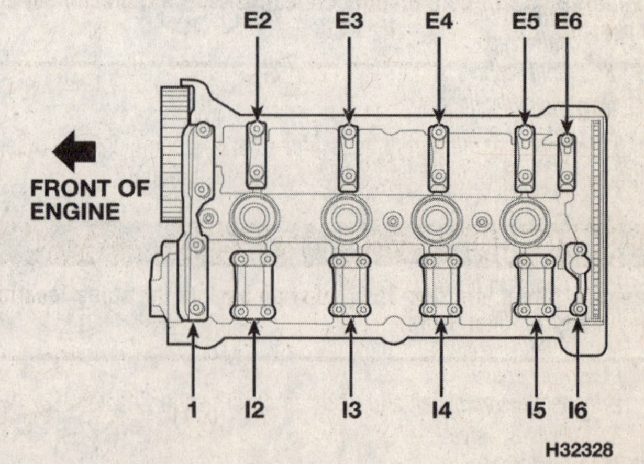

6.7 On 1.8L engines, the camshaft bearing caps should be marked as shown with a number and letter stamp or a marker to ensure correct reinstallation

6.9 On 2.0L engines, simply mark the bearing caps from 1 to 5 starting at the front (timing belt end of the engine)

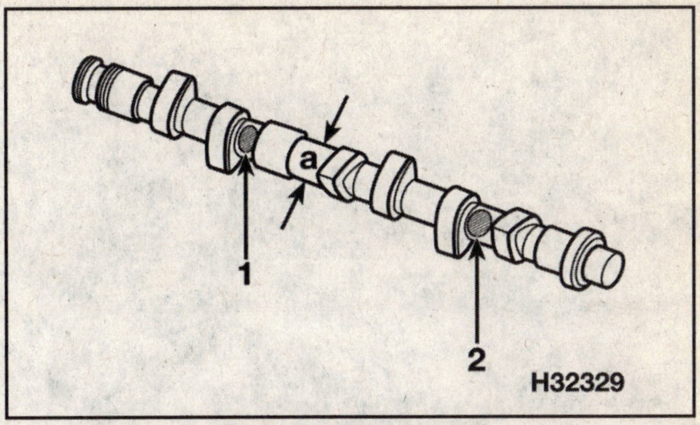

6.10 Camshaft identification markings - 2.0L engine

6.11a The lifters can be removed from the cylinder head with a magnet . . .

6.11b . . . and stored in individually marked plastic bags or a divided box as shown - be sure to keep them in order with their respective valves

CAUTION:

Keep the caps in order. They must go back in the same location they were removed from. After the camshaft has been removed, make a note of the camshaft identification markings (see illustration). This will help identify the camshaft if a replacement is necessary.

All engines

11 Remove the lifters from the cylinder head, keeping them in order with their respective valve and cylinder (see illustrations).

CAUTION:

Keep the lifters in order. They must go back in the same location they were removed from.

12 Inspect the camshaft and lifters as described in Chapter 2C.

INSTALLATION

Refer to illustrations 6.14, 6.15, 6.19, 6.21, 6.22 and 6.24

13 Apply clean engine oil onto the sides of the hydraulic lifters, and install them into position in their bores in the cylinder head. Push them down until they contact the valves, then lubricate the camshaft lobe contact surfaces.

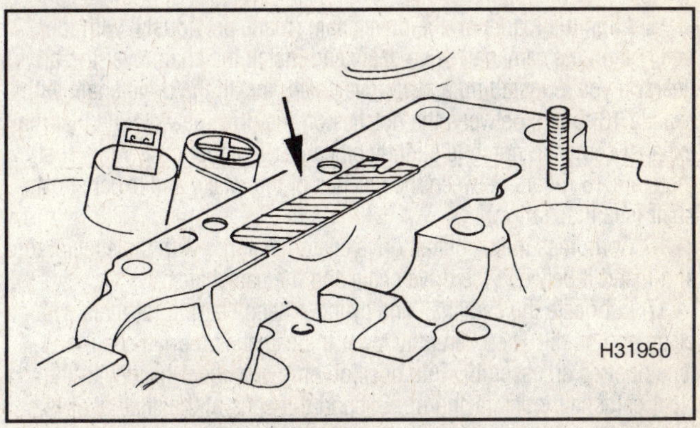

6.14 When installing the drive chain tensioner gasket, apply RTV sealant at the shaded area - 1.8L engine

1.8L engine

14 Clean the mating surfaces of the drive chain tensioner and the cylinder head and install a new drive chain tensioner gasket (see illustration).

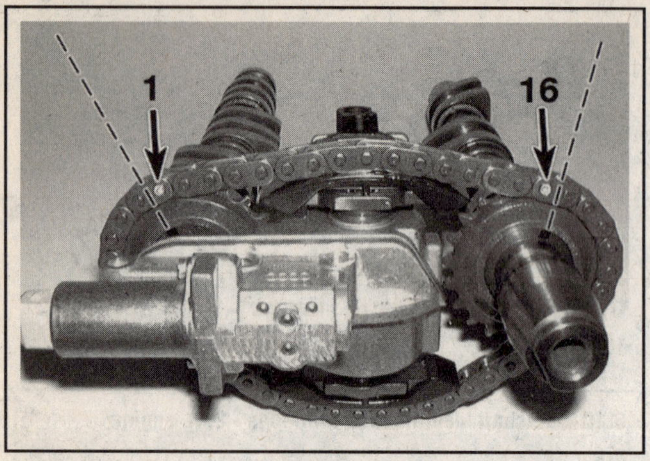

6.15 On 1.8L engines, install the drive chain over the camshaft drive gears with 16 rollers between the notches

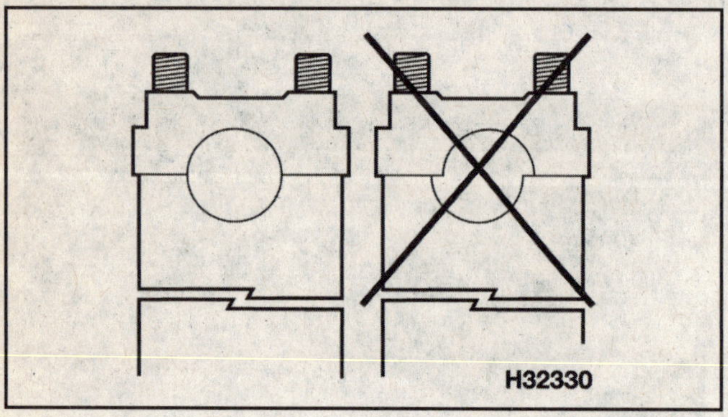

6.21 The camshaft bearing cap mounting holes are drilled off-center - be sure they're installed on the correct journal and with the arrows (made earlier) facing towards the timing belt end of the engine

15 Align the marks on the drive chain (made previously) with the notches on the camshaft drive gears and install the chain over the drive gears. If you're installing a new drive chain, install the drive chain with exactly 16 rollers between the notches on the drive gears. Note that the exhaust camshaft notch is slightly off center. In either case verify that there are 16 rollers between the notches of the intake and exhaust camshaft (see illustration).

16 Compress the camshaft drive chain tensioner with the special tool and insert it between the drive chain and the camshafts.

17 Lubricate the camshaft and cylinder head bearing journals with clean engine oil. Then carefully lower the camshafts, drive chain and tensioner as an assembly into position on the cylinder head with the No.1 camshaft lobes facing up. Support the ends of the shaft as it is inserted, to avoid damaging the lobes and journals.

18 Install the drive chain tensioner over the dowels on the cylinder head and tighten the bolts to the torque listed in this Chapter's Specifications.

19 Install the camshaft bearing caps in the reverse order of removal (see Step 7). Be sure to apply a small amount of RTV sealant to the mating surface of the front bearing cap before installing it (see illustration). After the bearing caps have been tightened, remove the drive chain tensioning tool from the tensioner. Reconfirm that there are 16 rollers between the notches of the intake and exhaust camshaft and that the notches align with the arrows on the caps and that the No.1 camshaft lobes face up.

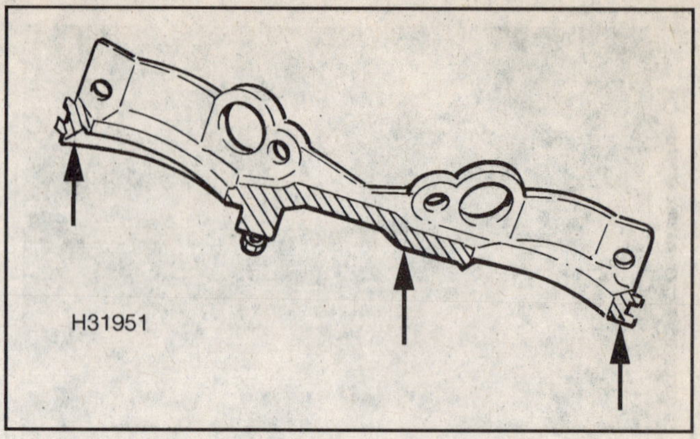

6.19 Apply a small amount of RTV sealant to the No.1 bearing cap at the shaded areas - 1.8L engine

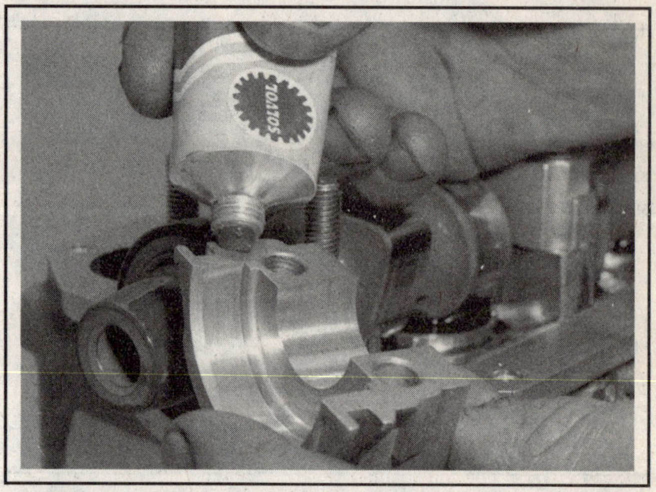

6.22 Apply a small amount of RTV sealant to the mating surface of the No. 1 bearing cap

2.0L engine

20 Lubricate the camshaft and cylinder head bearing journals with clean engine oil. Then carefully lower the camshaft into position on the cylinder head with the No.1 camshaft lobes facing up (180 degrees from the cylinder head mating surface). Support the ends of the shaft as it is inserted, to avoid damaging the lobes and journals.

21 Oil the upper surfaces of the camshaft bearing journals, then Install the No. 2 and 4 bearing caps over the camshaft and tighten the retaining nuts alternately and diagonally to the specified torque.

➡Note: The camshaft bearing caps are drilled off center (see illustration), make sure they're installed on the correct journal and with the arrows made in Step 9 facing towards the timing belt end of the engine.

22 Apply a small amount of RTV sealant to the mating surface of No. 1 bearing cap and install it, along with cap Nos. 3 and 5 over the camshaft (see illustration). Tighten the nuts to the specified torque starting with the center cap and working towards the end caps.

All engines

23 Clean the oil seal housing bore(s) and lubricate the lip of a new camshaft oil seal with clean engine oil and locate it over the end of the camshaft. Slide the seal along the camshaft until it locates squarely in the housing bore.

6.24 Gently drive the new seal into place with the spring side facing the engine

24 Using a socket with an outside diameter slightly smaller than the outside diameter of the seal, carefully drive the new seal into place with a hammer (see illustration). Make sure it's installed squarely and driven in to the same depth as the original. If a socket isn't available, a short section of pipe will also work.

➡Note: On 1.8L engines, make sure to install both camshaft oil seals, one for the intake camshaft and one for the exhaust camshaft.

25 Install the camshaft sprocket and on 1.8L engines, the camshaft position sensor reluctor ring, conical washer and retaining bolt. Tighten the bolts to the torque listed in this Chapter's Specifications.

26 Install the timing belt (see Section 5). When installing the timing belt, make sure the crankshaft is at TDC for the No. 1 cylinder and the camshaft sprocket mark is aligned with the rear timing cover.

27 The remainder of installation is the reverse of removal.

⁂ CAUTION:

If new lifters were used, wait at least 30 minutes before starting the vehicle to allow the lifters to bleed down. Failure to do so will result in serious engine damage.

7 Valve springs, retainers and seals - replacement

➧ Refer to illustrations 7.4, 7.8, 7.10, 7.15a, 7.15b, 7.15c, 7.16 and 7.17

➡Note: Broken valve springs and defective valve stem seals can be replaced without removing the cylinder heads. Two special tools and a compressed air source are normally required to perform this operation, so read through this Section carefully and rent or buy the tools before beginning the job.

1 Remove the valve cover referring to Section 4. Then refer to Section 6 and remove the camshaft and lifters.

2 Remove the spark plug from the cylinder which has the defective component. If all of the valve stem seals are being replaced, all of the spark plugs should be removed.

3 Turn the crankshaft until the piston in the affected cylinder is at top dead center (TDC) on the compression stroke (see Section 3 for instructions). If you're replacing all of the valve stem seals, begin with cylinder number one and work on the valves for one cylinder at a time. Move from cylinder-to-cylinder following the firing order sequence (see the Specifications listed at the end of this Chapter).

4 Thread an adapter into the spark plug hole (see illustration) and connect an air hose from a compressed air source to it. Most auto parts stores can supply the air hose adapter.

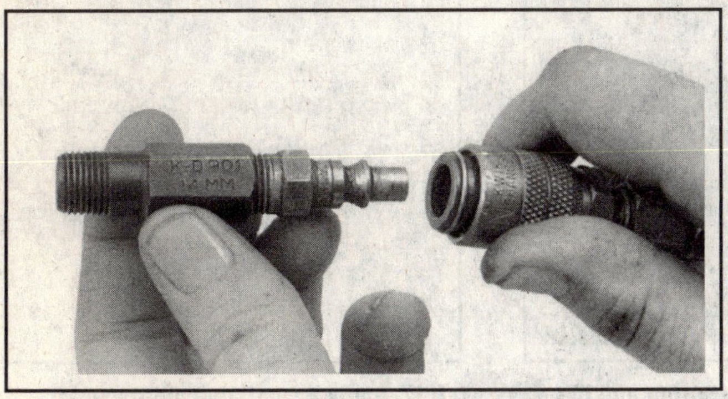

7.4 This is what the typical air hose adapter that threads into the spark plug hole looks like - they're commonly available at auto parts stores

➡Note: Many cylinder compression gauges utilize a screw-in fitting that may work with your air hose quick-disconnect fitting.

5 Apply compressed air to the cylinder. The valves should be held in place by the air pressure.

⁂ WARNING:

If the cylinder isn't exactly at TDC, air pressure may force the piston down, causing the engine to quickly rotate. DO NOT leave a wrench on the crankshaft drive sprocket bolt or you may be injured by the tool.

6 Stuff shop rags into the cylinder head holes around the valves to prevent parts and tools from falling into the engine.

7 Using a socket and a hammer gently tap on the top of each valve spring retainer several times. This will break the bond between the valve keeper and the spring retainer and allow the keeper to separate from the valve spring retainer as the valve spring is compressed.

8 Use a valve spring compressor to compress the spring. Remove the keepers with small needle-nose pliers or a magnet (see illustration).

7.8 While the valve spring tool is compressing the spring, remove the keepers with a small magnet or pliers

7.10 The old valve stem seals can be removed with a pair of needle-nose pliers

7.15a Install the protective plastic sleeve over the valve end face to avoid damage to the valve seal as the seal is installed

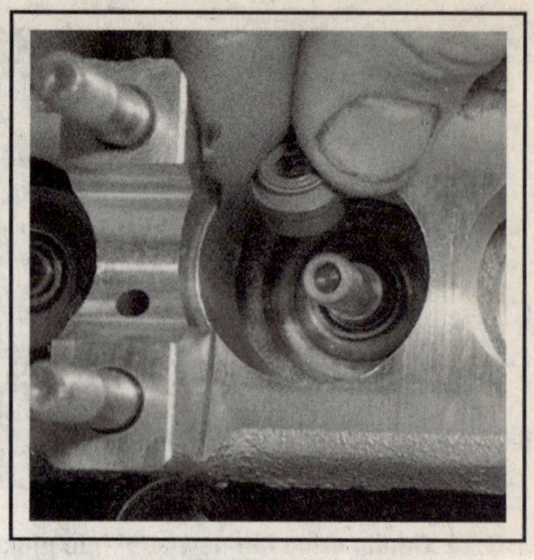

7.15b Push a new valve stem seal over the valve and down to the top of the guide, then remove the plastic installation tool

➡Note: Several different types of tools are available for compressing the valve springs with the head in place. Be sure to purchase or rent the "Import type" that bolts to the top of the cylinder head. This type uses a support bar across the cylinder head for leverage as the valve spring is compressed. The lack of clearance surrounding the valve springs on these engines prohibits typical types of valve spring compressors from being used.

9 Remove the valve spring and retainer.

➡Note: If air pressure fails to retain the valve in the closed position during this operation, the valve face or seat may be damaged. If so, the cylinder head will have to be removed for repair.

10 Remove the old valve stem seals, noting differences between the intake and exhaust seals (see illustration).

11 Wrap a rubber band or tape around the top of the valve stem so the valve won't fall into the combustion chamber, then release the air pressure.

12 Inspect the valve stem for damage. Rotate the valve in the guide and check the end for eccentric movement, which would indicate that the valve is bent.

13 Move the valve up-and-down in the guide and make sure it doesn't bind. If the valve stem binds, either the valve is bent or the guide is damaged. In either case, the head will have to be removed for repair.

14 Reapply air pressure to the cylinder to retain the valve in the closed position, then remove the tape or rubber band from the valve stem.

15 Lubricate the valve stem with engine oil and install a new seal on the valve guide (see illustrations).

16 Install the valve spring and the spring retainer in position over the valve (see illustration).

17 Compress the valve spring and carefully position the keepers in the groove. Apply a small dab of grease to the inside of each keeper to hold it in place (see illustration).

18 Remove the pressure from the spring tool and make sure the keepers are seated.

19 Disconnect the air hose and remove the adapter from the spark plug hole.

20 Install the camshaft, lifters, timing belt and the valve cover by referring to the appropriate Sections.

21 Install the spark plug(s) and hook up the wire(s).

22 Start and run the engine, then check for oil leaks and unusual sounds coming from the valve cover area.

7.15c Gently tap the new seal in place on the guide with a socket

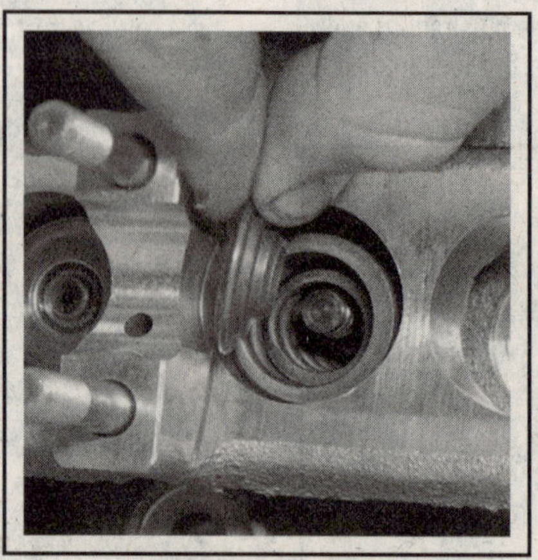

7.16 Install the valve spring and the retainer over the valve

7.17 Apply a small dab of grease to each keeper as shown here before installation - it'll hold them in place on the valve stem as the spring is released

8 Intake manifold - removal and installation

⁂ WARNING:

Wait until the engine is completely cool before beginning this procedure.

1 Remove the engine cover (see illustration 1.1).

1.8L ENGINE

➧ Refer to illustrations 8.2 and 8.9

2 Remove the upper intercooler hose and the EVAP hose from the throttle body (see illustration).

3 Disconnect the overrun solenoid and the fuel pressure regulator vacuum hoses from the intake manifold.

4 Label and detach the electrical connectors from the throttle control valve and the intake air temperature sensor.

5 Relieve the fuel system pressure and remove the fuel rail and injectors (see Chapter 4A).

6 Partially drain the engine coolant and remove the upper coolant pipe from the engine (see Chapter 3).

7 Remove the secondary air injection pump motor and support brackets (see Chapter 6A).

8 Remove the intake manifold support brace.

9 Remove the mounting nuts/bolts (see illustration), then detach the manifold and gasket from the engine.

10 Use a scraper to remove all traces of old gasket material and sealant from the manifold and cylinder head, then clean the mating surfaces with lacquer thinner or acetone. If the gasket was leaking, have the manifold checked for warpage at an automotive machine shop and resurfaced if necessary.

11 Install a new gasket, then position the manifold on the head and install the nuts/bolts.

12 Tighten the nuts/bolts in three or four equal steps to the torque listed in this Chapter's Specifications. Work from the center out towards the ends to avoid warping the manifold.

13 Install the remaining parts in the reverse order of removal.

14 Before starting the engine, check the throttle linkage for smooth operation.

15 Check the coolant and add some, if necessary, to bring it to the appropriate level. Run the engine and check for coolant and vacuum leaks.

16 Road test the vehicle and check for proper operation of all accessories, including the cruise control system.

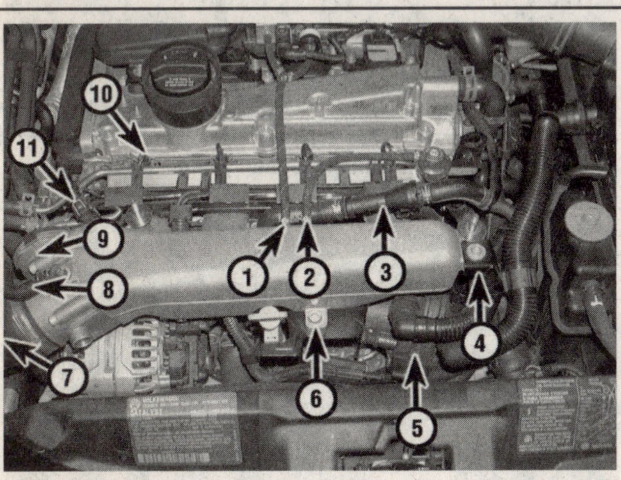

8.2 On 1.8L engines, disconnect the following components to allow removal of the intake manifold

1 *Vacuum hose (overrun solenoid)*
2 *Vacuum hose (fuel pressure regulator)*
3 *Crankcase ventilation hose*
4 *Air injection pump mounting bracket*
5 *Secondary air injection pump motor*
6 *Air injection pump mounting bracket*
7 *Upper intercooler hose*
8 *EVAP hose*
9 *Throttle control valve*
10 *Fuel rail and injectors*
11 *Intake air temperature sensor*

2.0L ENGINE

Upper intake manifold (plenum)

➧ Refer to illustrations 8.19a, 8.19b, 8.20, 8.21a and 8.21b

17 Clamp-off the coolant hoses that connect to the throttle body (see illustration 8.19a).

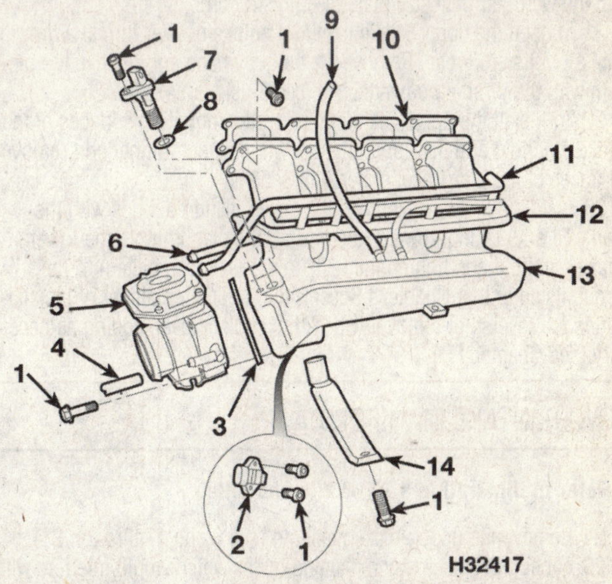

8.9 Exploded view of the intake manifold and related components - 1.8L engine

1 *Bolt*
2 *Throttle position sensor*
3 *Gasket*
4 *Connecting hose*
5 *Throttle valve control module*
6 *Fuel supply line/hose*
7 *Intake air temperature sensor*
8 *O-ring*
9 *Vacuum hose*
10 *Gasket*
11 *Fuel return line/hose*
12 *Fuel rail, fuel injector and fuel pressure regulator*
13 *Intake manifold*
14 *Support bracket*

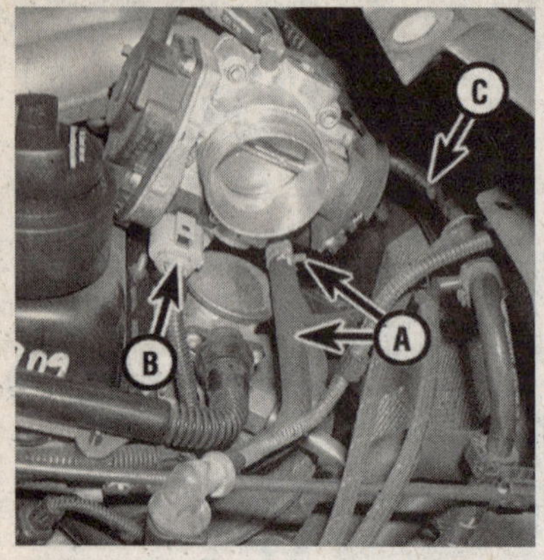

8.19a Remove the coolant hoses (A) and the electrical connector (B) and the brake booster vacuum hose (C) from the throttle body - 2.0L engine

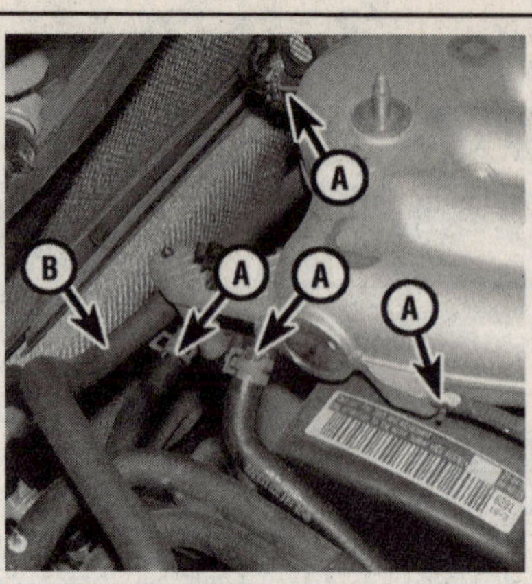

8.19b Detach the vacuum hoses (A) and the coolant reservoir hose (B) at the rear of the upper intake manifold - 2.0L engine

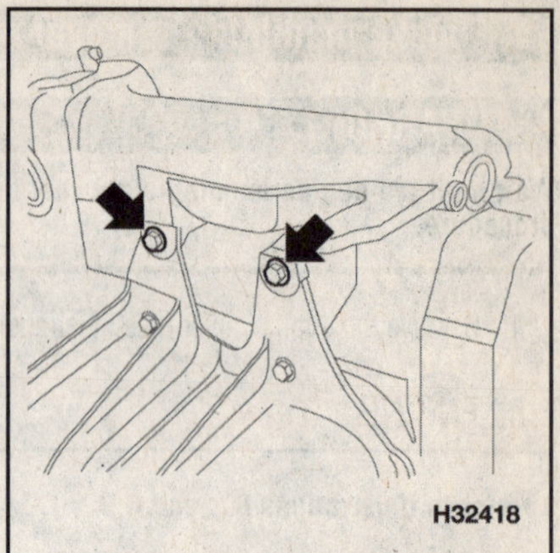

8.20 Intake manifold to warm air deflector plate mounting bolts

➡Note: The coolant hose leading to the coolant reservoir is easier to clamp-off near the right corner of the upper intake manifold, rather than where it connects to the throttle body.

18 Refer to Chapter 4A and remove the air intake duct and the accelerator cable.

19 Label and disconnect the hoses and electrical connectors attached to the plenum and throttle body (see illustrations). Be prepared for some coolant leakage at the throttle body.

20 Remove the bolts at the rear of the upper intake manifold securing it to the warm air deflector plate (see illustration).

21 Loosen the upper intake manifold bolts a quarter turn at a time until all bolts are loose. Remove the bolts by hand and remove the plenum with the throttle body attached (see illustrations).

22 To install the upper manifold, clean the mounting surfaces of the intake manifold and the upper plenum with lacquer thinner and remove all traces of the old gasket material or sealant.

23 Install the new gasket over the intake manifold studs with the marks (if any) facing upward, then install the plenum onto the lower intake manifold and tighten the bolts in a criss cross pattern to the torque listed in this Chapter's Specifications. The remainder of installation is the reverse of removal. Check the coolant level, adding as necessary (see Chapter 1).

LOWER INTAKE MANIFOLD

➧ Refer to illustration 8.27, 8.29a and 8.29b

24 Remove the upper intake manifold (see Steps 17 through 21).

25 Label and detach any remaining hoses which would interfere with the removal of the lower intake manifold.

26 Refer to Chapter 4A and relieve the fuel pressure. Remove the fuel rail and injectors from the lower intake manifold.

27 Remove the lower intake manifold support bracket and the oil dipstick tube from the engine (see illustration).

28 Remove the secondary air injection pump and bracket from the manifold (see Chapter 6A).

29 Loosen the manifold mounting bolts/nuts in 1/4-turn increments until they can be removed by hand starting at the center and moving towards the ends (see illustrations).

30 The manifold will probably be stuck to the cylinder heads and force may be required to break the gasket seal.

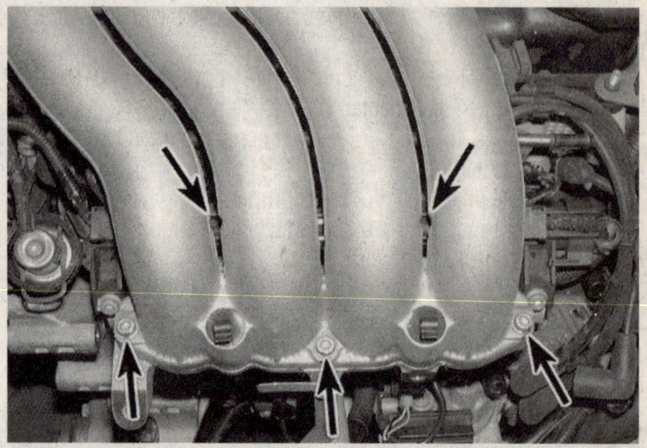

8.21a Upper intake manifold mounting bolts - 2.0L engine

8.21b Remove the upper intake manifold with the throttle body attached

⁂ CAUTION:

Don't pry between the manifold and the heads or damage to the gasket sealing surfaces may occur, leading to vacuum leaks.

8.27 Remove the support bracket (A) and the oil dipstick tube (B)

31 Carefully use a scraper to remove all traces of old gasket material and sealant from the manifold and cylinder heads, then clean the mating surfaces with lacquer thinner or acetone.

32 Install new gaskets, then position the lower manifold on the engine. Make sure the gaskets and manifolds are aligned over the studs in the cylinder heads and install the nuts.

33 Starting at the center and working towards the ends, tighten the nuts/bolts, in several steps, to the torque listed in this Chapter's Specifications.

34 The remainder of the installation is the reverse of the removal procedure. Check the coolant level, adding as necessary (see Chapter 1). Run the engine and check for fuel, coolant and vacuum leaks.

8.29a Lower intake manifold (upper) mounting bolts/nuts - 2.0L engine

8.29b Lower intake manifold (lower) mounting bolts - 2.0L engine

9 Exhaust manifold - removal and installation

Refer to illustrations 9.10, 9.11, 9.12, 9.13a and 9.13b

WARNING:

The engine must be completely cool before beginning this procedure.

REMOVAL

1 Remove the engine cover (see illustration 1.1).

2 Remove the air intake duct (see Chapter 4).

3 Remove the cowl cover (see Chapter 11).

4 On 1.8L engines, refer to Chapter 4A and remove the turbocharger and all of its related components, then proceed to Step 13.

5 On 2.0L engines, remove the upper intake manifold (see Section 8).

6 Remove the hoses from the secondary air injection control valve (see Chapter 6A).

7 Raise the front of the vehicle and support it securely on jackstands.

8 Disconnect the oxygen sensor electrical connectors and detach the wiring harness for the front O2 sensor from the retaining bracket on the manifold.

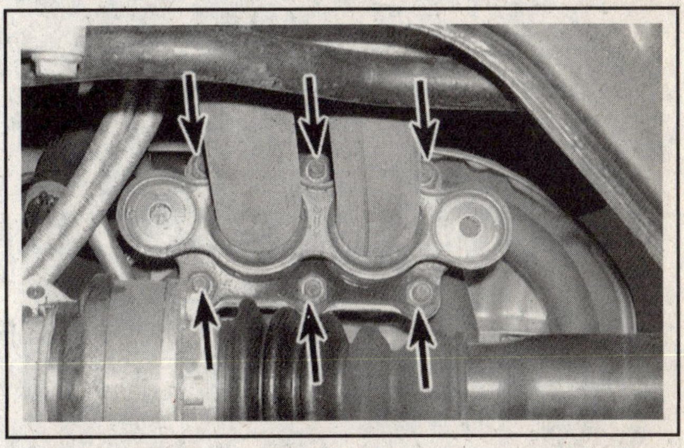

9.10 Exhaust pipe-to-exhaust manifold mounting nuts (upper arrows) - lower arrows show location of the exhaust manifold support brace - 2.0L engine

9 Apply penetrating oil to the exhaust manifold mounting nuts/bolts.

10 Disconnect the exhaust pipe from the exhaust manifold, then

9.11 Remove the air injection pipe (arrow) from the exhaust manifold and the air control valve on the warm air deflector plate - 2.0L engine

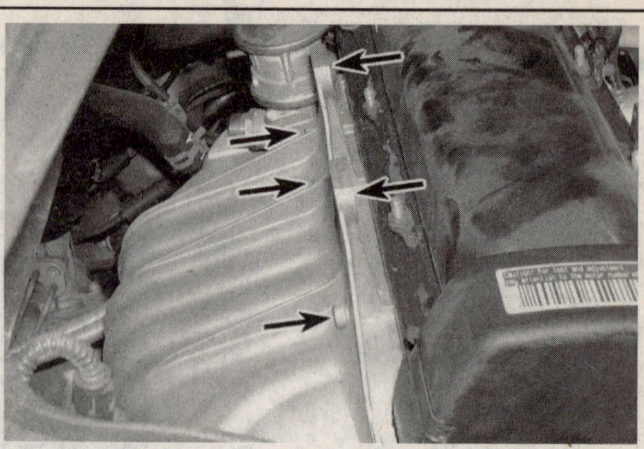

9.12 The warm air deflector plate is mounted to the cylinder head (left arrows) and the intake manifold (right arrows) - 2.0L engine

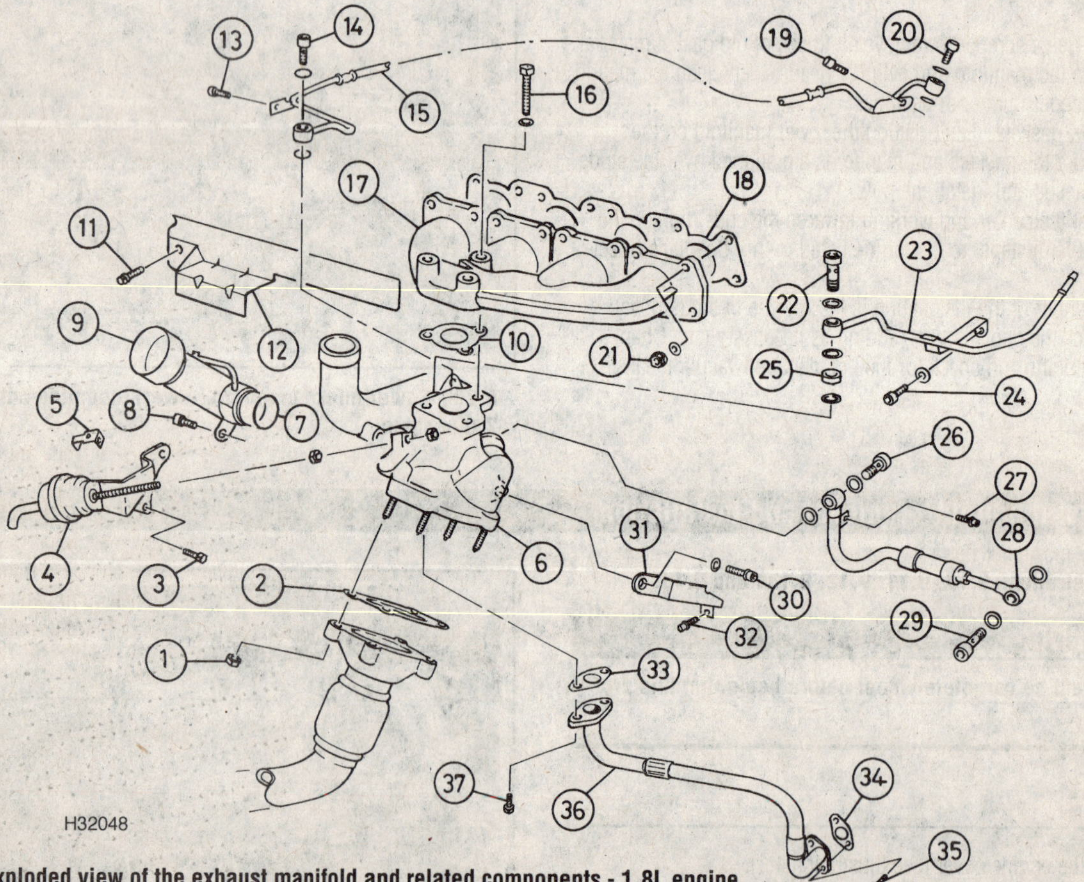

9.13a Exploded view of the exhaust manifold and related components - 1.8L engine

1 *Exhaust downpipe nuts*
2 *Exhaust downpipe gasket*
3 *Wastegate bracket bolt*
4 *Wastegate*
5 *Circlip*
6 *Turbocharger*
7 *O-ring*
8 *Mounting bolt*
9 *Air intake pipe stub*
10 *Gasket*
11 *Heat shield mounting bolt*
12 *Heat shield*
13 *Mounting bolt*
14 *Banjo bolt*
15 *Oil supply pipe*
16 *Turbocharger mounting bolts*
17 *Exhaust manifold*
18 *Manifold gasket*
19 *Mounting bolt*
20 *Banjo bolt*
21 *Manifold mounting nuts*
22 *Banjo bolt*
23 *Coolant return pipe*
24 *Mounting bolt*
25 *Spacer sleeve*
26 *Banjo bolt*
27 *Mounting bolt*
28 *Coolant supply pipe*
29 *Banjo bolt*
30 *Turbocharger-to-support bracket bolt*
31 *Turbocharger support bracket*
32 *Support bracket bolt*
33 *Gasket*
34 *Gasket*
35 *Oil return pipe flange bolt*
36 *Oil return pipe*
37 *Oil return pipe flange bolt*

remove the front exhaust pipe from the vehicle (see illustration). Remove the exhaust manifold support brace.

11 Disconnect the air injection pipe union nuts from the fittings on the exhaust manifold and the secondary air injection control valve (see illustration).

12 Remove the warm air deflector plate from the engine (see illustration).

13 Remove the nuts/bolts and detach the manifold and gasket (see illustrations).

INSTALLATION

14 Use a scraper to remove all traces of old gasket material and carbon deposits from the manifold and cylinder head mating surfaces. If the gasket was leaking, have the manifold checked for warpage at an automotive machine shop and resurfaced if necessary.

15 Position a new gasket over the cylinder head studs.

16 Install the manifold and thread the mounting nuts/bolts into place. Make sure to use hi-temp anti-seize compound on the exhaust manifold fasteners.

17 Working from the center out, tighten the nuts/bolts to the torque listed in this Chapter's Specifications in three or four equal steps.

9.13b Exhaust manifold upper mounting nuts (arrows) - 2.0L engine (lower nuts not visible in this photo)

18 Reinstall the remaining parts in the reverse order of removal.

19 Run the engine and check for exhaust leaks.

10 Cylinder head - removal and installation

➡Note: The cylinder head can be removed with the intake and exhaust manifold attached.

REMOVAL

➧ Refer to illustrations 10.9, 10.10, 10.12a and 10.12b

1 Drain the engine coolant (see Chapter 1).

2 Refer to Chapter 11 and remove the hood and the cowl cover.

3 Refer to Section 5 and remove the timing belt. Follow Steps 1 through 16. After the timing belt has been removed, reinstall the passenger side engine mount to support the engine during the removal and installation of the cylinder head.

4 On 1.8L engines, refer to Chapter 4A and remove the turbocharger and all its related components. On 2.0L engines, disconnect the front exhaust pipe from the exhaust manifold and remove the exhaust manifold support bracket (see Section 9).

5 Remove the valve cover (see Section 4).

6 Remove the secondary air injection pump and mounting bracket (see Chapter 6A).

7 Also referring to Chapter 4A, disconnect and remove the fuel supply and return lines from the fuel rail.

8 Unplug all electrical connectors and vacuum hoses from the cylinder head, labeling each wiring connector or hose to aid the installation process.

9 Refer to Chapter 3 and loosen the hose clamps and disconnect the radiator hoses from the ports on the cylinder head. Remove the coolant outlet flange from the end of the cylinder head (see illustration).

10 On 2.0L engines, loosen and withdraw the upper retaining screw from the timing belt rear cover (see illustration). Also remove the spark plug wires. On 1.8L engines, remove the locking pin and the threaded stud, nut and washer (tensioner compressing tool) from the timing belt tensioner.

10.9 Remove the coolant outlet flange (arrow) from the end of the cylinder head

10.10 On 2.0L engines, remove the upper bolt (arrow) from the rear timing belt cover

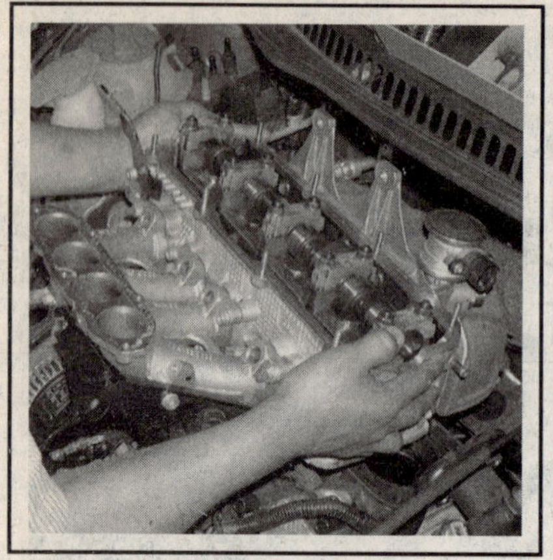

10.12a Lift the cylinder head off the engine with the manifolds attached

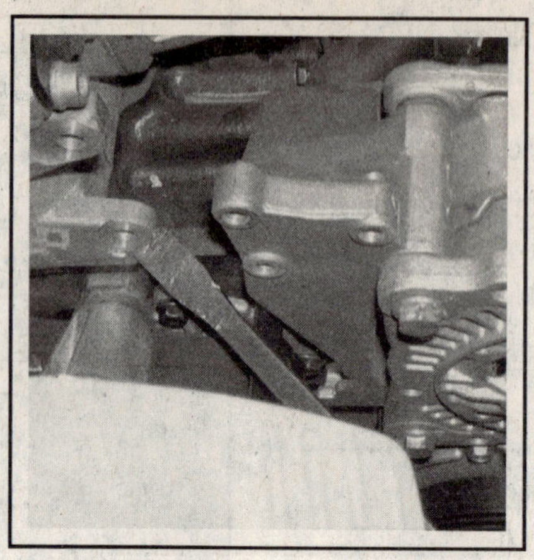

10.12b If the cylinder head is stuck, it may be necessary to pry upward on the casting protrusion to dislodge the head from the block

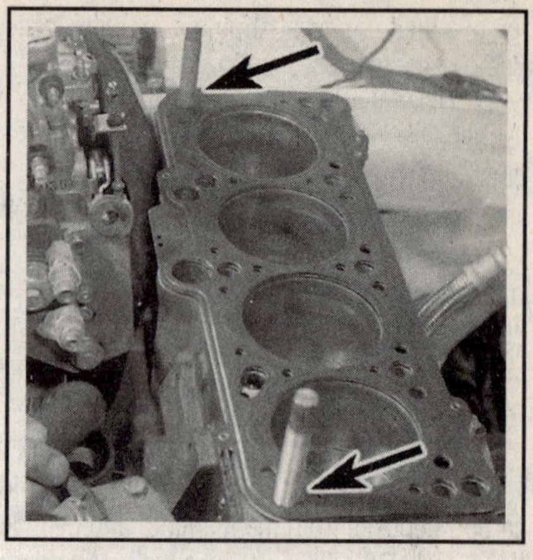

10.22 Two of the old head bolts (arrows) can be used as cylinder head alignment dowels

➡Note: This is the tool that was fabricated in Step 14 of Section 5 during the timing belt removal procedure.

11 Working in the reverse of the sequence shown in illustration 10.26a, progressively loosen the cylinder head bolts, by half a turn at a time, until all bolts can be unscrewed by hand. Discard the bolts - new ones must be installed on reassembly.

12 Check that nothing remains connected to the cylinder head, then lift the head away from the cylinder block; seek assistance if possible, as it is very heavy, especially when being removed with the manifolds (see illustration). If resistance is felt, carefully pry the cylinder head upward, beyond the gasket surface, at a casting protrusion (see illustration).

13 Remove the gasket from the top of the block. Do not discard the gasket - it will be needed for identification purposes.

14 If the cylinder head is to be disassembled for service, separate the manifold(s) as described in Sections 8 and 9. Disregard the steps that do not apply since the cylinder head is already removed from the vehicle, then proceed to Chapter 2C for overhaul procedures. Be sure to reinstall the manifolds back onto the cylinder head before installing the cylinder head on the vehicle.

INSTALLATION

➧ Refer to illustrations 10.22, 10.24, 10.26a and 10.26b

15 The mating faces of the cylinder head and cylinder block must be perfectly clean before installing the head. Use a hard plastic or wood scraper to remove all traces of gasket and carbon; also clean the piston crowns. Take particular care during the cleaning operations, as aluminum alloy is easily damaged. Also, make sure that the carbon is not allowed to enter the oil and water passages - this is particularly important for the lubrication system, as carbon could block the oil supply to the engine's components. Using adhesive tape and paper, seal the water, oil and bolt holes in the cylinder block.

16 Check the mating surfaces of the cylinder block and the cylinder head for nicks, deep scratches and other damage. If slight, they may be removed carefully with abrasive paper.

17 If warpage of the cylinder head gasket surface is suspected, use a straight-edge to check it for distortion, but note that head machining will not be possible - refer to Chapter 2C.

18 Clean out the cylinder head bolt holes using a suitable tap. Be sure they're clean and dry before installation of the head bolts.

19 It is possible for the piston crowns to strike and damage the valve heads if the camshaft is rotated with the timing belt removed and the crankshaft set to TDC. For this reason, the crankshaft must be set to a position other than TDC on No. 1 cylinder before the cylinder head is reinstalled. Use a wrench and socket on the crankshaft pulley center bolt to turn the crankshaft in the opposite direction of rotation (counterclockwise), until all four pistons are positioned halfway down their bores - approximately 90-degrees before TDC.

20 If the cylinder head has been resurfaced, make sure the valve seats have been reworked by the same amount to allow the correct piston to valve clearance before installing the cylinder head. See Chapter 2C for further information.

21 Cut off the heads from two of the old cylinder head bolts to use as alignment dowels during cylinder head installation. Also cut a slot in the end of the each bolt, big enough for a screwdriver blade, so that the alignment dowels can be removed after the cylinder head is installed. A simple hand held hacksaw can be used to fabricate the alignment dowels.

22 Install the alignment dowels in the outer rear holes of the cylinder block and position the new head gasket on the cylinder block, engaging it with the locating dowels (see illustration). Ensure that the manufacturer's "TOP" and part number markings are face up.

23 With the help of an assistant, place the cylinder head and manifolds centrally on the cylinder block, ensuring that the locating dowels engage with the recesses in the cylinder head. Check that the head gasket is correctly seated before allowing the full weight of the cylinder head to rest upon it.

➡Note: If the cylinder head had been disassembled for repair, be sure the camshaft(s) are reinstalled on the cylinder head with the No.1 cylinder camshaft lobes pointing upward.

24 Oil the threads and the underside of the bolt heads, then carefully enter each bolt into its relevant hole and screw them in hand tight (see illustration). Be sure to use NEW cylinder head bolts, as the old bolts are stretch type fasteners that will not provide the correct torque readings if reused.

25 Unscrew the homemade alignment dowels, using a flat-bladed screwdriver and install the remaining two bolts hand tight.

26 Working progressively and in the sequence shown (see illustra-

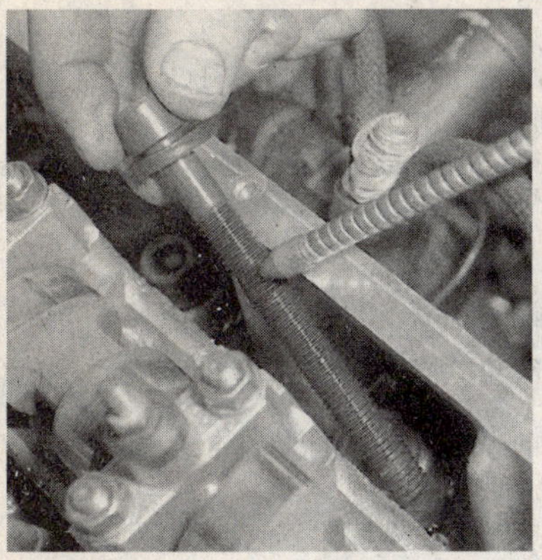

10.24 Apply oil to the threads of the cylinder head bolts before installing them

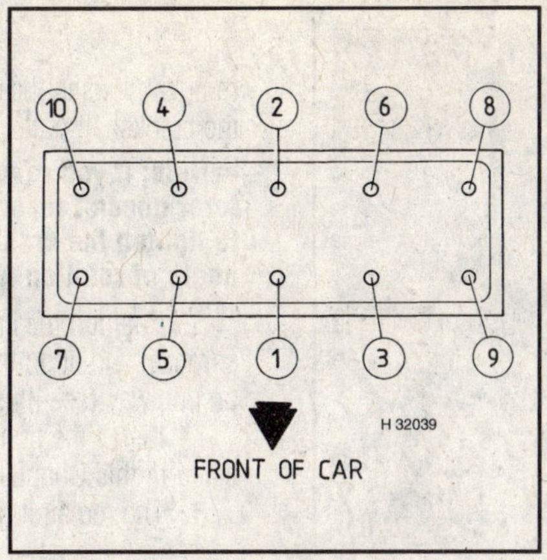

10.26a Cylinder head bolt TIGHTENING sequence

10.26b Using an angle measurement gauge during the final stages of tightening

tion), tighten the cylinder head bolts in three steps to the torque and angle of rotation listed in this Chapter's Specifications.

➡Note: It is recommended that an angle-measuring gauge be used during the final stages of the tightening, to ensure accuracy (see illustration). If a gauge is not available, use white paint to make alignment marks between the bolt head and cylinder head prior to tightening; the marks can then be used to check the bolt has been rotated through the correct angle during tightening.

27 On 2.0L engines, install the rear timing belt cover upper bolt and tighten it securely.

28 Rotate the crankshaft in the normal direction of rotation (clockwise) 90 degrees to TDC. Be sure the alignment mark on the crankshaft pulley aligns with the mark on the front cover. Refer to Section 3, if necessary.

29 Support the engine from underneath with a jack and a block of wood, then remove the right (passenger side) engine mount. which was reinstalled to support the engine as the cylinder head was removed.

30 Install the timing belt tensioner and the camshaft sprocket on the engine if removed. On 2.0L engines, be sure the tensioner is seated properly. On 1.8L engines, compress the timing belt tensioner and lock it in place (see Section 5).

31 Install and adjust the timing belt as described in Section 5.

32 The remainder of the installation is the reverse of removal.

33 Change the engine oil and coolant (see Chapter 1). Run the engine and check for leaks.

11 Crankshaft pulley - removal and installation

➧ Refer to illustrations 11.4, 11.5, and 11.6

1 With the parking brake applied and the shifter in Park (automatic) or in gear (manual), loosen the wheel bolts on the right front wheel, then raise the front of the vehicle and support it securely on jackstands.

2 Remove the right front wheel and the right splash shield from the wheelwell.

3 Remove the drivebelt (see Chapter 1).

4 Remove the bolts from the front of the crankshaft pulley and detach it from the engine (see illustration).

5 If you're removing the crankshaft pulley for other procedures in this manual, such as crankshaft front oil seal removal or oil pump

11.4 Crankshaft pulley retaining bolts (arrows)

11.5 If you're removing the crankshaft pulley to access the front oil seal or the oil pump drive chain, hold the crankshaft pulley with a chain or strap wrench and loosen the crankshaft drive sprocket retaining bolt first (wrap the pulley with a piece of old drivebelt)

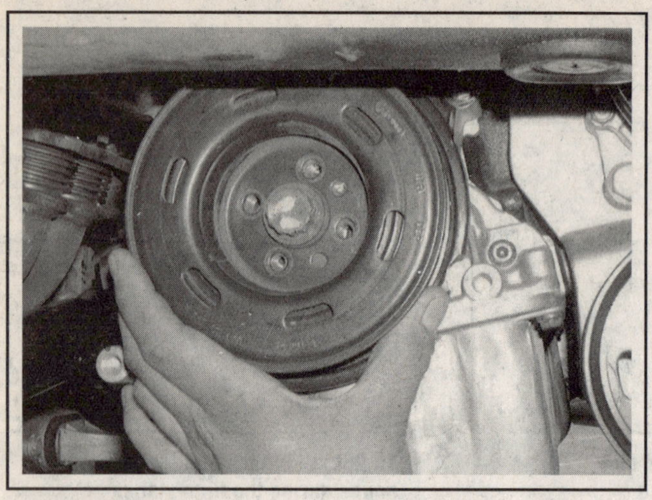

11.6 Align the crankshaft pulley mounting holes and install the bolts

removal, loosen the drive sprocket retaining bolt first, before removing the crankshaft pulley (see illustration).

➡**Note: If you remove this bolt, obtain a new one (the manufacturer doesn't recommend re-using it). Upon installation, be sure to tighten the crankshaft drive sprocket bolt to the torque and angle of rotation listed in this Chapter's Specifications.**

6 Position the crankshaft pulley/balancer on the crankshaft drive sprocket and align the mounting holes. Note that the pulley can only go on one way (see illustration).

7 Install the pulley mounting bolts and tighten them to the torque listed in this Chapter's Specifications.

8 The remaining installation steps are the reverse of removal.

12 Crankshaft front oil seal and housing - replacement

➧ **Refer to illustrations 12.2a, 12.2b, 12.4, 12.5 and 12.7**

1 Remove the timing belt and crankshaft sprocket (see Section 5).

2 Note how far the seal is recessed in the bore, then carefully pry it out of the front cover with a screwdriver or seal removal tool (see illustration). Don't scratch the housing bore or damage the crankshaft in the process (if the crankshaft is damaged, the new seal will end up leaking).

➡**Note: If a seal removal tool is unavailable, you can thread two self tapping screws (180 degrees apart from one another) into the front seal to pry the seal out (see illustration).**

3 Clean the bore in the housing and coat the outer edge of the new seal with engine oil or multi-purpose grease. Apply multi-purpose grease to the seal lip.

4 Using a socket with an outside diameter slightly smaller than the outside diameter of the seal, carefully drive the new seal into place with a hammer (see illustration). Make sure it's installed squarely and driven in to the same depth as the original. If a socket isn't available, a short section of large diameter pipe will also work. Check the seal after installation to make sure the spring didn't pop out of place.

5 If the front oil seal housing needs to be removed for access to other components such as to the oil pump drive chain, simply loosen the housing mounting bolts and remove the housing from the engine (see illustration). The front oil seal housing can be removed with or without the front oil seal. In some instances the front oil seal removal and installation is easier with the front housing removed, since the seal can be placed on a workbench and driven straight in and out of the bore with no special tools or adapters.

6 Before installing the front cover, make sure the mating surfaces of the cover, the cylinder block and the oil pan rail are perfectly clean. Use a hard plastic or wood scraper to remove all traces of gasket material. Take particular care when cleaning the front cover, as aluminum alloy is easily damaged.

7 Apply a 3/16-inch (5 mm) bead of RTV sealant to the cover rear sealing flange (see illustration). Also apply a 3/16-inch (5 mm) bead of RTV sealant to the oil pan flange and install the front cover onto the engine.

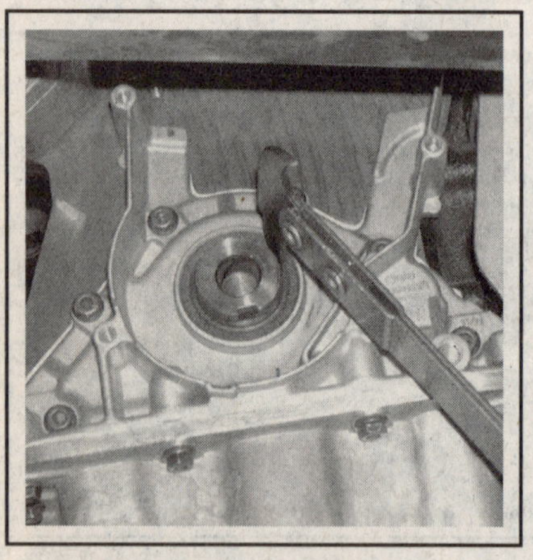

12.2a Pry the seal out very carefully with a seal removal tool or screwdriver, being careful not to nick or gouge the seal bore or the crankshaft

12.2b If a seal removal tool is unavailable, the front seal can also be removed with self tapping screws (as shown) to pry the seal out

12.4 Lubricate the seal lip and drive the new crankshaft seal into place with a seal driver or a large socket and a hammer

12.5 Front oil seal housing bolts (arrows) (2.0L engine shown, all others similar)

➡**Note: Be sure to lubricate the oil seal lip before installing the front cover onto the engine. This will aid the installation process and prevent dry start ups, which may damage the seal and lead to future oil leaks.**

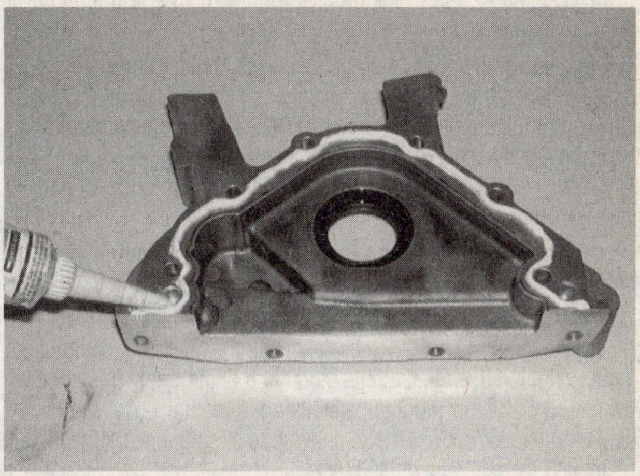

12.7 Apply a 3/16-inch (5 mm) bead of RTV sealant to the rear of the front oil seal housing as shown

8 Tighten the front oil seal housing bolts in several steps to the torque listed in this Chapter's Specifications.

9 Reinstall the crankshaft sprocket and timing belt (see Section 5).

10 Run the engine and check for oil leaks at the front seal.

13 Oil pan - removal and installation

Refer to illustrations 13.5, 13.6 and 13.11

REMOVAL

1 Set the parking brake and block the rear wheels.

2 Raise the front of the vehicle and support it securely on jackstands.

3 Remove the splash shields under the engine, if equipped.

➡**Note: On later models, there is a protective cover over the oil pan. Remove the three fasteners and the cover.**

4 Drain the engine oil (see Chapter 1). Remove the oil dipstick.

5 Remove the bellhousing-to-oil pan bolts (see illustration). On 1.8L engines, remove the oil return tube from the oil pan and the turbocharger (see Chapter 4A).

6 On 2.0L models so equipped, disconnect the connector for the oil level sensor. Remove the bolts and detach the oil pan. If it's stuck, carefully pry it loose with a small screwdriver or putty knife (see illustration). Don't damage the mating surfaces of the pan or the block or oil leaks could develop.

➡**Note: Some later models have a two-piece oil pan. On these models, only the bottom half of the oil pan must be removed unless the engine is being overhauled.**

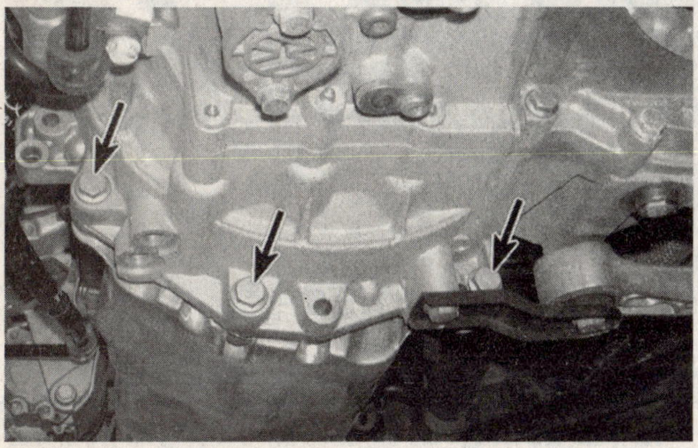

13.5 Oil pan-to-transaxle bolts (arrows)

INSTALLATION

7 Use a scraper to remove all traces of old sealant from the block and oil pan. Clean the mating surfaces with lacquer thinner or acetone.

8 Make sure the threaded bolt holes in the block are clean.

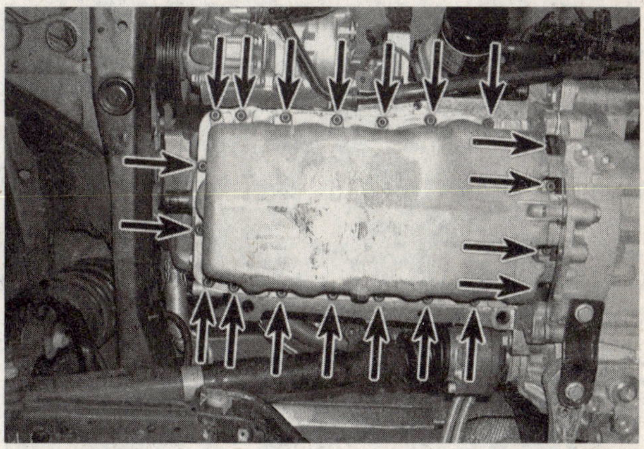

13.6 Oil pan retaining bolts (arrows) (2.0L engine shown, all others similar)

9 Check the oil pan flange for distortion, particularly around the bolt holes. Remove any nicks or burs as necessary.

10 Inspect the oil pump pick-up tube assembly for cracks and a blocked strainer. If the pick-up was removed, clean it thoroughly and install it now, using a new O-ring or gasket. Tighten the nuts/bolts to the torque listed in this Chapter's Specifications.

11 Apply a 3/16-inch (5 mm) bead of RTV sealant to the oil pan flange (see illustration).

➡**Note: The oil pan must be installed within 5 minutes once the sealant has been applied.**

12 Carefully position the oil pan on the engine block and install the oil pan-to-engine block bolts loosely. Install the oil pan to bellhousing bolts and tighten them just a little more than finger tight. This should draw the oil pan flush with the bellhousing. If the transaxle is not installed in the vehicle use a straightedge to align the rear surface of the oil pan with the rear face of the block.

13 Working from the center out, tighten the oil pan-to-engine block bolts to the torque listed in this Chapter's Specifications in three or four steps.

14 Tighten the oil pan-to-bellhousing bolts to the torque listed in this Chapter's Specifications. Connect the oil level indicator electrical connector, if equipped.

15 The remainder of installation is the reverse of removal. Be sure to wait one hour before adding oil to allow the sealant to properly cure.

16 Run the engine and check for oil pressure and leaks.

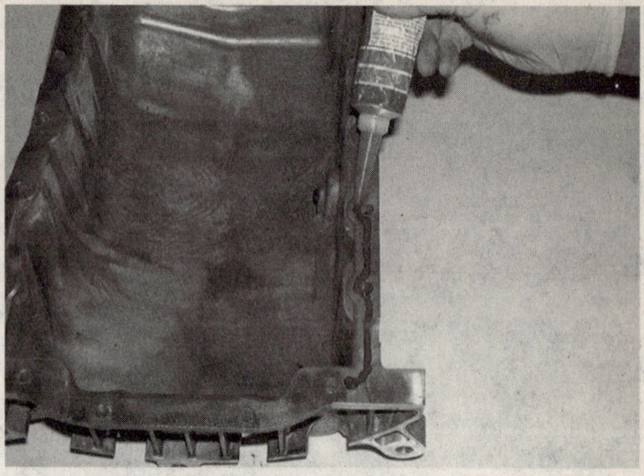

13.11 Apply a 3/16-inch (5 mm) bead of RTV sealant as shown to the oil pan sealing flange

14 Oil pump - removal, inspection and installation

REMOVAL

➧ **Refer to illustrations 14.5, 14.6, 14.9 and 14.10**

1 Remove the timing belt and the crankshaft drive sprocket (see Section 5).

2 After the timing belt has been removed, reinstall the passenger side engine mount to support the engine during the removal and installation of the oil pan and pump.

3 Remove the oil pan (see Section 13).

4 Remove the crankshaft front oil seal housing (see Section 12).

5 Remove the oil pump drive chain tensioner (see illustration).

➡**Note: On 2000 and later 2.0L engines, the oil pump is mounted in a case carrying the engine's balance shaft assembly, which bolts to the bottom of the engine block. To remove the oil pump on these models, first remove the center bolt securing the triangular-shaped chain guide/tensioner assembly to the block, which will provide access to the oil pump sprocket and pump.**

6 Mark the face of the oil pump driven sprocket and chain so they can be installed the same way (see illustration).

➡**Note: Installing the oil pump drive chain in a different direction from which it was originally installed will accelerate wear and cause premature failure.**

7 Wedge a screwdriver in one of the holes of the oil pump driven sprocket to hold the sprocket from turning as the driven sprocket retaining bolt is loosened, then remove the oil pump driven sprocket retain-

14.5 Remove the oil pump drive chain tensioner

14.6 Mark the face of the oil pump drive chain and the driven sprocket so the chain can be installed in the same position and direction of travel

14.9 Oil pump pick-up tube mounting bolts (arrows)

14.10 Oil pump mounting bolts (arrows)

ing bolt.

8 Remove the oil pump drive chain and driven sprocket from the engine.

9 Detach the two retaining bolts and separate the oil pump pick-up tube from the oil pump body. Lift out the pick-up tube and the O-ring (see illustration).

10 Remove the mounting bolts and separate the oil pump and the oil pan baffle from the engine (see illustration).

INSPECTION

11 Clean all components with solvent, then inspect them for wear and damage.

12 If damage or wear is noted, replacement of the entire oil pump assembly is recommended.

INSTALLATION

Refer to illustrations 14.17 and 14.20

13 Use a scraper to remove all traces of sealant and old gasket material from the pump case and engine block, then clean the mating surfaces with lacquer thinner or acetone.

14 Place the oil pump on the engine block over the dowel pins and install the two bolts facing the front of the engine block.

14.17 Always replace the pick-up tube O-ring (arrow)

15 Position the oil pan baffle plate in place and install the remaining mounting bolt.

16 Tighten the bolts to the torque listed in this Chapter's Specifications in several steps. Follow a criss-cross pattern to avoid warping the case.

17 Using a new O-ring, install the oil pick-up tube assembly (see illustration). Tighten the fasteners to the torque listed in this Chapter's Specifications.

18 Reinstall the oil pump drive chain and the driven sprocket in the original direction from which it was removed. Be sure to tighten the driven sprocket bolt to the correct torque Specifications.

19 Install the drive chain tensioner and bolt loosely on the engine block.

20 Engage the tab on the tensioner spring on the inside lip of the engine block (see illustration). Allow the tensioner to apply spring tension to the drive chain and tighten the tensioner retaining bolt.

21 Reinstall the remaining parts in the reverse order of removal. Be sure to install the crankshaft front oil seal housing to the engine block before installing the oil pan.

22 Add oil, start the engine and check for oil pressure and leaks.

23 Recheck the engine oil level.

14.20 Engage the tab on the drive chain tensioner spring on the inside lip of the engine block

15 Flywheel/driveplate - removal and installation

➧ **Refer to illustrations 15.3, 15.5 and 15.10**

⁂ CAUTION:

The manufacturer recommends replacing the flywheel/driveplate bolts with new ones whenever they are removed.

REMOVAL

1 Raise the vehicle and support it securely on jackstands, then refer to Chapter 7 and remove the transaxle. If it's leaking, now would be a very good time to replace the front pump seal/O-ring (automatic transaxle) or input shaft seal (manual transaxle).

2 Remove the pressure plate and clutch disc (Chapter 8) (manual transaxle equipped vehicles). Now is a good time to check/replace the clutch components.

3 Use a center punch or paint to make alignment marks on the flywheel/driveplate and crankshaft to ensure correct alignment during reinstallation (see illustration).

4 Remove the bolts that secure the flywheel/driveplate to the crankshaft. If the crankshaft turns, wedge a screwdriver in the ring gear teeth to jam the flywheel.

5 Remove the flywheel/driveplate from the crankshaft. Since the flywheel is fairly heavy, be sure to support it while removing the last bolt. Automatic transaxle equipped vehicles have spacers on both sides of the driveplate (see illustration). Keep them with the driveplate.

INSTALLATION

6 Clean the flywheel to remove grease and oil. Inspect the surface for cracks, rivet grooves, burned areas and score marks. Light scoring can be removed with emery cloth. Check for cracked and broken ring gear teeth. Lay the flywheel on a flat surface and use a straightedge to check for warpage.

7 Clean and inspect the mating surfaces of the flywheel/driveplate and the crankshaft. If the crankshaft rear seal is leaking, replace it before reinstalling the flywheel/driveplate.

8 Position the flywheel/driveplate and spacer (if used) against the crankshaft. Be sure to align the marks made during removal. Note that some engines have an alignment dowel or staggered bolt holes to ensure correct installation. Before installing the bolts, apply thread locking compound to the threads.

9 Wedge a screwdriver in the ring gear teeth to keep the flywheel/driveplate from turning and tighten the bolts to the torque listed in this Chapter's Specifications. Follow a criss-cross pattern and work up to the final torque in three or four steps.

10 On vehicles equipped with automatic transaxles, measure the installed height at three equal places around the driveplate and compare the average measurement to this Chapter's Specifications (see illustration). If the measurement is incorrect, the driveplate must be removed and shimmed to the proper height.

11 The remainder of installation is the reverse of the removal procedure.

15.3 Mark the flywheel/driveplate and the crankshaft so they can be reassembled in the same relative positions

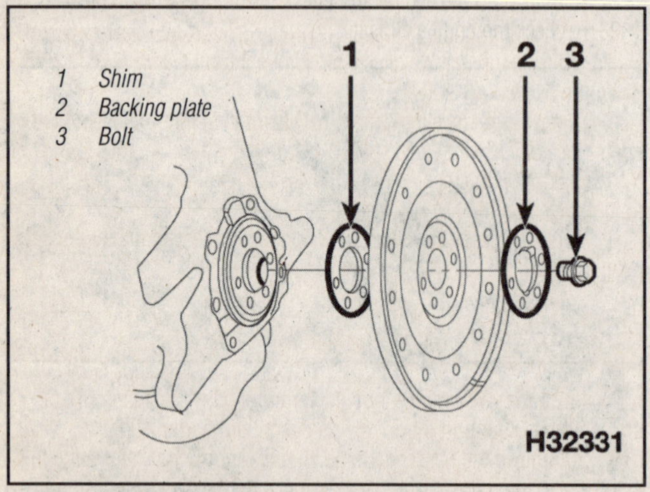

15.5 On vehicles with an automatic transaxle, there is a spacer plate on each side of the driveplate - mark each plate as it is removed so can be installed back in the same position

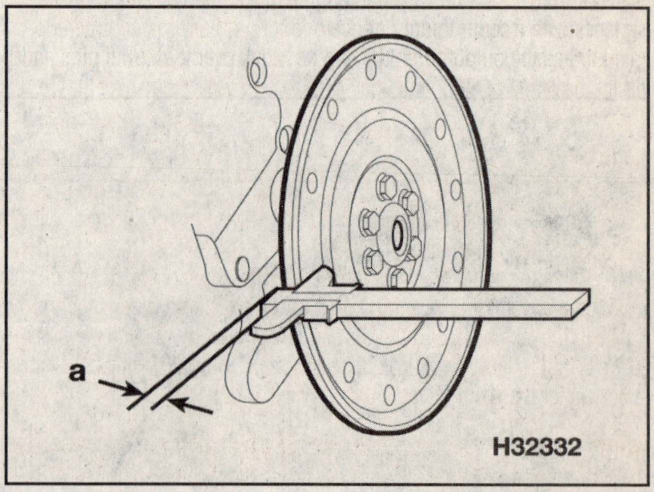

15.10 The driveplate installed height can be measured with a machinist's ruler

16 Rear main oil seal - replacement

Refer to illustrations 16.3 and 16.5

1 Remove the flywheel or driveplate (see Section 15).

2 The rear oil seal and housing are an integral part which must be removed and replaced together as a unit, however the rear seal and housing can be replaced without removing the oil pan.

3 Remove the rear oil seal housing mounting bolts and remove the housing from the engine (see illustration).

4 Before installing a new rear seal and housing, make sure the mating surfaces of the cover, the cylinder block and the oil pan rail are perfectly clean. Use a hard plastic or wood scraper to remove all traces of gasket material. Take particular care when cleaning the front cover, as aluminum alloy is easily damaged.

5 Apply a 3/16-inch (5 mm) bead of RTV sealant to the rear of the housing flange and to the oil pan sealing surface (see illustration). install the rear seal housing over the dowels onto the engine.

Note: Be sure to lubricate the oil seal lip before installing the housing onto the engine. This will aid the installation process and prevent dry start ups, which may damage the seal and lead to future oil leaks.

6 Tighten the rear oil seal housing bolts in several steps to the torque listed in this Chapter's Specifications. Be sure to wait one hour before starting the engine to allow the sealant to properly cure.

7 The remaining steps are the reverse of removal.

16.3 Crankshaft rear oil seal housing mounting bolts (arrows)

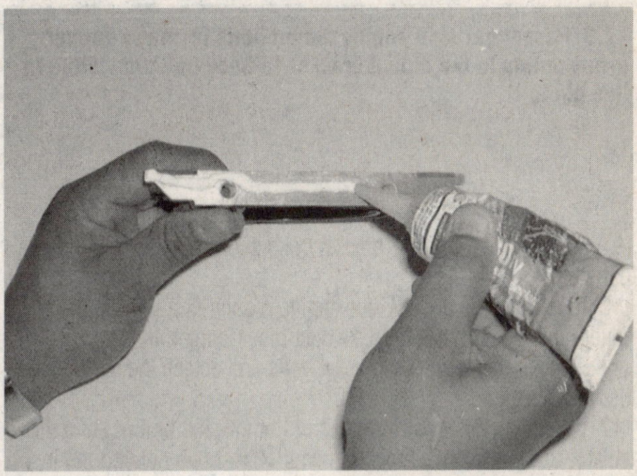

16.5 Apply a 3/16-inch (5 mm) bead of RTV sealant to the real oil seal housing sealing surfaces

17 Powertrain mounts - check and replacement

Refer to illustrations 17.9, 17.11 and 17.12

1 Engine mounts seldom require attention, but broken or deteriorated mounts should be replaced immediately or the added strain placed on the driveline components may cause damage or wear.

CHECK

2 During the check, the engine must be raised slightly to remove the weight from the mounts.

3 Raise the vehicle and support it securely on jackstands, then position a jack under the engine oil pan. Place a large block of wood between the jack head and the oil pan, then carefully raise the engine just enough to take the weight off the mounts. Do not position the wood block under the drain plug.

WARNING:

DO NOT place any part of your body under the engine when it's supported only by a jack!

4 Check the mounts to see if the rubber is cracked, hardened or separated from the metal plates. Sometimes the rubber will split right down the center.

5 Check for relative movement between the mount plates and the engine or frame (use a large screwdriver or pry bar to attempt to move the mounts). If movement is noted, lower the engine and tighten the mount fasteners.

6 Rubber preservative should be applied to the mounts to slow deterioration.

REPLACEMENT

7 Raise the vehicle and support it securely on jackstands (if not already done). Support the engine as described in Step 3.

8 To remove the passenger side engine mount, first relieve the fuel system pressure (see Chapter 4A). Remove the fuel and vapor lines from the fuel distribution block/leak detection pump and the coolant expansion tank from the passenger side of the engine compartment (see illustration 5.3). The expansion tank can be detached from the inner fenderwell and positioned aside without disconnecting the hoses.

17.9 Passenger side engine mount bolts (arrows) - upper arrow points to the mount bracket-to-body bolt not visible in this photo

17.11 Driver's side engine mount bolts (arrows) - upper arrow points to the mount bracket-to-body bolt not visible in this photo

9 Remove the mount-to-bracket bolts and detach the mount (see illustration).

10 To remove the driver's side engine mount, first remove the air cleaner housing and the air intake duct (see Chapter 4A).

11 Remove the mount-to-bracket bolts and detach the mount (see illustration).

12 To remove the lower torque strut, remove the bolt holding the insulator to the transaxle bracket, then the two bolts holding the insulator to the chassis (see illustration).

13 Installation is the reverse of removal. Always install NEW mounting bolts and be sure to tighten them securely.

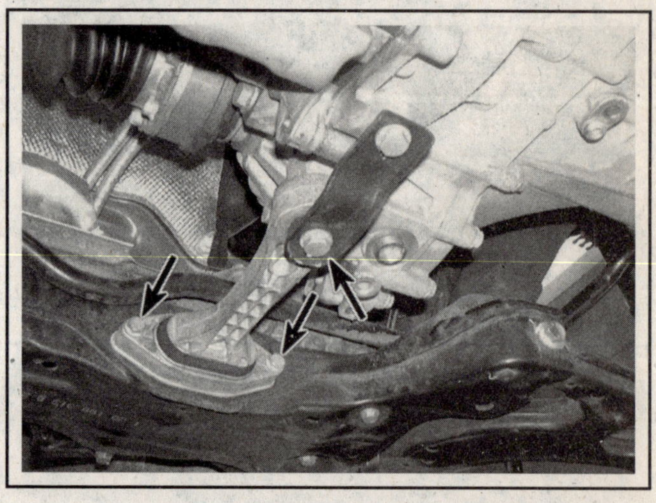

17.12 Lower torque strut mounting bolts (arrows)

Specifications

General

Displacement	
Turbo	109 cubic inches (1.8 liters)
Non-turbo	121 cubic inches (2.0 liters)
Bore and stroke	
1.8 liter	3.189 x 3.401 inches (81.01 x 86.38 mm)
2.0 liter	3.248 x 3.653 inches (82.51 x 92.78 mm)
Cylinder numbers (drivebelt end-to-transaxle end)	1-2-3-4
Firing order	1-3-4-2

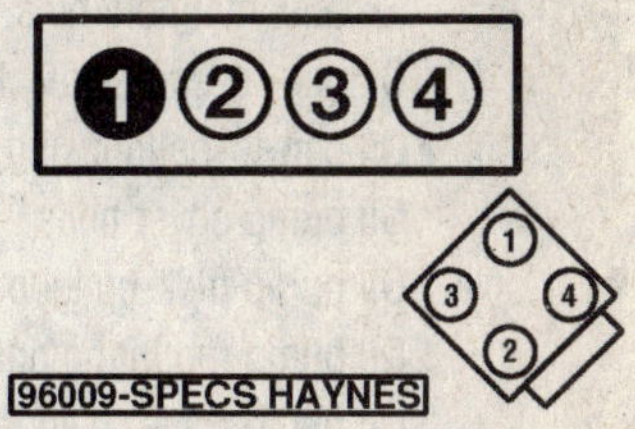

Cylinder numbering for 1.8L and 2.0L engines - coil terminal locations are for 2.0L engines only

Camshaft

Endplay	
1.8 liter	0.0080 inch (0.2 mm)
2.0 liter	0.006 inch (0.15 mm)
Journal diameter	N/A
Journal oil clearance (in cylinder head)	0.004 inch (0.1 mm)
Lobe lift	
Intake	N/A
Exhaust	N/A
Runout	0.0004 inch (0.01 mm)

Driveplate

Installed height	0.77 to 0.83 inch (19.5 - 21.1 mm)

Torque specifications

Torque specifications	Nm	Ft-lbs (unless otherwise indicated)
Camshaft sprocket bolt		
1.8L engine	65	48
2.0L engine	100	74
Camshaft bearing cap bolts/nuts		
1.8L engine	10	84 in-lbs
2.0L engine	20	15
Camshaft drive chain tensioner	10	84 in-lbs
Crankshaft drive sprocket bolt*		
Step one	90	66
Step two	Tighten an additional 90 degrees	
Crankshaft pulley bolts		
Bolts marked 8.8	25	18
Bolts marked 10.9	40	29
Crankshaft rear oil seal housing	15	132 in-lbs
Crankshaft front oil seal housing	15	132 in-lbs
Cylinder head bolts* (in sequence - see illustration 10.26a)		
Step one	40	30
Step two	Tighten an additional 90 degrees	
Step three	Tighten an additional 90 degrees	
Flywheel/driveplate bolts*		
Step one	60	44
Step two	Tighten an additional 90 degrees	

Torque specifications (continued)	Nm	Ft-lbs (unless otherwise indicated)
Intake manifold bolts		
1.8L engine	10	84 in-lbs
2.0L engine	20	15
Exhaust manifold nuts	24	18
Oil pan-to-engine block bolts	15	132 in-lbs
Oil pan-to-bellhousing bolts	45	33
Oil pump cover bolts	10	84 in-lbs
Oil pump pick-up tube	15	132 in-lbs
Oil pump mounting bolt	15	132 in-lbs
Oil pump drive chain tensioner bolt	15	132 in-lbs
Oil pump driven sprocket bolt	24	18
Timing belt cover-to-block bolts	10	84 in-lbs
Timing belt tensioner		
1.8L engine		
Damper-to-engine block bolts	20	15
Tensioner roller-to-cylinder head retaining bolt	27	20
2.0L engine		
Tensioner roller retaining nut	20	15
Valve cover-to-cylinder head nuts	10	84 in-lbs
Engine mount		
Mount-to-frame bolts*		
Step one	40	30
Step two	Tighten an additional 90 degrees	
Mount-to-engine or transaxle bolts*		
Models through 2004		
Step 1	60	44
Step 2	Tighten an additional 90 degrees	
2005 models	100	74
Mount support strap-to-body bolt	24	18
Lower torque strut		
Strut-to-chassis bolts		
Models through 2004		
Step 1	20	15
Step 2	Tighten an additional 90 degrees	
2005		
Step 1	25	19
Step 2	Tighten an additional 90 degrees	
Strut-to-bracket bolt		
Models through 2004	40	30
2005	50	37
Strut bracket-to bellhousing bolt*		
Models through 2004		
Step 1	20	15
Step 2		Tighten an additional 90 degrees
2005	50	37

** Replace with new bolt(s)*

2B

DIESEL ENGINE

Reference to other Chapters

CHECK ENGINE light on - See Chapter 6
Crankshaft pulley - removal and installation - See Chapter 2A
Cylinder compression check - See Chapter 2C
Drivebelt check, adjustment and replacement - See Chapter 1
Engine - removal and installation - See Chapter 2C
Engine oil and filter change - See Chapter 1
Engine overhaul - general information - See Chapter 2C
Flywheel/driveplate - removal and installation - See Chapter 2A
Oil pan - removal and installation - See Chapter 2A
Oil pump - removal, inspection and installation - See Chapter 2A
Powertrain mounts - check and replacement - See Chapter 2A
Rear main oil seal - removal and installation - See Chapter 2A
Spark plug replacement - See Chapter 1
Water pump - removal and installation - See Chapter 3

1 General information

✲✲ CAUTION 1:

Avoid disconnecting the battery whenever possible! Disconnecting the battery can cause driveability problems that require a special scan tool to remedy. See Chapter 5, Section 1 for the use of an auxiliary power source before disconnecting the battery.

✲✲ CAUTION 2:

These models are equipped with an anti-theft radio. Before performing a procedure that requires disconnecting the battery, make sure you have the proper activation code.

➡Note: The engine cover must removed before performing many of the procedures in this Chapter (see illustration 1.1 in Chapter 2A).

This Part of Chapter 2 is devoted to in-vehicle repair procedures for the 1.9L turbo ECOdiesel in-line four cylinder engine. This engine utilizes a cast-iron engine block with an aluminum cylinder head. The aluminum cylinder head is equipped with pressed-in valve guides, hardened valve seats and houses the single overhead camshaft, which is driven from the crankshaft by a timing belt. Hydraulic lifters are used to actuate the valves. The oil pump is mounted at the front of the engine and is driven by a chain from the crankshaft.

All information concerning engine removal and installation and engine block and cylinder head overhaul can be found in Part C of this Chapter.

The following repair procedures are based on the assumption that the engine is installed in the vehicle. If the engine has been removed from the vehicle and mounted on a stand, many of the steps outlined in this Part of Chapter 2 will not apply.

The Specifications included in this Part of Chapter 2 apply only to the procedures contained in this Part. Part C of Chapter 2 contains the Specifications necessary for cylinder head and engine block rebuilding.

2 Repair operations possible with the engine in the vehicle

Many major repair operations can be accomplished without removing the engine from the vehicle.

Clean the engine compartment and the exterior of the engine with some type of degreaser before any work is done. It will make the job easier and help keep dirt out of the internal areas of the engine.

Depending on the components involved, it may be helpful to remove the hood to improve access to the engine as repairs are performed (refer to Chapter 11 if necessary). Cover the fenders to prevent damage to the paint. Special pads are available, but an old bedspread or blanket will also work.

If intake, exhaust, oil or coolant leaks develop, indicating a need for gasket or seal replacement, the repairs can generally be made with the engine in the vehicle. The intake and exhaust manifold gaskets, oil pan gasket, crankshaft oil seals and cylinder head gasket are all accessible with the engine in place.

Exterior engine components, such as the intake and exhaust manifolds, the oil pan, the oil pump, the water pump, the starter motor, the alternator and the fuel system components can be removed for repair with the engine in place.

Since the cylinder head can be removed without pulling the engine, camshaft and valve component servicing can also be accomplished with the engine in the vehicle. Replacement of the timing belt and pulleys is also possible with the engine in the vehicle.

In extreme cases caused by a lack of necessary equipment, repair or replacement of piston rings, pistons, connecting rods and rod bearings is possible with the engine in the vehicle. However, this practice is not recommended because of the cleaning and preparation work that must be done to the components involved.

3 Top Dead Center (TDC) for number one piston - locating

➧ Refer to illustrations 3.4, 3.6a and 3.6b

1 Top Dead Center (TDC) is the highest point in the cylinder that each piston reaches as it travels up-and-down when the crankshaft turns. Each piston reaches TDC on the compression stroke and again on the exhaust stroke, but TDC generally refers to piston position on the compression stroke. The timing marks on the flywheel (manual transaxle) or the torque converter (automatic transaxle) are referenced to the number one piston at TDC.

2 Positioning the piston(s) at TDC is an essential part of procedures such as timing belt and sprocket replacement and fuel injection pump timing.

3 In order to bring any piston to TDC, the crankshaft must be turned using one of the methods outlined below. When looking at the timing belt end of the engine, normal crankshaft rotation is clockwise.

✲✲ WARNING:

Before beginning this procedure, be sure to place the transmission in Neutral, set the parking brake and remove the ignition key.

a) The preferred method is to turn the crankshaft with a large socket and breaker bar attached to the large bolt threaded into the center of the crankshaft pulley.

b) A remote starter switch, which may save some time, can also be used. Attach the switch leads to the S (switch) and B (battery) terminals on the starter motor. Once the piston is close to TDC, use a socket and breaker bar as described in the previous paragraph.

c) If an assistant is available to turn the ignition switch to the Start position in short bursts, you can get the piston close to TDC without a remote starter switch. Use a socket and breaker bar as described in Paragraph a) to complete the procedure.

4 Disable the fuel injection system by disconnecting the primary electrical connectors at the fuel cut off valve (see Chapter 4B). Remove the inspection plug at the top of the transaxle bellhousing (see illustration).

5 Remove the upper engine cover and the glow plugs (see Chapter 5B) and install a compression gauge in the number one cylinder glow plug hole.

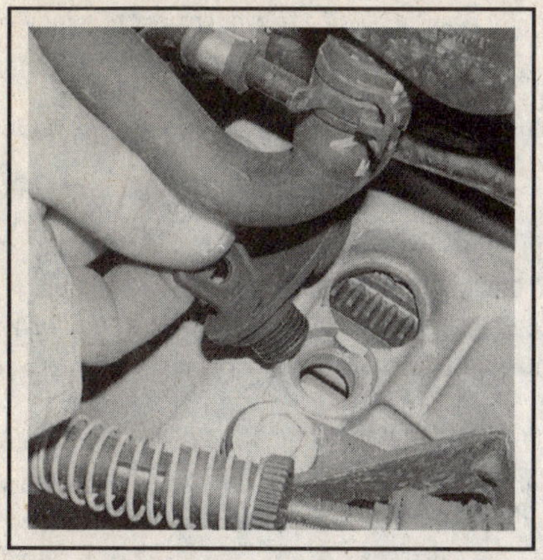

3.4 Remove the inspection cover from the top of the transaxle bellhousing

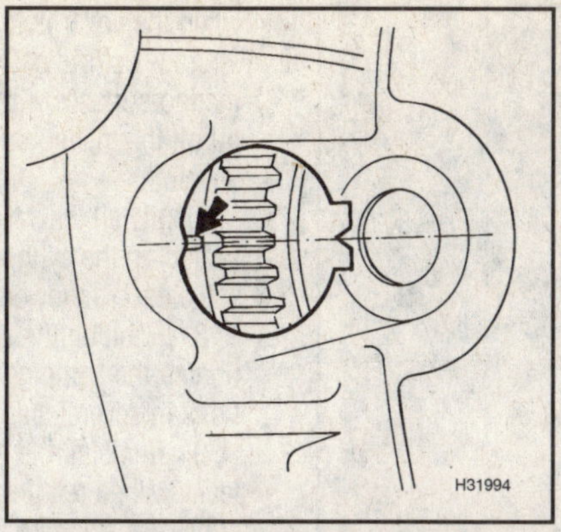

3.6a On vehicles equipped with manual transaxles, align the notch on the flywheel with the pointer on the transaxle case

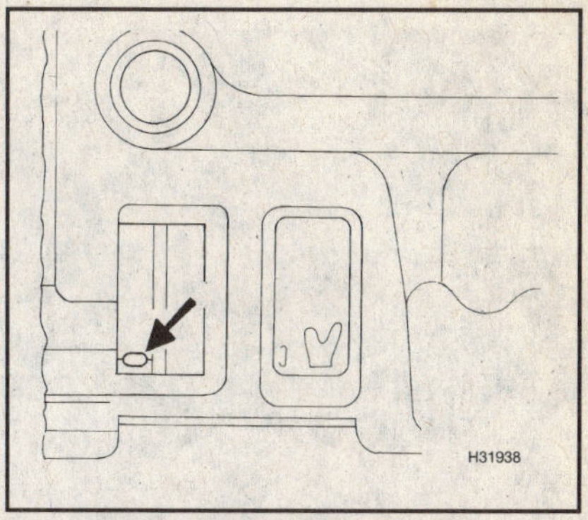

3.6b On vehicles equipped with automatic transaxles, align the mark on the torque converter with the lower edge of the window opening on the transaxle case

⁂ WARNING:

Be sure to use a diesel compression gauge that can handle at least 500 psi (35 kg/cm2).

Turn the crankshaft clockwise with a socket and breaker bar as described above.

6 When the piston approaches TDC, compression will be noted on the compression gauge. Continue turning the crankshaft until the timing mark on the flywheel (manual transaxle) or the torque converter (automatic transaxle) is aligned with the mark on the transaxle case (see illustrations). At this point, the number one cylinder is at TDC on the compression stroke. If the marks are aligned but there was no compression, the piston was on the exhaust stroke; continue rotating the crankshaft 360-degrees (1-turn) and line-up the marks.

➡Note: If a diesel compression gauge is not available, you can simply place your finger over the glow plug hole and feel for compression as the engine is rotated. Once compression at the No.1 glow plug hole is noted the remainder of the Step is the same.

7 After the number one piston has been positioned at TDC on the compression stroke, TDC for any of the remaining cylinders can be located by turning the crankshaft 180 degrees and following the firing order (refer to the *Specifications*). Rotating the engine 180 degrees past TDC #1 will put the engine at TDC compression for cylinder #3.

4 Valve cover - removal and installation

REMOVAL

➧ Refer to illustration 4.3

1 Remove the engine cover (see illustration 1.1 in Chapter 2A).

2 Remove the air intake duct from the air cleaner housing and position it aside (see Chapter 4B).

3 Detach the breather hose from the breather valve, then remove the breather valve from the valve cover (see illustration).

4 Remove the retaining nuts and detach the valve cover from the cylinder head.

5 If the cover is stuck to the head, bump the end with a block of wood and a hammer to jar it loose. If that doesn't work, try to slip a flexible putty knife between the head and cover to break the seal. Be careful not to damage the gasket.

⁂ CAUTION:

Don't pry at the cover-to-head joint or damage to the sealing surfaces may occur, leading to oil leaks after the cover is reinstalled.

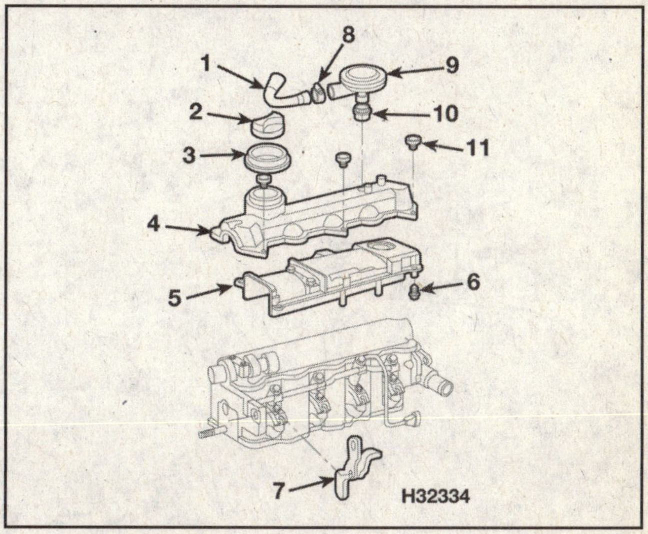

4.3 Valve cover details

1	*Breather hose*	6	*Oil deflector shield retaining bolt*	9	*Breather valve*
2	*Oil filler cap*	7	*Engine lifting bracket*	10	*Breather valve grommet*
3	*Oil filler cap seal*	8	*Clamp*	11	*Valve cover retaining nut*
4	*Valve cover*				
5	*Oil deflector shield*				

INSTALLATION

➧ Refer to illustration 4.7

6 The mating surfaces of the cylinder head and cover must be clean

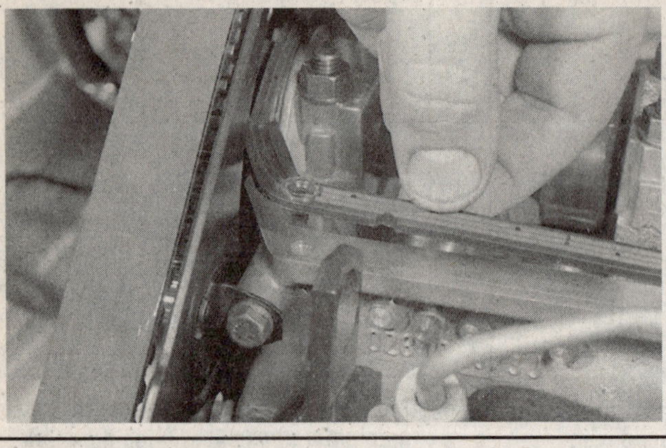

4.7 Ensure that the valve cover gasket is correctly seated on the cylinder head

when the cover is installed. If necessary, use a gasket scraper to remove all traces of old gasket material from the cylinder head, then clean the mating surfaces with lacquer thinner or acetone. If there's residue or oil on the mating surfaces when the cover is installed, oil leaks may develop.

7 Install the valve cover gasket. Ensure that the gasket is correctly seated on the cylinder head, and take care to avoid displacing it as the valve cover is lowered into position (see illustration).

8 Position the valve cover over the studs on the cylinder head, then tighten the retaining nuts to the torque listed in this Chapter's Specifications in several steps.

9 Reinstall the remaining parts, run the engine and check for oil leaks. When refitting hoses that were originally secured with crimp-type clips, use standard worm-drive clips in their place.

5 Timing belt and sprockets - removal, inspection and installation

GENERAL INFORMATION

1 The primary function of the timing belt is to drive the camshaft, but it is also used to drive the fuel injection pump and the water pump. Should the belt slip or break in service, the valve timing will be disturbed and piston-to-valve contact may occur, resulting in serious engine damage.

2 For this reason, it is important that the timing belt is tensioned correctly, and inspected regularly for signs of wear or deterioration.

3 Do not loosen the fuel injection pump sprocket bolts at any time during this procedure or the injection pump timing will be lost. Injection pump timing requires a specialized scan tool to reset the injection timing (see Chapter 4B).

REMOVAL

Refer to illustrations 5.7, 5.15a, 5.15b, 5.15c, 5.15d and 5.19

4 Remove the engine cover (see illustration 1.1 in Chapter 2A).

5 Refer to Chapter 1 and remove the fuel filter and the fuel filter mounting bracket.

6 Remove the upper hoses and the pipes leading from the intercooler to the intake manifold also referring to Chapter 4B.

7 Unclip and remove the upper timing belt cover from the engine (see illustration).

8 Remove the valve cover (see Section 4).

9 Remove the brake booster vacuum pump from the cylinder head

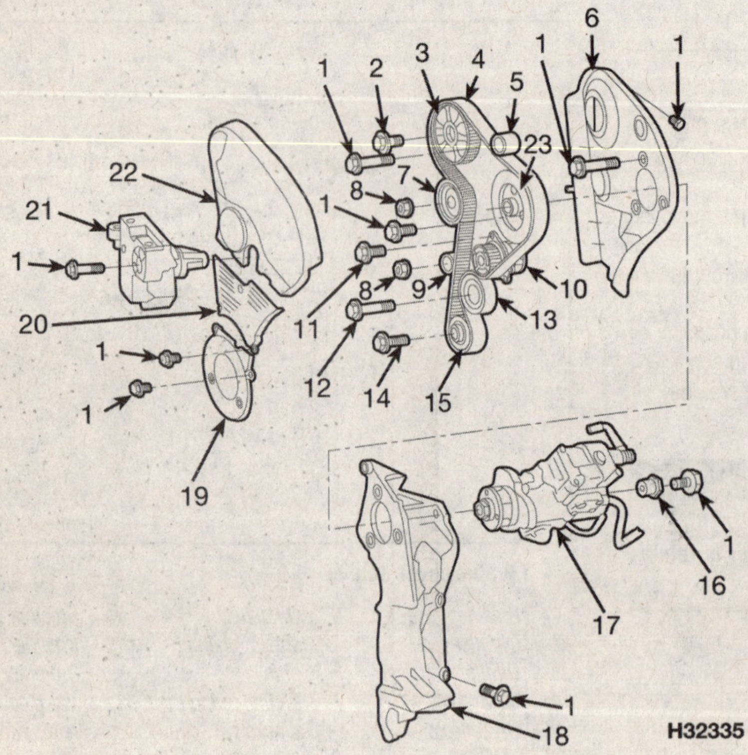

5.7 Exploded view of the timing belt and related components

1 *Bolt*
2 *Camshaft sprocket bolt*
3 *Camshaft sprocket*
4 *Timing belt*
5 *Idler pulley*
6 *Rear timing cover*
7 *Timing belt tensioner*
8 *Nut*
9 *Idler pulley*
10 *Water pump*
11 *Water pump bolt*
12 *Idler wheel bolt*
13 *Idler pulley*
14 *Crankshaft drive sprocket bolt*
15 *Crankshaft drive sprocket*
16 *Injection pump bushing*
17 *Injection pump*
18 *Alternator/power steering pump bracket*
19 *Lower timing belt cover*
20 *Center timing belt cover*
21 *Engine mount support bracket*
22 *Upper timing belt cover*
23 *Injection pump sprocket*

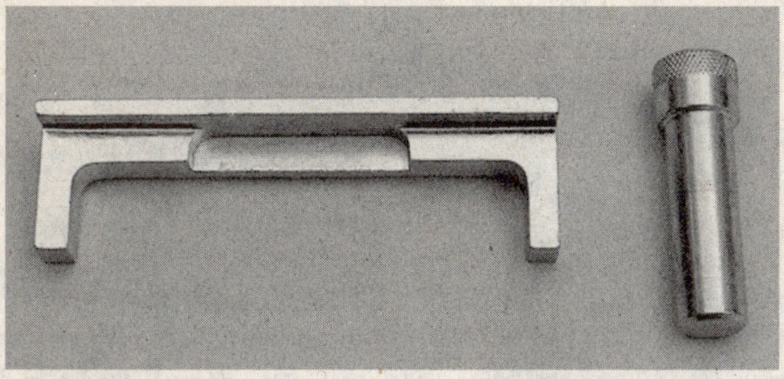

5.15a Engine locking tools

5.15b Engage the locking bar with the slot in the camshaft

(see Chapter 9).

10 Block the rear wheels and set the parking brake.

11 Loosen the wheel bolts on the right front wheel and raise the vehicle. Support the front of the vehicle securely on jackstands and remove the right front wheel.

12 Remove the engine under cover and the lower fender apron to allow access to the bottom of the engine.

13 Remove the lower hoses and the pipes leading from the intercooler to the turbocharger (see Chapter 4B, if necessary).

14 Remove the accessory drivebelt (see Chapter 1).

15 Rotate the engine in the normal direction of rotation (clockwise) until the No.1 cylinder is positioned at TDC (see Section 3). Then lock the engine in the TDC position as follows:

a) *To lock the engine in the TDC position, the camshaft (not the sprocket) and fuel injection pump sprocket must be secured in a reference position, using special locking tools. Improvised tools may be fabricated, but due to the exact measurements and machining involved, it is strongly recommended that a kit of locking tools is either borrowed or rented from a VW dealer, or purchased from a reputable tool manufacturer (see illustration).*

b) *Engage the edge of the locking bar with the slot in the end of the camshaft (see illustration).*

c) *With the locking bar still inserted, turn the camshaft slightly (by turning the crankshaft clockwise, as before), so that the locking bar rocks to one side, allowing one end of the bar to contact the cylinder head surface. At the other side of the locking bar, measure the gap between the end of the bar and the cylinder head using a feeler gauge.*

d) *Turn the camshaft back slightly, then pull out the feeler gauge. The idea now is to level the locking bar by inserting two feeler gauges, each with a thickness equal to half the originally measured gap, on either side of the locking bar and the cylinder head. This centers the camshaft, and sets the valve timing in reference position (see illustration).*

e) *Insert the locking pin through the fuel injection pump sprocket alignment hole, and thread it into the support bracket behind the sprocket. This locks the fuel injection pump in a reference position so that the timing belt can be removed without disturbing the fuel injection pump timing. (see illustration).*

16 Use a strap wrench to hold the crankshaft pulley from rotating. Loosen the crankshaft drive sprocket retaining bolt and the crankshaft pulley bolts, then remove the crankshaft pulley (see Chapter 2A). After the bolts are loosened, verify that the crankshaft has not moved from TDC.

➡Note: Loosening the drive sprocket bolt is only required if the crankshaft drive sprocket is expected to be removed. It is not typically necessary to remove the drive sprocket when you're simply replacing a timing belt, but it will need to be removed if you're replacing the crankshaft front oil seal or housing.

17 Support the engine from underneath with a jack and a block of wood, then remove the right (passenger side) engine mount (see Chapter 2A) and the engine mount support bracket from the front of the engine (see illustration 5.7).

18 Detach the retaining screws from the center and lower timing belt covers and remove the covers.

19 If you plan to re-use the timing belt, apply match marks on the

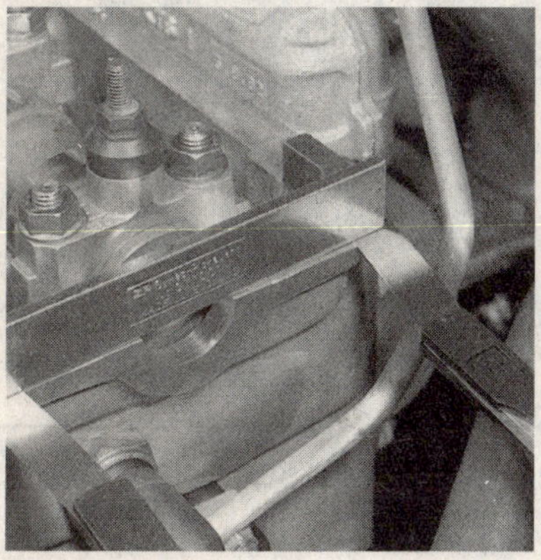

5.15c Camshaft centered and locked using locking bar and feeler gauges

5.15d Injection pump sprocket locked using locking pin (arrow)

5.19 If you intend to re-use the timing belt, apply directional marks on the belt and the rear timing cover

sprocket and belt and an arrow indicating direction of travel on the belt (see illustration).

20 Loosen the timing belt tensioner retaining nut to release the tension on the belt and remove the timing belt from the engine, taking care to avoid twisting or kinking it excessively.

21 If you're removing the upper part of the belt only, for camshaft seal replacement or cylinder head removal, it isn't necessary to detach the belt from the crankshaft sprocket. If the sprocket is worn or damaged, or if you need to replace the crankshaft front oil seal, remove the drive sprocket retaining bolt which was loosened in Step 16 and detach the crankshaft sprocket from the crankshaft.

INSPECTION

Refer to illustration 5.23

CAUTION:

Do not bend, twist or turn the timing belt inside out. Do not allow it to come in contact with oil, coolant or fuel. Do not turn the crankshaft or camshaft more than a few degrees (if necessary for tooth alignment) while the timing belt is removed.

22 Spin the idler pulleys and the timing belt tensioner and check the bearings for smooth operation and excessive play. Also inspect the remaining timing belt sprockets for any obvious damage. Replace all worn parts as necessary.

Note: Do not loosen the fuel injection pump sprocket bolts at any time during this procedure or the injection pump timing will be lost. Injection pump timing requires a specialized scan tool to reset the injection timing (see Chapter 4B).

23 Examine the belt for evidence of contamination by coolant or lubricant. If this is the case, find the source of the contamination before progressing any further. Check the belt for signs of wear or damage, particularly around the leading edges of the belt teeth (see illustration).

CAUTION:

If the belt appears to be in good condition and can be re-used, it is essential that it is reinstalled the same way around, otherwise accelerated wear will result, leading to premature failure.

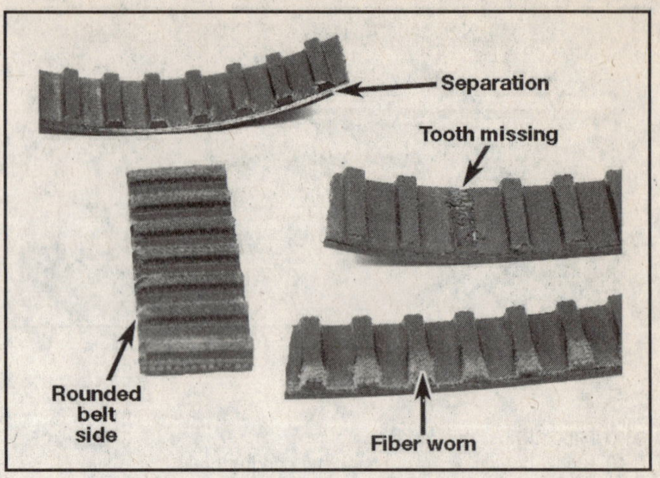

5.23 Check the timing belt for cracked and missing teeth - wear on one side of the belt indicates sprocket misalignment problems

24 Replace the belt if its condition is in doubt; the cost of belt replacement is negligible compared with potential cost of the engine repairs, should the belt fail in service. Similarly, if the belt is known to have covered more than 36,000 miles, it is prudent to replace it regardless of condition, as a precautionary measure.

INSTALLATION

Refer to illustrations 5.25, 5.26 and 5.30

25 Ensure that the crankshaft is still set to TDC on No. 1 cylinder, as described in Section 3. If the any of the timing sprockets or idler pulleys were removed for inspection or needed replacement, install them back onto the engine now. If the timing belt tensioner was removed, be sure the rear tab on the roller is properly positioned in the rear timing cover and install the retaining nut hand tight (see illustration).

26 Loosen the camshaft sprocket bolt by half a turn. Release the sprocket from the camshaft taper mounting by carefully tapping it with a pin punch, inserted through the hole provided in the timing belt inner cover (see illustration).

27 Loop the timing belt loosely under the crankshaft sprocket.

5.25 The timing belt tensioner is properly seated when the tab on the tensioning roller is engaged in the hole on the rear timing cover

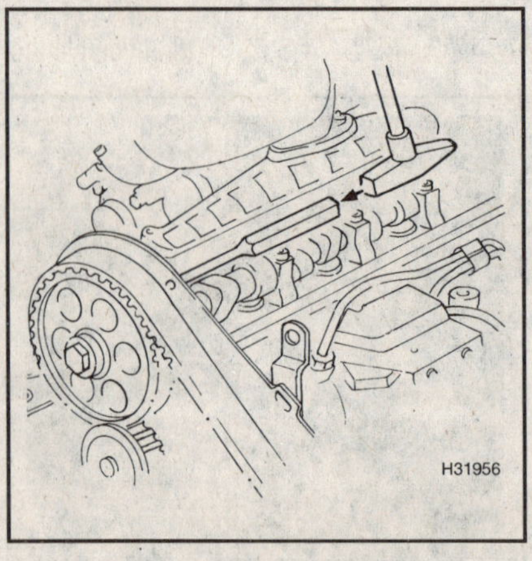

5.26 Releasing the camshaft sprocket from the taper using a pin punch

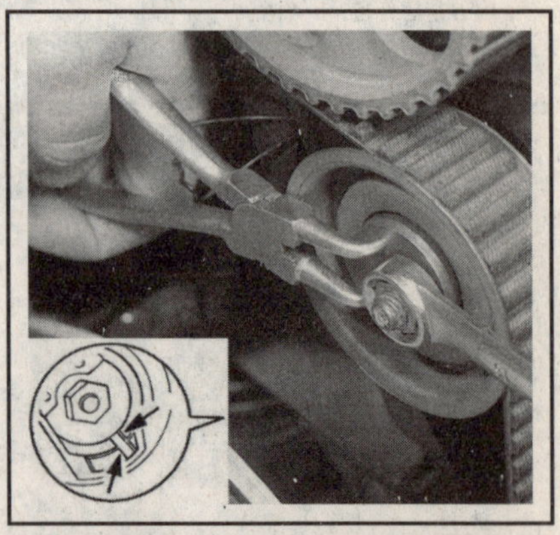

5.30 Tensioning the timing belt using a pair of snap-ring pliers in the belt tensioner - the belt is tensioned correctly when the notch on the front of the tensioner is aligned with the raised mark

⁂ CAUTION:

Observe the direction of rotation markings on the belt.

28 Engage the timing belt teeth with the crankshaft sprocket, then maneuver it into position over the idler pulley, water pump and the injection pump sprockets. Ensure the belt teeth seat correctly on the sprockets, then install the belt around the camshaft sprocket and the timing belt tensioner.

➡Note: Slight adjustments to the position of the camshaft sprocket may be necessary to achieve this.

29 Ensure that the 'front run' of the belt is taut and all the slack should be in the section of the belt that passes over the tensioner roller.

30 Tension the belt by turning the eccentrically-mounted tensioner clockwise until the lower marks align; two holes are provided in the side of the tensioner hub for this purpose - a pair of sturdy right-angled snap-ring pliers is a suitable substitute for the correct VW tool (see illustration). After the tensioner marks are aligned properly, tighten the tensioner locknut to the specified torque.

31 At this point, double check to make sure that the crankshaft is still set to TDC on No. 1 cylinder (see Section 3).

32 Tighten the camshaft sprocket retaining bolt to the torque listed in this Chapter's Specifications.

33 Remove the camshaft locking bar and the locking pin from the fuel injection pump sprocket.

34 Rotate the crankshaft through two complete revolutions. Reset the engine to TDC on No. 1 cylinder, with reference to Section 3 and check that the fuel injection pump sprocket locking pin can be inserted. Re-check the timing belt tension and adjust it, if necessary.

35 The remainder of the installation is the reverse of removal.

➡Note: If the CHECK ENGINE light comes on after performing this procedure it will be necessary to take the vehicle to a dealer service department or other qualified repair shop to have the injection pump timing reset with the proper scan tool. See Chapter 4B for further information.

6 Camshaft and lifters - removal and installation

➡Note: The camshaft and lifters should always be thoroughly inspected before installation and camshaft endplay should always be checked prior to camshaft removal. Although the hydraulic lifters are self adjusting and require no periodic service, there is an in-vehicle procedure for checking excessively noisy hydraulic lifters. Refer to Chapter 2C for the camshaft and lifter inspection procedures.

REMOVAL

➧ Refer to illustrations 6.4, 6.5 and 6.6

1 Remove the engine cover (see illustration 1.1 in Chapter 2A).
2 Remove the valve cover (see Section 4).
3 Remove the timing belt and camshaft sprocket (see Section 5).
4 Mark the camshaft bearing caps from 1 to 5, starting with the No.1 cap at the timing belt end. Also mark arrows indicating the front of the engine (see illustration). Loosen the No. 1, 3 and 5 camshaft bearing caps in two or three steps, then loosen the No. 2 and 4 bearing caps. Be sure to loosen the nuts alternately and evenly.
5 Remove the bearing caps and camshaft.

6.4 The camshaft bearing caps should be marked with a number and letter stamp or a marker to ensure correct reinstallation

⁂ CAUTION:

Keep the caps in order. They must go back in the same location they were removed from. After the camshaft has been removed, make a note of the camshaft identification markings (see illustration). This will help identify the camshaft if a replacement is necessary.

6 Remove the lifters from the cylinder head, keeping them in order with their respective valve and cylinder (see illustration).

⁂ CAUTION:

Keep the lifters in order. They must go back in the same location they were removed from.

7 Inspect the camshaft and lifters as described in Chapter 2C. The camshaft will have to be lifted slightly to clear the bearing journals on the cylinder head, then tilted to the rear to clear the rear timing cover. Be sure to discard the camshaft front oil seal - a new one must be used upon reassembly.

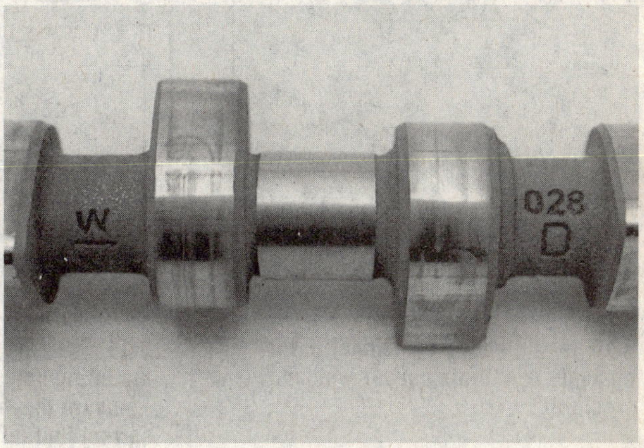

6.5 Camshaft identification markings

6.6 The lifters can be stored in individually marked plastic bags or a divided box as shown

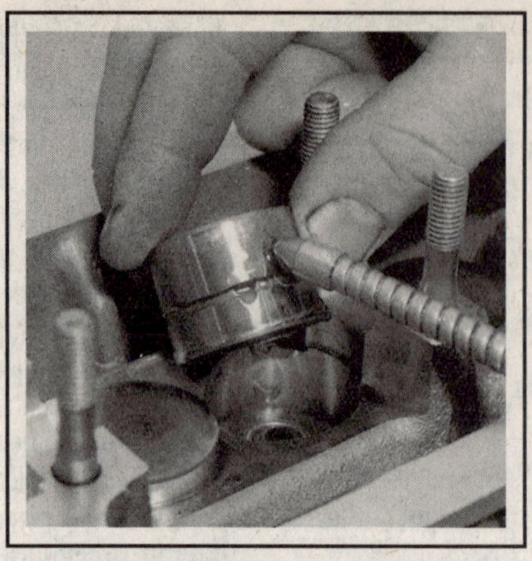

6.8 Install the lifters into their bores in the cylinder head

6.9 Lubricate the camshaft bearings with clean engine oil

INSTALLATION

➧ **Refer to illustrations 6.8, 6.9, 6.10, 6.11 and 6.12**

8 Apply clean engine oil onto the sides of the hydraulic lifters, and install them into position in their bores in the cylinder head. Push them down until they contact the valves, then lubricate the camshaft lobe contact surfaces (see illustration).

9 Lubricate the camshaft and cylinder head bearing journals with clean engine oil (see illustration). Then carefully lower the camshaft into position on the cylinder head with the No.1 camshaft lobes facing up (180 degrees from the cylinder head mating surface). Support the ends of the shaft as it is inserted, to avoid damaging the lobes and journals.

10 Lubricate the lip of a new camshaft oil seal with clean engine oil and locate it over the end of the camshaft. Slide the seal along the camshaft until it locates in the lower half of its housing in the cylinder head (see illustration).

11 Oil the upper surfaces of the camshaft bearing journals, then install the No. 2 and 4 bearing caps over the camshaft and tighten the retaining nuts alternately and diagonally to the specified torque.

➡**Note: The camshaft bearing caps are drilled off center (see illustration); be sure they're installed on the correct journal and with the arrows made in Step 4 facing towards the timing belt end of the engine.**

12 Apply a small amount of RTV sealant to the mating surfaces of caps Nos. 1 and 5 then install them, together with cap No. 3, over the camshaft and tighten the nuts to the specified torque (see illustration).

13 Loosely install the camshaft sprocket and install the timing belt (see Section 5), then tighten the sprocket bolt to the torque listed in this Chapter's Specifications.

14 The remainder of installation is the reverse of removal.

⁂ CAUTION:

If new lifters were used, wait at least 30 minutes before starting the vehicle to allow the lifters to bleed down. Failure to do so will result in serious engine damage.

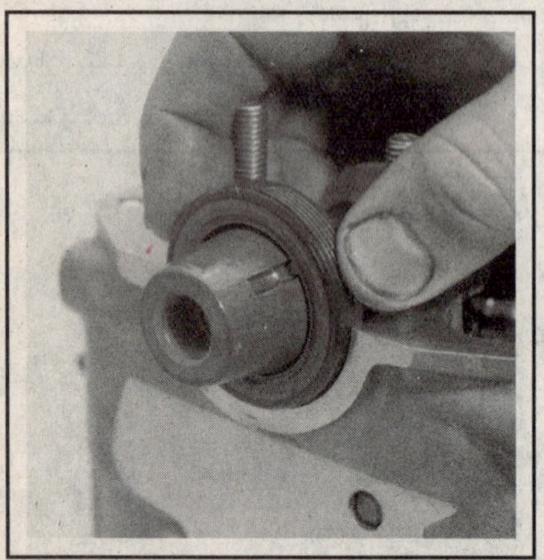

6.10 Installing the camshaft front oil seal - rear timing cover removed for clarity

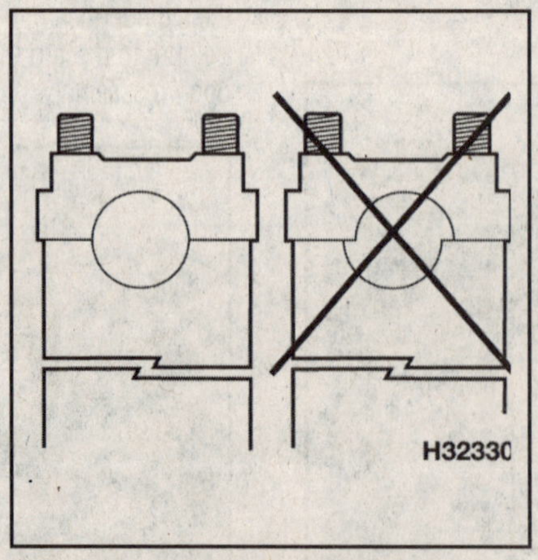

6.11 The camshaft bearing cap mounting holes are drilled off-center - be sure they're installed on the correct journal and with the arrows (made earlier) facing towards the timing belt end of the engine

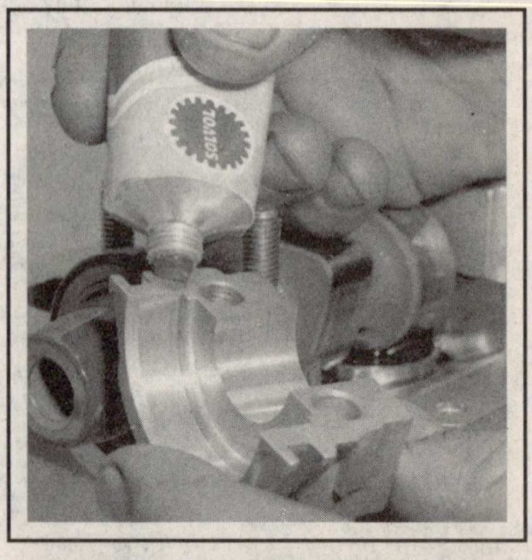

6.12 Apply a small amount of RTV sealant to the mating surfaces of the No. 1 and 5 bearing caps

7 Valve springs, retainers and seals - replacement

Refer to illustrations 7.7, 7.12a, 7.12b, 7.12c, 7.13 and 7.14

Note: Broken valve springs and defective valve stem seals can be replaced without removing the cylinder heads. Some special tools are normally required to perform this operation, so read through this Section carefully and rent or buy the tools before beginning the job.

1 Remove the valve cover referring to Section 4.

2 Refer to Section 6 and remove the camshaft and lifters. Remove the glow plugs (see Chapter 4B).

3 Turn the crankshaft until the piston in the affected cylinder is at Top Dead Center (TDC) on the compression stroke (see Section 3 for instructions). If you're replacing all of the valve stem seals, begin with cylinder number one and work on the valves for one cylinder at a time. Move from cylinder-to-cylinder following the firing order sequence (see the Specifications listed at the end of this Chapter).

4 A compressed air source is not necessary on diesel engines, since valve to piston clearance is smaller than normal compared to a typical gasoline engine. The valves will be held in place by the piston crown when the cylinder is at TDC.

5 Stuff shop rags into the cylinder head holes around the valves to prevent parts and tools from falling into the engine.

6 Using a socket and a hammer, gently tap on the top of each valve spring retainer several times. This will break the bond between the valve keeper and the spring retainer and allow the keeper to separate from the valve spring retainer as the valve spring is compressed.

7 Use a valve spring compressor to compress the spring. Remove the keepers with small needle-nose pliers or a magnet (see illustration).

Note: Several different types of tools are available for compressing the valve springs with the head in place. Be sure to purchase or rent the "Import type" that bolts to the top of the cylinder head. This type uses a support bar across the cylinder head for leverage as the valve spring is compressed. The lack of clearance surrounding the valve springs on these engines prohibits typical types of valve spring compressors from being used.

8 Remove the valve spring and retainer.

9 Remove the old valve stem seals, noting the differences between the intake and exhaust seals, if any.

10 Inspect the valve stem for damage. Rotate the valve in the guide and check the end for eccentric movement, which would indicate that the valve is bent.

11 Try to move the valve up-and-down in the guide and make sure it doesn't bind. If the valve stem binds, either the valve is bent or the guide is damaged. In either case, the head will have to be removed for repair.

12 Lubricate the valve stem with engine oil and install a new seal on the valve guide (see illustrations).

13 Install the valve spring and the spring retainer in position over the

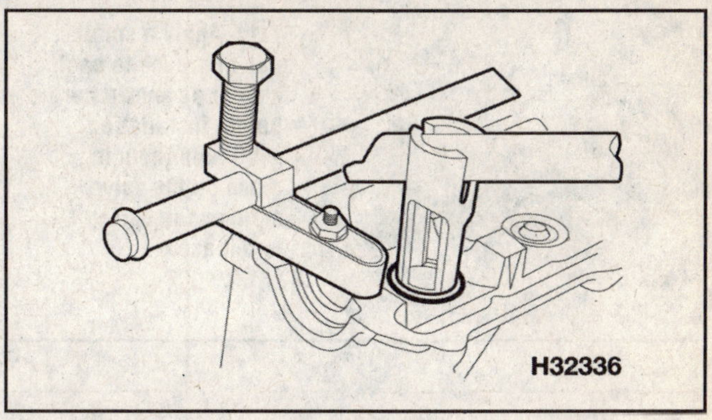

7.7 Compress the valve spring enough to release the valve stem keepers and lift them out with a magnet or needle-nose pliers

7.12a Install the protective plastic sleeve over the valve stem to avoid damage to the valve seal as the seal is installed

7.12b Push a new valve stem seal down over the valve to the top of the guide, then remove the plastic installation tool

7.12c Use a deep socket to press on the oil seal down over the valve guide

7.13 Install the valve spring and the retainer over the valve

7.14 Apply a small dab of grease to each keeper as shown here before installation - it'll hold them in place on the valve stem as the spring is released

valve (see illustration).

14 Compress the valve spring and carefully position the keepers in the groove. Apply a small dab of grease to the inside of each keeper to hold it in place (see illustration).

15 Remove the pressure from the spring tool and make sure the keepers are seated.

16 Install the camshaft, lifters, timing belt and the valve cover by referring to the appropriate Sections.

17 Install the glow plugs and hook up the electrical connector(s).

18 Start and run the engine, then check for oil leaks and unusual sounds coming from the valve cover area.

8 Intake manifold - removal and installation

Refer to illustration 8.9

WARNING:

Wait until the engine is completely cool before beginning this procedure.

REMOVAL

1 Remove the engine cover (see illustration 1.1 in Chapter 2A).

2 Remove the upper hoses and the pipes leading from the intercooler to the intake manifold (see Chapter 4B).

3 Remove the air intake duct and pipes from the air cleaner to the

1 *Exhaust manifold*
2 *Intake manifold*
3 *From intercooler*
4 *Intake manifold gasket*
5 *Mounting nut/bolt*
6 *Exhaust manifold gasket*
7 *Heat shield mounting bracket*
8 *Washer*
9 *Heat shield*
10 *Turbocharger*
11 *Wastegate*
12 *From air cleaner*
13 *Gasket*
14 *Seal*
15 *Oil return pipe*
16 *Banjo bolt*
17 *Flange bolt*
18 *Turbo support bracket bolt to engine*
19 *Turbo support bracket*
20 *Turbo support bracket bolt*
21 *Exhaust pipe gasket*
22 *Exhaust front pipe*
23 *Connection*
24 *Oil supply pipe mounting bolt*
25 *Oil supply pipe*

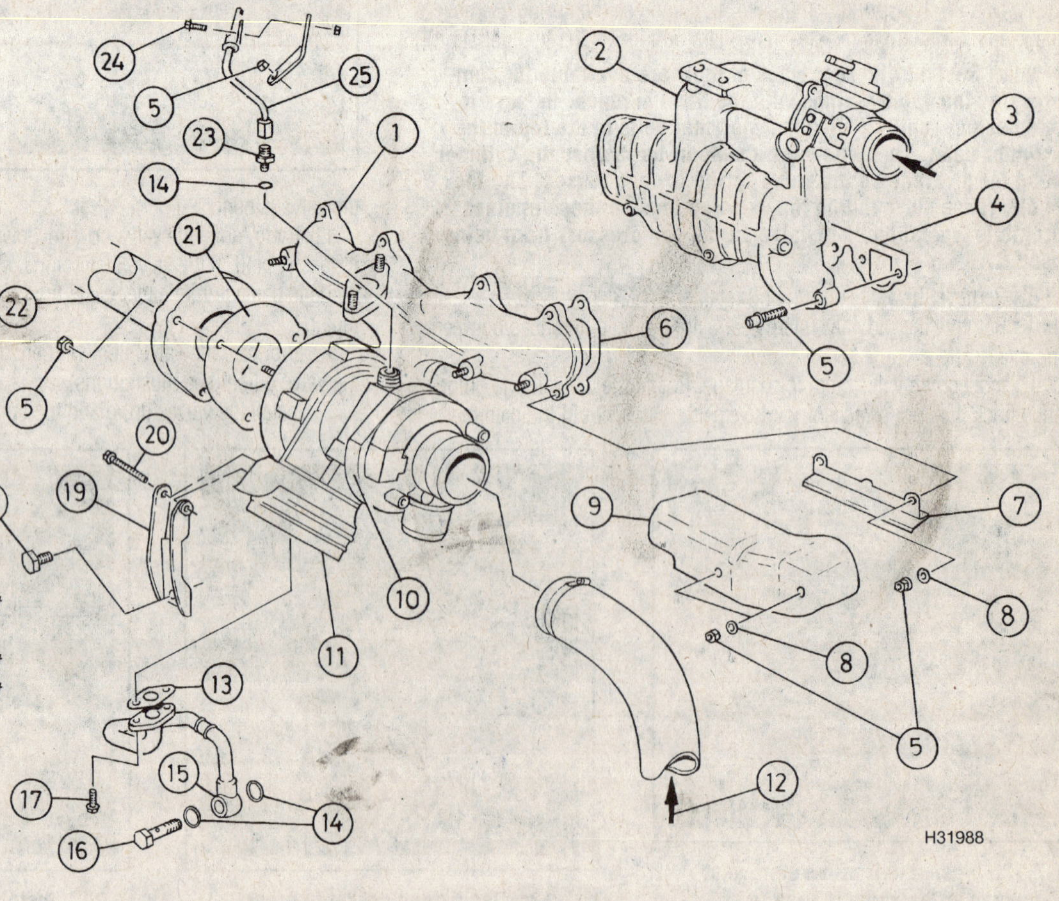

8.9 Exploded view of the intake and exhaust manifolds components

turbocharger, also referring to Chapter 4B.

4 Remove the crankcase ventilation valve from the top of the valve cover (see Section 4).

5 Remove windshield wiper arms and the cowl cover (see Chapters 11 and 12).

6 Disconnect the vacuum hose from the EGR valve. Then disconnect the vacuum hose and wire harness from the air shut off valve solenoid at the back of the intake manifold.

7 Remove the EGR cooler and pipes (see Chapter 6B).

8 Label and detach any remaining electrical connectors or hoses connected to the intake manifold.

9 Remove the mounting nuts/bolts (see illustration), then detach the manifold and gasket from the engine.

INSTALLATION

10 Use a scraper to remove all traces of old gasket material and sealant from the manifold and cylinder head, then clean the mating surfaces with lacquer thinner or acetone. Keep in mind that the intake manifold and the cylinder head are made of aluminum, so aggressive scraping is not suggested. If the gasket was leaking, have the manifold checked for warpage at an automotive machine shop and resurfaced if necessary.

11 Install a new gasket, then position the manifold on the head and install the nuts/bolts.

12 Tighten the nuts/bolts in several equal steps to the torque listed in this Chapter's Specifications. Work from the center out towards the ends to avoid warping the manifold.

13 Install the remaining parts in the reverse order of removal.

14 Check the coolant and add some, if necessary, to bring it to the appropriate level (see Chapter 1). Run the engine and check for coolant and intake air leaks.

15 Road test the vehicle and check for proper operation of all accessories, including the cruise control system.

9 Exhaust manifold - removal and installation

Refer to illustration 9.6

WARNING:

The engine must be completely cool before beginning this procedure.

1 The exhaust manifold and the turbocharger are integrated together and must be replaced as a unit. Follow the procedures outlined in Chapter 4B on turbocharger replacement to remove and disconnect the hoses and related components leading to the turbocharger, then proceed with the following Steps to remove and install the exhaust manifold/turbocharger unit.

2 Apply penetrating oil to the exhaust manifold mounting nuts/bolts.

3 Remove the mounting nuts/bolts and lower the exhaust manifold/turbocharger from the engine (see illustration 8.9).

4 Remove the exhaust manifold gaskets from the studs on the cylinder head.

5 Use a scraper to remove all traces of old gasket material and carbon deposits from the manifold and cylinder head mating surfaces. If the gasket was leaking, have the manifold checked for warpage at an automotive machine shop.

6 Position the new gaskets over the cylinder head studs.

Note: Be sure to position the exhaust manifold gaskets with the "notched" end facing up, otherwise they will obstruct the intake manifold gasket (see illustration).

7 Install the exhaust manifold/turbocharger and thread the mounting nuts/bolts into place.

8 Working from the center out, tighten the nuts/bolts to the torque listed in this Chapter's Specifications in several equal steps.

9 Refer to back to Chapter 4B and install the remaining parts in the reverse order of removal.

10 Run the engine and check for exhaust leaks.

11 Check the coolant level, adding as necessary (see Chapter 1).

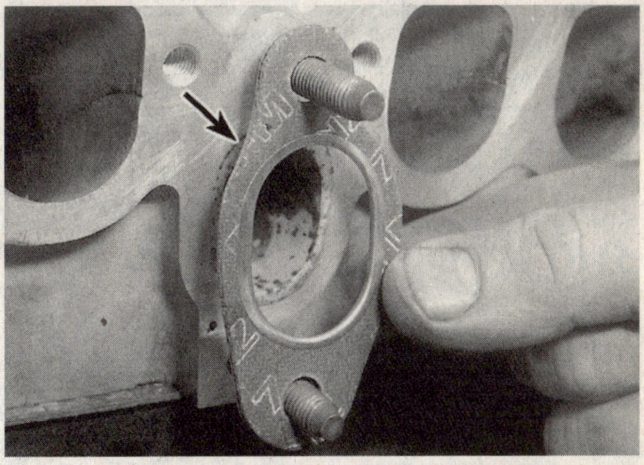

9.6 The exhaust manifold gaskets must be installed with the "notched end" (arrow) facing up

10 Cylinder head - removal and installation

WARNING:

The engine must be completely cool before beginning this procedure.

Note: The cylinder head can be removed with the intake and exhaust manifold/turbocharger attached.

REMOVAL

Refer to illustrations 10.10 and 10.12

1 Drain the engine coolant (see Chapter 1).

2 Refer to Chapter 11 and remove the hood and the cowl cover.

3 Refer to Section 5 and remove the timing belt. Follow Steps 1 through 20, then proceed to Step 26. After driving the camshaft

10.10 Remove the upper bolt securing the rear timing cover to the cylinder head

sprocket off the taper on the camshaft, remove the camshaft sprocket and the timing belt tensioner from the engine. After the timing belt, the timing belt tensioner and the camshaft sprocket have been removed, reinstall the passenger side engine mount to support the engine during the removal and installation of the cylinder head.

4 Disconnect the vacuum hose from the EGR valve. Then disconnect the vacuum hose and wire harness from the air shut off valve solenoid at the back of the intake manifold.

5 Remove the EGR cooler and pipes (see Chapter 6B).

6 Refer to Chapter 4B and remove all the related components from the turbocharger: the front exhaust pipe, the wastegate vacuum hose, the turbocharger support bracket, the air intake duct and pipes from the turbocharger to the air cleaner housing, the oil supply pipe, the oil return pipe and the heat shield.

7 Also referring to Chapter 4B disconnect and remove the fuel supply and return lines from the injectors and the injection pump head. Be sure to remove all the lines together as an assembly.

8 Unplug all electrical connectors, including the glow plug electrical bus connector from the cylinder head, labeling each wire or connector to aid the installation process.

9 Refer to Chapter 3 and loosen the clips and disconnect the radiator hoses from the ports on the cylinder head. Loosen the clips and disconnect the coolant reservoir hose and the heater intake and outlet coolant hoses from the ports on the cylinder head. Also disconnect the electrical connector from the coolant temperature sensor.

10 Loosen and withdraw the upper retaining screw from the timing belt rear cover (see illustration).

11 Working in the reverse of the sequence shown in illustration 10.27a, progressively loosen the cylinder head bolts, by half a turn at a time, until all bolts can be unscrewed by hand. Discard the bolts - new ones must be installed on reassembly.

12 Check that nothing remains connected to the cylinder head, then lift the head away from the cylinder block; seek assistance if possible, as it is very heavy, especially when being removed with the manifolds (see illustration).

13 Remove the gasket from the top of the block. Do not discard the gasket - it will be needed for identification purposes.

14 If the cylinder head is to be disassembled for service, separate the manifold(s) as described in Sections 8 and 9. Disregard the steps that do not apply since the cylinder head is already removed from the vehicle, then proceed to Chapter 2C for overhaul procedures.

INSTALLATION

Refer to illustrations 10.20, 10.23, 10.25, 10.27a and 10.27b

15 The mating faces of the cylinder head and cylinder block must be perfectly clean before installing the head. Use a hard plastic or wood scraper to remove all traces of gasket and carbon; also clean the piston

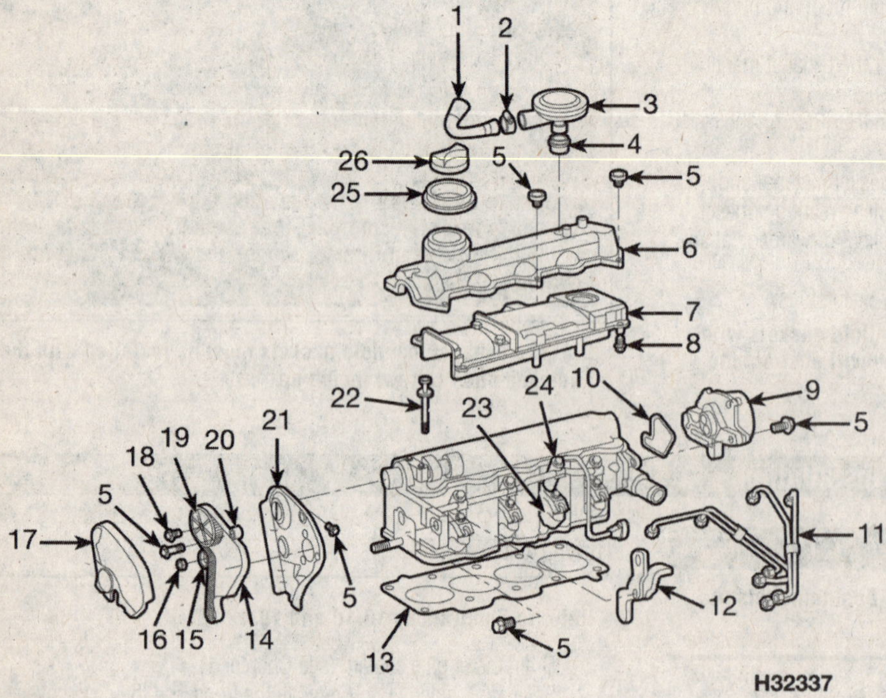

10.12 Exploded view of the cylinder head and related components

1 *Crankcase breather hose*
2 *Clamp*
3 *Crankcase breather valve*
4 *Sealing grommet*
5 *Bolt*
6 *Valve cover*
7 *Oil deflector*
8 *Bolt*
9 *Vacuum pump*
10 *Vacuum pump gasket*
11 *Fuel injector lines*
12 *Engine lifting bracket*
13 *Cylinder head gasket*
14 *Timing belt*
15 *Timing belt tensioner*
16 *Nut*
17 *Upper timing belt cover*
18 *Camshaft sprocket bolt*
19 *Camshaft sprocket*
20 *Idler wheel*
21 *Rear timing belt cover*
22 *Cylinder head bolt*
23 *Glow plug*
24 *Fuel injector*
25 *Sealing grommet*
26 *Oil filler cap*

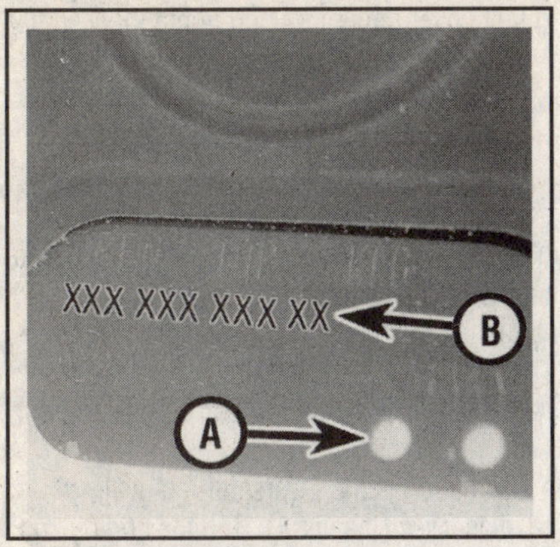

10.20 Cylinder head gasket identification markings - the punched holes (A) determine the thickness of the gasket and (B) is the part number

10.23 Two of the old head bolts (arrows) can be used as cylinder head alignment dowels

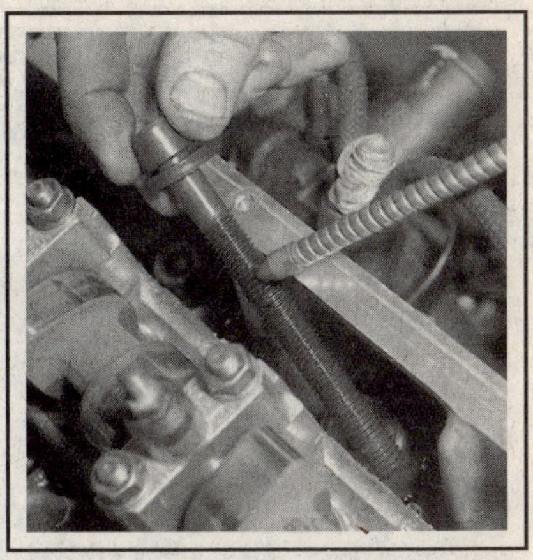

10.25 Apply oil to the threads of the cylinder head bolts before installing them

crowns. Take particular care during the cleaning operations, as aluminum alloy is easily damaged. Also, make sure that the carbon is not allowed to enter the oil and water passages - this is particularly important for the lubrication system, as carbon could block the oil supply to the engine's components. Using adhesive tape and paper, seal the water, oil and bolt holes in the cylinder block.

16 Check the mating surfaces of the cylinder block and the cylinder head for nicks, deep scratches and other damage. If slight, they may be removed carefully with abrasive paper.

17 If warpage of the cylinder head gasket surface is suspected, use a straight-edge to check it for distortion, but note that head machining will not be possible - refer to Chapter 2C.

18 Clean out the cylinder head bolt holes using a suitable tap. Be sure they are clean and dry before installation of the head bolts.

19 On diesel engines it is possible for the piston crowns to strike and damage the valve heads if the camshaft is rotated with the timing belt removed and the crankshaft set to TDC. For this reason, the crankshaft must be set to a position other than TDC on No. 1 cylinder before the cylinder head is reinstalled. Use a wrench and socket on the crankshaft pulley center bolt to turn the crankshaft in the opposite direction of rotation (counterclockwise), until all four pistons are positioned halfway down their bores (approximately 90-degrees before TDC).

20 Examine the old cylinder head gasket for manufacturer's identification markings. These will either be in the form of punched holes or a part number, on the edge of the gasket (see illustration). Unless new pistons have been installed, the new cylinder head gasket must be the same type as the old one.

21 If new pistons have been installed as part of an engine overhaul, before purchasing the new cylinder head gasket, refer to Chapter 2C and measure the piston projection. Purchase a new gasket according to the results of the measurement (also see Chapter 2C Specifications).

22 Cut off the heads from two of the old cylinder head bolts to use as alignment dowels during cylinder head installation. Also cut a slot in the end of the each bolt, big enough for a screwdriver blade so that the alignment dowels can be removed after the cylinder head is installed. A simple hand-held hacksaw can be used to fabricate the alignment dowels.

23 Install the alignment dowels in the outer rear holes of the cylinder block and position the new head gasket on the cylinder block, engaging it with the locating dowels (see illustration). Ensure that the manufacturer's "TOP" and part number markings are face up.

24 With the help of an assistant, place the cylinder head and manifolds centrally on the cylinder block, ensuring that the locating dowels engage with the recesses in the cylinder head. Check that the head gasket is correctly seated before allowing the full weight of the cylinder head to rest upon it.

➡Note: If the cylinder head had been disassembled for repair, be sure the camshaft locking tool is reinstalled on the cylinder head before installing the cylinder head onto the block. The locking tool should be installed as described in Section 5 with the No.1 cylinder camshaft lobes pointing upward.

25 Oil the threads and the underside of the bolt heads, then carefully guide each bolt into its relevant hole and screw them in hand tight (see illustration). Be sure to use NEW cylinder head bolts, as the old bolts are stretch-type fasteners that will not obtain the correct torque readings if reused.

26 Unscrew the homemade alignment dowels using a flat-bladed screwdriver and install the remaining two bolts hand-tight.

27 Working progressively and in the sequence shown (see illustration), tighten the cylinder head bolts in four steps to the torque and angle of rotation listed in this Chapter's Specifications.

➡Note: It is recommended that an angle measuring gauge be used during the final stages of the tightening, to ensure accuracy (see illustration). If a gauge is not available, use white paint to make alignment marks between the bolt head and cylinder head prior to tightening; the marks can then be used to check the bolt has been rotated through the correct angle during tightening.

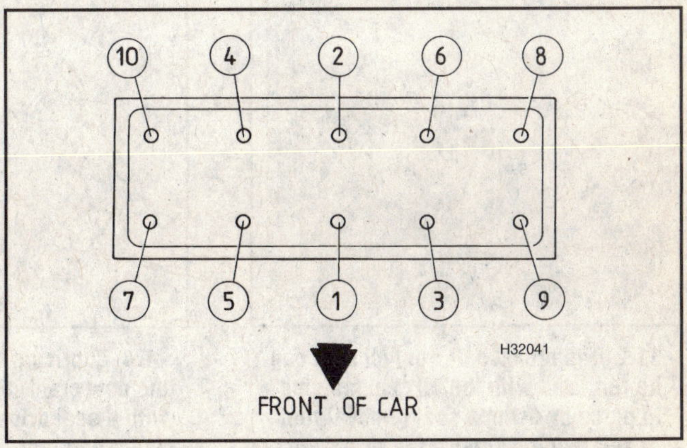

10.27a Cylinder head bolt TIGHTENING sequence

10.27b Using an angle measurement gauge during the final stages of tightening

28 Install the rear timing belt cover upper bolt and tighten it securely.

29 Rotate the crankshaft in the normal direction of rotation (clockwise) 90 degrees to TDC. Be sure the alignment marks on the flywheel or torque converter align with the mark on the bellhousing. Refer to Section 3, if necessary.

30 Support the engine from underneath with a jack and a block of wood, then remove the right (passenger side) engine mount. which was reinstalled to support the engine as the cylinder head was removed.

31 Install the timing belt tensioner and the camshaft sprocket on the engine loosely. Be sure the tensioner is seated properly in the rear timing cover hole, then install and adjust the timing belt as described in Section 5.

32 The remainder of the installation is the reverse of removal. Change the engine oil and coolant (see Chapter 1). Run the engine and check for leaks.

➡Note: If the CHECK ENGINE light comes on after performing this procedure it will be necessary to take the vehicle to a dealer service department or other qualified repair shop to have the injection pump timing reset with the proper scan tool. See Chapter 4B for further information.

11 Crankshaft front oil seal and housing - replacement

➧ Refer to illustrations 11.2, 11.4, 11.5 and 11.7

1 Remove the timing belt and crankshaft sprocket (see Section 5).

2 Note how far the seal is recessed in the bore, then carefully pry it out of the front cover with a screwdriver or seal removal tool. Don't scratch the housing bore or damage the crankshaft in the process (if the crankshaft is damaged, the new seal will end up leaking).

➡Note: If a seal removal tool is unavailable, you can thread two self-tapping screws (180 degrees apart from one another) into the front seal to pry the seal out (see illustration).

3 Clean the bore in the housing and coat the outer edge of the new seal with engine oil or multi-purpose grease. Apply multi-purpose grease to the seal lip.

4 Using a socket with an outside diameter slightly smaller than the outside diameter of the seal, carefully drive the new seal into place with a hammer (see illustration). Make sure it's installed squarely and driven in to the same depth as the original. If a socket isn't available, a short section of large diameter pipe will also work. Check the seal after installation to make sure the spring didn't pop out of place.

5 If the front oil seal housing needs to be removed for access to other components such as the oil pump drive chain, simply loosen the housing mounting bolts and remove the housing from the engine. The front oil seal housing can be removed with or without the front oil seal. In some instances the front oil seal removal and installation is easier with the front housing removed, since the seal can be placed on a workbench and driven straight in and out of the bore with no special tools or adapters (see illustration).

6 Before installing the front cover, make sure the mating surfaces of the cover, the cylinder block and the oil pan rail are perfectly clean. Remove all traces of gasket material. Be careful, as aluminum alloy is

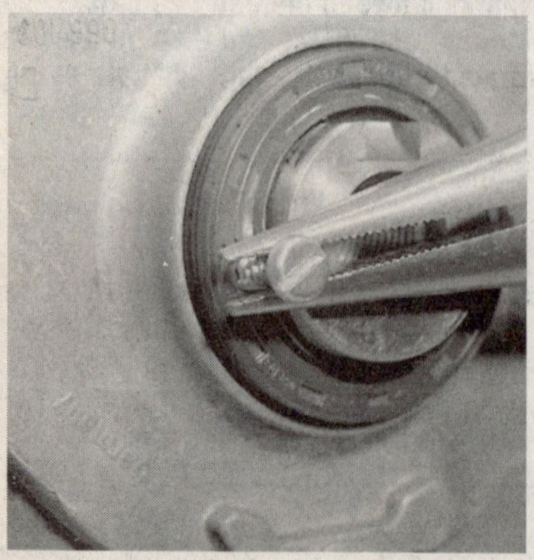

11.2 The crankshaft front oil seal can be removed with the oil seal housing in place by using a seal removal tool or self tapping screws (as shown) to pry the seal out

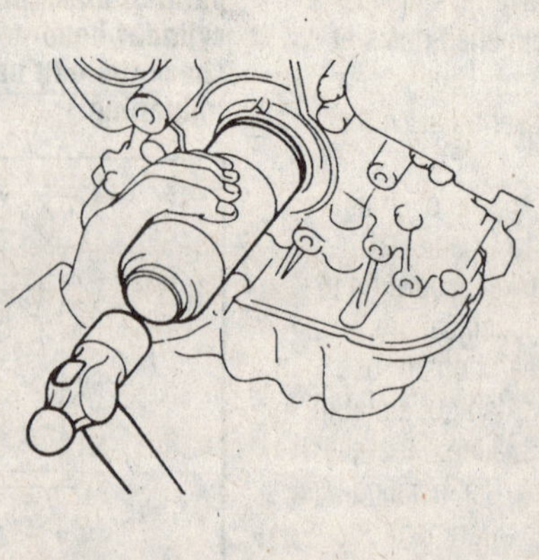

11.4 Lubricate the seal lip and drive the new crankshaft seal into place with a seal driver or a large socket and a hammer

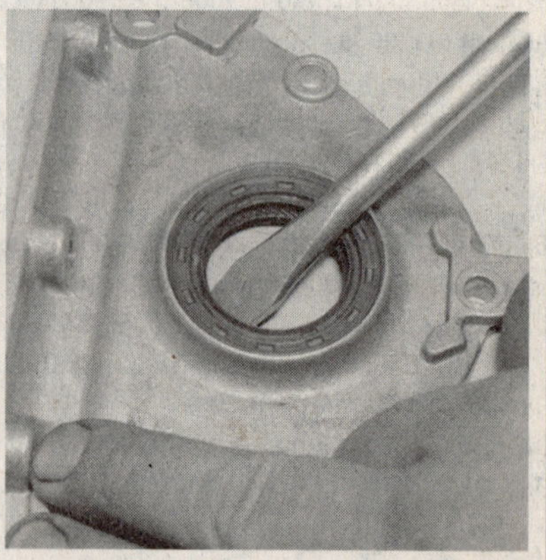

11.5 With the front oil seal housing removed, the oil seal can be pried out using conventional tools

easily damaged.

7 Apply a 3/16-inch (5 mm) bead of RTV sealant to the cover rear sealing flange (see illustration). Also apply a 3/16-inch (5 mm) bead of RTV sealant to the oil pan flange and install the front cover onto the engine.

➡Note: Be sure to lubricate the oil seal lip before installing the front cover onto the engine. This will aid the installation process and prevent dry start ups, which may damage the seal and lead to future oil leaks.

8 Tighten the front oil seal housing bolts in several steps to the torque listed in this Chapter's Specifications.

9 Reinstall the crankshaft sprocket and timing belt (see Section 5).

10 Run the engine and check for oil leaks at the front seal.

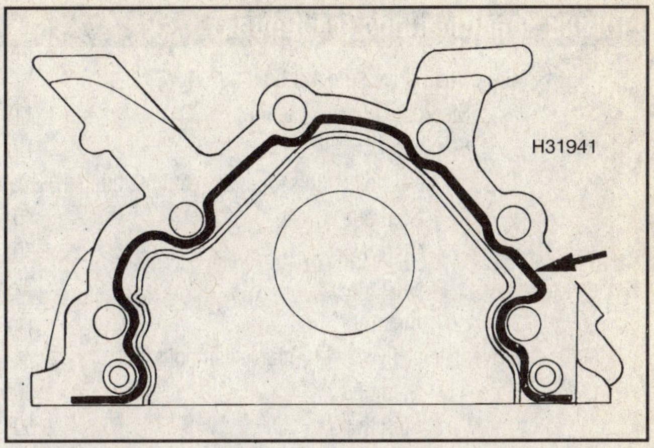

11.7 Apply a 3/16-inch (5 mm) bead of RTV sealant to the rear of the front oil seal housing as shown

Specifications

General

Displacement	116 cubic inches (1.9 liters)
Bore and stroke	3.130 x 3.759 inches (79.50 x 95.47 mm)
Cylinder numbers (drivebelt end-to-transaxle end)	1-2-3-4
Firing order	1-3-4-2

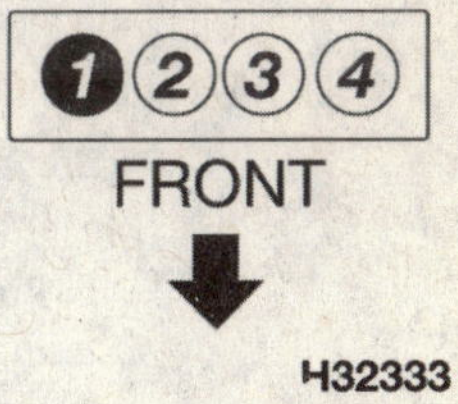

Cylinder numbering for 1.9L diesel engine

Camshaft

Endplay	0.0059 inch (0.15 mm)
Journal diameter	N/A
Journal oil clearance (in cylinder head)	0.0043 inch (0.110 mm)
Lobe lift	
Intake	N/A
Exhaust	N/A
Runout	0.0004 inch (0.010 mm)

Driveplate

Driveplate-to-cylinder block clearance	0.768 to 0.830 inch (19.5 to 21.10 mm)

Torque specifications

Torque specifications	Nm	Ft-lbs (unless otherwise indicated)
Camshaft sprocket bolt	45	33
Camshaft bearing cap nuts	20	15
Crankshaft front oil seal housing	15	132 in-lbs
Crankshaft rear oil seal housing	15	132 in-lbs
Crankshaft pulley/vibration damper bolts*		
Step one	10	84 in-lbs
Step two	Tighten an additional 90 degrees	

Torque specifications (continued)	Nm	Ft-lbs (unless otherwise indicated)
Crankshaft drive sprocket bolt*		
Step one	119	88
Step two	Tighten an additional 90 degrees	
Cylinder head bolts* (in sequence see illustration 10.27a)		
Step one	40	30
Step two	60	44
Step three	Tighten an additional 90 degrees	
Step four	Tighten an additional 90 degrees	
Driveplate or flywheel-to-crankshaft bolts*		
Step one	30	22
Step two	60	44
Step three	Tighten an additional 90 degrees	
Exhaust manifold nuts	24	18
Intake manifold bolts	24	18
Idler pulley bolts		
Upper pulley bolt	20	15
Center pulley nut	21	16
Lower pulley bolt*		
Step one	40	30
Step two	Tighten an additional 90 degrees	
Oil pan mounting bolts	15	132 in-lbs
Oil pump driven sprocket bolt	24	18
Oil pump drive chain tensioner bolt	15	132 in-lbs
Oil pump mounting bolts	15	132 in-lbs
Oil pump pick-up tube bolts	15	132 in-lbs
Powertrain mounts		
Mount-to-body bolts*		
Step one	40	30
Step two	Tighten an additional 90 degrees	
Mount bracket-to-body bolts	24	18
Mount bracket-to-cylinder block bolts (passenger side)*		
Step one	60	44
Step two	Tighten an additional 90 degrees	
Mount-to-transaxle bracket bolts (driver's side)*		
Step one	60	44
Step two	Tighten an additional 90 degrees	
Lower torque strut-to-subframe bolts*		
Step one	20	15
Step two	Tighten an additional 90 degrees	
Lower torque strut-to-bracket bolt*	40	30
Lower torque strut bracket-to-transaxle bolt*		
Step one	40	30
Step two	Tighten an additional 90 degrees	
Rear timing belt cover-to-cylinder head bolts	30	22
Timing belt covers-to-block bolts	10	84 in-lbs
Timing belt tensioner locknut	20	15
Valve cover-to-cylinder head nuts	10	84 in-lbs

** Replace with new bolt(s)*

2C

GENERAL ENGINE OVERAHUL PROCEDURES

Section

Reference to other Chapters

CHECK ENGINE light - See Chapter 6A or 6B

1 General information - engine overhaul

Included in this portion of Chapter 2 are the general overhaul procedures for the cylinder head and internal engine components.

The information ranges from advice concerning preparation for an overhaul and the purchase of replacement parts to detailed, step-by-step procedures covering removal and installation of internal engine components and the inspection of parts.

The following Sections have been written based on the assumption that the engine has been removed from the vehicle. For information concerning in-vehicle engine repair, as well as removal and installation of the external components necessary for the overhaul, see Chapter 2A (gasoline engines) or Chapter 2B (diesel engines), and Section 8 of this Chapter.

The Specifications included in this Part are only those necessary for the inspection and overhaul procedures which follow. Refer to Chapter 2, Part A or Part B for additional Specifications.

It's not always easy to determine when, or if, an engine should be completely overhauled, as a number of factors must be considered.

High mileage is not necessarily an indication that an overhaul is needed, while low mileage doesn't preclude the need for an overhaul. Frequency of servicing is probably the most important consideration. An engine that's had regular and frequent oil and filter changes, as well as other required maintenance, will most likely give many thousands of miles of reliable service. Conversely, a neglected engine may require an overhaul very early in its life.

Excessive oil consumption is an indication that piston rings, valve seals and/or valve guides are in need of attention. Make sure that oil leaks aren't responsible before deciding that the rings and/or guides are bad. Perform a cylinder compression check to determine the extent of the work required (see Section 3). Also check the vacuum readings under various conditions (see Section 4).

Loss of power, rough running, knocking or metallic engine noises, excessive valve train noise and high fuel consumption rates may also point to the need for an overhaul, especially if they're all present at the same time. If a complete tune-up doesn't remedy the situation, major mechanical work is the only solution.

An engine overhaul involves restoring the internal parts to the specifications of a new engine. During an overhaul, the piston rings are replaced and the cylinder walls are reconditioned (re-bored and/or honed). If a re-bore is done by an automotive machine shop, new oversize pistons will also be installed. The main bearings and connecting rod bearings are generally replaced with new ones and, if necessary, the crankshaft may be reground to restore the journals. Generally, the valves are serviced as well, since they're usually in less-than-perfect condition at this point. While the engine is being overhauled, other components, such as the starter and alternator, can be rebuilt as well. The end result should be a like new engine that will give many trouble free miles.

➡**Note: Critical cooling system components such as the hoses, drivebelts, thermostat and water pump should be replaced with new parts when an engine is overhauled. The radiator should be checked carefully to ensure that it isn't clogged or leaking (see Chapter 3). If you purchase a rebuilt engine or short block, some rebuilders will not warranty their engines unless the radiator has been professionally flushed. Also, we don't recommend overhauling the oil pump - always install a new one when an engine is rebuilt.**

Before beginning the engine overhaul, read through the entire procedure to familiarize yourself with the scope and requirements of the job. Overhauling an engine isn't difficult, but it is time-consuming. Plan on the vehicle being tied up for a minimum of two weeks, especially if parts must be taken to an automotive machine shop for repair or reconditioning. Check on availability of parts and make sure that any necessary special tools and equipment are obtained in advance. Most work can be done with typical hand tools, although a number of precision measuring tools are required for inspecting parts to determine if they must be replaced. Often an automotive machine shop will handle the inspection of parts and offer advice concerning reconditioning and replacement.

➡**Note: Always wait until the engine has been completely disassembled and all components, especially the engine block, have been inspected before deciding what service and repair operations must be performed by an automotive machine shop. Since the block's condition will be the major factor to consider when determining whether to overhaul the original engine or buy a rebuilt one, never purchase parts or have machine work done on other components until the block has been thoroughly inspected. As a general rule, time is the primary cost of an overhaul, so it doesn't pay to install worn or substandard parts.**

As a final note, to ensure maximum life and minimum trouble from a rebuilt engine, everything must be assembled with care in a spotlessly-clean environment.

2 Oil pressure check

➧ Refer to illustration 2.2

1 Low engine oil pressure can be a sign of an engine in need of rebuilding. A "low oil pressure" indicator (often called an "idiot light") is not a test of the oiling system. Such indicators only come on when the oil pressure is dangerously low. Even a factory oil pressure gauge in the instrument panel is only a relative indication, although much better for driver information than a warning light. A better test is with a mechanical (not electrical) oil pressure gauge. When used in conjunction with an accurate tachometer, an engine's oil pressure performance can be compared to the manufacturers Specifications.

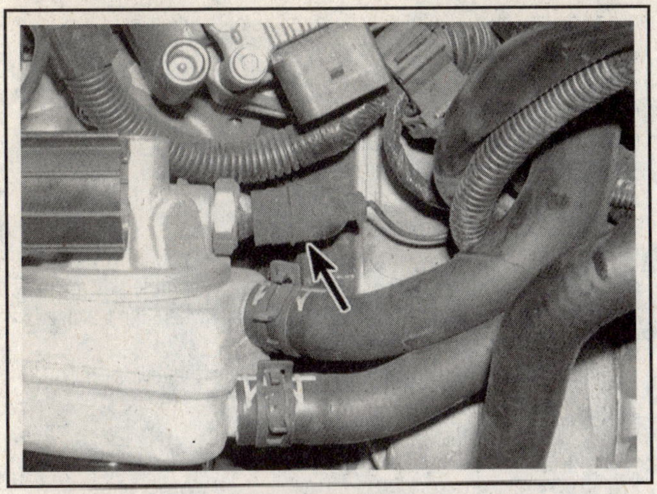

2.2 The oil pressure sending unit (arrow) is mounted on the oil filter adapter housing which is located on the (front) side of the engine block (2.0L engine shown, all others similar)

2 Find the oil pressure indicator sending unit (see illustration).

3 Remove the oil pressure sending unit and install a fitting which will allow you to directly connect your hand-held, mechanical oil pressure gauge. Use Teflon tape or sealant on the threads of the adapter and the fitting on the end of your gauge's hose.

4 Connect an accurate tachometer to the engine, according to the tachometer manufacturer's instructions.

5 Check the oil pressure with the engine running (full operating temperature) at the specified engine speed, and compare it to this Chapter's Specifications. If it's extremely low, the bearings and/or oil pump are probably worn out.

3 Cylinder compression check

Refer to illustration 3.6

1 A compression check will tell you what mechanical condition the upper end (pistons, rings, valves, head gaskets) of the engine is in. Specifically, it can tell you if the compression is down due to leakage caused by worn piston rings, defective valves and seats or a blown head gasket.

Note: The engine must be at normal operating temperature and the battery must be fully charged for this check.

2 Begin by cleaning the area around the spark plugs before you remove them. Compressed air should be used, if available, otherwise a small brush will work. The idea is to prevent dirt from getting into the cylinders as the compression check is being done.

3 Remove all of the spark plugs (see Chapter 1) or glow plugs from the engine.

4 Block the throttle wide open.

5 On gasoline engines, disable the fuel and ignition systems by disconnecting the primary electrical connectors at the ignition coil pack/modules (see Chapter 5) and the electrical connectors at the fuel injectors (see Chapter 4A). On diesel engines, disable the fuel injection system by unplugging the electrical connector to the fuel shut off valve (see Chapter 4B).

6 Install the compression gauge in the number one spark plug hole (see illustration).

WARNING:

If you are checking the compression on a diesel engine, make sure the gauge is capable of reading pressures up to 500 psi (35 kg/cm2).

7 Crank the engine over at least seven compression strokes and watch the gauge. The compression should build up quickly in a healthy engine. Low compression on the first stroke, followed by gradually increasing pressure on successive strokes, indicates worn piston rings. A low compression reading on the first stroke, which doesn't build up during successive strokes, indicates leaking valves or a blown head

3.6 A compression gauge with a threaded fitting for the spark plug hole is preferred over the type that requires hand pressure to maintain the seal - be sure to open the throttle valve as far as possible during the compression check

gasket (a cracked head could also be the cause). Deposits on the undersides of the valve heads can also cause low compression. Record the highest gauge reading obtained.

8 Repeat the procedure for the remaining cylinders, turning the engine over for the same length of time for each cylinder, and compare the results to this Chapter's Specifications.

GASOLINE ENGINES ONLY

WARNING:

DO NOT perform this step on a diesel engine. Checking compression on a diesel engine with oil in the cylinder could cause it to fire, which would damage the gauge and possibly the engine, and could cause injury.

9 If the readings are below normal, add some engine oil (about three squirts from a plunger-type oil can) to each cylinder, through the spark plug hole, and repeat the test.

10 If the compression increases after the oil is added, the piston rings are definitely worn. If the compression doesn't increase significantly, the leakage is occurring at the valves or head gasket. Leakage past the valves may be caused by burned valve seats and/or faces or warped, cracked or bent valves.

ALL ENGINES

11 If two adjacent cylinders have equally low compression, there's a strong possibility the head gasket between them is blown. The appearance of coolant in the combustion chambers or the crankcase would verify this condition.

12 If one cylinder is about 20-percent lower than the others, and the engine has a slightly rough idle, a worn exhaust lobe on the camshaft could be the cause.

13 If the compression is unusually high, the combustion chambers are probably coated with carbon deposits. If that's the case, the cylinder heads should be removed and decarbonized.

14 If compression is way down or varies greatly between cylinders, it would be a good idea to have a leak-down test performed by an automotive repair shop. This test will pinpoint exactly where the leakage is occurring and how severe it is.

15 Installation of the remaining components is the reverse of removal.

4 Vacuum gauge diagnostic checks

➧ Refer to illustrations 4.4 and 4.6

➡Note: This procedure does not apply to diesel engines.

1 A vacuum gauge provides valuable information about what is going on in the engine at a low cost. You can check for worn rings or cylinder walls, leaking head or intake manifold gaskets, incorrect carburetor adjustments, restricted exhaust, stuck or burned valves, weak valve springs, improper ignition or valve timing and ignition problems.

2 Unfortunately, vacuum gauge readings are easy to misinterpret, so they should be used in conjunction with other tests to confirm the diagnosis.

3 Both the gauge readings and the rate of needle movement are important for accurate interpretation. Most gauges measure vacuum in inches of mercury (in-Hg). As vacuum increases (or atmospheric pressure decreases), the reading will increase. Also, for every 1,000-foot increase in elevation above sea level, the gauge readings will decrease about one inch of mercury.

4 Connect the vacuum gauge directly to intake manifold vacuum, not to ported vacuum (see illustration). Be sure no hoses are left disconnected during the test or false readings will result.

5 Before you begin the test, allow the engine to warm up completely. Block the wheels and set the parking brake. With the transmission in Park, start the engine and allow it to run at normal idle speed.

6 Read the vacuum gauge; an average, healthy engine should normally produce about 17 to 22 inches of vacuum with a fairly steady needle. Refer to the following vacuum gauge readings and what they indicate about the engine's condition (see illustration).

7 A low, steady reading usually indicates a leaking gasket between the intake manifold and carburetor or throttle body, a leaky vacuum hose, late ignition timing or incorrect camshaft timing. Eliminate all other possible causes, utilizing the tests provided in this Chapter before you remove the timing belt cover to check the timing marks.

8 If the reading is three to eight inches below normal and it fluctuates at that low reading, suspect an intake manifold gasket leak at an intake port.

9 If the needle has regular drops of about two to four inches at a steady rate, the valves are probably leaking. Perform a compression or leak-down test to confirm this.

10 An irregular drop or down-flick of the needle can be caused by a sticking valve or an ignition misfire. Perform a compression or leak-down test and read the spark plugs.

4.4 A simple vacuum gauge can be very handy in diagnosing engine condition and performance

11 A rapid vibration of about four inches-Hg vibration at idle combined with exhaust smoke indicates worn valve guides. Perform a leak-down test to confirm this. If the rapid vibration occurs with an increase in engine speed, check for a leaking intake manifold gasket or head gasket, weak valve springs, burned valves or ignition misfire.

12 A slight fluctuation, say one inch up and down, may mean ignition problems. Check all the usual tune-up items and, if necessary, run the engine on an ignition analyzer.

13 If there is a large fluctuation, perform a compression or leak-down test to look for a weak or dead cylinder or a blown head gasket.

14 If the needle moves slowly through a wide range, check for a clogged PCV system, throttle body or intake manifold gasket leaks.

15 Check for a slow return after revving the engine by quickly snapping the throttle open until the engine reaches about 2,500 rpm and let it shut. Normally the reading should drop to near zero, rise above normal idle reading (about 5 in-Hg over) and then return to the previous idle reading. If the vacuum returns slowly and doesn't peak when the throttle is snapped shut, the rings may be worn. If there is a long delay, look for a restricted exhaust system (often the muffler or catalytic converter). An easy way to check this is to temporarily disconnect the exhaust ahead of the suspected part and re-test.

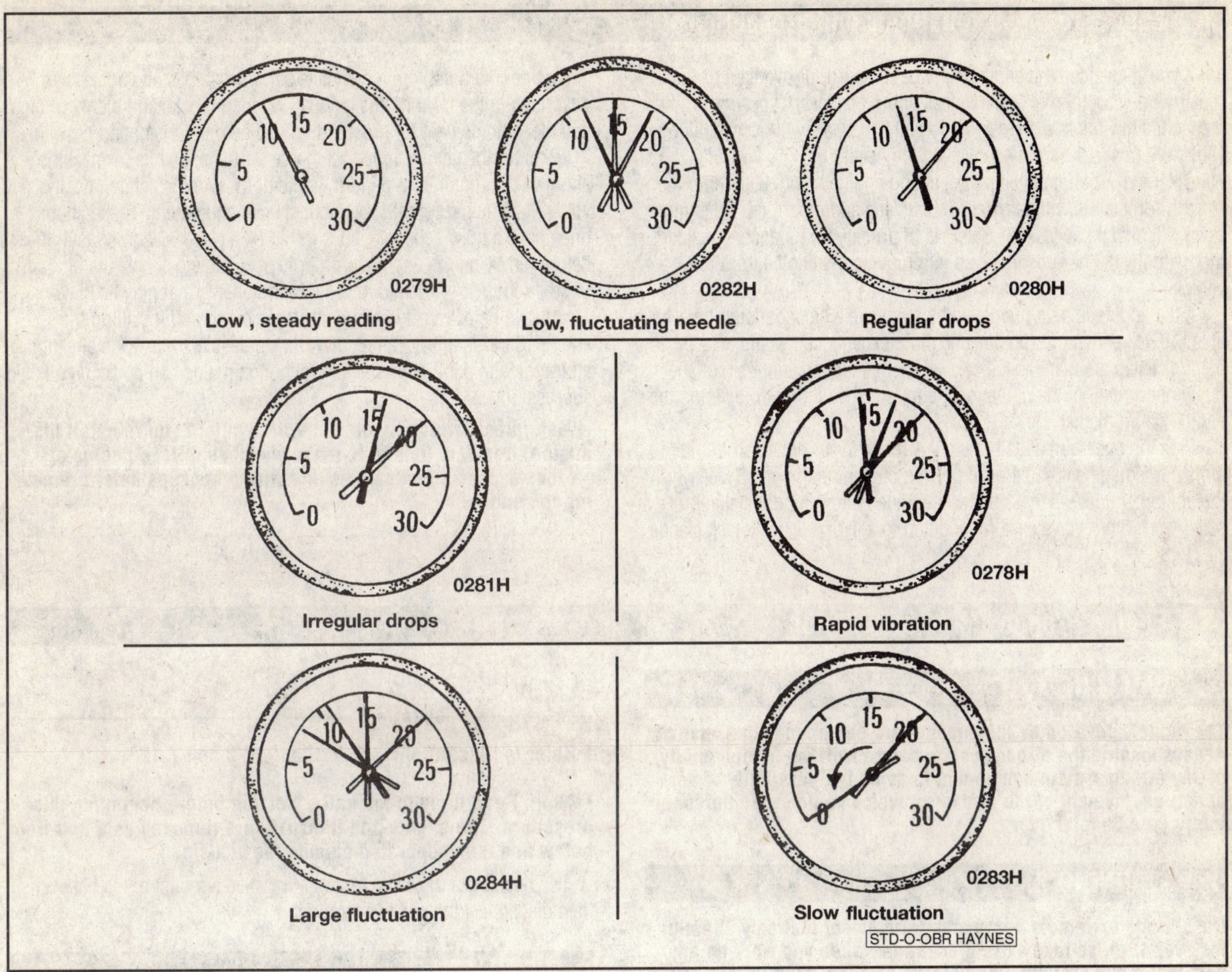

4.6 Typical vacuum gauge diagnostic readings

5 Engine rebuilding alternatives

The home mechanic is faced with a number of options when performing an engine overhaul. The decision to replace the engine block, piston/connecting rod assemblies and crankshaft depends on a number of factors, with the number one consideration being the condition of the block. Other considerations are cost, access to machine shop facilities, parts availability, time required to complete the project and the extent of prior mechanical experience.

Some of the rebuilding alternatives include:

Individual parts - If the inspection procedures reveal the engine block and most engine components are in reusable condition, purchasing individual parts may be the most economical alternative. The block, crankshaft and piston/connecting rod assemblies should all be inspected carefully. Even if the block shows little wear, the cylinder bores should be surface-honed.

Short-block - A short-block consists of an engine block with a crankshaft and piston/connecting rod assemblies already installed. All new bearings are incorporated and all clearances will be correct. The existing camshaft, valve train components, cylinder head and external parts can be bolted to the short block with little or no machine shop work necessary.

Long-block - A long-block consists of a short block plus an oil pump, oil pan, cylinder head, valve cover, camshaft and valve train components, timing sprockets and a timing belt. All components are installed with new bearings, seals and gaskets incorporated throughout. The installation of manifolds and external parts is all that's necessary.

Give careful thought to which alternative is best for you and discuss the situation with local automotive machine shops, auto parts dealers and experienced rebuilders before ordering or purchasing replacement parts.

6 Engine removal - methods and precautions

If you've decided the engine must be removed for overhaul or major repair work, several preliminary steps should be taken. Locating a suitable place to work is extremely important. Adequate work space, along with storage space for the vehicle, will be needed.

Cleaning the engine compartment and engine before beginning the removal procedure will help keep tools clean and organized. An engine hoist will also be necessary. Safety is of primary importance, considering the potential hazards involved in removing the engine from this vehicle.

If the engine is being removed by a novice, a helper should be available. Advice and aid from someone more experienced would also be helpful. There are many instances when one person cannot simultaneously perform all of the operations required when lifting or lowering the engine out of the vehicle.

Plan the operation ahead of time. Arrange for or obtain all of the tools and equipment you'll need prior to beginning the job. Some of the equipment necessary to perform engine removal and installation safely and with relative ease (in addition to a hydraulic jack, jack stands and an engine hoist) are a complete set of wrenches and sockets as described in the front of this manual, wooden blocks and plenty of rags and cleaning solvent for mopping up spilled oil, coolant and gasoline.

Plan for the vehicle to be out of use for quite a while. A machine shop will be required to perform some of the work which the do-it-yourselfer can't accomplish without special equipment. These shops often have a busy schedule, so it would be a good idea to consult them before removing the engine in order to accurately estimate the amount of time required to rebuild or repair components that may need work.

Always be extremely careful when removing and installing the engine. Serious injury can result from careless actions. Plan ahead, take your time and a job of this nature, although major, can be accomplished successfully.

➡Note: Because it may be some time before you reinstall the engine, it is very helpful to make sketches or take photos of various accessory mountings and wiring hookups before removing the engine.

7 Engine - removal and installation

⁂ WARNING 1:

The models covered by this manual are equipped with airbags. Always disable the airbag system before working in the vicinity of any airbag system component to avoid the possibility of accidental deployment of the airbag(s), which could cause personal injury (see Chapter 12).

⁂ WARNING 2:

Gasoline is extremely flammable (and diesel fuel only slightly less volatile), so take extra precautions when you work on any part of the fuel system. Don't smoke or allow open flames or bare light bulbs near the work area, and don't work in a garage where a gas-type appliance (such as a water heater or a clothes dryer) is present. Since gasoline and diesel fuel is carcinogenic, wear latex gloves when there's a possibility of being exposed to fuel, and, if you spill any fuel on your skin, rinse it off immediately with soap and water. Mop up any spills immediately and do not store fuel-soaked rags where they could ignite. The fuel system on gasoline engines is under constant pressure, so, if any fuel lines are to be disconnected, the fuel pressure in the system must be relieved first (see Chapter 4A). When you perform any kind of work on the fuel system, wear safety glasses and have a Class B type fire extinguisher on hand.

⁂ WARNING 3:

The air conditioning system is under high pressure - have a dealer service department or service station evacuate the system and recover the refrigerant before disconnecting any of the hoses or fittings.

⁂ CAUTION:

When disassembling the air intake system on turbocharged vehicles, ensure that no foreign material can get into the turbo air intake port. Cover the opening with a sheet of plastic and a rubber band. The turbocharger compressor blades could be severely damaged if debris is allowed to enter.

REMOVAL

➧ Refer to illustrations 7.6, 7.17a, 7.17b and 7.19

➡Note: Read through the entire Section before beginning this procedure. The engine and transaxle are removed as a unit from below and then separated outside the vehicle.

1 Disconnect the cable from the negative terminal of the battery, then disconnect the positive terminal.

⁂ CAUTION:

On models equipped with an anti-theft audio system, be sure to have the correct activation code before performing any procedure which requires disconnecting the battery (see the front of this manual).

2 If you're removing a gasoline engine, relieve the fuel system pressure (see Chapter 4A).

3 Place protective covers on the fenders and cowl and remove the hood (see Chapter 11).

4 Remove the air cleaner assembly (see Chapter 4A). Remove the battery and its mounting bracket (see Chapter 5). On turbocharged models, also disconnect the intercooler hoses (see Chapter 4A or 4B).

5 Raise the vehicle and support it securely on jackstands. Drain the cooling system, transaxle and engine oil and remove the drivebelts (see Chapter 1).

6 Clearly label, then disconnect all vacuum lines, coolant and emissions hoses, wiring harness connectors, ground straps and fuel lines. Masking tape and/or a touch up paint applicator work well for marking items (see illustration). Take instant photos or sketch the locations of components and brackets.

7 Remove the cooling fans, and disconnect the radiator hoses and heater hoses (see Chapter 3).

8 Disconnect the fuel lines from the fuel rail or the injection pump (see Chapter 4A or 4B). Plug or cap all open fittings.

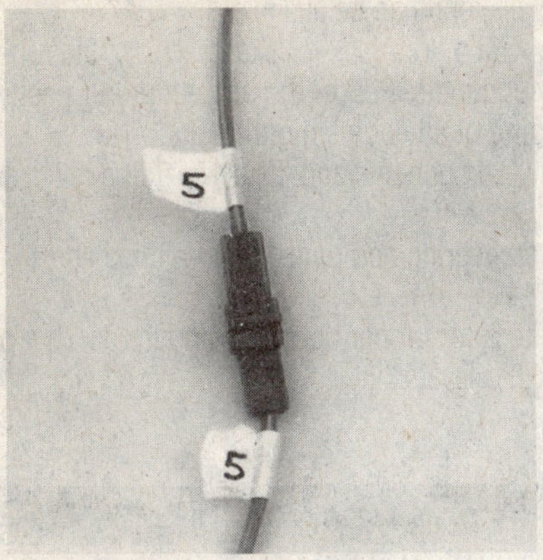

7.6 Label each wire before unplugging the connector

7.17a Attach the chain or sling to the lifting eyes - all engines are equipped with at least one

7.17b On some engines, it will be necessary to attach the sling or chain to the casting protrusion on the cylinder head

9 Refer to Chapter 3 and unbolt and set aside the air conditioning compressor, without disconnecting the refrigerant lines.

10 Disconnect the throttle cable (if equipped) from the engine (see Chapter 4A).

11 Unbolt the power steering pump. Tie the pump aside without disconnecting the hoses (see Chapter 10). Remove the alternator (see Chapter 5).

12 Remove the secondary air injection pump and hoses from the engine if equipped (see Chapter 6A).

13 Label and disconnect the main engine electrical harnesses from the starter and the engine.

14 On 2.0L engines, it may be helpful to remove the upper intake manifold to make engine removal easier (see Chapter 2A). Be sure to label and disconnect all hoses, connectors, and any ground straps.

15 Remove the front section of the exhaust system (see Chapter 4A).

16 Remove the driveaxles and the driveaxle boot protective cover if equipped (see Chapter 8). Disconnect the electrical connectors and the shift linkage from the transaxle (see Chapter 7A or 7B).

17 Attach a lifting sling or chain to the lifting eye (if equipped) on the engine (see illustration). If two lifting eyes are not provided, attach the lifting sling or chain to a safe place such as the side of the cylinder head (see illustration). Position a hoist and connect the sling to it. Take up the slack until there is slight tension on the hoist.

18 Recheck to be sure nothing except the mounts are still connecting the engine/transaxle to the vehicle. Disconnect anything still remaining.

19 Support the transaxle with a floor jack. Place a block of wood on the jack head to prevent damage to the transaxle. Raise the engine enough to remove the engine mount bolts (see illustration).

⁂ WARNING:

Do not place any part of your body under the engine/transaxle when it's supported only by a hoist or other lifting device.

20 Slowly lower the engine/transaxle out of the vehicle.

21 Once the powertrain is on the floor, disconnect the engine lifting hoist and have an assistant help you carefully rock the engine transaxle back until it is laying flat on the floor so it can be slid out from under the vehicle.

7.19 Raise the engine enough to remove the engine mounts, then remove the jack supporting the transaxle and lower the engine/transaxle to the floor

➡Note: A sheet of old hardboard or paneling between the engine and floor makes moving the powertrain easier. A helper will be needed to move the powertrain.

22 Separate the engine from the transaxle (see Chapter 7A or 7B).

23 Place the engine on the floor or remove the flywheel/driveplate and mount the engine on an engine stand.

INSTALLATION

24 Check the engine/transaxle mounts. If they're worn or damaged, replace them.

25 On automatic transaxle equipped models, inspect the converter seal and bushing, and apply a dab of grease to the nose of the converter and to the seal lips. If the vehicle is equipped with a manual transaxle, inspect the clutch components (see Chapter 8).

26 Carefully guide the transaxle into place, following the procedure outlined in Chapter 7A or 7B.

CAUTION:

Do not use the bolts to force the engine and transaxle into alignment. It may crack or damage major components.

27 Install the engine-to-transaxle bolts and tighten them securely.
28 Slide the engine/transaxle over a sheet of hardboard or paneling until it is in the approximate position under the vehicle, then tilt it upright.
29 Roll the engine hoist into position, attach the sling or chain in a position that will allow a good balance, and slowly raise the powertrain until the mounts at the transaxle end can be attached.
30 Support the transaxle with a floor jack for extra security, then reinstall the right-side engine mounts. Follow the procedure in Chapter 2, Part A for the final tightening of all engine mount bolts.
31 Reinstall the remaining components and fasteners in the reverse order of removal.
32 Add coolant, oil, power steering and transmission fluids as needed (see Chapter 1).
33 Run the engine and check for proper operation and leaks. Shut off the engine and recheck the fluid levels.

8 Engine overhaul - disassembly sequence

1 It's much easier to disassemble and work on the engine if it's mounted on a portable engine stand. A stand can often be rented quite cheaply from an equipment rental yard. Before it's mounted on a stand, the flywheel/driveplate should be removed from the engine.
2 If a stand isn't available, it's possible to disassemble the engine with it blocked up on the floor. Be extra careful not to tip or drop the engine when working without a stand.
3 If you're going to obtain a rebuilt engine, all external components must come off first, to be transferred to the replacement engine, just as they will if you're doing a complete engine overhaul yourself. These include:

Alternator mounting brackets
Emissions control components
Ignition coil/module assembly, spark plug wires and spark plugs (gasoline engines)
Glow plug/preheating system components (diesel engines)
Valve cover
Timing belt covers and timing belt
Water pump
Thermostat and housing cover and coolant supply tubes
Fuel system components
Turbocharger (if equipped)
Intake/exhaust manifolds
Camshaft and crankshaft position sensors
Oil filter and adapter housing
Oil dipstick and tube
Flywheel/driveplate

➡Note: When removing the external components from the engine, pay close attention to details that may be helpful or important during installation. Note the installed position of gaskets, seals, spacers, pins, brackets, washers, bolts and other small items.

4 If you're obtaining a short-block, then the cylinder heads, oil pan and oil pump will have to be removed as well. See *Engine rebuilding alternatives* for additional information regarding the different possibilities to be considered.
5 If you're planning a complete overhaul, the engine must be disassembled and the internal components removed in the following general order:

Camshaft and lifters
Cylinder head
Oil pan
Front oil seal housing
Oil pump and pick-up tube
Piston/connecting rod assemblies
Rear main oil seal housing
Crankshaft and main bearings

6 Before beginning the disassembly and overhaul procedures, make sure the following items are available. Also, refer to *Engine overhaul - reassembly* sequence for a list of tools and materials needed for engine reassembly.

Common hand tools
Small cardboard boxes or plastic bags for storing parts
Gasket scraper
Ridge reamer
Engine balancer puller
Micrometers
Telescoping gauges
Dial indicator set
Valve spring compressor
Cylinder surfacing hone
Piston ring groove-cleaning tool
Electric drill motor
Tap and die set
Wire brushes
Oil gallery brushes
Cleaning solvent

9 Cylinder head - disassembly

Refer to illustrations 9.2 and 9.3

➡Note: New and rebuilt cylinder heads are commonly available for most engines at dealerships and auto parts stores. Due to the fact that some specialized tools are necessary for the disassembly and inspection procedures, and replacement parts aren't always readily available, it may be more practical and economical for the home mechanic to purchase a replacement head rather than taking the time to disassemble, inspect and recondition the original.

1 Cylinder head disassembly involves removal of the intake and exhaust valves and related components. It is already assumed that the camshaft(s) and lifters are removed from the cylinder head. If they're not already removed, label the parts and store them separately so they can be reinstalled in their original locations.

9.2 A small plastic bag, with an appropriate label, can be used to store the valve train components so they can be kept together and reinstalled in the original positions

9.3 You'll need a valve spring compressor with a special adapter to compress the spring and allow removal of the keepers from the valve stem

2 Before the valves are removed, arrange to label and store them, along with their related components, so they can be kept separate and reinstalled in their original locations (see illustration).

3 Compress the springs on the first valve with a spring compressor and remove the keepers (see illustration). Carefully release the valve spring compressor and remove the retainer, the spring and the spring seat (if used).

4 Pull the valve out of the head, then remove the oil seal from the guide. If the valve binds in the guide (won't pull through), push it back into the head and deburr the area around the keeper groove with a fine file or whetstone.

5 Repeat the procedure for the remaining valves. Remember to keep all the parts for each valve together so they can be reinstalled in the same locations.

6 Once the valves and related components have been removed and stored in an organized manner, the heads should be thoroughly cleaned and inspected. If a complete engine overhaul is being done, finish the engine disassembly procedures before beginning the cylinder head cleaning and inspection process.

10 Cylinder head - cleaning and inspection

1 Thorough cleaning of the cylinder head and related valve train components, followed by a detailed inspection, will enable you to decide how much valve service work must be done during the engine overhaul.

➡Note: If the engine was severely overheated, the cylinder head is probably warped (see Step 12).

CLEANING

2 Scrape all traces of old gasket material and sealant off the head gasket, intake manifold and exhaust manifold mating surfaces. Be very careful not to gouge the cylinder head. Special gasket-removal solvents that soften gaskets and make removal much easier are available at auto parts stores.

3 Remove all built-up scale from the coolant passages.

4 Run a stiff wire brush through the various holes to remove deposits that may have formed in them.

5 Run an appropriate-size tap into each of the threaded holes to remove corrosion and thread sealant that may be present. If compressed air is available, use it to clear the holes of debris produced by this operation.

⁂ WARNING:

Wear eye protection when using compressed air!

6 Clean the exhaust manifold and intake manifold stud threads with a wire brush.

7 Clean the cylinder head with solvent and dry it thoroughly.

8 Compressed air will speed the drying process and ensure that all holes and oil passages are clean.

➡Note: Decarbonizing chemicals are available and may prove very useful when cleaning cylinder heads and valve train components. They're very caustic and should be used with caution. Be sure to follow the instructions on the container.

9 Clean all the valve springs, keepers and retainers with solvent and dry them thoroughly. Do the components from one valve at a time to avoid mixing up the parts.

10 Scrape off any heavy deposits that may have formed on the valves, then use a motorized wire brush to remove deposits from the valve heads and stems. Again, make sure the valves don't get mixed up.

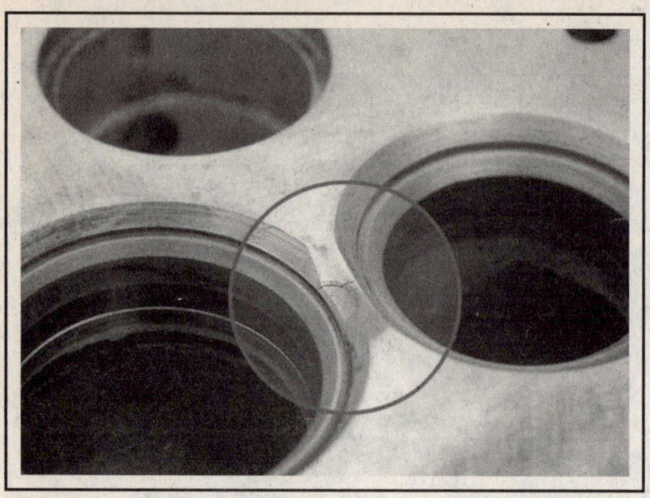

10.11 Check for cracks between the valve seats

10.12a Check the cylinder head gasket surface for warpage by trying to slip a feeler gauge under the straightedge (see this Chapter's Specifications for the maximum warpage allowed and use a feeler gauge of that thickness)

INSPECTION

➡Note: Be sure to perform all of the following inspection procedures before concluding machine shop work is required. Make a list of the items that need attention.

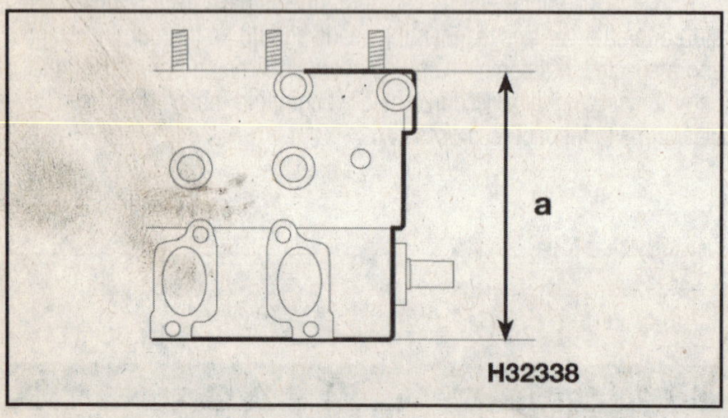

10.12b The cylinder head minimum height dimension is measured from the valve cover rail to the deck surface

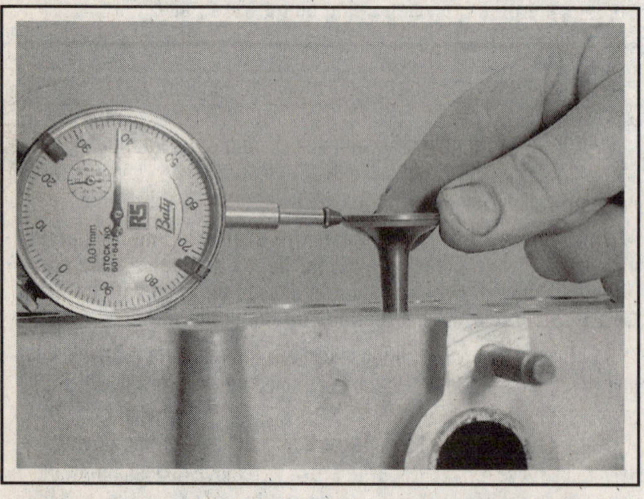

10.14 A dial indicator can be used to determine the valve stem-to-guide clearance - measure the maximum deflection of the valve in its guide

Cylinder head

➧ Refer to illustrations 10.11, 10.12a, 10.12b and 10.14

11 Inspect the head very carefully for cracks, evidence of coolant leakage and other damage. If cracks are found, check with an automotive machine shop concerning repair. If repair isn't possible, a new cylinder head must be obtained (see illustration).

12 Using a straightedge and feeler gauge, check the head gasket mating surface for warpage, then compare the measurement against this Chapter's Specifications (see illustration). If the warpage exceeds the limit on gasoline engines, it can be resurfaced at an automotive machine shop as long as the cylinder head is within the specified minimum height listed in this Chapter's Specifications (see illustration).

➡Note: If the head is resurfaced on gasoline engines, it will be necessary to rework the valve seats the same amount which was removed from the cylinder head or piston-to-valve clearance will be compromised, which may lead to severe engine damage. If the cylinder head warpage exceeds the limit on diesel engines, it must be replaced.

13 Examine the valve seats in each of the combustion chambers. If they're pitted, cracked or burned, the head will require valve service that's beyond the scope of the home mechanic.

14 Check the valve stem-to-guide clearance by measuring the lateral movement of the valve stem with a dial indicator attached securely to the head (see illustration). Install the valve into the guide until the stem is flush with the top of the guide. The total valve stem movement indicated by the gauge needle must be compared to the specifications in this Chapter. After this is done, if there's still some doubt regarding the condition of the valve guides, they should be checked by an automotive machine shop (the cost should be minimal).

Valves

➧ Refer to illustrations 10.15 and 10.16

15 Carefully inspect each valve face for uneven wear, deformation, cracks, pits and burned areas. Check the valve stem for scuffing and galling and the neck for cracks. Rotate the valve and check for any obvious indication that it's bent. Look for pits and excessive wear on the end

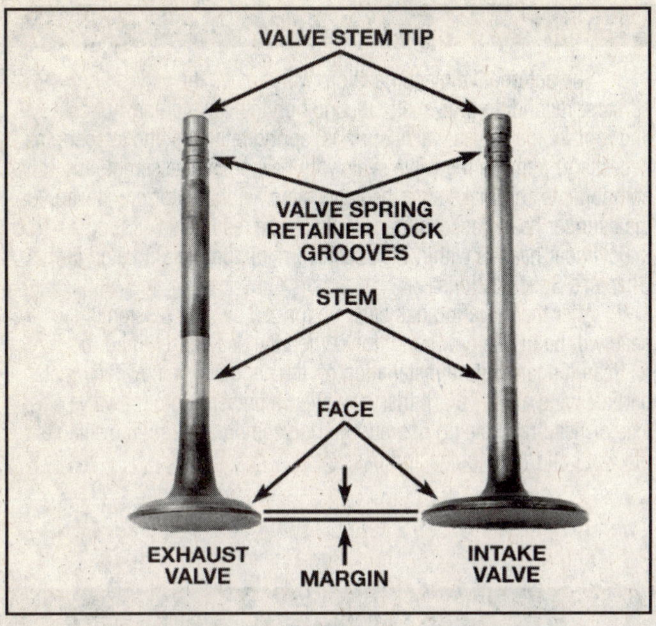

10.15 Check for valve wear at the points shown here

of the stem. The presence of any of these conditions (see illustration) indicates the need to consult an automotive machine shop.

➡**Note: The manufacturer recommends the valves be replaced, if refacing is necessary.**

16 Also measure the stem diameter at several points along their lengths (see illustration). Taper should not exceed the limit listed in this Chapter's Specifications.

Valve components

➧ **Refer to illustrations 10.17 and 10.18**

17 Check each valve spring for wear (on the ends) and pits. Measure the free length of each intake valve spring and compare them with one another (see illustration). Any springs that are shorter have sagged and shouldn't be re-used. Now repeat this check on the exhaust valve springs. If, in either check, any of the springs measures shorter than another (intake-to-intake, exhaust-to-exhaust) replace all of the springs as a set. The tension of all springs should be checked with a special fixture before deciding they're suitable for use in a rebuilt engine (take the springs to an automotive machine shop for this check).

➡**Note: If the engine has accumulated many miles, it's a good idea to replace all of the springs as a matter of course.**

18 Stand each spring on a flat surface and check it for squareness (see illustration). If any of the springs are distorted or sagged, replace all of them with new parts.

19 Check the spring retainers, spring seats and the keepers for obvious wear and cracks. Any questionable parts should be replaced with new ones, as extensive damage will occur if they fail during engine operation.

Camshaft(s) and lifters

20 Refer to Section 21 of this Chapter for the camshaft and lifter inspection procedures. Be sure to inspect the camshaft bearing journals on the cylinder head before the head is sent to a machine shop to have the valves serviced. If the journals are gouged or scored the cylinder head will have to be replaced regardless of the condition of the valves and related components.

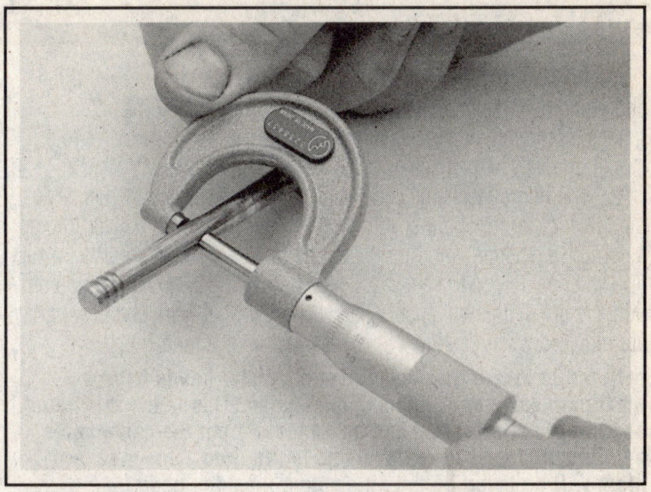

10.16 Measure the diameter of the valve stems at several points

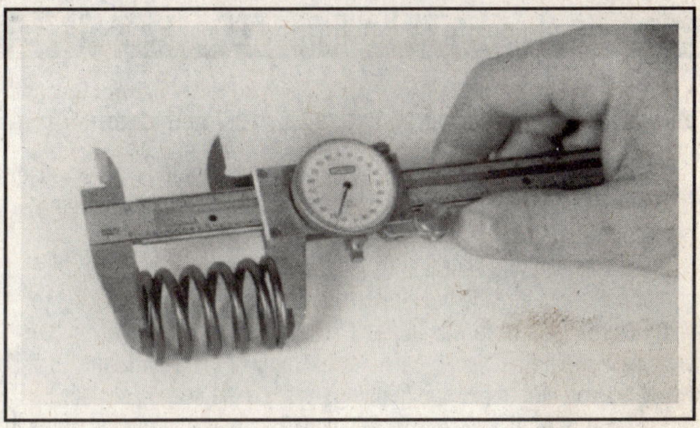

10.17 Measure the free length of each valve spring with a dial or vernier caliper

All components

21 If the inspection process indicates the valve components are in generally poor condition and worn beyond the limits specified, which is usually the case in an engine that's being overhauled, reassemble the valves in the cylinder head (see Section 11 for valve servicing recommendations).

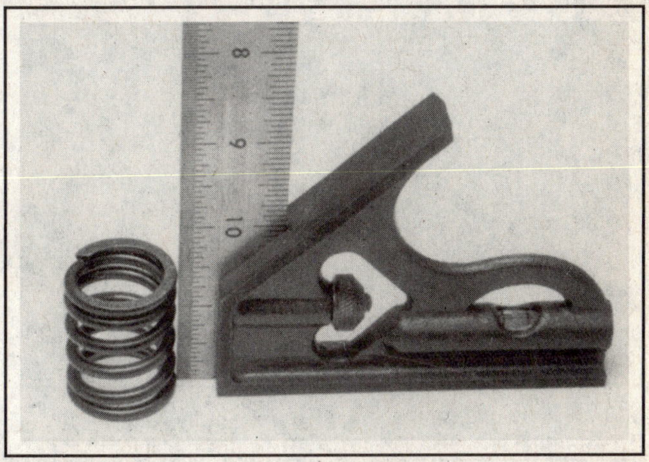

10.18 Check each valve spring for squareness

11 Valves - servicing

1 Because of the complex nature of the job and the special tools and equipment needed, servicing of the valve seats and the valve guides, commonly known as a valve job, should be done by a professional (the valves themselves aren't serviceable).

2 The home mechanic can remove and disassemble the head, do the initial cleaning and inspection, then reassemble and deliver it to an automotive machine shop for the actual service work. Doing the inspection will enable you to see what condition the head and valvetrain components are in and will ensure that you know what work and new parts are required when dealing with an automotive machine shop.

➡Note: Be aware that Volkswagen cylinder heads have a maximum valve seat refacing dimension. This is the maximum amount of material that can be removed from the valve seats before cylinder head replacement is required. This measurement will be taken by the automotive machine shop.

3 The automotive machine shop, will remove the valves and springs, recondition the seats, recondition the valve guides, check and replace the valves, valve springs, spring retainers and keepers (as necessary), replace the valve seals with new ones, reassemble the valve components and make sure the valve stem height is correct. If warped, the cylinder head gasket surface will also be resurfaced as long as the cylinder head is within the specified minimum height listed in this Chapter's Specifications.

4 After the valve job has been performed by a professional, the head will be in like new condition. When the head is returned, be sure to clean it again before installation on the engine to remove any metal particles and abrasive grit that may still be present from the valve service or head resurfacing operations. Use compressed air, if available, to blow out all the oil holes and passages.

12 Cylinder head - reassembly

➧ Refer to illustrations 12.2, 12.3, 12.6, 12.7 and 12.9

1 Regardless of whether or not the head was sent to an automotive repair shop for valve servicing, make sure it's clean before beginning reassembly.

2 If the head was sent out for valve servicing, the valves and related components will already be in place. Begin the reassembly procedure with Step 8. If the head was not sent out for service, the valves, at the very least should be lapped before reassembly of the cylinder head. Apply a small amount of fine grinding paste on the sealing surface (valve face) of each valve and install them into their appropriate guide in the cylinder head. Attach a valve lapping tool to the valve head. Using a back and forth rotating motion grind the valve head into its seat. Periodically lift the valve and rotate it to redistribute the grinding paste (see illustration). After the lapping process has been completed for each valve it will be necessary to clean the valves and seats of all lapping compound. Be sure to mark the valves before they're removed so they can be installed back into the same valve guide on reassembly.

3 Beginning at one end of the head, lubricate and install the first valve. Apply moly-base grease or clean engine oil to the valve stem (see illustration).

4 Install the spring seat and shims, if originally installed, before the valve seals.

5 Install new seals on each of the valve guides. Gently tap each seal into place until it's completely seated on the guide. Many seal sets come with a plastic installer, but use hand pressure. Do not hammer on the seals or they could be driven down too far and subsequently leak. Don't twist or cock the seals during installation or they won't seal properly on the valve stems.

6 The valve components (see illustration) may be installed in the following order:

Valves
Valve spring seat
Valve stem seals
Valve spring shims (if any)
Valve springs
Retainers
Keepers

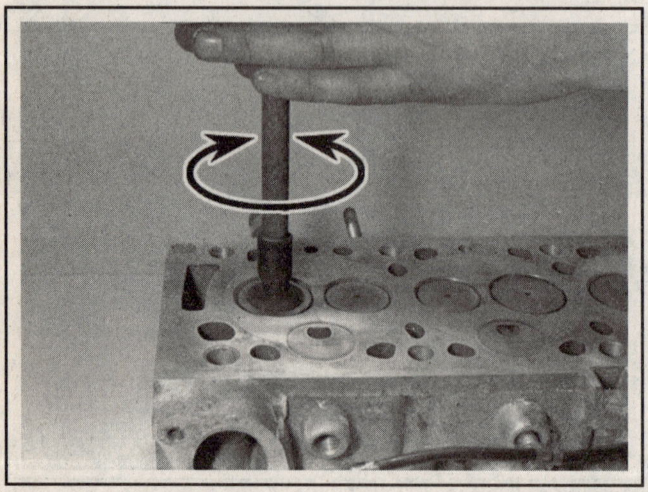

12.2 Lapping in the valves

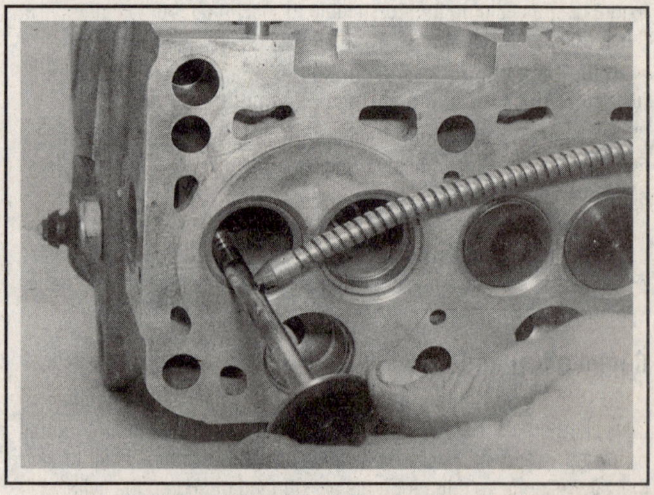

12.3 Lubricate the valve stem with clean engine oil before installing it into the guide

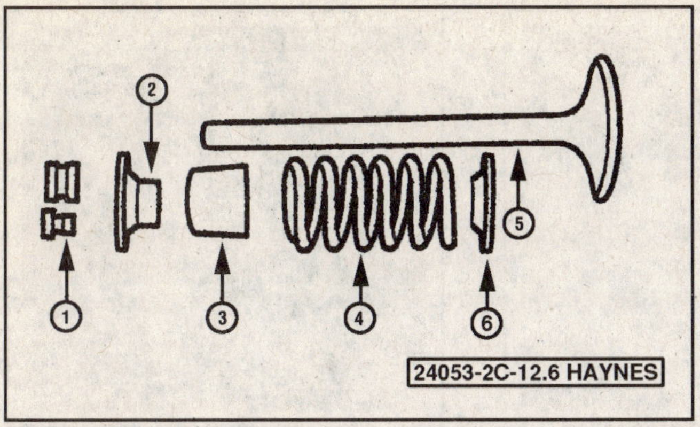

12.6 Typical valve components

1	*Keepers*	4	*Spring*
2	*Retainer*	5	*Valve*
3	*Oil seal*	6	*Valve spring seat*

7 Compress the springs with a valve spring compressor and carefully install the keepers in the groove, then slowly release the compressor and make sure the keepers seat properly. Apply a small dab of grease to each keeper to hold it in place if necessary (see illustration). Tap the valve stem tips with a plastic hammer to seat the keepers, if necessary.

8 Repeat the procedure for the remaining valves. Be sure to return the components to their original locations - don't mix them up!

9 Check the installed valve height with a straightedge and a dial or venier caliper. If the head was sent out for service work, the installed height should be correct (but don't automatically assume it is). The measurement is taken from valve cover rail to the top of each valve stem (see illustration). If the height is less than specified in this Chapter, the valve seats have been reworked past their limits and will not allow proper operation of the hydraulic valve lifters. Valve seats that have been reworked past their limits must be replaced with new ones or a new cylinder head is required.

12.7 Apply a small dab of grease to each keeper as shown here before installation - it'll hold them in place on the valve stem as the spring is released

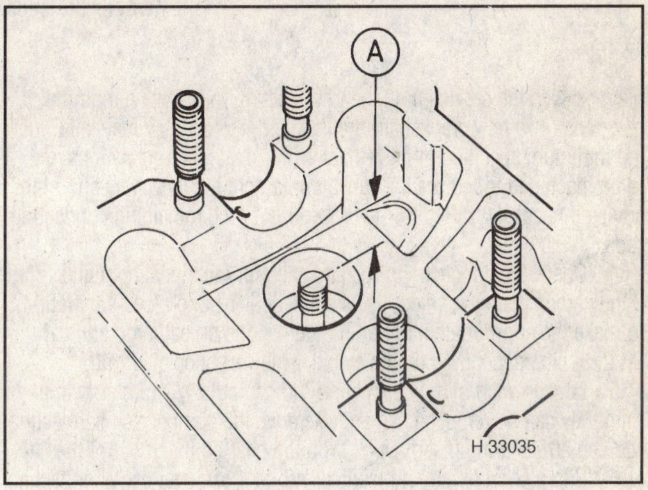

12.9 The valve installed height is measured from the valve cover rail to the top of the valve stem

13 Pistons and connecting rods - removal

Refer to illustrations 13.1, 13.3, 13.4 and 13.6

Note: Prior to removing the piston/connecting rod assemblies, remove the cylinder head, the oil pan, the oil pump drive chain, the oil pump and baffle by referring to the appropriate Sections in Chapter 2 Part A or B.

1 Use your fingernail to feel if a ridge has formed at the upper limit of ring travel (about 1/4-inch down from the top of each cylinder). If carbon deposits or cylinder wear have produced ridges, they must be completely removed with a special tool (see illustration). Follow the manufacturer's instructions provided with the tool. Failure to remove the ridges before attempting to remove the piston/connecting rod assemblies may result in piston breakage.

2 After the cylinder ridges have been removed, turn the engine upside-down so the crankshaft is facing up.

3 Before the connecting rods are removed, check the endplay (side clearance) with feeler gauges. Slide them between the first connecting rod and the crankshaft throw until the play is removed (see illustration). The endplay is equal to the thickness of the feeler gauge(s). If the end-

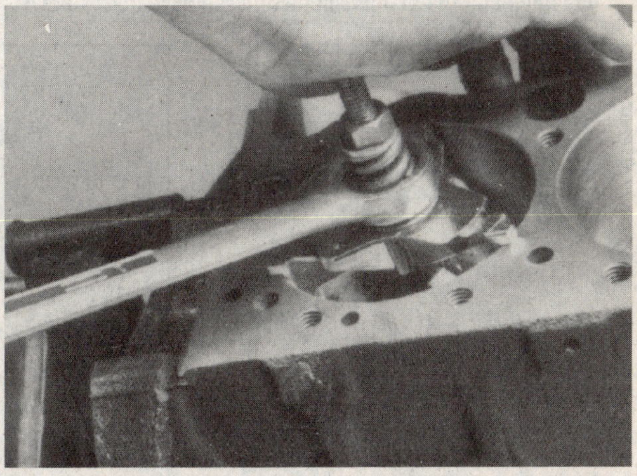

13.1 A ridge reamer is required to remove the ridge from the top of each cylinder - do this before removing the pistons!

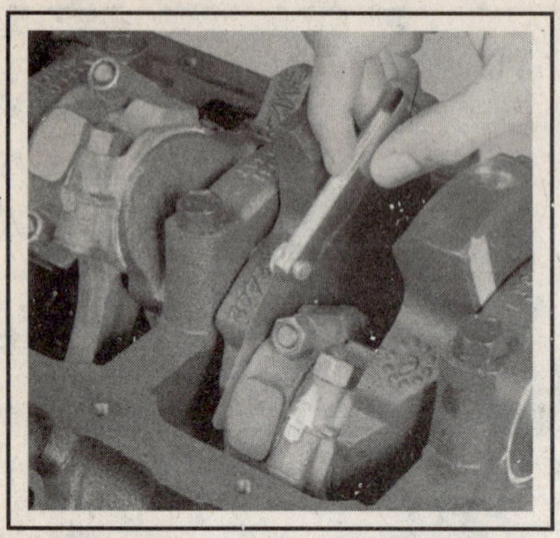

13.3 Check the connecting rod side clearance with a feeler gauge as shown

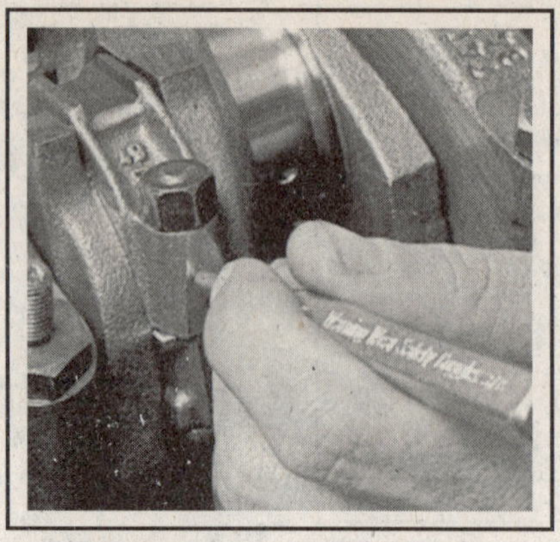

13.4 Mark the rod bearing caps in order from the front of the engine to the rear (one mark for the front cap, two for the second one and so on)

13.6 To prevent damage to the crankshaft journals and cylinder walls, slip sections of rubber or plastic hose over the rod bolts before removing the pistons/rods

play exceeds the service limit, new connecting rods will be required. If new rods (or a new crankshaft) are installed, the endplay may fall under the minimum specified in this Chapter (if it does, the rods will have to be machined to restore it - consult an automotive machine shop for advice if necessary). Repeat the procedure for the remaining connecting rods.

4 Check the connecting rods and caps for identification marks. If they aren't plainly marked, use a small center-punch (see illustration) to make the appropriate number of indentations on each rod and cap (1, 2, 3, etc., depending on the cylinder they're associated with).

5 Loosen each of the connecting rod cap nuts 1/2-turn at a time until they can be removed by hand. Remove the number one connecting rod cap and bearing insert. Don't drop the bearing insert out of the cap.

6 Slip a short length of plastic or rubber hose over each connecting rod cap bolt to protect the crankshaft journal and cylinder wall as the piston is removed (see illustration).

7 Remove the bearing insert and push the connecting rod/piston assembly out through the top of the engine. Use a wooden or plastic hammer handle to push on the upper bearing surface in the connecting rod. If resistance is felt, double-check to make sure all of the ridge was removed from the cylinder.

8 Repeat the procedure for the remaining cylinders.

9 After removal, reassemble the connecting rod caps and bearing inserts in their respective connecting rods and install the cap nuts finger tight. Leaving the old bearing inserts in place until reassembly will help prevent the connecting rod bearing surfaces from being accidentally nicked or gouged.

10 Don't separate the pistons from the connecting rods.

14 Crankshaft - removal

➧ Refer to illustrations 14.1, 14.3 and 14.4

➡Note: The crankshaft can be removed only after the engine has been removed from the vehicle. It's assumed the flywheel/driveplate, timing belt, oil pan, oil pump, the front and rear oil seal housings and the piston/connecting rod assemblies have already been removed.

1 Before the crankshaft is removed, check the endplay. Mount a dial indicator with the stem in line with the crankshaft and touching the snout of the crank (see illustration).

2 Push the crankshaft all the way to the rear and zero the dial indi-

14.1 Measuring crankshaft endplay using a dial indicator

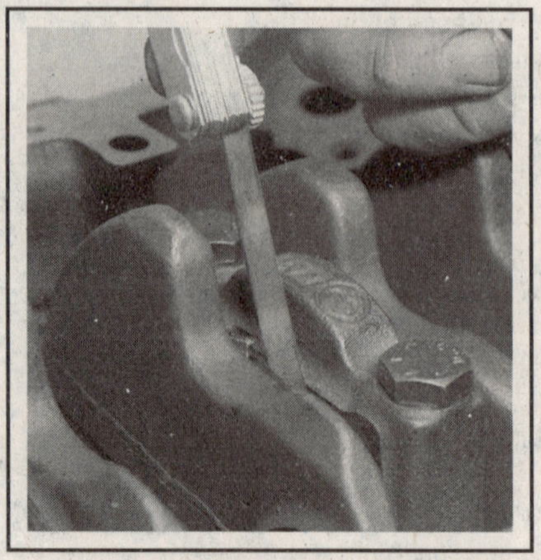

14.3 Checking crankshaft endplay with a feeler gauge

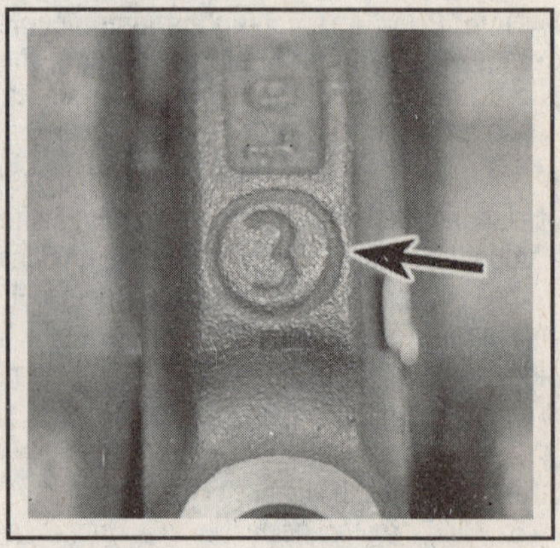

14.4 Manufacturer's identification markings on the main bearing caps (arrow)

cator. Next, pry the crankshaft to the front as far as possible and check the reading on the dial indicator. The distance it moves is the endplay. If it's greater than listed in this Chapter's Specifications, check the crankshaft thrust surfaces for wear. If no wear is evident, new main bearings should correct the endplay.

3 If a dial indicator isn't available, feeler gauges can be used. Gently pry or push the crankshaft all the way to the front of the engine. Slip feeler gauges between the crankshaft and the front face of the thrust main bearing to determine the clearance (see illustration).

➡**Note: The thrust bearing is located at the number three main bearing cap on all engines.**

4 Check the main bearing caps to see if they're marked to indicate their locations. They should be numbered consecutively from the front of the engine to the rear (see illustration). If they aren't, mark them with number stamping dies or a center-punch. Main bearing caps generally have a cast-in arrow, which points to the front of the engine. Loosen the main bearing cap bolts 1/4-turn at a time each, until they can be removed by hand. Note if any stud bolts are used and make sure they're returned to their original locations when the crankshaft is reinstalled.

5 Gently tap the caps with a soft-face hammer, then separate them from the engine block. If necessary, use the bolts as levers to remove the caps. Try not to drop the bearing inserts if they come out with the caps.

6 Carefully lift the crankshaft straight out of the engine. It may be a good idea to have an assistant available, since the crankshaft is quite heavy. Be careful not to damage the reluctor ring for the crankshaft position sensor. With the bearing inserts in place in the engine block and main bearing caps, return the caps to their respective locations on the engine block and tighten the bolts finger tight.

15 Engine block - cleaning

➧ **Refer to illustrations 15.3, 15.4a, 15.4b, 15.4c and 15.8**

1 Remove the main bearing caps and separate the bearing inserts from the caps and the engine block. Tag the bearings, indicating which cylinder they were removed from and whether they were in the cap or the block, then set them aside.

2 Using a gasket scraper, remove all traces of gasket material from the engine block. Be very careful not to nick or gouge the gasket sealing surfaces.

3 Remove all of the covers and threaded oil gallery plugs from the block. The plugs are usually very tight - they may have to be drilled out and the holes retapped. Use new plugs when the engine is reassembled. Where applicable, it will also be necessary to remove the oil spray jets from the engine block (see illustration).

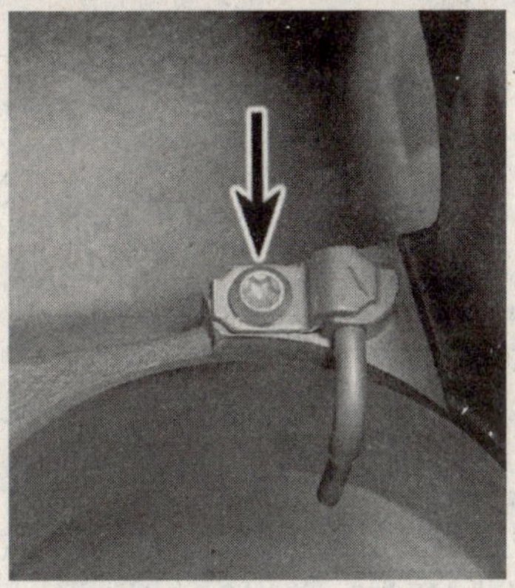

15.3 On 1.8L and 1.9L engines, it will be necessary to remove the oil spray nozzles from the block before removing the piston and connecting rod assemblies

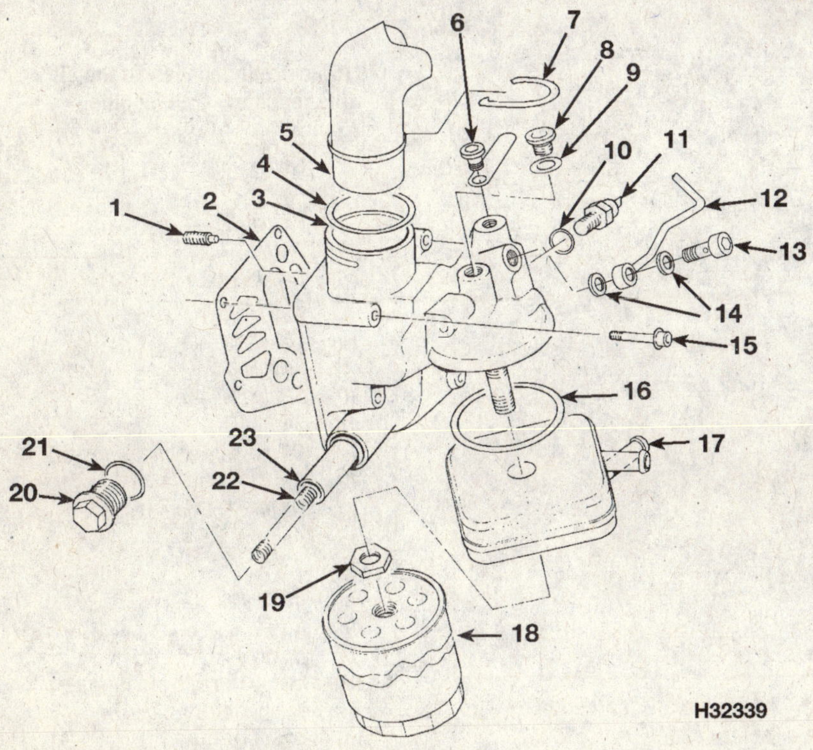

15.4a Exploded view of the oil filter adapter - 1.8L engine

1 *Check valve*
2 *Gasket*
3 *Oil filter adapter*
4 *O-ring*
5 *Pipe*
6 *Bolt*
7 *Retaining clip*
8 *Sealing plug*
9 *Seal*
10 *Seal*
11 *Oil pressure switch*
12 *Oil supply pipe*
13 *Banjo bolt*
14 *Gasket*
15 *Bolt*
16 *Gasket*
17 *Oil cooler*
18 *Oil filter*
19 *Nut*
20 *Sealing plug*
21 *Seal*
22 *Spring*
23 *Oil pressure relief valve*

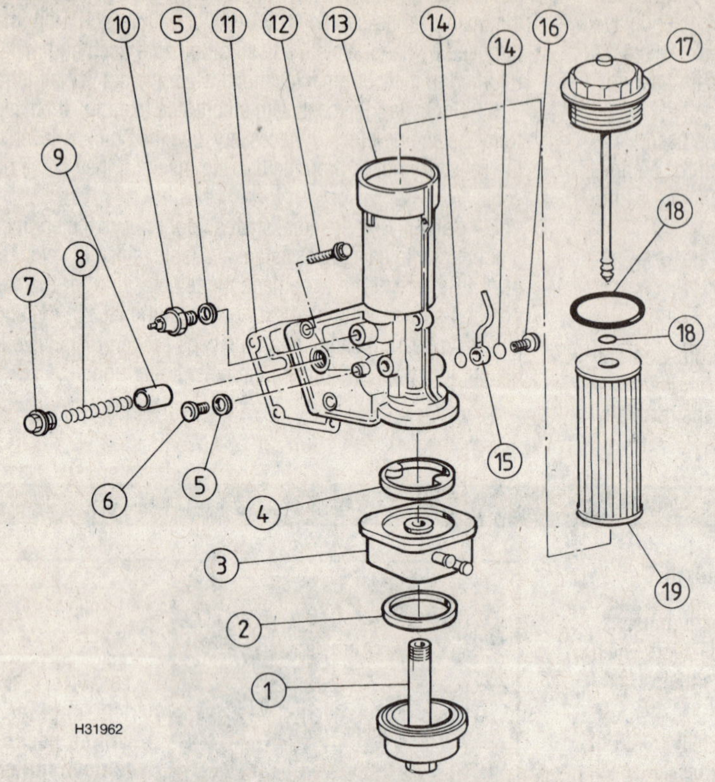

15.4b Exploded view of the oil filter adapter - 1.9L engine

1 *Oil cooler securing plate*
2 *O-ring*
3 *Oil cooler*
4 *O-ring*
5 *Washer*
6 *Sealing plug*
7 *Oil pressure relief valve plug*
8 *Spring*
9 *Piston*
10 *Oil pressure warning light switch*
11 *Gasket*
12 *Mounting bolt*
13 *Oil filter housing*
14 *Seal*
15 *Oil supply pipe to turbo*
16 *Banjo bolt - turbo (or sealing plug)*
17 *Oil filter cover*
18 *O-ring*
19 *Oil filter*

4 Remove the oil filter adapter from the engine. Disassemble the components from the adapter housing and inspect them for wear and damage. Look for knicks and scoring especially on the pressure relief valve piston (see illustrations). Clean the components and the oil passages in the housing thoroughly with solvent, then dry them with compressed air. If in doubt about the condition of the adapter housing and its components, replace it with a new one.

5 If the engine is extremely dirty, it should be taken to an automotive machine shop to be cleaned.

6 After the block is returned, clean all oil holes and oil galleries one

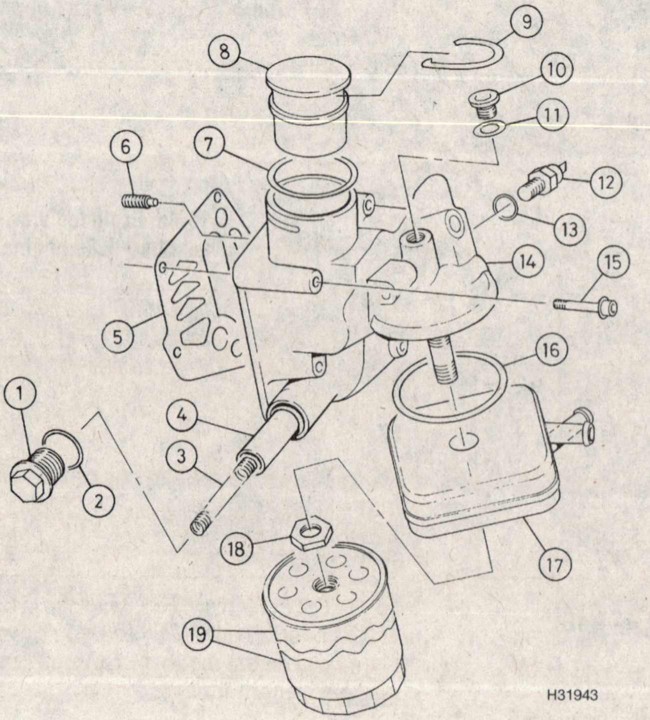

15.4c Exploded view of the oil filter adapter - 2.0L engine

1 *Oil pressure relief valve plug*
2 *Sealing ring*
3 *Spring*
4 *Piston*
5 *Gasket*
6 *Check valve*
7 *Seal*
8 *Sealing cap*
9 *Retaining clip*
10 *Sealing plug*
11 *Seal*
12 *Oil pressure warning light switch*
13 *Seal*
14 *Oil filter housing*
15 *Bolt*
16 *Seal*
17 *Oil cooler*
18 *Nut*
19 *Oil filter*

more time. Brushes specifically designed for this purpose are available at most auto parts stores. Flush the passages with warm water until the water runs clear, dry the block thoroughly and wipe all machined surfaces with a light, rust preventive oil. If you have access to compressed air, use it to speed the drying process and blow out all the oil holes and galleries.

WARNING:

Wear eye protection when using compressed air!

7 If the block isn't extremely dirty or sludged up, you can do an adequate cleaning job with hot soapy water and a stiff brush. Take plenty of time and do a thorough job. Regardless of the cleaning method used, be sure to clean all oil holes and galleries very thoroughly, dry the block completely and coat all machined surfaces with light oil.

8 The threaded holes in the block must be clean to ensure accurate torque readings during reassembly. Run the proper size tap into each of the holes to remove rust, corrosion, thread sealant or sludge and restore damaged threads (see illustration). If possible, use compressed air to clear the holes of debris produced by this operation. Now is a good time to clean the threads on the head bolts and the main bearing cap bolts as well.

9 Reinstall the main bearing caps and tighten the bolts finger tight.

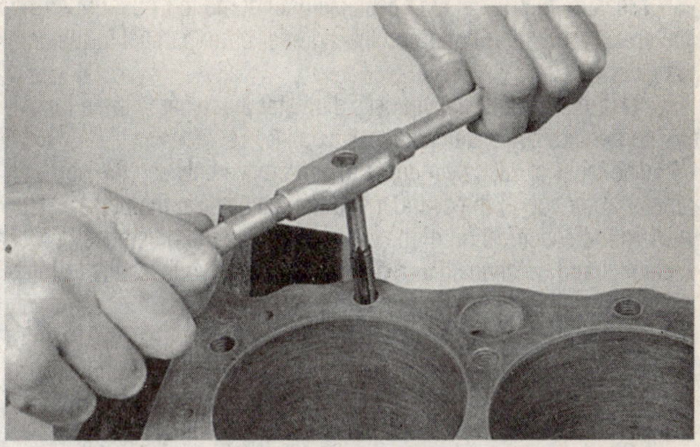

15.8 All bolt holes in the block - particularly the main bearing cap and head bolt holes - should be cleaned and restored with a tap (be sure to remove debris from the holes after this is done)

10 Apply non-hardening sealant (such as Permatex no. 2 or Teflon pipe sealant) to the new oil gallery plugs and thread them into the holes in the block. Make sure they're tightened securely.

11 If the engine isn't going to be reassembled right away, cover it with a large plastic trash bag to keep it clean.

16 Engine block - inspection

Refer to illustrations 16.4a, 16.4b and 16.4c

Note: The manufacturer recommends checking the block deck for warpage and the main bearing bore concentricity and alignment. Since special measuring tools are needed, the checks should be done by an automotive machine shop.

1 Before the block is inspected, it should be cleaned as described in Section 15.

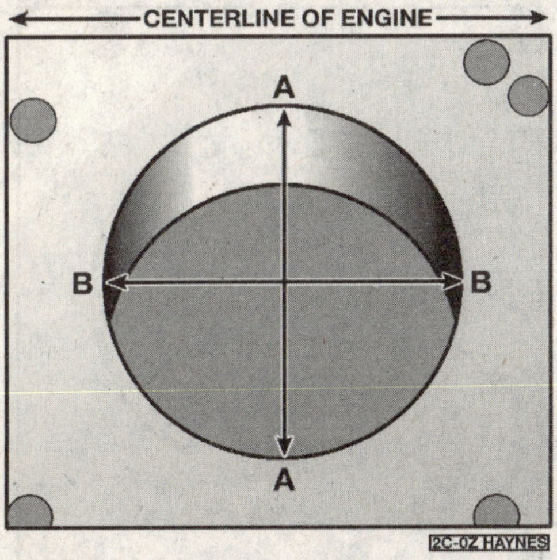

16.4a Measure the diameter of each cylinder at a right angle to the engine centerline (A), and parallel to the engine centerline (B) - out-of-round is the difference between A and B; taper is the difference between the diameter at the top of the cylinder and the diameter at the bottom of the cylinder

2 Visually check the block for cracks, rust and corrosion. Look for stripped threads in the threaded holes. It's also a good idea to have the block checked for hidden cracks by an automotive machine shop that has the special equipment to do this type of work. If defects are found, have the block repaired, if possible, or replaced.

3 Check the cylinder bores for scuffing and scoring.

Note: The following checks should not be made with the engine block mounted on a stand - the cylinders will be distorted and the measurements will be inaccurate.

4 Check the cylinders for taper and out-of-round conditions as follows (see illustrations):

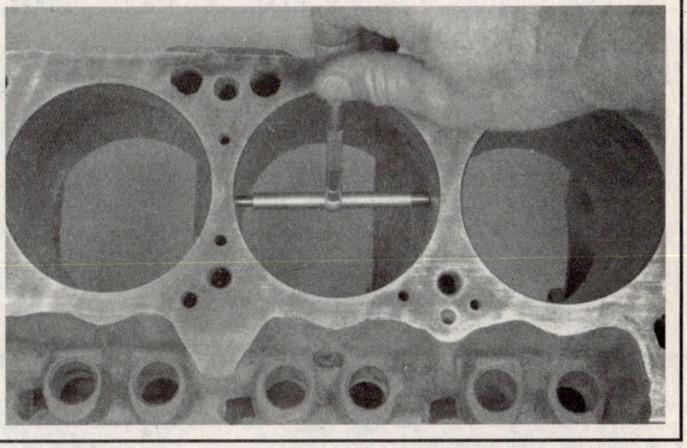

16.4b The ability to "feel" when the telescoping gauge is at the correct point will be developed over time, so work slowly and repeat the check until you're satisfied the bore measurement is accurate

5 Measure the diameter of each cylinder at the top (just under the ridge area), center and bottom of the cylinder bore, parallel to the crankshaft axis.

6 Next, measure each cylinder's diameter at the same three locations perpendicular to the crankshaft axis.

7 The taper of each cylinder is the difference between the bore diameter at the top of the cylinder and the diameter at the bottom. The out-of-round specification of the cylinder bore is the difference between the parallel and perpendicular readings. Compare your results to this Chapter's Specifications.

8 If the cylinder walls are badly scuffed or scored, or if they're out-of-round or tapered beyond the limits given in this Chapter's Specifications, have the engine block rebored and honed at an automotive machine shop.

9 If a rebore is done, oversize pistons and rings will be required.

10 Using a precision straightedge and feeler gauge, check the block deck (the surface the cylinder heads mate with) for distortion as you did with the cylinder heads (see Section 10). If it's distorted beyond the specified limit, the block decks can be resurfaced by an automotive machine shop, but is not recommended on these engines.

11 If the cylinders are in reasonably good condition and not worn to the outside of the limits, and if the piston-to-cylinder clearances can be maintained properly, they don't have to be rebored. Honing is all that's necessary (see Section 17).

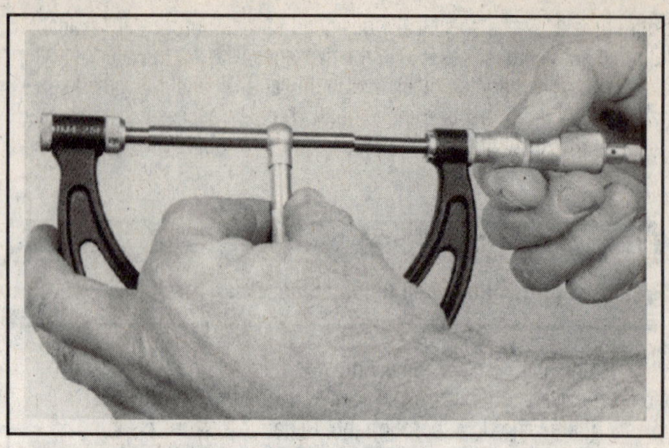

16.4c The gauge is then measured with a micrometer to determine the bore size

17 Cylinder honing

Refer to illustrations 17.3a and 17.3b

1 Prior to engine reassembly, the cylinder bores must be honed so the new piston rings will seat correctly and provide the best possible combustion chamber seal.

Note: If you don't have the tools or don't want to tackle the honing operation, most automotive machine shops will do it for a reasonable fee.

2 Before honing the cylinders, install the main bearing caps and tighten the bolts to the torque listed in this Chapter's Specifications.

3 Two types of cylinder hones are commonly available - the flex hone or "bottle brush" type and the more traditional surfacing hone with spring-loaded stones. Both will do the job, but for the less experienced mechanic the "bottle brush" hone will probably be easier to use. You'll also need some honing oil (kerosene will work if honing oil isn't available), rags and an electric drill motor. Proceed as follows:

17.3a A "bottle brush" hone will produce a better cross hatch pattern when using a drill motor to hone the cylinders

a) *Mount the hone in the drill motor, compress the stones and slip it into the first cylinder (see illustration). Be sure to wear safety goggles or a face shield!*

b) *Lubricate the cylinder with plenty of honing oil, turn on the drill and move the hone up-and-down in the cylinder at a pace that will produce a fine crosshatch pattern on the cylinder walls, and with the drill square and centered with the bore. Ideally, the crosshatch lines should intersect at approximately a 45-60-degree angle (see illustration). Be sure to use plenty of lubricant and don't take off any more material than is absolutely necessary to produce the desired finish.*

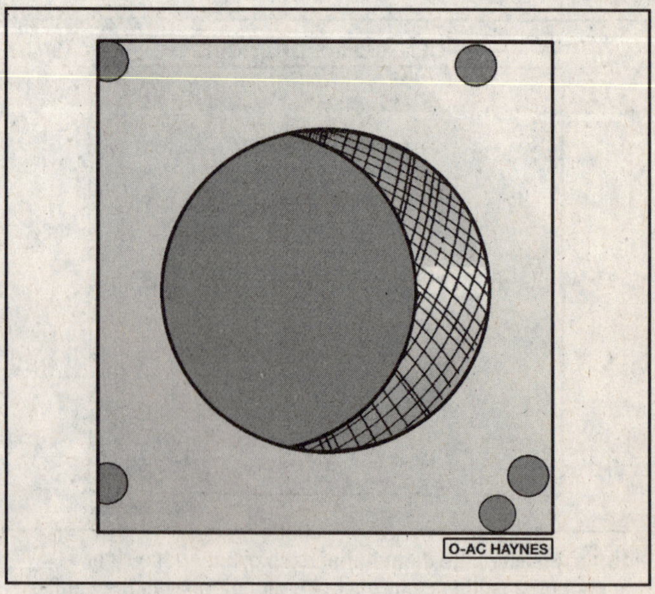

17.3b The cylinder hone should leave a smooth, crosshatch pattern with the lines intersecting at approximately a 60-degree angle

➡Note: Piston ring manufacturers may specify a different cross-hatch angle - read and follow any instructions included with the new rings.

c) *Don't withdraw the hone from the cylinder while it's running. Instead, shut off the drill and continue moving the hone up-and-down in the cylinder until it comes to a complete stop, then compress the stones and withdraw the hone. If you're using a "bottle brush" type hone, stop the drill motor, then turn the chuck in the normal direction of rotation while withdrawing the hone from the cylinder.*

d) *Wipe the oil out of the cylinder and repeat the procedure for the remaining cylinders.*

4 After the honing job is complete, chamfer the top edges of the cylinder bores with a small file so the rings won't catch when the pistons are installed. Be very careful not to nick the cylinder walls with the end of the file.

5 The entire engine block must be washed again very thoroughly with warm, soapy water to remove all traces of the abrasive grit produced during the honing operation.

➡Note: The bores can be considered clean when a lint-free white cloth - dampened with clean engine oil - used to wipe them out doesn't pick up any more honing residue, which will show up as gray areas on the cloth. Be sure to run a brush through all oil holes and galleries and flush them with running water.

6 After rinsing, dry the block and apply a coat of light rust preventive oil to all machined surfaces. Wrap the block in a plastic trash bag to keep it clean and set it aside until reassembly.

18 Pistons and connecting rods - inspection

➧ Refer to illustrations 18.4a, 18.4b, 18.10, 18.11, 18.13a and 18.13b

1 Before the inspection process can be carried out, the piston/connecting rod assemblies must be cleaned and the original piston rings removed from the pistons.

➡Note: Always use new piston rings when the engine is reassembled.

2 Using a piston ring installation tool, carefully remove the rings from the pistons. Be careful not to nick or gouge the pistons in the process.

3 Scrape all traces of carbon from the top of the piston. A hand-held wire brush or a piece of fine emery cloth can be used once the majority of the deposits have been scraped away. Do not, under any circumstances, use a wire brush mounted in a drill motor to remove deposits from the pistons. The piston material is soft and may be eroded away by the wire brush.

4 Use a piston ring groove-cleaning tool to remove carbon deposits from the ring grooves. If a tool isn't available, a piece broken off the old ring will do the job. Be very careful to remove only the carbon deposits - don't remove any metal and do not nick or scratch the sides of the ring grooves (see illustrations).

5 Once the deposits have been removed, clean the piston/rod assemblies with solvent and dry them with compressed air (if available).

⁂ WARNING:

Wear eye protection. Make sure the oil return holes in the back sides of the ring grooves are clear.

6 If the pistons and cylinder walls aren't damaged or worn excessively, and if the engine block isn't rebored, new pistons won't be necessary. Normal piston wear appears as even vertical wear on the piston thrust surfaces and slight looseness of the top ring in its groove. New piston rings, however, should always be used when an engine is rebuilt.

7 Carefully inspect each piston for cracks around the skirt, at the pin bosses and at the ring lands.

8 Look for scoring and scuffing on the thrust faces of the skirt, holes in the piston crown and burned areas at the edge of the crown. If the skirt is scored or scuffed, the engine may have been suffering from overheating and/or abnormal combustion, which caused excessively high operating temperatures. The cooling and lubrication systems should be checked thoroughly. A hole in the piston crown is an indication that abnormal combustion (preignition) was occurring. Burned

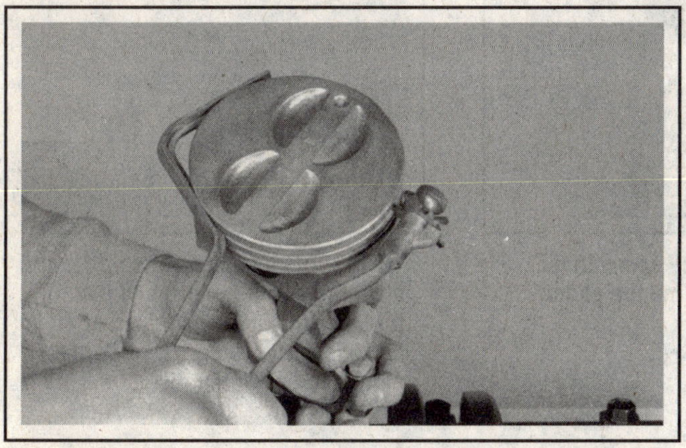

18.4a The piston ring grooves can be cleaned with a special tool, as shown here . . .

18.4b . . . or a section of broken ring

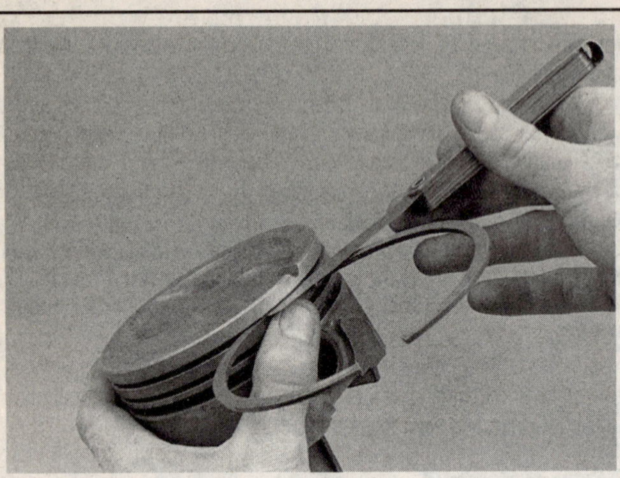

18.10 Check the ring side clearance with a feeler gauge at several points around the groove

areas at the edge of the piston crown are usually evidence of spark knock (detonation). If any of the above problems exist, the causes must be corrected or the damage will occur again. The causes may include intake air leaks, incorrect fuel/air mixture, low octane fuel, ignition timing and EGR system malfunctions.

9 Corrosion of the piston, in the form of small pits, indicates coolant is leaking into the combustion chamber and/or the crankcase. Again, the cause must be corrected or the problem may persist in the rebuilt engine.

10 Measure the piston ring side clearance by laying a new piston ring in each ring groove and slipping a feeler gauge in beside it (see illustration). Check the clearance at three or four locations around each groove. Be sure to use the correct ring for each groove - they are different. If the side clearance is greater than specified in this Chapter, new pistons will have to be used.

11 Check the piston-to-bore clearance by measuring the bore (see Section 16) and the piston diameter. Make sure the pistons and bores are correctly matched. Measure the piston across the skirt, at a 90-degree angle to the piston pin (see illustration). The measurement must be taken at a specific point to be accurate. The pistons are measured 1/4-inch (6.3 mm) from the bottom of the skirt, at right angles to the piston pin. Measure the cylinder bore at three equal places-in the bore (top, middle and bottom) and use the average dimension for comparison with the piston measurement.

12 Subtract the piston diameter from the bore diameter to obtain the clearance. If it's greater than specified, the block will have to be rebored and new pistons and rings installed. If the pistons are graphite coated (dark gray in color) an 0.0008 inch (0.02 mm) tighter clearance is acceptable as the piston coating will eventually wear off though normal use.

13 Check the piston pin to the rod and piston clearances by twisting the piston and rod in opposite directions. Any noticeable play indicates excessive wear, which must be corrected (see illustrations). The piston/connecting rod assemblies should be taken to an automotive machine shop to have the pistons and rods re-sized and new pins installed.

14 If the pistons must be removed from the connecting rods for any reason, they should be taken to an automotive machine shop to also have the connecting rods checked for bend and twist, since automotive machine shops have special equipment for this purpose.

➡Note: Unless new pistons and/or connecting rods are being installed it is not necessary to completely disassemble the pistons from the connecting rods.

15 Check the connecting rods for cracks and other damage. Temporarily remove the rod caps, lift out the old bearing inserts, wipe the rod and cap bearing surfaces clean and inspect them for nicks, gouges and scratches. After checking the rods, replace the old bearings, slip the caps into place and tighten the nuts finger tight.

➡Note: If the engine is being rebuilt because of a connecting rod knock, be sure to install new or remanufactured connecting rods.

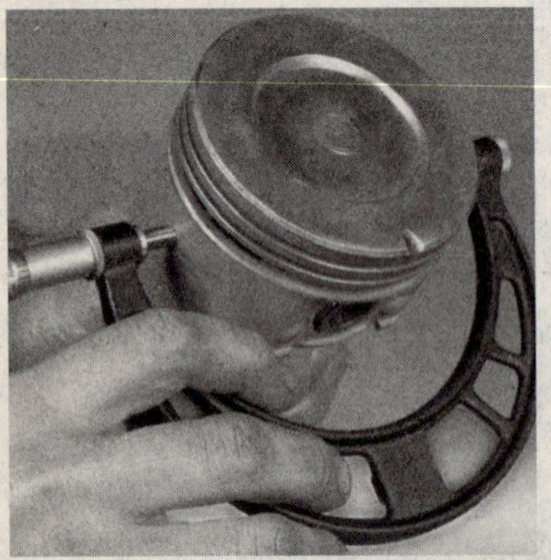

18.11 Measure the piston diameter at a 90-degree angle to the piston pin and at the specified distance from the bottom of the piston skirt

18.13a Insert a small screwdriver into the slot and pry out the piston pin retaining clips

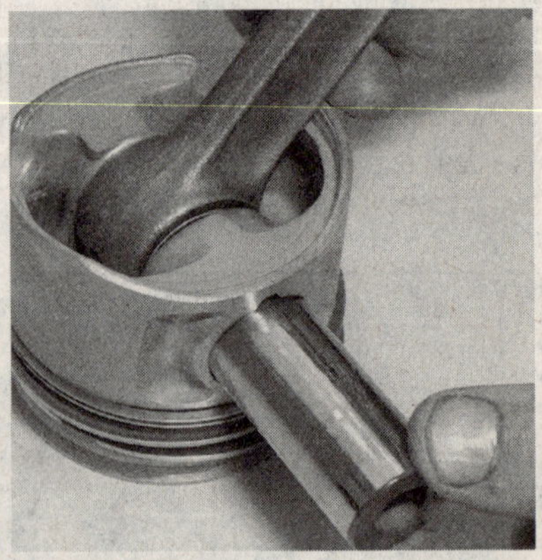

18.13b The piston pin should require a slight push to remove and install it - any excessive movement or looseness will require replacement of the rod and/or the piston and pin assembly

19 Crankshaft - inspection

Refer to illustrations 19.1, 19.2, 19.5 and 19.7

1 Remove all burrs from the crankshaft oil holes with a stone, file or scraper (see illustration).

2 Clean the crankshaft with solvent and dry it with compressed air (if available).

WARNING:

Wear eye protection when using compressed air.

Be sure to clean the oil holes with a stiff brush (see illustration) and flush them with solvent.

3 Check the main and connecting rod bearing journals for uneven wear, scoring, pits and cracks.

4 Check the rest of the crankshaft for cracks and other damage. It should be Magnafluxed to reveal hidden cracks - an automotive machine shop will handle the procedure.

5 Using a micrometer, measure the diameter of the main and connecting rod journals and compare the results to this Chapter's Specifications (see illustration). By measuring the diameter at a number of points around each journal's circumference, you'll be able to determine whether or not the journal is out-of-round. Take the measurement at each end of the journal, near the crank throws, to determine if the journal is tapered.

6 If the crankshaft journals are damaged, tapered, out-of-round or worn beyond the limits given in the Specifications, have the crankshaft reground by an automotive machine shop. Be sure to use the correct-size bearing inserts if the crankshaft is reconditioned.

7 Check the oil seal journals at each end of the crankshaft for wear and damage. If the seal has worn a groove in the journal, or if it's nicked or scratched (see illustration), the new seal may leak when the engine is reassembled. In some cases, an automotive machine shop may be able to repair the journal by pressing on a thin sleeve. If repair isn't feasible, a new or different crankshaft should be installed.

8 Examine the main and rod bearing inserts (see Section 20). Also inspect the crankshaft reluctor ring at the rear of the crankshaft for knicks and damage. Damage to this component may result in severe driveabilty problems.

19.1 The oil holes should be chamfered so sharp edges don't gouge or scratch the new bearings

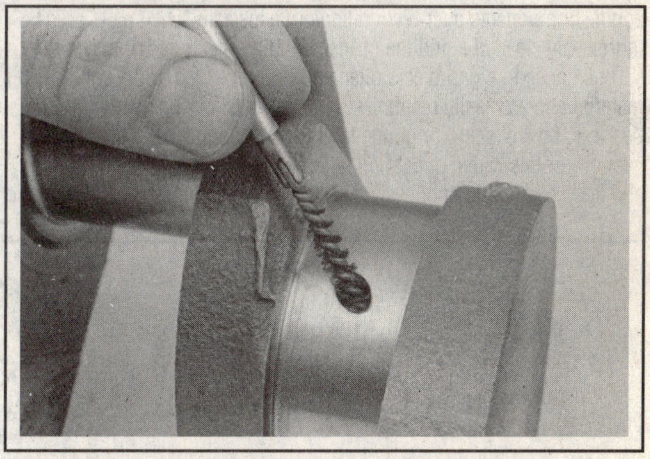

19.2 Use a wire or stiff plastic bristle brush to clean the oil passages in the crankshaft

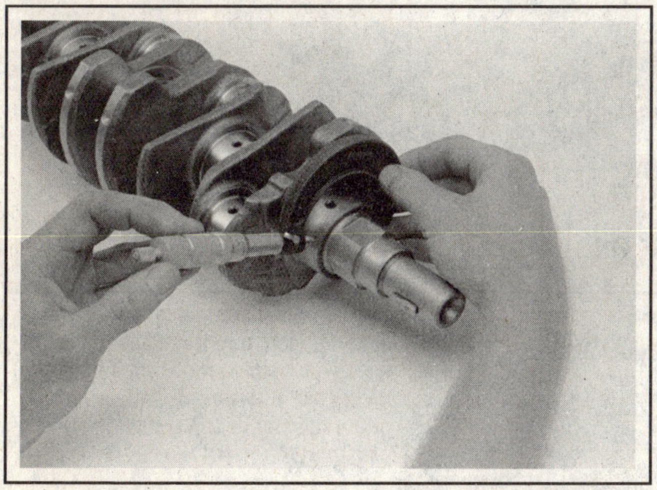

19.5 Measure the diameter of each crankshaft journal at several points to detect taper and out-of-round conditions

19.7 If the seals have worn grooves in the crankshaft journals, or if the seal contact surfaces are nicked or scratched, the new seals will leak

20 Main and connecting rod bearings - inspection and selection

Refer to illustration 20.1

1 Even though the main and connecting rod bearings should be replaced with new ones during the engine overhaul, the old bearings should be retained for close examination, as they may reveal valuable information about the condition of the engine (see illustration).

2 Bearing failure occurs because of lack of lubrication, the presence of dirt or other foreign particles, overloading the engine and corrosion. Regardless of the cause of bearing failure, it must be corrected before the engine is reassembled to prevent it from happening again.

3 When examining the bearings, remove them from the engine block, the main bearing caps, the connecting rods and the rod caps and lay them out on a clean surface in the same general position as their location in the engine. This will enable you to match any bearing problems with the corresponding crankshaft journal.

4 Dirt and other foreign particles get into the engine in a variety of ways. It may be left in the engine during assembly, or it may pass through filters or the PCV system. It may get into the oil, and from there into the bearings. Metal chips from machining operations and normal engine wear are often present. Abrasives are sometimes left in engine components after reconditioning, especially when parts aren't thoroughly cleaned using the proper cleaning methods. Whatever the source, these foreign objects often end up embedded in the soft bearing material and are easily recognized. Large particles won't embed in the bearing and will score or gouge the bearing and journal. The best prevention for this cause of bearing failure is to clean all parts thoroughly and keep everything spotlessly clean during engine assembly. Frequent and regular engine oil and filter changes are also recommended.

5 Lack of lubrication (or lubrication breakdown) has a number of interrelated causes. Excessive heat (which thins the oil), overloading (which squeezes the oil from the bearing face) and oil leakage or throw off (from excessive bearing clearances, worn oil pump or high engine speeds) all contribute to lubrication breakdown. Blocked oil passages, which usually are the result of misaligned oil holes in a bearing shell, will also oil starve a bearing and destroy it. When lack of lubrication is the cause of bearing failure, the bearing material is wiped or extruded from the steel backing of the bearing. Temperatures may increase to the point where the steel backing turns blue from overheating.

6 Driving habits can have a definite effect on bearing life. Low speed operation in too high a gear (lugging the engine) puts very high loads on bearings, which tends to squeeze out the oil film. These loads cause the bearings to flex, which produces fine cracks in the bearing face (fatigue failure). Eventually the bearing material will loosen in pieces and tear away from the steel backing. Short trip driving leads to corrosion of bearings because insufficient engine heat is produced to drive off the condensed water and corrosive gases. These products collect in the engine oil, forming acid and sludge. As the oil is carried to the engine bearings, the acid attacks and corrodes the bearing material.

7 Incorrect bearing installation during engine assembly will lead to bearing failure as well. Tight-fitting bearings leave insufficient oil clearance and will result in oil starvation. Dirt or foreign particles trapped behind a bearing insert result in high spots on the bearing which lead to failure.

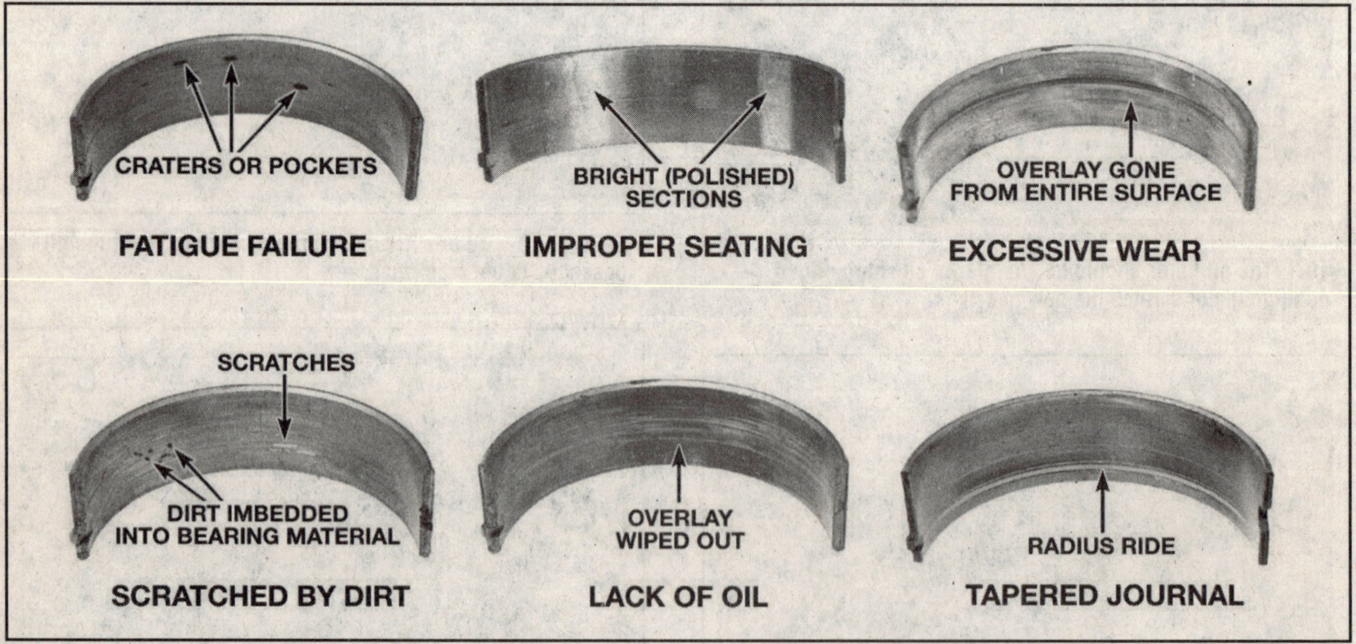

20.1 Typical bearing failures

21 Camshaft, lifters and bearings - inspection

➧ Refer to illustrations 21.1, 21.2, 21.4, 21.5, 21.6a, 21.6b and 21.9

1 Visually check the camshaft bearing surfaces for pitting, score marks, galling and abnormal wear. If the bearing surfaces are damaged, the cylinder head will have to be replaced (see illustration).

2 Measure the outside diameter of each camshaft bearing journal and record your measurements (see illustration). Compare them to the journal outside diameter specified in this Chapter, then measure the inside diameter of each corresponding camshaft bearing and record the measurements. Subtract each cam journal outside diameter from its respective cam bearing bore inside diameter to determine the oil clearance for each bearing. Compare the results to the specified journal-to-bearing clearance. If any of the measurements fall outside the standard specified wear limits in this Chapter, either the camshaft or the cylinder head, or both, must be replaced.

3 Check camshaft runout by placing the camshaft back into the cylinder head and set up a dial indicator on the center journal. Zero the dial indicator. Turn the camshaft slowly and note the dial indicator readings. Record your readings and compare them with the specified runout in this Chapter. If the measured runout exceeds the runout specified in this Chapter, replace the camshaft.

4 Check the camshaft endplay by placing a dial indicator with the stem in line with the camshaft and touching the snout (see illustration). Push the camshaft all the way to the rear and zero the dial indicator. Next, pry the camshaft to the front as far as possible and check the reading on the dial indicator. The distance it moves is the endplay. If it's greater than the Specifications listed in Chapter 2A or 2B, check the bearing caps for wear. If the bearing caps are worn the cylinder head must be replaced.

5 Compare the camshaft lobe height by measuring each lobe with a micrometer (see illustration). Measure each of the intake lobes and write the measurements and relative positions down on a piece of paper. Then measure each of the exhaust lobes and record the measurements and relative positions also. This will let you compare all of the intake lobes to one another and all of the exhaust lobes to one another. If the difference between the lobes exceeds 0.005 inch the camshaft should be replaced. Do not compare intake lobe heights to exhaust lobe heights as lobe lift may be different. Only compare intake lobes-to-intake lobes and exhaust lobes-to exhaust lobes for this comparison.

21.1 Inspect the camshaft bearing surfaces in the cylinder head for pits, score marks and abnormal wear - if damage is noted, the cylinder head must be replaced

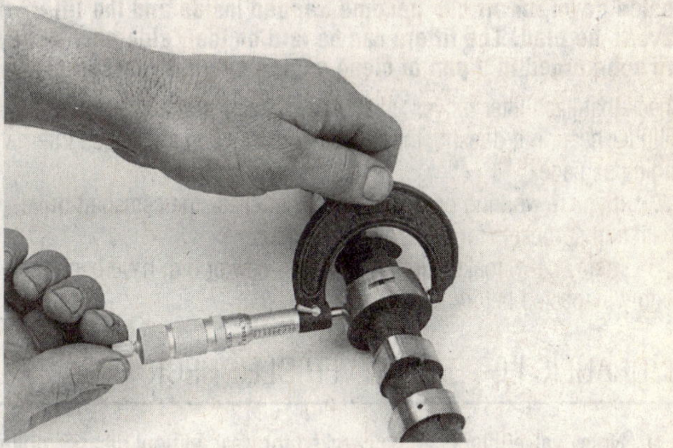

21.2 Measure the outside diameter of each camshaft journal and the inside diameter of each bearing surface on the cylinder head to determine the oil clearance measurement

21.4 Checking camshaft endplay with a dial indicator

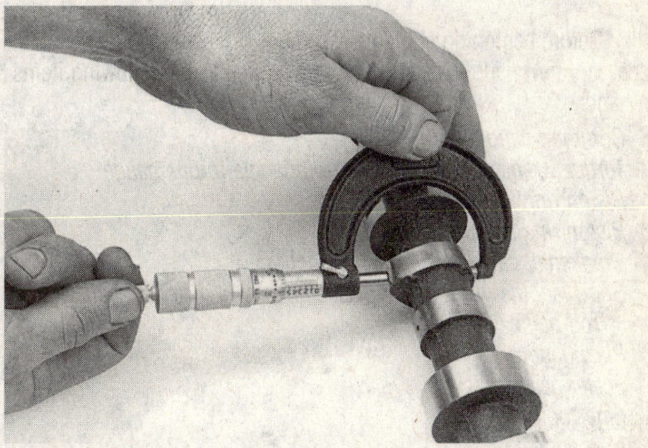

21.5 Measuring the camshaft lobe height with a micrometer - make sure you move the micrometer to get the highest reading (top of cam lobe)

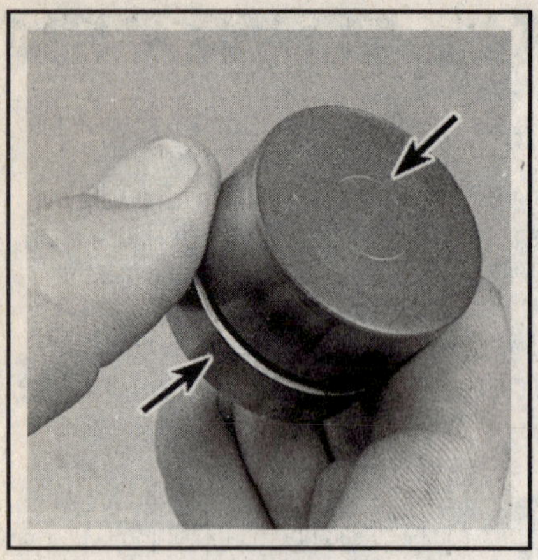

21.6a Inspect the indicated areas (arrows) of each valve lifter

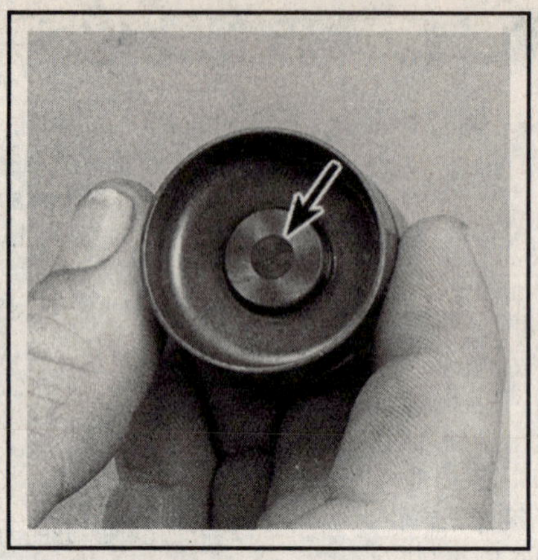

21.6b Also check the valve stem contact area of the lifters

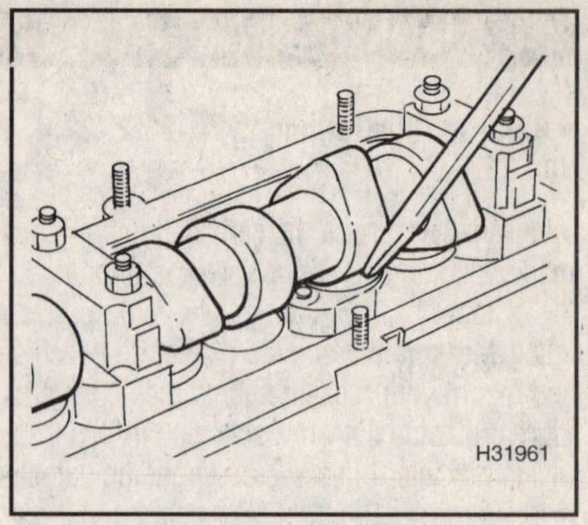

21.9 When checking noisy lifters with the engine in vehicle, it may be necessary to bleed the lifter down by depressing it with a wooden or plastic tool

6 Inspect the contact and sliding surfaces of each lifter for wear and scratches (see illustrations).

➡Note: If the lifter pad is worn, it's a good idea to check the corresponding camshaft. Do not lay the lifters on their side or upside down, or air can become trapped inside and the lifter will have to be bled. The lifters can be laid on their side only if they are submerged in a pan of clean engine oil until reassembly.

Check that each lifter moves up and down freely in its bore on the cylinder head. If it doesn't, the valve may stick open and cause internal engine damage.

7 If you're working on a DOHC engine, check the camshaft drive chain and sprockets for signs of wear.

8 In any case, make sure all the parts, new or old, have been thoroughly inspected before reassembly.

HYDRAULIC LIFTERS - IN VEHICLE CHECK

9 Noisy valve lifters can be checked for wear without disassembling the engine by following the procedure outlined below:

a) Run the engine until it reaches normal operating temperature.
b) Remove the valve cover (see Chapter 2A or 2B).
c) Rotate the engine by hand until the No.1 piston is located at TDC (see Chapter 2A or 2B).
d) Insert a feeler gauge between the camshaft lobe and the lifter to measure the clearance. If the clearance exceeds 0.008 inch (0.2 mm) (gasoline engine) or 0.004 inch (0.1 mm) (diesel engine) the lifter and/or the camshaft lobe has worn beyond it limits.
e) If no clearance exists, depress the lifter to let it bleed down and check the clearance again (see illustration).
f) Lifter clearance on the remaining cylinders can be checked by following the firing order sequence and positioning each of the remaining pistons at TDC.

➡Note: Lifter clearance can also be checked on any lifter whose cam lobe is pointing upward.

g) If the clearance is beyond the maximum allowed, inspect the camshaft as described in Step 5.
h) If the camshaft is OK, the lifters are faulty and must be replaced.

22 Engine overhaul - reassembly sequence

1 Before beginning engine reassembly, make sure you have all the necessary new parts, gaskets and seals as well as the following items on hand:

Common hand tools
Torque wrench (1/2-inch drive) with angle-torque gauge
Piston ring installation tool
Piston ring compressor
Crankshaft balancer installation tool
Short lengths of rubber or plastic hose to fit over connecting rod bolts
Plastigage
Feeler gauges
Fine-tooth file
New engine oil
Engine assembly lube or moly-base grease
Gasket sealant
Thread locking compound

2 In order to save time and avoid problems, engine reassembly must be done in the following general order:

Crankshaft and main bearings
Piston/connecting rod assemblies
Oil spray nozzles (if equipped)
Oil pump, baffle and oil pump pick-up tube
Oil pump drive chain and tensioner
Oil filter adapter housing
Front and rear main oil seal housings
Cylinder head with lifters and camshaft(s)
Fuel injection pump (diesel engine)
Timing belt
Oil pan
Valve cover
Flywheel/driveplate
Intake and exhaust manifolds
Vacuum pump (diesel engines only)

23 Piston rings - installation

Refer to illustrations 23.3, 23.4, 23.5, 23.9a, 23.9b and 23.12

1 Before installing the new piston rings, the ring end gaps must be checked. It's assumed the piston ring side clearance has been checked and verified correct (see Section 18).

2 Lay out the piston/connecting rod assemblies and the new ring sets so the ring sets will be matched with the same piston and cylinder during the end gap measurement and engine assembly.

3 Insert the top (number one) ring into the first cylinder and square it up with the cylinder walls by pushing it in with the top of the piston (see illustration). The ring should be near the bottom of the cylinder, at the lower limit of ring travel.

4 To measure the end gap, slip feeler gauges between the ends of the ring until a gauge equal to the gap width is found (see illustration). The feeler gauge should slide between the ring ends with a slight amount of drag. Compare the measurement to this Chapter's Specifications. If the gap is larger or smaller than specified, double-check to make sure you have the correct rings before proceeding.

5 If the gap is too small, it must be enlarged or the ring ends may come in contact with each other during engine operation, which can cause serious engine damage. The end gap can be increased by filing the ring ends very carefully with a fine file. Mount the file in a vise equipped with soft jaws, slip the ring over the file with the ends contacting the file teeth and slowly move the ring to remove material from the ends. When performing this operation, file only from the outside in (see illustration).

Note: When you have the end gap correct, remove any burrs from the filed ends of the rings with a whetstone.

6 Excess end gap isn't critical unless it's greater than 0.040-inch (1.0 mm). Again, double-check to make sure you have the correct rings for the engine. If the engine block has been bored oversize, necessitating oversize pistons, matching oversize rings are required.

7 Repeat the procedure for each ring that will be installed in the first cylinder and for each ring in the remaining cylinders. Remember to keep rings, pistons and cylinders matched up.

8 Once the ring end gaps have been checked/corrected, the rings can be installed on the pistons.

9 The oil control ring (lowest one on the piston) is usually installed first. Some piston ring manufactures supply one-piece oil rings - others

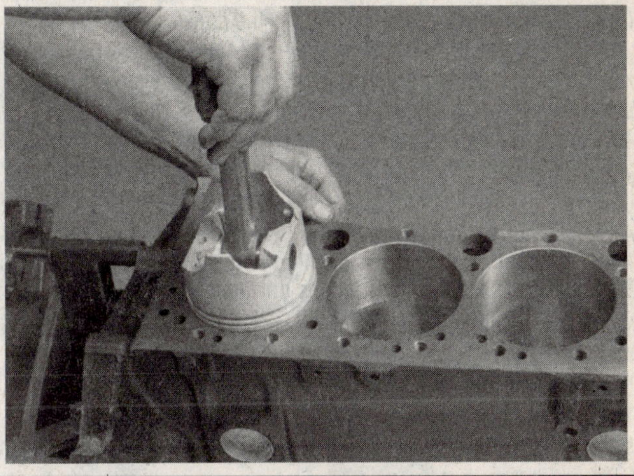

23.3 When checking piston ring end gap, the ring must be square in the cylinder bore (this is done by pushing the ring down with the top of a piston as shown)

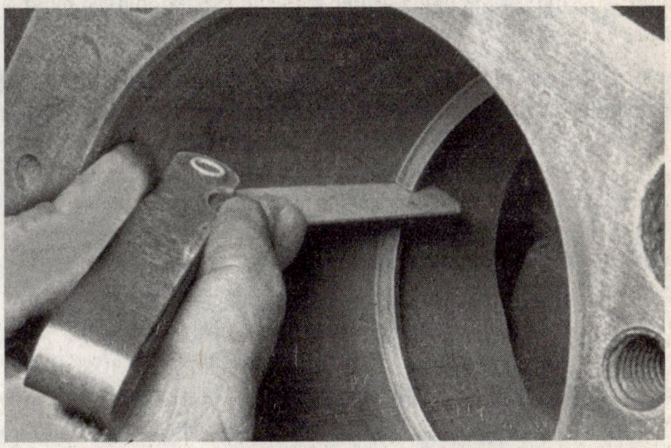

23.4 With the ring square in the cylinder, measure the end gap with a feeler gauge

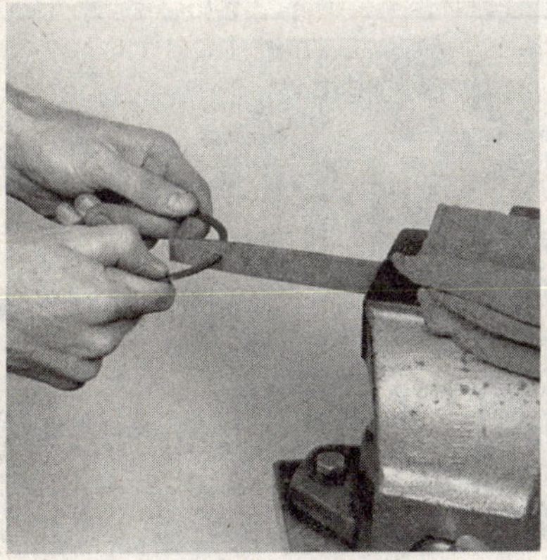

23.5 If the end gap is too small, clamp a file in a vise and file the ring ends (from the outside in only) to enlarge the gap slightly

23.9a Installing the spacer/expander in the oil control ring groove

23.9b DO NOT use a piston ring installation tool when installing the oil ring side rails

may supply three-piece oil rings. One-piece rings can be installed as shown in illustration 23.12. If you're installing three-piece oil rings, slip the spacer/expander into the groove (see illustration). If an anti-rotation tang is used, make sure it's inserted into the drilled hole in the ring groove. Next, install the lower side rail. Don't use a piston ring installation tool on the oil ring side rails, as they may be damaged. Instead, place one end of the side rail into the groove between the spacer/expander and the ring land, hold it firmly in place and slide a finger around the piston while pushing the rail into the groove (see illustration). Next, install the upper side rail in the same manner.

➡Note: Some engines may have a two piece oil ring. If so, follow the installation instructions that come with the piston rings if they differ from the instructions outlined here.

10 After the three oil ring components have been installed, check to make sure both the upper and lower side rails can be turned smoothly in the ring groove.

11 The number two (middle) ring is installed next. It's usually stamped with a mark, which must face up, toward the top of the piston.

➡Note: Always follow the instructions printed on the ring package or box - different manufacturers may require different approaches. Don't mix up the top and middle rings, as they have different cross-sections.

12 Use a piston ring installation tool and make sure the identification mark is facing the top of the piston, then slip the ring into the middle groove on the piston (see illustration). Don't expand the ring any more than necessary to slide it over the piston.

23.12 Installing the compression rings with a ring expander - the "TOP" mark (arrow) must face up

13 Install the number one (top) ring in the same manner. Make sure the mark is facing up. Be careful not to confuse the number one and number two rings.

14 Repeat the procedure for the remaining pistons and rings.

24 Crankshaft - installation and main bearing oil clearance check

1 Crankshaft installation is the first step in engine reassembly. It's assumed at this point that the engine block and crankshaft have been cleaned, inspected and repaired or reconditioned.

2 Position the engine with the bottom facing up.

3 Remove the main bearing cap bolts and lift out the caps. Lay them out in the proper order to ensure correct installation.

4 If they're still in place, remove the original bearing inserts from the block and the main bearing caps. Wipe the bearing surfaces of the block and caps with a clean, lint-free cloth. They must be kept spotlessly clean.

MAIN BEARING OIL CLEARANCE CHECK

➧ Refer to illustrations 24.5, 24.6, 24.11 and 24.15

➡Note: Don't touch the faces of the new bearing inserts with your fingers. Oil and acids from your skin can etch the bearings.

5 Clean the back sides of the new main bearing inserts and lay one in each main bearing saddle in the block. If one of the bearing inserts from each set has a large groove in it, make sure the grooved insert is installed in the block (see illustration). Lay the other bearing from each set in the corresponding main bearing cap. Make sure the tab on the bearing insert fits into the recess in the block or cap, neither higher than the cap's edge nor lower.

⁂ CAUTION:

The oil holes in the block must line up with the oil holes in the bearing inserts. Do not hammer the bearing into place and don't nick or gouge the bearing faces. No lubrication should be used at this time.

6 The four thrustwashers must be installed on each side of the number three main journal and the main cap (see illustration).

7 Clean the faces of the bearings in the block and the crankshaft main bearing journals with a clean, lint-free cloth.

8 Check or clean the oil holes in the crankshaft, as any dirt here can go only one way - straight through the new bearings.

9 Once you're certain the crankshaft is clean, carefully lay it in position in the main bearings.

10 Before the crankshaft can be permanently installed, the main bearing oil clearance must be checked.

11 Cut several pieces of the appropriate size Plastigage (they should be slightly shorter than the width of the main bearings) and place one piece on each crankshaft main bearing journal, parallel with the journal axis (see illustration).

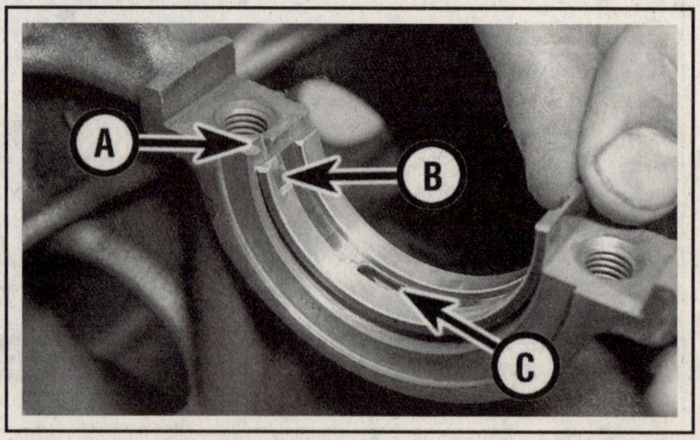

24.5 Bearing shell correctly installed

A Recess in bearing saddle
B Lug on bearing shell
C Oil hole

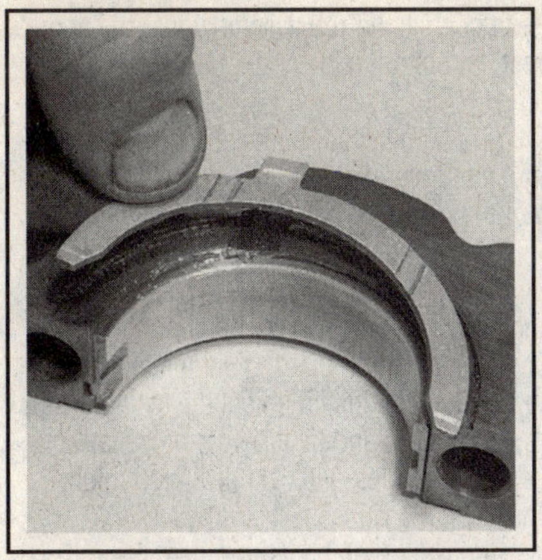

24.6 Installing the thrustwashers on the No. 3 main bearing cap

24.11 Lay the Plastigage strips (arrow) on the main bearing journals, parallel to the crankshaft centerline

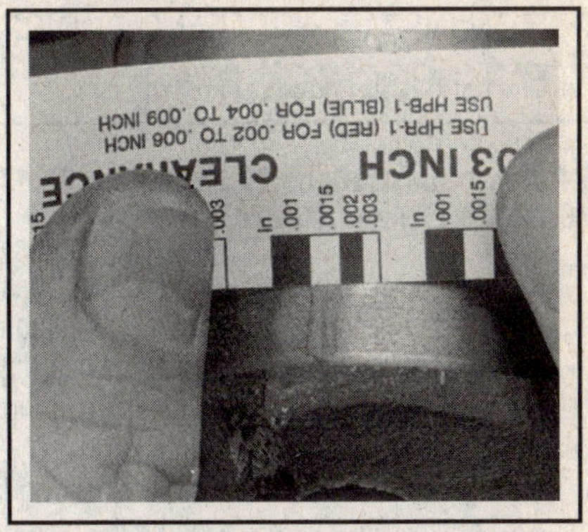

24.15 Measuring the width of the crushed Plastigage to determine the main bearing oil clearance (be sure to use the correct scale - standard and metric ones are included)

12 Clean the faces of the bearings in the caps and install the caps in their original locations (don't mix them up) with the arrows pointing toward the front of the engine. Don't disturb the Plastigage.

13 Starting with the center main and working out toward the ends, tighten the main bearing cap bolts to the torque listed in this Chapter's Specifications in three steps. Don't rotate the crankshaft at any time during this operation, and do not tighten one cap completely - tighten all caps equally. Before tightening, the main caps should be seated using light taps with a brass or plastic mallet.

14 Remove the bolts/studs and carefully lift off the main bearing caps. Keep them in order. Don't disturb the Plastigage or rotate the crankshaft. If any of the main bearing caps are difficult to remove, tap them gently from side-to-side with a soft-face hammer to loosen them.

15 Compare the width of the crushed Plastigage on each journal to the scale printed on the Plastigage envelope to obtain the main bearing oil clearance (see illustration). Check the Specifications to make sure it's correct.

16 If the clearance is not as specified, the bearing inserts may be the wrong size (which means different ones will be required). Before deciding different inserts are needed, make sure no dirt or oil was between the bearing inserts and the caps or block when the clearance was measured. If the Plastigage was wider at one end than the other, the journal may be tapered (see Section 19).

17 Carefully scrape all traces of the Plastigage material off the main bearing journals and/or the bearing faces. Use your fingernail or the edge of a credit card - don't nick or scratch the bearing faces.

FINAL CRANKSHAFT INSTALLATION

18 Carefully lift the crankshaft out of the engine.

19 Clean the bearing faces in the block, then apply a thin, uniform layer of moly-base grease or engine assembly lube to each of the bearing surfaces. Be sure to coat the thrust faces as well as the journal face of the thrust bearing.

20 Make sure the crankshaft journals are clean, then lay the crankshaft back in place in the block.

21 Clean the faces of the bearings in the caps, then apply lubricant to them.

22 Install the caps in their original locations with the arrows (made earlier) pointing toward the front of the engine.

23 With all caps in place and bolts just started, tap the ends of the crankshaft forward and backward with a lead or brass hammer to line up the main bearing and crankshaft thrust surfaces.

24 Following the procedures outlined in Step 13, retighten all main bearing cap bolts to the torque listed in this Chapter's Specifications, starting with the center main and working out toward the ends.

25 Rotate the crankshaft a number of times by hand to check for any obvious binding.

26 The final step is to check the crankshaft endplay with feeler gauges or a dial indicator as described in Section 14. The endplay should be correct if the crankshaft thrust faces aren't worn or damaged and new bearings have been installed.

25 Rear main oil seal - replacement

All models are equipped with a one piece rear main oil seal and housing. The crankshaft must be installed first and the main bearing caps bolted in place before the seal and housing can be installed on the engine block. Refer to Chapter 2A for the rear main seal replacement procedure. Disregard the Steps that do not apply since the engine is out of the vehicle and the oil pan is not installed.

26 Pistons and connecting rods - installation and rod bearing oil clearance check

1 Before installing the piston/connecting rod assemblies, the cylinder walls must be perfectly clean, the top edge of each cylinder must be chamfered, and the crankshaft must be in place.

2 Remove the cap from the end of the number one connecting rod (check the marks made during removal). Remove the original bearing inserts and wipe the bearing surfaces of the connecting rod and cap with a clean, lint-free cloth. They must be kept spotlessly clean.

PISTON INSTALLATION AND ROD BEARING OIL CLEARANCE CHECK

Refer to illustrations 26.5a, 26.5b, 26.11, 26.13, 26.17, 26.22, 26.27a and 26.27b

3 Clean the back side of the new upper bearing insert, then lay it in place in the connecting rod. Make sure the tab on the bearing fits into the recess in the rod. Don't hammer the bearing insert into place and be very careful not to nick or gouge the bearing face. Don't lubricate the bearing at this time.

4 Clean the back side of the other bearing insert and install it in the rod cap. Again, make sure the tab on the bearing fits into the recess in the cap, and don't apply any lubricant. It's critically important that the mating surfaces of the bearing and connecting rod are perfectly clean and oil free when they're assembled.

5 Stagger the piston ring gaps around the piston (see illustrations).

6 Slip a section of plastic or rubber hose over each connecting rod cap bolt.

7 Lubricate the piston and rings with clean engine oil and attach a piston ring compressor to the piston. Leave the skirt protruding about 1/4-inch (6 mm) to guide the piston into the cylinder. The rings must be compressed until they're flush with the piston.

8 Rotate the crankshaft until the number one connecting rod journal is at BDC (bottom dead center) and apply a coat of engine oil to the cylinder walls.

9 With the mark or notch on top of the piston facing the front of the engine, gently insert the piston/connecting rod assembly into the number one cylinder bore and rest the bottom edge of the ring compressor on the engine block.

10 Tap the top edge of the ring compressor to make sure it's contacting the block around its entire circumference.

11 Gently tap on the top of the piston with the end of a wooden or plastic hammer handle (see illustration) while guiding the end of the connecting rod into place on the crankshaft journal. The piston rings may try to pop out of the ring compressor just before entering the cylinder bore, so keep some pressure down on the ring compressor. Work slowly, and if any resistance is felt as the piston enters the cylinder, stop immediately. Find out what's hanging up and fix it before proceeding. Do not, for any reason, force the piston into the cylinder - you might break a ring and/or the piston.

12 Once the piston/connecting rod assembly is installed, the connecting rod bearing oil clearance must be checked before the rod cap is permanently bolted in place.

13 Cut a piece of the appropriate size Plastigage slightly shorter than the width of the connecting rod bearing and lay it in place on the number one connecting rod journal, parallel with the journal axis (see illustration).

14 Clean the connecting rod cap bearing face, remove the protective hoses from the connecting rod bolts and install the rod cap. Make sure the mating mark on the cap is on the same side as the mark on the connecting rod.

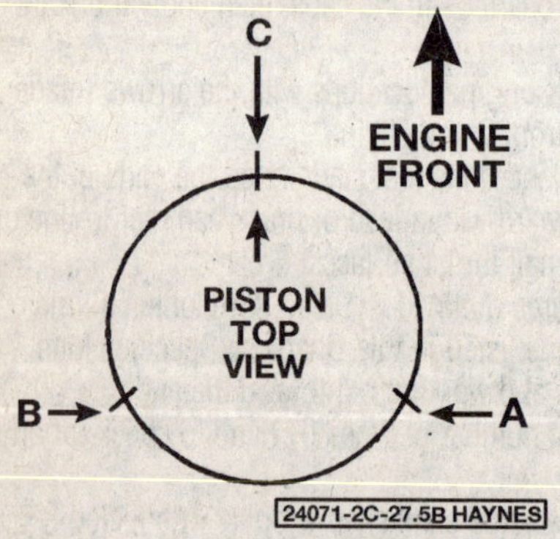

26.5a Ring end gap positions

A *Top compression ring gap*
B *Second compression ring gap*
C *Oil ring gap*

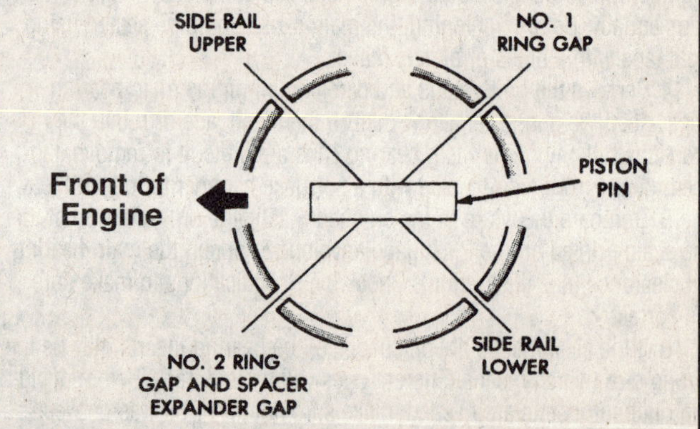

26.5b Ring end gap positions (with three-piece oil rings)

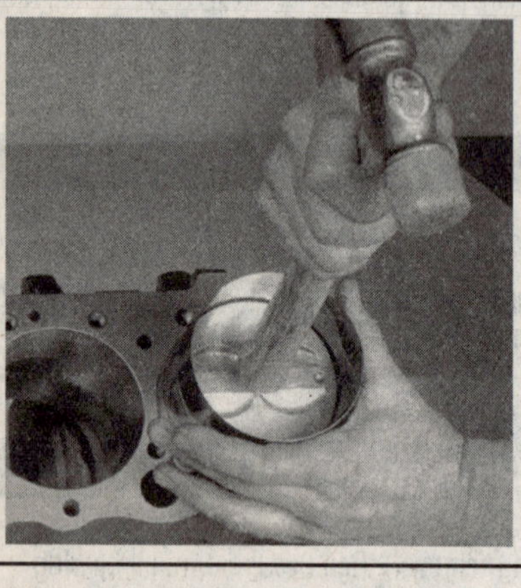

26.11 Drive the piston into the cylinder bore with the end of a wooden or plastic hammer handle

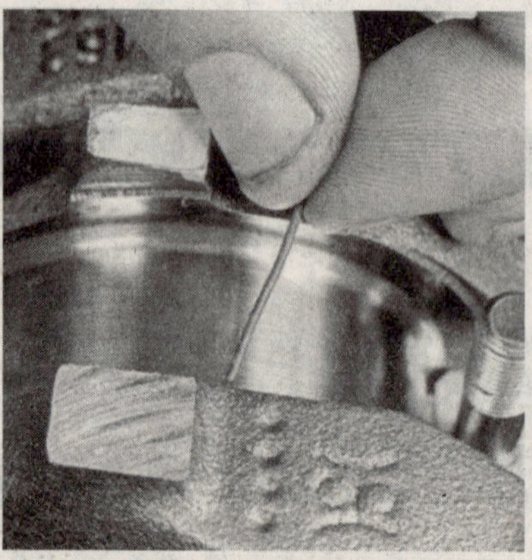

26.13 Lay the Plastigage strips on each rod bearing journal, parallel to the crankshaft centerline

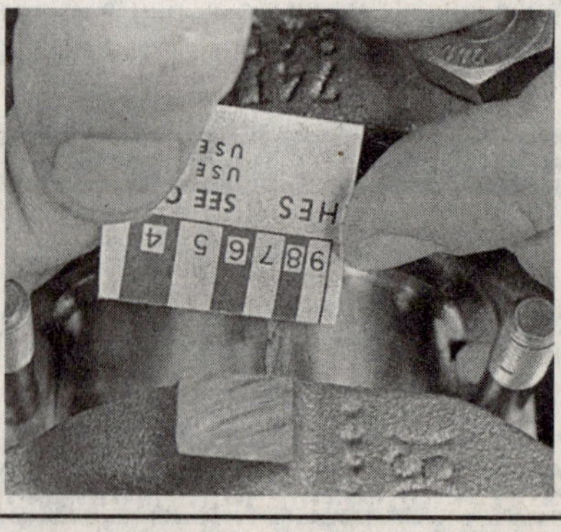

26.17 Measuring the width of the crushed Plastigage to determine the rod bearing oil clearance (be sure to use the correct scale - standard and metric ones are included)

15 Install the nuts and tighten them to the torque listed in this Chapter's Specifications. Work up to it in three steps.

➡Note: Use a thin-wall socket to avoid erroneous torque readings that can result if the socket is wedged between the rod cap and nut. If the socket tends to wedge itself between the nut and the cap, lift up on it slightly until it no longer contacts the cap. Do not rotate the crankshaft at any time during this operation.

16 Remove the nuts and detach the rod cap, being very careful not to disturb the Plastigage.

17 Compare the width of the crushed Plastigage to the scale printed on the Plastigage envelope to obtain the oil clearance (see illustration). Compare it to this Chapter's Specifications to make sure the clearance is correct.

18 If the clearance is not as specified, the bearing inserts may be the wrong size (which means different ones will be required). Before deciding different inserts are needed, make sure no dirt or oil was between the bearing inserts and the connecting rod or cap when the clearance was measured. Also, recheck the journal diameter. If the Plastigage was wider at one end than the other, the journal may be tapered (see Section 19).

FINAL CONNECTING ROD INSTALLATION

19 Carefully scrape all traces of the Plastigage material off the rod journal and/or bearing face. Be very careful not to scratch the bearing - use your fingernail or the edge of a credit card.

20 Make sure the bearing faces are perfectly clean, then apply a uniform layer of clean moly-base grease or engine assembly lube to both of them. You'll have to push the piston into the cylinder to expose the face of the bearing insert in the connecting rod - be sure to slip the protective hoses over the rod bolts first.

21 Slide the connecting rod back into place on the journal, remove the protective hoses from the rod cap bolts, install the rod cap and tighten the nuts to the torque listed in this Chapter's Specifications. Again, work up to the torque in three steps.

22 Repeat the entire procedure for the remaining pistons/connecting rods. On diesel engines it will be necessary to double check the orientation of the valve relief pockets to make sure the pistons are installed correctly (see illustration).

23 The important points to remember are:

a) *Keep the back sides of the bearing inserts and the insides of the connecting rods and caps perfectly clean when assembling them.*
b) *Make sure you have the correct piston/rod assembly for each cylinder.*
c) *The arrow or mark on the piston must face the front of the engine.*
d) *Lubricate the cylinder walls with clean oil.*
e) *Lubricate the bearing faces when installing the rod caps after the oil clearance has been checked.*

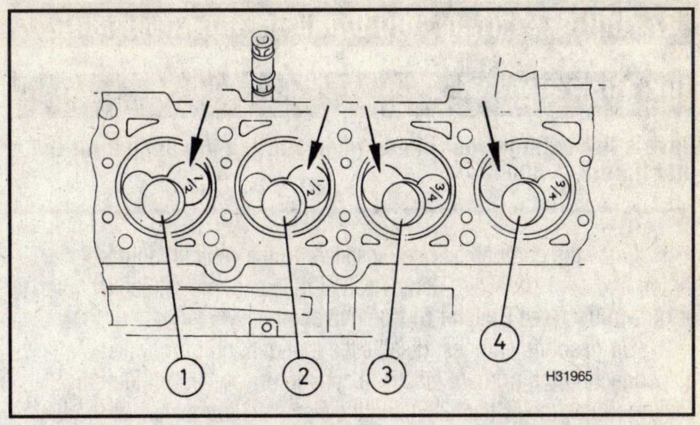

26.22 On diesel engines, the intake valve reliefs (arrows) should point towards the transaxle on the No. 1 and 2 cylinders and point towards the timing belt on the No. 3 and 4 cylinders

24 After all the piston/connecting rod assemblies have been properly installed, rotate the crankshaft a number of times by hand to check for any obvious binding.

25 As a final step, the connecting rod endplay must be checked (see Section 13).

26 Compare the measured endplay to this Chapter's Specifications to make sure it's correct. If it was correct before disassembly and the original crankshaft and rods were reinstalled, it should still be right. If new rods or a new crankshaft were installed, the endplay may be inadequate. If so, the rods will have to be removed and taken to an automotive machine shop for re-sizing.

27 On diesel engines, it will be necessary to check the piston projection to help determine the correct thickness of head gasket to be used. Rotate the engine so that the No.1 piston is located at TDC. Using a dial indicator or a depth micrometer measure the distance that the piston protrudes past the deck surface of the block (see illustration). Repeat the measuring process on the remaining three cylinders and record the highest reading. Three different thickness head gaskets are available for diesel engines depending on the amount the piston protrudes above the deck surface. Compare the highest reading to the gasket selection chart below to select the proper head gasket thickness (see illustration).

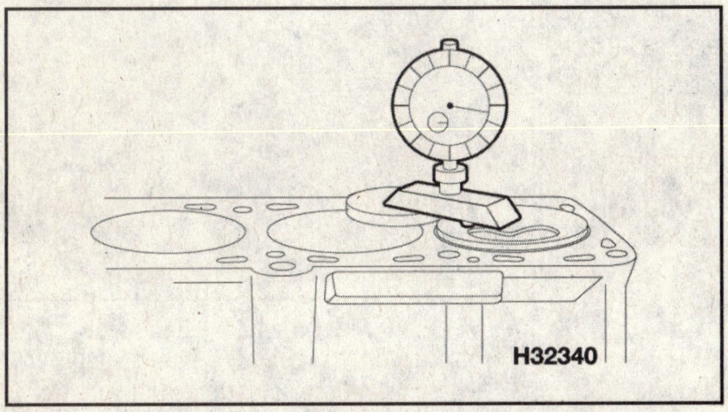

26.27a Measuring piston projection with a dial indicator

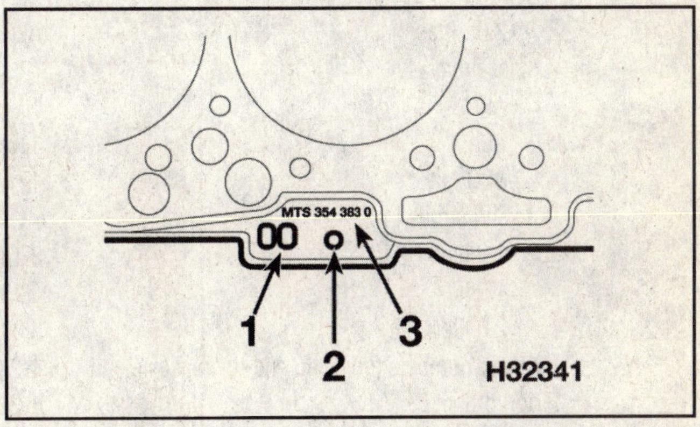

26.27b Head gasket identification marks - diesel engine

1 *Production control code*
2 *Identification holes*
3 *Part number*

27 Initial start-up and break-in after overhaul

⁂ WARNING:

Have a fire extinguisher handy when starting the engine for the first time.

1 Once the engine has been installed in the vehicle, double-check the oil and coolant levels. Remove all of the spark plugs (see Chapter 1) or glow plugs (see Chapter 5) from the engine.

2 On gasoline engines, disable the fuel and ignition systems by disconnecting the primary electrical connectors at the ignition coil pack/modules (see Chapter 5) and the electrical connectors at the fuel injectors (see Chapter 4A). On diesel engines, disable the fuel injection system by unplugging the electrical connector to the fuel shut off valve (see Chapter 4B).

3 Install the spark plugs or glow plugs, hook up the spark plug wires or electrical connectors for the glow plugs. Restore the fuel and ignition system functions.

4 Start the engine. It may take a few moments for the fuel system to build up pressure, but the engine should start without a great deal of effort.

➡Note: If the engine keeps backfiring, recheck the valve timing and spark plug wire routing.

5 After the engine starts, it should be allowed to warm up to normal operating temperature. While the engine is warming up, make a thorough check for fuel, oil and coolant leaks.

➡Note: On diesel engines, it may be necessary to reset the injection pump timing with a scan tool before the engine will start (see Chapter 4B).

6 Shut the engine off and recheck the engine oil and coolant levels.

7 Drive the vehicle to an area with no traffic, accelerate from 30 to 50 mph, then allow the vehicle to slow to 30 mph with the throttle closed. Repeat the procedure 10 or 12 times. This will load the piston rings and cause them to seat properly against the cylinder walls. Check again for oil and coolant leaks.

8 Drive the vehicle gently for the first 500 miles (no sustained high speeds) and keep a constant check on the oil level. It isn't unusual for an engine to use oil during the break-in period.

9 At approximately 500 to 600 miles, change the oil and filter.

10 For the next few hundred miles, drive the vehicle normally. Don't pamper it or abuse it.

11 After 2000 miles, change the oil and filter again and consider the engine broken in.

Specifications

General

VIN engine code	
D	1.8L (turbo-gasoline)
F	1.9L (turbo-diesel)
B	2.0L (gasoline)
Engine designation	
1.8L (turbo-gasoline)	APH
1.9L (turbo-diesel)	ALH
2.0L (gasoline)	AEG
Displacement	
APH	109 cubic inches (1.8 liters)
ALH	116 cubic inches (1.9 liters)
AEG	121 cubic inches (2.0 liters)
Bore and stroke	
1.8 liter	3.189 x 3.401 inches (81.0 x 86.4 mm)
1.9 liter	3.130 x 3.759 inches (79.9 x 95.5 mm)
2.0 liter	3.248 x 3.653 inches (82.5 x 92.8 mm)
Cylinder numbers (drivebelt end-to-transaxle end)	1-2-3-4
Firing order	1-3-4-2

General (continued)	
Cylinder compression pressure	
Gasoline engines	
Minimum	110 psi (7.7 kg/cm2)
Maximum variation between cylinders	30-percent from the highest reading
Diesel engine	
Minimum	279 psi (19.53 kg/cm2)
Maximum variation between cylinders	20-percent from the highest reading
Oil pressure	
Gasoline engines	
At idle	29 psi (2.03 kg/cm2) minimum
At 2,000 rpm	44 to 66 psi (3.08 to 4.62 kg/cm2)
Diesel engine	
At idle	N/A
At 2,000 rpm	29 psi (2.03 kg/cm2) minimum
Maximum oil pressure	
All engines	103 psi (7.21 kg/cm2)
Cylinder head	
Warpage limit	0.004 inch (0.10 mm)
Cylinder head height (minimum)	
Gasoline engines	
1.8 liter	5.480 inches (139.20 mm)
2.0 liter	5.220 inches (132.58 mm)
Diesel engine	
1.9 liter	Cylinder head resurfacing not permissible
Valves and related components	
Valve face angle	
All engines	45-degrees
Valve seat angle	
All engines	45-degrees
Valve margin width	
All engines	N/A
Valve stem diameter	
Intake valves	
1.8 liter	0.235 inch (5.963 mm)
1.9 liter	0.274 inch (6.963 mm)
2.0 liter	0.272 inch (6.92 mm)
Exhaust valves	
1.8 liter	0.234 inch (5.943 mm)
1.9 liter	0.273 inch (6.943 mm)
2.0 liter	0.272 inch (6.92 mm)
Valve stem-to-guide clearance (maximum deflection)	
Intake valves	
1.8 liter	0.031 inch (0.8 mm)
1.9 liter	0.051 inch (1.3 mm)
2.0 liter	0.039 inch (1.0 mm)
Exhaust valves	
1.8 liter	0.031 inch (0.8 mm)
1.9 liter	0.051 inch (1.3 mm)
2.0 liter	0.051 inch (1.3 mm)

Valves and related components (continued)

Valve spring	
Free length	N/A
Installed height	N/A
Valve installed height (minimum dimension)*	
Intake valves	
1.8 liter	
Outer valves	1.339 inch (34.0 mm)
Inner valves	1.327 inch (33.7 mm)
1.9 liter	1.409 inch (35.8 mm)
2.0 liter	1.331 inch (33.8 mm)
Exhaust valves	
1.8 liter	1.354 inch (34.4 mm)
1.9 liter	1.421 inch (36.1 mm)
2.0 liter	1.343 inch (34.1 mm)

**Measured from tip of valve stem to top of the valve cover rail.*

Crankshaft

Endplay	
Gasoline engines	
Standard	0.0028 to 0.0091 inch (0.07 to 0.23 mm)
Service limit	0.0118 inch (0.30 mm)
Diesel engine	
Standard	0.0028 to 0.0067 inch (0.07 to 0.17 mm)
Service limit	0.0146 inch (0.37 mm)
Runout N/A	
Main bearing journal diameters	
All engines	
Standard	2.1260 inches (54.0 mm)
1st undersize	2.1161 inches (53.75 mm)
2nd undersize	2.1063 inches (53.50 mm)
3rd undersize	2.0965 inches (53.25 mm)
Tolerance	-0.0008 to -0.0014 inch (-0.022 to -0.037 mm)
Out-of-round limit	0.0002 inch (0.005 mm)
Taper limit	0.0003 inch (0.007 mm)
Main bearing oil clearance	
Gasoline engines	
Standard	0.0004 to 0.0016 inch (0.01 to 0.04 mm)
Service limit	0.0059 inch (0.15 mm)
Diesel engine	
Standard	0.0012 to 0.0031 inch (0.03 to 0.08 mm)
Service limit	0.0067 inch (0.17 mm)

Connecting rods

Connecting rod bearing journal diameters	
All engines	
Standard	1.8819 inches (47.80 mm)
1st undersize	1.8720 inches (47.55 mm)
2nd undersize	1.8622 inches (47.30 mm)
3rd undersize	1.8524 inches (47.05 mm)

Tolerance	-0.0008 to -0.0016 inch (-0.022 to -0.042 mm)
Out-of-round limit	0.0002 inch (0.005 mm)
Taper limit	0.0003 inch (0.007 mm)
Connecting rod bearing oil clearance	
Gasoline engines	
Standard	0.0004 to 0.0024 inch (0.01 to 0.06 mm)
Service limit	0.0047 inch (0.12 mm)
Diesel engine	
Standard	0.0004 to 0.0020 inch (0.01 to 0.05 mm)
Service limit	0.0031 inch (0.08 mm)
Connecting rod side clearance (endplay)	0.0020 to 0.0122 inch (0.05 to 0.31 mm)

Engine block

Cylinder bore diameter	
1.8 liter	
Standard	3.189 inches (81.01 mm)
1st oversize	3.209 inches (81.51 mm)
1.9 liter	
Standard	3.1303 inches (79.51 mm)
1st oversize	3.1401 inches (79.76 mm)
2nd oversize	3.1500 inches (80.01 mm)
2.0 liter	
Standard	3.2484 inches (82.51 mm)
1st oversize	3.2681 inches (83.01 mm)
Out-of-round limit	
1.8 liter	0.0016 inch (0.04 mm)
1.9 liter	0.0039 inch (0.10 mm)
2.0 liter	0.0031 inch (0.08 mm)
Taper limit	
1.8 liter	0.0016 inch (0.04 mm)
1.9 liter	0.0039 inch (0.10 mm)
2.0 liter	0.0031 inch (0.08 mm)
Block deck warpage limit	0.004 inch (0.10 mm)

Pistons and rings

Piston diameter	
1.8 liter	
Standard*	3.1875 inches (80.965 mm)
1st oversize*	3.2073 inches (81.465 mm)
1.9 liter	
Standard	3.1287 inches (79.47 mm)
1st oversize	3.1386 inches (79.72 mm)
2nd oversize	3.1484 inches (79.97 mm)
2.0 liter	
Standard*	3.2466 inches (82.465 mm)
1st oversize	3.2663 inches (82.965 mm)
Piston ring end gap	
1.8 liter	
Top compression ring	0.0059 to 0.0157 inch (0.15 to 0.40 mm
Second compression ring	0.0059 to 0.0157 inch (0.15 to 0.40 mm
Oil control ring	0.0098 to 0.0197 inch (0.25 to 0.50 mm)

Piston ring end gap (continued)	
1.9 liter	
Top compression ring	0.0079 to 0.0157 inch (0.20 to 0.40 mm)
Second compression ring	0.0079 to 0.0157 inch (0.20 yo 0.40 mm)
Oil control ring	0.0098 to 0.0197 inch (0.25 to 0.50 mm)
2.0 liter	
Top compression ring	0.0079 to 0.0157 inch (0.20 to 0.40 mm)
Second compression ring	0.0079 to 0.0157 inch (0.20 to 0.40 mm)
Oil control ring	0.0098 to 0.0197 inch (0.25 to 0.50 mm)
Piston ring side clearance	
1.8 liter	
Top compression ring	0.0007 to 0.0027 inch (0.02 to 0.07 mm)
Second compression ring	0.0007 to 0.0027 inch (0.02 to 0.07 mm)
Oil control ring	0.0007 to 0.0024 inch (0.02 to 0.06 mm)
1.9 liter	
Top compression ring	0.0024 to 0.0035 inch (0.06 to 0.09 mm)
Second compression ring	0.0020 to 0.0031 inch (0.05 to 0.08 mm)
Oil control ring	0.0012 to 0.0024 inch (0.03 to 0.06 mm)
2.0 liter	
Top compression ring	0.0024 to 0.0035 inch (0.06 to 0.09 mm)
Second compression ring	0.0024 to 0.0035 inch (0.06 to 0.09 mm)
Oil control ring	0.0012 to 0.0024 inch (0.03 to 0.06 mm)

** Dimension without graphite coating (the graphite coating, which is dark gray, will add approximately 0.0008 inch [0.02 mm] to the diameter).*

Torque specifications**	Nm	Ft-lbs (unless other wise indicated)
Main bearing cap bolts (always replace)		
Step 1	65	48
Step 2	Tighten an additional 90-degrees	
Connecting rod cap nuts (always replace)		
Step 1	30	22
Step 2	Tighten an additional 90-degrees	
Oil spray nozzles	27	20

****Note: Refer to Chapter 2A or 2B for additional torque specifications.**

3

COOLING, HEATING AND AIR CONDITIONING SYSTEMS

Section

Reference to other Chapters

Engine oil cooler - See Chapter 2C

1 General information

All vehicles covered by this manual employ a pressurized engine cooling system with thermostatically controlled coolant circulation. Coolant is drawn from the radiator by an impeller-type water pump mounted at the front of the block. The coolant is then circulated through the engine block and the cylinder head before it's redirected back into the radiator.

A wax pellet type thermostat is located in the thermostat housing on the engine. During warm up, the closed thermostat prevents coolant from circulating through the radiator. When the engine reaches normal operating temperature, the thermostat opens and allows hot coolant to travel through the radiator, where it is cooled before returning to the engine.

The cooling system is pressurized by a spring-loaded expansion tank cap, which, by maintaining pressure, increases the boiling point of the coolant. If the coolant temperature goes above this increased boiling point, the extra pressure in the system forces the reservoir cap valve off its seat and allows the coolant to escape through the overflow tube into the expansion tank. When the system cools, the excess coolant is automatically drawn from the reservoir tank back into the radiator.

The expansion tank serves as both the point at which fresh coolant is added to the cooling system to maintain the proper fluid level and as a holding tank for overheated coolant.

The heating system works by directing air through the heater core mounted in the dash and then to the interior of the vehicle by a system of ducts. Temperature is controlled by mixing heated air with fresh air, using a system of doors in the ducts, and a blower motor.

Air conditioning is an optional accessory, consisting of an evaporator core located under the dash, a condenser in front of the radiator, a receiver/drier in the engine compartment and a belt-driven compressor mounted at the front of the engine.

Diesel engines are equipped with coolant glow plugs to help heat the engine during cold weather. These glow plugs are tested and replaced in the same manner as the engine glow plugs, refer to Chapter 5 for more information. Diesel engines are also equipped with an EGR cooler that is mounted at the rear of the engine between the intake manifold and the exhaust manifold/turbocharger, refer to Chapter 6B for more information on this component.

2 Antifreeze - general information

Refer to illustration 2.4

WARNING:

Do not allow antifreeze to come in contact with your skin or painted surfaces of the vehicle. Rinse off spills immediately with plenty of water. Antifreeze is highly toxic if ingested. Never leave antifreeze lying around in an open container or in puddles on the floor; children and pets are attracted by it's sweet smell and may drink it. Check with local authorities about disposing of used antifreeze. Many communities have collection centers which will see that antifreeze is disposed of safely. Never dump used antifreeze on the ground or pour it into drains.

2.4 An inexpensive hydrometer can be used to test the condition of your coolant

CAUTION:

The manufacturer recommends using only silicate and phosphate-free coolant for these systems. Silicate and phosphate-free antifreeze is red in color. Never mix green-colored ethylene glycol antifreeze and red-colored silicate and phosphate-free coolant because doing so will destroy the efficiency of the antifreeze.

The cooling system should be filled with the proper antifreeze solution which will prevent freezing down to at least -20-degrees F (even lower in cold climates). It also provides protection against corrosion and increases the coolant boiling point.

The cooling system should be drained, flushed and refilled at least every other year (see Chapter 1). The use of antifreeze solutions for periods of longer than two years is likely to cause damage and encourage the formation of rust and scale in the system. However, these models are filled with a new, long-life silicate and phosphate-free coolant which the manufacturer claims is good for the lifetime of the vehicle.

Before adding antifreeze to the system, check all hose connections. Antifreeze can leak through very minute openings.

The exact mixture of antifreeze to water which you should use depends on the relative weather conditions. The mixture should contain at least 50-percent antifreeze, but should never contain more than 70-percent antifreeze. Consult the mixture ratio chart on the antifreeze container before adding coolant. Hydrometers are available at most auto parts stores to test the coolant (see illustration). Always use antifreeze which meets the vehicle manufacturer's specifications.

3 Thermostat - check and replacement

WARNING:

The engine must be completely cool when this procedure is performed.

CAUTION:

Don't drive the vehicle without a thermostat! The computer may stay in open loop mode and emissions and fuel economy will suffer.

CHECK

1 Before assuming the thermostat is to blame for a cooling system problem, check the coolant level, drivebelt tension (see Chapter 1) and temperature gauge (or light) operation.

2 If the engine seems to be taking a long time to warm up (based on heater output or temperature gauge operation), the thermostat is probably stuck open. Replace the thermostat with a new one.

3 If the engine runs hot, use your hand to check the temperature of the upper radiator hose. If the hose isn't hot, but the engine is, the thermostat is probably stuck closed, preventing the coolant inside the engine from escaping to the radiator. Replace the thermostat.

4 If the upper radiator hose is hot, it means the coolant is flowing and the thermostat is open. Consult the *Troubleshooting* Section at the front of this manual for cooling system diagnosis.

REPLACEMENT

Refer to illustrations 3.8, 3.11 and 3.14

5 Drain the coolant from the radiator (see Chapter 1). If the coolant is relatively new or in good condition, save it and reuse it. If it is to be replaced, see Section 2 for **cautions** about proper handling of used antifreeze.

6 Follow the lower radiator hose to the engine to locate the thermostat housing cover. The thermostat is located on the side of the engine facing the radiator.

7 Squeeze the tabs on the hose clamp and pull the hose clamp back over the hose.

8 Detach the hose from the thermostat housing cover (see illustration). If the hose sticks, grasp it near the end with a pair of adjustable pliers and twist it to break the seal, then pull it off. If the hose is old or deteriorated, cut it off and install a new one.

3.8 Lower radiator hose clamp at the thermostat housing cover (2.0L engine shown, all others similar)

A Radiator hose clamp
B Thermostat housing cover bolts

9 If the outer surface of the cover fitting that mates with the hose is deteriorated (corroded, pitted, etc.) it may be damaged further by hose removal. If it is, the thermostat housing cover will have to be replaced.

10 Remove the bolts/nuts and detach the thermostat cover. If the cover is stuck, tap it with a soft-face hammer to jar it loose. Be prepared for some coolant to spill as the gasket seal is broken.

11 Note how it's installed (which end is facing up), then remove the thermostat and the cover O-ring (see illustration).

12 Clean the mating surfaces of the engine block and the thermostat housing cover.

13 Install the thermostat and make sure the correct end faces out - the spring is directed toward the engine. On diesel engines install the thermostat and rotate it 1/4 turn clockwise.

14 Install a new O-ring over the thermostat and reattach the thermostat housing cover to the engine block (see illustration). Tighten the bolts to the torque listed in this Chapter's Specifications. Now may be a good time to check and replace the hoses and clamps (see Chapter 1).

15 The remaining steps are the reverse of the removal procedure.

16 Refer to Chapter 1 and refill the system, then run the engine and check carefully for leaks.

17 Repeat steps 1 through 4 to be sure the repairs corrected the previous problem(s).

3.11 Remove the thermostat noting which end faces out

3.14 Correct installation of the thermostat and the cover O-ring

4 Engine cooling fan and circuit - check and component replacement

WARNING:

Keep hands, tools and clothing away from the fan. To avoid injury or damage DO NOT operate the engine with a damaged fan. Do not attempt to repair fan blades - replace a damaged fan with a new one.

CHECK

Refer to illustration 4.1

1 To test a fan motor, unplug the electrical connector at the motor and use fused jumper wires to connect battery power and ground directly to the fan (see illustration). If the fan doesn't operate, replace the motor.

2 If the motor tests OK, check the cooling fan thermo switch, located at the bottom of the radiator on the left hand side.

3 Remove the electrical connector from the cooling fan thermo switch and apply a fused jumper wire between the red wire and the red/white wire with the ignition switch in the On position. Both fans should run at
Low speed.

4 Next, apply the fused jumper wire between the red wire and the red/yellow wire with the ignition switch in the On position. Both fans should run at high speed.

5 If the fans run as described in Steps 3 and 4, it indicates that the cooling fans and circuit are operating properly and the thermo switch is faulty.

6 If the fans do not operate with the thermo switch bypassed check for voltage at the red wire of the thermo switch electrical connector. There should be battery voltage.

7 If voltage is not present, check the No. 3 and the No. 8 fuses in the engine compartment fuse holder and the No.16 fuse in the passenger compartment fuse box. If the fuses are OK, the problem lies in the wiring harness or the fan control module.

8 Refer to the wiring schematics at the end of Chapter 12 and check the wiring for open or short circuits. The fan control module can only be diagnosed as faulty through process of elimination.

REPLACEMENT

Refer to illustration 4.10

WARNING:

Keep hands, tools and clothing away from the fan. To avoid injury or damage DO NOT operate the engine with a damaged fan. Do not attempt to repair fan blades - replace a damaged fan with a new one.

9 Raise the vehicle and support it securely on jackstands. Remove the lower engine cover and disconnect the electrical connector(s) from the cooling fan(s).

10 Detach the wiring harness clips and remove the cooling fan to shroud mounting bolts (see illustration). Pull the fan assembly outward slightly to dislodge it from the fan shroud, then guide the fan assembly out of engine compartment from the bottom, making sure that all wiring clips are disconnected. Be careful not to contact the radiator cooling fins.

11 If the fan blades or the fan motor are damaged, they can be replaced by removing the fan blade from the fan motor, then removing the fan motor from the fan housing (see illustration 4.10).

12 Installation is the reverse of removal.

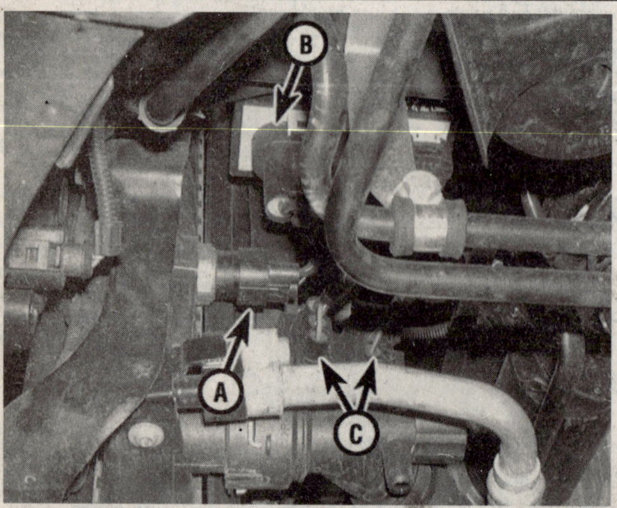

4.1 Location of the cooling fan thermo switch (A), the fan control module (B) and the cooling fan electrical connectors (C)

4.10 Cooling fan to shroud mounting bolts (A), the fan motor mounting bolts (B) and the wiring harness retaining clip (C)

5 Radiator and expansion tank - removal and installation

WARNING 1:

The air conditioning system is under high pressure. DO NOT loosen any fittings or remove any components until after the system has been discharged. Air conditioning refrigerant should be properly discharged into an EPA-approved container at a dealership service department or an automotive air conditioning repair facility. Always wear eye protection when disconnecting air conditioning system fittings.

WARNING 2:

The engine must be completely cool when this procedure is performed.

RADIATOR

Refer to illustrations 5.3, 5.5, 5.8, 5.9 and 5.11

1 Drain the cooling system as described in Chapter 1. Refer to the coolant **Warning** in Section 2.

2 Remove the front fenders and bumper cover assembly (see Chapter 11).

3 Disconnect the upper coolant hoses from the radiator (see illustration).

4 Raise the vehicle and support it securely on jackstands.

5 Disconnect the lower coolant hoses and the electrical connector from the fan thermo switch. Also disconnect the cooling fan electrical connectors (see illustration).

6 Remove the drive belt tensioner, if necessary.

7 Disconnect the retaining clamps for the air conditioning lines.

8 Detach the lock carrier panel and the radiator mounting bolts (see illustration).

9 Remove the condenser mounting bolts (if equipped) and separate the condenser from the radiator without disconnecting the air conditioning lines (see illustration).

10 Remove the radiator and cooling fans as an assembly from the vehicle.

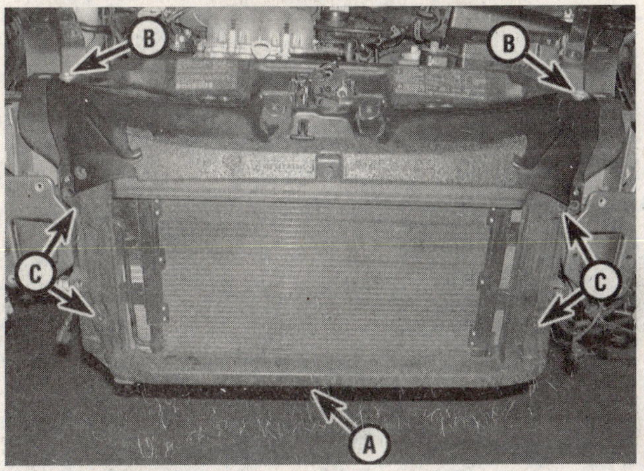

5.8 To remove the lock carrier panel (A), detach the upper bolts (B) and the radiator mounting bolts (C)

5.3 Removing the upper radiator hose

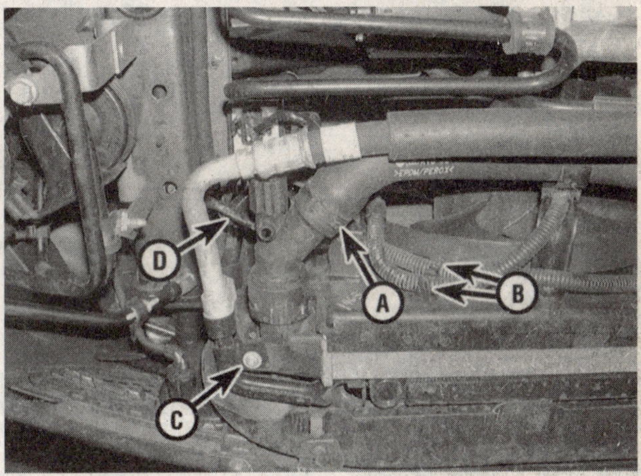

5.5 Remove the following components from the bottom of the radiator

A Lower radiator hose
B Cooling fan electrical connectors
C Air conditioning line retaining clamp
D Cooling fan thermo switch electrical connector

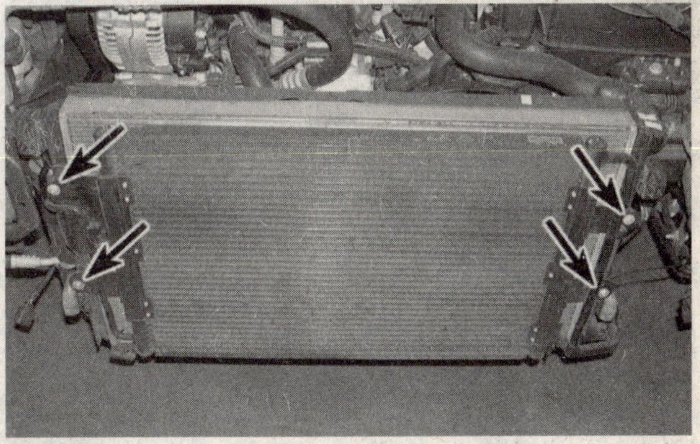

5.9 Condenser-to-radiator mounting bolts (arrows)

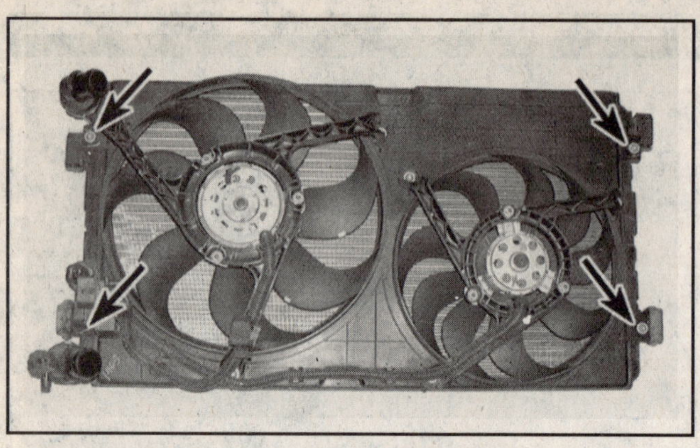

5.11 Cooling fan shroud mounting bolts (arrows)

11 Remove the cooling fans and shroud from the radiator (see illustration).

12 Prior to installation of the radiator, replace any damaged hose clamps and/or radiator hoses. If leaks have been noticed or there have been cooling problems, have the radiator cleaned and tested at a radiator shop.

13 Radiator installation is the reverse of removal.

14 After installation, fill the system with the proper mixture of antifreeze, bleed the air from the cooling system as described in Chapter 1.

EXPANSION TANK

Refer to illustration 5.17

15 Drain the cooling system as described in Chapter 1 until the expansion tank is empty. Refer to the coolant **Warning** in Section 2.

5.17 Detach the coolant hoses (A), the mounting bolts (B) and the electrical connector (C).

16 Remove the coolant recovery hoses from the expansion tank.

17 Detach the reservoir mounting bolts, then lift the expansion tank from the notch on the inner fenderwell and disconnect the coolant level sensor connector (see illustration).

18 Remove the expansion tank from the engine compartment.

19 Prior to installation, make sure the reservoir is clean and free of debris which could be drawn into the radiator (wash it with soapy water and a brush if necessary, then rinse thoroughly).

20 Installation is the reverse of removal.

6 Water pump - removal and installation

Refer to illustrations 6.6 and 6.8

WARNING:

Wait until the engine is completely cool before starting this procedure.

1 Raise the vehicle and support it securely on jackstands.

2 Drain the coolant (see Chapter 1).

3 Remove the serpentine drivebelt (see Chapter 1).

4 Remove the timing belt (see Chapter 2A or 2B).

5 Remove any idler pulleys that would interfere with the removal of

6.6 Water pump retaining bolts (arrows)

6.8 Seat the water pump O-ring (arrow) firmly in its groove

the water pump.

6 Remove the water pump mounting bolts from the engine block and the water pump (see illustration). On 2.0L engines, this will include removing the rear timing belt cover as well.

7 Clean the mating surfaces on the water pump and the engine block. Wipe the mating surfaces with a rag saturated with lacquer thinner or acetone.

8 First, install the water pump O-ring to the water pump on the bench, then install the water pump onto the engine block (see illustration). Apply a small amount of coolant to the O-ring to ease installation into the block.

9 Install the water pump to engine block bolts and tighten them to the torque listed in this Chapter's Specifications.

10 The remainder of the installation procedure is the reverse of removal. Be sure to properly install the timing belt. Add coolant to the specified level (see Chapter 1). Start the engine and check for the proper coolant level and the water pump and hoses for leaks. Bleed the cooling system of air as described in Chapter 1.

7 Coolant temperature gauge sending unit - check and replacement

CHECK

1 The coolant temperature indicator system Is composed of a temperature gauge or warning light mounted in the dash and a coolant temperature sensor mounted on the engine. This coolant temperature sensor doubles as an information sensor for the fuel and emissions systems (see Chapter 6) and as a sending unit for the temperature gauge.

2 If an overheating indication occurs, check the coolant level in the system and then make sure the wiring between the gauge and the sending unit is secure and all fuses are intact.

3 Check the operation of the coolant temperature sensor (see Chapter 6A). If the sensor is defective, replace it with a new part of the same specification.

4 If the coolant temperature sensor is good, have the temperature gauge checked by a dealer service department. This test will require a scan tool to access the information as it is processed by the Engine Control Module.

REPLACEMENT

5 Refer to Chapter 6 for the engine coolant temperature sensor replacement procedure.

8 Blower motor and circuit - check

Refer to illustrations 8.3 and 8.4

WARNING:

These models have airbags. Always disconnect the negative battery cable and wait two minutes before working in the vicinity of the impact sensors, steering column or instrument panel to avoid the possibility of accidental deployment of the airbag, which could cause personal injury (see Chapter 12).

1 Check the fuses and all connections in the circuit for looseness and corrosion. Make sure the battery is fully charged.

2 Place the transaxle in Park (automatic) or Neutral (manual) and set the parking brake securely. Turn the ignition switch to the Run position, it isn't necessary to start the vehicle.

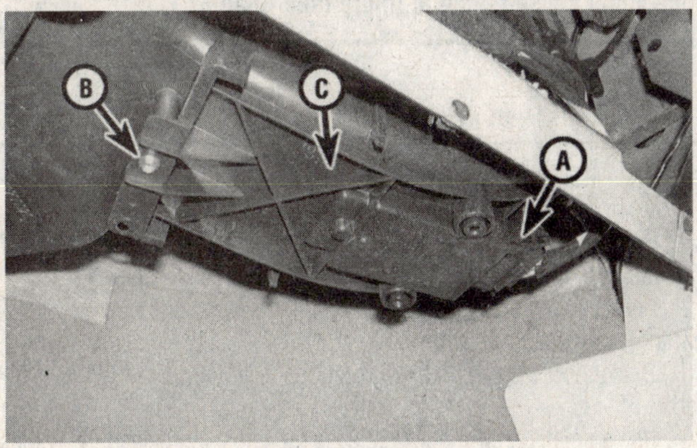

8.3 Disconnect the resistor electrical connector (A), then remove the cover mounting screws (B) and remove the cover (C)

3 Remove the cover (below the glove box) for access to the blower motor. It will be necessary to disconnect the resistor connector, then remove the cover and reconnect the resistor connector for this test (see illustration).

4 Backprobe the blower motor electrical connector and connect a voltmeter to the two terminals in the blower motor connector (see illustration).

Note: Refer to Chapter 12 for additional information on backprobing a connector.

5 Move the blower switch through each of its positions and note the voltage readings. Changes in voltage indicate that the motor speeds will also vary as the switch is moved to the different positions.

8.4 Connect a voltmeter to the heater blower motor connector by backprobing the connector, operate the blower motor, and check the running voltage at each blower switch position

6 If there is voltage present, but the blower motor does not operate, the blower motor is probably faulty. Disconnect the blower motor connector, then hook one side of the blower motor terminals to a chassis ground and the other to a fused source of battery voltage. If the blower doesn't operate, it is faulty.

7 If there was no voltage present at the blower motor at one or more speeds, and the motor itself tested OK, check the blower motor resistor.

8 Disconnect the electrical connector from the blower motor resistor (see illustration 8.3). With the ignition On, check for voltage at each of the terminals in the connector as the blower speed switch is moved to the different positions. If the voltmeter does not respond correctly to the switch and the blower is known to be good then the resistor is probably faulty. If there is no voltage present from the switch, then the switch, control panel or related wiring is probably faulty.

9 Follow the blower motor ground wire from the motor to the chassis and check the ground terminal for continuity to ground against the chassis metal.

9 Blower motor - removal and installation

9.3 Position the lower section of the instrument panel aside and lower the blower motor from the housing

➧ Refer to illustration 9.3

⁂ WARNING:

These models have airbags. Always disconnect the negative battery cable and wait two minutes before working in the vicinity of the impact sensors, steering column or instrument panel to avoid the possibility of accidental deployment of the airbag, which could cause personal injury (see Chapter 12).

1 Remove the glove box and the glove box support brace (see Chapter 11).

2 Remove the lower cover and disconnect the electrical connector from the blower motor (see illustration 8.4).

3 Pull the lower section of the instrument panel outward and unclip the blower motor from the blower housing (see illustration).

4 Pull the blower motor and fan straight down to remove it from the vehicle.

5 Installation is the reverse of removal.

10 Heater and air conditioning control assembly - removal and installation

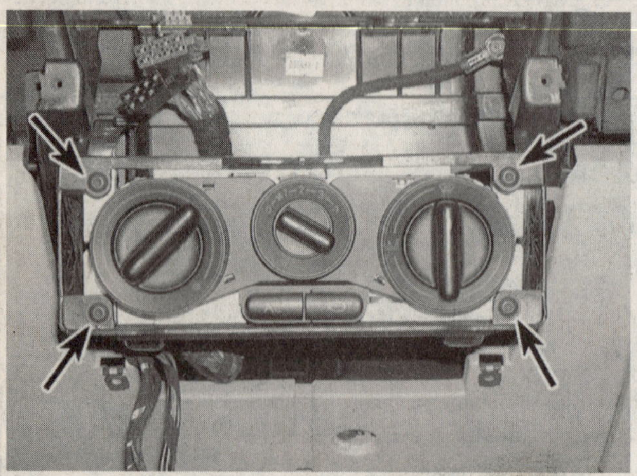

10.2 Remove the screws (arrows) retaining the heater/air conditioning control assembly to the instrument panel

⁂ WARNING:

These models have airbags. Always disconnect the negative battery cable and wait two minutes before working in the vicinity of the impact sensors, steering column or instrument panel to avoid the possibility of accidental deployment of the airbag, which could cause personal injury (see Chapter 12).

REMOVAL

➧ Refer to illustrations 10.2 and 10.4

1 Remove the center instrument panel bezel to allow access to the heater/air conditioning control mounting screws (see Chapter 11).

2 Remove the control assembly retaining screws and pull the unit from the dash (see illustration).

3 Disconnect the electrical connections from the rear of the control head.

4 Use a small screwdriver to release the clips and detach the cables from the actuating arms (see illustration). Note the color and location of the cables as the cables are removed.

10.4 Pull the control assembly outward and disconnect the electrical connectors, then detach the cable retaining clamps (arrows)

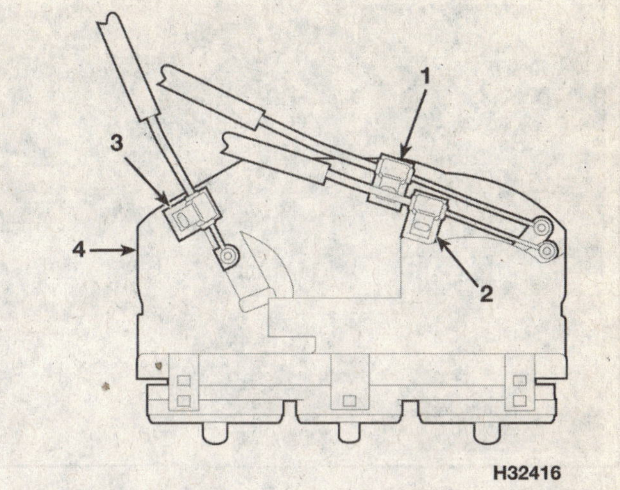

10.5 Heater control cable installation details

1 *Air distribution control knob (yellow cable to center duct)*
2 *Air distribution control knob (green cable to defroster duct)*
3 *Temperature control knob (beige cable to temperature duct)*
4 *Control head*

INSTALLATION

Refer to illustration 10.5

5 To install the control assembly, attach the cables to the actuating arms first, then snap the cable retaining clip in place.

Note: When reconnecting cables to the control assembly, be sure to attach the correct color cable with the correct actuating arm (see illustration).

6 The remainder of the installation is the reverse the removal.

11 Heater core - removal and installation

WARNING 1:

These models have airbags. Always disconnect the negative battery cable and wait two minutes before working in the vicinity of the impact sensors, steering column or instrument panel to avoid the possibility of accidental deployment of the airbag, which could cause personal injury (see Chapter 12).

WARNING 2:

The air conditioning system is under high pressure. DO NOT loosen any fittings or remove any components until after the system has been discharged. Air conditioning refrigerant should be properly discharged into an EPA-approved container at a dealership service department or an automotive air conditioning facility. Always wear eye protection when disconnecting air conditioning system fittings.

REMOVAL

Refer to illustrations 11.4, 11.5a, 11.5b, 11.6, 11.8, 11.9, 11.10, 11.11a and 11. 11b

1 Have the air conditioning system discharged by a dealership service department or an automotive air conditioning facility.

2 Drain the cooling system (see Chapter 1).

3 Remove the instrument panel and cross beam from the passenger compartment. Refer to Chapter 11 and read through the entire instrument panel removal procedure before attempting to remove it. The instrument panel removal procedure is quite lengthy and can be particularly difficult for a beginner.

4 Once the instrument panel cross beam is removed from the vehicle, detach the cross beam support braces and the lower air duct (see illustration).

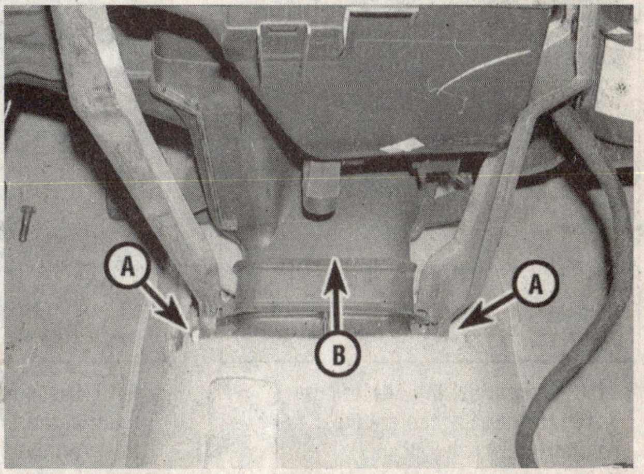

11.4 Cross beam support brackets (A) and the lower air duct (B)

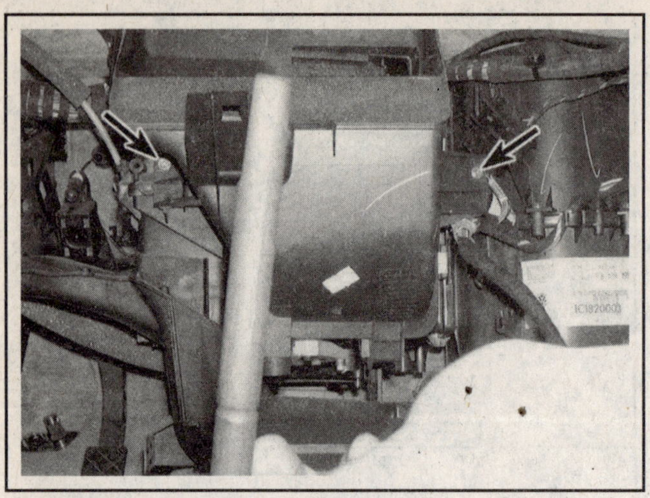

11.5a Detach the center air duct from the heater/air conditioning unit

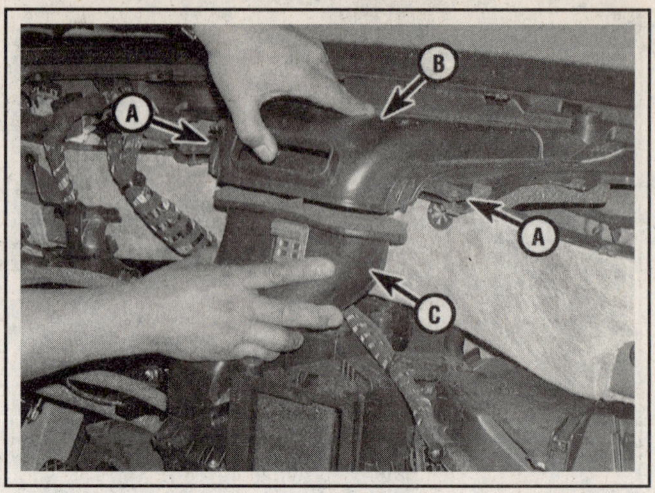

11.5b Remove the retaining screws (A), then lift the upper duct (B) to remove the middle duct (C)

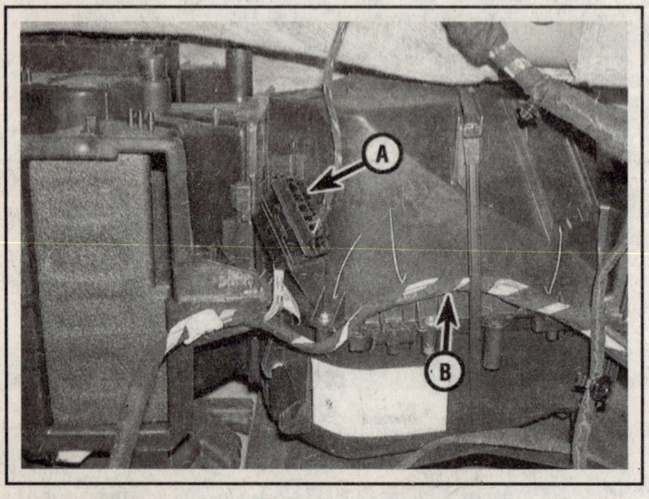

11.6 Disconnect the wiring harness connectors (A) and position the main wiring harness (B) aside

5 Remove the upper air ducts from the center of the heater/air conditioning unit (see illustrations).

6 Disconnect all electrical connectors from the heater/air conditioning unit and position the wiring harness aside (see illustration). Refer to Section 9 to disconnect the wires from the blower motor.

7 Working in the engine compartment, remove the air intake duct (see Chapter 4A), the engine cover (see Chapter 2A) and the cowl cover (see Chapter 11).

8 Disconnect the heater hoses at the heater core inlet and outlet on the engine side of the firewall (see illustration) and plug the open fittings. If the hoses are stuck to the pipes, cut them off and replace them with new ones upon installation.

9 Remove the air conditioning lines from the evaporator core fittings at the firewall. Also remove the mounting nuts securing the heater/air conditioning unit to the passenger side of the firewall (see illustration).

10 Remove the mounting nuts securing the heater/air conditioning unit to the driver's side of the firewall (see illustration). Once the heating/air conditioning unit is unbolted from the firewall it can be removed from the vehicle and set on a workbench.

11.8 Disconnect the heater core hoses (arrows) at the engine compartment firewall

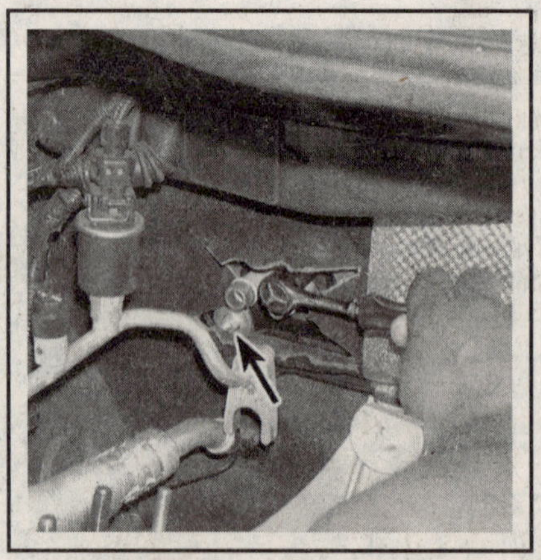

11.9 Disconnect the air conditioning lines at the firewall (if equipped), then remove the two mounting nuts on the passenger side of the engine compartment

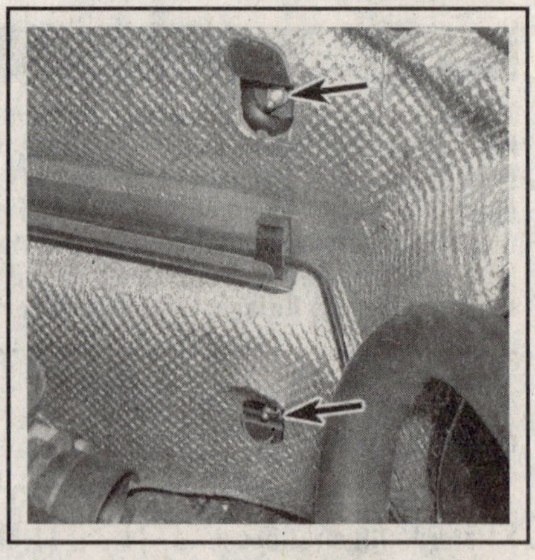

11.10 Heater/air conditioning unit mounting nuts - driver's side

11.11a Detach the heater core retaining screws (arrows) . . .

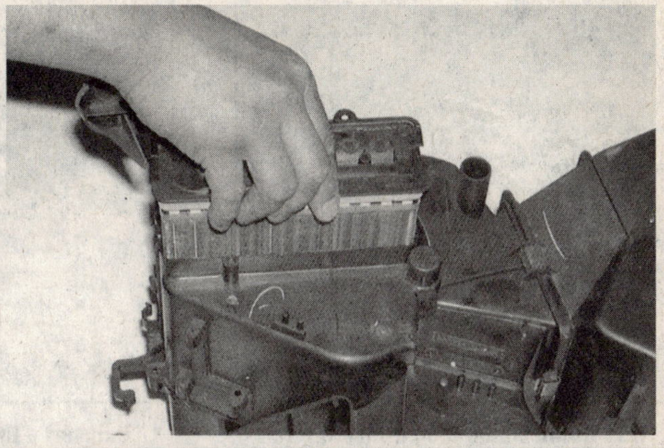

11.11b . . . and remove the heater core from the housing

11 Remove the heater core retaining screws and carefully remove the heater core from the heating/air conditioning unit (see illustrations).

INSTALLATION

12 Installation is the reverse of removal.

➡Note: When reinstalling the heater core, make sure any original insulating/sealing materials are in place around the heater core pipes and around the core.

13 Refill the cooling system (see Chapter 1).

14 Start the engine and check for proper operation.

12 Air conditioning and heating system - check and maintenance

AIR CONDITIONING SYSTEM

Refer to illustration 12.1

WARNING:

The air conditioning system is under high pressure. Do not loosen any hose fittings or remove any components until after the system has been discharged. Air conditioning refrigerant should be properly discharged into an EPA-approved recovery/recycling unit at a dealer service department or an automotive air conditioning repair facility. Always wear eye protection when disconnecting air conditioning system fittings.

CAUTION 1:

All models covered by this manual use environmentally friendly R-134a. This refrigerant (and its appropriate refrigerant oils) are not compatible R-12 refrigerant system components and must never be mixed or the components will be damaged.

CAUTION 2:

When replacing entire components, additional refrigerant oil should be added equal to the amount that is removed with the component being replaced. Be sure to read the can before adding any oil to the system, to make sure it is compatible with the R-134a system.

1 The following maintenance checks should be performed on a regular basis to ensure that the air conditioning continues to operate at peak efficiency.

a) *Inspect the condition of the compressor drivebelt. If it is worn or deteriorated, replace it (see Chapter 1).*
b) *Check the drivebelt tension and, if necessary, adjust it (see Chapter 1).*
c) *Inspect the system hoses. Look for cracks, bubbles, hardening and deterioration. Inspect the hoses and all fittings for oil bubbles or seepage. If there is any evidence of wear, damage or leakage, replace the hose(s).*
d) *Inspect the condenser fins for leaves, bugs and any other foreign material that may have embedded itself in the fins. Use a "fin comb" or compressed air to remove debris from the condenser.*
e) *Make sure the system has the correct refrigerant charge.*
f) *If you hear water sloshing around in the dash area or have water dripping on the carpet, check the evaporator housing drain tube (see illustration) and insert a piece of wire into the opening to check for blockage.*

2 It's a good idea to operate the system for about ten minutes at least once a month. This is particularly important during the winter months because long term non-use can cause hardening, and subsequent failure, of the seals. Note that using the Defrost function operates the compressor.

3 If the air conditioning system is not working properly, proceed to Step 6 and perform the general checks outlined below.

4 Because of the complexity of the air conditioning system and the special equipment necessary to service it, in-depth troubleshooting and repairs beyond checking the refrigerant charge and the compressor clutch operation are not included in this manual. However, simple checks and component replacement procedures are provided in this Chapter.

5 The most common cause of poor cooling is simply a low system refrigerant charge. If a noticeable drop in system cooling ability occurs, one of the following quick checks will help you determine whether the refrigerant level is low. Should the system lose its cooling ability, the following procedure will help you pinpoint the cause.

12.1 Remove the plug from the firewall heat shield and check that the evaporator housing drain flap (arrow) is clear of any blockage - the view here is from below the engine looking up

12.9 Insert a thermometer in the center duct while operating the air conditioning system - the output air should be 35-40 degrees F less than the ambient temperature, depending on humidity (but not lower than 40-degrees F)

12.18 A basic charging kit for 134a systems is available at most auto parts stores - it must say 134a (not R-12) and so should the can of refrigerant

Check

Refer to illustration 12.9

6 Warm the engine up to normal operating temperature.

7 Place the air conditioning temperature selector at the coldest setting and put the blower at the highest setting. Open the doors (to make sure the air conditioning system doesn't cycle off as soon as it cools the passenger compartment).

8 After the system reaches operating temperature, feel the two pipes connected to the evaporator at the firewall.

9 The pipe (thinner tubing) leading from the condenser outlet to the evaporator should be cold, and the evaporator outlet line (the thicker tubing that leads back to the compressor) should be slightly colder (3 to 10 degrees F colder). If the evaporator outlet is considerably warmer than the inlet, the system needs a charge. Insert a thermometer in the center air distribution duct (see illustration) while operating the air conditioning system at its maximum setting - the temperature of the output air should be 35 to 40 degrees F below the ambient air temperature (down to approximately 40 degrees F). If the ambient (outside) air temperature is very high, say 110 degrees F, the duct air temperature may be as high as 60 degrees F, but generally the air conditioning is 35 to 40 degrees F cooler than the ambient air.

10 If the air isn't as cold as it used to be, the system probably needs a charge.

11 If the air warm and the system doesn't seem to be operating properly check the operation of the compressor clutch.

12 Have an assistant switch the air conditioning On while you observe the front of the compressor. The clutch will make an audible click and the center of the clutch should rotate.

13 If the clutch does not operate, check the appropriate fuses. Inspect the fuses in the interior fuse panel.

14 If the clutch doesn't respond, disconnect the clutch connector at the compressor and check for battery voltage at the compressor clutch connector. There should be battery voltage with the air conditioning switched On.

15 Check for continuity to ground on the black wire terminal of the compressor clutch connector.

16 If power and ground are available and the clutch doesn't operate when connected, the compressor clutch is defective.

17 Further inspection or testing of the system is beyond the scope of the home mechanic and should be left to a professional.

Adding refrigerant

Refer to illustrations 12.18, 12.21 and 12.22

CAUTION:

Make sure any refrigerant, refrigerant oil or replacement component your purchase is designated as compatible with environmentally friendly R-134a systems.

18 Purchase a R-134a automotive charging kit at an auto parts store (see illustration). A charging kit includes a 12-ounce can of refrigerant, a tap valve and a short section of hose that can be attached between the tap valve and the system low side service valve. Because one can of refrigerant may not be sufficient to bring the system charge up to the proper level, it's a good idea to buy an additional can.

WARNING:

Never add more than two cans of refrigerant to the system.

19 Hook up the charging kit by following the manufacturer's instructions.

WARNING:

DO NOT hook the charging kit hose to the system high side! The fittings on the charging kit are designed to fit only on the low side of the system.

20 Back off the valve handle on the charging kit and screw the kit onto the refrigerant can, making sure first that the O-ring or rubber seal inside the threaded portion of the kit is in place.

WARNING:

Wear protective eyewear when dealing with pressurized refrigerant cans.

21 Remove the dust cap from the low-side charging and attach the quick-connect fitting on the kit hose (see illustration).

12.21 Attach the refrigerant kit to the low-side charging port (arrow) - it's near the front of the vehicle - the cap should be marked with an "L"

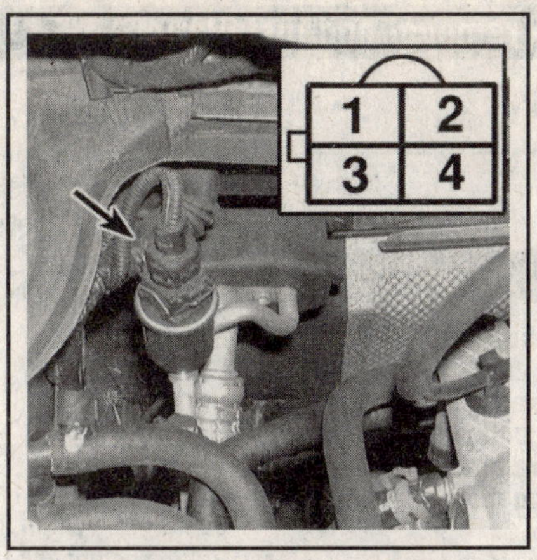

12.22 To engage the compressor, remove the pressure switch connector (arrow) and use a jumper wire to bridge terminals 1 and 2

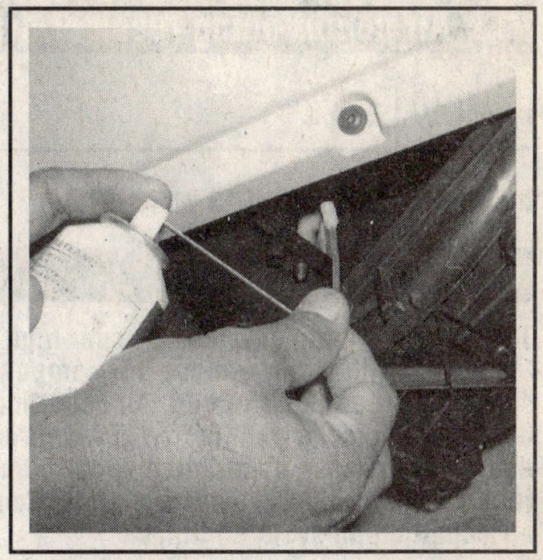

12.32 With the blower motor resistor removed, spray the disinfectant at the evaporator core

22 Warm up the engine and turn On the air conditioning. Keep the charging kit hose away from the fan and other moving parts.

➡Note: The charging process requires the compressor to be running. If the clutch cycles off, you can put the air conditioning switch on High and leave the car doors open to keep the clutch on and compressor working or use a jumper wire across the terminals of the pressure switch to keep the compressor engaged (see illustration).

23 Turn the valve handle on the kit until the stem pierces the can, then back the handle out to release the refrigerant. You should be able to hear the rush of gas. Add refrigerant to the low side of the system, keeping the can upright at all times, but shaking it occasionally. Allow stabilization time between each addition.

➡Note: The charging process will go faster if you wrap the can with a hot-water-soaked shop rag to keep the can from freezing up.

24 If you have an accurate thermometer, you can place it in the center air conditioning duct inside the vehicle and keep track of the output air temperature (see illustration 12.9). A charged system that is working properly should cool down to approximately 40-degrees F. If the ambient (outside) air temperature is very high, say 110 degrees F, the duct air temperature may be as high as 60 degrees F, but generally the air conditioning is 35-40 degrees F cooler than the ambient air.

25 When the can is empty, turn the valve handle to the closed position and release the connection from the low-side port. Replace the dust cap.

26 Remove the charging kit from the can and store the kit for future use with the piercing valve in the UP position, to prevent inadvertently piercing the can on the next use.

HEATING SYSTEMS

27 If the carpet under the heater core is damp, or if antifreeze vapor or steam is coming through the vents, the heater core is leaking. Remove it (see Section 11) and install a new unit (most radiator shops will not repair a leaking heater core).

28 If the air coming out of the heater vents isn't hot, the problem could stem from any of the following causes:

a) *The thermostat is stuck open, preventing the engine coolant from warming up enough to carry heat to the heater core. Replace the thermostat (see Section 3).*
b) *There is a blockage in the system, preventing the flow of coolant through the heater core. Feel both heater hoses at the firewall. They should be hot. If one of them is cold, there is an obstruction in one of the hoses or in the heater core, or the heater control valve is shut. Detach the hoses and back flush the heater core with a water hose. If the heater core is clear but circulation is impeded, remove the two hoses and flush them out with a water hose.*
c) *If flushing fails to remove the blockage from the heater core, the core must be replaced (see Section 11).*

ELIMINATING AIR CONDITIONING ODORS

➧ Refer to illustration 12.32

29 Unpleasant odors that often develop in air conditioning systems are caused by the growth of a fungus, usually on the surface of the evaporator core. The warm, humid environment there is a perfect breeding ground for mildew to develop.

30 The evaporator core on most vehicles is difficult to access, and factory dealerships have a lengthy, expensive process for eliminating the fungus by opening up the evaporator case and using a powerful disinfectant and rinse on the core until the fungus is gone. You can service your own system at home, but it takes something much stronger than basic household germ-killers or deodorizers.

31 Aerosol disinfectants for automotive air conditioning systems are available in most auto parts stores, but remember when shopping for them that the most effective treatments are also the most expensive. The basic procedure for using these sprays is to start by running the system in the RECIRC mode for ten minutes with the blower on its highest speed. Use the highest heat mode to dry out the system and keep the compressor from engaging by disconnecting the wiring connector at the compressor (see Section 14).

32 The disinfectant can usually comes with a long spray hose. Remove the blower motor resistor (see Section 8), point the nozzle inside the hole and to the left towards the evaporator core, and spray according to the manufacturer's recommendations (see illustration). Try to cover the whole surface of the evaporator core, by aiming the spray up, down and sideways. Follow the manufacturer's recommendations for the length of spray and waiting time between applications.

33 Once the evaporator has been cleaned, the best way to prevent the mildew from coming back again is to make sure your evaporator housing drain tube is clear (see illustration 12.1).

13 Air conditioning receiver/drier - removal and installation

REMOVAL

➧ **Refer to illustrations 13.3 and 13.4**

⁂ WARNING:

The air conditioning system is under high pressure. DO NOT loosen any fittings or remove any components until after the system has been discharged. Air conditioning refrigerant should be properly discharged into an EPA-approved container at a dealership service department or an automotive air conditioning repair facility. Always wear eye protection when disconnecting air conditioning system fittings.

1 Have the air conditioning system discharged (see **Warning** above).
2 Raise the vehicle and support it securely on jackstands.
3 Disconnect the refrigerant outlet line from the bottom of the receiver/drier (see illustration). Cap or plug the open line immediately to prevent the entry of dirt or moisture.
4 Disconnect the refrigerant inlet line from the top of the receiver/drier (see illustration), using an Allen wrench. Cap or plug the open line immediately to prevent the entry of dirt or moisture.
5 Loosen the clamp bolt (see illustration 13.4) on the mounting bracket and slide the receiver/drier assembly up and out of the engine compartment.

INSTALLATION

6 If you are replacing the receiver/drier with a new one, add 1/2 ounce (13.5 cc) of fresh refrigerant oil to the new unit (oil must be R-134a compatible).
7 Place the new receiver/drier into position in the bracket.
8 Install the inlet and outlet lines, using clean refrigerant oil on the new O-rings. Tighten the mounting bolts securely.
9 Have the system evacuated, recharged and leak tested by a dealership service department or an automotive air conditioning repair facility.

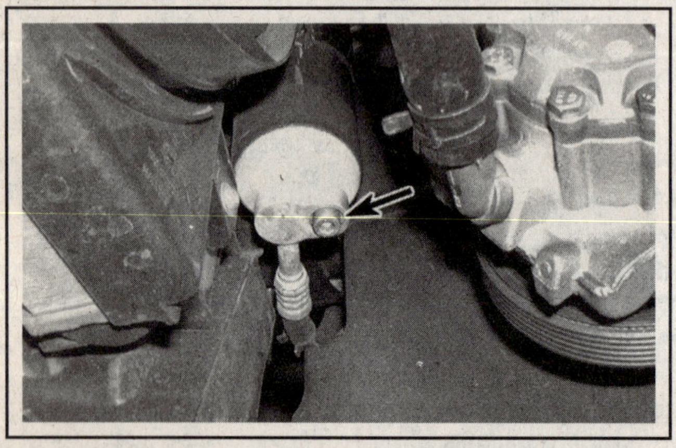

13.3 Disconnect the outlet line - view here is from below at front of the vehicle

13.4 Remove the inlet line (A) and the mounting bracket bolt (B) - view here is from above in the engine compartment

14 Air conditioning compressor - removal and installation

➡**Note: Whenever the compressor is replaced because of internal damage, the expansion valve should also be replaced (see Section 16).**

REMOVAL

➧ **Refer to illustrations 14.7 and 14.8**

⁂ WARNING:

The air conditioning system is under high pressure. DO NOT loosen any fittings or remove any components until after the system has been discharged. Air conditioning refrigerant should be properly discharged into an EPA-approved container at a dealership service department or an automotive air conditioning repair facility. Always wear eye protection when disconnecting air conditioning system fittings.

➡**Note: The receiver/drier (see Section 13) should be replaced whenever the compressor is replaced.**

1 Have the air conditioning system discharged (see **Warning** above).
2 Raise the vehicle and support it securely on jackstands.
3 Remove the splash shield from below the engine (if equipped).
4 Clean the compressor thoroughly around the refrigerant line fittings.
5 Remove the serpentine drivebelt (see Chapter 1).
6 Disconnect the electrical connector from the air conditioning compressor.
7 Disconnect the suction and discharge lines from the compressor. Both lines are mounted to the back of the compressor by one bolt for each line. Plug the open fittings to prevent the entry of dirt and moisture, and discard the seals between the plate and compressor (see illustration).
8 Remove the compressor mounting bolts (see illustration). Detach

the compressor from the mounting bracket and remove the compressor from the engine compartment.

INSTALLATION

9 If a new compressor is being installed, pour the oil from the old compressor into a graduated container and add that exact amount of new refrigerant oil to the new compressor. Also follow any directions included with the new compressor.

➡**Note: Some replacement compressors come with refrigerant oil in them. Follow the directions with the compressor regarding the draining of excess oil prior to installation.**

⁂ CAUTION:

The oil used must be labeled as compatible with R-134a refrigerant systems.

10 Installation is the reverse of the disassembly. When installing the line fitting bolt to the compressor, use new seals lubricated with clean refrigerant oil, and tighten the bolt securely.

11 Have the system evacuated, recharged and leak tested by a dealership service department or an automotive air conditioning repair facility.

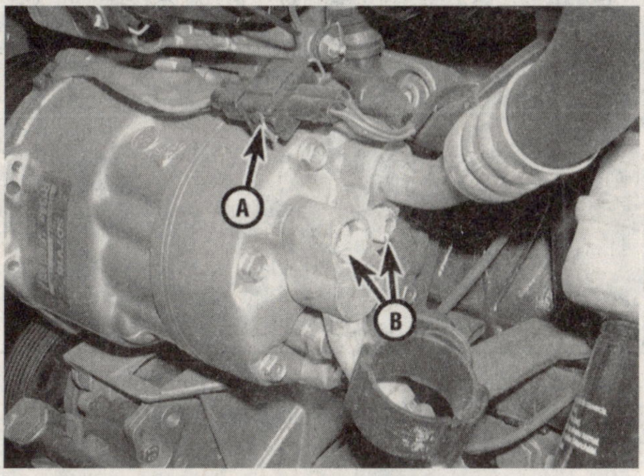

14.7 Disconnect the electrical connector (A) and the retaining bolts (B) securing the refrigerant lines to the back of the compressor

14.8 Compressor mounting bolts (arrows)

15 Air conditioning condenser - removal and installation

➧ **Refer to illustration 15.5**

⁂ WARNING:

The air conditioning system is under high pressure. DO NOT loosen any fittings or remove any components until after the system has been discharged. Air conditioning refrigerant should be properly discharged into an EPA-approved container at a dealership service department or an automotive air conditioning repair facility. Always wear eye protection when disconnecting air conditioning system fittings.

1 Have the air conditioning system discharged (see **Warning** above).

2 Raise the vehicle and support it securely on jackstands.

3 Remove the splash shield from below the engine (if equipped).

4 Remove the front fenders and bumper cover assembly (see Chapter 11).

5 Disconnect the condenser inlet and outlet lines (see illustration). The condenser outlet line fastens to the top of the receiver/drier (see illustration 13.4).

6 Detach the lock carrier panel (see illustration 5.8).

7 Remove the condenser mounting bolts (see illustration 5.9) and separate the condenser from the radiator.

8 Installation is the reverse of removal. Always use new O-rings on air conditioning system fittings. If you are replacing the condenser with a new one, add 1/2 ounce (13.5 cc) of fresh refrigerant oil to the new unit (oil must be R-134a compatible).

9 Have the system evacuated, recharged and leak tested by a dealership service department or an automotive air conditioning repair facility.

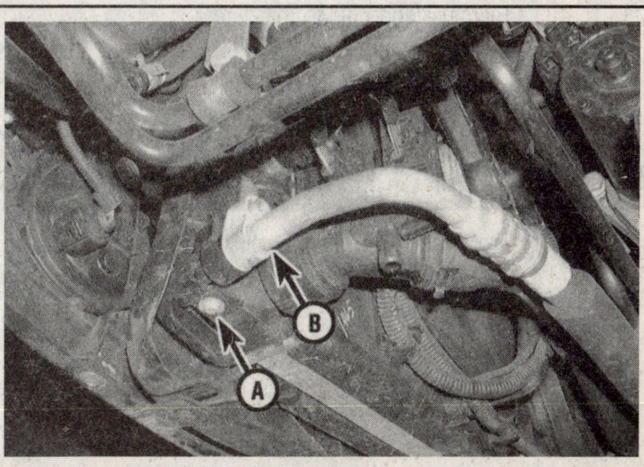

15.5 Condenser inlet line retaining bracket (A) and fitting (B)

16 Air conditioning expansion valve - removal and installation

Refer to illustration 16.3

WARNING:

The air conditioning system is under high pressure. DO NOT loosen any fittings or remove any components until after the system has been discharged. Air conditioning refrigerant should be properly discharged into an EPA-approved container at a dealership service department or an automotive air conditioning repair facility. Always wear eye protection when disconnecting air conditioning system fittings.

1 Have the air conditioning system discharged and the refrigerant recovered (see **Warning** above).

2 Remove the engine cover.

3 Disconnect the evaporator line fitting at the firewall (see illustration). Cap or plug the open lines immediately to prevent the entry of dirt or moisture.

4 Detach the expansion valve retaining bolt. Remove the expansion valve and O-rings. Cap or plug the open lines immediately to prevent the entry of dirt or moisture into the evaporator core.

5 Installation is the reverse of removal. Always use a new O-ring when installing the expansion valve.

6 Retighten the refrigerant lines, then have the system evacuated, recharged and leak tested by a dealership service department or an automotive air conditioning repair facility.

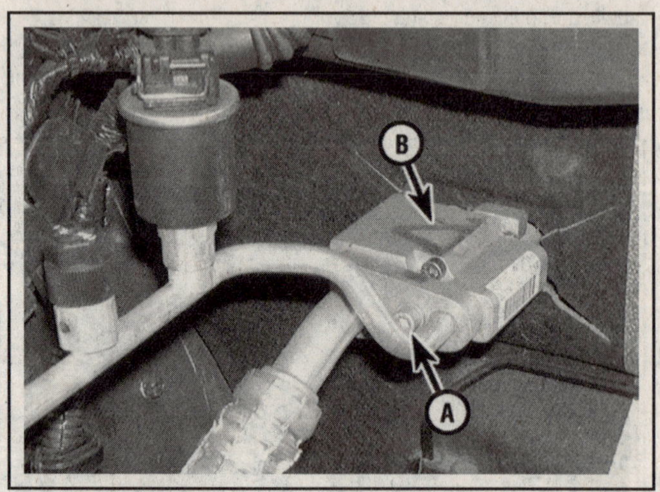

16.3 Working in the engine compartment, disconnect the refrigerant lines (A) and the expansion valve (B)

Specifications

General

Coolant capacity	See Chapter 1
Expansion tank pressure cap rating	15 psi
Thermostat opening temperature	185-degrees F (85-degrees C)
Refrigerant type	R-134a
Refrigerant capacity	24.7 ounces (700 grams)

Torque specifications

Thermostat housing nuts/bolts		
Through 2004 models	10	84 in-lbs
2005	15	141 in-lbs
Coolant outlet pipe-to-cylinder head	10 Nm	84 in-lbs
Water pump attaching bolts	15 Nm	132 in-lbs

4A

FUEL AND EXHAUST SYSTEMS: GASOLINE ENGINES

Section

Reference to other Chapters

Air filter replacement - See Chapter 1
CHECK ENGINE light - See Chapter 6
Exhaust system check - See Chapter 1
Fuel system check - See Chapter 1

1 General information

Refer to illustration 1.1

WARNING:

Gasoline is extremely flammable, so take extra precautions when you work on any part of the fuel system. Don't smoke or allow open flames or bare light bulbs near the work area, and don't work in a garage where a gas-type appliance (such as a water heater or a clothes dryer) is present. Since gasoline is carcinogenic, wear latex gloves when there's a possibility of being exposed to fuel, and, if you spill any fuel on your skin, rinse it off immediately with soap and water. Mop up any spills immediately and do not store fuel-soaked rags where they could ignite. The fuel system is under constant pressure, so, if any fuel lines are to be disconnected, the fuel pressure in the system must be relieved first. When you perform any kind of work on the fuel system, wear safety glasses and have a Class B type fire extinguisher on hand.

All gasoline models covered by this manual are equipped with a Bosch Motronic fuel injection system (see illustration). This system uses timed impulses to sequentially inject the fuel directly into the intake ports of each cylinder. The injectors are controlled by the Engine Control Module (ECM). The ECM monitors various engine parameters and delivers the exact amount of fuel, in the correct sequence, into the intake ports. This Chapter's information pertains to the air and fuel delivery components of the system only. Refer to Section 12 for additional general information regarding the fuel injection system. Refer to Chapter 6A for information regarding the electronic control system.

All models are equipped with an electric fuel pump, mounted in the fuel tank. Access to the fuel pump is provided through an access hole under the rear seat cushion. The fuel level sending unit is an integral component of the fuel pump module and it must be removed from the fuel tank in the same manner.

The exhaust system consists of an exhaust manifold, a catalytic converter, an exhaust pipe and a muffler. Each of these components is replaceable. For further information regarding the catalytic converter, refer to Chapter 6A.

1.8L models are equipped with a turbocharger and intercooler. The turbocharger increases power by using an exhaust gas driven turbine to pressurize the intake charge before it enters the combustion chambers. The amount of intake manifold pressure (boost) is regulated by an exhaust by-pass valve (wastegate). The wastegate is controlled by the ECM. The heated compressed air is routed through an air-to-air radiator (intercooler). The intercooler removes excess heat from the compressed air, increasing its density and allowing for more boost pressure.

1.1 Typical fuel system components (2.0L gasoline engine shown)

1 *Fuel feed and return lines*
2 *Throttle body/throttle control module*
3 *Fuel rail and injectors*
4 *Air filter housing*
5 *Accelerator cable*
6 *Fuel pressure regulator*

2 Fuel pressure relief procedure

➧ Refer to illustration 2.3

⁂ WARNING:

See the Warning in Section 1.

➡Note: After the fuel pressure has been relieved, it's a good idea to lay a shop towel over any fuel connection to be disassembled, to absorb the residual fuel that may leak out when servicing the fuel system.

1 Before servicing any fuel system component, you must relieve the fuel pressure to minimize the risk of fire or personal injury.

2 Remove the fuel filler cap - this will relieve any pressure built up in the tank.

3 Remove the fuel pump fuse (no. 28) from the fuse box (see illustration).

4 Attempt to start the engine. The engine should immediately stall. Continue to crank the engine for approximately three seconds.

5 Turn the ignition Off and remove the key.

6 Place shop towels around the fuel fitting to be disconnected to absorb any residual fuel that may spill out.

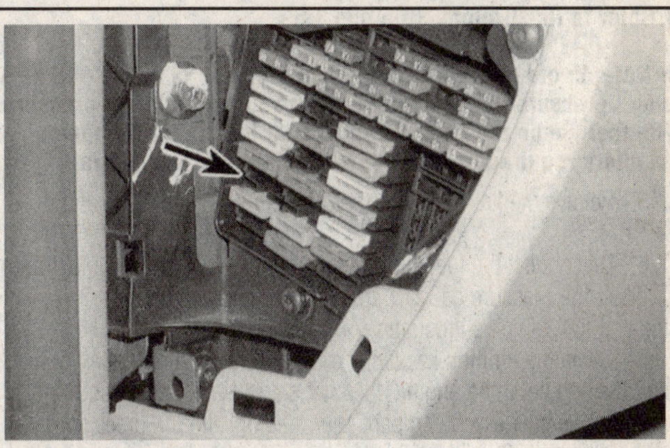

2.3 To disable the fuel pump, remove fuse no. 28 (arrow) from the fuse box

7 After completing the service or repair, install the fuel pump fuse.

3 Fuel pump/fuel pressure - check

⁂ WARNING:

See the Warning in Section 1.

PRELIMINARY CHECK

➧ Refer to illustration 3.3

1 If you suspect insufficient fuel delivery, check the following items first:

a) Check the battery and make sure it's fully charged (see Chapter 5).
b) Check the fuel pump fuse (no. 28).
c) Check the fuel filter for restriction.
d) Inspect all fuel lines to ensure that the problem is not simply a leak in a line.

2 Place the transmission in Park (automatic) or Neutral (manual) and apply the parking brake. Have an assistant cycle the ignition key On and Off several times (or attempt to start the engine, if the engine does not start) while you listen for the sound of the fuel pump operating inside the fuel tank. Remove the rear seat cushion and the fuel pump access cover and listen at, or feel the top of the fuel pump module, if necessary. You should hear a "whirring" sound indicating the fuel pump is operating. If the fuel pump is operating, proceed to the pressure check.

3 If there is no sound, remove the fuel pump access cover and disconnect the fuel pump electrical connector. Connect a test light or voltmeter to terminals 1 and 4 of the fuel pump harness connector (see illustration). Cycle the ignition key On and Off several times (or attempt to start the engine, if the engine does not start) - battery voltage should be indicated. If battery voltage is not indicated, check the fuel pump circuit, referring to Chapter 12 and the wiring diagrams. Check the related fuses, the fuel pump relay and the related wiring to ensure power is reaching the fuel pump connector. Check the ground circuit for continuity.

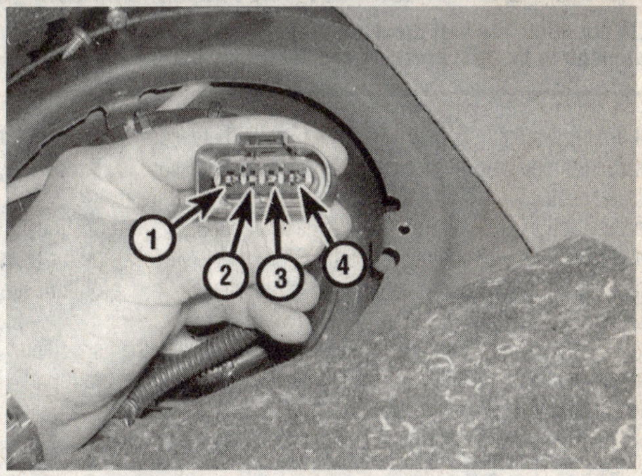

3.3 Check for battery power across terminals 1 and 4 of the fuel pump connector with the ignition key On

4 If the power and ground circuits are good and the fuel pump does not operate, remove the fuel pump and check for open circuits in the fuel pump module wiring and connectors. If the wiring and connectors are good, replace the fuel pump (see Section 7).

PRESSURE CHECK

Refer to illustrations 3.6a and 3.6b

Note: In order to perform the fuel pressure test, you will need a fuel pressure gauge capable of measuring high fuel pressure. The fuel gauge must be equipped with the proper fittings or adapters required to attach it to the fuel line and fuel rail.

5 Relieve the fuel pressure (see Section 2). Remove the engine cover.

6 Disconnect the fuel supply line from the fuel rail (see illustration). Connect the pressure gauge with a T-fitting and adapter hose to the fuel line and fuel rail (see illustration).

7 Cycle the ignition key On and Off several times (or attempt to start the engine, if the engine does not start). Note the pressure indicated on the gauge and compare your reading with the pressure listed in this Chapter's Specifications.

8 If the fuel pressure is lower than specified, pinch-off the fuel return line.

CAUTION:

Use special pliers designed specifically for pinching a rubber fuel line (available at most auto parts stores). Use of any other type pliers may damage the fuel line.

Cycle the ignition key On and Off several times and note the fuel pressure.

CAUTION:

Do not allow the fuel pressure to rise above 65 psi (448 kPa) or damage to the fuel pressure regulator may occur.

If the fuel pressure is now above the specified pressure, replace the fuel pressure regulator (see Section 15). If the fuel pressure is still lower than specified, check the fuel lines and the fuel filter for restrictions. If no restriction is found, remove the fuel pump module (see Section 7) and check the fuel strainer for restrictions, check the fuel pipe for leaks and check the fuel pump wiring for high resistance. If no problems are found, replace the fuel pump.

9 If the fuel pressure recorded in Step 7 is higher than specified, check the fuel return line for restrictions. If no restrictions are found, replace the fuel pressure regulator (see Section 15).

10 If the fuel pressure is within specifications, start the engine.

WARNING:

Make sure the fuel pressure gauge hose is positioned away from the engine drivebelt before starting the engine.

With the engine running, the fuel pressure should be 3 to 10 psi (21 to 69 kPa) below the pressure recorded in Step 7. If it isn't, remove the vacuum hose from the fuel pressure regulator and verify there is 12 to 14 in-Hg (305 to 356 mm-Hg) of vacuum present at the hose. If vacuum is not present at the hose, check the hose for a restriction or a break. If vacuum is present, reconnect the hose to the fuel pressure regulator. If the fuel pressure regulator does not decrease the fuel pressure with vacuum applied, replace the fuel pressure regulator.

3.6a To test the fuel pressure, remove the fuel supply line from the fuel rail . . .

11 Now check the fuel system hold pressure. Turn the engine off and monitor the fuel pressure for ten minutes; the fuel pressure should not drop more than 10 psi (69 kPa) within ten minutes. If it does, there is a leak in the fuel line, a fuel injector is leaking, the fuel pump check valve is defective or the fuel pressure regulator is defective. To determine the source of the leak, cycle the ignition key On and Off several times to obtain the highest fuel pressure reading, then immediately pinch-off the fuel supply hose between the fuel gauge T-fitting and the fuel rail. If the pressure drops below 10 psi (69 kPa) within ten minutes, the main fuel line is leaking or the fuel pump is defective. If the pressure holds, remove the clamp from the supply line, pressurize the system and clamp off the return line. If the pressure drops below 10 psi (69 kPa) within ten minutes, an injector is probably leaking (or the fuel rail is leaking, but such a leak should be very apparent). If the pressure holds, remove the clamp from the return line. If the pressure now drops, the fuel pressure regulator is defective.

3.6b . . . and attach a fuel pressure gauge between the fuel line and fuel rail with a T-fitting

4 Fuel lines and fittings - repair and replacement

Refer to illustrations 4.2 and 4.9

WARNING:

See the Warning in Section 1.

1 Always relieve the fuel pressure before servicing fuel lines or fittings (see Section 2).

2 Special flexible fuel supply, return and vapor lines extend from the fuel tank to the engine compartment. The lines are color coded and secured to the underbody with a plastic retainer (see illustration). Rubber hose completes the connection from the engine compartment junction block to the fuel rail. All fuel lines must be occasionally inspected for leaks or damage.

3 If evidence of contamination is found in the system or fuel filter during disassembly, the line should be disconnected and blown out. Check the fuel strainer on the fuel pump for damage and deterioration.

4 Don't route fuel line or hose within four inches of any part of the exhaust system or within ten inches of the catalytic converter. Fuel line must never be allowed to chafe against the engine, body or frame. A minimum of 1/4-inch clearance must be maintained around a fuel line.

5 Because fuel lines used on fuel-injected vehicles are under high pressure, they require special consideration.

6 In the event of fuel line damage, it is necessary to replace the damaged flexible lines with factory replacement parts. Others may fail from the high pressures of this system.

7 When replacing a fuel line, remove all fasteners attaching the fuel line to the vehicle body and route the new line exactly as originally installed.

8 When replacing rubber hose, always use hose specifically designated as fuel hose and replace the hose clamp with a new one.

9 If the fuel line is equipped with a quick-connect fitting, clean any debris from around the fitting and twist the fitting back-and-forth to loosen the seal. Press in on the tab to disconnect the fitting and carefully remove the fuel line from the vehicle (see illustration).

CAUTION:

The quick-connect fittings are not serviced separately. Do not attempt to repair these types of fuel lines in the event the fitting or line becomes damaged. Replace the entire fuel line as an assembly.

10 Installation is the reverse of removal with the following additions:

a) *Clean the quick-connect fittings with a lint-free cloth and apply clean engine oil to the fittings.*

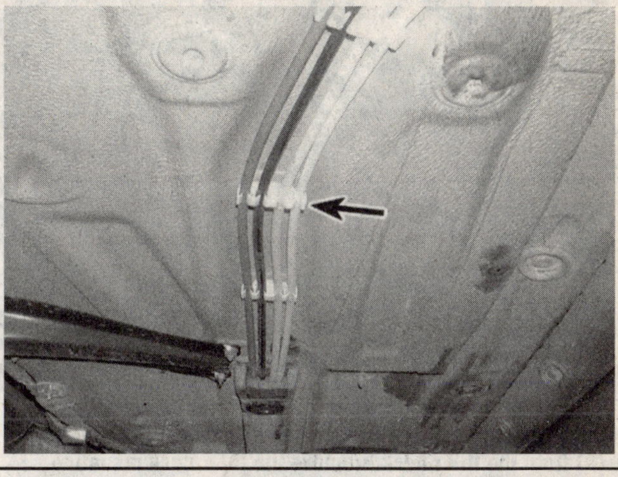

4.2 The fuel lines are secured to the underbody with plastic retainers - the fuel lines are color coded; the white lines are EVAP system vapor lines, the black line is the fuel supply line and the blue line is the fuel return line

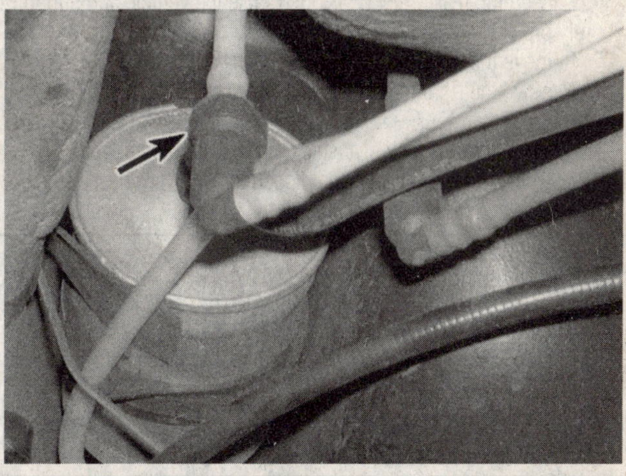

4.9 To disconnect a quick-connect fitting, press the tab in (arrow) and pull the lines apart

b) *After connecting a quick-connect fitting, check the integrity of the connection by attempting to pull the lines apart.*

c) *Cycle the ignition key On and Off several times and check for leaks at the fitting, before starting the engine.*

5 Fuel tank - removal and installation

Refer to illustrations 5.10, 5.11, 5.12 and 5.13

WARNING:

See the Warning in Section 1.

1 Remove the fuel tank filler cap to relieve fuel tank pressure.

2 Relieve the fuel system pressure (see Section 2).

3 Using a siphoning kit (available at most auto parts stores), siphon the fuel into an approved gasoline container.

WARNING:

Do not start the siphoning action by mouth!

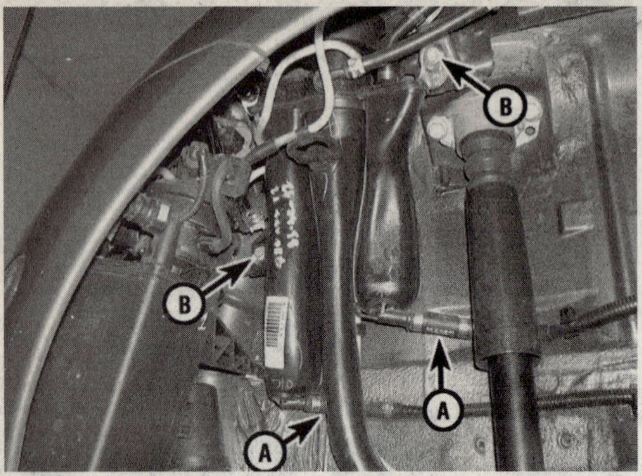

5.10 Loosen the hose clamps and disconnect the vent hoses (A) from the filler neck - remove the filler neck mounting bolts (B)

5.11 Disconnect the fuel supply line quick-connect fitting from the fuel filter

4 Remove the rear seat cushion (see Chapter 11).

5 Remove the fuel pump access cover and disconnect the fuel pump electrical connector (see Section 3).

6 Loosen the rear wheel bolts, then raise the vehicle and support it securely on jackstands.

7 Remove the rear axle (see Chapter 10).

8 Remove the right rear fenderwell liner (see Chapter 11, Section 13, Step 3).

9 Remove the fuel tank heat shield and cover, if equipped.

10 Loosen the hose clamps and disconnect the vent hoses from the fuel tank filler neck (see illustration). Remove the fuel filler neck mounting bolts. Detach the overflow valve from the bracket on the inner fender.

11 Disconnect the fuel supply line from the fuel filter (see illustration). Disconnect the return line from the fuel line at the body (see Section 4).

12 Disconnect the EVAP hose from the canister (see illustration). Remove the screw from the fender brace and position the EVAP hose outside the brace. Detach the vapor line from the clip on the filler neck.

13 Position a transmission jack under the fuel tank and support the tank. Remove the fuel tank strap bolts and remove the straps (see illustration).

14 Lower the jack and remove the tank from the vehicle.

15 Installation is the reverse of removal.

5.12 Squeeze the tabs and disconnect the EVAP hose (A) from the canister, remove the fender brace screw (B) and move the EVAP hose outside the brace - detach the vapor line from the clip (C) at the filler neck

5.13 Remove the fuel tank strap bolts (arrows) and remove the straps

6 Fuel tank cleaning and repair - general information

1 The fuel tanks installed in the vehicles covered by this manual are not repairable. If the fuel tank becomes damaged, it must be replaced.

2 Cleaning the fuel tank (due to fuel contamination) should be performed by a professional with the proper training to carry out this critical and potentially dangerous work. Even after cleaning and flushing, explosive fumes may remain inside the fuel tank.

3 If the fuel tank is removed from the vehicle, it should not be placed in an area where sparks or open flames could ignite the fumes coming out of the tank. Be especially careful inside a garage where a gas-type appliance is located.

7 Fuel pump - removal and installation

Refer to illustrations 7.2, 7.3, 7.5 and 7.6

WARNING:

See the Warning in Section 1.

1 Relieve the fuel system pressure (see Section 2).

2 Remove the rear seat cushion (see Chapter 11). Remove the fuel pump access cover (see illustration).

3 Disconnect the fuel pump electrical connector. Referring to Section 4, disconnect the fuel supply and return lines from the fuel pump module (see illustration).

Note: The fuel supply line is identified with a black mark. The fuel return line is identified with a blue mark.

4 Apply alignment marks on the body and fuel pump module flange so the fuel pump module can be installed in the original position.

5 Loosen the fuel pump module retaining ring (see illustration).

6 Remove the fuel pump module from the tank (see illustration). Angle the assembly slightly to avoid damaging the fuel level sending unit float.

WARNING:

Some fuel may remain in the module reservoir and spill as the module is removed. Have several shop towels ready and a drain pan nearby to place the module in.

7 The electric fuel pump is not serviced separately. In the event of failure, the complete assembly must be replaced. Transfer the fuel level sending unit to the new fuel pump module assembly, if necessary (see Section 8).

8 Clean the fuel tank sealing surface and install a new seal on the fuel pump module.

9 Install the fuel pump module, aligning the marks made in Step 4.

10 Press the fuel pump module down until seated and tighten the retaining ring.

11 The remainder of installation is the reverse of removal.

7.2 Remove the screws (arrows) and the fuel pump access cover

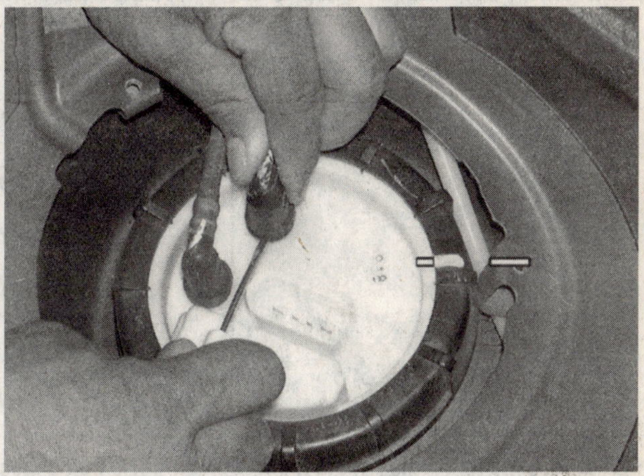

7.3 Press the tab in and disconnect the fuel supply line and the fuel return line from the fuel pump module - note the alignment marks on the fuel pump flange, retaining ring and body panel

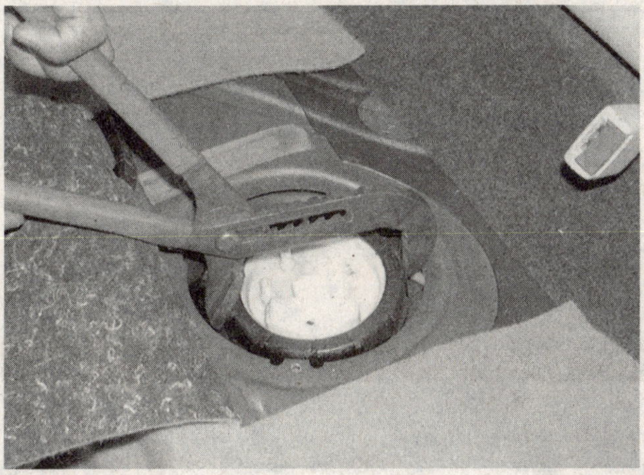

7.5 Using an appropriate tool, loosen the fuel pump module retaining ring

7.6 Carefully remove the fuel pump module from the tank and drain the fuel from the reservoir

8 Fuel level sending unit - check and replacement

WARNING:

See the Warning in Section 1.

CHECK

Refer to illustration 8.2

1 Remove the fuel pump module (see Section 7).

2 Connect the probes of an ohmmeter to the two fuel level sensor terminals (2 and 3) of the fuel pump module electrical connector (see illustration).

3 Position the float in the down (empty) position and note the reading on the ohmmeter.

4 Move the float up to the full position while watching the meter.

5 If the fuel level sending unit resistance does not change smoothly as the float travels from empty to full, replace the fuel level sending unit assembly.

REPLACEMENT

Refer to illustrations 8.7 and 8.8

6 Remove the fuel pump module (see Section 7).

7 Disconnect the two fuel level sending unit wire terminals from the module connector terminals under the flange (see illustration). Detach the wires from the retaining clips, noting the routing of the wiring for reinstallation.

8 Using a small screwdriver, pry the retaining tabs up and slide the fuel level sending unit off the module (see illustration).

9 Install the fuel level sending unit onto the fuel pump module. Slide the unit up until the retaining tabs snap into place. Connect the two wire terminals to the module connector. Route the wires as originally installed and secure them to the retaining clips.

10 The remainder of installation is the reverse of removal.

8.2 Check the resistance of the fuel level sending unit across terminals 2 and 3

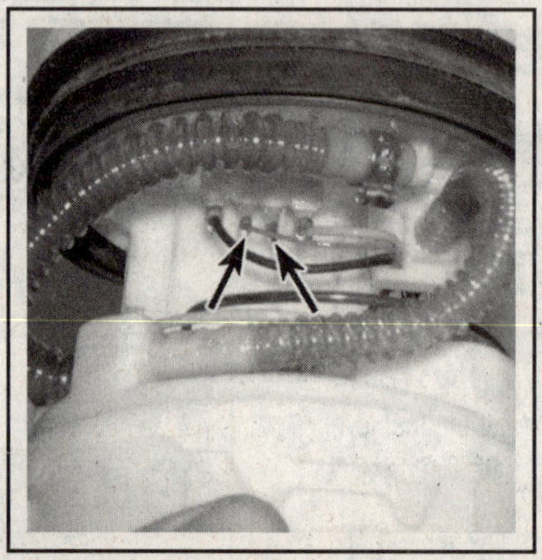

8.7 Disconnect the two fuel level sending unit wire terminals from the fuel pump module

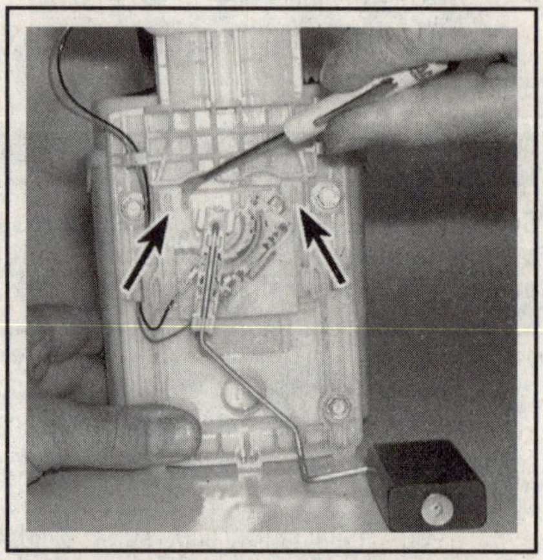

8.8 Pry the retaining tabs up and slide the fuel level sending unit off the module

9 Air filter housing - removal and installation

Refer to illustrations 9.1, 9.3 and 9.5

1 Remove intake duct resonator screw and the air filter cover screws (see illustration).

2 Lift the back of the cover up and detach the cover from the retainers along the front edge.

3 Disconnect the electrical connector from the Mass Airflow sensor. Loosen the hose clamp and detach the mass airflow sensor from the air intake duct (see illustration). Remove the cover with the mass airflow sensor attached.

4 Remove air filter element.

5 Disconnect the secondary air injection pump hose from the air filter housing. Remove the mounting bolts, pull the housing up and detach the duct from the inner fender (see illustration). Remove the assembly from the engine compartment.

6 If necessary, disconnect the electrical connector from the crankcase ventilation heater, detach the breather hoses and separate the air intake duct from the throttle body.

7 Installation is the reverse of removal.

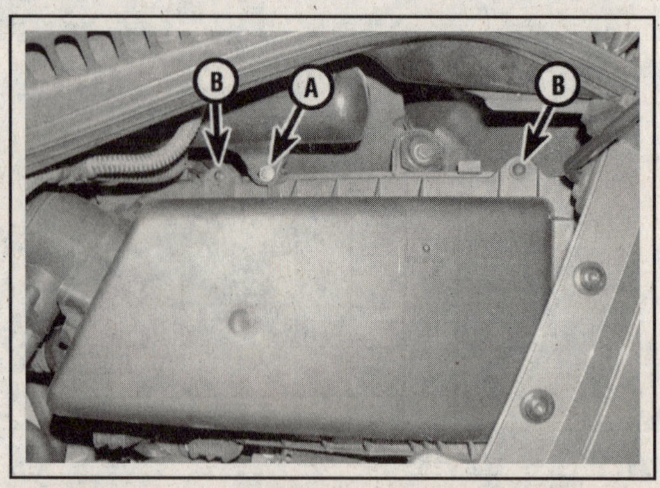

9.1 Remove the intake duct resonator screw (A) and the air filter cover screws (B) - screw locations vary with model year

9.3 Disconnect the electrical connector from the mass airflow sensor (A), squeeze the tabs on the hose clamp together (B) and separate the Mass Airflow sensor from the air intake duct

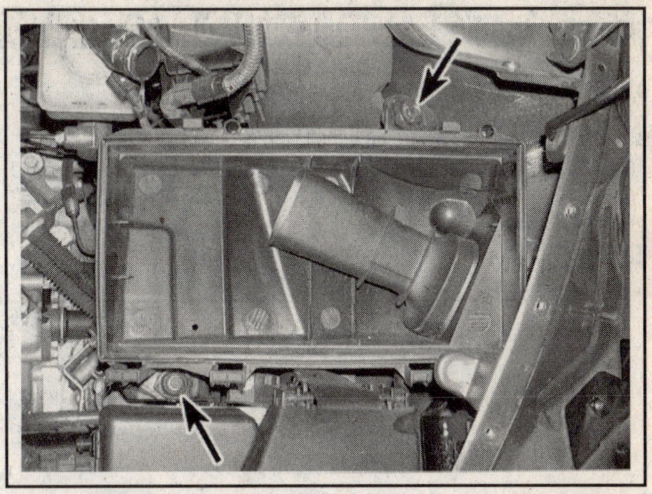

9.5 Remove the air filter housing mounting screws (arrows)

10 Accelerator cable (2.0L engine) - replacement

Refer to illustrations 10.2 10.3 and 10.4

Note: Some 2000 and later models have Electronic Power Control, in which the throttle control is operated by an electric motor on the throttle body. There is no mechanical cable in the system (see Section 11).

1 Remove the engine cover.

2 Rotate the throttle lever and separate the accelerator cable end from the throttle lever (see illustration).

3 Detach the cable grommet from the bracket (see illustration). Detach the cable from the cable retainers on the intake manifold.

4 Remove the trim panel from under the dash and detach the cable from the accelerator pedal (see illustration).

5 Push the cable and the firewall grommet through the firewall and into the engine compartment.

6 Remove the cable from the engine compartment.

7 Installation is the reverse of removal, but before installing the cable grommet on the bracket, adjust the cable as follows:

a) Have an assistant fully depress the accelerator pedal.

b) Remove the retaining clip.

c) Adjust the cable by rotating the grommet on the cable housing until the slot on the grommet is aligned with the bracket and throttle plate is at wide open throttle.

d) Install the cable grommet onto the bracket.

e) Install the retaining clip.

f) Have your assistant release and depress the accelerator pedal several times. Check the throttle lever and make sure it contacts both the closed throttle and wide open throttle stops, if it doesn't, readjust the cable.

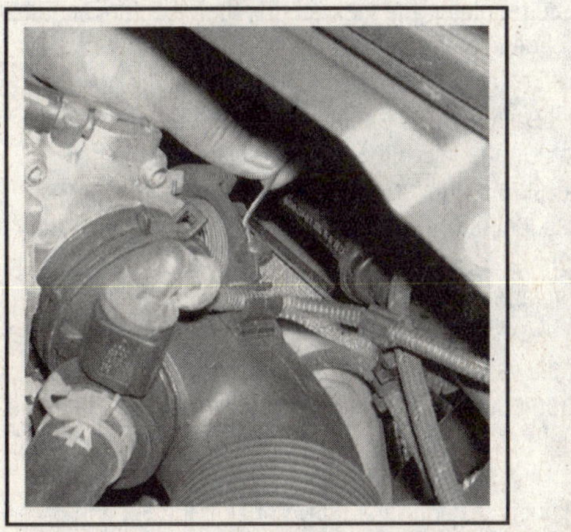

10.2 Rotate the throttle lever and pass the cable through the slot in the lever

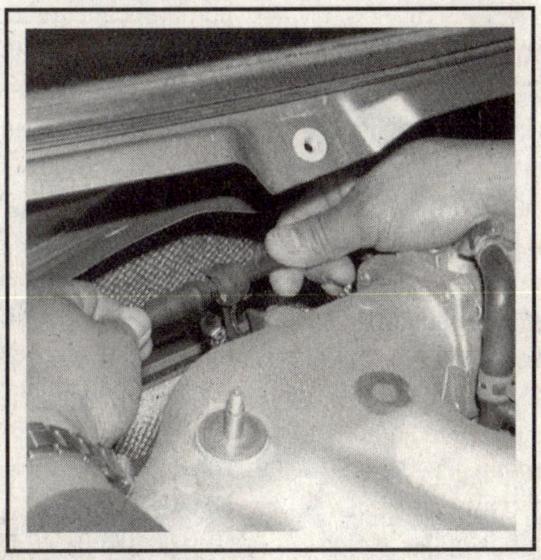

10.3 Remove the cable grommet from the bracket

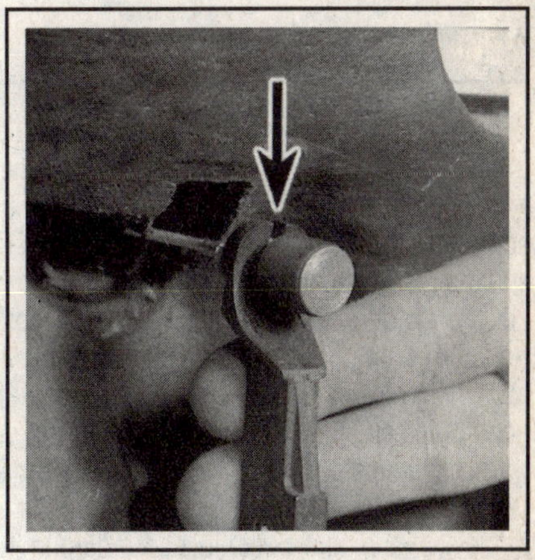

10.4 Pull the accelerator cable retainer out of the pedal and slide the cable through the slot (arrow)

11 Electronic accelerator pedal module (1.8L turbo engine) - replacement

Refer to illustration 11.4

1 1.8L turbo models are equipped with an electronically controlled accelerator system, called Electronic Power Control (EPC). The system consists of the accelerator pedal module, the throttle control module and the ECM. The system does not use the traditional accelerator cable. The ECM controls the throttle position based on the voltage signal received from the accelerator pedal module. Refer to Chapter 6A for more information on the electronic accelerator system.

2 Remove the insulation panel in the driver's footwell.

3 Disconnect the electrical connector from the pedal module.

4 Remove the mounting nuts and remove the module from the footwell (see illustration).

5 Installation is the reverse of removal.

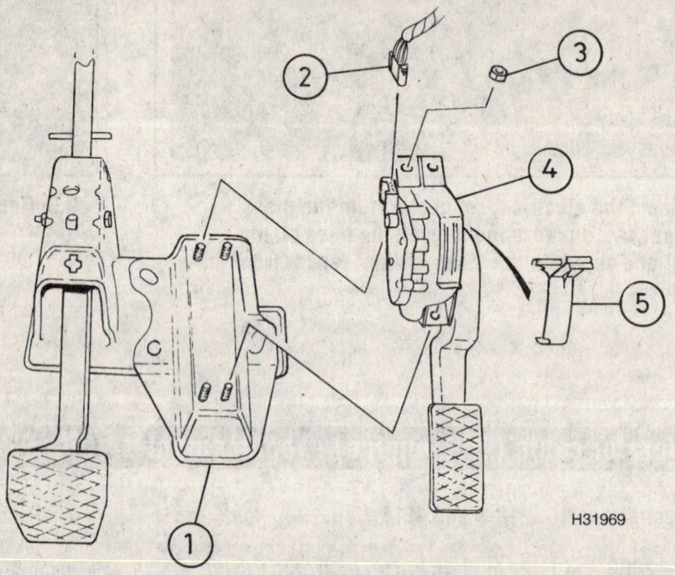

11.4 Electronic pedal module and related components (1.8L turbo)

1 *Pedal bracket*
2 *Pedal module electrical connector*
3 *Mounting nut*
4 *Accelerator pedal position sensor*
5 *Footwell panel bracket*

12 Fuel injection system - general information

The fuel injection system consists of three sub-systems: air intake, engine control and fuel delivery. The system uses an Engine Control Module (ECM) along with the sensors (coolant temperature sensor, throttle position sensor, mass airflow sensor, oxygen sensor, etc.) to determine the proper air/fuel ratio under all operating conditions.

The fuel injection system and the engine control system are closely linked in function and design. For additional information, refer to Chapter 6A.

AIR INTAKE SYSTEM

The air intake system consists of the air filter, the air intake ducts, the throttle body, the air intake plenum and the intake manifold.

When the engine is idling, the air/fuel ratio is controlled by the idle control system, which consists of the Engine Control Module (ECM) and the throttle position actuator. The throttle position actuator is an electric motor contained within the throttle control module and controlled by the ECM. The ECM commands the throttle position actuator to open or close the throttle plate depending upon the running conditions of the engine (air conditioning system, power steering, cold and warm running etc.). The ECM receives information from the sensors (vehicle speed, coolant temperature, air conditioning, power steering mode etc.) and adjusts the idle according to the demands of the engine and driver. Refer to Chapter 6A for information on the throttle control module.

EMISSIONS AND ENGINE CONTROL SYSTEM

The emissions and engine control system is described in detail in Chapter 6A.

FUEL DELIVERY SYSTEM

The fuel delivery system consists of these components: the fuel pump, the fuel pressure regulator, the fuel rail, the fuel injectors and the associated hoses and lines.

The fuel pump is an electric type located in the fuel tank. Fuel is drawn through an inlet screen into the pump, flows through the one-way valve, passes through the fuel filter and is delivered to the fuel rail and injectors. The pressure regulator maintains a constant fuel pressure to the injectors. Excess fuel is routed back to the fuel tank through the fuel pressure regulator.

The injectors are solenoid-actuated pintle type consisting of a solenoid, plunger, needle valve and housing. When current is applied to the solenoid coil, the needle valve raises and pressurized fuel sprays out the nozzle. The injection quantity is determined by the length of time the valve is open (the length of time during which current is supplied to the solenoid coils).

The fuel pump relay is located in the relay panel at the left end of the instrument panel. The fuel pump relay connects battery voltage to the fuel pump. The ECM controls the fuel pump relay. If the ECM senses there is NO signal from the engine speed sensor (as with the engine not running or cranking), the ECM will de-energize the relay.

13 Fuel injection system - check

➧ Refer to illustrations 13.7, 13.8, 13.9 and 13.10

➡Note: The following procedure is based on the assumption that the fuel pressure is adequate (see Section 3).

1 Check all electrical connectors that are related to the system. Check the ground wire connections for tightness. The main engine ground point is located on the transaxle case. Other ground points in the engine compartment are located on the left side of the engine compartment under the battery tray and on the left side of the engine compartment behind the headlight. Loose connectors and poor grounds can cause many problems that resemble more serious malfunctions.

2 Check to see that the battery is fully charged, as the control unit and sensors depend on an accurate supply voltage in order to properly meter the fuel.

3 Check the air filter element - a dirty or partially blocked filter will severely impede performance and economy (see Chapter 1).

4 Check the related fuses. If a blown fuse is found, replace it and see if it blows again. If it does, search for a wire shorted to ground in the harness.

5 Check the air intake duct to the intake manifold for leaks, which will result in an excessively lean mixture. Also check the condition of all vacuum hoses connected to the intake manifold and/or throttle body.

6 Remove the air intake duct from the throttle body and check for dirt, carbon or other residue build-up. If it's dirty, clean it with carburetor cleaner spray, a toothbrush and a shop towel.

7 With the engine running, place an automotive stethoscope against each injector, one at a time, and listen for a clicking sound, indicating operation (see illustration). If you don't have a stethoscope, place the tip of a screwdriver against the injector and listen through the handle. If you hear the injectors operating, the electrical circuits are functioning, but the injectors may be dirty or fouled from carbon deposits - commercial cleaning products may help or they may require replacement. If one or more injectors are not operating, proceed with the injector check.

8 Turn the engine Off. Disconnect any one of the injector connectors and check for battery voltage at terminal no. 1 of the injector harness connector with the ignition key On (see illustration). If battery voltage is not present, check the fuse, fuel pump relay and related wiring (see Chapter 12).

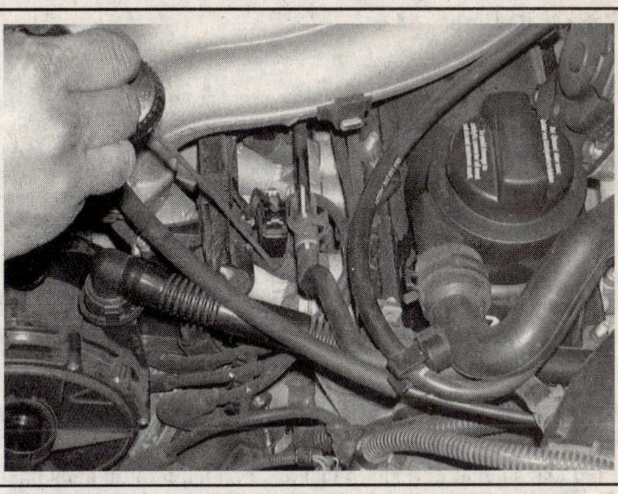

13.7 Use a stethoscope to determine if the injectors are working properly - they should make a steady clicking sound that rises and falls with engine speed changes

9 Install an injector test light ("noid" light) into the disconnected injector electrical connector (see illustration). Crank the engine over. Confirm that the light flashes. This tests the ECM control of the injectors. If the light does not flash, have the ECM checked at a dealer service department or other properly equipped repair facility. Test each injector connector, if necessary.

10 Disconnect the injector electrical connectors and measure the resistance of each injector (see illustration). Compare the measurements with the resistance value listed in this Chapter's Specifications. Replace any injector whose resistance value does not fall within specifications.

11 The remainder of the engine control system checks can be found in Chapter 6A.

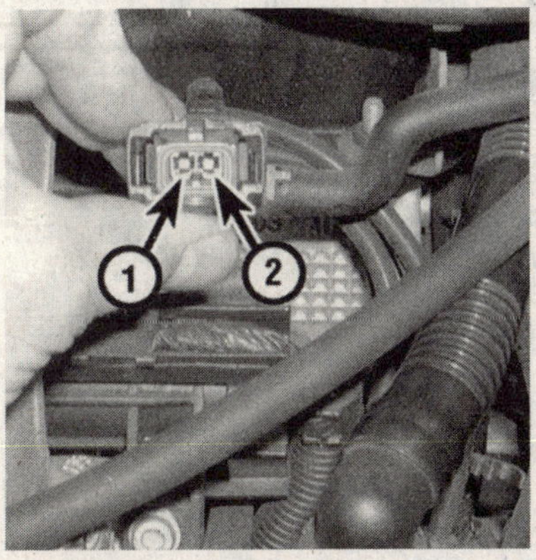

13.8 Fuel injector harness connector terminal identification

1 *12-volt supply*
2 *Injector control*

13.9 Install the "noid" light (available at most auto parts stores) into each injector electrical connector and confirm that it blinks when the engine is cranking

13.10 Measure the resistance of each injector across the two terminals of the injector

14 Throttle body - removal and installation

➧ Refer to illustration 14.6

1 Remove the engine cover.
2 Remove the air intake duct and air filter cover (see Section 9).
3 Disconnect the electrical connector from the throttle control module.
4 Detach the hoses from the throttle body.
5 On 2.0L models, detach the accelerator cable from the throttle body (see Section 10).
6 Remove the mounting bolts/nuts and remove the throttle body and gasket (see illustration).
7 Remove all traces of old gasket material from the throttle body and intake manifold and install a new gasket.

⁂ CAUTION:

Do not use solvent or a sharp tool to clean the throttle body gasket surface or damage to the throttle body may occur.

8 Install the throttle body and tighten the bolts to the torque listed in this Chapter's Specifications.
9 The remainder of installation is the reverse of removal.

14.6 Throttle body mounting bolts (arrows)

15 Fuel pressure regulator - replacement

➧ Refer to illustration 15.4 and 15.5

⁂ WARNING:

See the Warning in Section 1.

1 Relieve the fuel system pressure (see Section 2).
2 Remove the engine cover.
3 Disconnect the vacuum hose from the port on the regulator.

15.4 Remove the fuel pressure regulator retaining clip (arrow) and withdraw the regulator from the fuel rail

4 Remove the fuel pressure regulator retaining clip and withdraw the fuel pressure regulator from the fuel rail (see illustration).
5 Be sure to replace the O-ring seals, lubricating them with a light film of engine oil (see illustration).

➡Note: The small O-ring may remain in the fuel rail; recover and replace it.

6 Press the fuel pressure regulator into the fuel rail until fully seated and install the retaining clip.
7 The remainder of installation is the reverse of removal.

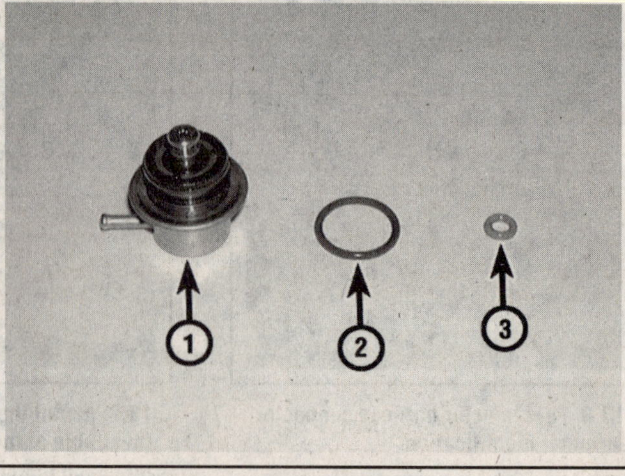

15.5 Fuel pressure regulator details

1 Fuel pressure regulator
2 Large O-ring
3 Small O-ring

16 Fuel rail and injectors - removal and installation

⁂ WARNING:

See the Warning in Section 1.

REMOVAL

➧ Refer to illustrations 16.5a, 16.5b, 16.6, 16.7, 16.8a and 16.8b

1 Relieve the fuel pressure (see Section 2).

2 Remove the engine cover.

3 On 2.0L models, remove the upper intake manifold (see Chapter 2A). Cover the lower intake manifold ports to prevent any foreign objects from entering the engine.

4 Disconnect the vacuum hose from the fuel pressure regulator. Clearly label and remove any other vacuum hoses or electrical wiring that will interfere with the fuel rail removal.

5 Disconnect the fuel inlet and return lines from the fuel rail (see illustration). On 2.0L models, disconnect the vapor line from the fuel rail (see illustration).

6 Disconnect the fuel injector electrical connectors. Detach the wiring harness retainers from the fuel rail and position the harness aside (see illustration).

➡Note: Apply a numbered tag to each connector with the corresponding cylinder number.

16.5b On 2.0L models, disconnect the vapor line (arrow) from the fuel rail

16.7 Remove the bolts (arrows) securing the fuel rail to the intake manifold

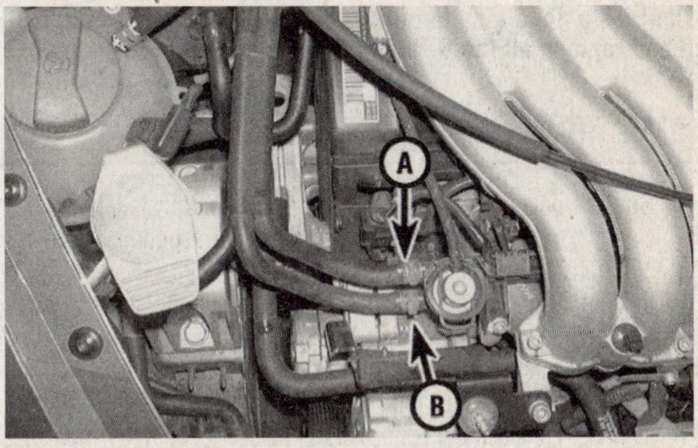

16.5a Disconnect the fuel supply line (A) and return line (B) from the fuel rail

7 Clean any debris from around the injectors. Remove the fuel rail mounting bolts (see illustration). Gently rock the fuel rail and injectors to loosen the injectors. Remove the fuel rail and fuel injectors as an assembly.

8 Remove the retaining clip and remove the injector(s) from the fuel

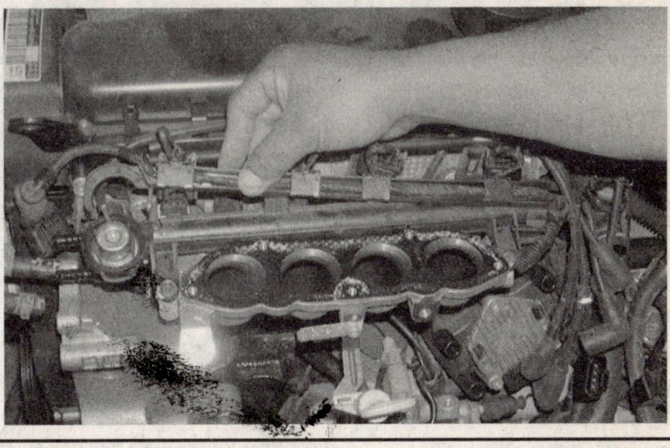

16.6 Disconnect the electrical connectors from the fuel injectors, detach the wiring harness retainers and position the wiring harness aside

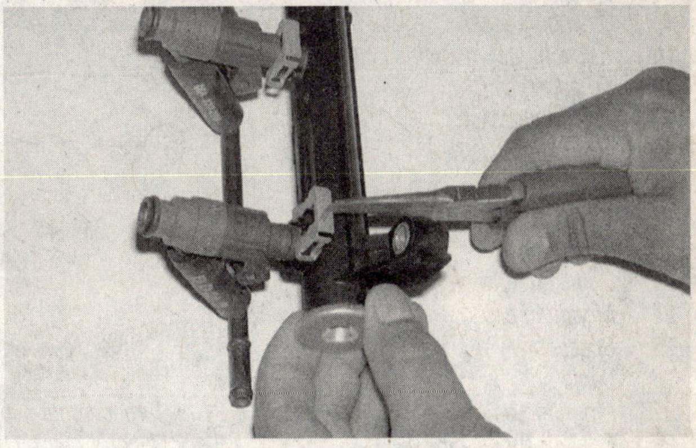

16.8a Remove the injector retaining clip and pull the injector off the fuel rail

rail assembly (see illustrations). Remove and discard the O-rings and seals.

➡Note: Whether you're replacing an injector or a leaking O-ring, it's a good idea to remove all the injectors from the fuel rail and replace all the O-rings.

INSTALLATION

9 Coat the new O-rings with clean engine oil and install them on the injector(s), then insert each injector into its corresponding bore in the fuel rail. Install the injector retaining clip.

10 Install the injector and fuel rail assembly on the intake manifold. Fully seat the injectors, then tighten the fuel rail mounting nuts to the torque listed in this Chapter's Specifications.

11 Connect the fuel lines and make sure they're securely installed.

12 Connect the electrical connectors to each injector, referring to the numbered tags.

13 The remainder of installation is the reverse of removal.

14 After the injector/fuel rail assembly installation is complete, turn the ignition switch to On, but don't operate the starter (this activates the fuel pump for about two seconds, which builds up fuel pressure in the fuel lines and the fuel rail). Repeat this about two or three times, then check the fuel lines, fuel rail and injectors for fuel leakage.

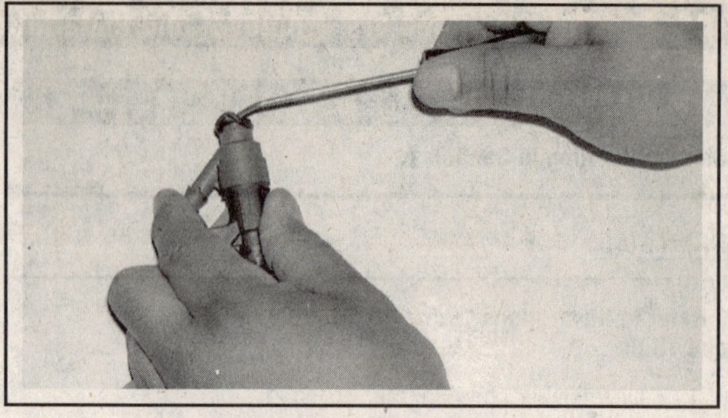

16.8b Carefully remove the O-rings from the injector

17 Turbocharger and intercooler (1.8L turbo engine) - check and replacement

CHECK

1 The turbocharger is a precision component which can be severely damaged by a lack of lubrication or from foreign material entering the air intake duct. Turbocharger failure may be indicated by poor engine performance, blue/gray exhaust smoke or unusual noises from the

17.8 Turbocharger and related components

1 *Exhaust downpipe nuts*
2 *Exhaust downpipe gasket*
3 *Wastegate bracket bolt*
4 *Wastegate*
5 *Circlip*
6 *Turbocharger*
7 *O-ring*
8 *Mounting bolt*
9 *Air inlet pipe stub*
10 *Gasket*
11 *Heat shield mounting bolt*
12 *Heat shield*
13 *Mounting bolt*
14 *Banjo bolt*
15 *Oil supply pipe*
16 *Turbocharger mounting bolts*
17 *Exhaust manifold*
18 *Manifold gasket*
19 *Mounting bolt*
20 *Banjo bolt*
21 *Manifold mounting nuts*
22 *Banjo bolt*
23 *Coolant return pipe*
24 *Mounting bolt*
25 *Spacer sleeve*
26 *Banjo bolt*
27 *Mounting bolt*
28 *Coolant supply pipe*
29 *Banjo bolt*
30 *Turbocharger-to-support bracket bolt*
31 *Turbocharger support bracket*
32 *Support bracket bolt*
33 *Gasket*
34 *Gasket*
35 *Oil return pipe flange bolt*
36 *Oil return pipe*
37 *Oil return pipe flange bolt*
38 *Wastegate setting/locknuts*

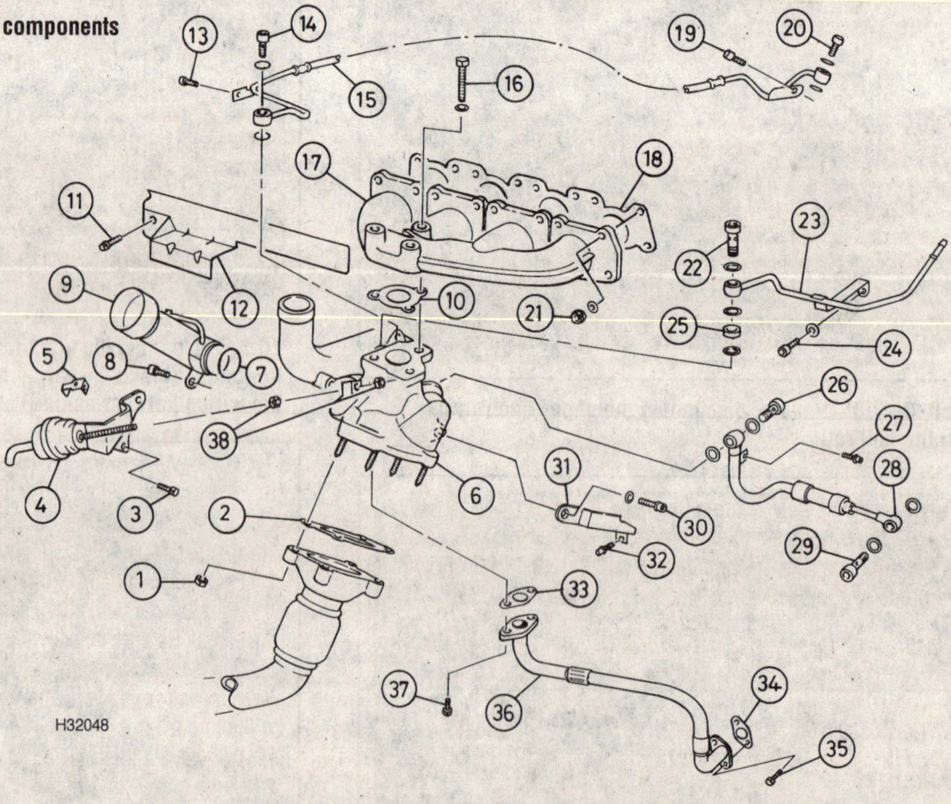

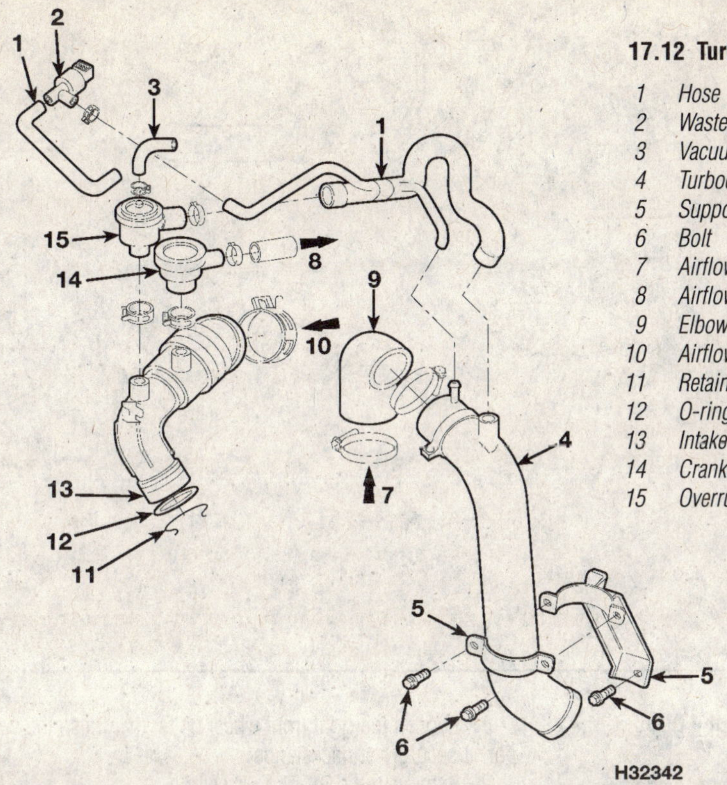

17.12 Turbocharger connecting pipes

1 *Hose*
2 *Wastegate regulator valve*
3 *Vacuum hose*
4 *Turbocharger outlet pipe*
5 *Support bracket*
6 *Bolt*
7 *Airflow from turbocharger*
8 *Airflow from crankcase ventilation housing*
9 *Elbow*
10 *Airflow from air filter*
11 *Retaining ring*
12 *O-ring*
13 *Intake hose*
14 *Crankcase pressure regulating valve*
15 *Overrun recirculation valve*

turbocharger. If a turbocharger failure is suspected, check the following areas:

a) *Check the intake air duct for looseness or damage. Make sure there are no restrictions in the air intake system, dirty air filter element or damaged intercooler.*
b) *Check the system vacuum hoses for restrictions or damage.*
c) *Check the system wiring for damage and electrical connectors for looseness or corrosion.*
d) *Make sure the wastegate actuator linkage is not binding.*
e) *Check the exhaust system for damage or restrictions.*
f) *Check the lubricating oil supply and drainback lines for damage or restrictions.*
g) *Check the coolant supply and return lines for damage and restrictions.*
h) *If the turbocharger requires replacement due to failure, be sure to change the engine oil and filter (see Chapter 1).*

2 Complete diagnosis of the turbocharger and control system require special techniques and equipment. If the previous checks fail to identify the problem, take the vehicle to a dealership service department or other properly equipped repair facility for diagnosis.

REPLACEMENT

Refer to illustrations 17.8, 17.12, 17.20, 17.24 and 17.26

Turbocharger

WARNING:

Wait until the engine is completely cool before beginning this procedure.

3 Drain the cooling system (see Chapter 1).

4 Remove the engine cover.

5 Remove the cowl panel (see Chapter 11).

6 Raise the vehicle and support it securely on jackstands.

7 Remove the engine compartment undercover.

8 Remove the nuts securing the exhaust pipe to the turbocharger (see illustration). Remove the exhaust support bolts, as necessary and lower the exhaust pipe.

9 Disconnect the vacuum hose from the wastegate actuator.

10 Remove the turbocharger support bracket.

11 Remove the bolts and disconnect the oil return pipe from the turbocharger.

12 Disconnect the hoses from the recirculation valve and crankcase breather valve on the air intake duct. Loosen the intake duct clamp at the air filter and separate the duct from the air filter assembly. Remove the retaining clip, disconnect the air intake duct from the turbocharger intake and remove the duct (see illustration).

13 Loosen the clamps, remove the bracket bolts and remove the turbocharger outlet pipe and elbow.

14 Loosen the coolant return pipe union bolt at the turbocharger. Remove the bracket bolt and disconnect the coolant return pipe from the turbocharger. Be sure to recover the sealing washers and spacer.

15 Loosen the oil supply union bolt from the turbocharger. Remove the bracket bolt from the heat shield and disconnect the oil supply pipe from the turbocharger.

16 Remove the heat shield.

17 Remove the turbocharger-to-exhaust manifold bolts. Lower the turbocharger, tilt it to the side, loosen the coolant supply pipe union bolt and disconnect the coolant supply pipe from the turbocharger.

18 Remove the turbocharger from the engine compartment.

17.20 Intercooler and related components

1. *Air deflector*
2. *Intercooler*
3. *Grommet*
4. *Bolt*
5. *Intercooler-to-intake manifold pipe*
6. *Grommet*
7. *Hose*
8. *Nut*
9. *Turbocharger-to-intercooler pipe*
10. *Bolt*
11. *Support bracket*
12. *Grommet*
13. *Hose*
14. *Bolt*
15. *Bolt*
16. *Boost pressure sensor*
17. *O-ring*

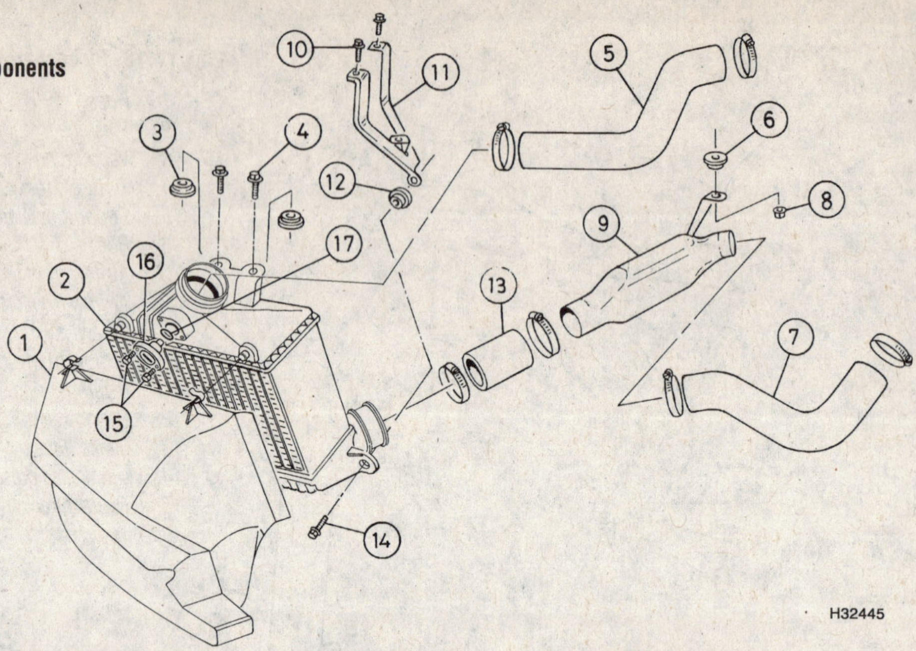

➡Note: The manufacturer recommends replacing the turbocharger/exhaust manifold as a unit only.

19 Installation is the reverse of removal with the following additions:

a) *Replace all gaskets, seals, union bolt washers and self-locking nuts.*

b) *Tighten the turbocharger mounting bolts to the torque listed in this Chapter's Specifications.*

c) *Change the oil and filter (see Chapter 1).*

d) *Before starting the engine, remove the fuel pump fuse from the fuse box and crank the engine over until oil pressure builds.*

Intercooler

20 The intercooler is located in the lower section of the right front fender ahead of the front wheel (see illustration).

21 Loosen the right front wheel bolts, then raise the vehicle and support it securely on jackstands.

22 Remove the right front wheel.

23 Remove the retainers and remove the right front fenderwell liner (see Chapter 11, Section 13, Step 3).

24 Remove the right turn signal lamp housing (see Chapter 12). Working through the turn signal opening, disconnect the boost pressure sensor electrical connector, loosen the hose clamp and disconnect the outlet air duct from the intercooler (see illustration).

25 Loosen the hose clamps and disconnect the air inlet duct from the intercooler.

26 Remove the mounting bolts and remove the intercooler and air deflector from the wheel well (see illustration). Detach the air deflector from the intercooler.

27 Installation is the reverse of removal.

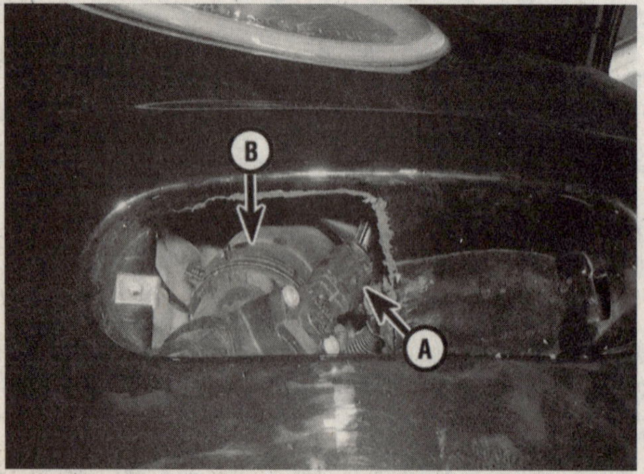

17.24 Working through the turn signal lamp opening, disconnect the boost pressure sensor electrical connector (A) and the outlet duct hose clamp (B)

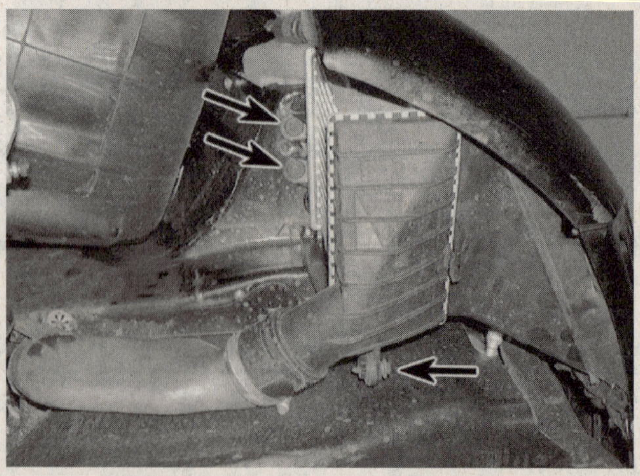

17.26 Remove the mounting bolts (arrows) and remove the intercooler and air deflector from the wheelwell

18 Exhaust system servicing - general information

⁂ WARNING:

Inspection and repair of exhaust system components should be done only after enough time has elapsed after driving the vehicle to allow the system components to cool completely. Also, when working under the vehicle, make sure it is securely supported on jackstands.

1 The exhaust system consists of the exhaust manifold, catalytic converter, muffler, resonators, the tailpipe and all connecting pipes, brackets, hangers and clamps. The exhaust system is attached to the body with mounting brackets and rubber hangers. If any of the parts are improperly installed, excessive noise and vibration will be transmitted to the body.

MUFFLER AND PIPES

➧ Refer to illustrations 18.2a and 18.2b

2 Conduct regular inspections of the exhaust system to keep it safe and quiet. Look for any damaged or bent parts, open seams, holes, loose connections, excessive corrosion or other defects which could allow exhaust fumes to enter the vehicle (see illustrations). Also check the catalytic converter when you inspect the exhaust system (see below). Deteriorated exhaust system components should not be repaired; they should be replaced with new parts.

3 If the exhaust system components are extremely corroded or rusted together, welding equipment will probably be required to remove them. The convenient way to accomplish this is to have a muffler repair shop remove the corroded sections with a cutting torch. If, however, you want to save money by doing it yourself (and you don't have a welding outfit with a cutting torch), simply cut the exhaust pipes with a hacksaw at the separation point. If you do decide to tackle the job at home, be sure to wear safety goggles to protect your eyes from metal chips and work gloves to protect your hands.

4 Here are some simple guidelines to follow when repairing the exhaust system:

a) *Work from the back to the front when removing exhaust system components.*
b) *Apply penetrating oil to the exhaust system component fasteners to make them easier to remove.*
c) *Use new gaskets, hangers and clamps when installing exhaust system components.*
d) *Apply anti-seize compound to the threads of all exhaust system fasteners during reassembly.*
e) *Be sure to allow sufficient clearance between newly installed parts and all points on the underbody to avoid overheating the floor pan and possibly damaging the interior carpet and insulation. Pay particularly close attention to the catalytic converter and heat shield.*

18.2a Exhaust system details - front section (typical)

1 *Bolt*
2 *Bolt*
3 *Fitting nut*
4 *Nut*
5 *Pipe*
6 *Bolt*
7 *Union*
8 *Connector*
9 *Oxygen sensor (pre-catalyst)*
10 *Oxygen sensor (post catalyst)*
11 *Connector*
12 *Exhaust flow to muffler*
13 *Catalytic converter*
14 *Guide tube*
15 *Nut*
16 *Support*
17 *Gasket*
18 *Exhaust manifold*
19 *Heat shield*

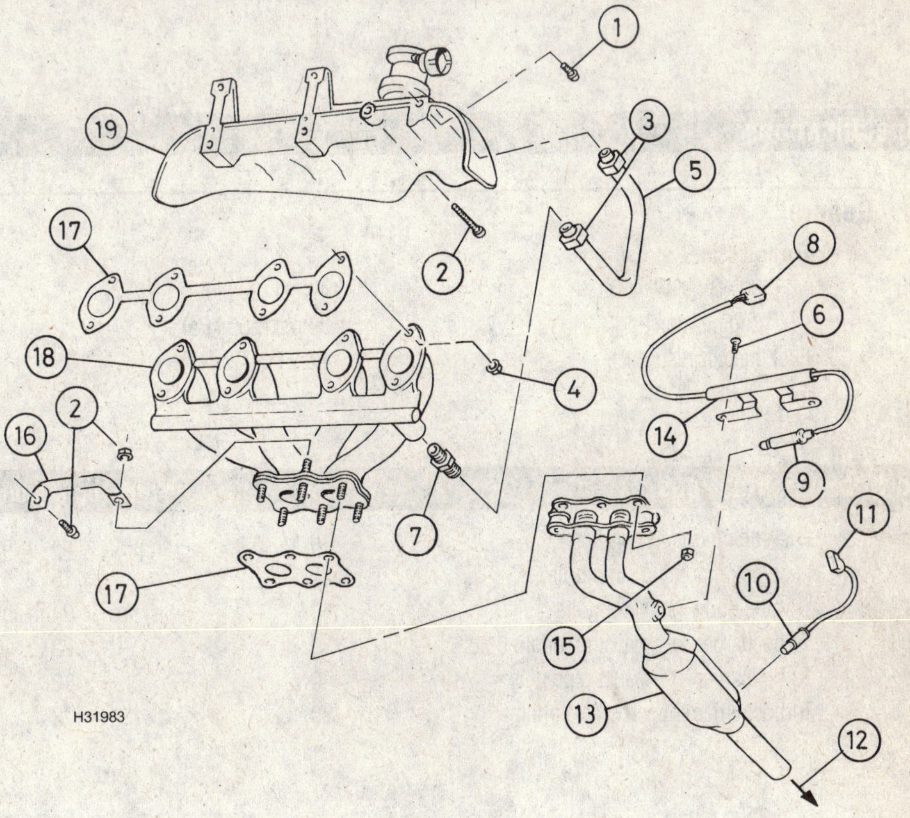

18.2b Exhaust system details - rear section

1 *Exhaust flow from catalytic converter*
2 *Clamp*
3 *Bolt*
4 *Mount*
5 *Resonator*
6 *Separation point*
7 *Mount*
8 *Bracket*
9 *Muffler*
10 *Mount*
11 *Nut*
12 *Crossmember*
13 *Nut*
14 *Washer*

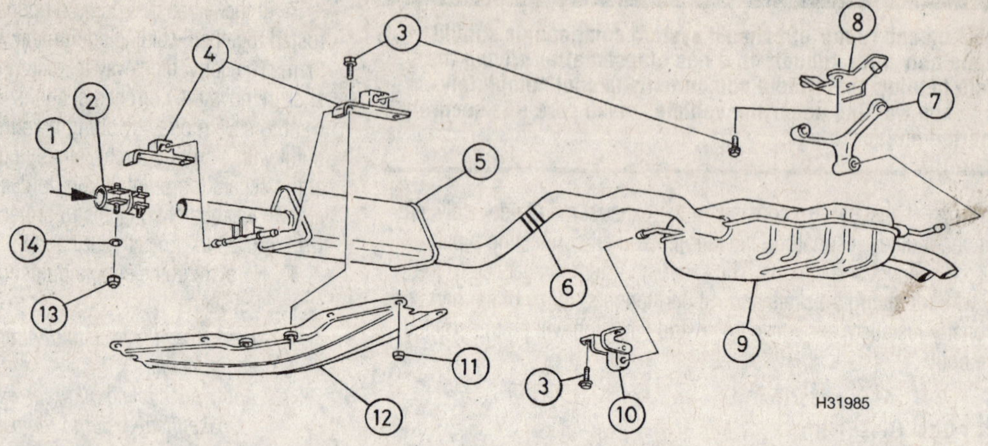

CATALYTIC CONVERTER

WARNING:

The converter gets very hot during operation. Make sure it has cooled down before you touch it.

➡Note: See Chapter 6 for additional information on the catalytic converter.

5 Periodically inspect the heat shield for cracks, dents and loose or missing fasteners.

6 Inspect the converter for cracks or other damage.

7 If the catalytic converter requires replacement, refer to Chapter 6A.

Specifications

General

Fuel pressure	
Key On, engine Off	44 psi (303 kPa)
Engine running	39 psi (269 kPa)
Fuel injector resistance	12.0 to 17.0 ohms

Torque specifications

Torque specifications	Nm	Ft-lbs (unless otherwise indicated)
Fuel rail mounting bolts	10	84 in-lbs
Fuel tank mounting strap bolts	25	18
Throttle body mounting bolts	10	84 in-lbs
Turbocharger-to-exhaust manifold bolts	30	22
Turbocharger-to-exhaust pipe nuts	40	30
Turbocharger support bracket bolt	30	22

4B

FUEL AND EXHAUST SYSTEMS: DIESEL ENGINE

Section

Reference to other Chapters

Air filter housing - removal and installation - See Chapter 4A
Air filter replacement - See Chapter 1
CHECK ENGINE light - See Chapter 6
Exhaust system check - See Chapter 1
Exhaust system servicing - general information - See Chapter 4A
Fuel level sending unit - check and replacement - See Chapter 4A
Fuel lines and fittings - repair and replacement - See Chapter 4A
Fuel system check - See Chapter 1
Fuel tank - removal and installation - See Chapter 4A
Fuel tank cleaning and repair - general information - See Chapter 4A

1 General information

WARNING 1:

Diesel fuel isn't as volatile as gasoline, but it is flammable, so take extra precautions when you work on any part of the fuel system. Don't smoke or allow open flames or bare light bulbs near the work area. Don't work in a garage or other enclosed space where there is a gas-type appliance (such as a water heater or clothes dryer). Avoid direct skin contact with diesel fuel - wear protective clothing, safety glasses and gloves when handling fuel system components and have a Class B fire extinguisher on hand. Ensure that the work area is well ventilated to prevent the build up of diesel fuel vapor.

WARNING 2:

Fuel injectors operate at extremely high pressures and the jet of fuel produced at the nozzle is capable of piercing skin, with potentially fatal results. When working with pressurized injectors, take great care to avoid exposing any part of the body to the fuel spray. It is recommended that any pressure testing of the fuel system components should be carried out by a diesel fuel systems specialist.

CAUTION:

Under no circumstances should diesel fuel be allowed to come into contact with coolant hoses, wiring or rubber components - wipe off accidental spillage immediately. Hoses that have been contaminated with fuel for an extended period should be replaced. Diesel fuel systems are particularly sensitive to contamination from dirt, air and water. Pay particular attention to cleanliness when working on any part of the fuel system, to prevent the entry of dirt. Thoroughly clean the area around fuel unions before disconnecting them. Store dismantled components in sealed containers to prevent contamination and the formation of condensation. Only use lint-free cloths and clean fuel for component cleansing. Avoid using compressed air when cleaning components in place.

Diesel models covered by this manual are equipped with a Turbocharged Direct Injection (TDI) fuel injection system. This system uses an electronically controlled mechanical fuel injection pump and injector nozzles to sequentially inject the fuel directly into the combustion chamber of each cylinder. The fuel injection pump is controlled by the Engine Control Module (ECM). The ECM monitors various engine parameters and delivers the exact amount of fuel, in the correct sequence, to each cylinder. This Chapter's information pertains to the air and fuel delivery components of the system only. Refer to Chapter 6B for information regarding the electronic engine control system.

The exhaust system consists of an exhaust manifold, a catalytic converter, an exhaust pipe and a muffler. Each of these components is replaceable. For further information regarding the catalytic converter, refer to Chapter 6A.

Diesel models are equipped with a turbocharger and intercooler. The turbocharger increases power by using an exhaust gas driven turbine to pressurize the intake charge before it enters the combustion chambers. The amount of intake manifold pressure (boost) is regulated by an exhaust by-pass valve (wastegate). The wastegate regulator valve is controlled by the ECM. The heated compressed air is routed through an air-to-air radiator (intercooler). The intercooler removes excess heat from the compressed air, increasing its density and allowing for more boost pressure.

2 Electronic accelerator system - general information

1.9L TDI models are equipped with an electronically controlled accelerator system. The system consists of the throttle position sensor (located at the throttle pedal), the fuel injection pump and the ECM. The system does not use the traditional accelerator cable.

The engine speed of a diesel engine is governed by the amount of fuel injected into each cylinder, the more fuel injected, the faster the engine runs. The fuel injection pump contains an electronic component known as the fuel quantity adjuster. The fuel quantity adjuster varies the amount of fuel delivered to the fuel injectors. The ECM controls the fuel quantity adjuster based on the voltage signal received from the throttle position sensor. Refer to Chapter 6B for more information on the throttle position sensor and the fuel quantity adjuster.

3 Fuel shut-off valve - check and replacement

WARNING:

Observe the precautions in Section 1 before working on any component of the fuel system.

CHECK

1 The fuel shut-off valve shuts fuel off from the fuel injection pump when the ignition key is turned off, stopping the engine from running. The ECM controls the fuel shut-off valve, supplying battery power to the valve when the ignition key is operated. A faulty fuel shut-off valve would not allow the engine to start, or conversely, not allow the engine to shut down when the ignition key is switched off. The fuel shut-off valve is located on the fuel injection pump (see illustration 8.3).

2 To check the fuel shut-off valve, connect a 12-volt test light or voltmeter to the electrical terminal on the valve. Have an assistant turn the ignition key on - voltage should be indicated on the light or meter and you should hear the valve click.

3 If voltage is not indicated, check the wiring from the valve to the ECM. If the wiring is good, have the ECM checked at a dealer service department or other properly equipped repair facility.

4 If voltage is indicated but the valve did not click, the valve is probably defective. Remove the valve, spring and plunger and check the plunger for freedom of movement in the valve. If the plunger binds in the valve, replace the valve. Also, check the bore in the injection pump for foreign material that may cause the valve to bind in the injection pump.

REPLACEMENT

5 Clean the area around the fuel shut-off valve and place a shop towel around the valve to absorb the fuel.

6 Disconnect the electrical terminal from the valve.

7 Remove the valve from the fuel injection pump.

8 Install the valve with a new O-ring. Tighten the valve to the torque listed in this Chapter's Specifications and clean up the fuel.

9 Connect the electrical terminal to the valve.

4 Air shut-off valve - check and replacement

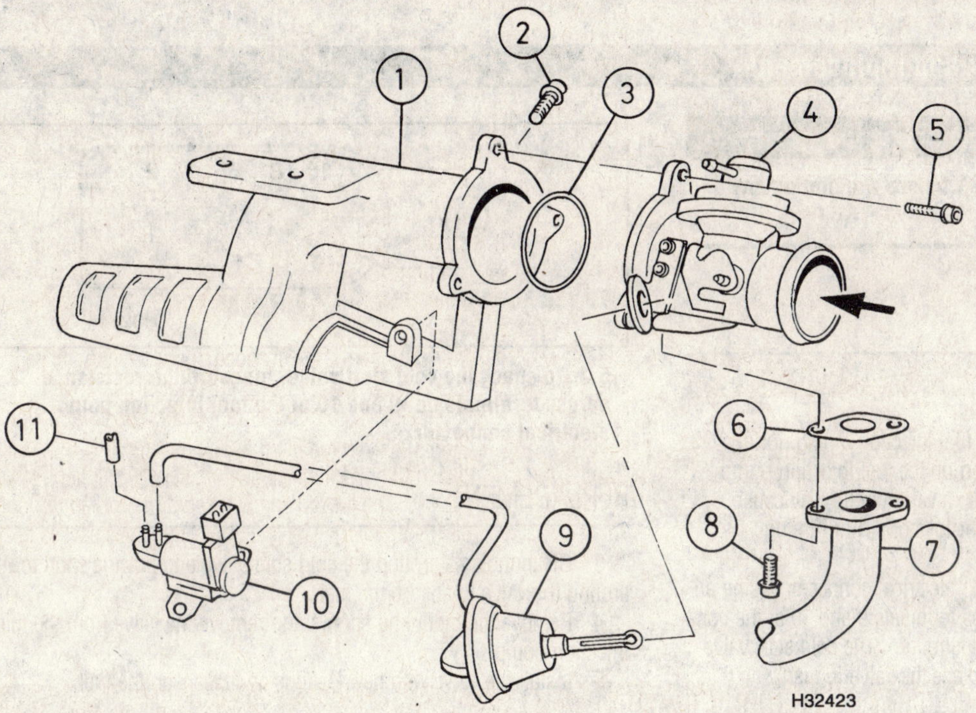

4.1 Air shut-off valve components

1 *Intake manifold*
2 *Bolt*
3 *O-ring*
4 *Intake manifold inlet flange (with EGR valve and air shut-off valve flap)*
5 *Bolt*
6 *Gasket*
7 *EGR pipe*
8 *Bolt*
9 *Air shut-off valve actuator*
10 *Air shut-off valve vacuum valve/solenoid*
11 *Vacuum hose*

CHECK

➧ **Refer to illustrations 4.1 and 4.3**

1 The air shut-off valve closes off the intake air passage when the engine is shut down. Shutting off the air reduces the shock associated with shutting down a diesel engine. The air shut-off valve is comprised of three components; the valve flap in the intake manifold flange, the valve actuator and the vacuum valve/solenoid (see illustration). The ECM controls the vacuum valve/solenoid. When the ignition key is turned off the ECM energizes the vacuum valve/solenoid, vacuum is applied to the valve actuator, opening the flap in the intake manifold flange. After approximately three seconds from shutting the engine down, the ECM de-energizes the vacuum valve/solenoid, vacuum is vented and the air valve opens.

2 If a problem is suspected with the air shut-off valve, check the vacuum supply to the vacuum valve/solenoid. Vacuum should be present with the engine running. Check the actuator and linkage for binding. Disconnect the vacuum hose from the actuator and apply vacuum to the actuator with a hand-held vacuum pump. Make sure the actuator opens and closes the valve flap.

3 Disconnect the electrical connector from the vacuum valve/solenoid. Using an ohmmeter, measure the resistance across the two terminals of the valve/solenoid (see illustration). The valve/solenoid resistance should be 25 to 45 ohms; if it isn't, replace the vacuum valve/solenoid.

REPLACEMENT

Air valve flap

➡**Note: The air valve flap is incorporated into the intake manifold flange along with the EGR valve. If the air valve flap is defective, the entire flange must be replaced.**

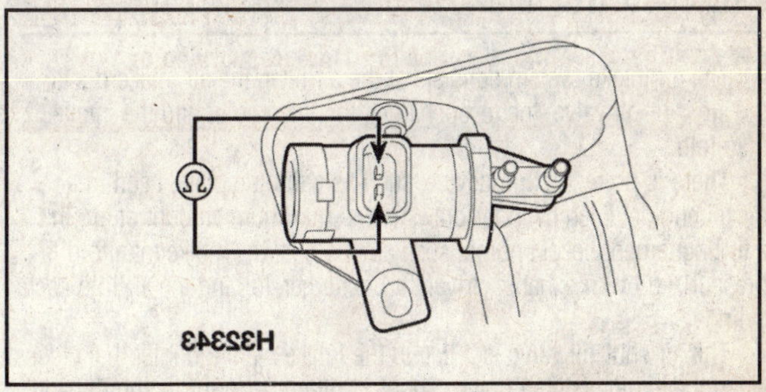

4.3 Measure the resistance across the two terminals of the air shut-off valve vacuum valve/solenoid

4 Disconnect the vacuum hose from the EGR valve.

5 Disconnect the linkage from the air valve actuator.

6 Remove the intake hose from the flange.

7 Remove the bolts and disconnect the EGR pipe from the flange.

8 Remove the flange mounting bolts and remove the flange.

9 Installation is the reverse of removal, using a new O-ring seal and EGR pipe gasket.

Air valve actuator

10 Disconnect the vacuum hose from the actuator.

11 Disconnect the actuator linkage from the air valve flap.

12 Remove the actuator from the flange.

13 Installation is the reverse of removal.

Vacuum valve/solenoid

14 Disconnect the electrical connector from the vacuum valve/solenoid.

15 Disconnect the vacuum hoses from the vacuum valve/solenoid.

16 Remove the mounting bolts and remove the vacuum valve/solenoid from the intake manifold.

17 Installation is the reverse of removal.

5 Cold start valve - check and replacement

⁂ WARNING:

Observe the precautions in Sec-tion 1 before working on any component of the fuel system.

CHECK

➧ Refer to illustration 5.2

1 The cold start valve advances the fuel injection timing during cold starts. The cold start valve is located on the fuel injection pump (see illustration 8.3). A problem with the valve or circuit may cause unstable engine idle (timing advanced) or loss of power (timing retarded).

2 Disconnect the fuel injection pump electrical connector. Using an ohmmeter, measure the resistance across terminals 9 and 10 at the connector (fuel injection pump side) (see illustration). The cold start valve resistance should be 12 to 20 ohms, replace the valve if it isn't.

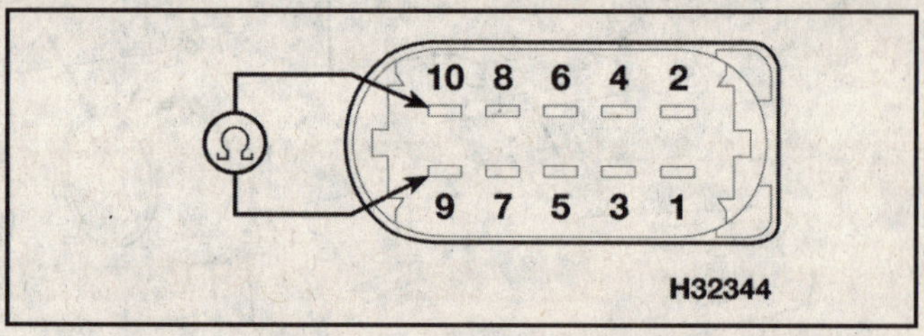

5.2 To check the cold start valve, measure the resistance across terminals no. 9 and 10 of the fuel injection pump electrical connector

REPLACEMENT

3 Clean the area around the cold start valve and place a shop towel around the valve to absorb the fuel.

4 Remove the mounting screws and remove the valve from the fuel injection pump.

5 Install the valve with new O-rings and clean-up the fuel.

6 Fuel injection system - general information

The fuel injection system consists of three sub-systems: air intake, engine control and fuel delivery. The system uses an Engine Control Module (ECM) along with the sensors (coolant temperature sensor, throttle position sensor, mass airflow sensor, engine speed sensor, etc.) to determine the proper fuel delivery rate under all operating conditions.

The fuel injection system and the engine control system are closely linked in function and design. For additional information, refer to Chapter 6B.

AIR INTAKE SYSTEM

The air intake system consists of the air filter, the air intake ducts, the air shut-off valve, the turbocharger, the intercooler and the intake manifold.

There is no restriction device, such as a throttle plate, used in a diesel engine. Each cylinder draws in the maximum amount of air. In a turbocharged diesel engine, such as this one, the intake manifold is pressurized by an exhaust-driven turbocharger, forcing the air into each cylinder.

The air shut-off valve is a flap in the intake manifold inlet that closes when the engine is shut down. Shutting-off the air prior to shutting-off the fuel reduces the shock associated with diesel engine shut down.

EMISSIONS AND ENGINE CONTROL SYSTEM

The emissions and engine control system is described in detail in Chapter 6B.

FUEL DELIVERY SYSTEM

The fuel delivery system consists of the fuel injection pump, the fuel injectors and the high pressure lines connecting the two.

Fuel is drawn from the fuel tank and through the fuel filter by the fuel injection pump - no other fuel pump is used. The fuel injection pump pressurizes the fuel and delivers the correct amount of fuel to the fuel injectors in firing order sequence. The fuel injection pump is a mechanical pump with electronic controls. The ECM precisely controls the amount of fuel injected and the injection timing. The fuel injectors are connected to the fuel injection pump with special high-pressure fuel lines.

The injectors are mechanical two-stage spring loaded injectors. When the pressure reaches a preset value (3190 to 3355 psi - 220 to 230 bar) the fuel pressure overcomes the spring pressure and fuel is sprayed out of the nozzle. The injector tip contains 5-ports, allowing the fuel to spray out the nozzle in a more lateral direction promoting a more efficient combustion process.

7 Fuel delivery system - check

WARNING:

Observe the precautions in Section 1 before working on any component of the fuel system.

1 If the engine is hard to start or unable to start and you suspect a fuel delivery problem, perform the following test to verify fuel is reaching the fuel injectors.

2 Check for fuel leaks at the injection pump, fuel lines, fittings and injectors. If a leak is detected it must be repaired before proceeding.

3 Clean the area around the number one fuel injector line fitting. Using a flare-nut wrench, loosen the fitting approximately one-half turn.

4 Wrap a shop towel around the fitting to absorb the fuel and crank the engine with the starter. If a steady stream of fuel is discharged from the injector fitting, the fuel delivery system is functioning properly.

5 If no fuel is observed at the injector fitting, check the following items:

a) Check for fuel in the tank.
b) Check the fuel shut-off valve (see Section 3).
c) Check the fuel lines from the fuel tank to the fuel injection pump for an air leak or restriction.
d) Check the fuel filter for a restriction.
e) Check for an engine mechanical problem, such as a broken timing belt.

6 If all the items listed are good, have the fuel injection pump checked at a dealer service department or other properly equipped repair facility.

7 Tighten the fuel injector line fitting and clean-up the fuel.

8 Fuel injection pump - removal and installation

WARNING:

Observe the precautions in Section 1 before working on any component of the fuel system.

➡Note 1: Special tools are required to perform this procedure. Obtain the tools before beginning work.

➡Note 2: After pump installation, final fuel injection pump timing must be checked with the engine running using the manufacturer's scan tool (or equivalent) and adjusted, if necessary. If the proper scan tool is not available, complete the procedure and drive the vehicle to a dealer service department or other properly equipped repair facility for the final injection pump timing procedure.

REMOVAL

Refer to illustrations 8.3, 8.10, 8.13, 8.14 and 8.15

1 Disconnect the fuel injection pump electrical connector and remove the wire terminal from the fuel shut-off valve.

2 Remove the fuel supply hose from the fuel injection pump. Cap the fitting of the pump and plug the hose.

3 Loosen the fuel line fittings at the fuel injectors and fuel injection pump (see illustration). Remove the fuel lines as an assembly. Be very careful not to bend a high pressure fuel line.

4 Cap the fittings on the fuel injection pump and cap all the injectors to prevent contamination.

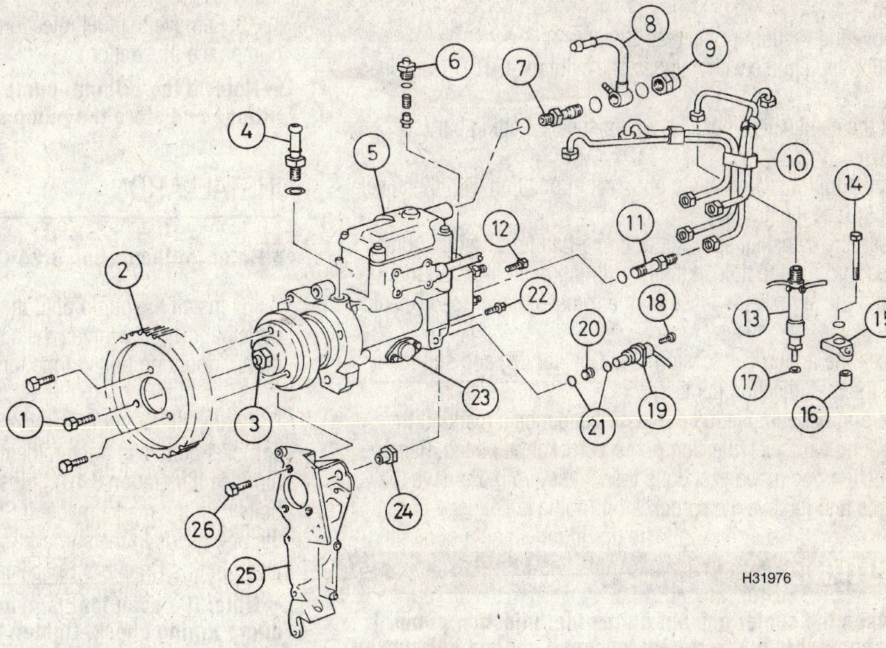

8.3 Fuel injection pump and fuel injector components

1 *Bolt*
2 *Fuel injection pump sprocket*
3 *Nut*
4 *Fuel supply fitting*
5 *Fuel injection pump*
6 *Fuel shut-off valve*
7 *Fuel return line union*
8 *Fuel return line fitting*
9 *Fuel return line union nut*
10 *Fuel injector lines*
11 *Fuel injector line union*
12 *Bolt*
13 *Fuel injector*
14 *Bolt*
15 *Retainer*
16 *Spacer*
17 *Copper sealing washer*
18 *Screw*
19 *Cold start valve*
20 *Strainer*
21 *O-ring*
22 *Bolt*
23 *Timing cover*
24 *Sleeve*
25 *Mounting bracket*
26 *Bolt*

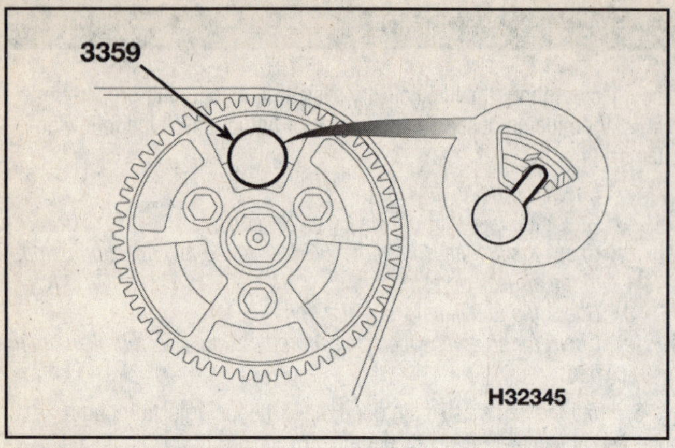

8.10 Insert the fuel injection pump sprocket alignment pin through the slot in the sprocket hub

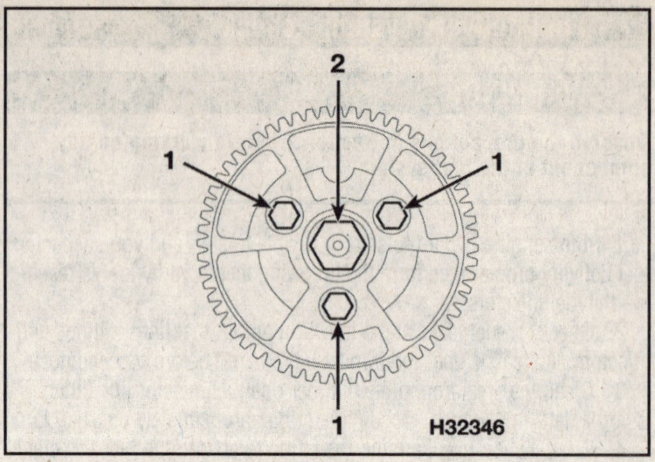

8.13 Remove the fuel injection pump sprocket bolts (1) - DO NOT loosen the center hub nut (2)

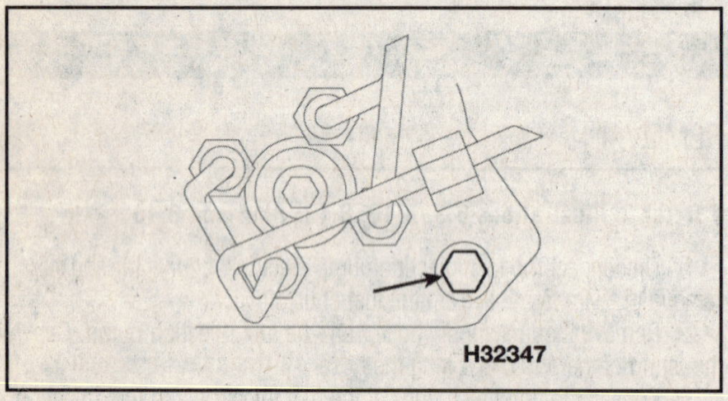

8.14 Remove the fuel injection pump support bracket bolt

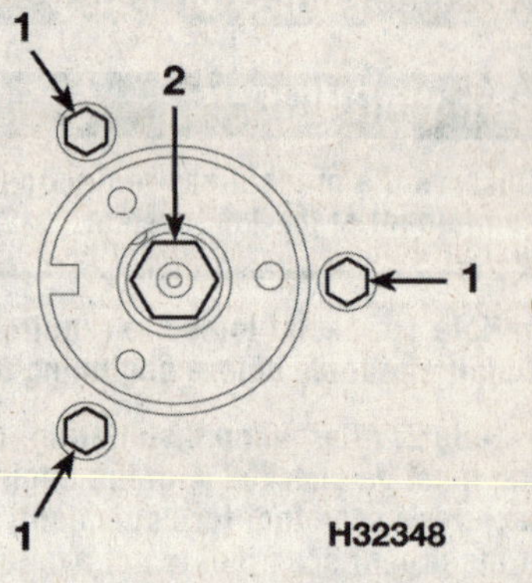

8.15 Remove the fuel injection pump mounting bolts (1) - DO NOT loosen the center hub nut (2)

5 Remove the intercooler-to-intake manifold air duct (see Section 10).

6 Remove the valve cover and the timing belt upper cover (see Chapter 2B).

7 Remove the vacuum pump (see Chapter 9).

8 Position the engine with number one cylinder at TDC (see Chapter 2B).

9 Lock the camshaft in position with a special setting bar (see Chapter 2B).

10 Lock the fuel injection pump sprocket in position with the special alignment pin (see illustration).

11 Loosen the camshaft sprocket bolt one-half turn. Using a brass drift inserted through the hole in the rear timing belt cover, tap the sprocket until the sprocket is loose on the camshaft taper (see Chapter 2B).

12 Loosen the timing belt tension (see Chapter 2B) and slip the timing belt off the fuel injection pump sprocket.

13 If the original pump is to be reinstalled, scribe a mark on the sprocket around each fuel injection pump sprocket bolt head. Remove the fuel injection pump sprocket bolts (see illustration). Remove the alignment pin and remove the sprocket.

⁂ CAUTION:

Do not loosen the center hub nut on the fuel injection pump. If the center hub nut is inadvertently loosened and the hub moved, the fuel injection pump must be sent to a qualified fuel injection repair facility for recalibration.

14 Remove the fuel injection pump support bracket bolt (see illustration).

15 Remove the fuel injection pump mounting bolts and remove the pump (see illustration).

➡Note: If the original pump is to be reinstalled, cap all fuel line fittings and store the pump so the fuel will not drain out.

INSTALLATION

➧ Refer to illustration 8.24

16 Install the pump onto the engine and tighten the pump mounting bolts hand-tight. Tighten the support bracket bolt first, then tighten the pump mounting bolts to the torque listed in this Chapter's Specifications.

17 Install the pump sprocket onto the pump hub. Install new sprocket bolts and tighten them hand-tight. Install the pump alignment pin (see illustration 8.10). If installing the original pump align the pump bolts with the marks made in Step 13. If installing a new pump, center the bolt heads in the sprocket slots. Tighten the bolts to the torque listed in this Chapter's Specifications.

➡Note: If a scan tool is available for the final fuel injection pump timing check, tighten the bolts to the first step (15 ft-lbs) at this time. Tighten the bolts to the final torque after adjusting the fuel injection pump timing.

18 Make sure the engine is positioned with number one cylinder at

TDC, the camshaft is locked in position and the camshaft sprocket is loose on the shaft. Install and tension the timing belt (see Chapter 2B).

19 Tighten the camshaft sprocket bolt to the torque listed in the Chapter 2B Specifications.

20 Remove the fuel injection pump sprocket alignment pin and the camshaft setting bar. Using a breaker bar and socket on the crankshaft pulley bolt, rotate the engine two complete revolutions and recheck the timing belt tension.

21 Install the fuel injection lines and the fuel supply line. Connect the electrical connectors to the pump.

22 Install the valve cover and the vacuum pump. Install the timing belt upper cover (unless a scan tool is available to check the injection pump timing).

23 Install the intake air duct.

24 If installing a new fuel injection pump, remove the fuel return line port fitting and connect a hand-held vacuum pump to the fuel injection pump using clear section of hose and the appropriate adapter (see illustration). Using the vacuum pump, draw fuel into the fuel injection pump until fuel is visible exiting the fuel injection pump. Do not draw fuel into a plastic vacuum pump or the vacuum pump may be damaged.

CAUTION:

Do not attempt to start the engine without priming a dry pump first or damage to the fuel injection pump will occur.

25 Remove the vacuum pump and adapter and install the fuel return line.

26 Start the engine and check for fuel leaks at the fuel lines.

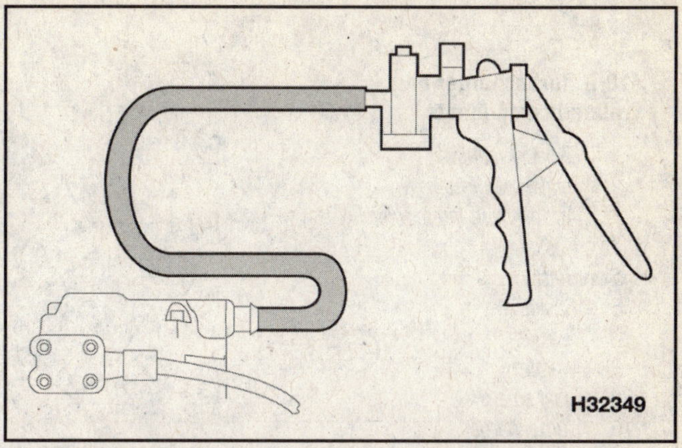

8.24 If the fuel injection pump is dry, connect a hand-held vacuum pump to the return line fitting and draw fuel into the fuel injection pump

27 If a scan tool is available, connect the scan tool, start the engine and check and adjust the fuel injection pump timing following the instructions provided with the scan tool. After completing the timing procedure, install the timing belt upper cover.

28 If a scan tool is not available, drive the vehicle to a dealer service department or other properly equipped repair facility and have the fuel injection pump timing checked and adjusted.

Note: It will be necessary for the repair facility to replace the fuel injection pump sprocket bolts, if timing adjustment is required.

9 Fuel injectors - removal and installation

WARNING:

Observe the precautions in Section 1 before working on any component of the fuel system.

REMOVAL

1 Loosen the fuel line fittings at the fuel injectors and fuel injection pump. Remove the fuel lines as an assembly. Be very careful not to bend a high pressure fuel line.

2 Remove the fuel return hoses from the fuel injectors.

3 Remove the fuel injector mounting bolt, retainer and spacer and withdraw the fuel injector from the cylinder head (see illustration 8.3). Disconnect the electrical connector from the fuel injector needle lift sensor prior to removing injector number three.

INSTALLATION

4 Remove the copper sealing washer from the injector tip. If the washer is not on the injector, retrieve it from the cylinder head. Install a new copper washer.

5 Install the injector into the cylinder head. Press the injector in until fully seated.

6 Install the retainer and mounting bolt. Tighten the bolt to the torque listed in this Chapter's Specifications.

7 Install the high pressure lines and tighten the fitting nuts. Install the fuel return lines and connect the fuel injector lift sensor.

8 Crank the engine over with the starter and check for fuel leaks. It is not necessary to bleed the air from the injectors on this system.

10 Turbocharger and intercooler - check and replacement

CHECK

1 The turbocharger is a precision component which can be severely damaged by a lack of lubrication or from foreign material entering the air intake duct. Turbocharger failure may be indicated by poor engine performance, blue/gray exhaust smoke or unusual noises from the turbocharger. If a turbocharger failure is suspected, check the following areas:

a) *Check the intake air duct for looseness or damage. Make sure there are no restrictions in the air intake system, dirty air filter element or damaged intercooler.*
b) *Check the system vacuum hoses for restrictions or damage.*
c) *Check the system wiring for damage and electrical connectors for looseness or corrosion.*

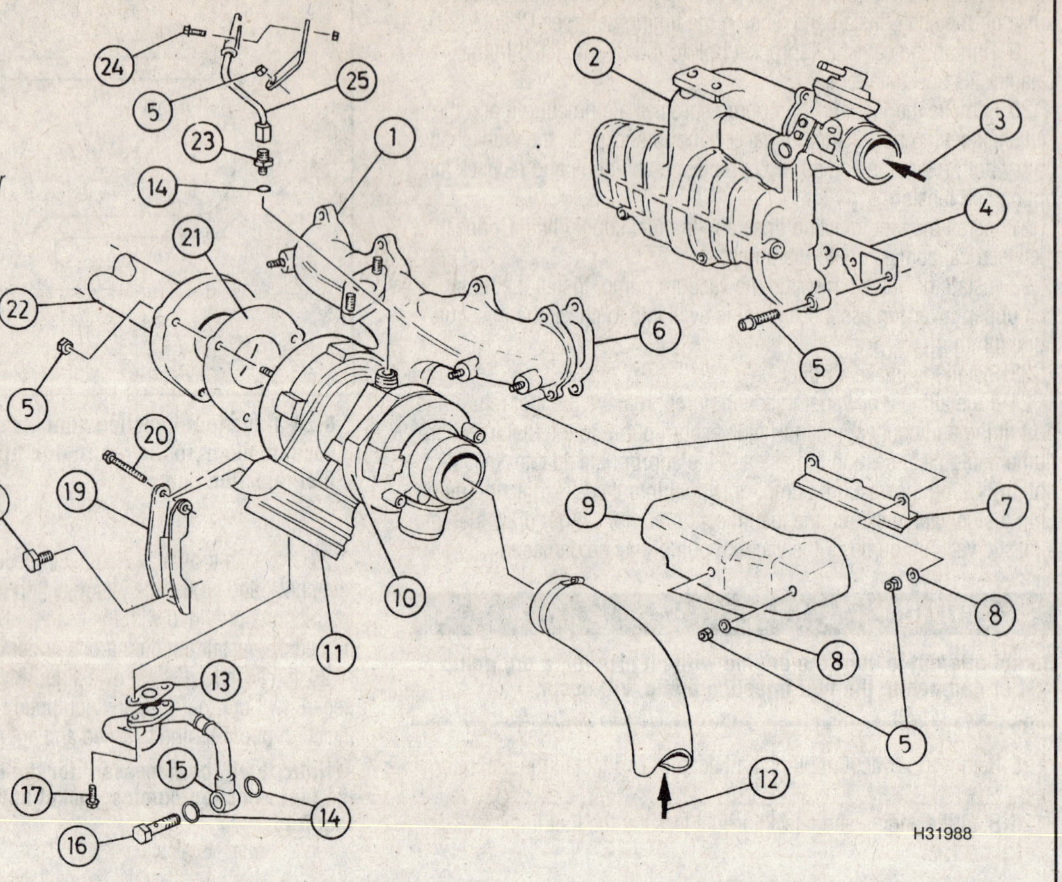

10.9 Turbocharger and related components

1 *Exhaust manifold*
2 *Intake manifold*
3 *Airflow from intercooler*
4 *Gasket*
5 *Bolt*
6 *Gasket*
7 *Bracket*
8 *Washer*
9 *Heat shield*
10 *Turbocharger*
11 *Wastegate actuator*
12 *Airflow from air filter*
13 *Gasket*
14 *O-ring*
15 *Oil return pipe*
16 *Union bolt*
17 *Bolt*
18 *Bolt*
19 *Bracket*
20 *Bolt*
21 *Gasket*
22 *Exhaust pipe*
23 *Fitting*
24 *Bolt*
25 *Oil supply line*

- d) *Make sure the wastegate actuator linkage is not binding.*
- e) *Check the exhaust system for damage or restrictions.*
- f) *Check the lubricating oil supply and drainback lines for damage or restrictions.*
- g) *If the turbocharger requires replacement due to failure, be sure to change the engine oil and filter (see Chapter 1).*

2 Complete diagnosis of the turbocharger and control system require special techniques and equipment. If the previous checks fail to identify the problem, take the vehicle to a dealership service department or other properly equipped repair facility for diagnosis.

REPLACEMENT

➧ **Refer to illustrations 10.9 and 10.16**

Turbocharger

⁂ WARNING:

Wait until the engine is completely cool before beginning this procedure.

➡**Note: The turbocharger and exhaust manifold are integrated together and replaced as a unit.**

3 Remove the engine cover. Remove the cowl panel (see Chapter 11).

4 Raise the vehicle and support it securely on jackstands.

5 Remove the engine compartment undercover. Drain the cooling system (see Chapter 1).

6 Remove the right-side driveaxle (see Chapter 8).

7 Remove the nuts securing the exhaust pipe to the turbocharger. Remove the front portion of the exhaust pipe.

8 Disconnect the vacuum hose from the wastegate actuator.

9 Remove the turbocharger support bracket (see illustration).

10 Remove the bolts and disconnect the oil return pipe from the turbocharger.

11 Loosen the intake duct clamp at the air filter and separate the duct from the air filter assembly. Remove the retaining clip, disconnect the air intake duct from the turbocharger intake and remove the duct.

12 Loosen the oil supply union bolt from the turbocharger. Remove the bracket bolt from the heat shield and disconnect the oil supply pipe from the turbocharger.

13 Remove the heat shield. Remove the EGR cooler (see Chapter 6B).

14 Remove the exhaust manifold bolts (see Chapter 2B) and remove the turbocharger and exhaust manifold from the engine.

15 Installation is the reverse of removal with the following additions:

- a) *Replace all gaskets, seals, union bolt washers and self-locking nuts.*
- b) *Tighten the turbocharger mounting bolts to the torque listed in this Chapter's Specifications.*
- c) *Change the engine oil and filter (see Chapter 1).*
- d) *Refill the cooling system (see Chapter 1).*
- e) *Before starting the engine, disconnect the electrical lead from the fuel shut-off valve and crank the engine over until oil pressure builds.*

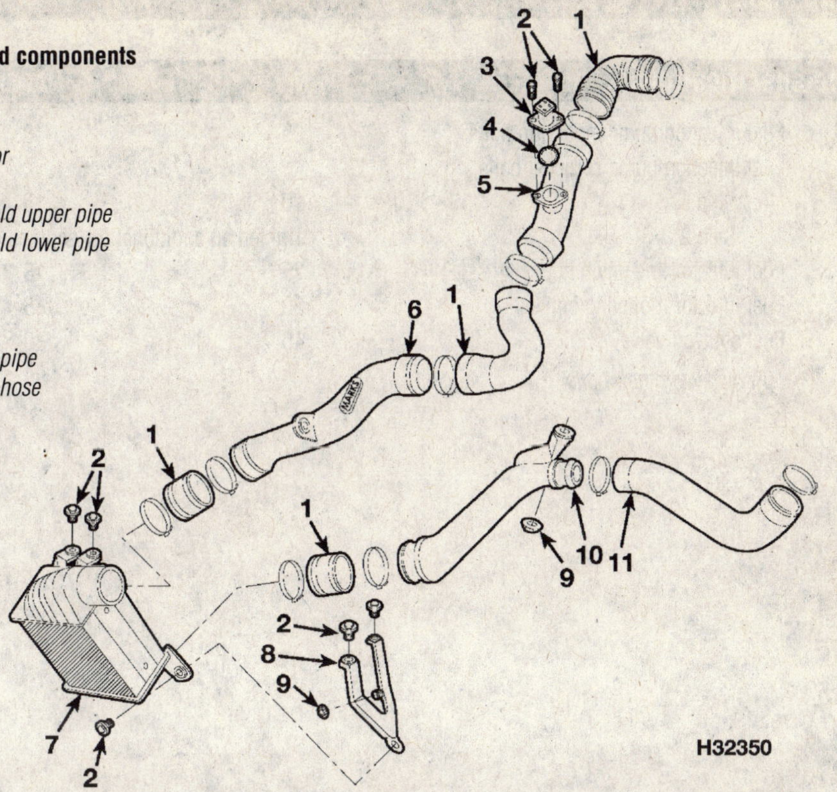

10.16 Intercooler and related components

1. *Hose*
2. *Bolt*
3. *Intake air temperature sensor*
4. *O-ring*
5. *Intercooler-to-intake manifold upper pipe*
6. *Intercooler-to-intake manifold lower pipe*
7. *Intercooler*
8. *Bracket*
9. *Nut*
10. *Intercooler-to-turbocharger pipe*
11. *Intercooler-to-turbocharger hose*

Intercooler

16 The intercooler is located in the lower section of the right front fender, ahead of the front wheel (see illustration).

17 Raise the vehicle and support it securely on jackstands.

18 Remove the engine compartment undercover.

19 Remove the right front wheel.

20 Remove the retainers and remove the right front fenderwell liner (see Chapter 11, Section 13, Step 3).

21 Loosen the hose clamps and disconnect the inlet and outlet ducts from the intercooler.

22 Remove the mounting bolts and remove the intercooler and air deflector from the wheelwell. Detach the air deflector from the intercooler.

23 Installation is the reverse of removal.

Specifications

Torque specifications	Nm	Ft-lbs
Fuel injection pump mounting bolts	25	18
Fuel injection pump sprocket bolts*		
Step 1	20	15
Step 2	Tighten an additional 90-degrees	
Fuel injector line union nut	25	18
Fuel injector mounting bolt	20	15
Fuel shut-off valve	40	30

** Replace with new bolts*

5 ENGINE ELECTRICAL SYSTEMS

Section

Reference to other Chapters

Battery check, maintenance and charging - See Chapter 1
CHECK ENGINE light - See Chapter 6
Drivebelt check, adjustment and replacement - See Chapter 1
Spark plug replacement - See Chapter 1

1 General information, precautions and battery disconnection

GENERAL INFORMATION

➧ **Refer to illustration 1.1**

The engine electrical systems include all ignition, charging and starting components (see Illustration). Because of their engine-related functions, these components are discussed separately from body electrical devices such as the lights, the instruments, etc. (which are included in Chapter 12).

PRECAUTIONS

Always observe the following precautions when working on the electrical system:

a) *Be extremely careful when servicing engine electrical components. They are easily damaged if checked, connected or handled improperly.*
b) *Never leave the ignition switched on for long periods of time when the engine is not running.*
c) *Never disconnect the battery cables while the engine is running.*
d) *Maintain correct polarity when connecting battery cables from another vehicle during jump starting - see the "Booster battery (jump) starting" section at the front of this manual.*
e) *Always disconnect the negative battery cable from the battery before working on the electrical system, but read the following battery disconnection procedure first.*

It's also a good idea to review the safety-related information regarding the engine electrical systems located in the "Safety first!" section at the front of this manual, before beginning any operation included in this Chapter.

BATTERY DISCONNECTION

Several systems on the vehicle require battery power to be available at all times, either to ensure their continued operation (such as the radio, alarm system, power door locks, windows, etc.) or to maintain control unit memories (such as that in the engine management system's Engine Control Module [ECM]) which would be lost if the battery were to be disconnected. Therefore, whenever the battery is to be disconnected, first note the following to ensure that there are no unforeseen consequences of this action:

a) *These models are equipped with an anti-theft radio. Before performing a procedure that requires disconnecting the battery, make sure you have the proper activation code.*

1.1 Typical engine electrical system components

1 *Spark plug wires*
2 *Battery cable*
3 *Underhood fuse box*
4 *Battery (inside housing)*
5 *Ignition coils*
6 *Alternator*

b) *The engine management system's ECM will lose the information stored in its memory when the battery is disconnected. This includes idling and operating values, any fault codes detected and system monitors required for emissions testing. Whenever the battery is disconnected, the information relating to idle speed control and other operating values will have to be re-programmed into the unit's memory using a scan tool (see Chapter 6A).*

c) *On any vehicle with power door locks, it is a wise precaution to remove the key from the ignition and to keep it with you, so that it does not get locked inside if the power door locks should engage accidentally when the battery is reconnected!*

Devices known as "memory-savers" can be used to avoid some of the above problems. Precise details vary according to the device used. Typically, it is plugged into the cigarette lighter and is connected by its own wires to a spare battery; the vehicle's own battery is then disconnected from the electrical system, leaving the "memory-saver" to pass sufficient current to maintain audio unit security codes and ECM memory values, and also to run permanently live circuits such as the clock and radio memory, all the while isolating the battery in the event of a short-circuit occurring while work is carried out.

WARNING 1:

Some of these devices allow a considerable amount of current to pass, which can mean that many of the vehicle's systems are still operational when the main battery is disconnected. If a "memory-saver" is used, ensure that the circuit concerned is actually "dead" before carrying out any work on it!

WARNING 2:

If work is to be performed around any of the airbag system components, the battery must be disconnected. If a memory-saver device is used, power will be supplied to the airbag and personal injury may result if the airbag is accidentally deployed.

The battery on these vehicles is located under a cover at the left front corner of the engine compartment. To disconnect the battery for service procedures requiring power to be cut from the vehicle, lift the battery cover lid, peel back the insulator, loosen the negative cable clamp nut and detach the negative cable from the negative battery post (see Section 3). Isolate the cable end to prevent it from accidentally coming into contact with the battery post.

2 Battery - emergency jump starting

Refer to the *Booster battery (jump) starting* procedure at the front of this manual.

3 Battery - check and replacement

WARNING:

Hydrogen gas is produced by the battery, so keep open flames and lighted cigarettes away from it at all times. Always wear eye protection when working around a battery. Rinse off spilled electrolyte immediately with large amounts of water.

CAUTION 1:

These models are equipped with an anti-theft radio. before performing a procedure that requires disconnecting the battery, make sure you have the activation code.

CAUTION 2:

Disconnecting the battery can cause driveability problems that require a scan tool to remedy. See Section 1 for the use of an auxiliary voltage input device before disconnecting the battery.

CHECK

Refer to illustrations 3.1a, 3.1b and 3.1c –

1 A battery cannot be accurately tested until it is at or near a fully charged state. Disconnect the negative battery cable from the battery and perform the following tests:

a) ***Battery state of charge test*** - *Visually inspect the indicator eye (if equipped) on the top of the battery. If the indicator eye is dark in color, charge the battery as described in Chapter 1. If the battery is equipped with removable caps, check the battery electrolyte. The electrolyte level should be above the upper edge of the plates. If the level is low, add distilled water. DO NOT OVERFILL. The excess electrolyte may spill over during periods of heavy charging. Test the specific gravity of the electrolyte using a hydrometer (see illustration). Remove the caps and extract a sample of the electrolyte and observe the float inside the barrel of the hydrometer. Follow the instructions from the tool manufacturer and determine the specific gravity of the electrolyte for each cell. A fully charged battery will indicate approximately 1.270 (green zone) at 68-degrees F (20-degrees C). If the specific gravity of the electrolyte is low (red zone), charge the battery as described in Chapter 1.*

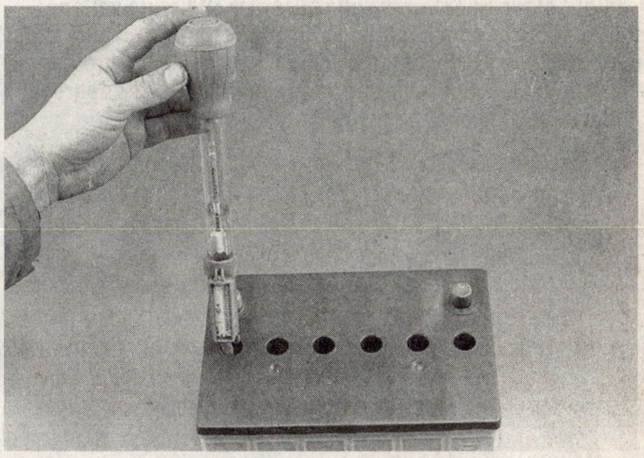

3.1a Use a battery hydrometer to draw electrolyte from the battery cell - this hydrometer is equipped with a thermometer to make temperature corrections

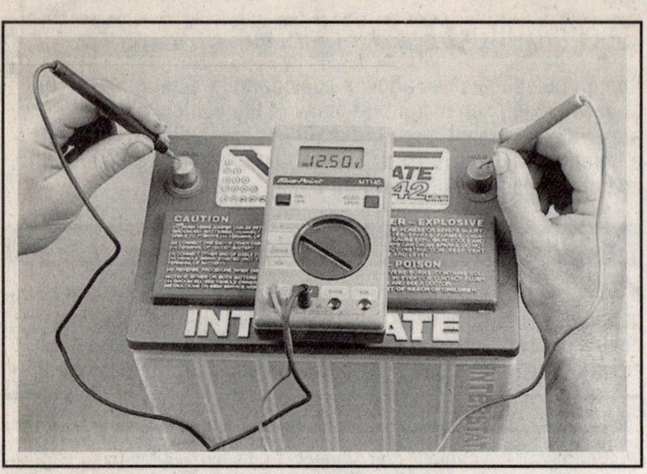

3.1b To test the open circuit voltage of the battery, connect the black probe of the voltmeter to the negative terminal and the red probe to the positive terminal of the battery - a fully charged battery should indicate approximately 12.5 volts depending on the outside air temperature

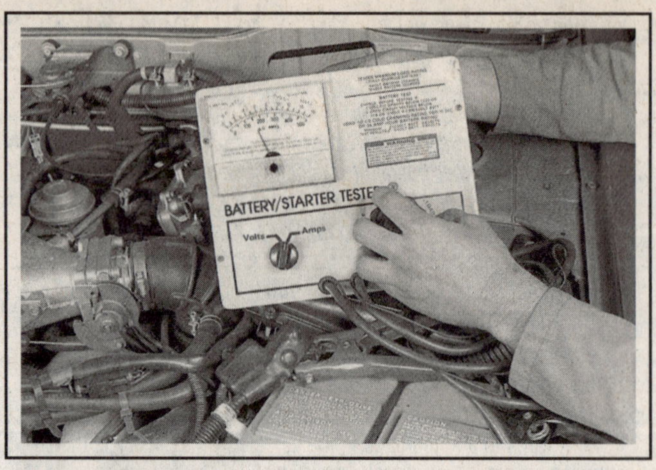

3.1c Some battery load testers are equipped with an ammeter which enables the battery load to be precisely dialed in, as shown - less expensive testers have a load switch and a voltmeter only

b) ***Open circuit voltage test*** *- Using a digital voltmeter, perform an open circuit voltage test (see illustration). Connect the negative probe of the voltmeter to the negative battery post and the positive probe to the positive battery post. The battery voltage should be greater than 12.5 volts. If the battery is less than the specified voltage, charge the battery before proceeding to the next test. Do not proceed with the battery load test until the battery is fully charged.*

c) ***Battery load test*** *- An accurate check of the battery condition can only be performed with a load tester (available at most auto parts stores). This test evaluates the ability of the battery to operate the starter and other accessories during periods of heavy amperage draw (load). Install a special battery load testing tool onto the battery terminals (see illustration). Load test the battery according to the tool manufacturer's instructions. This tool utilizes a carbon pile to increase the load demand (amperage draw) on the battery. Maintain the load on the battery for 15 seconds and observe that the battery voltage does not drop below 9.6 volts. If the battery condition is weak or defective, the tool will indicate this condition immediately.*

➡Note: Cold temperatures will cause the minimum voltage requirements to drop slightly. Follow the chart given in the tool manufacturer's instructions to compensate for cold climates. Minimum load voltage for freezing temperatures (32 degrees F/0-degrees C) should be approximately 9.1 volts.

d) ***Battery drain test*** *- This test will indicate whether there's a constant drain on the vehicle's electrical system that can cause the battery to discharge. Make sure all accessories are turned Off. If the vehicle has an underhood light, verify it's working properly, then disconnect it. Connect one lead of a digital ammeter to the disconnected negative battery cable clamp and the other lead to the negative battery post. A drain of approximately 100 milliamps or less is considered normal (due to the engine control computer, digital clocks, digital radios and other components which normally cause a key-off battery drain). An excessive drain (approximately 500 milliamps or more) will cause the battery to discharge. The problem circuit or component can be located by removing the fuses, one at a time, until the excessive drain stops and normal drain is indicated on the meter.*

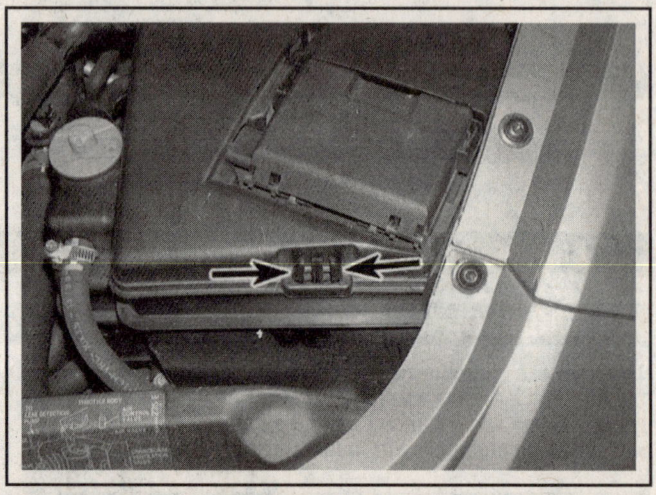

3.2 Squeeze the latch and lift the battery cover lid

REPLACEMENT

➧ Refer to illustrations 3.2, 3.4, 3.6 and 3.7

⁂ CAUTION:

Always disconnect the negative cable first and hook it up last or the battery may be shorted by the tool being used to loosen the cable clamps.

2 Squeeze the latch and lift the battery cover lid (see illustration). Peel back the top insulator.

3 Loosen the cable clamp nut and remove the negative battery cable from the negative battery post. Isolate the cable end to prevent it from accidentally coming into contact with the battery post.

4 Open the fuse box cover, remove the nut and disconnect the positive cable from the fuse box (see illustration). Carefully remove the fuse box cover and wiring harness protector. Position the fuse box and the battery cover lid aside.

5 Loosen the cable clamp nut and remove the positive battery cable from the positive battery post.

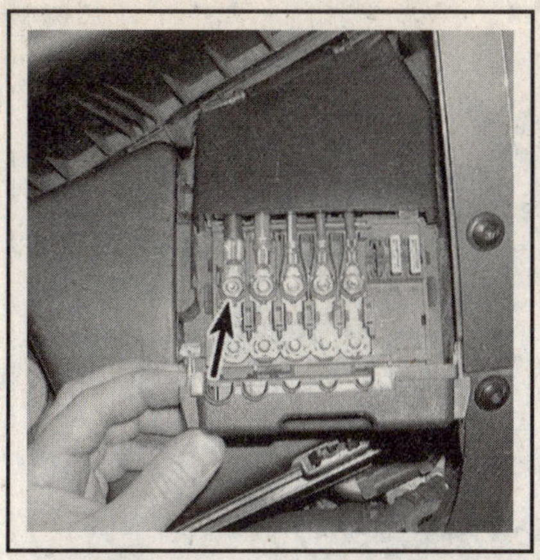

3.4 Open the fuse box cover and disconnect the positive cable from the fuse box

3.6 Remove the power steering reservoir bolt (A) and the battery cover bolt (B)

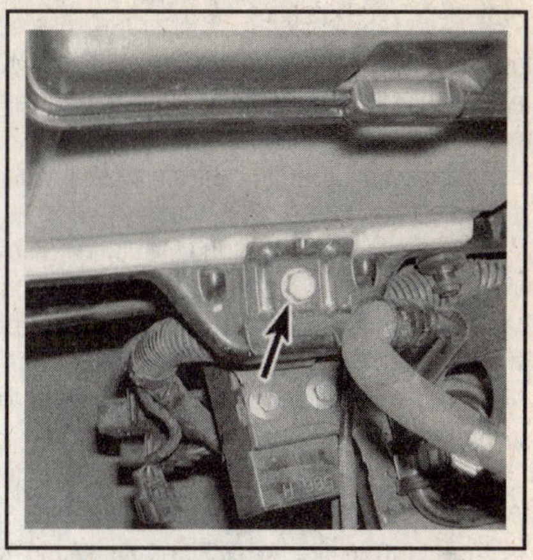

3.7 Remove the battery hold-down clamp bolt (arrow)

6 Remove the power steering reservoir mounting bolt and the battery cover bolt (see illustration). Position the reservoir aside (without disconnecting the hoses). Detach the battery cover from the locating tabs and remove the battery cover.

7 Remove the battery hold-down clamp (see illustration).

8 Remove the battery insulator and lift out the battery. Be careful - it's heavy.

➡Note: Battery straps and handlers are available at most auto parts stores for a reasonable price. They make it easier to remove and carry the battery.

9 While the battery is out, inspect the battery tray for corrosion. If corrosion exists, clean the deposits with a mixture of baking soda and water to prevent further corrosion. Flush the area with plenty of clean water and dry thoroughly.

10 If you are replacing the battery, make sure you replace it with a battery with the identical dimensions, amperage rating, cold cranking rating, etc.

11 When installing the battery, make sure the center notch in the battery foot is aligned with the hold-down clamp hole in the battery tray. Install the hold-down clamp and tighten the bolt to the torque listed in this Chapter's Specifications. Do not over-tighten the bolt.

12 The remainder of installation is the reverse of removal.

4 Battery cables - replacement

➧ Refer to illustrations 4.4a and 4.4b

⁂ CAUTION 1:

These models are equipped with an anti-theft radio. Before performing a procedure that requires disconnecting the battery, make sure you have the activation code.

⁂ CAUTION 2:

Disconnecting the battery can cause driveability problems that require a scan tool to remedy. See Section 1 for the use of an auxiliary voltage input device before disconnecting the battery.

1 Periodically inspect the entire length of each battery cable for damage, cracked or burned insulation and corrosion. Poor battery cable connections can cause starting problems and decreased engine performance.

2 Check the cable-to-terminal connections at the ends of the cables for cracks, loose wire strands and corrosion. The presence of white, fluffy deposits under the insulation at the cable terminal connection is a sign that the cable is corroded and should be replaced. Check the terminals for distortion, missing mounting bolts and corrosion.

3 When removing the cables, always disconnect the negative cable from the negative battery post first and hook it up last or the battery may be shorted by the tool used to loosen the cable clamps. Even if only the positive cable is being replaced, be sure to disconnect the negative cable from the negative battery post first (see Chapter 1 for further information regarding battery cable maintenance).

4 Disconnect the old cables from the battery, then disconnect them from the opposite end. Detach the cables from the starter solenoid, underhood fuse box and ground terminals, as necessary (see illustrations). Note the routing of each cable to ensure correct installation.

4.4a The positive cable is connected to the starter

5 If you are replacing either or both of the battery cables, take them with you when buying new cables. It is vitally important that you replace the cables with identical parts. Cables have characteristics that make them easy to identify: Positive cables are usually red and larger in cross-section; ground cables are usually black and smaller in cross-section.

6 Clean the threads of the starter solenoid or ground connection with a wire brush to remove rust and corrosion. Apply a light coat of battery terminal corrosion inhibitor or petroleum jelly to the threads to prevent future corrosion.

7 Attach the cable to the terminal and tighten the mounting nut/bolt securely.

8 Before connecting a new cable to the battery, make sure that it reaches the battery post without having to be stretched.

9 After installing the cables, connect the negative cable to the negative battery post.

4.4b The negative battery cable is connected to a transaxle bolt stud

5 Ignition system - general information

All gasoline engine models are equipped with a distributorless ignition system. The ignition system consists of the battery, ignition coils, spark plugs, camshaft position sensor, engine speed sensor and the Engine Control Module (ECM). The ECM controls the ignition timing and spark advance characteristics for the engine. The ignition timing is not adjustable.

The engine speed sensor and camshaft position sensor generate pulses that are input to the Engine Control Module. The ECM determines piston position and engine speed from these two sensors. The ECM calculates injector sequence and ignition timing from the piston position. Refer to Chapter 6A for testing and replacement procedures for the engine speed sensor and camshaft position sensor.

The 2.0L engine utilizes a coil pack consisting of two ignition coils and a power unit. This type of ignition system uses a "waste spark" method of spark distribution. Each cylinder is paired with its opposing cylinder in the firing order (1-4, 2-3) so one cylinder under compression fires simultaneously with its opposing cylinder, where the piston is on the exhaust stroke. Since the cylinder on the exhaust stroke requires very little of the available voltage to fire its plug, most of the voltage is used to fire the plug of the cylinder on the compression stroke. In a conventional ignition system, one end of the ignition coil secondary winding is connected to engine ground. In a waste spark system, neither end of the secondary winding is grounded - instead, one end of the coil secondary winding is directly attached to the spark plug and the other end is attached to the spark plug of the companion cylinder.

1.8L turbo models utilize an individual ignition coil/power unit for each cylinder. The unit is positioned directly over each spark plug. The ECM fires each coil sequentially in the firing order sequence.

The ECM controls the ignition system by opening and closing the primary ignition coil control circuit. The computerized ignition system provides complete control of the ignition timing by determining the optimum timing in response to engine speed, coolant temperature, throttle position and engine load. These parameters are relayed to the ECM by the camshaft position sensor, engine speed sensor, throttle position sensor, coolant temperature sensor and mass airflow sensor. Refer to Chapter 6A for additional information on the various sensors.

The ignition system is also integrated with a knock sensor system. The system uses two knock sensors in conjunction with the ECM to control spark timing. The knock sensor system allows the engine to use maximum spark advance without spark knock, which improves driveability and fuel economy.

6 Ignition system - check

➧ Refer to illustrations 6.2, 6.4 and 6.5

⁂ WARNING:

Because of the high voltage generated by the ignition system, extreme care should be taken whenever an operation is performed involving ignition components. This not only includes the ignition coil, but related components and test equipment.

1 If a malfunction occurs and the vehicle won't start, do not immediately assume that the ignition system is causing the problem. First, check the following items:

a) Make sure the battery cable clamps, where they connect to the battery, are clean and tight.

b) Test the condition of the battery (see Section 3). If it does not pass all the tests, replace it with a new battery.

c) Check the external ignition coil wiring and connections.

d) Check the related fuses inside the fuse box (see Chapter 12). If they're burned, determine the cause and repair the circuit.

2 If the engine turns over but won't start, make sure there is sufficient secondary ignition voltage to fire the spark plug. On 2.0L models, disconnect a spark plug wire from one of the spark plugs and attach a calibrated ignition system tester (available at most auto parts stores)

to the spark plug boot. Connect the clip on the tester to a bolt or metal bracket on the engine (see illustration). On 1.8L turbo models, remove an ignition coil (see Section 7) and attach the calibrated ignition system tester to the spark plug boot. Reconnect the electrical connector to the coil and clip the tester to a good ground. Crank the engine and watch the end of the tester to see if a bright blue, well-defined spark occurs (weak spark or intermittent spark is the same as no spark).

3 If spark occurs, sufficient voltage is reaching the plug to fire it (repeat the check at the remaining spark plug wires or ignition coils to verify that the spark plug wires, connectors and ignition coils are good). If the ignition system is operating properly the problem lies elsewhere; i.e. a mechanical or fuel system problem. However, the plugs themselves may be fouled, so remove and check them as described in Chapter 1.

4 If no spark occurs, remove the spark plug wire or boot from the suspected ignition coil and check the terminals for damage. Using an ohmmeter, check the wire or boot for an open or high resistance (compare your measurement with the values listed in this Chapter's Specifications). On 2.0L models, disconnect the spark plug wires from all the ignition coil towers and check the ignition coil secondary resistance across each pair of coil towers (see illustration). Compare your measurement with the values listed in this Chapter's Specifications. Replace the ignition coil assembly if either coil is not within specifications.

5 Check for battery voltage to the ignition coil with the ignition key On (engine not running). Disconnect the coil electrical connector and check for power at the indicated terminals of the coil connector (see illustration):

a) On 2.0L models connect a voltmeter to terminals 2 and 4.
b) On 1.8L Turbo models, connect a voltmeter to terminals 1 and 2; then 1 and 4.

6 Battery voltage should be available with the ignition key On. If there is no battery voltage present, check the wiring and/or circuit between the fuse box and ignition coil (don't forget to check the fuses). Also check the ground circuit for continuity.

➡Note: Refer to the wiring diagrams at the end of Chapter 12 for wire color identification for testing and additional information on the circuits.

7 If battery voltage is available to the ignition coil, check the ignition coil control circuits as follows (see illustration 6.5):

⁂ CAUTION:

Use only an LED test light to avoid damaging the ECM.

a) Remove fuse no. 32 from the fuse box (this disables the fuel injectors so the engine will not start or flood).

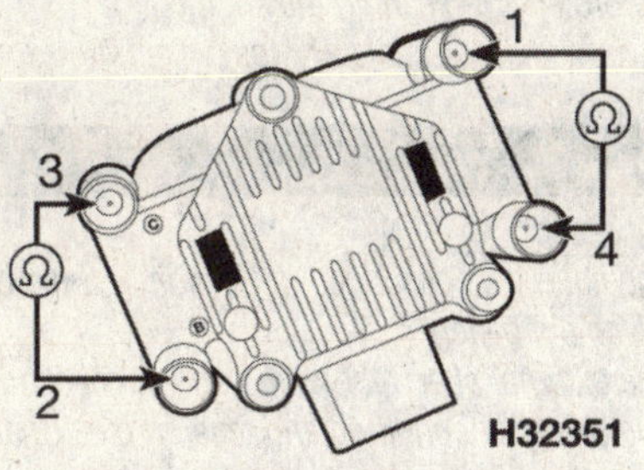

6.4 On 2.0L models, measure the ignition coil secondary resistance across each pair of coil towers

6.2 To use a calibrated ignition tester, disconnect a wire from a spark plug, connect the tester to the spark plug boot, clip the tester to a convenient ground and crank the engine over - if there's enough power to fire the plug, bright blue sparks will be visible between the electrode tip and the tester body (weak sparks or intermittent sparks are the same as no sparks)

b) On 2.0L models, connect an LED test light to terminals 1 and 4 of the coil harness connector. Perform the remainder of the test, then connect the LED test light to terminals 3 and 4 and perform the test.
c) On 1.8L Turbo models, connect an LED test light to terminals 2 and 3 of the coil harness connector.
d) Crank the engine with the starter and confirm that the LED test light flashes as the engine rotates. A flashing LED test light indicates the ECM, camshaft position sensor and engine speed sensor are functioning properly.

8 If a control signal is not present at the ignition coil, refer to Chapter 6A and check the camshaft position sensor and the engine speed sensor. If the sensors are good and there is no control signal, have the ECM checked by a dealer service department or other qualified repair shop.

9 If battery voltage and a control signal exist at the ignition coil and there is no spark, replace the ignition coil.

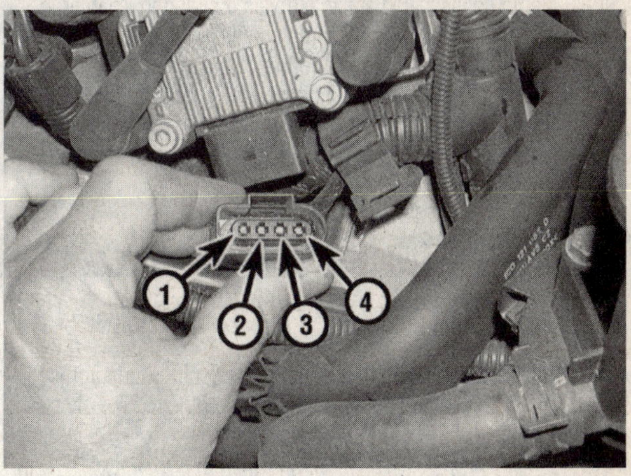

6.5 Ignition coil electrical connector terminal identification

7 Ignition coil(s) - replacement

Refer to illustrations 7.3 and 7.4

1 Remove the engine cover.

2 On 2.0L models, remove the secondary air injection pump (see Chapter 6).

3 On 2.0L models, disconnect the electrical connector from the ignition coil. Label and detach the spark plug wires. Remove the mounting bolts and remove the coil assembly (see illustration).

4 On 1.8L turbo models, disconnect the electrical connector from the ignition coil. Remove the ignition coil mounting screws and pull the coil straight up and out of the cylinder head (see illustration).

5 Installation is the reverse of removal.

7.3 Ignition coil pack mounting screws - 2.0L models

7.4 On 1.8L turbo models, remove the ignition coil mounting screws (arrows) and pull the coil straight up

8 Charging system - general information and precautions

The charging system includes the alternator, a charge indicator light, the battery and the wiring between all the components. The battery supplies electrical power for the vehicle electrical system. The charging system maintains the battery in a fully charged condition. The alternator generates DC voltage to charge the battery and is driven by a serpentine drivebelt at the front of the engine.

The alternator voltage regulator regulates the alternator voltage output. The voltage regulator limits the charging voltage to a maximum preset value. This prevents overcharging the battery. The voltage regulator and brush assembly may be replaced in the event of failure.

The charging system doesn't ordinarily require periodic maintenance. However, the drivebelt, battery, battery cables, wiring and connections should be inspected at the intervals outlined in Chapter 1.

The dashboard warning light should come ON when the ignition key is turned to ON, but it should go off immediately after the engine is started. If it remains on, there is a malfunction in the charging system. Some vehicles are also equipped with a voltmeter. If the voltmeter indicates abnormally high or low voltage, check the charging system (see Section 10).

Be very careful when making electrical circuit connections to a vehicle equipped with an alternator and note the following:

a) *When reconnecting wires to the alternator from the battery, be sure to note the polarity.*
b) *Before using arc welding equipment to repair any part of the vehicle, disconnect the wires from the alternator and the battery terminals.*
c) *Never start the engine with a battery charger connected.*
d) *Always disconnect both battery cables before using a battery charger.*
e) *The alternator is turned by an engine drivebelt which could cause serious injury if your hands, hair or clothes become entangled in it with the engine running.*
f) *Because the alternator is connected directly to the battery, it could arc or cause a fire if overloaded or shorted out.*
g) *Wrap a plastic bag over the alternator and secure it with rubber bands before steam-cleaning the engine.*

9 Charging system - check

Refer to illustration 9.6

1 If a malfunction occurs in the charging circuit, do not immediately assume that the alternator is causing the problem. First check the following items:

a) *The battery cables where they connect to the battery. Make sure the connections are clean and tight.*
b) *Check the battery as described in Section 3. If the battery is defective, replace the battery.*
c) *Check the external alternator wiring and connections.*
d) *Check the drivebelt condition and tension (see Chapter 1).*
e) *Check the alternator mounting bolts for tightness.*
f) *Run the engine and check the alternator for abnormal noise.*

2 The charging system warning light on the instrument cluster

should illuminate when the ignition key is switched on and go off when the engine is running.

3 If the warning light does not illuminate when the ignition key is switched on, switch the ignition off and disconnect the wiring connector from the alternator (do not disconnect the large output wire). Connect the black wire terminal in the harness connector to a good engine ground point using a jumper wire and switch the ignition on. The warning light should illuminate - if it doesn't, there is an open circuit in the black wire between the alternator and the instrument cluster, or the instrument cluster is defective.

4 If the warning light illuminates with the connector grounded but not when connected to the alternator, remove the voltage regulator brush assembly and check the brushes (see Section 11). Measure the brush length and compare it with the value listed in this Chapter's Specifications. Check the brush holder and springs for damage. If the brushes are worn or defective, replace the voltage regulator/brush assembly. If the brushes are good, replace the alternator.

5 If the warning light is illuminated with the engine running, switch the engine off and disconnect the wiring connector from the alternator (do not disconnect the large output wire). Switch the ignition on and check the warning light. If the warning light is still on, the black wire from the alternator to the instrument cluster is grounded or the instrument cluster is defective. If the warning light went out, switch the ignition off, reconnect the connector and proceed with the charging system check.

6 Lift up the battery cover lid and connect a voltmeter to the positive and negative battery terminals (see Illustration). Check the battery voltage with the engine off. It should be approximately 12.4 to 12.6 volts if the battery is fully charged.

7 Start the engine and check the battery voltage again. It should now be greater than the voltage recorded in Step 2, but not more than 14.5 volts. Turn On all the vehicle accessories (air conditioning, rear window defogger, blower motor, etc.) and increase the engine speed to 2000 rpm - the voltage should not drop below the voltage recorded in Step 2.

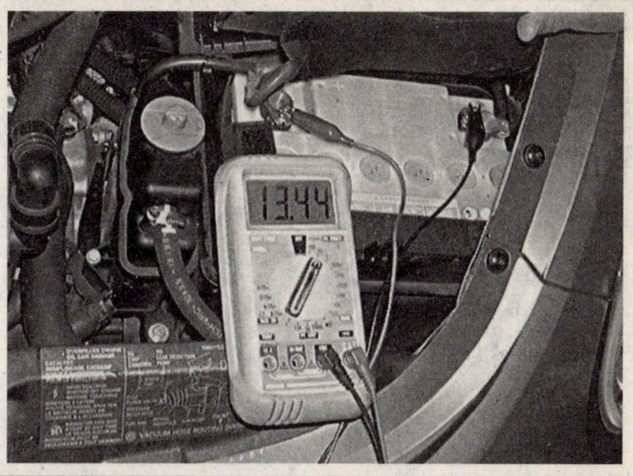

9.6 To measure battery voltage, attach the voltmeter leads to the remote battery terminals (engine OFF) - to measure charging voltage, start the engine

8 If the indicated voltage is greater than the specified charging voltage, replace the voltage regulator (see Section 11).

9 If the indicated voltage reading is less than the specified charging voltage, the alternator is probably defective. Have the charging system checked at a dealer service department or other properly equipped repair facility.

➡Note: Many auto parts stores will bench test an alternator off the vehicle. Refer to your local auto parts store regarding their policy, many will perform this service free of charge.

10 Alternator - removal and installation

➧ Refer to illustrations 10.4 and 10.5

⁂ CAUTION 1:

These models are equipped with an anti-theft radio. Before performing a procedure that requires disconnecting the battery, make sure you have the activation code.

⁂ CAUTION 2:

Disconnecting the battery can cause driveability problems that require a scan tool to remedy. See Section 1 for the use of an auxiliary voltage input device before disconnecting the battery.

1 Disconnect the negative battery cable from the battery.
2 Remove the engine cover.
3 Remove the alternator drivebelt (see Chapter 1).
4 Disconnect the output wire and the electrical connector from the alternator (see illustration).

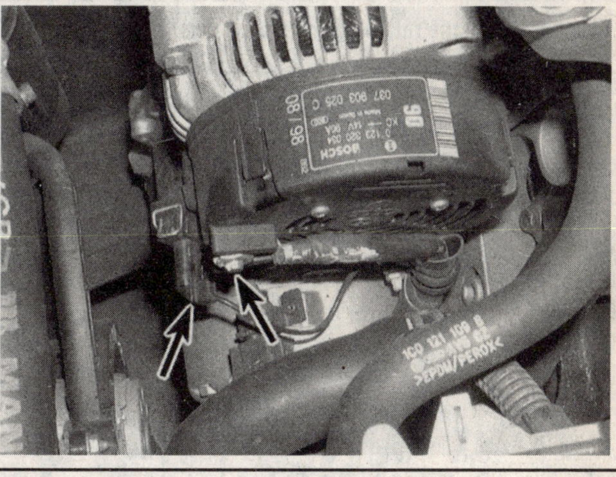

10.4 Disconnect the alternator electrical connections (arrows)

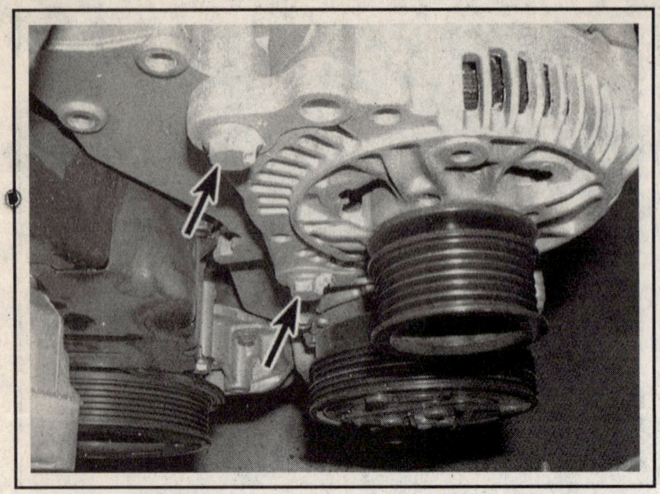

10.5 Remove the alternator mounting bolts (arrows)

5 Remove the mounting bolts and remove the alternator from the engine (see illustration).

6 If you are replacing the alternator, take the old one with you when purchasing a replacement unit. Make sure the new/rebuilt unit looks identical to the old alternator. Look at the terminals - they should be the same in number, size and location as the terminals on the old alternator. Finally, look at the identification numbers - they will be stamped into the housing or printed on a tag attached to the housing. Make sure the numbers are the same on both alternators.

7 Many new/rebuilt alternators do not have a pulley installed, so you may have to switch the pulley from the old unit to the new/rebuilt one. When buying an alternator, find out the shop's policy regarding pulleys; some shops will perform this service free of charge.

8 Installation is the reverse of removal. Tighten the mounting bolts to the torque listed in this Chapter's Specifications.

9 Install the drivebelt (see Chapter 1).

10 Check the charging voltage to verify proper operation of the alternator (see Section 9).

11 Voltage regulator and brushes - replacement

Refer to illustrations 11.3a, 11.3b, 11.3c, 11.4a, 11.4b, 11.5 and 11.6

1 Remove the alternator (see Section 10).

2 Place the alternator on a clean work surface, with the pulley facing down.

3 Remove the retaining screws, pry open the clips and remove the plastic cover from the rear of the alternator (see illustrations).

4 Remove the voltage regulator/brush assembly screws and remove the unit from the alternator (see illustrations).

5 Measure the brush contact free length and compare your measurement with the value listed in this Chapter's specifications (see illus-

11.3a Remove the retaining screws (arrows) . . .

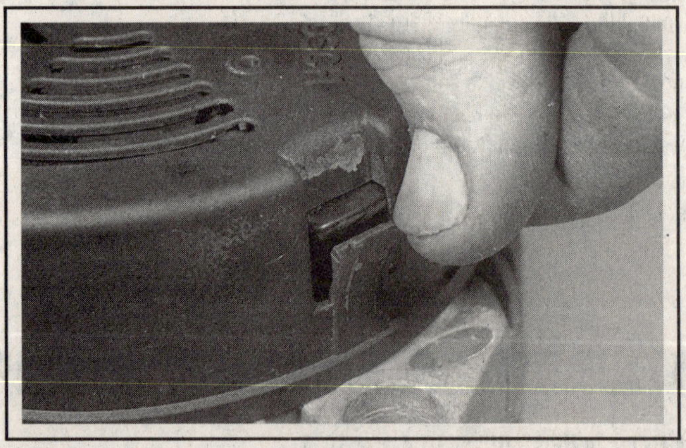

11.3b . . . then pry open the clips . . .

11.3c . . . and remove the plastic cover from the rear of the alternator

11.4a Unscrew the voltage regulator/ brush assembly screws . . .

11.4b . . . and remove the assembly from the alternator

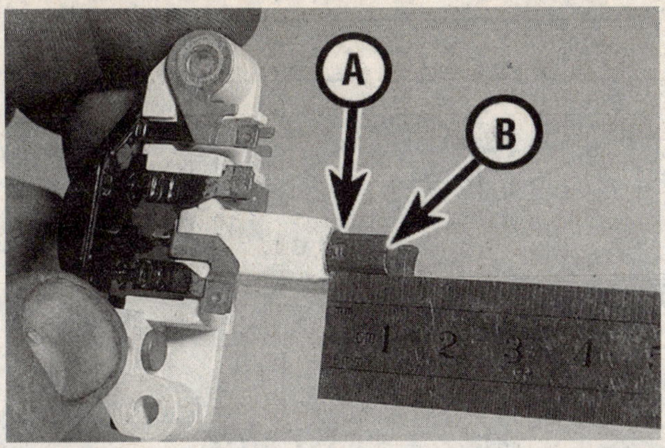

11.5 Measure the free length of the brush contacts - take the measurement from the manufacturers emblem (A) etched in the side of the brush contact, to the shallowest part of the curved end face of the brush (B)

tration). Replace the unit if the brushes are worn below the minimum value.

6 Inspect the brush contact surface of the slip rings (see illustration). Minor imperfections may be cleaned with crocus cloth. If they are excessively worn, burnt or pitted, replace the alternator.

11.6 Inspect the surfaces of the slip rings (arrows), at the end of the alternator shaft

7 Installation is the reverse of removal.

8 Install the alternator and check the charging voltage as described in Section 9.

12 Starting system - general information and precautions

The sole function of the starting system is to turn over the engine quickly enough to allow it to start. The starting system consists of the battery, starter circuit, starter motor assembly and the wiring connecting the components.

The starter motor assembly is bolted to the front of the transaxle bellhousing.

When the ignition key is turned to the START position, the starter solenoid is actuated through the starter control circuit which includes a starter relay located in the relay panel. The starter solenoid then connects the battery to the starter motor. The battery supplies the electrical energy to the starter motor, which does the actual work of cranking the engine.

Always observe the following precautions when working on the starting system:

- a) *Excessive cranking of the starter motor can overheat it and cause serious damage. Never operate the starter motor for more than 15 seconds at a time without pausing to allow it to cool for at least two minutes.*
- b) *The starter is connected directly to the battery and could arc or cause a fire if mishandled, overloaded or shorted.*
- c) *Always detach the cable from the negative terminal of the battery before working on the starting system (see Section 1).*

13 Starter motor and circuit - check

Refer to illustration 13.4

1 If a malfunction occurs in the starting circuit, do not immediately assume that the starter is causing the problem. First, check the following items:

- a) *Make sure the battery cable clamps, where they connect to the battery, are clean and tight.*
- b) *Check the condition of the battery cables (see Section 4). Replace any defective battery cables with new parts.*
- c) *Test the condition of the battery (see Section 3). If it does not pass all the tests, replace it with a new battery.*
- d) *Check the starter motor wiring and connections.*
- e) *Check the starter motor mounting bolts for tightness.*
- f) *Check the related fuses in the fuse box (see Chapter 12). If they're blown, determine the cause and repair the circuit.*
- g) *Check the ignition switch circuit for correct operation (see Chapter 12).*
- h) *Check the starter relay (see Chapter 12).*
- i) *Check the operation of the Park/Neutral position switch (see Chapter 7B) or the clutch start switch (see Chapter 8). These systems must operate correctly to provide battery voltage to the starter solenoid.*

2 If the starter does not activate when the ignition switch is turned to the start position, check for battery voltage to the starter solenoid. This will determine if the solenoid is receiving the correct voltage from the ignition switch. Install a 12-volt test light or a voltmeter to the starter solenoid terminal. While an assistant turns the ignition switch to the start position, observe the test light or voltmeter. The test light should shine brightly or battery voltage should be indicated on the voltmeter. If voltage is not available to the starter solenoid, refer to the wiring diagrams in Chapter 12 and check the fuses, switches and starter relay in series with the starting system. If voltage is available but there is no movement from the starter motor, remove the starter from the engine (see Section 14) and bench test the starter (see Step 4).

3 If the starter turns over slowly, check the starter cranking voltage and the current draw from the battery. This test must be performed with

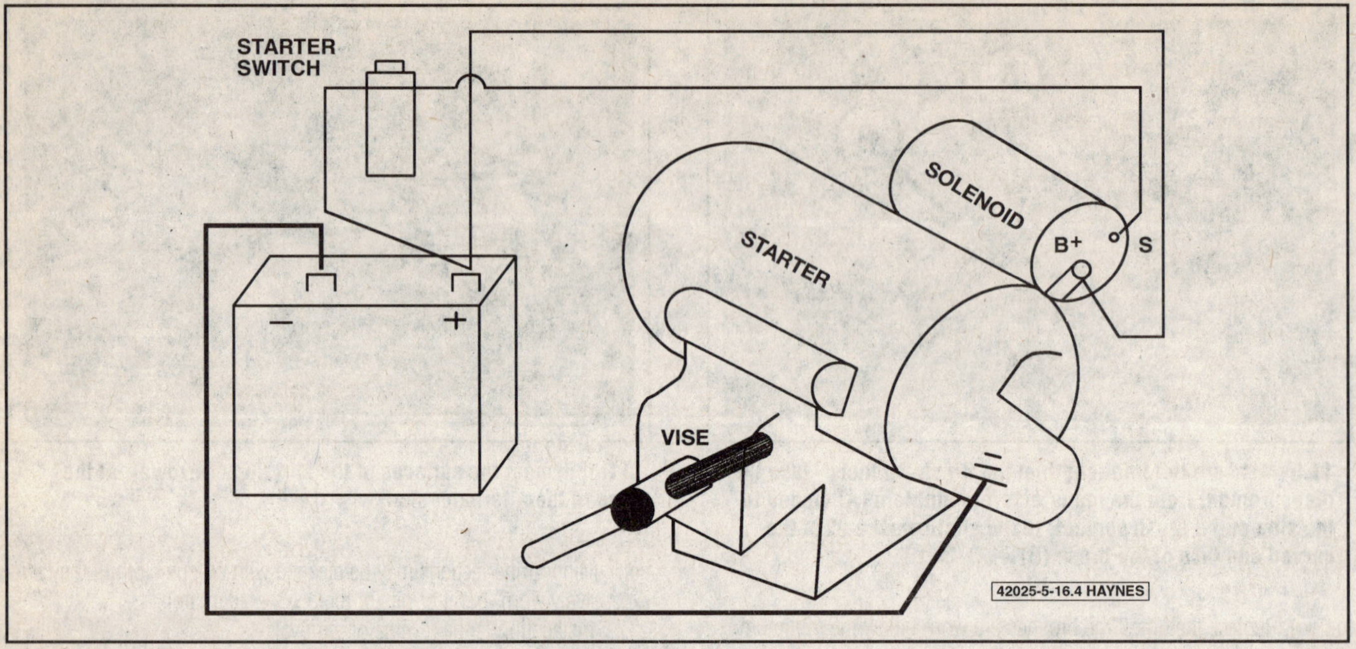

13.4 Starter motor bench testing details

the starter assembly on the engine. Crank the engine over (for 10 seconds or less) and observe the battery voltage. It should not drop below 8.5 volts. Also, observe the current draw using an ammeter. Typically a starter amperage draw should not exceed 200 amps. If the starter motor amperage draw is excessive, have it tested by a dealer service department or other qualified repair shop. There are several conditions that may affect the starter cranking potential. The battery must be in good condition and the battery cold-cranking rating must not be underrated for the particular application. Be sure to check the battery specifications carefully. The battery terminals and cables must be clean and not corroded. Also, in cases of extreme cold temperatures, make sure the battery and/or engine block is warmed before performing the tests.

4 If the starter is receiving voltage but does not activate, remove and check the starter motor assembly on the bench. Most likely the starter motor or solenoid is defective. In some rare cases, the engine may be seized so be sure to try and rotate the crankshaft pulley (see Chapter 2A or 2B) before proceeding. With the starter assembly mounted in a vise on the bench, install one jumper cable from the positive terminal of a test battery to the B+ terminal on the starter. Install another jumper cable from the negative terminal of the battery to the body of the starter (see illustration). Install a starter switch and apply battery voltage to the solenoid S terminal (for 10 seconds or less) and observe the solenoid plunger, shift lever and overrunning clutch extend and rotate the pinion drive. If the pinion drive extends but does not rotate, the solenoid is operating but the starter motor is defective. If there is no movement but the solenoid clicks, the solenoid and/or the starter motor is defective. If the solenoid plunger extends and rotates the pinion drive, the starter assembly is operating properly.

14 Starter motor - removal and installation

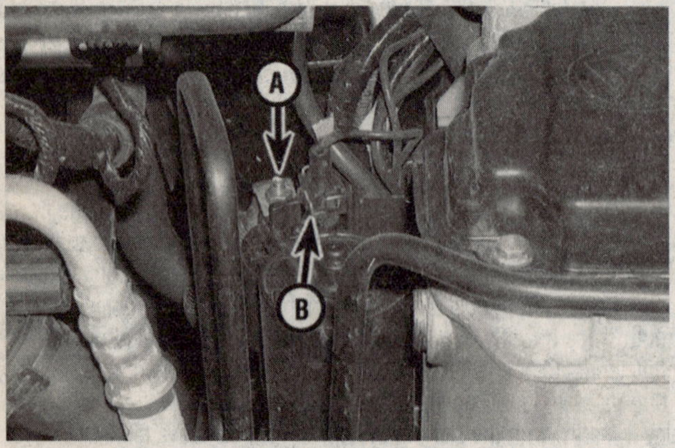

14.4 Remove the nut (A) and disconnect the battery cable from the starter motor - disconnect the connector (B) from the starter solenoid

➧ Refer to illustrations 14.4 and 14.5

✱✱ CAUTION 1:

These models are equipped with an anti-theft radio. Before performing a procedure that requires disconnecting the battery, make sure you have the activation code.

✱✱ CAUTION 2:

Disconnecting the battery can cause driveability problems that require a scan tool to remedy. See Section 1 for the use of an auxiliary voltage input device before disconnecting the battery.

1 Disconnect the negative battery cable from the battery.
2 Raise the vehicle and support it securely on jackstands.
3 Remove the splash pan from under the engine.

➡Note: On some models, the starter is accessible from above if the battery and battery tray are removed (see Section 3). Unbolt the clamp securing the power steering fluid line and set the line aside for starter access.

4 Disconnect the wires from the terminals on the starter motor solenoid (see illustration).

5 Remove the starter mounting bolts (see illustration).

6 Withdraw the starter from the transaxle bellhousing.

7 Installation is the reverse of removal.

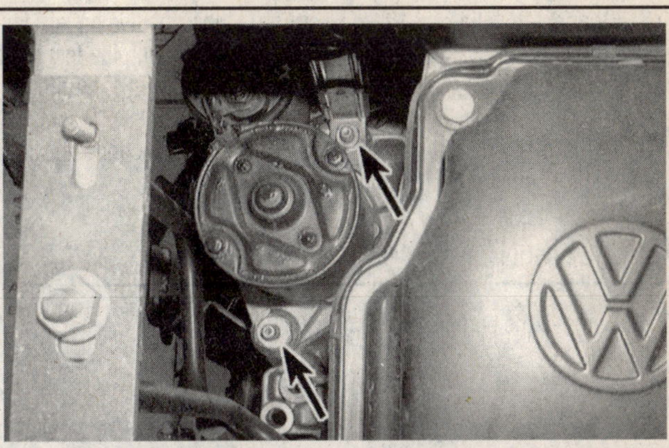

14.5 Remove the starter mounting bolts (arrows)

15 Glow plug system (diesel engine) - general information

To assist cold starting, diesel engine models are equipped with a pre-heating system, which comprises four glow plugs, a glow plug relay, a dash-mounted warning lamp, the Engine Control Module (ECM) and the associated electrical wiring.

The glow plugs are miniature electric heating elements, encapsulated in a metal case with a probe at one end and electrical connection at the other. Each combustion chamber has a glow plug threaded into it. When the glow plug is energized, the air in the combustion chamber is heated, allowing optimum combustion temperature to be achieved more readily before fuel is injected into the cylinder.

The duration of the pre-heating period is governed by the Engine Control Module, which monitors the temperature of the engine via the coolant temperature sensor and alters the pre-heating time to suit the conditions.

A dash-mounted warning lamp informs the driver that pre-heating is taking place. The lamp extinguishes when sufficient pre-heating has taken place to allow the engine to be started, but power will still be supplied to the glow plugs for a further period until the engine is started. If no attempt is made to start the engine, the power supply to the glow plugs is switched off to prevent battery drain and glow plug burn-out.

After the engine has been started, the glow plugs continue to operate for a further period of time. This helps to improve fuel combustion while the engine is warming up, resulting in quieter, smoother running and reduced exhaust emissions.

The Check Engine warning lamp will illuminate during normal driving if a pre-heating system malfunction occurs and a diagnostic trouble code will be stored in the ECM memory. Refer to Chapter 6B for additional information on the On-Board Diagnostic system.

16 Glow plugs - check and replacement

CHECK

➧ Refer to illustration 16.5

1 Remove the engine cover.

2 Disconnect the electrical connector from the engine coolant temperature sensor (see Chapter 6B).

➡Note: This will allow the glow plugs to be energized, regardless of engine temperature.

3 Carefully disconnect the electrical connector for each glow plug and position the bus connector so the electrical terminals will not contact any engine component.

4 Connect a 12-volt test light or voltmeter to one of the glow plug terminals and switch the ignition key on. Battery voltage should be indicated on the meter for approximately 20 seconds. If no voltage is indicated, check the glow plug fuses and glow plug relay (see Chapter 12).

➡Note: The glow plug fuses are located in the fuse box on the battery cover. The glow plug relay is located in the relay panel under the left-side of the instrument panel. Test each glow plug terminal, then switch the ignition key off.

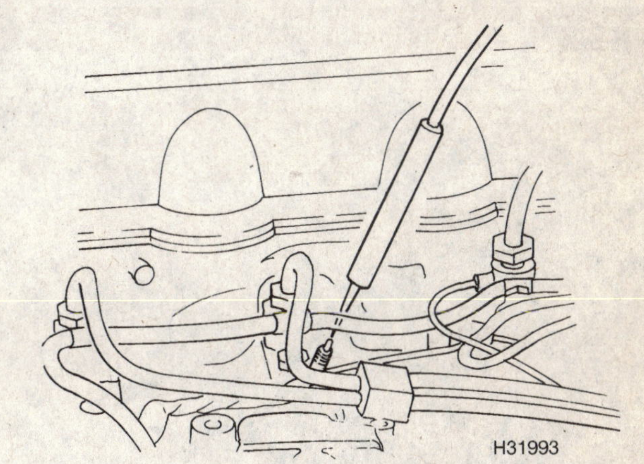

16.5 To test a glow plug, connect a 12-volt test light to the positive battery terminal, disconnect the electrical connector and touch the glow plug terminal with the test light - if the test light illuminates, the glow plug is good

5 Connect the clip of a 12-volt test light to the positive battery terminal. Touch the electrical terminal on the glow plug with the test light tip (see illustration). If the glow plug is good, the test light will illuminate.

6 Replace any defective glow plugs. Install the bus connector and connect the engine coolant temperature sensor.

REPLACEMENT

7 Remove the engine cover.

8 Carefully disconnect the electrical terminal for each glow plug and position the bus connector aside.

9 Using a deep socket, remove the glow plugs from the cylinder head.

10 Installation is the reverse of removal. Tighten the glow plugs to the torque listed in this Chapter's Specifications.

Specifications

General

Alternator brush wear limit	0.19 inch (5.0 mm)
Battery voltage	
Engine off	12.0 to 12.5 volts
Engine running	13.5 to 14.5 volts
Ignition coil secondary resistance (2.0L engine)	4,000 to 6,000 ohms
Spark plug wire resistance (2.0L engine)	4,000 to 6,000 ohms
Spark plug boot resistance (1.8L engine)	2,000 ohms

Torque specifications	Nm	Ft-lbs (unless otherwise indicated)
Alternator mounting bolts	25 to 30	18 to 22
Battery hold-down clamp bolt	22	16
Glow plugs (diesel engine)	15	132 in-lbs
Starter mounting bolts	60	44

6A

EMISSIONS AND ENGINE CONTROL SYSTEMS: GASOLINE ENGINES

Section

1 General information

Refer to illustrations 1.1a, 1.1b, 1.7a and 1.7b

To prevent pollution of the atmosphere from incompletely burned and evaporating gases, and to maintain good driveability and fuel economy, a number of emission control systems are incorporated (see illustrations). They include the:

Electronic engine control system
Crankcase ventilation system
Evaporative emissions control system
Secondary air injection system
Catalytic converter

All of these systems are linked, directly or indirectly, to the emission control system.

The Sections in this Chapter include general descriptions, checking procedures within the scope of the home mechanic (when possible) and component replacement procedures for each of the systems listed above.

Before assuming that an emissions control system is malfunctioning, check the fuel and ignition systems carefully. The diagnosis of some emission control devices requires specialized tools, equipment and training. If checking and servicing become too difficult or if a procedure is beyond your ability, consult a dealer service department or other properly equipped repair facility. Remember, the most frequent cause of emissions problems is simply a loose or broken vacuum hose or wire, so always check the hose and wiring connections first.

This doesn't mean, however, that emission control systems are particularly difficult to maintain and repair. You can quickly and easily perform many checks and service procedures at home with common test equipment and hand tools.

Note: Because of a Federally mandated warranty which covers the emission control system components, check with your dealer about warranty coverage before working on any emissions-related systems. Once the warranty has expired, you may wish to perform some of the component checks and/or replacement procedures in this Chapter to save money.

Pay close attention to any special precautions outlined in this Chapter. It should be noted that the illustrations of the various systems may not exactly match the system installed on the vehicle you're working on because of changes made by the manufacturer during production or from year-to-year.

A Vehicle Emissions Control Information (VECI) label is located in the engine compartment (see illustrations). This label contains important emissions specifications and adjustment information, as well as a vacuum hose schematic with emissions components identified. When servicing the engine or emissions systems, the VECI label in your particular vehicle should always be checked for up-to-date information.

1.1a Typical emission and engine control system components - 2.0L models

1 *Crankcase ventilation hose*
2 *Secondary air injection solenoid valve*
3 *Mass airflow sensor/intake air temperature sensor*
4 *Engine coolant temperature sensor*
5 *Secondary air injection pump*
6 *Knock sensor*
7 *Camshaft position sensor*
8 *Throttle control module*

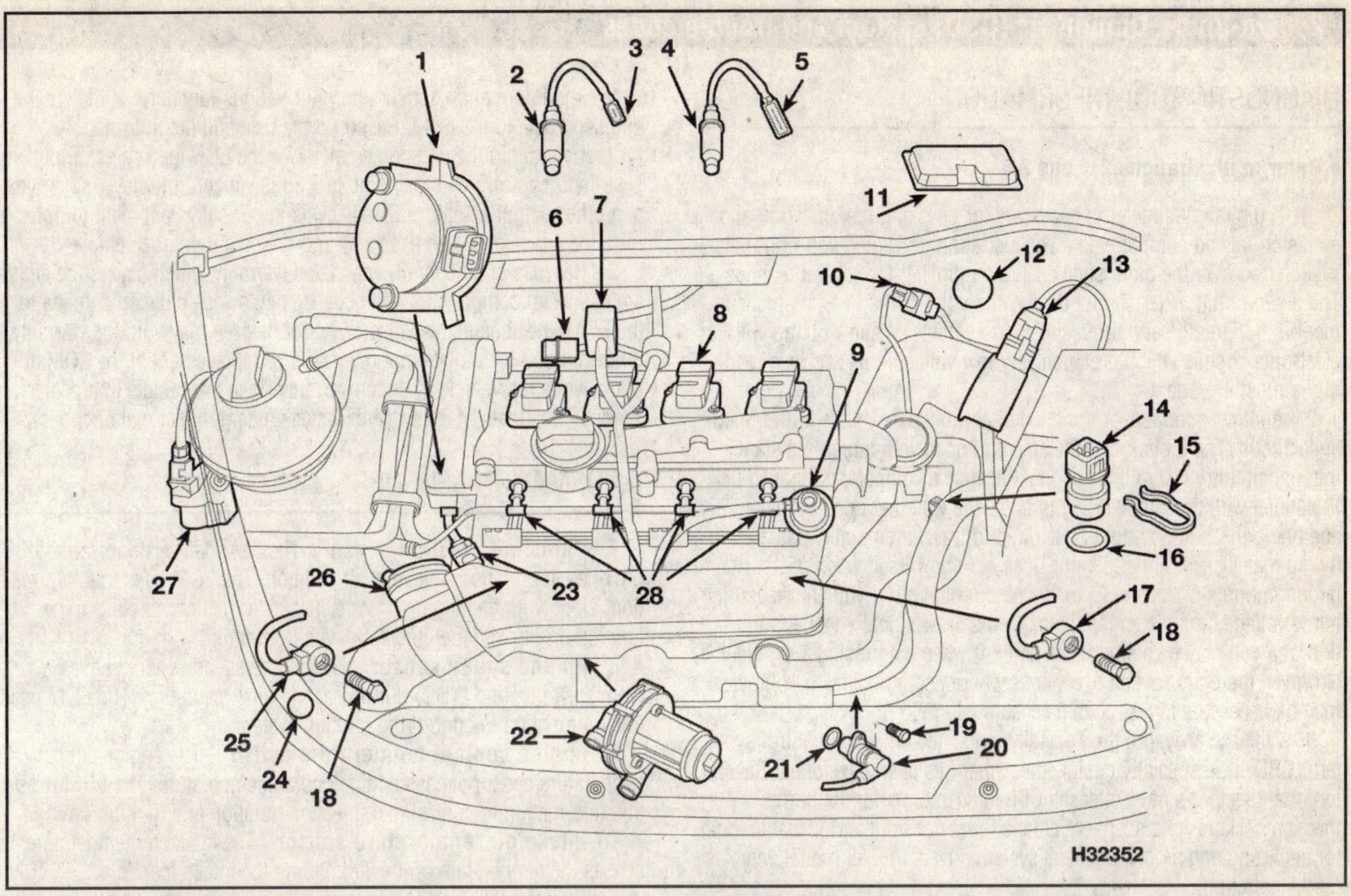

1.1b Typical emission and engine control system components - 1.8L turbo models

1. *Camshaft position sensor*
2. *Oxygen sensor - pre catalyst*
3. *Oxygen sensor connector - pre catalyst*
4. *Oxygen sensor - post catalyst*
5. *Oxygen sensor connector - post catalyst*
6. *Secondary air injection solenoid valve*
7. *Overrun recirculation valve*
8. *Ignition coil*
9. *Fuel pressure regulator*
10. *Wastegate regulator valve*
11. *Engine Control Module*
12. *Secondary air injection pump relay*
13. *Mass airflow sensor*
14. *Engine coolant temperature sensor*
15. *Retaining clip*
16. *O-ring*
17. *Knock sensor*
18. *Bolt*
19. *Bolt*
20. *Engine speed sensor*
21. *O-ring*
22. *Secondary air injection pump*
23. *Intake air temperature sensor*
24. *Boost pressure sensor*
25. *Knock sensor*
26. *Throttle control module*
27. *EVAP purge valve*
28. *Fuel injector*

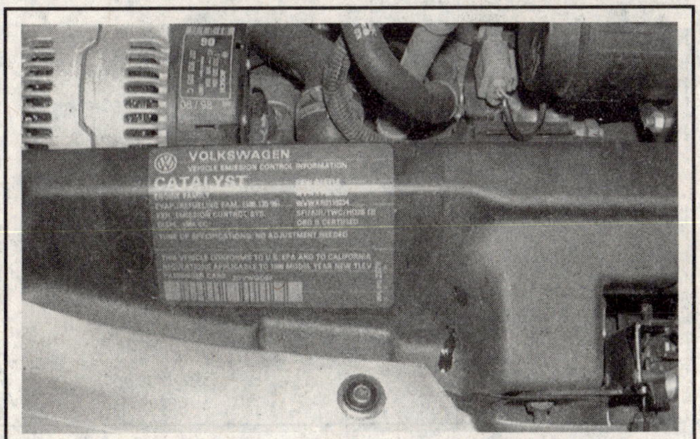

1.7a The Vehicle Emission Control Information (VECI) label is located in the engine compartment and contains information on the emission devices on your vehicle

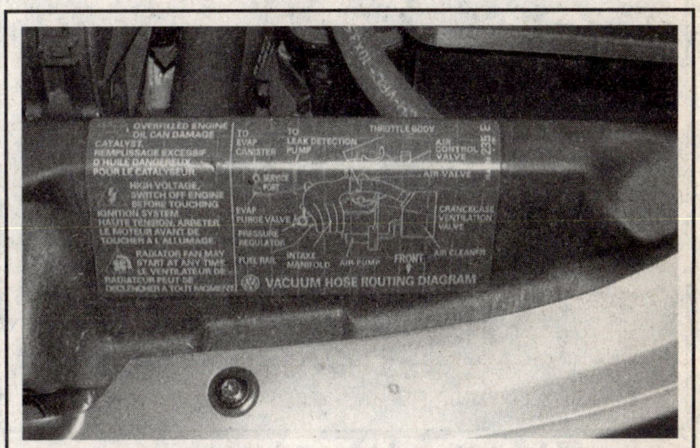

1.7b Vacuum hose routing diagram of the emissions system vacuum hoses

2 On-Board Diagnostic (OBD) system and trouble codes

DIAGNOSTIC TOOL INFORMATION

➧ **Refer to illustrations 2.1 and 2.3**

1 A digital multimeter is necessary for checking fuel injection and emission related components (see illustration). A digital volt-ohmmeter is preferred over the older style analog multimeter for several reasons. The analog multimeter cannot display the volts-ohms or amps measurement in hundredths and thousandths increments. When working with electronic circuits which are often very low voltage, this accurate reading is most important.

2 Another good reason for the digital multimeter is the high impedance circuit. The digital multimeter is equipped with a high resistance internal circuitry (10 million ohms). Because a voltmeter is hooked up in parallel with the circuit when testing, it is vital that none of the voltage being measured should be allowed to travel the parallel path set up by the meter itself. This dilemma does not show itself when measuring larger amounts of voltage (9 to 12 volt circuits) but if you are measuring a low voltage circuit such as the oxygen sensor signal voltage, a fraction of a volt may be a significant amount when diagnosing a problem. However, there are several exceptions where using an analog voltmeter may be necessary to test certain sensors.

3 With the arrival of the Federally mandated emission control system (OBD-II), a specially designed scanner has been developed. Several tool manufacturers have released OBD-II scan tools for the home mechanic. Hand-held scanners are the most powerful and versatile tools for analyzing engine management systems used on later model vehicles (see illustration).

ON-BOARD DIAGNOSTIC SYSTEM GENERAL DESCRIPTION

4 All models described in this manual are equipped with the second generation On-Board Diagnostic (OBD-II) system. The system consists of an on-board computer, known as the Engine Control Module (ECM), information sensors and output actuators.

5 The information sensors monitor various functions of the engine and send data to the ECM. Based on the data and the information programmed into the computer's memory, the ECM generates output signals to control various engine functions via control relays, solenoids and other output actuators. The ECM is specifically calibrated to optimize the emissions, fuel economy and driveability of the vehicle.

6 Because of a Federally mandated warranty which covers the emissions system components and because any owner-induced damage to the ECM, the sensors and/or the control devices may void the warranty, it isn't a good idea to attempt diagnosis or replacement of the ECM at home while the vehicle is under warranty. Take the vehicle to a dealer service department if the ECM or a system component malfunctions.

INFORMATION SENSORS

7 **Camshaft position sensor** - The camshaft position sensor provides information on camshaft position. The ECM uses this information, along with the engine speed sensor information, to control the ignition timing and fuel injection synchronization.

8 **Engine speed sensor** - The engine speed sensor senses crankshaft position (TDC) during each engine revolution. The ECM uses this information to control the ignition system.

9 **Engine coolant temperature sensor** - The engine coolant temperature sensor senses engine coolant temperature. The ECM uses this information to control fuel injection duration and ignition timing.

10 **Intake air temperature sensor** - The intake air temperature senses the temperature of the air entering the intake manifold. The ECM uses this information to control fuel injection duration.

11 **Knock sensor** - The knock sensor is a piezoelectric element that detects the sound of engine detonation, or "pinging." The ECM uses the input signal from the knock sensor to recognize detonation and retard spark advance to avoid engine damage.

12 **Mass airflow sensor** - The mass airflow sensor measures the amount of air passing through the sensor body and ultimately entering the engine. The ECM uses this information to control fuel delivery.

13 **Oxygen sensor** - The oxygen sensors generate a voltage signal that varies with the difference between the oxygen content of the exhaust

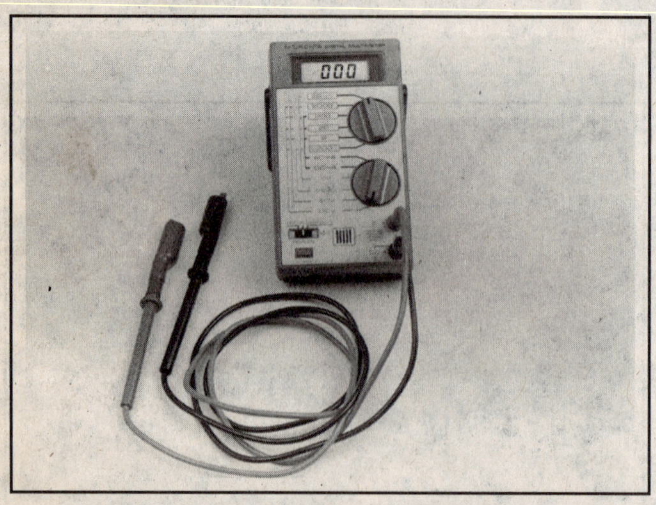

2.1 Digital multimeters can be used for testing all types of circuits; because of their high impedance, they are much more accurate than analog meters for measuring low-voltage computer circuits

2.3 Scanners like these from Actron and AutoXray are powerful diagnostic aids - they can tell you just about anything you want to know about your engine management system

and the oxygen in the surrounding air. The ECM uses this information to determine if the fuel system is running rich or lean.

14 **Throttle position sensor** - The throttle position sensor senses throttle movement and position. This signal enables the ECM to determine when the throttle is closed, in a cruise position, or wide open. The ECM uses this information to control fuel delivery and ignition timing. The throttle position sensor is a component of the throttle control module.

15 **Closed throttle position switch** - The closed throttle position switch signals the ECM when the throttle is in the fully closed position. The ECM uses this information to control the engine idle speed. The closed throttle position switch is a component of the throttle control module.

16 **Vehicle speed sensor** - The vehicle speed sensor provides information to the ECM to indicate vehicle speed.

17 **Miscellaneous ECM inputs** - In addition to the various sensors, the ECM monitors various switches and circuits to determine vehicle operating conditions. The switches and circuits include:

a) Accelerator pedal position (1.8L turbo engine)
b) Air conditioning system
c) Antilock brake system
d) Barometric pressure sensor (inside ECM)
e) Battery voltage
f) Brake switch
g) Clutch pedal switch
h) Cruise control system
i) EVAP system
j) Park/neutral position switch
k) Power steering pressure switch
l) Turbocharger boost pressure (1.8L turbo engine)
m) Sensor signal and ground circuits
n) Transaxle control module

OUTPUT ACTUATORS

18 **Check Engine light** - The ECM will illuminate the Check Engine light if a malfunction in the electronic engine control system occurs.

19 **EVAP canister purge valve solenoid** - The evaporative emission canister purge valve solenoid is operated by the ECM to purge the fuel vapor canister and route fuel vapor to the intake manifold for combustion.

20 **EVAP system leak detection pump** - The EVAP system is equipped with a self-diagnostic leak detection system. The ECM controls the operation of the leak detection pump.

21 **Secondary air injection pump and vacuum valve/solenoid** - The ECM operates the secondary air injection pump and opens the vacuum valve to inject fresh air into the exhaust stream, lowering emission levels under certain operating conditions.

22 **Fuel injectors** - The ECM opens the fuel injectors individually in firing order sequence. The ECM also controls the time the injector is held open (pulse width). The pulse width of the injector (measured in milliseconds) determines the amount of fuel delivered. For more information on the fuel delivery system and the fuel injectors, including injector replacement, refer to Chapter 4A.

23 **Fuel pump relay** - The fuel pump relay is activated by the ECM with the ignition switch in the Start or Run position. When the ignition switch is turned on, the relay is activated to supply initial line pressure to the system. For more information on fuel pump check and replacement, refer to Chapter 4A.

24 **Throttle valve actuator (2.0L engine)** - The throttle valve actuator controls the throttle plate at idle position. The more the throttle plate is opened, the higher the idle speed. The throttle plate opening and the resulting idle speed is controlled by the ECM. The throttle valve actuator is a component of the throttle control module.

25 **Throttle valve actuator (1.8L turbo engine)** - The throttle valve actuator controls the throttle plate at all engine speeds. The more the throttle plate is opened, the higher the engine speed. The throttle plate opening and the resulting engine speed is controlled by the ECM. The throttle valve actuator is a component of the throttle control module.

26 **Ignition coils** - The ECM controls spark delivery and ignition timing depending on engine operation conditions. Refer to Chapter 5 for more information on the ignition system.

27 **Oxygen sensor heaters** - Each oxygen sensor is equipped with a heating element. Heating the oxygen sensor allows it to reach operating temperature quickly. The ECM controls the oxygen sensor heaters.

28 **Turbocharger boost control (1.8L turbo engine)** - The ECM monitors intake manifold pressure and controls the turbocharger wastegate with the boost pressure control valve. The engine control system calculates the engine torque needed depending on driver demand and engine operating conditions, the ECM will then adjust the boost pressure to meet the demands.

OBTAINING DIAGNOSTIC TROUBLE CODES

Refer to illustration 2.30

Note: The diagnostic trouble codes on all models can only be extracted from the Engine Control Module (ECM) using a specialized scan tool. Have the vehicle diagnosed by a dealer service department or other qualified automotive repair facility if the proper scan tool is not available.

29 The ECM will illuminate the CHECK ENGINE light (also known as the Malfunction Indicator Lamp) on the dash if it recognizes a fault in the system. The light will remain illuminated until the problem is repaired and the code is cleared or the ECM does not detect any malfunction for several consecutive drive cycles.

30 The diagnostic codes for the On-Board Diagnostic (OBD) system can only be extracted from the ECM using a scan tool. The scan tool is programmed to interface with the OBD system by plugging into the diagnostic connector (see illustration). When used, the scan tool has

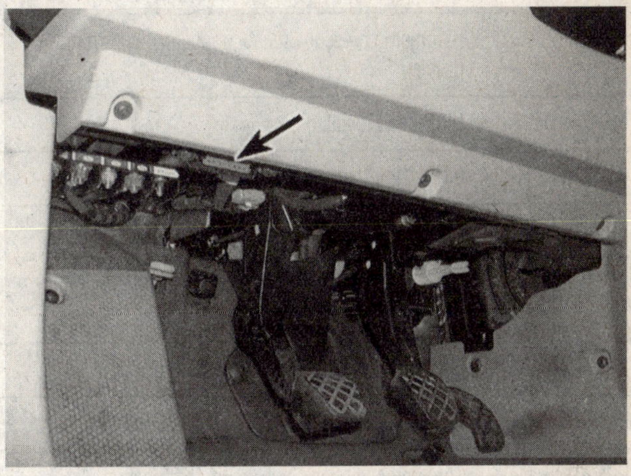

2.30 The diagnostic connector is typically located under the instrument panel

the ability to diagnose in-depth driveability problems and it allows freeze frame data to be retrieved from the ECM stored memory. Freeze frame data is an OBD II ECM feature that records all related sensor and actuator activity on the ECM data stream whenever an engine control or emissions fault is detected and a trouble code is set. This ability to look at the circuit conditions and values when the malfunction occurs provides a valuable tool when trying to diagnose intermittent driveability problems. If the tool is not available and intermittent driveability problems exist, have the vehicle checked at a dealer service department or other qualified repair shop.

CLEARING DIAGNOSTIC TROUBLE CODES

31 After the system has been repaired, the codes must be cleared from the ECM memory using a scan tool. Do not attempt to clear the codes by disconnecting battery power. If battery power is disconnected from the ECM, the ECM will lose the current engine operating parameters and driveability will suffer until the ECM is programmed with a scan tool.

32 Always clear the codes from the ECM before starting the engine after a new electronic emission control component is installed onto the engine. The ECM stores the operating parameters of each sensor. The ECM may set a trouble code if a new sensor is allowed to operate before the parameters from the old sensor have been erased.

DIAGNOSTIC TROUBLE CODE IDENTIFICATION

33 The accompanying list of diagnostic trouble codes is a compilation of all the codes that may be encountered using a generic scan tool. Additional trouble codes may be available with the use of the manufacturer specific scan tool. Not all codes pertain to all models and not all codes will illuminate the Check Engine light when set. All models require a scan tool to access the diagnostic trouble codes.

TROUBLE CODES

Code	Code Identification
P0011-P0013	Intake cam position sensor
P0016	Crankshaft sensor circuit, correlations fault
P0030-P0032	Oxygen sensor heater circuit
P0036-P0038	Oxygen sensor circuit, sensor 2
P0042-P0044	Oxygen sensor circuit, sensor 3
P0068	MAP/MAF, TPS circuit
P0097	IAT circuit, low
P0098	IAT circuit, high
P0101	MAF circuit, range
P0102	Mass air flow sensor circuit, low input
P0103	Mass air flow sensor circuit, high input
P0106	Barometric pressure sensor, range or performance problem
P0107-P0108	MAP circuit, voltage
P0112	Intake air temperature circuit, low input
P0113	Intake air temperature circuit, high input
P0116	Engine coolant temperature circuit, range or performance problem
P0117	Engine coolant temperature circuit, low input
P0118	Engine coolant temperature circuit, high input
P0120	Throttle position sensor circuit, malfunction

Code	Code Identification
P0121	Throttle position sensor circuit, range or performance problem
P0122	Throttle position sensor circuit, low input
P0123	Throttle position sensor circuit, high input
P0125	Insufficient coolant temperature for closed loop fuel control
P0130	Oxygen sensor circuit malfunction (pre-converter sensor)
P0131	Oxygen sensor circuit, low voltage (pre-converter sensor)
P0132	Oxygen sensor circuit, high voltage (pre-converter sensor)
P0133	Oxygen sensor circuit, slow response (pre-converter sensor)
P0134	Oxygen sensor circuit - no activity detected (pre-converter sensor)
P0135	Oxygen sensor heater circuit malfunction (pre-converter sensor)
P0136	Oxygen sensor circuit malfunction (post-converter sensor)
P0137	Oxygen sensor circuit, low voltage (post-converter sensor)
P0138	Oxygen sensor circuit, high voltage (post-converter sensor)
P0139	Oxygen sensor circuit, slow response (post-converter sensor)

Code	Code Identification
P0140	Oxygen sensor circuit - no activity detected (post-converter sensor)
P0141	Oxygen sensor heater circuit malfunction (post-converter sensor)
P0142	Oxygen sensor circuit, sensor 2, slow response
P0143	Oxygen sensor circuit, sensor 3, low voltage
P0144	Oxygen sensor circuit, sensor 3, high voltage
P0145	Oxygen sensor circuit, sensor 3, slow response
P0146	Oxygen sensor circuit, sensor 3, no response
P0147	Oxygen sensor circuit, sensor 3, heater circuit
P0170	Fuel trim malfunction
P0171	System too lean
P0172	System too rich
P0201	Injector circuit, cylinder 1, open
P0202	Injector circuit, cylinder 2, open
Code	Code Identification
P0203	Injector circuit, cylinder 3, open
P0204	Injector circuit, cylinder 4, open
P0219	Engine overspeed
P0221-P0230	Throttle pedal position sensor
P0234	Turbocharger overboost condition
P0235-P0246	Turbocharger boost sensor circuit, range or performance problem
P0237	Turbocharger boost sensor circuit, low input
P0238	Turbocharger boost sensor circuit, high input
P0261	Injector circuit, cylinder 1, current low
P0262	Injector circuit, cylinder 1, current high
P0264	Injector circuit, cylinder 2, current low
P0265	Injector circuit, cylinder 2, current high
P0267	Injector circuit, cylinder 3, current low
P0268	Injector circuit, cylinder 3, current high
P0270	Injector circuit, cylinder 4, current low
P0271	Injector circuit, cylinder 4, current low
P0299	Turbocharger, low boost

Code	Code Identification
P0300	Random/multiple cylinder misfire detected
P0301	Cylinder no. 1 misfire detected
P0302	Cylinder no. 2 misfire detected
P0303	Cylinder no. 3 misfire detected
P0304	Cylinder no. 4 misfire detected
P0318	Rough road sensor, circuit A
P0321	Engine speed sensor, range or performance problem
P0322	Engine speed sensor, no input
P0324	Knock sensor, system error
P0327	Knock sensor no. 1 circuit, low input
P0328	Knock sensor no. 1 circuit, high input
P0332	Knock sensor no. 2 circuit, low input
P0333	Knock sensor no.2 circuit, high input
P0340	Camshaft position sensor, A circuit
P0341	Camshaft position sensor circuit, range or performance problem
P0342	Camshaft position sensor circuit, low input
P0343	Camshaft position sensor circuit, high input
P0351	Ignition coil, A circuit
P0352	Ignition coil, B circuit
P0353	Ignition coil, C circuit
P0354	Ignition coil, D circuit
P0411	Secondary air injection system, incorrect flow
P0412-P0418	Secondary air injection system, switching valve circuit
P0420	Catalyst system efficiency below threshold
P0422	Catalyst system efficiency below threshold
P0440	Evaporative emission control system malfunction
P0441	Evaporative emission control system, incorrect purge flow
P0442	Evaporative emission control system, small leak detected
P0445	Evaporative emission control system, large leak detected
P0458-P0459	Evaporative emission control system, purge valve circuit

Code	Code Identification
P0501	Vehicle speed sensor circuit, range or performance problem
P0506	Idle control system, rpm lower than expected
P0507	Idle control system, rpm higher than expected
P0510	Closed throttle position switch malfunction
P0532	Air conditioning refrigerant pressure sensor circuit, low input
P0533	Air conditioning refrigerant pressure sensor circuit, high input
P0560	System voltage malfunction
P0562	System voltage low
P0563	System voltage high
P0571	Brake switch malfunction
P0600	Engine Control Module, communications error
P0601-P0602	Engine Control Module, programming error
P0604	Engine Control Module, memory error (RAM)
P0605	Engine Control Module, memory error (ROM)
P0606 and P0613	Transmission Control Module, programming error
P0614	ECM/TCM, incompatibility
P0627-P0629	Fuel pump, circuit A
P0638	Throttle control actuator, range
P0686-P0688	Engine Control Module, power relay control circuit
P0700	Transmission control system, module
P0702	Transmission control system, electrical
P0704	Transmission clutch circuit, input
P0705	Transmission range sensor, circuit
P0706	Transmission range sensor, performance
P0710-P0714	Transmission fluid temp sensor, performance or circuit
P0715-P0717	Transmission turbine speed sensor, performance or circuit
P0720-P0722	Transmission output speed sensor, performance or circuit
P0725-P0727	Engine speed input, circuit or performance

Code	Code Identification
P0729-P0735	Automatic transmission, incorrect gear ratio
P0740	Torque converter clutch circuit, open
P0743	Torque converter clutch circuit, electrical
P0746-P0749	Transmission pressure control solenoid
P0751-P0753	Transmission shift solenoid A
P0756-P0758	Transmission shift solenoid B
P0761-P0763	Transmission shift solenoid C
P0766-P0768	Transmission shift solenoid D
P0711-P0773	Transmission shift solenoid E
P0776-P0778	Transmission pressure control solenoid B
P0785	Transmission shift timing solenoid
P0796-P0798	Transmission pressure control solenoid C
P0811	Excessive clutch slippage
P0845-P0849	Transmission fluid pressure sensor, circuit A
P0840-P0844	Transmission fluid pressure sensor, circuit B
P0863-P0865	Transmission control module, circuit
P0884-P0892	Transmission Control module, power relay
P1102-P01103	Oxygen sensor heater circuit, short to B+ (pre-converter sensor)
P1105	Oxygen sensor heater circuit, short to B+ (post-converter sensor)
P1111	Oxygen sensor control, lean
P1111	Oxygen sensor control, rich
P1113-P01113	Oxygen sensor circuit, incorrect resistance (pre-converter sensor)
P1115	Oxygen sensor heater circuit, short to ground (pre-converter sensor)
P1116	Oxygen sensor heater circuit, open (pre-converter sensor)
P1117	Oxygen sensor heater circuit, short to ground (post-converter sensor)
P1118	Oxygen sensor heater circuit, open (post-converter sensor)
P1127	Long term fuel trim too rich
P1128	Long term fuel trim too lean

Code	Code Identification
P1136	Fuel trim too rich at idle
P1137	Fuel trim too lean at idle
P1142-P01143	Load detection, value out of range
P1149	Oxygen sensor control, value out of range
P1151	Long-term fuel trim, lean
P1152-P01166	Long-term fuel trim, rich
P1171	Throttle actuator position sensor range or performance problem
P1172	Throttle actuator position sensor, signal low
P1173	Throttle actuator position sensor, signal high
P1176	Oxygen sensor, correction limit attained
P1196	Oxygen sensor heater electrical malfunction (pre-converter sensor)
P1198	Oxygen sensor heater electrical malfunction (post-converter sensor)
P1201	Fuel injector circuit, Cylinder 1, open
P1202	Fuel injector circuit, Cylinder 2, open
P1203	Fuel injector circuit, Cylinder 3, open
P1204	Fuel injector circuit, Cylinder 4, open
P1213	Fuel injector circuit, short to B+ (cylinder no. 1)
P1214	Fuel injector circuit, short to B+ (cylinder no. 2)
P1215	Fuel injector circuit, short to B+ (cylinder no. 3)
P1216	Fuel injector circuit, short to B+ (cylinder no. 4)
P1225	Fuel injector circuit, short to ground (cylinder no. 1)
P1226	Fuel injector circuit, short to ground (cylinder no. 2)
P1227	Fuel injector circuit, short to ground (cylinder no. 3)
P1228	Fuel injector circuit, short to ground (cylinder no. 4)
P1237	Fuel injector circuit, open (cylinder no. 1)
P1238	Fuel injector circuit, open (cylinder no. 2)
P1239	Fuel injector circuit, open (cylinder no. 3)
P1240	Fuel injector circuit, open (cylinder no. 4)
P1255-P1256	ECT circuit, open or short

Code	Code Identification
P1287	Turbocharger bypass valve, open
P1288	Turbocharger bypass valve, short to B+
P1289	Turbocharger bypass valve, Short to ground
P1295	Turbocharger bypass, faulty input
P1296	Turbocharger cooling system, malfunction
P1297	Turbocharger throttle valve, loose hose, pressure drop
P1300	Misfire detected, fuel problem
P1325	Knock sensor limit attained (cylinder no. 1)
P1326	Knock sensor limit attained (cylinder no. 2)
P1327	Knock sensor limit attained (cylinder no. 3)
P1328	Knock sensor limit attained (cylinder no. 4)
P1335	Engine torque control adaptation at limit
P1336	Engine torque control adaptation at limit
P1338	Camshaft position sensor, open or short
P1340	Camshaft position sensor out of sequence
P1355	Ignition control circuit, open (cylinder no. 1)
P1356	Ignition control circuit, short to B+ (cylinder no. 1)
P1357	Ignition control circuit, short to ground (cylinder no. 1)
Code	Code Identification
P1358	Ignition control circuit, open (cylinder no. 2)
P1359	Ignition control circuit, short to B+ (cylinder no. 2)
P1360	Ignition control circuit, short to ground (cylinder no. 2)
P1361	Ignition control circuit, open (cylinder no. 3)
P1362	Ignition control circuit, short to B+ (cylinder no. 3)
P1363	Ignition control circuit, short to ground (cylinder no. 3)
P1364	Ignition control circuit, open (cylinder no. 4)
P1365	Ignition control circuit, short to B+ (cylinder no. 4)
P1366	Ignition control circuit, short to ground (cylinder no. 4)

Code	Code Identification
P1386	Engine Control Module malfunction
P1387	Engine Control Module malfunction
P1388	Engine Control Module malfunction
P1409	EVAP system, purge control valve circuit
P1410	Fuel tank vent valve, short to B+
P1420	Secondary air injection valve circuit, malfunction
P1421	Secondary air injection valve circuit, short to ground
P1422	Secondary air injection valve circuit, Short to B+
P1424	Secondary air injection system, leak detected
P1425	Fuel tank vent valve circuit, short to ground
P1426	Fuel tank vent valve circuit, open
P1432	Secondary air injection valve open
P1433	Secondary air injection system, pump relay circuit open
P1434	Secondary air injection system, pump relay circuit short to B+
P1435	Secondary air injection system, pump relay circuit short to ground
P1436	Secondary air injection system, pump relay circuit, electrical fault
P1450	Secondary air injection system, pump relay circuit short to B+
P1451	Secondary air injection system, pump relay circuit short to ground
P1452	Secondary air injection system, pump relay circuit open
P1470	Leak detection pump, electrical fault
P1471	EVAP system leak detection pump circuit, short to B+
P1472	EVAP system leak detection pump circuit, short to ground
P1473	EVAP system leak detection pump circuit, open
P1475	EVAP system leak detection pump vent system malfunction
P1476	EVAP system leak detection pump, pressure low
P1477	EVAP system leak detection pump vent system malfunction
P1478	EVAP system leak detection pump, breather blocked

Code	Code Identification
P1500	Fuel pump relay circuit malfunction
P1501	Fuel pump relay circuit, short to ground
P1502	Fuel pump relay circuit, short to B+
P1517	ECM, main relay voltage supply
P1539	Clutch pedal switch signal
P1541	Fuel pump relay circuit, open
P1542	Throttle actuator sensor, range or performance problem
P1543	Throttle actuator sensor, signal low
P1544	Throttle actuator sensor, signal high
P1545	Throttle valve actuator malfunction
P1546	Turbocharger boost pressure control valve circuit, short to B+
P1547	Turbocharger boost pressure control valve circuit, short to ground
P1548	Turbocharger boost pressure control valve circuit, open
P1550	Turbocharger boost pressure control valve, pressure deviation
P1555	Turbocharger boost pressure control valve, upper limit exceeded
P1556	Turbocharger boost pressure control valve, positive deviation
P1557	Turbocharger boost pressure control valve, negative deviation
P1558	Throttle valve actuator, electrical malfunction
P1559	Throttle control module to Engine Control Module adaptation error
P1560	Engine overspeed condition
P1564	Throttle control valve, voltage low
P1565	Throttle valve actuator, lower limit not attained
P1568	Throttle valve actuator, mechanical malfunction
P1569	Cruise control switch signal
P1580	Throttle control valve, drive malfunction
P1582	Idle adaptation at limit
P1592	Charge air pressure sensor
P1602	Engine Control Module, power supply low voltage
P1603	Engine Control Module, failed self-check

Code	Code Identification
P1604	Engine Control Module malfunction
P1606	ABS control unit, road surface information fault
P1609	ABS control unit, crash shut-off triggered
P1610	ABS control unit, memory error
P1612	Engine Control Module, incorrect programming
P1624	ABS control unit, fault lamp active
P1626	Engine Control Module, no manual transaxle data
P1630	Accelerator pedal position sensor no. 1, signal low
P1631	Accelerator pedal position sensor no.1, signal high
P1633	Accelerator pedal position sensor no.2, signal low
P1634	Accelerator pedal position sensor no.2, signal high
P1635	Data bus, no drivetrain message
P1639	Accelerator pedal position sensor, range or performance problem
P1640	Engine Control Module, ROM error
P1648	Engine Control Module, no powertrain data
P1649	Engine Control Module, no ABS data
P1650	Engine Control Module, no instrument cluster data
P1676	Electronic accelerator pedal warning light circuit malfunction
P1677	Electronic accelerator pedal warning light circuit, short to B+
P1678	Electronic accelerator pedal warning light circuit, short to ground
P1679	Electronic accelerator pedal warning light circuit, open
P1681	Engine Control Module, programming not finished
P1690	MIL circuit, electrical fault
P1691	Check Engine light circuit, open
P1692	Check Engine light circuit, short to ground
P1693	Check Engine light circuit, short to B+
P1778	Automatic transmission, valve 7, electrical fault
P1780	Automatic transmission, valve 7, torque retraction fault
P1823	Automatic transmission, valve 3, pressure control
P1828	Automatic transmission, valve 4, pressure control
P1847	ABS control unit, valve 3, memory fault
P1851	Engine Control Module, no ABS data
P1853	Engine Control Module, no ABS data
P1854	Data bus, software fault
P1855	Data bus, hardware fault
P1854	Data bus, software fault
P1857	ECM, load signal fault
P1861	TPS sensor, fault from ECM
P1866	TCM data bus, missing information

3 Engine Control Module (ECM) - removal and installation

Refer to illustrations 3.3 and 3.4

CAUTION:

Avoid static electricity damage to the Engine Control Module (ECM) by grounding yourself to the body of the vehicle before touching the ECM and using a special anti-static pad to store the ECM on, once it is removed.

Note 1: Anytime the ECM is replaced with a new unit it must be reprogrammed by a dealership service department with special equipment. The following procedure pertains to removal and installation of the original ECM only. If the ECM must be replaced with a new unit, take the vehicle to a dealership service department.

Note 2: Anytime battery power is disconnected from the ECM, stored operating parameters will be lost from the ECM, which will cause various driveability problems until the ECM can be reset with a scan tool.

1 Disconnect the cable from the negative battery terminal.

CAUTION:

These models are equipped with an anti-theft radio. Before performing a procedure that requires disconnecting the battery, make sure you have the activation code.

2 The ECM is located under the plenum cover, at the base of the windshield on the drivers side. Remove the plenum panel on the driver's side (see Chapter 11). Remove the metal plenum close-out plate under

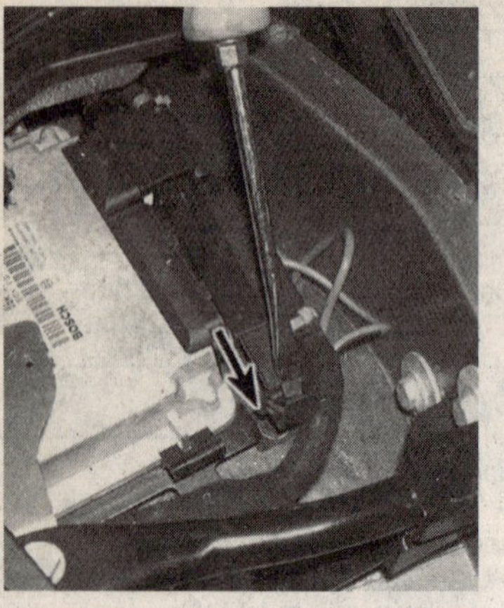

3.3 To disconnect the electrical connectors from the ECM, pry the locking tab out

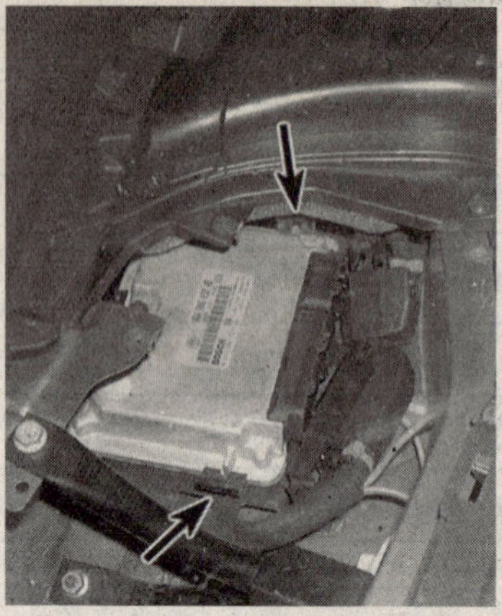

3.4 Carefully pry off the ECM retaining clips (arrows)

the plenum panel.

3 Disconnect the electrical connectors from the ECM (see illustration).

4 Carefully pry off the ECM retainer clips (see illustration). Remove the ECM from the plenum.

5 Installation is the reverse of removal.

4 Throttle control module - check and replacement

1 The throttle control module is connected to the end of the throttle shaft on the throttle body. The throttle control module contains four major components of the engine control system; the throttle position sensor, the closed throttle switch, the throttle valve actuator and a feedback sensor for the throttle valve actuator. The ECM uses the throttle position sensor to control fuel delivery based on driver demand. On 2.0L models, the throttle valve actuator controls the idle speed and cruise control functions. 1.8L turbo models are not equipped with a throttle cable, the throttle valve actuator controls the throttle plate under all driving conditions. The throttle control module is calibrated to the throttle body during manufacture, therefore as the throttle control module requires replacement, the complete throttle body assembly must be replaced.

CHECK

Refer to illustrations 4.2, 4.4 and 4.5

Note: Performing the following test will set a diagnostic trouble code and illuminate the Check Engine light. Clear the diagnostic trouble code after performing the tests and making the necessary repairs (see Section 2).

2 The throttle control module is located on the side of the throttle body (see illustration). Check the terminals in the connector and the wires leading to the sensor for looseness and breaks. Repair as required.

3 A scan tool is required for complete diagnosis of the throttle control module (see Section 2), but the following checks may identify a definite failure with the throttle control module or circuit.

4 Disconnect the electrical connector from the throttle control module and check the voltage supply and ground circuits (see illustration). On 2.0L models, connect a voltmeter to terminals 2 and 5 - a minimum of 4.5 volts should be present. Connect a voltmeter to terminals 2 and 6 - a minimum of 9.0 volts should be present.

5 On 1.8L turbo models through 2003, connect a voltmeter to terminals 2 and 6 - a minimum of 4.5 volts should be present (see illustration).On 2004 and 2005 models, check between 2 and 4 (vehicles with cruise control) or between 4 and 7 (without cruise control).

4.2 The throttle control module (arrow) is located on the side of the throttle body

4.4 Disconnect the electrical connector from the throttle control module and check the voltage supply and ground circuits at the harness connector

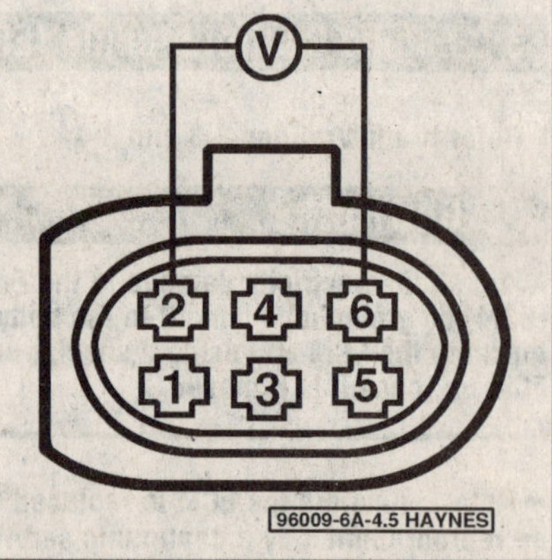

4.5 Throttle control module electrical connector terminal identification (harness-side) - 1.8L turbo models

6 If the specified voltage is not present, check the power and ground circuits.

➡Note: Refer to the wiring diagrams at the end of Chapter 12 for additional information of the circuits.

7 Using an ohmmeter, measure the resistance between terminals 2 and 8 (2.0L) or terminals 3 and 5 (1.8L) on the throttle control module. On 2.0L models the resistance should be 3.0 to 200 ohms. On 1.8L turbo models the resistance should be 1.0 to 5.0 ohms. If the resistance is not as specified, replace the throttle control module.

REPLACEMENT

CAUTION:

If the throttle control module is replaced, the ECM must be programmed with a scan tool to accept the new throttle control module.

8 The throttle control module is calibrated to the throttle body during manufacture, therefore if the throttle control module requires replacement, replace the throttle body assembly (see Chapter 4A).

5 Mass airflow sensor - check and replacement

1 The mass airflow sensor measures the amount of air passing through the sensor body and ultimately entering the engine through the throttle body. The ECM uses this information to control fuel delivery - the more air entering the engine (acceleration), the more fuel required.

CHECK

➧ Refer to illustrations 5.2 and 5.3

➡Note: Performing the following test will set a diagnostic trouble code and illuminate the Check Engine light. Clear the diagnostic trouble code after performing the tests and making the necessary repairs (see Section 2).

2 The mass airflow sensor is located in the air intake duct attached to the air filter housing cover (see illustration). A scan tool is necessary to check the output of the mass airflow sensor (see Section 2). The scan tool displays the sensor output in grams per second. With the engine idling at normal operating temperature, the display should read approximately 2.8 to 5.6 grams per second (2.0L) or 1.0 to 10.0 grams per second (1.8L). When the engine is accelerated the values should rise quickly and remain steady at a steady engine speed.

3 Before checking the mass airflow sensor operation, check the voltage supply circuit. Disconnect the electrical connector from the mass airflow sensor and connect the positive lead of a voltmeter to terminal no. 2 and the negative lead to a good engine ground point (see illustration). Turn the ignition key On - the voltage should read approximately 12.0 volts. If the voltage is incorrect, check the circuit from the mass airflow sensor to the fuse box (don't forget to check the fuses first). If the power circuit is good, check the mass airflow sensor operation with a scan tool. If the mass airflow sensor does not respond as described in Step 2, replace the mass airflow sensor.

➡Note: Before condemning the mass airflow sensor, check the air intake duct for leaks between the sensor and the intake manifold. A reading below the specified value may be caused by an air leak.

REPLACEMENT

➧ Refer to illustration 5.6

4 Disconnect the electrical connector from the mass airflow sensor.
5 Remove the air filter housing cover (see Chapter 4A).
6 Remove the screws retaining the mass airflow sensor to the air filter cover and remove the sensor (see illustration).

CAUTION:

Handle the mass airflow sensor with care. Damage to this sensor will affect the operation of the entire fuel injection system.

7 Installation is the reverse of removal.

5.2 The mass airflow sensor is located in the air intake duct attached to the air filter housing cover

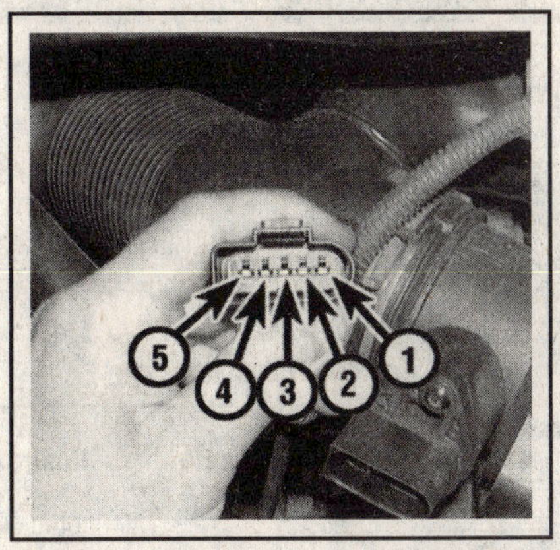

5.3 Disconnect the electrical connector from the mass airflow sensor and check the voltage supply and ground circuits at the harness connector

5.6 Remove the retaining screws (arrows) and remove the mass airflow sensor

6 Intake air temperature sensor - check and replacement

1 The intake air temperature sensor is a thermistor (a resistor which varies the value of its resistance in accordance with temperature changes). The change in the resistance values will directly affect the voltage signal from the sensor to the ECM. As the sensor temperature INCREASES, the resistance values will DECREASE. As the sensor temperature DECREASES, the resistance values will INCREASE.

CHECK

➧ Refer to illustrations 6.2, 6.3a and 6.3b

2 On 2.0L models, the intake air temperature sensor is incorporated into the mass airflow sensor. On 1.8L turbo models, the intake air temperature sensor is located on the intake manifold near the throttle body (see illustration). Check the terminals in the connector and the wires leading to the sensor for looseness and breaks. Repair as required.

3 With the ignition switch OFF, disconnect the electrical connector from the mass airflow sensor (2.0L) or intake air temperature sensor (1.8L) (see illustrations). Using an ohmmeter, measure the resistance between the two intake air temperature sensor terminals on the sensor while it is completely cold - at 86-degrees F (30-degrees C) the resistance should be 1.5 to 2.0 K-ohms. Reconnect the electrical connector to the sensor, start the engine and warm it up until it reaches operating temperature. Disconnect the connector and check the resistance again - at 176-degrees F (80-degrees C) the resistance should be 275 to 375 ohms. Compare your measurements to the resistance chart (see illustration). If the sensor resistance test results are incorrect, replace the intake air temperature sensor.

REPLACEMENT

4 On 2.0L models, replace the mass airflow sensor (see Section 5).

5 On 1.8L Turbo models, disconnect the electrical connector from the sensor, remove the mounting bolt and withdraw the sensor from the intake manifold.

6 Installation is the reverse of removal.

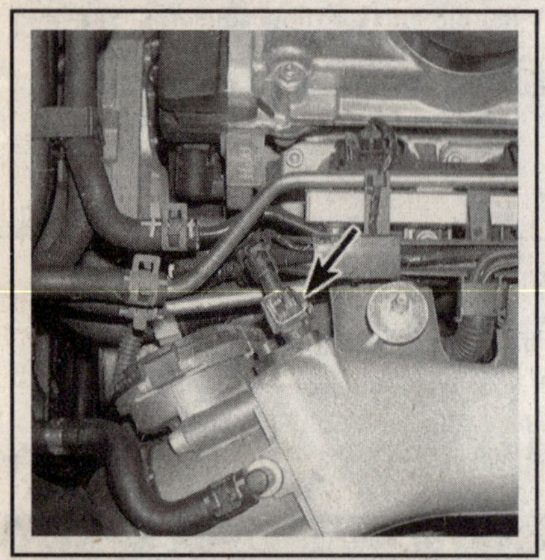

6.2 Intake air temperature sensor location - 1.8L Turbo models

6.3a On 2.0L models, disconnect the electrical connector from the mass airflow sensor and measure the intake air temperature resistance across terminal no. 1 and 3

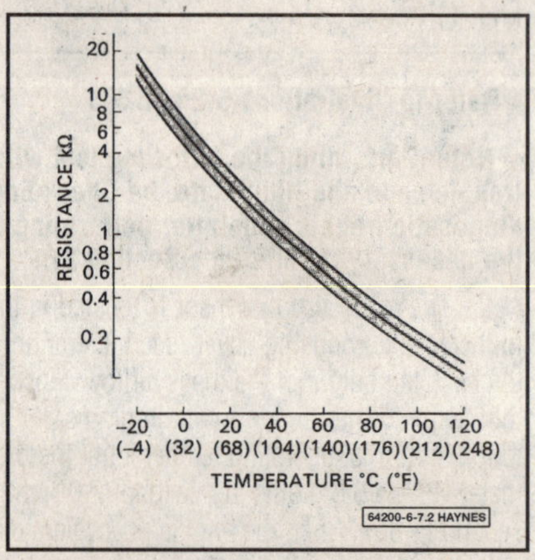

6.3b Intake air temperature sensor and engine coolant temperature sensor approximate temperature vs. resistance values

7 Engine coolant temperature sensor - check and replacement

7.2a Engine coolant temperature sensor location - 2.0L models

1 The engine coolant temperature sensor is a thermistor (a resistor which varies the value of its resistance in accordance with temperature changes). The change in the resistance values will directly affect the voltage signal from the sensor to the ECM. As the sensor temperature INCREASES, the resistance values will DECREASE. As the sensor temperature DECREASES, the resistance values will INCREASE.

CHECK

➧ Refer to illustrations 7.2a, 7.2b and 7.3

2 The engine coolant temperature sensor is located in the coolant pipe near the coolant outlet (see illustrations). Check the terminals in the connector and the wires leading to the sensor for looseness and breaks. Repair as required.

3 With the ignition switch OFF, disconnect the electrical connec-

7.2b Engine coolant temperature sensor location - 1.8L turbo models

tor from the engine coolant temperature sensor. Using an ohmmeter, measure the resistance between terminals 1 and 3 on the sensor while it is completely cold (see illustration) - at 86-degrees F (30-degrees C) the resistance should be 1.5 to 2.0 K-ohms. Reconnect the electrical connector to the sensor, start the engine and warm it up until it reaches operating temperature (176-degrees F, 80-degrees C) the resistance should be 275 to 375 ohms. Compare your measurements to the resistance chart (see illustration 6.3b). If the sensor resistance test results are incorrect, replace the engine coolant temperature sensor.

➡Note: A more accurate check may be performed by removing the sensor and suspending the tip of the sensor in a container of water. Heat the water on the stove while you monitor the resistance of the sensor.

7.3 Measure the resistance of the engine coolant air temperature sensor across terminals 1 and 3

REPLACEMENT

⁂ WARNING:

Wait until the engine is completely cool before beginning this procedure.

4 Partially drain the cooling system (see Chapter 1).

5 Disconnect the electrical connector from the sensor.

6 Carefully pry the retaining clip out and withdraw the sensor from the coolant pipe.

7 Before installing the new sensor, replace the O-ring.

8 Installation is the reverse of removal.

8 Engine speed sensor - check and replacement

1 The engine sensor provides the ECM with a crankshaft position signal. The ECM uses the signal to determine a crankshaft reference point (Top Dead Center) and calculate engine speed (RPM). The signal is also used by the On-board Diagnostic system for misfire detection. The ignition system will not operate if the ECM does not receive an engine speed sensor input.

8.2 Engine speed sensor electrical connector location (arrow)

CHECK

➧ Refer to illustrations 8.2 and 8.3

2 The engine speed sensor is located on the side of the engine block near the oil filter. The sensor is equipped with a pigtail lead and a remote connector (see illustration). Check the terminals in the connector and the wires leading to the sensor for looseness and breaks. Repair as required.

3 Disconnect the electrical connector from the engine speed sensor.

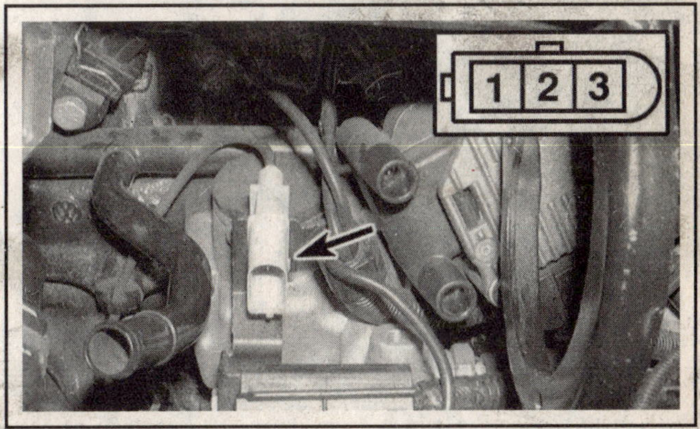

8.3 Measure the resistance of the engine speed sensor across terminals 2 and 3

8.7 Remove the engine speed sensor mounting bolt

Using an ohmmeter, measure the resistance of the sensor across terminals 2 and 3 (see illustration). The resistance should be 480 to 1000 ohms (2.0L) or 730 to 1000 ohms (1.8L).

4 Check for continuity between terminals 1 and 2, then 1 and 3. No continuity (infinity) should be indicated.

5 If the engine speed sensor fails either test, replace the sensor.

REPLACEMENT

Refer to illustration 8.7

6 Disconnect the engine speed sensor electrical connector.

7 Remove the engine speed sensor mounting bolt and withdraw the sensor from the engine block (see illustration).

8 Replace the O-ring and lightly lubricate it with clean engine oil.

9 Installation is the reverse of removal.

9 Camshaft position sensor - check and replacement

1 The camshaft position sensor, in conjunction with the crankshaft position sensor, determines the ignition timing and fuel injection synchronization on each cylinder. The sensor is a magnetic pick-up device triggered from a reluctor wheel on the camshaft sprocket (2.0L) or camshaft (1.8L).

CHECK

Refer to illustrations 9.2a, 9.2b, 9.3 and 9.4

Note: Performing the following test will set a diagnostic trouble code and illuminate the Check Engine light. Clear the diagnostic trouble code after performing the tests and making the necessary repairs (see Section 2).

2 The camshaft position sensor is located at the front of the cylinder head under the timing belt cover. On 2.0L models, the camshaft position sensor is located behind the camshaft sprocket and is equipped with a pigtail lead and a remote connector (see illustration). On 1.8L turbo models, the camshaft position sensor is located at the front of the intake camshaft (see illustration). Check the terminals in the connector and the wires leading to the sensor for looseness and breaks. Repair as required.

3 Before checking the camshaft position sensor, check the voltage supply and ground circuits from the ECM. Disconnect the electrical connector from the camshaft position sensor and connect a voltmeter to terminals 1 and 3 at the harness connector (see illustration). Turn the ignition key On - the voltage should read approximately 5.0 volts. If the voltage is incorrect, check the wiring from the camshaft position sensor to the ECM. If the circuits are good, have the ECM checked at a dealer service department or other properly equipped repair facility.

4 To check the camshaft position sensor operation, reconnect the connector to the camshaft position sensor and using a suitable probe, backprobe terminal no. 2 of the camshaft position sensor connector (see illustration) (see Chapter 12 for additional information on how to backprobe a connector). Connect the positive lead of a voltmeter to the probe and the negative lead to a good engine ground point. Turn the ignition key On. Rotate the engine slowly with a breaker bar and socket attached to the crankshaft pulley center bolt while watching the meter. The voltage should fluctuate between zero volts and 10.0 volts as the vanes in the reluctor wheel pass the sensor. If the test results are incorrect, replace the camshaft position sensor.

9.2a Camshaft position sensor electrical connector location - 2.0L models

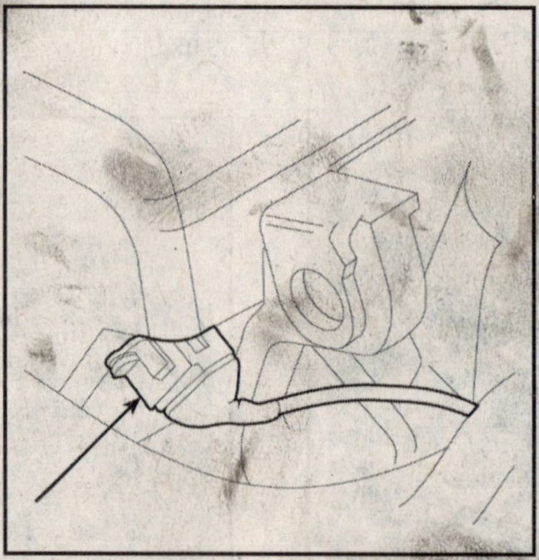

9.2b Camshaft position sensor electrical connector location - 1.8L turbo models

9.3 Disconnect the electrical connector from the camshaft position sensor and check the voltage supply and ground circuits at the harness connector

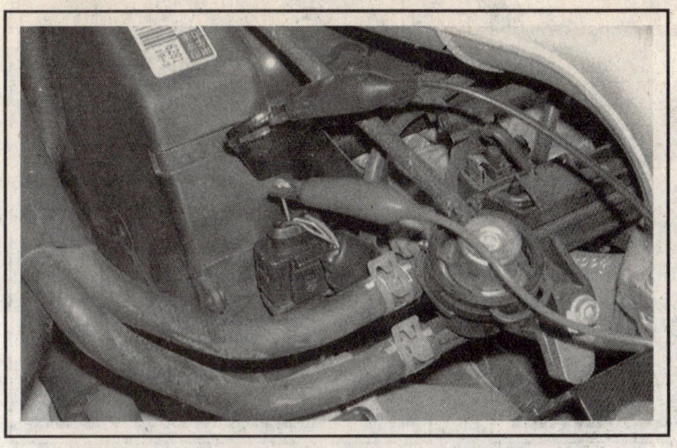

9.4 To check the camshaft position sensor, backprobe terminal no. 2 of the sensor connector with a voltmeter

➡Note: Rotate the engine slowly through at least two complete revolutions. Removing the spark plugs from the engine will make the crankshaft much easier to turn.

⁂ WARNING:

If you remove the spark plugs, unplug the primary (low voltage) electrical connector(s) from the ignition coil(s) to disable the ignition system.

REPLACEMENT

➧ Refer to illustration 9.7

2.0L models

5 Remove the timing belt and the camshaft sprocket (see Chapter 2A).

9.7 On 2.0L models, remove the timing belt rear cover bolts (A) and the camshaft position sensor mounting bolt (B)

6 Remove the timing belt rear cover. Disconnect the camshaft position sensor electrical connector and detach the electrical connector from the bracket.

7 Remove the camshaft position sensor mounting bolt and remove the sensor from the cylinder head (see illustration).

8 Installation is the reverse of removal.

1.8L turbo models

9 Disconnect the camshaft position sensor electrical connector.

10 Remove the timing belt upper cover (see Chapter 2A).

11 Remove the camshaft position sensor mounting bolts and remove the sensor from the cylinder head.

12 Installation is the reverse of removal.

10 Oxygen sensor - check and replacement

➡Note: All models are equipped with two oxygen sensors; one pre-converter oxygen sensor and one post-converter oxygen sensor.

1 The oxygen in the exhaust reacts with the elements inside the oxygen sensor to produce a voltage output that varies from 0.1 volt (high oxygen, lean mixture) to 0.9 volt (low oxygen, rich mixture). The pre-converter oxygen sensor (mounted in the exhaust system before the catalytic converter) provides a feedback signal to the ECM that indicates the amount of oxygen remaining in the exhaust gas after combustion. The ECM monitors this variable voltage continuously to determine the required fuel injector pulse width and to control the engine air/fuel ratio. A mixture ratio of 14.7 parts air to 1 part fuel is the ideal ratio for minimum exhaust emissions, as well as the best combination of fuel economy and engine performance. Based on oxygen sensor signals, the ECM tries to maintain this air/fuel ratio of 14.7:1 at all times.

2 The post-converter oxygen sensor (mounted in the exhaust system after the catalytic converter) has no effect on ECM control of the air/fuel ratio. However, the post-converter sensor is identical to the pre-converter sensor and operates in the same way. The ECM uses the post-converter signal to monitor the efficiency of the catalytic converter. A post-converter oxygen sensor will produce a slower fluctuating voltage signal that reflects the lower oxygen content in the post-catalyst exhaust.

3 An oxygen sensor produces no voltage when it is below its normal operating temperature of about 600-degrees F. During this warm-up period, the ECM operates in an open-loop fuel control mode. It does not use the oxygen sensor signal as a feedback indication of residual oxygen in the exhaust. Instead, the ECM controls fuel metering based on the inputs of other sensors and its own programs.

4 Proper operation of an oxygen sensor depends on four conditions:

a) ***Electrical*** - *The low voltages generated by the sensor require good, clean connections which should be checked whenever a sensor problem is suspected or indicated.*

b) ***Outside air supply*** - *The sensor needs air circulation to the internal portion of the sensor. Whenever the sensor is installed, make sure the air passages are not restricted.*

c) ***Proper operating temperature*** - *The ECM will not react to the sensor signal until the sensor reaches approximately 600-degrees F. This factor must be considered when evaluating the performance of the sensor.*

d) ***Unleaded fuel*** - *Unleaded fuel is essential for proper operation of the sensor.*

5 The ECM can detect several different oxygen sensor problems and set diagnostic trouble codes to indicate the specific fault (see Sec-

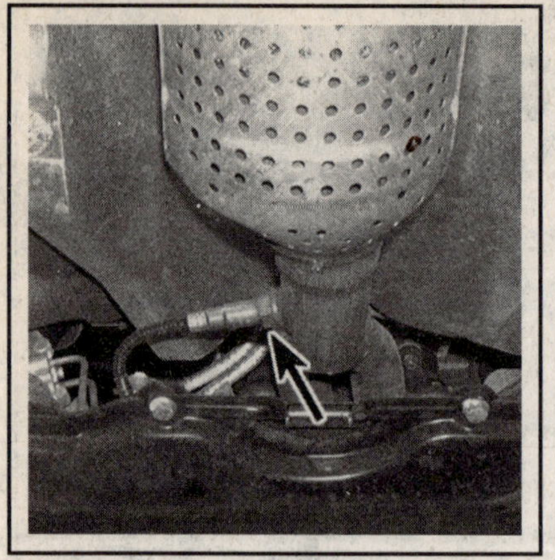

10.6a The pre-converter oxygen sensor is located in the exhaust pipe before the catalytic converter

10.6b The post-converter oxygen sensor is located in the exhaust pipe after the catalytic converter

10.6c The electrical connectors for the oxygen sensors are located under a cover

tion 2). When an oxygen sensor fault occurs, the ECM will disregard the oxygen sensor signal voltage and revert to open-loop fuel control as described previously.

CHECK

➧ **Refer to illustrations 10.6a, 10.6b, 10.6c and 10.8**

CAUTION:

The oxygen sensor is very sensitive to excessive circuit loads and circuit damage of any kind. For safest testing, disconnect the oxygen sensor connector, install jumper wires between the two connectors and connect your voltmeter to the jumper wires. If jumper wires aren't available, carefully backprobe the wires in the connector shell with suitable probes (such as T-pins). Do not puncture the oxygen sensor wires or try to backprobe the sensor itself. Use only a digital voltmeter to test an oxygen sensor.

➡**Note: Performing the following test will set a diagnostic trouble code and illuminate the Check Engine light. Clear the diagnostic trouble code after performing the tests and making the necessary repairs (see Section 2).**

6 The oxygen sensor connectors are located under a cover on the vehicle under-body (see illustrations). Connect a voltmeter to the black wire and the brown wire at the oxygen sensor connector. Turn the ignition ON but do not start the engine. The meter should read approximately 400 to 450 millivolts (0.40 to 0.45 volt). If it doesn't, trace and repair the circuit from the sensor to the ECM.

➡**Note: Refer to the wiring diagrams at the end of Chapter for additional information on the oxygen sensor circuits.**

7 Start the engine and let it warm up to normal operating temperature; again check the oxygen sensor signal voltage.

a) Voltage from a pre-converter sensor should range from 100 to 900 millivolts (0.1 to 0.9 volt) and switch actively between high and low readings.

b) Voltage from a post-converter sensor should also read between 100 to 900 millivolts (0.1 to 0.9 volt) but it should not switch actively. The post-converter oxygen sensor voltage may stay toward the center of its range (about 400 millivolts) or stay for relatively longer periods of time at the upper or lower limits of the range.

8 Check the battery voltage supply and ground circuits to the oxygen sensor heater. Disconnect the electrical connector and connect a voltmeter to terminals 1 and 2 of the sensor harness connector (see illustration). Turn the ignition ON, the meter should read approximately 12 volts. If battery voltage is not present, check the power and ground circuits to the sensor (don't forget to check the fuses first).

9 Allow the oxygen sensor to cool and check the oxygen sensor heater for an open circuit. With the connector disconnected. Connect an ohmmeter to the two oxygen sensor heater terminals of the connector (oxygen sensor side). The oxygen sensor pigtail is generally not color coded, but the heater wires are usually the white wires. If an open circuit or excessive resistance is indicated, replace the oxygen sensor.

➡**Note: If the tests indicate that a sensor is good, and not the cause of a driveability problem or diagnostic trouble code, check the wiring harness and connectors between the sensor and the ECM for an open or short circuit. If no problems are found, have the vehicle checked by a dealer service department or other qualified repair shop.**

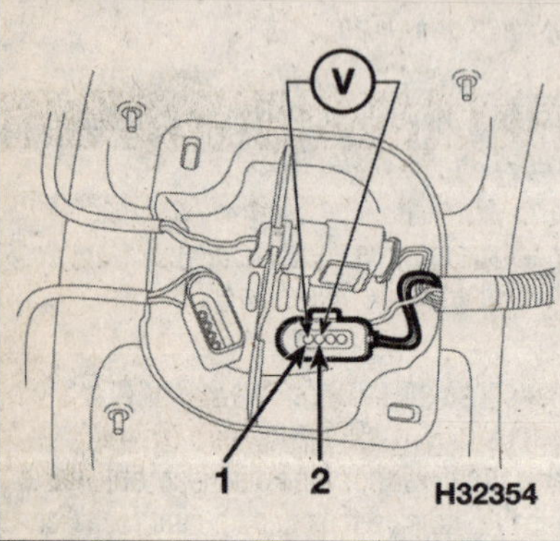

10.8 Check the battery voltage supply and ground circuits to the oxygen sensor heater

REPLACEMENT

➧ **Refer to illustration 10.13**

10 The exhaust pipe contracts when cool, and the oxygen sensor may be hard to loosen when the engine is cold. To make sensor removal easier, start and run the engine for a minute or two; then shut it off. Be careful not to burn yourself during the following procedure. Also observe these guidelines when replacing an oxygen sensor.

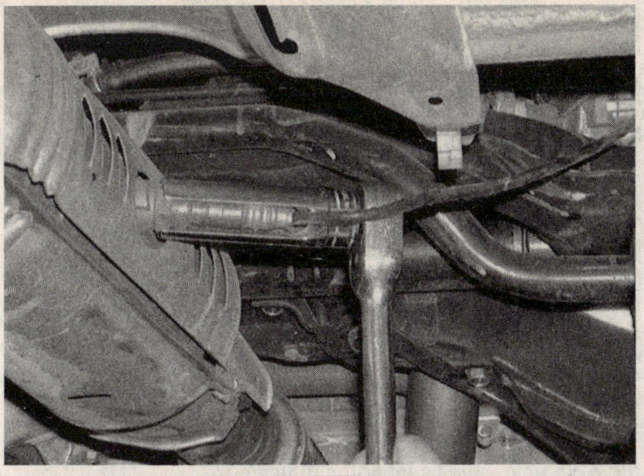

10.13 A special slotted socket, allowing clearance for the wiring harness, may be required for oxygen sensor removal (the tool is available at most auto parts stores)

- a) *The sensor has a permanently attached pigtail and electrical connector which should not be removed from the sensor. Damage or removal of the pigtail or electrical connector can harm operation of the sensor.*
- b) *Keep grease, dirt and other contaminants away from the electrical connector and the louvered end of the sensor.*
- c) *Do not use cleaning solvents of any kind on the oxygen sensor.*
- d) *Do not drop or roughly handle the sensor.*

11 Raise the vehicle and place it securely on jackstands.

12 Remove the cover and disconnect the electrical connector from the sensor.

13 Using a suitable wrench or specialized oxygen sensor socket, unscrew the sensor from the exhaust manifold (see illustration).

14 Anti-seize compound must be used on the threads of the sensor to aid future removal. The threads of most new sensors will be coated with this compound. If not, be sure to apply anti-seize compound before installing the sensor.

15 Install the sensor and tighten it securely.

16 Reconnect the electrical connector to the sensor, install the cover and lower the vehicle.

11 Knock sensor - check and replacement

1 The knock sensor detects abnormal vibration (spark knock or pinging) in the engine. The knock control system is designed to reduce spark knock during periods of heavy detonation. This allows the engine to use maximum spark advance to improve driveability. Knock sensors produce AC output voltage which increases with the severity of the knock. The signal is fed into the ECM and the timing is retarded to compensate for the severe detonation.

11.2 Knock sensor location (one of two sensors)

CHECK

➧ Refer to illustration 11.2

2 All models, are equipped with two knock sensors. They are located on the side of the engine block below the intake manifold (see illustration).

3 A scan tool is required to thoroughly check the knock sensor system, but the following check can be performed to identify a definite knock sensor failure.

4 Disconnect the electrical connector from the knock sensor. Using an ohmmeter, check for continuity across the two terminals of the knock sensor. No continuity (infinity) should be indicated on the meter, if continuity is indicated, replace the knock sensor.

REPLACEMENT

5 Disconnect the electrical connector from the knock sensor.

6 Remove the sensor from the engine block (see illustration 11.2).

7 Installation is the reverse of removal. Tighten the knock sensor bolt to 15 ft-lbs (20 Nm).

12 Vehicle speed sensor - check and replacement

1 The Vehicle Speed Sensor (VSS) is mounted on the transaxle. The sensor is triggered by a toothed rotor on the transaxle output shaft. As the output shaft rotates, the sensor produces a fluctuating voltage, the frequency of which is proportional to vehicle speed. The ECM uses the sensor input signal for several different engine and transmission control functions. The VSS signal also drives the speedometer on the instrument panel. A defective VSS can cause various driveability and transaxle problems.

CHECK

➧ Refer to illustration 12.2

2 Raise the vehicle and support it securely on jacks the vehicle speed sensor (see illustration). Check the connector and the wires leading to the sensor for Repair as required.

3 To check the VSS operation with a voltmeter, backprobe the white/blue wire terminal of the VSS connector using a suitable probe (see Chapter 12 for additional information on how to backprobe a connector). Connect the other lead of the voltmeter to a good ground and turn the ignition key On. Hold the right front tire steady and rotate the left front tire by hand while watching the voltmeter. The sensor should produce a fluctuating voltage from zero to approximately 5.0 volts.

REPLACEMENT

4 Raise the vehicle and support it securely on jackstands.
5 Disconnect the electrical connector from the VSS.
6 Remove mounting bolt and withdraw the VSS from the transaxle case.
7 Replace the sensor O-ring.
8 Installation is the reverse of removal.

12.2 Vehicle speed sensor location

13 Accelerator control system (1.8L turbo engine)

Models equipped with the 1.8L turbo engine utilize an electronic accelerator control system. The system does not use an accelerator cable, the throttle valve is operated by the throttle valve actuator contained within the throttle control module. The throttle valve actuator is controlled by the ECM. The accelerator pedal module contains a pedal position sensor. The pedal position signal is fed to the ECM and the ECM commands the throttle valve actuator to open the throttle accordingly.

With the engine Off and the ignition key On, the ECM opens the throttle valve in direct relationship to the accelerator pedal input. But when the engine is running, under load, the ECM operates the throttle valve independently of the accelerator pedal. The throttle valve may be opened significantly farther than the driver may be demanding. This allows the ECM to maintain the engine at peak operating efficiency.

The system operation is monitored by the On Board Diagnosis system, but uses a separate warning light. The Electronic Power Control (EPC) warning light is located near the top of the speedometer. The EPC light will illuminate with the ignition key on and should go out when the engine is started. If the EPC warning light remains on, or comes on, with the engine running, a problem with the system has been identified and a Diagnostic Trouble Code is retained in the ECM memory. If there is a complete failure with the system the ECM will limit the engine speed to approximately 1200 rpm.

A quick check of the system can be performed by removing the air intake duct and observing the throttle valve as an assistant depresses the accelerator pedal. When the accelerator pedal is depressed half-way, the throttle valve should open to half throttle. When the accelerator pedal is depressed fully, the throttle valve should open to full throttle. The manufacturer's scan tool is required to properly diagnose the system. If a fault is suspected with the system or if the EPC warning light is on, take the vehicle to a dealership service department as soon as possible. For additional information, refer to Chapter 4A for information on the electronic accelerator pedal module and Section 4 of this Chapter for information on the throttle control module.

14 Turbocharger boost control system (1.8L turbo engine)

Refer to illustration 14.1

The turbocharger boost control system consists of the boost pressure sensor, the boost pressure (wastegate) control valve, the recirculation air valve and solenoid, the ECM and the connecting hoses and wiring (see illustration).

The ECM monitors boost pressure sensor signal and controls the vacuum supplied to the turbocharger wastegate actuator with the boost pressure control valve. The engine control system calculates the engine torque needed depending on driver demand and engine operating conditions, the ECM will then adjust the boost pressure to meet the demands. The recirculation air valve returns air to the intake duct when throttle is closed minimizing turbo lag. The ECM controls the recirculation air valve with a vacuum valve/solenoid.

If a problem is suspected with the system, check the wiring, connectors, vacuum hoses and connecting pipes. Using a hand-held vacuum pump, apply vacuum to the wastegate actuator (with the engine not running) and check the actuator arm and lever for freedom of movement. Disconnect the electrical connector from the boost pressure control valve and using an ohmmeter, check the solenoid windings for an open circuit. The boost pressure control valve should measure approximately 25 to 35 ohms, if an open circuit or excessive resistance is indicated, replace the control valve.

A scan tool is required for complete diagnosis of the system control circuits. If a scan tool is not available, have the system checked at a dealer service department or other properly equipped repair facility.

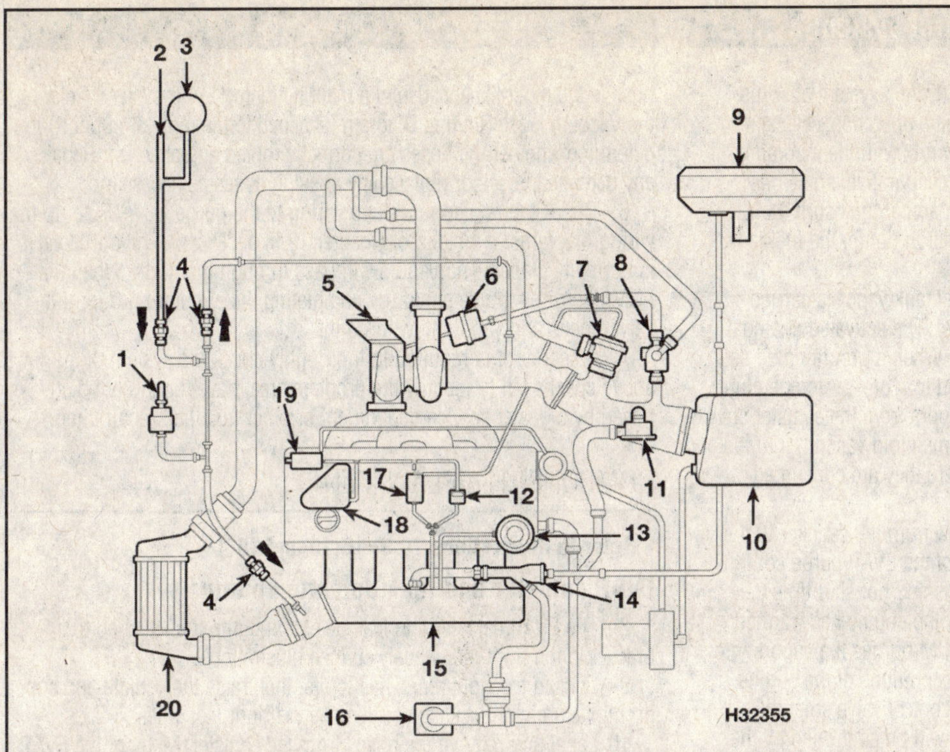

14.1 Turbocharger boost pressure control system

1 *EVAP purge valve*
2 *Hose*
3 *Vacuum reservoir*
4 *Check valve*
5 *Turbocharger*
6 *Wastegate actuator*
7 *Recirculation air valve*
8 *Boost pressure control valve*
9 *Power brake booster*
10 *Air filter assembly*
11 *Crankcase pressure regulator valve*
12 *Check valve*
13 *Fuel pressure regulator*
14 *Vacuum amplifier*
15 *Intake manifold*
16 *Crankcase ventilation housing*
17 *Recirculation valve solenoid*
18 *Vacuum reservoir*
19 *Secondary air injection valve*
20 *Intercooler*

15 Crankcase ventilation system

➧ Refer to illustrations 15.2a and 15.2b

1 When the engine is running, a certain amount of the gasses produced during combustion escapes past the piston rings into the crankcase as blow-by gasses. The crankcase ventilation system is designed to reduce the resulting hydrocarbon emissions (HC) by routing the gasses and vapors from the crankcase into the intake manifold and combustion chambers, where they are consumed during engine operation.

2 Crankcase vapors pass through a hose connected from the valve cover to the air intake duct (see illustration). The oil/air separator at the valve cover separates the oil suspended in the blow-by gases and allows the oil to drain back into the crankcase. The crankcase vapors are drawn from the oil/air separator through a hose connected to the air intake duct where they mix with the incoming air and are burned during the normal combustion process. A heating element is incorporated into the air intake duct to prevent icing in extremely cold weather. On 1.8L turbo models, a pressure regulating valve is installed in the breather hose (see illustration).

3 A plugged breather, valve or hose will cause excessive crankcase pressures resulting in oil leaks and sludge build-up in the crankcase. Check the components for restrictions and clean or replace the components as necessary. Be sure to check the basic mechanical condition of the engine before condemning the crankcase ventilation system (see Chapter 2C).

15.2a The crankcase ventilation system hose (arrow) runs from the valve cover to the intake duct

15.2b 1.8L turbo models are equipped with a crankcase ventilation pressure regulator valve (arrow)

16 Evaporative emissions control system

1 The fuel evaporative emissions control (EVAP) system absorbs fuel vapors from the fuel tank and, during engine operation, releases them into the engine intake system where they mix with the incoming air/fuel mixture. The main components of the evaporative emissions system are the canister (filled with activated charcoal to absorb fuel vapors), the purge control valve, the leak detection pump, the fuel tank and the vapor and purge lines.

2 After passing through a check valve, fuel tank vapor is carried through the vapor hose to the charcoal canister. The activated charcoal in the canister absorbs and stores the vapors. When a programmed set of conditions are met (engine running, warmed to a pre-set temperature, etc.), the ECM opens the purge valve. Fuel vapors from the canister are then drawn through the purge hose by intake manifold vacuum into the intake manifold and combustion chamber where they are consumed during normal engine operation.

3 The ECM regulates the rate of vapor flow from the canister to the intake manifold by controlling the duty cycle of the EVAP purge control valve solenoid. During cold running conditions and hot start time delay, the ECM does not energize the solenoid. After the engine has warmed up to the correct operating temperature, the ECM purges the vapors into the intake manifold according to the running conditions of the engine. The ECM will cycle (ON then OFF) the purge control valve solenoid about 5 to 10 times per second. The flow rate will be controlled by the pulse width, or length of time, the solenoid is allowed to be energized.

4 The EVAP system is equipped with a leak detection monitor system. The system is a self-diagnostic system designed to detect a leak in the EVAP system. Each time the engine is started cold, the PCM energizes the leak detection pump. The pump pressurizes the EVAP system then shuts off. The PCM is able to detect a leak if the pump continues to run, unable to pressurize the system. If a leak is detected, the PCM will trigger a diagnostic trouble code (see Section 2).

CHECK

➡Note: The evaporative emissions control system, like all emission control systems, is protected by a Federally-mandated warranty. The EVAP system probably won't fail during the service life of the vehicle; however, if it does, the hoses or charcoal canister are usually to blame.

5 Always check the hoses first. A disconnected, damaged or missing hose is the most likely cause of a malfunctioning EVAP system. Refer to the Vacuum Hose Routing Diagram (attached to the radiator support) to determine whether the hoses are correctly routed and attached. Repair any damaged hoses or replace any missing hoses as necessary.

6 Check the related fuses and wiring to the purge valve. Refer to the wiring diagrams at the end of Chapter 12, if necessary. The purge valve is normally closed - no vapors will pass through the ports. When the ECM energizes the solenoid (by completing the circuit to ground), the valve opens and vapors flow through.

7 A scan tool is required to thoroughly check the system. If the above checks fail to identify the problem area, have the system diagnosed by a dealer service department or other qualified repair shop.

COMPONENT REPLACEMENT

➧ Refer to illustrations 16.8, 16.10 and 16.13

EVAP canister and leak detection pump

8 The EVAP canister and leak detection pump are attached to a bracket in the right rear wheelwell (see illustration).

9 Loosen the right rear wheel bolts, then raise the vehicle and support it securely on jackstands. Remove the right rear wheel.

10 Label and remove the hoses from the canister. Many of the EVAP system hoses are equipped with quick-connect fittings (see illustration). Disconnect the fitting as follows:

a) Clean the area around the fitting.
b) Twist the fitting back-and-forth several times to loosen the seal.
c) Depress the locking tabs and pull the fitting straight off the nipple.

11 Remove the mounting bolts and remove the canister. Remove the leak detection pump from the canister, if necessary.

12 Installation is the reverse of removal.

Purge control valve

13 The purge control valve is mounted on a bracket above the right front wheelwell (see illustration).

14 Disconnect the electrical connector. Label and remove the hoses from the purge control valve.

15 Remove the mounting bolts and remove the purge control valve from the bracket.

16 Installation is the reverse of removal.

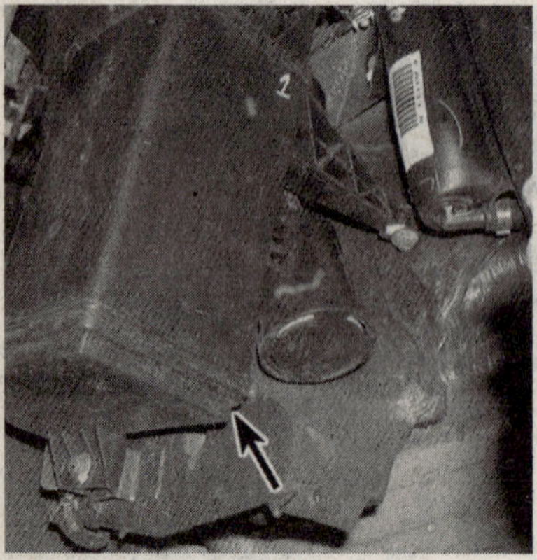

16.8 The EVAP canister and leak detection pump are mounted in the right rear wheelwell

16.10 To disconnect an EVAP system quick-connect fitting, squeeze the tabs together

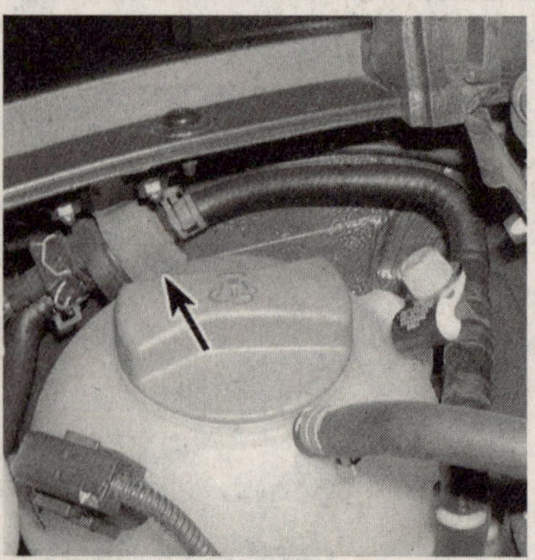

16.13 EVAP purge control valve location (arrow)

17 Secondary air injection system

1 The secondary air injection system is used to reduce tailpipe emissions on initial engine start-up. The system uses an electric motor/pump assembly, vacuum valve/solenoid, combination air shut-off/check valve and tubing to inject fresh air directly into the exhaust manifold. The fresh air (oxygen) reacts with the exhaust gas in the catalytic converter to reduce HC and CO levels. The air pump and solenoid are controlled by the ECM. During initial start-up, when the coolant temperature is between 15-degrees C/59-degrees F and 35-degrees C/95-degrees F, the ECM will energize the vacuum valve/solenoid, opening the check valve and operate the air pump for approximately one minute. During normal operation, the check valve is closed to prevent exhaust backflow into the system.

CHECK

2 Check the air pump hoses and the vacuum hoses. Repair any damaged hoses or replace any missing hoses as necessary. Check the vacuum source to the vacuum valve. Intake manifold vacuum should be present with the engine running.

3 Check the related fuses and wiring to the air pump and vacuum valve/solenoid. Refer to the wiring diagrams at the end of Chapter 12, if necessary. The vacuum valve is normally closed - no vacuum is applied to the check valve. When the ECM energizes the solenoid (by completing the circuit to ground), the valve opens, vacuum is applied to the check valve, the check valve opens and air flows through the tube into the exhaust pipe.

4 A scan tool is required to thoroughly check the system. If the above checks fail to identify the problem area, have the system diagnosed by a dealer service department or other qualified repair shop.

COMPONENT REPLACEMENT

➧ Refer to illustrations 17.7, 17.11 and 17.15

Air pump

5 Remove the engine cover.

6 Disconnect the electrical connector and remove the hoses from the air pump.

7 Remove the mounting bolts and remove the air pump assembly (see illustration).

8 Installation is the reverse of removal.

Vacuum valve

9 Disconnect the electrical connector from the vacuum valve.

10 Label and disconnect the vacuum hoses from the valve.

11 Remove the mounting nut and remove the vacuum valve (see illustration).

12 Installation is the reverse of removal.

Check valve (combination valve)

13 Remove the air intake duct.

14 Remove the vacuum hose and the air hose from the check valve.

15 Remove the mounting bolts and remove the check valve (see illustration).

16 Installation is the reverse of removal.

17.7 Remove the secondary air injection pump mounting bolts (arrows) (two of three shown)

17.11 Secondary air injection vacuum valve location (arrow)

17.15 Secondary air injection check valve location (arrow)

18 Catalytic converter

➡Note: Because of a Federally mandated warranty which covers emissions-related components such as the catalytic converter, check with a dealer service department before replacing the converter at your own expense.

1 The catalytic converter is an emission control device added to the exhaust system to reduce pollutants from the exhaust gas stream. A three-way (reduction) catalyst design is used. The catalytic coating on the three-way catalyst contains platinum and rhodium, which lowers the levels of oxides of nitrogen (NOx) as well as hydrocarbons (HC) and carbon monoxide (CO).

2 The test equipment for a catalytic converter is expensive and highly sophisticated. If you suspect that the converter on your vehicle is malfunctioning, take it to a dealer or authorized emissions inspection facility for diagnosis and repair.

18.6 Catalytic converter and related components

1 *Catalytic converter*
2 *Pre-converter oxygen sensor*
3 *Post-converter oxygen sensor*
4 *Oxygen sensor electrical connectors (under cover)*

18.8 Loosen the clamp bolts, slide the clamp forward (arrow) and separate the catalytic converter pipe from the exhaust system

CHECK

3 Whenever the vehicle is raised for servicing of underbody components, check the converter for leaks, corrosion, dents and other damage. Check the welds/flange bolts that attach the front and rear ends of the converter to the exhaust system. If damage is discovered, the converter should be replaced.

4 A catalytic converter may become plugged. The easiest way to check for a restricted converter is to use a vacuum gauge to diagnose the effect of a blocked exhaust on intake vacuum.

- *a) Connect a vacuum gauge to an intake manifold vacuum source.*
- *b) Warm the engine to operating temperature, place the transmission in Park and apply the parking brake.*
- *c) Note and record the vacuum reading at idle.*
- *d) Open the throttle until the engine speed is about 2000 rpm.*
- *e) Release the throttle quickly and record the vacuum reading.*
- *f) Perform the test three more times, recording the reading after each test.*
- *g) If the reading after the fourth test is more than one in-Hg lower than the reading recorded at idle, the catalytic converter, muffler or exhaust pipes may be plugged or restricted.*

REPLACEMENT

Refer to illustrations 18.6 and 18.8

Note: Refer to Section 18 in Chapter 4A for additional information.

5 Raise the vehicle and support it securely on jackstands.

6 Disconnect the electrical connectors from the oxygen sensors (see illustration).

7 Remove the exhaust pipe-to-exhaust manifold flange bolts and separate the exhaust pipe from the exhaust manifold (see Chapter 2A). Support the exhaust pipe.

8 Loosen the clamp bolts and detach the catalytic converter and header pipe from the exhaust system (see illustration).

9 Clean the carbon deposits from the mounting flanges and install new gaskets.

10 Installation is the reverse of removal.

6B

EMISSIONS AND ENGINE CONTROL SYSTEMS: DIESEL ENGINES

Section

Reference to other Chapters

Catalytic converter - See Chapter 6A
Engine Control Module (ECM) - removal and installation - See Chapter 6A
Vehicle speed sensor - check and replacement - See Chapter 6A

1 General information

Refer to illustration 1.1

To prevent pollution of the atmosphere from incompletely burned fuel and to maintain good driveability and fuel economy, a number of emission control systems are incorporated (see illustration). They include the:

Electronic engine control system
Crankcase ventilation system
Exhaust Gas Recirculation system
Catalytic converter

All of these systems are linked, directly or indirectly, to the emission control system.

The Sections in this Chapter include general descriptions, checking procedures within the scope of the home mechanic (when possible) and component replacement procedures for each of the systems listed above.

Before assuming that an emissions control system is malfunctioning, check the fuel system carefully. The diagnosis of some emission control devices requires specialized tools, equipment and training. If checking and servicing become too difficult or if a procedure is beyond your ability, consult a dealer service department or other properly equipped repair facility. Remember, the most frequent cause of emissions problems is simply a loose or broken vacuum hose or wire, so always check the hose and wiring connections first.

This doesn't mean, however, that emission control systems are particularly difficult to maintain and repair. You can quickly and easily perform many checks and do most of the regular maintenance at home with common hand tools.

Note: Because of a Federally mandated warranty which covers the emission control system components, check with your dealer about warranty coverage before working on any emissions-related systems. Once the warranty has expired, you may wish to perform some of the component checks and/or replacement procedures in this Chapter.

Pay close attention to any special precautions outlined in this Chapter. It should be noted that the illustrations of the various systems may not exactly match the system installed on the vehicle you're working on because of changes made by the manufacturer during production or from year-to-year.

A Vehicle Emissions Control Information (VECI) label is located in the engine compartment. This label contains important emissions specifications and adjustment information, as well as a vacuum hose schematic with emissions components identified. When servicing the engine or emissions systems, the VECI label in your particular vehicle should always be checked for up-to-date information.

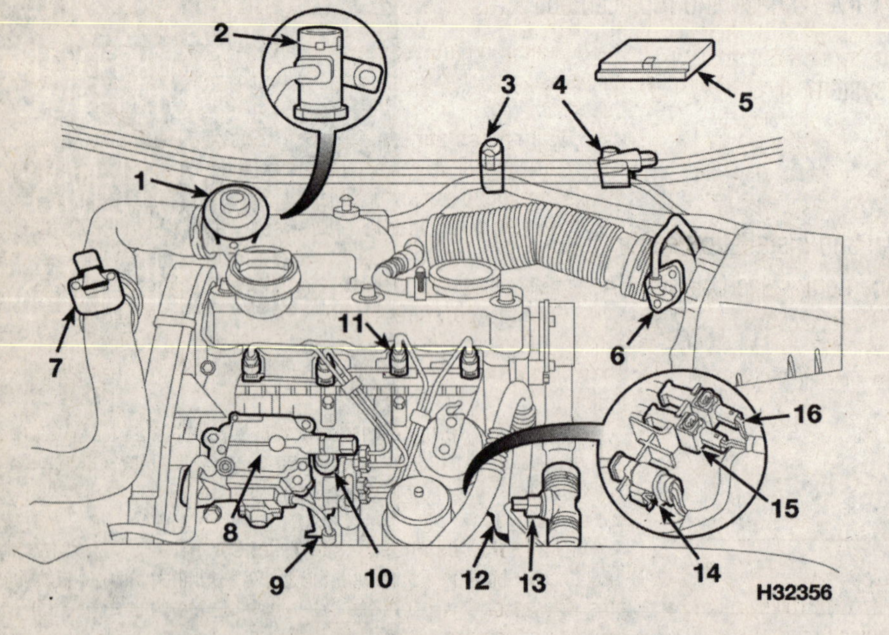

1.1 Typical emission and engine control system components - diesel engine

1. *EGR valve*
2. *Air shut-off valve vacuum valve*
3. *EGR vacuum regulator valve*
4. *Wastegate regulator valve*
5. *Engine Control Module*
6. *Mass airflow sensor*
7. *Manifold absolute pressure/intake air temperature sensor*
8. *Fuel injection pump quantity adjuster*
9. *Cold start valve*
10. *Fuel shut-off valve*
11. *Fuel injector needle lift sensor*
12. *Engine speed sensor*
13. *Engine coolant temperature sensor*
14. *Fuel injection pump electrical connector*
15. *Engine speed sensor electrical connector*
16. *Fuel injector needle lift sensor electrical connector*

2 On-Board Diagnostic (OBD) system and trouble codes

DIAGNOSTIC TOOL INFORMATION

Refer to illustrations 2.1 and 2.3

1 A digital multimeter is necessary for checking fuel injection and emission related components (see illustration). A digital volt-ohmmeter is preferred over the older style analog multimeter for several reasons. The analog multimeter cannot display the volts-ohms or amps measurement in hundredths and thousandths increments. When working with electronic circuits which are often very low voltage, this accurate read-

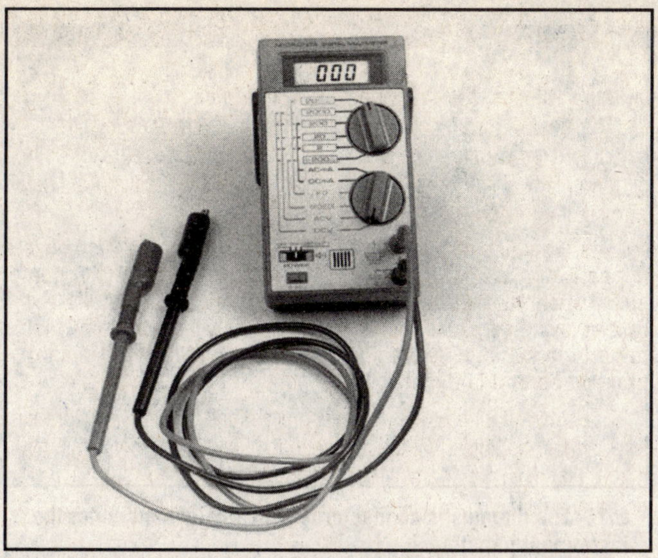

2.1 Digital multimeters can be used for testing all types of circuits; because of their high impedance, they are much more accurate than analog meters for measuring low-voltage computer circuits

ing is most important.

2 Another good reason for the digital multimeter is the high impedance circuit. The digital multimeter is equipped with a high resistance internal circuitry (10 million ohms). Because a voltmeter is hooked up in parallel with the circuit when testing, it is vital that none of the voltage being measured should be allowed to travel the parallel path set up by the meter itself. This dilemma does not show itself when measuring larger amounts of voltage (9 to 12 volt circuits) but if you are measuring a low voltage circuit such as sensor signal voltage, a fraction of a volt may be a significant amount when diagnosing a problem. However, there are several exceptions where using an analog voltmeter may be necessary to test certain sensors.

3 With the arrival of the Federally mandated emission control system (OBD-II), a specially designed scanner has been developed. Several tool manufacturers have released OBD-II scan tools for the home mechanic. Hand-held scanners are the most powerful and versatile tools for analyzing engine management systems used on later model vehicles (see illustration).

ON-BOARD DIAGNOSTIC SYSTEM GENERAL DESCRIPTION

4 All models described in this manual are equipped with the second generation On-Board Diagnostic (OBD-II) system. The system consists of an onboard computer, known as the Engine Control Module (ECM), information sensors and output actuators. The information sensors monitor various functions of the engine and send data to the ECM. Based on the data and the information programmed into the computer's memory, the ECM generates output signals to control various engine functions via control relays, solenoids and other output actuators. The ECM is specifically calibrated to optimize the emissions, fuel economy and driveability of the vehicle.

5 Because of a Federally mandated warranty which covers the emissions system components and because any owner-induced damage to the ECM, the sensors and/or the control devices may void the warranty, it isn't a good idea to attempt diagnosis or replacement of the ECM at home while the vehicle is under warranty. Take the vehicle to a dealer service department if the ECM or a system component malfunctions.

2.3 Scanners like these from Actron and AutoXray are powerful diagnostic aids - they can tell you just about anything you want to know about your engine management system

INFORMATION SENSORS

6 **Engine speed sensor** - The engine speed sensor senses crankshaft position (TDC) during each engine revolution. The ECM uses this information to control fuel injection quantity and timing.

7 **Engine coolant temperature sensor** - The engine coolant temperature sensor senses engine coolant temperature. The ECM uses this information to control fuel injection quantity and timing.

8 **Fuel injector needle lift sensor** - The ECM monitors the fuel injection timing with the signal from the fuel injection needle lift sensor. The sensor is located in fuel injector number three.

9 **Fuel temperature sensor** - The fuel temperature sensor senses the temperature of the fuel being delivered to the fuel injectors. The ECM uses this information to control fuel injection quantity and timing.

10 **Intake air temperature sensor** - The intake air temperature sensor senses the temperature of the air entering the intake manifold. The ECM uses this information to control fuel injection quantity and timing.

11 **Manifold absolute pressure sensor** - The ECM monitors the turbocharger boost pressure with the signal from the manifold absolute pressure sensor. The ECM uses this information to control the turbocharger wastegate regulator valve.

12 **Mass airflow sensor** - The mass airflow sensor measures the amount of air passing through the sensor body and ultimately entering the engine. The ECM uses this information to control fuel injection quantity and turbocharger boost pressure.

13 **Modulating piston displacement sensor** - The ECM monitors the fuel injection quantity with the modulating piston displacement sensor.

14 **Throttle position sensor** - The throttle position sensor senses throttle movement and position. The ECM uses this information to control fuel delivery and engine speed according to driver demand.

15 **Vehicle speed sensor** - The vehicle speed sensor provides information to the ECM to indicate vehicle speed.

16 **Miscellaneous ECM inputs** - In addition to the various sensors, the ECM monitors various switches and circuits to determine vehicle operating conditions. The switches and circuits include:

a) Air conditioning system
b) Antilock brake system
c) Barometric pressure sensor (inside ECM)
d) Battery voltage
e) Brake switch
f) Clutch pedal switch
g) Cruise control system
h) Park/neutral position switch
i) Power steering pressure switch
j) Sensor signal and ground circuits
k) Transaxle control module

OUTPUT ACTUATORS

17 **Air shut-off valve** - The ECM closes the air shut-off valve when the ignition key is turned off. Shutting off the air supply softens the shock of shutting-down a diesel engine.

18 **Check Engine light** - The ECM will illuminate the Check Engine light if a malfunction in the electronic engine control system occurs.

19 **Cold start valve** - The ECM energizes the cold start valve when the engine is started cold. The cold start valve advances the fuel injection timing to aid in cold starting. See Chapter 4B for more information regarding the cold start valve.

20 **EGR regulator valve** - The ECM controls the operation of the EGR valve by regulating the vacuum supply to the valve.

21 **Fuel shut-off valve** - The ECM controls the operation of the fuel shut-off valve. The fuel shut-off valve interrupts fuel flow to the fuel injection pump and is used to start and stop the engine from running.

22 **Fuel quantity adjuster** - The ECM precisely controls the fuel injection quantity (and resulting engine speed) based on signals received from the various engine sensors.

23 **Glow plugs** - The ECM controls the operation of the glow plug system. The glow plugs allow the engine to start easily in cold conditions.

24 **Turbocharger wastegate regulator valve** - The ECM monitors intake manifold pressure and controls the turbocharger wastegate with the wastegate regulator valve. The engine control system calculates the engine torque needed depending on driver demand and engine operating conditions, the ECM will then adjust the boost pressure to meet the demands.

OBTAINING DIAGNOSTIC TROUBLE CODES

Refer to illustration 2.26

Note: The diagnostic trouble codes on all models can only be extracted from the Engine Control Module (ECM) using a specialized scan tool. Have the vehicle diagnosed by a dealer service department or other qualified automotive repair facility if the proper scan tool is not available.

25 The ECM will illuminate the CHECK ENGINE light (also known as the Malfunction Indicator Lamp) on the dash if it recognizes a fault in the system. The light will remain illuminated until the problem is

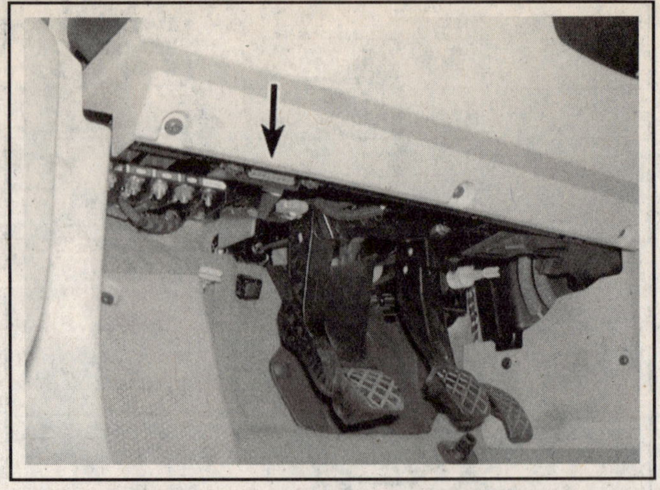

2.26 The diagnostic connector is typically located under the instrument panel

repaired and the code is cleared or the ECM does not detect any malfunction for several consecutive drive cycles.

26 The diagnostic codes for the On-Board Diagnostic (OBD) system can only be extracted from the ECM using a scan tool. The scan tool is programmed to interface with the OBD system by plugging into the diagnostic connector (see illustration). When used, the scan tool has the ability to diagnose in-depth driveability problems and it allows freeze frame data to be retrieved from the ECM stored memory. Freeze frame data is an OBD II ECM feature that records all related sensor and actuator activity on the ECM data stream whenever an engine control or emissions fault is detected and a trouble code is set. This ability to look at the circuit conditions and values when the malfunction occurs provides a valuable tool when trying to diagnose intermittent driveability problems. If the tool is not available and intermittent driveability problems exist, have the vehicle checked at a dealer service department or other qualified repair shop.

CLEARING DIAGNOSTIC TROUBLE CODES

27 After the system has been repaired, the codes must be cleared from the ECM memory using a scan tool. Do not attempt to clear the codes by disconnecting battery power. If battery power is disconnected from the ECM, the ECM will lose the current engine operating parameters and driveability will suffer until the ECM is programmed with a scan tool.

28 Always clear the codes from the ECM before starting the engine after a new electronic emission control component is installed onto the engine. The ECM stores the operating parameters of each sensor. The ECM may set a trouble code if a new sensor is allowed to operate before the parameters from the old sensor have been erased.

DIAGNOSTIC TROUBLE CODE IDENTIFICATION

29 The accompanying list of diagnostic trouble codes is a compilation of all the codes that may be encountered using a generic scan tool. Additional trouble codes may be available with the use of the manufacturer specific scan tool. Not all codes pertain to all models and not all codes will illuminate the Check Engine light when set. All models require a scan tool to access the diagnostic trouble codes.

Code	Code Identification
P0030	Ho2s, heater control circuit
P0100-P0103	Mass air flow sensor, faulty signal
P0105-P0108	MAP sensor, circuit or performance
P0112-P0113	IAT, circuit
P0116	Engine coolant temperature sensor, faulty signal
P0117-P0118	ECT, circuit
P0121	Throttle position sensor, faulty signal
P0123	Throttle position sensor circuit, short to B+
P0128	Coolant thermostat
P0135	O2 sensor, heater circuit
P0181-P0183	Fuel temperature sensor
P0200	Injector circuit, cylinder 1, open
P0201	Injector circuit, cylinder 2, open
P0202	Injector circuit, cylinder 3, open
P0203	Injector circuit, cylinder 4, open
P0216	Injector timing control circuit
P0225-P0228	Accelerator pedal position switch
P0230-P0232	Fuel pump circuit
P0234	Turbocharger overboost condition
P0237-P0238	Turbocharger boost solenoid, A circuit
P0243-P0246	Turbocharger wastegate solenoid
P0251-P0252	Injection pump metering control
P0263	Contribution or balance, cylinder 1
P0266	Contribution or balance, cylinder 2
P0269	Contribution or balance, cylinder 3
P0272	Contribution or balance, cylinder 4
P0299	Turbocharger underboost condition
P0300	Random/multiple cylinder misfire detected
P0301	Cylinder no. 1 misfire detected
P0302	Cylinder no. 2 misfire detected
P0303	Cylinder no. 3 misfire detected
P0304	Cylinder no. 4 misfire detected
P0321	Engine speed sensor, faulty signal
P0322	Engine speed sensor, no signal output
P0380-P0381	Glow plug monitor circuit

Code	Code Identification
P0401	EGR, insufficient flow
P0402	EGR, excess flow
P0403	EGR, control circuit
P0404	EGR, range, performance
P0501	Vehicle speed sensor circuit, faulty signal
P0560	System voltage malfunction
P0600	Serial communication, memory error
P0602	Control module, program error
P0604	Engine Control module, internal RAM error
P0605	Engine Control Module, internal ROM error
P0606	ECM/PCM, communication
P0607	Control module, performance
P0613	TCM, processor
P0614	ECM/TCM, incompatibility
P0642-P0645	Sensor reference, voltage A
P0652-P0653	Sensor reference, voltage B
P0670	Glow plug circuit
P0671	Glow plug circuit, cylinder 1
P0672	Glow plug circuit, cylinder 2
P0673	Glow plug circuit, cylinder 3
P0674	Glow plug circuit, cylinder 4
P0684	Glow plug control module, communication to ECM
P0700	Automatic transmission control system, MIL request
P0701	Automatic transmission control system, range
P0702	Automatic transmission control system, MIL electrical
P0705	Automatic transmission control system, range sensor
P0710-P0713	Transmission fluid temperature sensor, A circuit
P0715-P0717	Input turbine speed sensor
P0720-P0722	Output turbine speed sensor
P0725-P0727	Engine speed input
P0730-P0734	Automatic transmission, incorrect gear ratio
P0740-P0743	Torque converter clutch, circuit
P0746-P0749	Automatic transmission, pressure control solenoid A
P0751-P0753	Automatic transmission, shift solenoid A
P0756-P0758	Automatic transmission, shift solenoid B

Code	Code Identification
P0761-P0763	Automatic transmission, shift solenoid C
P0766-P0768	Automatic transmission, shift solenoid D
P0771-P0773	Automatic transmission, shift solenoid E
P0776-P0778	Automatic transmission, pressure control solenoid B
P0781	Automatic transmission, first-to-second shift
P0782	Automatic transmission, second-to-third shift
P0783	Automatic transmission, third-to-fourth shift
P0784	Automatic transmission, fourth-to-fifth shift
P0785	Automatic transmission, shift timing solenoid
P0791	Automatic transmission, intermediate shaft speed sensor
P0796-P0798	Automatic transmission, pressure control solenoid
P0811	Automatic transmission, clutch slippage
P0829	Automatic transmission, fifth-to-sixth shift
P0840-P0841	Automatic transmission, fluid pressure sensor switch A
P0845-P0846	Automatic transmission, fluid pressure sensor switch B
P0863-P0865	TCM, communication circuit
P0889-P0892	TCM, power relay circuit
P0914	Automatic transmission, gear shift position circuit
P0919	Automatic transmission, gear shift position error
P0929-P0931	Automatic transmission, gear shift lock control solenoid
P1144	Mass airflow sensor, short to ground or open circuit
P1145	Mass airflow sensor, short to B+
P1146	Mass airflow sensor, voltage supply circuit open or shorted
P1155	Manifold absolute pressure sensor, short to B+
P1156	Manifold absolute pressure sensor, short to ground or open circuit
P1157	Manifold absolute pressure sensor, voltage supply circuit shorted or open
P1160	Intake air temperature sensor, short to ground
P1161	Intake air temperature sensor, short to B+ or open circuit
P1162	Fuel temperature sensor, short to ground
P1163	Fuel temperature sensor, short to B+ or open circuit
P1245	Fuel injector needle lift sensor, short to ground
P1246	Fuel injector needle lift sensor, faulty signal
P1247	Fuel injector needle lift sensor, short to B+ or open circuit

Code	Code Identification
P1248	Fuel injection control malfunction
P1251	Cold start valve circuit, short to B+
P1252	Cold start valve circuit, short to ground or open circuit
P1255	Engine coolant temperature sensor, short to ground
P1256	Engine coolant temperature sensor, short to B+ or open circuit
P1354	Modulating piston displacement sensor circuit, open or shorted
P1402	EGR regulator valve circuit, short to B+
P1403	EGR system control malfunction
P1441	EGR regulator valve circuit, short to ground or open circuit
P1537	Fuel shut-off valve malfunction
P1538	Fuel shut-off valve circuit, short to ground or open circuit
P1540	Vehicle speed sensor, signal high
P1546	Turbocharger wastegate regulator valve circuit, short to B+
P1549	Turbocharger wastegate regulator valve circuit, short to ground or open circuit
P1550	Turbocharger boost pressure control
P1561	Fuel quantity adjuster control malfunction
P1562	Fuel quantity adjuster reaches upper limit
P1563	Fuel quantity adjuster reaches lower limit
P1612	Engine Control Module, incorrect programming
P1616	Glow plug indicator light, short to B+
P1617	Glow plug indicator light, short to ground or open circuit
P1618	Glow plug relay circuit, short to B+
P1619	Glow plug relay circuit, short to ground or open circuit
P1626	Engine Control Module, no communication with transaxle control module
P1632	Throttle position sensor circuit, voltage supply open or shorted
P1693	Check Engine light circuit, short to B+
P1694	Check Engine light circuit, short to ground or open circuit
P1851	Engine Control Module, no communication with ABS module
P1854	Engine Control Module, data bus failure

3 Throttle position sensor - check and replacement

1 The throttle position sensor is connected to the accelerator pedal. The throttle position sensor sends a varying voltage signal to the ECM as the accelerator is depressed. The ECM uses the throttle position sensor signal to control fuel delivery based on driver demand. These models are not equipped with an accelerator cable; the ECM controls the fuel quantity adjuster and corresponding engine speed under all driving conditions.

CHECK

➧ Refer to illustrations 3.3 and 3.4

2 The throttle control module is attached to the accelerator pedal bracket (see illustration). Check the terminals in the connector and the wires leading to the sensor for looseness and breaks. Repair as required.

3 A scan tool is required for complete diagnosis of the throttle control system (see Section 2), but the following checks may identify a definite failure with the throttle position sensor.

4 Disconnect the electrical connector from the throttle position sensor. Using an ohmmeter, measure the resistance between terminals 1 and 3, then 2 and 3 on the sensor (see illustration). With the throttle closed, the resistance should be 800 to 1400 ohms at each pair of terminals.

5 Measure the resistance between terminals 4 and 6. With the throttle closed, the resistance should be 800 to 1200 ohms. Depress the throttle fully, an open circuit (infinity) should now be indicated on the meter. On models equipped with an automatic transaxle, repeat this step on terminals 5 and 6, the same results should be indicated.

6 If the resistance is not as specified, replace the throttle position sensor.

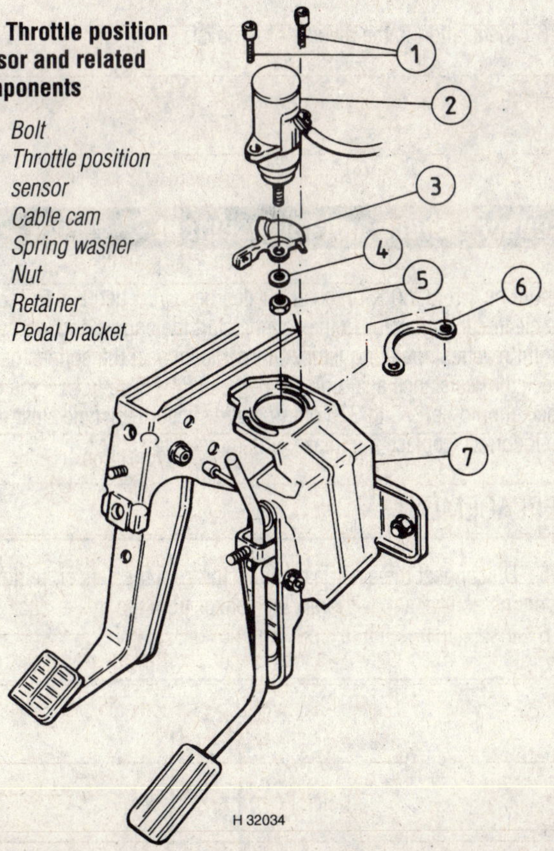

3.3 Throttle position sensor and related components

1 *Bolt*
2 *Throttle position sensor*
3 *Cable cam*
4 *Spring washer*
5 *Nut*
6 *Retainer*
7 *Pedal bracket*

REPLACEMENT

➧ Refer to illustration 3.10

7 Remove the insulation panel from the driver's footwell.

8 Disconnect the electrical connector from the throttle position sensor.

9 Remove the mounting screws, detach the cable cam from the accelerator pedal and remove the sensor.

10 Remove the nut and transfer the cable disc to the new sensor. Align the cable cam with the indicated notch on the sensor (see illustration).

11 Install the sensor, leaving the mounting screws loose.

12 Depress the accelerator pedal against the pedal stop. If the pedal stop cannot be reached, rotate the throttle position sensor in the bracket slots.

13 Tighten the throttle position sensor mounting screws and connect the electrical connector.

14 Install the insulation panel.

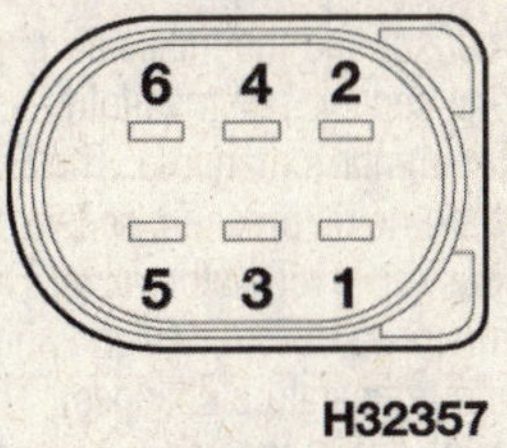

3.4 Throttle position sensor terminal identification

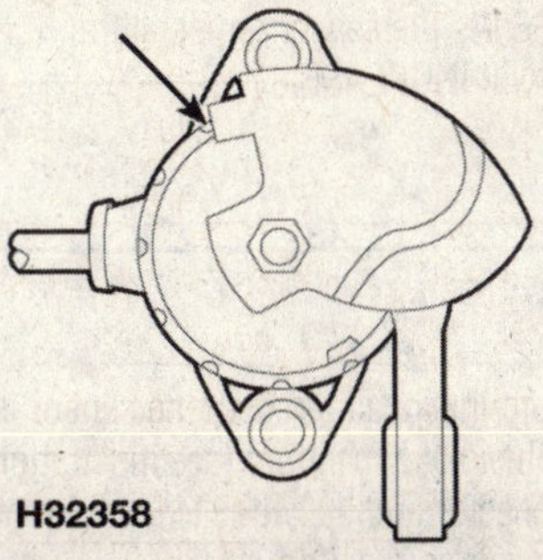

3.10 When installing the cable disc to the throttle position sensor, align the cable cam with notch shown

4 Mass airflow sensor - check and replacement

1 The mass airflow sensor measures the amount of air passing through the sensor body and ultimately entering the engine through the intake manifold. The ECM uses this information to fine tune fuel delivery - the more air entering the engine (acceleration), the more fuel required.

CHECK

➧ Refer to illustration 4.3

2 The mass airflow sensor is located on the air filter housing (see

illustration 1.1). A scan tool is necessary to check the output of the mass airflow sensor (see Section 2). The scan tool displays the sensor output in mg/H. With the engine idling at normal operating temperature, the display should read approximately 200 to 450 mg/H. When the engine is accelerated to 3000 rpm the values should rise above 800 mg/H.

3 Before checking the mass airflow sensor operation, check the voltage supply circuits. Disconnect the electrical connector from the mass airflow sensor and connect a voltmeter to terminals 2 and 3 (harness side) (see illustration). Turn the ignition key On - the voltage should read approximately 12.0 volts. If the voltage is incorrect, check the circuit from the mass airflow sensor to the fuse box (don't forget to check the fuses first). Connect the voltmeter to terminals 3 and 4 - the voltage should read approximately 5.0 volts. If the voltage is incorrect, check the circuit from the mass airflow sensor to the ECM. If the power circuits are good, check the mass airflow sensor operation with a scan tool. If the mass airflow sensor does not respond as described in Step 2, replace the mass airflow sensor.

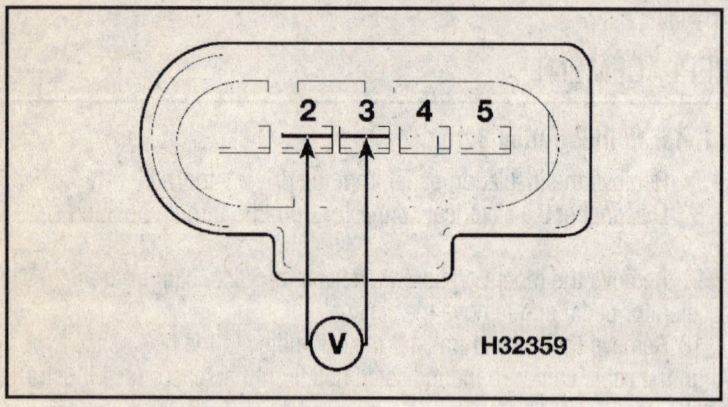

4.3 Mass airflow sensor electrical connector terminal identification (harness side)

➡Note: Before condemning the mass airflow sensor, check the air intake duct for leaks between the sensor and the intake manifold. A reading below the specified value may be caused by an air leak.

REPLACEMENT

4 Disconnect the electrical connector from the mass airflow sensor.

5 Remove the retaining screws and remove the sensor from the air intake duct.

CAUTION:

Handle the mass airflow sensor with care. Damage to this sensor will affect the operation of the engine control system.

6 Installation is the reverse of removal.

5 Manifold absolute pressure/intake air temperature sensor - check and replacement

1 The manifold absolute pressure sensor and the intake air temperature sensor are incorporated into one sensor. The manifold absolute pressure sensor senses the pressure in the intake manifold. The ECM compares this signal with the internal barometric pressure sensor to determine control of the turbocharger wastegate regulator valve. The intake air temperature sensor is a thermistor (a resistor which varies the value of its resistance in accordance with temperature changes). The change in the resistance values will directly affect the voltage signal from the sensor to the ECM. As the sensor temperature INCREASES, the resistance values will DECREASE. As the sensor temperature DECREASES, the resistance values will INCREASE.

CHECK

Refer to illustration 5.4

2 The manifold absolute pressure/intake air temperature sensor is located in the intake duct ahead of the intake manifold (see illustration 1.1). Check the terminals in the connector and the wires leading to the sensor for looseness and breaks. Repair as required.

3 Because much of the manifold absolute pressure sensor circuitry is located inside the ECM, a scan tool is required for diagnosis (see Section 2), but the intake air temperature sensor can be checked as follows.

4 With the ignition switch OFF, disconnect the electrical connector from the manifold absolute pressure/intake air temperature sensor. Using an ohmmeter, measure the resistance between terminals 1 and 2 on the sensor (see illustration). With the engine cool, the resistance should be 1.5 to 2.0 K-ohms at 30-degrees C/86-degrees F. Reconnect the electrical connector to the sensor, start the engine and warm it up until it reaches operating temperature, disconnect the connector and check the resistance again. At 80-degrees C/176-degrees F the resistance should be 275 to 375 ohms. If the sensor resistance test results are incorrect, replace the intake air temperature sensor.

REPLACEMENT

5 Disconnect the electrical connector from the sensor, remove the mounting screws and withdraw the sensor from the intake duct.

6 Installation is the reverse of removal.

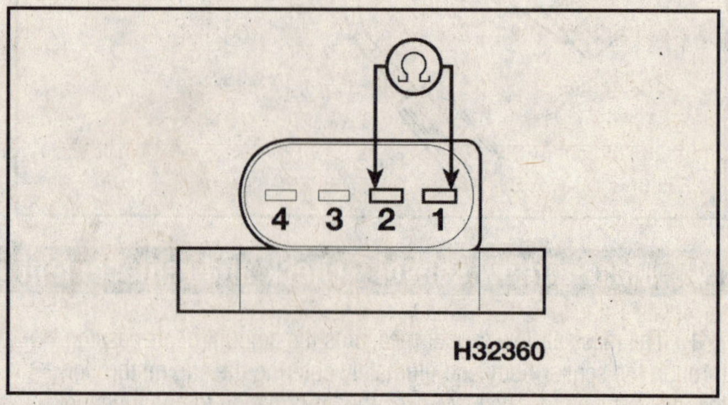

5.4 Manifold absolute pressure/intake air temperature sensor terminal identification

6 Engine coolant temperature sensor - check and replacement

1 The engine coolant temperature sensor is a thermistor (a resistor which varies the value of its resistance in accordance with temperature changes). The change in the resistance values will directly affect the voltage signal from the sensor to the ECM. As the sensor temperature INCREASES, the resistance values will DECREASE. As the sensor temperature DECREASES, the resistance values will INCREASE.

CHECK

➧ Refer to illustration 6.3

2 The engine coolant temperature sensor threads into a coolant pipe near the coolant outlet (see illustration 1.1). Check the terminals in the connector and the wires leading to the sensor for looseness and breaks. Repair as required.

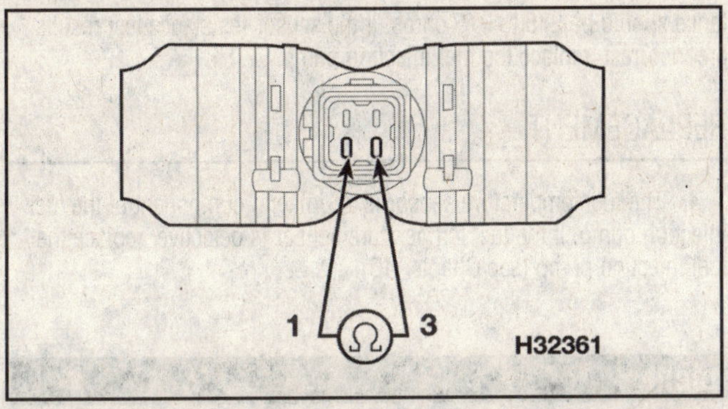

6.3 Measure the resistance of the engine coolant air temperature sensor across terminals 1 and 3

3 With the ignition switch OFF, disconnect the electrical connector from the engine coolant temperature sensor. Using an ohmmeter, measure the resistance between terminals 1 and 3 on the sensor (see illustration). With the engine cool (30-degrees C/86-degrees F), the resistance should be 1.5 to 2.0 K-ohms. Reconnect the electrical connector to the sensor, start the engine and warm it up until it reaches operating temperature. Disconnect the connector and check the resistance again. At 80-degrees C/176-degrees F the resistance should be 275 to 375 ohms. If the sensor resistance test results are incorrect, replace the engine coolant temperature sensor.

REPLACEMENT

⁂ WARNING:

Wait until the engine is completely cool before beginning this procedure.

4 Partially drain the cooling system (see Chapter 1).

5 Disconnect the electrical connector from the sensor and carefully unscrew the sensor.

6 Before installing the new sensor, wrap the threads with Teflon sealing tape to prevent leakage and thread corrosion.

7 Installation is the reverse of removal. Refill the cooling system (see Chapter 1).

7 Engine speed sensor - check and replacement

1 The engine sensor provides the ECM with a crankshaft position signal. The ECM uses the signal to determine a crankshaft reference point (Top Dead Center) and calculate engine speed (RPM). The signal is also used by the On-Board Diagnostic system for misfire detection. The fuel system will not operate if the ECM does not receive an engine speed sensor input.

CHECK

➧ Refer to illustration 7.3

2 The engine speed sensor is located on the side of the engine block near the oil filter. The sensor is equipped with a pigtail lead and a remote connector (see illustration 1.1). Check the terminals in the connector and the wires leading to the sensor for looseness and breaks. Repair as required.

3 Disconnect the electrical connector from the engine speed sensor. Using an ohmmeter, measure the resistance of the sensor across terminals 1 and 2 of the connector (sensor side) (see illustration). The resistance should be approximately 1000 to 1600 ohms.

4 If the engine speed sensor resistance is not as specified, replace the sensor.

REPLACEMENT

5 Disconnect the engine speed sensor electrical connector and remove the connector from the bracket.

6 Remove the engine speed sensor mounting bolt and withdraw the sensor from the engine block.

7 Replace the O-ring and lightly lubricate it with clean engine oil.

8 Installation is the reverse of removal.

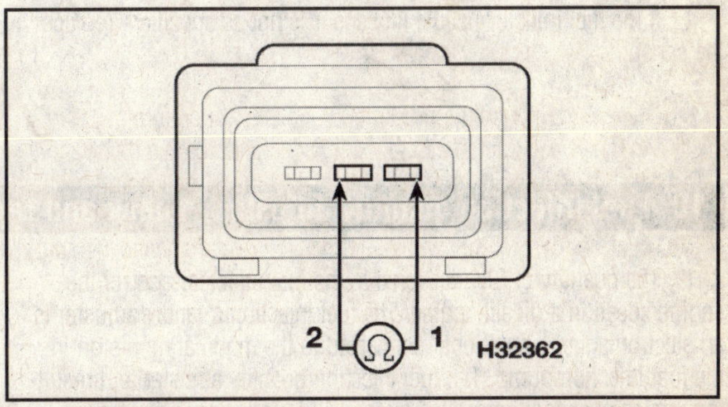

7.3 Measure the resistance of the engine speed sensor across terminals 1 and 2

8 Fuel temperature sensor

1 The fuel temperature sensor is a thermistor (a resistor which varies the value of its resistance in accordance with temperature changes). The change in the resistance values will directly affect the voltage signal from the sensor to the ECM. As the sensor temperature INCREASES, the resistance values will DECREASE. As the sensor temperature DECREASES, the resistance values will INCREASE.

CHECK

➧ Refer to illustration 8.3

2 The fuel temperature sensor is located in the fuel injection pump. Check the terminals in the connector and the wires leading to the fuel injection pump for looseness and breaks. Repair as required.

3 With the ignition switch OFF, disconnect the electrical connector from the fuel injection pump. Using an ohmmeter, measure the resistance between terminals 4 and 7 of the connector (fuel injection pump side) (see illustration). With the engine cool (30-degrees C/86-degrees F), the resistance should be 1.5 to 2.0 K-ohms. Reconnect the electrical connector to the fuel injection pump, start the engine and warm it up until it reaches operating temperature. Disconnect the connector and check the resistance again. At 80-degrees C/176-degrees F the resistance should be 275 to 375 ohms. If the sensor resistance test results are incorrect, replace the fuel injection pump.

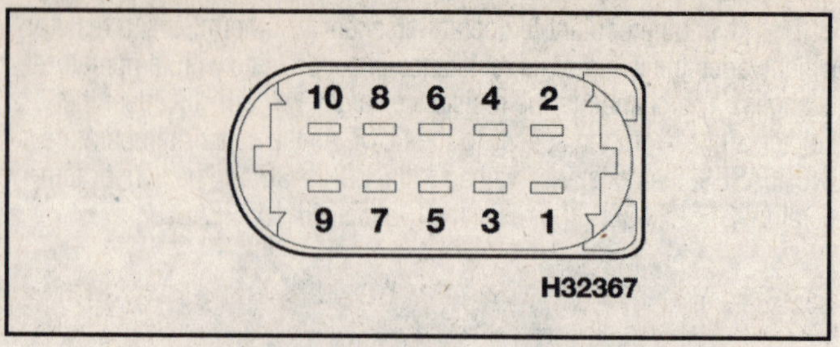

8.3 Fuel injection pump electrical connector terminal identification (fuel injection pump side)

REPLACEMENT

4 The fuel temperature sensor is an integral component of the fuel injection pump. If the fuel temperature sensor is defective, replace the fuel injection pump (see Chapter 4B).

9 Fuel injector needle lift sensor - check and replacement

1 The fuel injector needle lift sensor provides the ECM with a fuel injector needle position signal. The ECM uses the signal to determine fuel injection timing.

CHECK

➧ Refer to illustration 9.3

2 The fuel injector needle lift sensor is incorporated into the cylinder no. 3 fuel injector. The sensor is equipped with a pigtail lead and a remote connector (see illustration 1.1). Check the terminals in the connector and the wires leading to the sensor for looseness and breaks. Repair as required.

3 Disconnect the electrical connector from the fuel injector needle lift sensor. Using an ohmmeter, measure the resistance across the two terminals of the sensor connector (sensor side) (see illustration). The resistance should be 80 to 120 ohms.

4 If the fuel injector needle lift sensor is not as specified, replace the sensor.

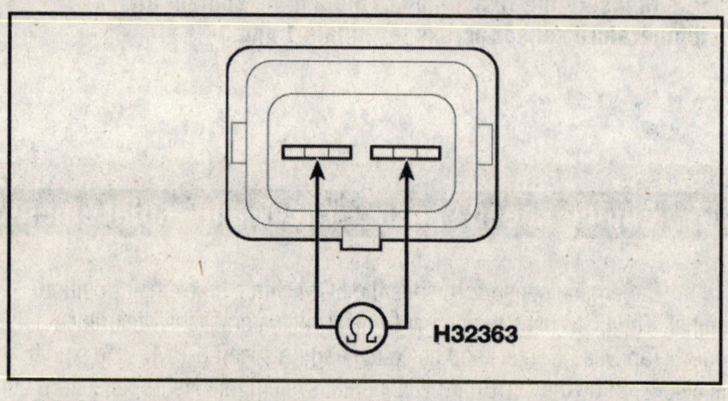

9.3 Measure the resistance of the fuel injector needle lift sensor across the two terminals of the sensor

REPLACEMENT

5 The fuel injector needle lift sensor is incorporated into the cylinder no. 3 fuel injector. If the fuel injector needle lift sensor is defective, replace the cylinder no. 3 fuel injector (see Chapter 4B).

10 Fuel injection quantity adjuster - check and replacement

1 The quantity of fuel delivered to the fuel injectors control the engine speed in a diesel engine. The fuel injection quantity adjuster is an electromagnetic positioner connected to the modulating piston in the fuel injection pump. The fuel injection quantity adjuster and resulting engine speed, is controlled by the ECM. The modulating piston is equipped with a feedback sensor, used by the ECM to fine tune the fuel injection quantity adjuster.

CHECK

2 The fuel injection quantity adjuster is located in the fuel injection pump. The fuel injection pump is equipped with a pigtail lead and a remote connector (see illustration 1.1). Check the terminals in the connector and the wires leading to the sensor for looseness and breaks. Repair as required.

3 Disconnect the electrical connector from the fuel injection pump. Using an ohmmeter, measure the resistance across the following terminals of the fuel injection pump connector (fuel injection pump side) (see illustration 8.3).

a) Terminals 1 and 3 - 4.9 to 7.5 ohms
b) Terminals 2 and 3 - 4.9 to 7.5 ohms
c) Terminals 5 and 6 - 0.5 to 2.5 ohms

4 If the resistance is not as specified, replace the fuel injection pump.

REPLACEMENT

5 The fuel injection quantity adjuster is an integral component of the fuel injection pump. If the fuel injection quantity adjuster is defective, replace the fuel injection pump (see Chapter 4B).

11 Turbocharger wastegate regulator valve - check and replacement

1 The ECM monitors the manifold absolute pressure sensor signal and controls the vacuum supplied to the turbocharger wastegate actuator with the wastegate regulator valve.

CHECK

Refer to illustration 11.3

2 The wastegate regulator valve is located at the driver's side of the engine compartment, under the cowl (see illustration 1.1). Check the terminals in the connector and the wires leading to the valve for looseness and breaks. Repair as required.

3 Disconnect the electrical connector from the wastegate regulator valve. Using an ohmmeter, measure the resistance of the valve across the two terminals (see illustration). The resistance should be approximately 25 to 45 ohms.

4 If the wastegate regulator valve resistance is not as specified, replace the valve.

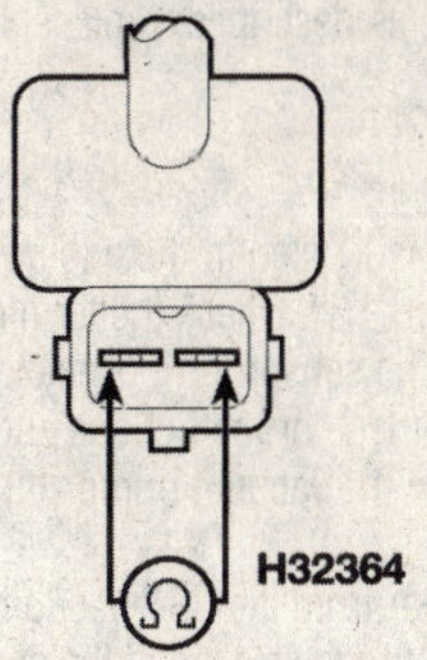

11.3 Measure the resistance of the wastegate regulator valve across the two terminals of the valve

REPLACEMENT

5 Disconnect the electrical connector and vacuum hoses from the valve.

6 Remove the mounting screw and remove the valve.

7 Installation is the reverse of removal.

12 Crankcase ventilation system

1 When the engine is running, a certain amount of the gasses produced during combustion escapes past the piston rings into the crankcase as blow-by gasses. The crankcase ventilation system is designed to reduce the resulting hydrocarbon emissions (HC) by routing the gasses and vapors from the crankcase into the intake manifold and combustion chambers, where they are consumed during engine operation.

2 Crankcase vapors pass through a hose connected from the valve cover to the air intake duct. The oil/air separator at the valve cover separates the oil suspended in the blow-by gases and allows the oil to drain back into the crankcase. The crankcase vapors are drawn from the oil/air separator through a hose connected to the air intake duct where they mix with the incoming air and are burned during the normal combustion process. A heating element is incorporated into the air intake duct to prevent icing in extremely cold weather.

3 A plugged breather, valve or hose will cause excessive crankcase pressures resulting in oil leaks and sludge build-up in the crankcase. Check the components for restrictions and clean or replace the components as necessary. Be sure to check the basic mechanical condition of the engine before condemning the crankcase ventilation system (see Chapter 2C).

13 Exhaust Gas Recirculation (EGR) system

1 The Exhaust Gas Recirculation (EGR) system is used to lower NOx (oxides of nitrogen) emission levels caused by high combustion temperatures. The EGR valve recirculates a small amount of exhaust gases into the intake manifold. The additional mixture lowers the temperature of combustion thereby reducing the formation of NOx compounds.

2 The EGR system consists of the EGR valve, an EGR gas cooler, the EGR vacuum regulator valve, the ECM and a series of connecting pipes and vacuum hoses. The EGR valve is incorporated into the intake manifold inlet flange. The EGR gas cooler is connected to the exhaust manifold and the EGR valve by a semi-flexible pipe. The EGR vacuum valve regulator valve is mounted under the cowl.

3 The ECM controls the EGR flow rate by energizing the EGR vacuum regulator valve solenoid coil. When the vacuum regulator valve is energized vacuum is applied the EGR valve, opening the EGR passage.

4 Exhaust gas from the exhaust manifold is cooled by the EGR gas cooler as the gas flows through the cooler. Cooling the exhaust gas further improves the efficiency of the EGR system.

CHECK

➧ **Refer to illustrations 13.7 and 13.10**

EGR valve

5 Remove the air intake duct from the intake manifold flange.

6 Disconnect the vacuum hose from the EGR valve and connect a hand-held vacuum pump to the vacuum port on the EGR valve.

7 Apply vacuum to the EGR valve while watching the diaphragm rod through the intake flange opening (see illustration). The EGR valve diaphragm should pull the rod up when vacuum is applied. Release the vacuum and the rod should move down to its original position.

8 If the EGR valve does not operate as described, replace the EGR valve/intake manifold flange.

EGR vacuum regulator valve

9 The EGR vacuum regulator valve is located in the engine compartment, under the cowl (see illustration 1.1). Check the terminals in the connector and the wires leading to the sensor for looseness and breaks. Check the vacuum hoses for damage or restrictions and make sure vacuum is present at the hose from the vacuum pump with the engine running. Repair as required.

10 Disconnect the electrical connector from the EGR vacuum regulator valve. Using an ohmmeter, measure the resistance of the sensor across the two terminals of the valve (see illustration). The resistance should be 14 to 20 ohms.

11 If the EGR vacuum regulator valve resistance is not as specified, replace the valve.

REPLACEMENT

➧ **Refer to illustration 13.15**

EGR valve

➡**Note: The EGR valve is incorporated into the intake manifold flange along with the air valve flap. If the EGR valve is defective, the entire flange must be replaced.**

12 Disconnect the vacuum hose from the EGR valve.

13 Disconnect the air valve actuator linkage from the air valve flap and remove the actuator.

14 Remove the intake hose from the flange.

15 Remove the bolts and disconnect the EGR pipe from the flange (see illustration).

16 Remove the flange mounting bolts and remove the flange.

17 Installation is the reverse of removal, using a new O-ring seal and EGR pipe gasket.

EGR vacuum regulator valve

18 Disconnect the electrical connector from the vacuum regulator valve.

19 Label and remove the vacuum hoses from the valve ports. Make a careful note of their orientation to aid installation later.

20 Remove the retaining screws and remove the valve.

21 Installation is a reversal of removal.

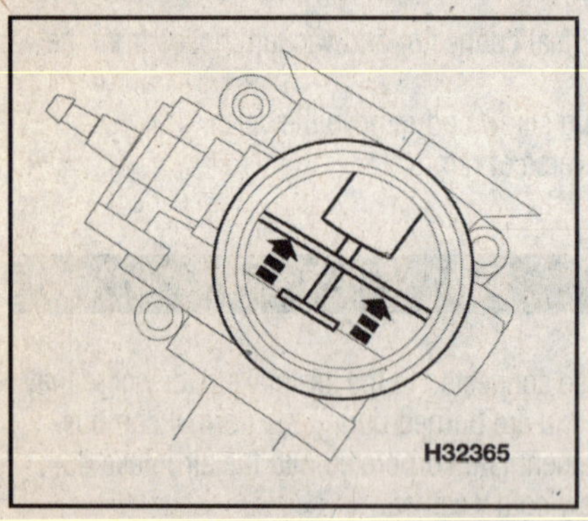

13.7 Apply vacuum to the EGR valve while watching the diaphragm rod through the intake flange opening - the EGR valve diaphragm should pull the rod up when vacuum is applied

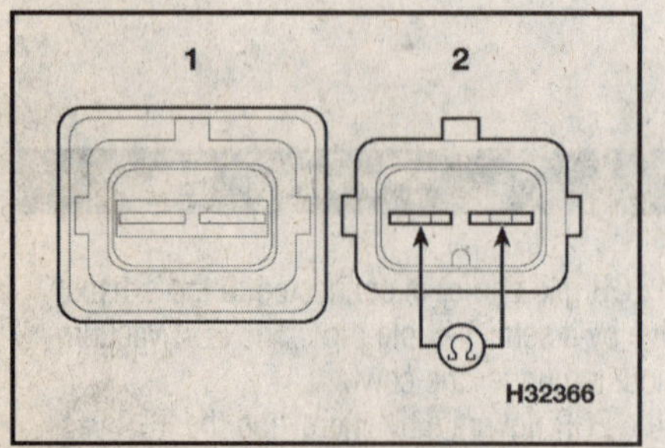

13.10 Disconnect the electrical connector (1) and measure the resistance of the EGR vacuum regulator valve across the two terminals of the valve (2)

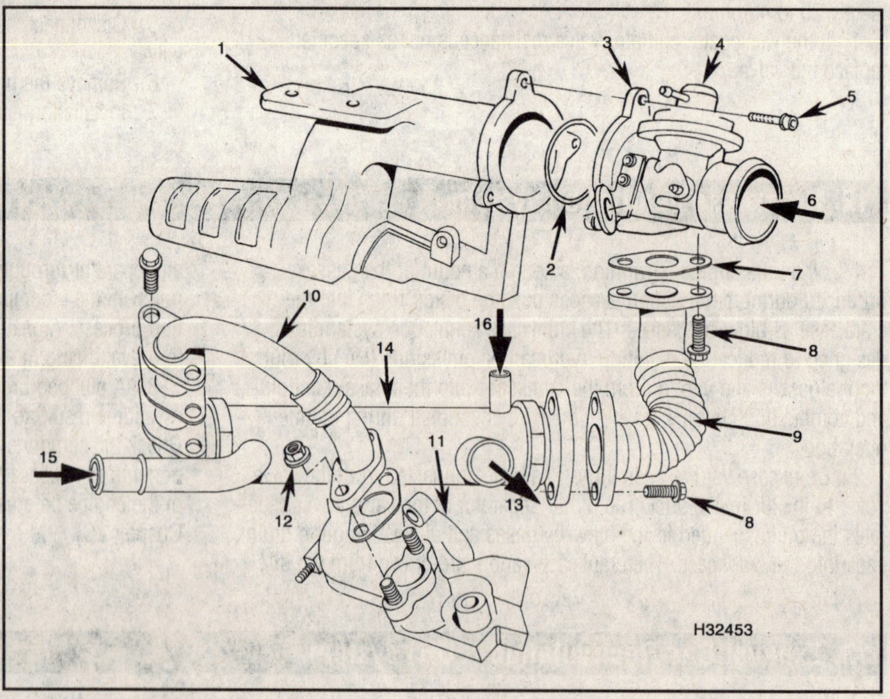

13.15 EGR system components

1 *Intake manifold*
2 *O-ring*
3 *Intake manifold inlet flange*
4 *EGR valve*
5 *Bolt*
6 *Intake air from air filter*
7 *Gasket*
8 *Bolt*
9 *EGR cooler-to-EGR valve pipe*
10 *Exhaust manifold-to-EGR cooler pipe*
11 *Exhaust manifold*
12 *Nut*
13 *Coolant flow to radiator*
14 *EGR gas cooler*
15 *Coolant flow from heater*
16 *Coolant flow from expansion tank*

7A MANUAL TRANSAXLE

Section

Reference to other Chapters

Manual transaxle lubricant change - See Chapter 1
Manual transaxle lubricant level check - See Chapter 1
Transaxle mount replacement - See Chapter 2A

1 General information

The vehicles covered by this manual have either a 5-speed or 6-speed manual transaxle or a 4-speed or 6-speed automatic transaxle. Information on the manual transaxle is included in this Part of Chapter 7. Service procedures for the automatic transaxle are contained in Chapter 7, Part B.

The manual transaxle is a compact, lightweight aluminum alloy housing containing both the transmission and differential assemblies.

Because of the complexity of the transaxle and the special tools needed to work on it, internal repair procedures for the manual transaxle are beyond the scope of this manual. The information in this Chapter is devoted to removal and installation procedures.

2 Shift cables and lever - removal, installation and adjustment

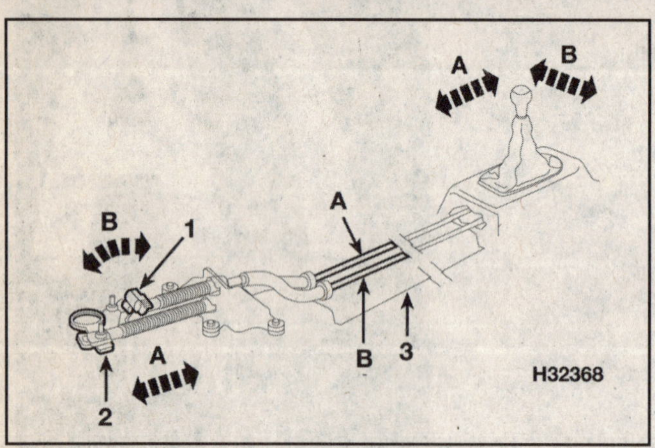

2.2a Operation details of the shift cables

1	*Relay lever/actuating arm*	A	*Gear selector cable*
2	*Gear selector lever*	B	*Gate selector cable*
3	*Heat shield*		

REMOVAL AND INSTALLATION

Cables

➧ Refer to illustrations 2.2a, 2.2b, 2.2c, 2.3a, 2.3b, 2.6, 2.7, 2.8a, 2.8b, 2.9 and 2.10

1 Remove the air filter housing (see Chapter 4A).

2 Remove the clips retaining the cables to the bracket on the transaxle (see illustrations).

3 Detach the cables from the levers on the transaxle. The gate selector cable oan be detached from its lever by pulling up on the plastic tab and sliding the cable eye off the pin (see illustrations); the gear selector cable is attached to its lever by a bolt.

4 Apply the parking brake. Raise the front of the vehicle and support it securely on jackstands.

5 Unbolt the exhaust pipe from the exhaust manifold (see Chapter 2A or 2B), then separate the front half of the exhaust system from the rear half at the clamp (see Chapter 6A, Catalytic converter).

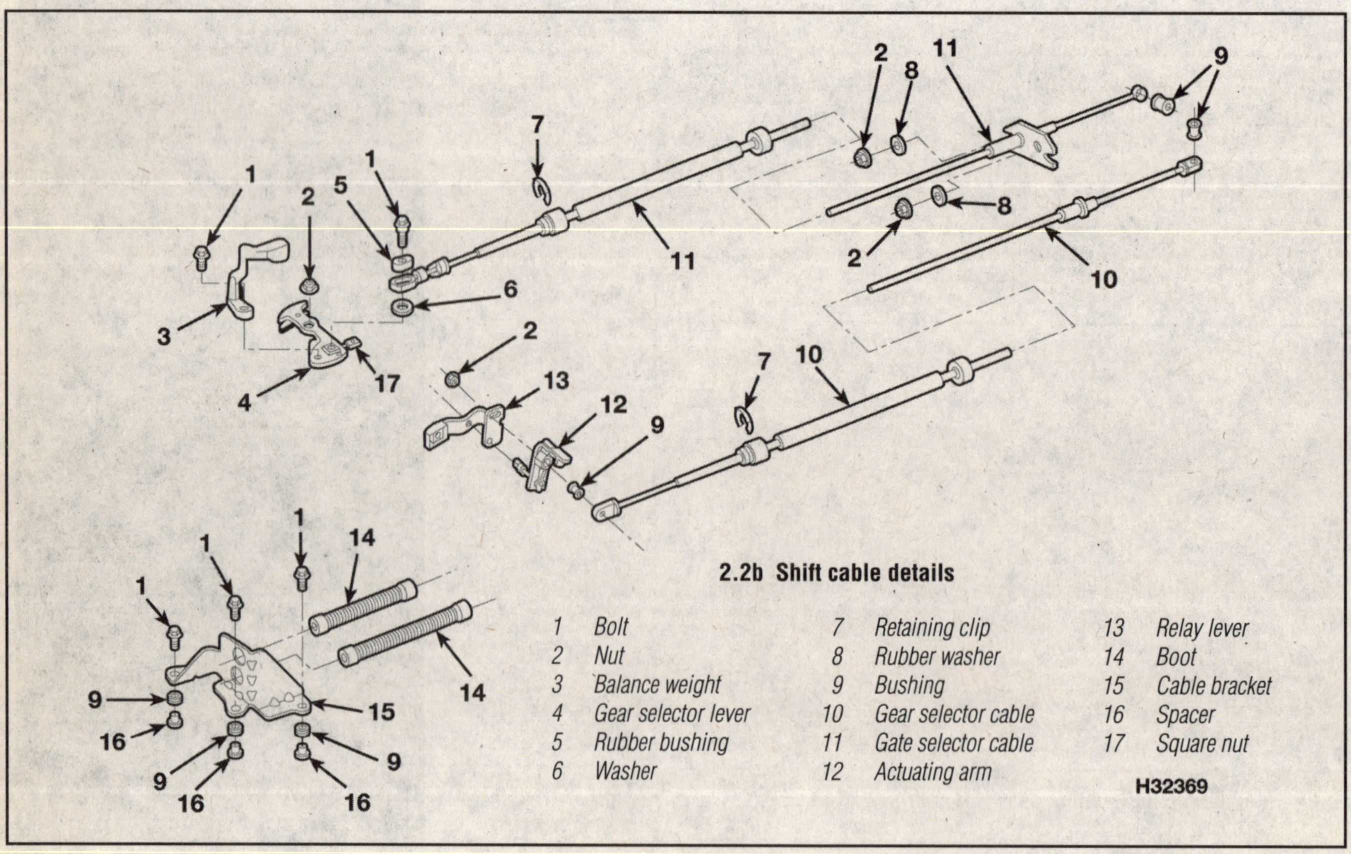

2.2b Shift cable details

1	*Bolt*	7	*Retaining clip*	13	*Relay lever*
2	*Nut*	8	*Rubber washer*	14	*Boot*
3	*Balance weight*	9	*Bushing*	15	*Cable bracket*
4	*Gear selector lever*	10	*Gear selector cable*	16	*Spacer*
5	*Rubber bushing*	11	*Gate selector cable*	17	*Square nut*
6	*Washer*	12	*Actuating arm*		

2.2c Remove the clips that retain the cables to the bracket

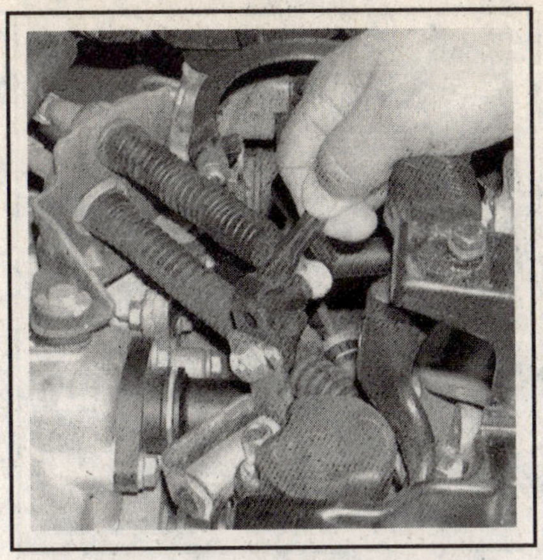

2.3a Pull up on this plastic tab and slide the gate selector cable off the post on the lever

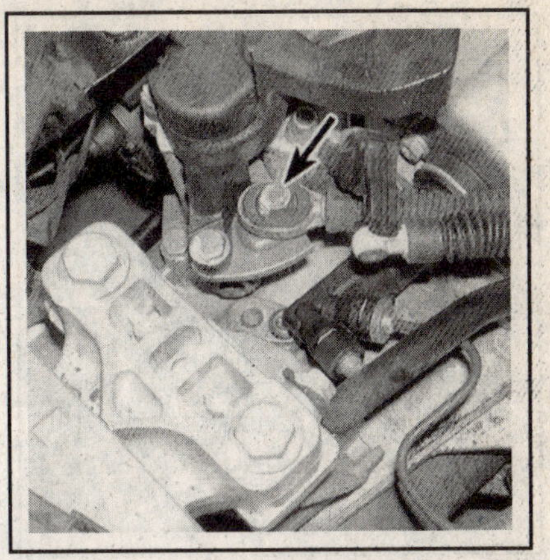

2.3b Remove this bolt (arrow) to detach the gear selector cable from its lever

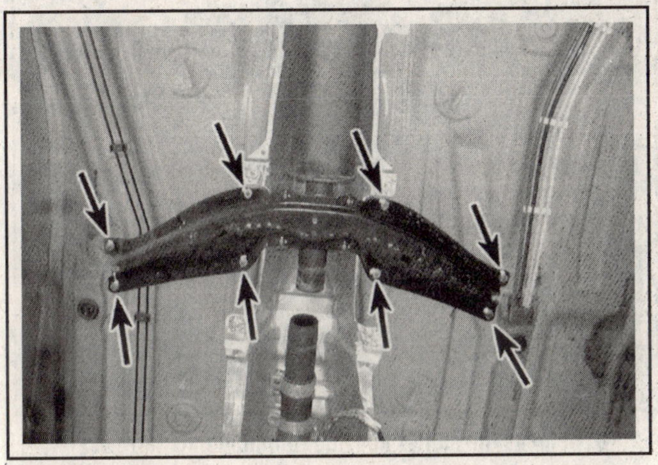

2.6 Unbolt the exhaust system crossmember and allow it to hang down

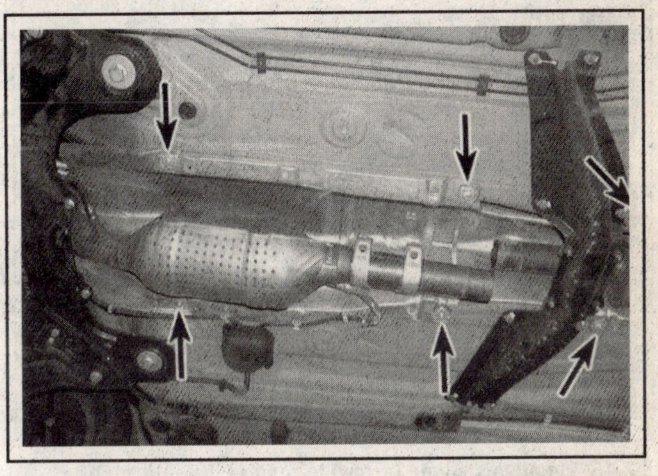

2.7 Remove the retainers (arrows) and guide the heat shields out towards the rear for access to the shift cables

6 Remove the crossmember that supports the rear portion of the exhaust system (see illustration).

7 Pry the retainers off their posts and remove the heat shields for access to the cables (see illustration).

8 Unscrew the bolts and, using a long prybar and a hammer, drive the cover plate off the bottom of the shifter housing (see illustrations).

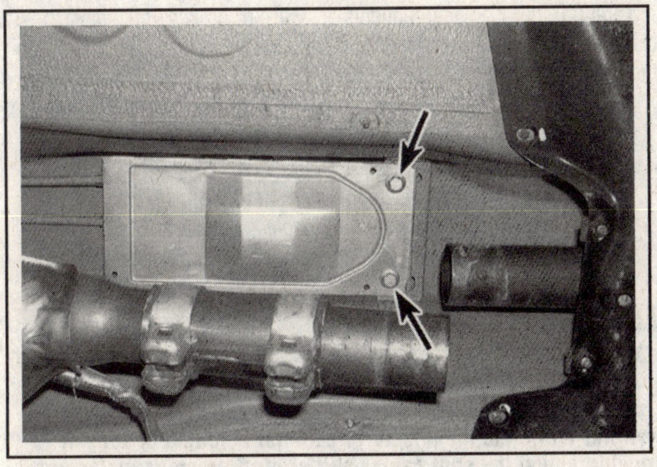

2.8a Remove the bolts at the rear of the cover plate . . .

2.8b . . . then knock the cover forward and off the shifter housing

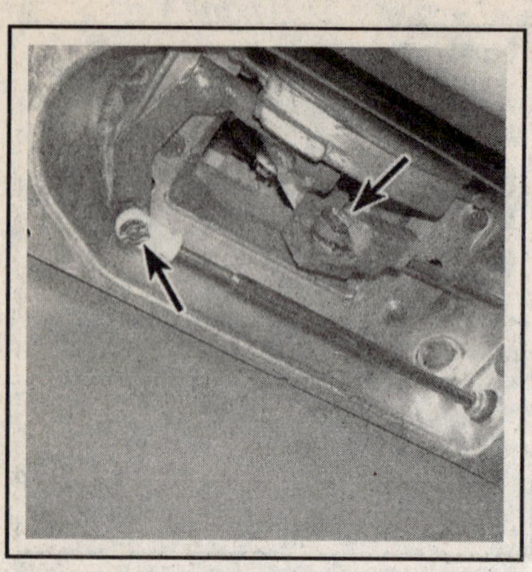

2.9 Pull out the spring clips and detach the cable ends from the levers . . .

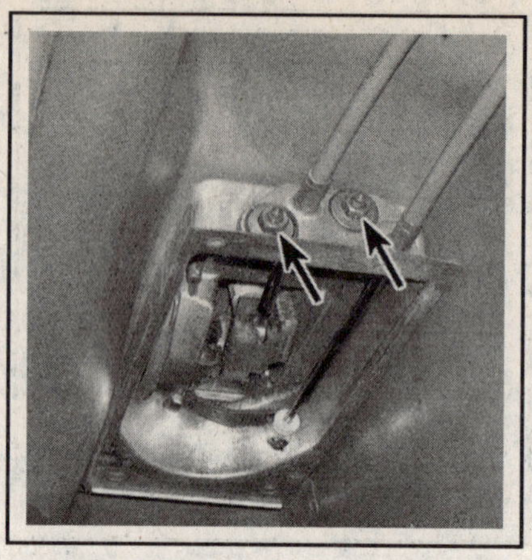

2.10 . . . then remove the nuts securing the cable bracket

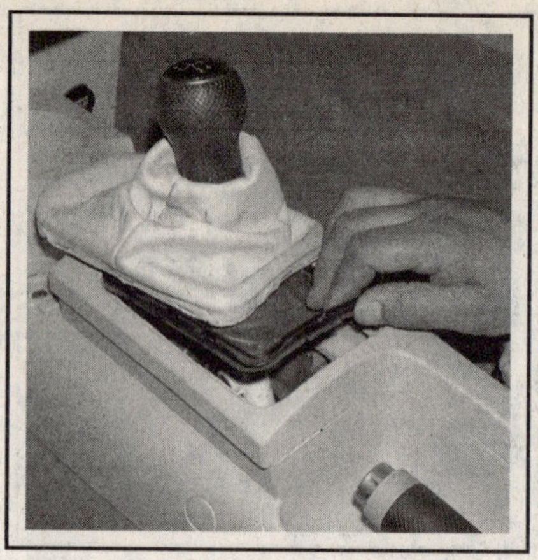

2.14 Pull up on the shift boot and the sealing boot to detach them (this can usually be done without the use of tools)

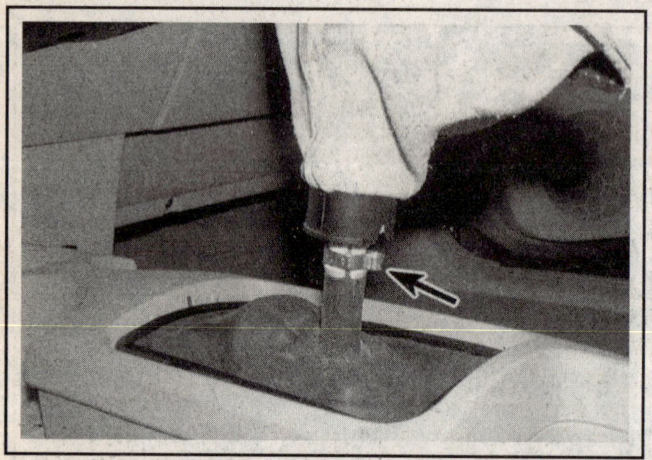

2.15 Cut off this clamp (arrow) to remove the shift knob

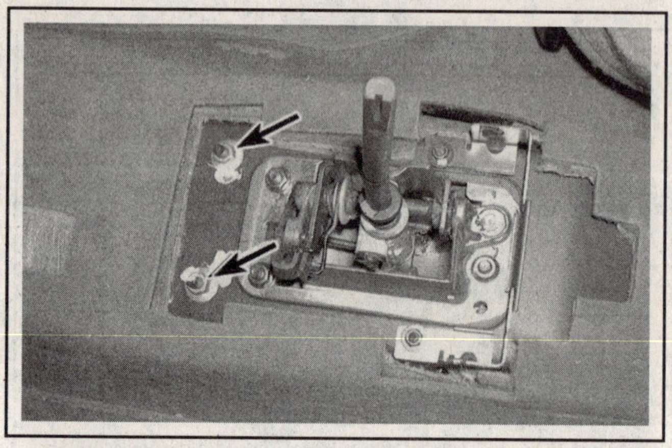

2.16 Remove the shifter housing retaining nuts (arrows)

CAUTION:

Make sure the prybar only contacts the cover plate, not the shifter housing.

9 Remove the spring clips and detach the cables from their pins on the gate selector and gear lever (see illustration).

10 Unscrew the nuts securing the cable bracket, then pull the cables out of the shifter housing (see illustration).

11 Guide the cables down between the engine and the firewall and remove them.

12 Installation is the reverse of removal, with the following points:

a) Lubricate the pins on the shift levers with multi-purpose grease.
b) Replace the gaskets on the shifter cover plate and the cable bracket if they are damaged.
c) Tighten the cable bracket nuts and the shift housing (cover plate) bolts to the torque values listed in this Chapter's Specifications.
d) Adjust the cables as described below.

Shift lever

Refer to illustrations 2.14, 2.15, 2.16 and 2.18

13 Perform Steps 4 through 10.

14 Working inside the vehicle, pull the shift lever boot and the sealing boot from the center console (see illustration).

15 Cut off the clamp and remove the shift knob and boot (see illustration).

16 Remove the nuts securing the shifter housing to the floor pan (see illustration).

17 Lower the shifter housing and lever through the floor pan.

18 The shift lever assembly can be disassembled and inspected (see illustration), but if any components are worn or broken you'll have to check on the availability of replacement parts - you may have to replace the entire mechanism.

19 Installation is the reverse of removal, with the following points:

a) Lubricate all contact surfaces with multi-purpose grease.
b) Replace any torn gaskets.
c) Tighten the shifter housing fasteners and the cable bracket fasteners to the torque listed in this Chapter's Specifications.
d) Adjust the shift cables if shift operation isn't smooth.

ADJUSTMENT

Refer to illustrations 2.21a and 2.21b

Note: While the manufacturer calls for a couple of special tools to adjust the shift cables/shifter mechanism, satisfactory results can usually be obtained with a little thought, patience and the help of an assistant.

2.18 Exploded view of the shifter assembly

1. *Lock washer*
2. *Bushing*
3. *Nut*
4. *Mounting plate*
5. *Fulcrum pin*
6. *Bolt*
7. *Gate selector bracket*
8. *Gate selector cable*
9. *Spring clip*
10. *Gear selector cable*
11. *Cap*
12. *Gate selector housing*
13. *Spring*
14. *Guide bushing*
15. *Snap-ring*
16. *Spacer sleeve*
17. *Gear lever guide*
18. *Damping washer*
19. *Shift lever*

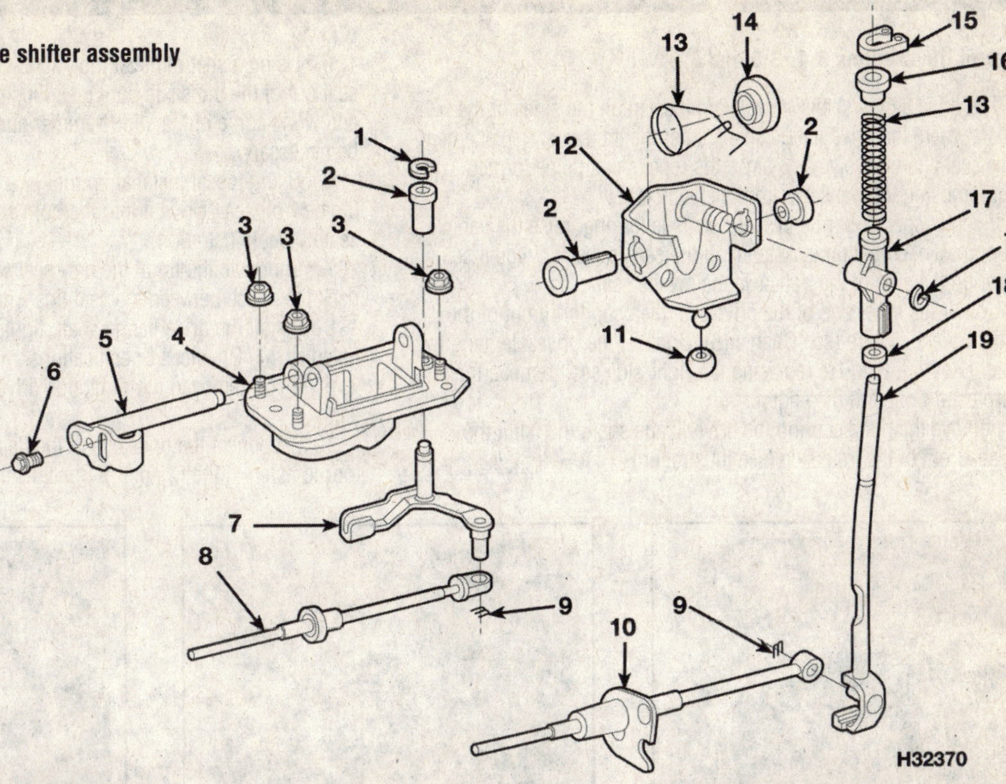

20 Remove the air filter housing (see Chapter 4A).

21 With the cables attached and the transaxle shift lever in the Neutral position (the car will be able to roll if it's in Neutral), check the position of the shifter inside the vehicle - it should be resting right in the middle, in the 3rd/4th gear plane. If it isn't, lift up on the shifter boot and the sealing boot and loosen the nut that holds the fulcrum pin to the mounting plate (see illustration) and, in the engine compartment, the nut holding the relay lever to the actuating arm (see illustration). Verify that the transaxle is in Neutral, have an assistant move the shift lever to the 3rd/4th gear plane, then tighten the bolt and the nut. This will adjust the gate selector cable.

22 Move the shifter back-and-forth several times between 3rd and 4th gears with the clutch pedal depressed - you should be able to feel an equal amount of movement in each direction. If the detent, or engagement, into one gear is more positive than the other, place the lever back into Neutral and have an assistant hold it there. Working in the engine compartment, loosen the bolt holding the gear selector cable to the gear selector lever (see illustration 2.3b). Make sure the transaxle is in Neutral, have your assistant hold the shift lever in Neutral, then tighten the bolt and check the operation of the shifter.

23 Repeat Steps 21 and 22 if necessary, then install the air filter housing and road test the vehicle.

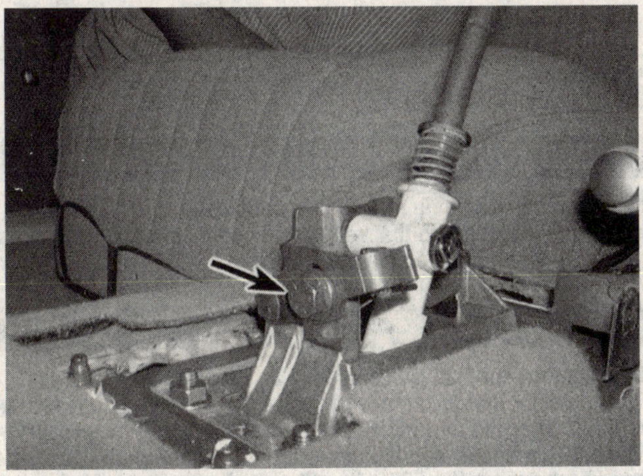

2.21a To adjust the gate selector cable, loosen this bolt at the shift lever mechanism (arrow) . . .

2.21b . . . and this nut at the transaxle - with the transaxle and shift lever in Neutral, move the shift lever into the 3rd/4th gear plane and tighten the bolt

3 Drive flange shaft oil seal - replacement

Refer to illustrations 3.4, 3.5 and 3.6

1 The drive flange shaft oil seals are located on the sides of the transaxle, where the drive flange shafts enter the transaxle. Replacement of these seals is relatively easy, since the repairs can be performed without removing the transaxle from the vehicle.

2 If you suspect that one of these seals is leaking, raise the vehicle and support it securely on jackstands. If the seal is leaking, you'll see lubricant on the side of the transaxle, below the seal.

3 Unbolt the inner end of the driveaxle, then support it out of the way with a piece of wire (see Chapter 8). Don't let the driveaxle hang by the outer CV joint. If you're replacing the right-side seal, remove the CV joint protector bolted to the engine.

4 Remove the bolt securing the drive flange shaft, then pull the flange shaft out of the transaxle (see illustration).

5 Using a seal removal tool, screwdriver or prybar, carefully pry the seal out of the transaxle bore (see illustration). If the seal is stubborn and won't come out, a slide hammer and real removal attachment may be necessary.

6 Using a seal installation tool or a large socket as a drift, install the new oil seal. Drive it into the bore squarely and make sure it's fully seated (see illustration).

7 Lubricate the lip of the new seal with multi-purpose grease. Also pack the space between the seal lips with the same grease.

8 Install the drive flange shaft, tightening the bolt to the torque listed in this Chapter's Specifications.

9 If you removed the right side driveaxle, install the CV joint protector.

10 Reconnect the driveaxle to the flange and tighten the bolts to the torque listed in the Chapter 8 Specifications.

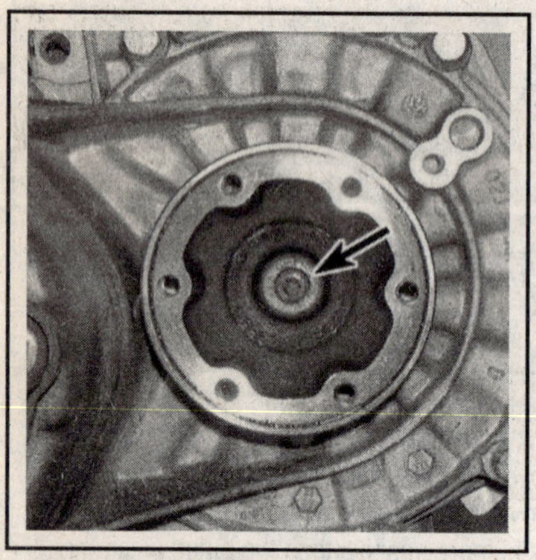

3.4 The drive flange shaft is retained by a single bolt (arrow)

3.5 Carefully pry out the oil seal with a seal removal tool or a screwdriver; make sure you don't damage the seal bore or the new seal may leak

3.6 Installing the drive flange shaft seal with a seal installation tool

4 Back-up light switch - check and replacement

CHECK

1 Turn the ignition key to the On position and move the shift lever to the Reverse position. The switch should close the back-up light circuit and turn on the back-up lights.

2 If it doesn't, check the back-up light fuse (see Chapter 12).

3 If the fuse is okay, verify that there's voltage available on the battery side of the switch (with the ignition turned to On) (see illustration 4.7 for the location of the switch).

4 If there's no voltage on the battery side of the switch, check the wire between the fuse and the switch (refer to the wiring diagrams at the back of this manual for the proper wire to check); if there is voltage, put the shift lever in Reverse and see if there's voltage on the other terminal of the switch.

5 If there's no voltage available, replace the switch; if there is voltage, check to see if the wire between the switch and the bulbs has an open somewhere.

6 Keep in mind that it is possible that both bulbs could be burned out, but not very likely.

REPLACEMENT

Refer to illustration 4.7

CAUTION 1:

These models are equipped with an anti-theft radio. Before performing a procedure that requires disconnecting the battery, make sure you have the activation code.

CAUTION 2:

Disconnecting the battery can cause driveability problems that require a scan tool to remedy. If the vehicle exhibits driveability problems after the battery is reconnected, it may be necessary to take it to a dealer service department or other qualified repair shop to reprogram the ECM. See Chapter 5, Section 1 for information on the use of an auxiliary voltage input device (memory saver) before disconnecting the battery.

7 The back-up light switch is located on the top of the transaxle, forward of the shift levers (see illustration).
8 Remove the air filter housing (see Chapter 4A).
9 Remove the battery (see Chapter 5).
10 Unplug the electrical connector from the switch.
11 Unscrew the mounting bolts and detach the switch from the transaxle.
12 Installation is the reverse of removal. Tighten the switch bolts securely.

4.7 The back-up light switch is located on the top of the transaxle, just forward of the shift levers - it's secured by two bolts (arrows) (transaxle removed for clarity)

5 Manual transaxle - removal and installation

Refer to illustrations 5.18a, 5.18b, 5.19a, 5.19b, 5.21, 5.24a and 5.24b

WARNING:

The manufacturer recommends replacing the transaxle mounting bracket-to-transaxle bolts and the transaxle mount-to-body bolts with new ones whenever they are removed.

CAUTION 1:

These models are equipped with an anti-theft radio. Before performing a procedure that requires disconnecting the battery, make sure you have the activation code.

CAUTION 2:

Disconnecting the battery can cause driveability problems that require a scan tool to remedy. If the vehicle exhibits driveability problems after the battery is reconnected, it may be necessary to take it to a dealer service department or other qualified repair shop to reprogram the ECM. See Chapter 5, Section 1 for information on the use of an auxiliary voltage input device (memory saver) before disconnecting the battery.

REMOVAL

1 Select a solid, level surface to park the vehicle upon. Give yourself enough space to move around it easily. Apply the parking brake and block the rear wheels.
2 Refer to Chapter 11 and remove the hood from its hinges.
3 Remove the battery (see Chapter 5).
4 Remove the air filter housing (see Chapter 4A).
5 Remove the windshield wiper arms (see Chapter 12), the cowl cover and plenum close-out panel (see Chapter 11).
6 Detach the shift cables from their levers on the transaxle (see Section 2), then unbolt the cable bracket from the top of the transaxle. Secure the cables and bracket out of the way.
7 Unplug the electrical connectors from the Vehicle Speed Sensor (see Chapter 6A) and the back-up light switch.
8 Unbolt the clutch release cylinder from the transaxle and tie it out of the way (see Chapter 8). Don't disconnect the hose.

CAUTION:

Don't depress the clutch pedal while the release cylinder is removed.

9 Unbolt the balance weight from the gear selector lever.
10 Support the engine from above with an engine support fixture, using the engine lifting points (see Chapter 2C; these are the same points that are used when attaching an engine hoist for engine removal). You can obtain one of these at most equipment rental yards.
11 Raise the front of the vehicle and support it securely on jackstands.
12 Remove the under-vehicle splash shield and the left-side inner fender panel.
13 Turn the steering wheel all the way to the left. Unbolt the inner ends of the driveaxles from the drive flanges (see Chapter 8). Support the driveaxles with pieces of wire, as high up as possible.
14 Remove the right (passenger's side) drive flange shaft (see Section 3).
15 Remove the starter motor (see Chapter 5).
16 Support the transaxle with a jack (preferably a special jack made for this purpose; these can be obtained at most equipment rental yards). Safety chains will help steady the transaxle on the jack.

5.18a Remove the engine-to-transaxle bolt (lower arrow) and the flywheel cover plate bolt (upper arrow) from the right side . . .

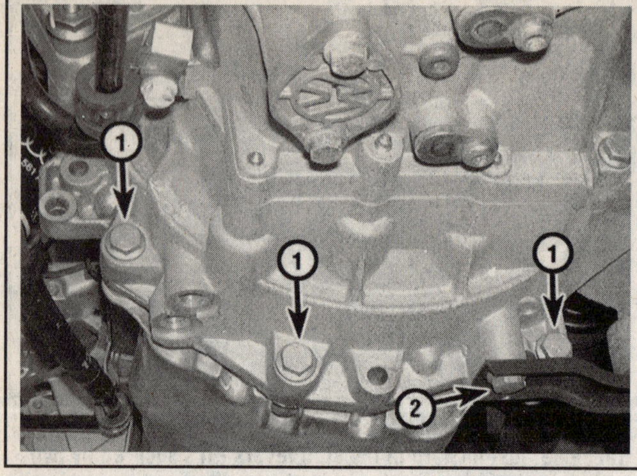

5.18b . . . and the transaxle-to-engine bolts from the left side (1); the support strut bolt (2) will already have been removed

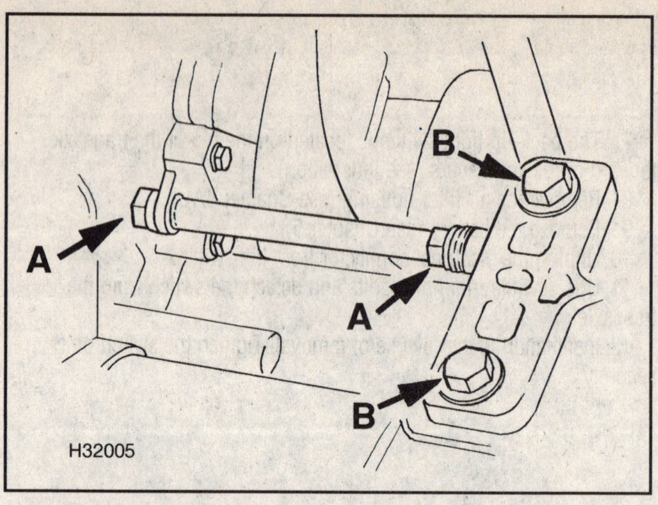

5.19a Working in the engine compartment, remove the support strut bolts (A) and the transaxle mount bolts (B) . . .

17 Remove the engine support strut (see Chapter 2A, Section 17).

18 Remove the transaxle mounting bolts that are accessible from underneath (see illustrations). Lay them out in order or mark them so they can be returned to their original locations.

19 Working from above, remove the left-side transaxle mount bolts (see illustrations). Also unbolt the support strut between the mount bracket and transaxle.

20 Lower the engine support bracket and the jack supporting the transaxle, simultaneously, then unbolt the transaxle mount bracket from the left side of the transaxle. Also unbolt the clips that secure the power steering line to the end of the transaxle and reposition the line.

21 Remove the transaxle mounting bolts that are accessible from above (see illustration). Lay them out in order or mark them so they can be returned to their original locations.

22 Check that nothing remains connected to the transaxle. Carefully pull the transaxle away from the engine, lowering the jack and engine support fixture, as necessary, to allow the transaxle to clear the left side of the vehicle.

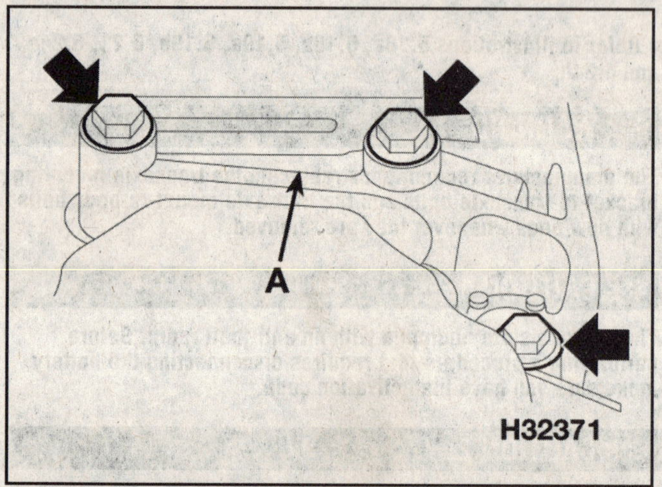

5.19b . . . then lower the engine/transaxle assembly and unbolt the transaxle mount bracket (A) and remove it

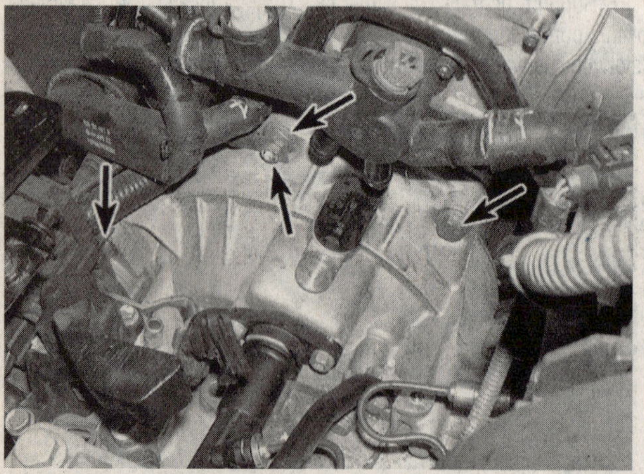

5.21 Transaxle upper mounting fasteners

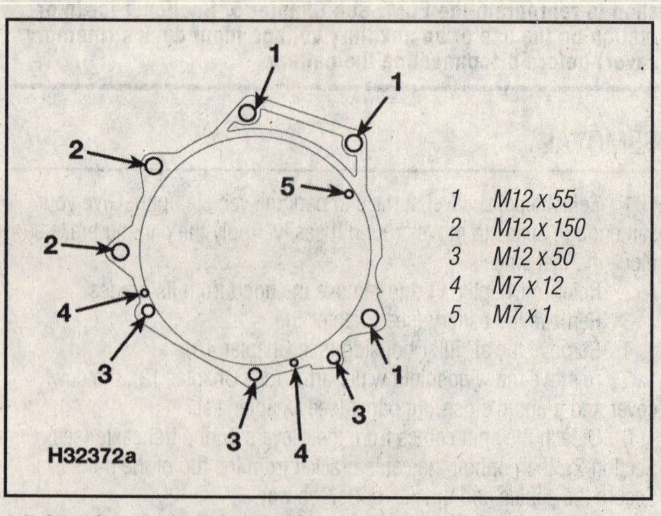

5.24a Several different lengths and sizes of bolts are used to secure the transaxle and engine together - they must be returned to their proper positions - 02J transaxle

⁂ CAUTION:

Don't allow the transaxle to hang by the input shaft.

23 Lower the transaxle and remove it from under the vehicle.

INSTALLATION

24 Installation of the transaxle is essentially a reversal of the removal procedure, but note the following points:

a) *Lubricate the splines of the input shaft with a thin film of high-temperature grease.*
b) *Tighten the transaxle mounting fasteners to the torque values listed in this Chapter's Specifications (see illustrations).*
c) *Refer to Chapter 2A and tighten the engine mounting bolts to the correct torque.*
d) *Refer to Chapter 8 and reconnect the driveaxles.*
e) *On completion, refer to Chapter 1 and check the transaxle lubricant level.*

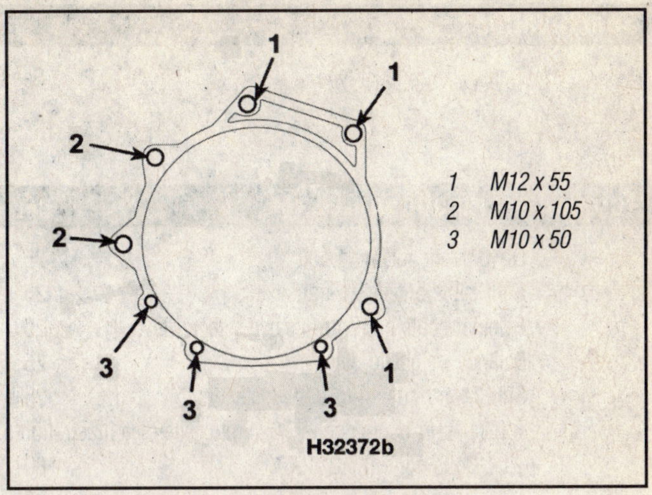

5.24b Bolt sizes and lengths - 02M transaxle

6 Manual transaxle overhaul - general information

1 Overhauling a manual transaxle is a difficult job for the do-it-yourselfer. It involves the disassembly and reassembly of many small parts. Numerous clearances must be precisely measured and, if necessary, changed with select fit spacers and snap-rings. As a result, if transaxle problems arise, it can be removed and installed by a competent do-it-yourselfer, but overhaul should be left to a transmission repair shop. Rebuilt transaxles may be available, check with your dealer parts department and auto parts stores. At any rate, the time and money involved in an overhaul is almost sure to exceed the cost of a rebuilt unit.

2 Nevertheless, it's not impossible for an inexperienced mechanic to rebuild a transaxle if the special tools are available and the job is done in a deliberate step-by-step manner so nothing is overlooked.

3 The tools necessary for an overhaul include internal and external snap-ring pliers, a bearing puller, a slide hammer, a set of pin punches, a dial indicator and possibly a hydraulic press. In addition, a large, sturdy workbench and a vise or transaxle stand will be required.

4 During disassembly of the transaxle, make careful notes of how each piece comes off, where it fits in relation to other pieces and what holds it in place. If you note how each part is installed before removing it, getting the transaxle back together again will be much easier.

5 Before taking the transaxle apart for repair, it will help if you have some idea what area of the transaxle is malfunctioning. Certain problems can be closely tied to specific areas in the transaxle, which can make component examination and replacement easier. Refer to the *Troubleshooting* section at the front of this manual for information regarding possible sources of trouble.

Specifications

Transaxle lubricant type	See Chapter 1

Torque specifications

	Nm	Ft-lbs (unless otherwise indicated)
Drive flange retaining bolt	25	18
Shift cable bracket-to-transaxle	25	18
Selector cable bolt-to-transaxle selector lever	25	18
Balance weight-to-selector lever	25	18
Shift lever housing-to-body fasteners	25	18
Transaxle-to-engine bolts - 02J transaxle (see illustration 5.24a)		
Bolts no. 1 (M12 x 55)	80	59
Bolts no. 2 (M12 x 150)	80	59
Bolts no. 3 (M12 x 50)	45	33
Bolts no. 4 (M7 x 12)	10	84 in-lbs
Bolt no. 5 (M7 x 12)	10	84 in-lbs
Transaxle-to-engine bolts - 02M transaxle (see illustration 5.24b)		
Bolts no. 1 (M12 x 55)	80	59
Bolts no. 2 (M10 x 105)	80	59
Bolts nó. 3 (M10 x 50)	40	30
Transaxle mounting bracket-to-transaxle bolts		
Step 1	40	30
Step 2	Tighten an additional 90-degrees (1/4-turn)	
Transaxle mount-to-body bolts (see illustration 5.19a)		
Bolts A	25	18
Bolts B		
Step 1	60	44
Step 2	Tighten an additional 90-degrees (1/4-turn)	

7B AUTOMATIC TRANSAXLE

Section

Reference to other Chapters

Automatic transaxle fluid and filter change - See Chapter 1
Automatic transaxle fluid level check - See Chapter 1
Engine mounts - check and replacement - See Chapter 2A

1 General information

The vehicles covered by this manual have either a 5-speed or 6-speed manual transaxle, a 4-speed (01M), 6-speed (09G) or 6-speed direct shift (02E) automatic transaxle. All information on the automatic transaxle is included in this Part of Chapter 7. Information for the manual transaxle can be found in Part A of this Chapter.

The overall operation of the transaxle is managed by a Transmission Control Module (TCM) and as a result there are no manual adjustments (aside from shift cable adjustment). Comprehensive fault diagnosis can therefore only be carried out using dedicated electronic test equipment.

If the transaxle requires major repair work, it should be taken to a dealer service department or an automotive transmission repair shop.

2 Diagnosis - general

Automatic transaxle malfunctions may be caused by five general conditions:

a) Poor engine performance
b) Improper adjustments
c) Hydraulic malfunctions
d) Mechanical malfunctions
e) Malfunctions in the computer or its signal network

Diagnosis of these problems should always begin with a check of the easily repaired items: fluid level and condition (see Chapter 1) and shift cable adjustment. Next, perform a road test to determine if the problem has been corrected or if more diagnosis is necessary. If the problem persists after the preliminary tests and corrections are completed, additional diagnosis should be done by a dealer service department or transmission repair shop. Refer to the *Troubleshooting Section* at the front of this manual for information on symptoms of transaxle problems.

PRELIMINARY CHECKS

1 Drive the vehicle to warm the transaxle to normal operating temperature.

2 Check the fluid level as described in Chapter 1, Section 22.

a) If the fluid level is unusually low, add enough fluid to bring the level within the designated capacity, then check for external leaks (see below).
b) If the fluid level is abnormally high, drain off the excess, then check the drained fluid for contamination by coolant. The presence of engine coolant in the automatic transmission fluid indicates that a failure has occurred in the transaxle fluid cooler (see Section 8).
c) If the fluid is foaming, drain it and refill the transaxle, then check for coolant in the fluid, or a high fluid level.

3 Check the engine idle speed.

➡Note: If the engine is malfunctioning, do not proceed with the preliminary checks until it has been repaired and runs normally.

4 Inspect the shift cable (see Section 3). Make sure that it's properly adjusted and that it operates smoothly.

FLUID LEAK DIAGNOSIS

5 Most fluid leaks are easy to locate visually. Repair usually consists of replacing a seal or gasket. If a leak is difficult to find, the following procedure may help.

6 Identify the fluid. Make sure it's transmission fluid and not engine oil or brake fluid (the automatic transmission fluid in these models is a transparent yellow color).

7 Try to pinpoint the source of the leak. Drive the vehicle several miles, then park it over a large sheet of cardboard. After a minute or two, you should be able to locate the leak by determining the source of the fluid dripping onto the cardboard.

8 Make a careful visual inspection of the suspected component and the area immediately around it. Pay particular attention to gasket mating surfaces. A mirror is often helpful for finding leaks in areas that are hard to see.

9 If the leak still cannot be found, clean the suspected area thoroughly with a degreaser or solvent, then dry it.

10 Drive the vehicle for several miles at normal operating temperature and varying speeds. After driving the vehicle, visually inspect the suspected component again.

11 Once the leak has been located, the cause must be determined before it can be properly repaired. If a gasket is replaced but the sealing flange is bent, the new gasket will not stop the leak. The bent flange must be straightened.

12 Before attempting to repair a leak, check to make sure that the following conditions are corrected or they may cause another leak.

➡Note: Some of the following conditions cannot be fixed without highly specialized tools and expertise. Such problems must be referred to a transmission shop or a dealer service department.

Gasket leaks

13 Check the pan periodically. Make sure the bolts are tight, no bolts are missing, the gasket is in good condition and the pan is flat (dents in the pan may indicate damage to the valve body inside).

14 If the pan gasket is leaking, the fluid level may be too high, the vent may be plugged, the pan bolts may be too tight, the pan sealing flange may be warped, the sealing surface of the transaxle housing may be damaged, the gasket may be damaged or the transaxle casting may be cracked or porous. If sealant instead of gasket material has been used to form a seal between the pan and the transaxle housing, it may be the wrong sealant.

Seal leaks

15 If a transaxle seal is leaking, the fluid level may be too high, the vent may be plugged, the seal bore may be damaged, the seal itself may be damaged or improperly installed, the surface of the shaft protruding through the seal may be damaged or a loose bearing may be causing excessive shaft movement.

16 Make sure the filler tube seal is in good condition and the tube is properly seated. If transmission fluid is evident, check the O-ring for damage.

Case leaks

17 If the case itself appears to be leaking, the casting is porous and will have to be repaired or replaced.

18 Make sure the oil cooler hose fittings are tight and in good condition.

Fluid comes out vent pipe or fill tube

19 If this condition occurs, the transaxle is overfilled, there is coolant in the fluid, the case is porous, the dipstick is incorrect, the vent is plugged or the drain-back holes are plugged.

3 Shift cable - removal, installation and adjustment

REMOVAL AND INSTALLATION

Refer to illustrations 3.2 and 3.6

1 Move the shift lever to the 'P' position. Remove the air filter housing (see Chapter 4).

2 At the transaxle end of the cable, pry the cable end off the selector lever, then remove the clip holding the cable to its bracket (see illustration).

3 Raise the vehicle and support it securely on jackstands. Apply the parking brake.

4 Refer to Chapter 7A, Section 2, and perform Steps 5 through 7.

5 Remove the cover from the bottom of the shift lever housing.

6 Pry the shift cable end off the shift lever pin, then remove the clip and pass the cable and casing through the housing (see illustration).

7 Guide the cable down between the engine and the firewall and remove it.

8 Installation is the reverse of removal, with the following points:

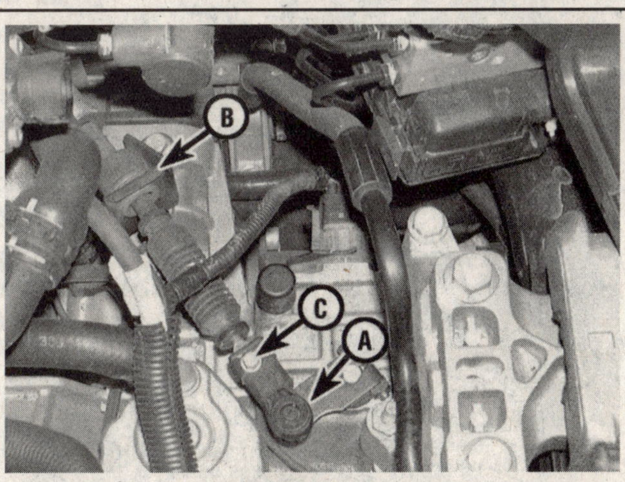

3.2 To detach the shift cable at the transaxle, pry the cable end off the selector lever (A), then remove the clip (B) and detach the cable from the bracket - (C) is the adjustment bolt

- *a) Lubricate the pins on the shift levers with multi-purpose grease.*
- *b) When connecting the cable end to the selector lever at the transaxle, squeeze it on with a pair of pliers - don't try to press it on just by pushing down on it, as the selector lever might get bent.*
- *c) Replace the gasket on the shifter cover plate if it is damaged*
- *d) Adjust the cables as described below.*

ADJUSTMENT

9 Move the selector lever to the 'P' position.

10 At the transaxle end of the cable, loosen the bolt at the ball socket of the cable end (see illustration 3.2). Push the selector shaft up against its end stop, corresponding to the 'P' position; try to turn the front wheels by hand - they should be locked. Now tighten the bolt to the torque listed in this Chapter's Specifications.

11 Verify the operation of the shift lever by shifting through all gear positions and checking that every gear can be selected smoothly and without delay.

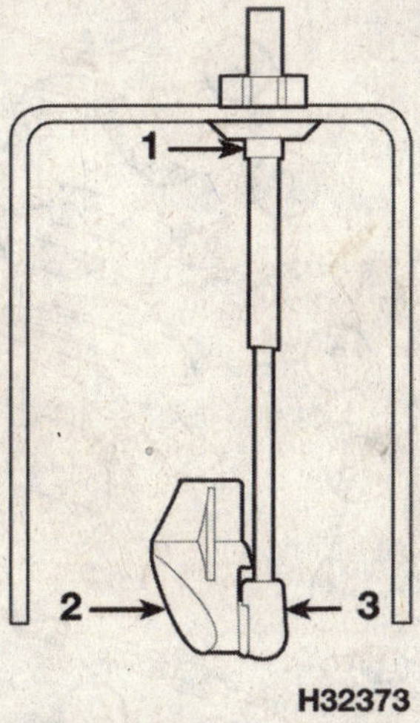

3.6 To detach the shift cable at the shift lever housing, pry the cable end (3) off the bottom of the shift lever (2), then remove the clip (1) and pass the cable through the housing

4 Shift lever assembly - removal and installation

Refer to illustrations 4.6 and 4.12

1 Refer to Section 3 and perform Steps 3 through 6. On models with 09G 6-speed transmission, position the shift lever in Reverse.

2 Working inside the vehicle, push down on the sleeve of the shift handle, then push in on the knob and pull the handle up and off the lever. The button can be retained in the In position by inserting a pin or drill bit in the hole at the left-forward side of the button, while the button is depressed.

3 Pry the covers off the left and right rear of the shift housing.

4 Refer to Chapter 11 and remove the center console.

5 Detach the shift interlock cable from the locking lever (see Section 5).

6 Detach all wire harness clips from the shift housing. Disconnect the electrical connectors from the shift lock solenoid and the gear position indicator (see illustration).

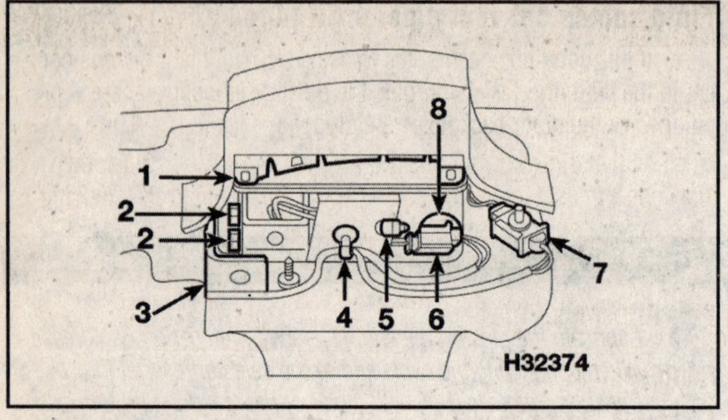

4.6 Shift lever details

1 *Frame*
2 *Shift lock solenoid bolts*
3 *Retaining bracket*
4 *Wiring harness retainer*
5 *Cable tie*
6 *Shift-lock solenoid electrical connector*
7 *Gear position indicator electrical connector*
8 *Connector retainer*

7 Pry the frame off at each corner, being careful not to crack it.

8 Place the shift lever in position 1 and unscrew the two retaining bracket nuts, then remove the bracket.

9 Remove the two bolts that retain the shift lock solenoid (see illustration 4.6).

10 Remove the two nuts that secure the front of the shifter assembly.

11 Carefully lower the assembly down through the floor.

12 The shift lever assembly can be disassembled and inspected (see illustration), but if any components are worn or broken you'll have to check on the availability of replacement parts - you may have to replace the entire mechanism.

13 Installation is the reverse of removal, with the following points:

a) *Adjust the shift cable (see Section 3).*
b) *Adjust the shift interlock cable (see Section 5).*
c) *Tighten the shift lever housing bolts securely.*

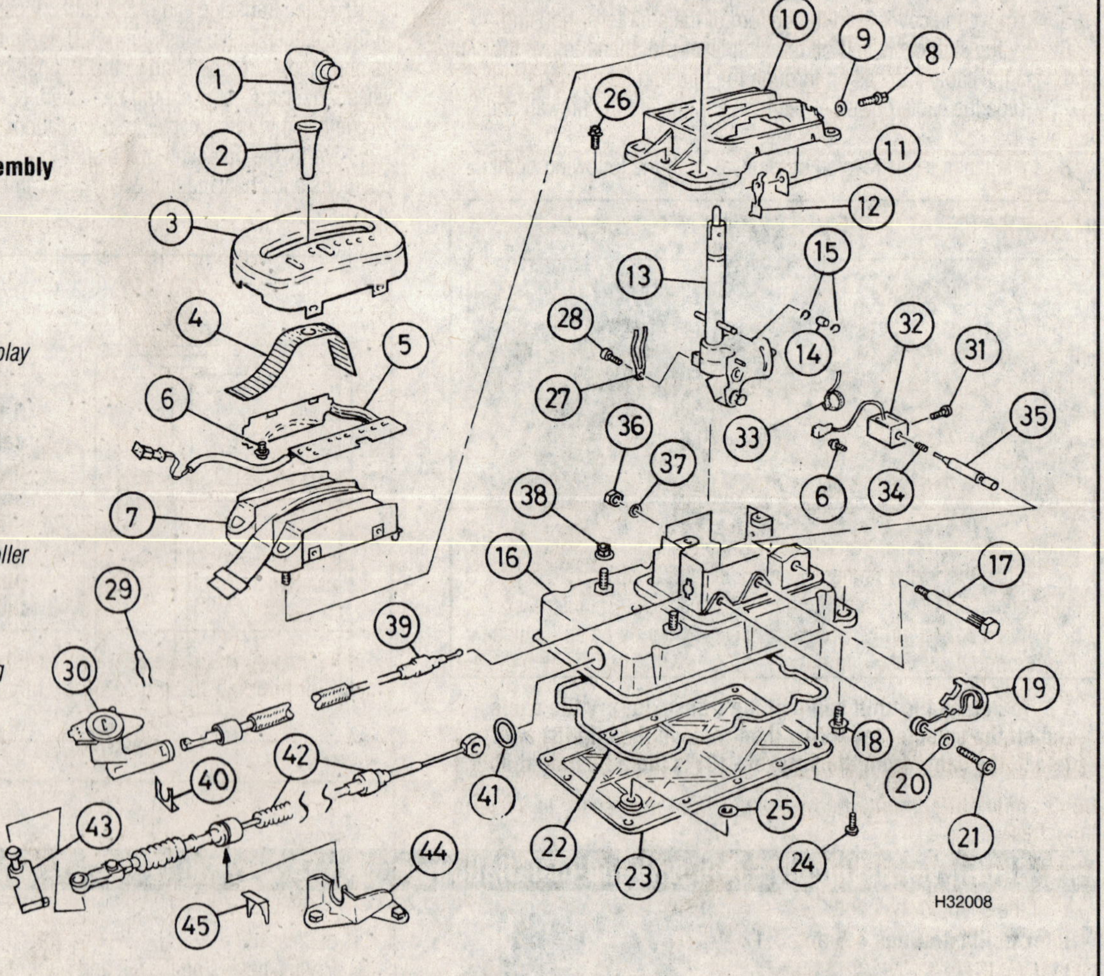

4.12 Shift lever assembly - exploded view

1 *Shifter handle*
2 *Sleeve*
3 *Cover*
4 *Cover strip*
5 *Gear indicator display*
6 *Retainer*
7 *Bracket*
8 *Bolt*
9 *Washer*
10 *Locking segment*
11 *Plate*
12 *Locating spring/roller*
13 *Shift lever*
14 *Roller*
15 *Circlip*
16 *Shift lever housing*
17 *Fulcrum pin*
18 *Bolt*
19 *Locking lever*
20 *Washer*
21 *Bolt*
22 *Gasket*
23 *Cover*
24 *Bolt*
25 *O-ring*
26 *Bolt*
27 *Contact spring*
28 *Bolt*
29 *Clip*
30 *Steering lock*
31 *Bolt*
32 *Shift lock solenoid*
33 *Cable tie*
34 *Spring*
35 *Locking pin*
36 *Nut*
37 *Washer*
38 *Nut*
39 *Key interlock cable*
40 *Clip*
41 *Gasket*
42 *Shift cable*
43 *Selector lever*
44 *Bracket*
45 *Clip*

5 Shift interlock system - description, check and component replacement

WARNING 1:

These models are equipped with airbags. Always disable the airbag system before working in the vicinity of any airbag system component to avoid the possibility of accidental deployment of the airbag(s), which could cause personal injury (see Chapter 12).

WARNING 2:

Do not use a memory saving device to preserve the ECM's memory when working on or near airbag system components.

CAUTION 1:

These models are equipped with an anti-theft radio. Before performing a procedure that requires disconnecting the battery, make sure you have the activation code.

CAUTION 2:

Disconnecting the battery can cause driveability problems that require a scan tool to remedy. If the vehicle exhibits driveability problems after the battery is reconnected, it may be necessary to take it to a dealer service department or other qualified repair shop to reprogram the ECM.

DESCRIPTION

1 The shift lock system prevents the shift lever from being shifted out of Park or Neutral until the brake pedal is applied and the button on the lever is pushed in. It also prevents the ignition key from being turned to the Lock position until the shift lever has been placed in the Park position.

SOLENOID CHECK

2 Remove the center console (see Chapter 11).

3 Follow the wiring harness from the shift lock solenoid back to the electrical connector, then unplug the connector (see illustration 4.6). Using a pair of jumper wires, momentarily apply battery voltage and ground to the solenoid terminals and verify that there's an audible "click."

CAUTION:

Don't apply battery voltage any longer than necessary to perform this check.

4 If the shift lock solenoid doesn't click when energized, replace it.

COMPONENT REPLACEMENT

Shift lock solenoid

5 Refer to Section 4 and remove the shift lever assembly. Remove the console (see Chapter 11).

6 Place the shifter in position 1, then wiggle the lever back and forth and work the solenoid and locking pin/spring out from the housing. On models with the 09G transaxle, disconnect the electrical connector at the printed circuit board on the left side of the shifter, then disconnect the solenoid connector. Access the solenoid by removing the cover on the underside of the floorpan.

7 Installation is the reverse of removal.

Key interlock cable

Replacement

Refer to illustrations 5.14, 5.15 and 5.16

8 Disconnect the cable from the negative terminal of the battery.

9 Remove the steering wheel (see Chapter 10).

10 Remove the steering column cover (see Chapter 11).

11 Remove the center console (see Chapter 11).

12 Turn the ignition key to the On position and place the shift lever in the Park position.

13 Remove the shift lever handle (see Section 4, Step 2).

14 Pull out the clip that retains the cable to the key lock cylinder housing, then detach the cable from the housing (see illustration).

15 Detach the interlock cable from the locking lever at the shifter housing (see illustration 4.12), then push in on the tabs and detach the cable casing from the shifter housing (see illustration).

16 Remove the cable from the instrument panel, noting how it's routed (see illustration). To ease installation of the new cable, it's a good idea to attach a length of wire to the end of the old cable; as you remove the old cable, the length of wire will occupy its place. Then attach the wire to the new cable and use it to help you pull the new cable into place.

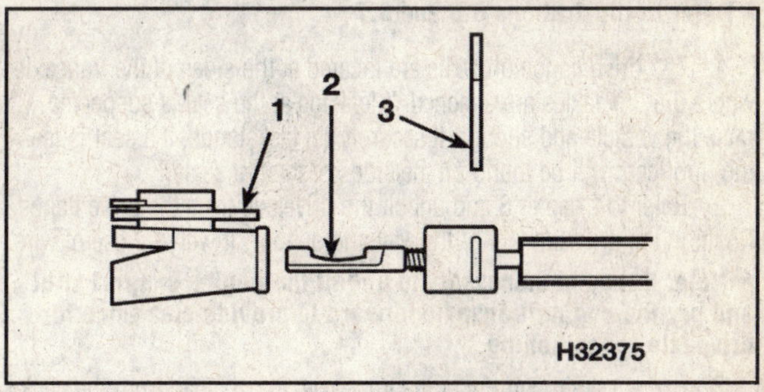

5.14 The key interlock cable (2) is retained to the key lock cylinder housing (1) with a spring clip (3)

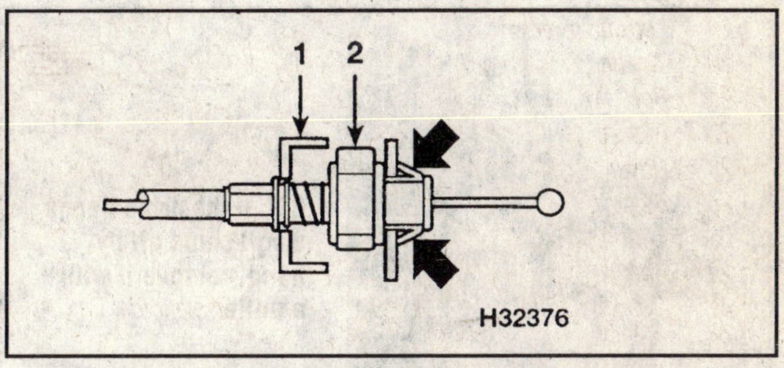

5.15 To detach the interlock cable from the shifter housing, depress the tabs (arrows)

1 Sliding sleeve *2 Adjustment clip*

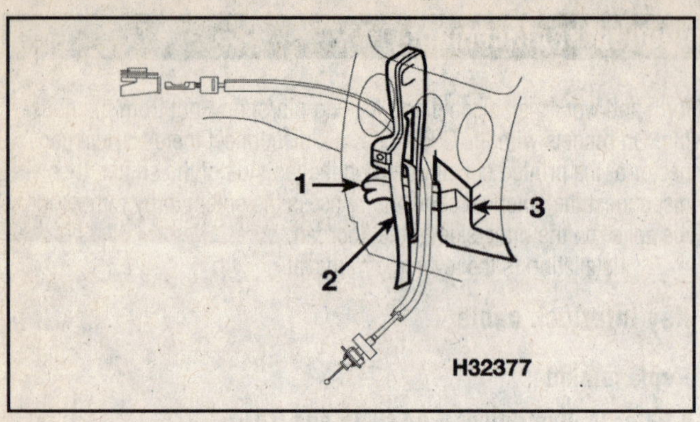

5.16 Key interlock cable routing details

1 Footwell vent
2 Instrument panel support
3 Heater unit

17 To install the cable, reverse the removal procedure, then adjust the cable as described in the next Step.

Adjustment

Refer to illustration 5.22

18 Remove the center console, if not already done (see Chapter 11).

19 Place the shift lever in the Park position, then turn the ignition key to Lock and remove it from the lock cylinder.

20 Move the sliding sleeve on the cable housing forward, then push up on the adjustment clip (see illustration 5.15).

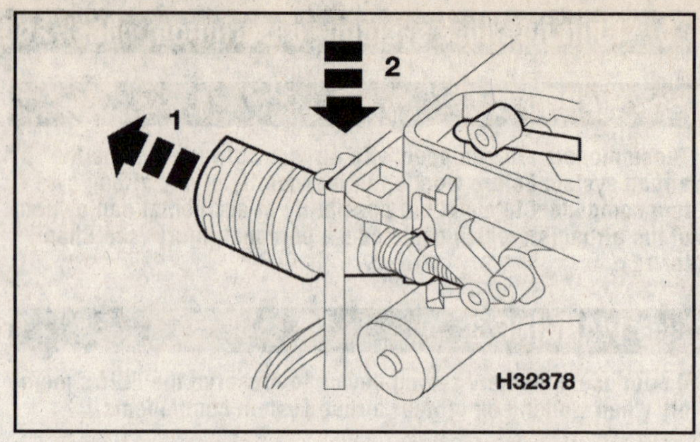

5.22 Place a 0.030-inch (0.8 mm) feeler gauge between the shift lever roller and the locking lever, then pull the cable casing in the direction of arrow (1) to remove any slack; finally, depress the adjustment clip (2)

21 Remove the bracket from the shift housing (see illustration 4.12).

22 Insert a 0.030-inch (0.8 mm) feeler gauge between the shift lever roller and the locking lever (see illustration), then pull lightly on the cable casing, away from the shifter housing. Now push the adjustment clip down and slide the sleeve over the clip.

Note: On models with the 09G transaxle, a factory adjustment pin is required to align the interlock cable.

6 Drive flange oil seal - replacement

Refer to illustrations 6.3 and 6.7

1 The drive flange oil seals are located at the sides of the transaxle, where the driveaxles are attached. If leakage at the seal is suspected, raise the vehicle and support it securely on jackstands. If a seal is leaking, lubricant will be found on the sides of the transaxle.

2 Refer to Chapter 8 and unbolt the driveaxle from the drive flange. Position the driveaxle out of the way and support it with a piece of wire.

Note: It may be necessary to unbolt the engine support strut and pry the engine/transaxle forward to provide clearance for driveaxle repositioning.

3 Place a drain pan under the transaxle. Pry the cap from the center of the flange, then remove the snap-ring. Remove the drive flange, using a puller, if necessary (see illustration).

4 Pry the oil seal out of the transaxle bore with a seal removal tool or a large screwdriver.

5 Pack the open side of the new seal with multi-purpose grease, then drive it into its bore using a seal installation tool or a large socket with an outside diameter slightly smaller than that of the seal. Make sure the seal enters the bore squarely, and is driven in until it is completely seated.

6 If a groove is worn in the drive flange where it contacts the seal, the drive flange must be replaced.

7 Lubricate the seal contact area of the flange with multi-purpose grease. Install the drive flange and install a new snap-ring. If the drive flange will not go on by hand, slight tapping with a hammer may seat

6.3 If the drive flange won't come off by hand, remove it with a puller

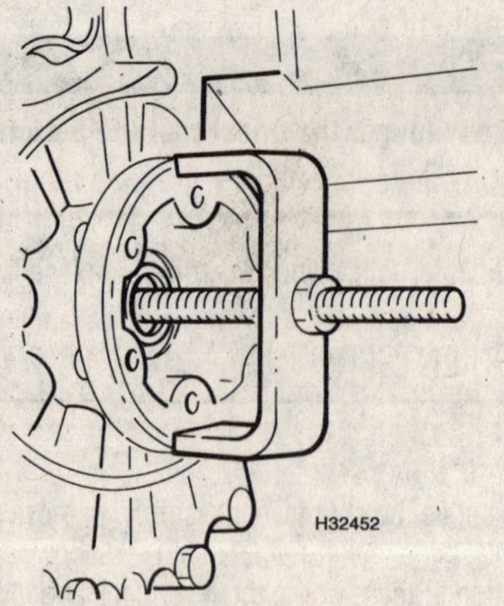

6.7 The drive flange may have to be pressed on using a setup like this

it. If not, it will have to be pressed on using a threaded rod, a piece of metal fabricated to contact the flange surface, and a nut (see illustration).

8 Reconnect the driveaxle to the drive flange, tightening the bolts to the torque listed in the Chapter 8 Specifications.

9 Check the differential lubricant, adding as necessary to bring it to the appropriate level (see Chapter 1).

7 Multi-function Transmission Range (TR) switch - check and replacement

Refer to illustration 7.3

CHECK

1 The Transmission Range (TR) switch is a solid state electronic device that works in conjunction with the Transaxle Control Module (TCM). Diagnosis must be performed with a special factory scan tool. For this reason, if the back-up lights don't work or the vehicle starts in any gear position other than Park or Neutral, have the problem diagnosed by a technician equipped with the proper tool.

REPLACEMENT

2 Remove the air filter housing (see Chapter 4A).

3 Disconnect the electrical connector from the switch (see illustration).

4 On early models, remove the bolt from the hold-down clamp, then pull the switch out of the case.

5 On later models, the range switch is serviced by removing the nut and washer securing the shift arm, then removing the two mounting bolts

6 Installation is the reverse of the removal procedure. On early models, be sure to install a new O-ring on the switch and lubricate it with clean automatic transmission fluid before installing. Tighten the hold-down bolt(s) securely.

7.3 On early models, the Transmission Range (TR) switch is located on the rear side of the transaxle - it is retained by a clamp and a bolt (arrow)

8 Automatic transaxle fluid cooler - removal and installation

Refer to illustrations 8.3a and 8.3b

WARNING:

Wait until the engine is completely cool before beginning this procedure.

CAUTION 1:

These models are equipped with an anti-theft radio. Before performing a procedure that requires disconnecting the battery, make sure you have the activation code.

CAUTION 2:

Disconnecting the battery can cause driveability problems that require a scan tool to remedy. If the vehicle exhibits driveability problems after the battery is reconnected, it may be necessary to take it to a dealer service department or other qualified repair shop to reprogram the ECM. See Chapter 5, Section 1 for information on the use of an auxiliary voltage input device (memory saver) before disconnecting the battery.

1 Remove the air filter housing (see Chapter 4A).

2 Remove the battery and battery tray (see Chapter 5).

3 Pinch off the coolant lines using locking pliers, then detach the lines from the cooler (see illustrations). Be prepared for coolant spillage.

8.3a Pinch off the coolant hoses to prevent excessive coolant loss, then squeeze the clamps (1) and slide them back on the hoses, detach the hoses, then unscrew the banjo bolts (2) - typical, later models have a round cooler with one mounting bolt

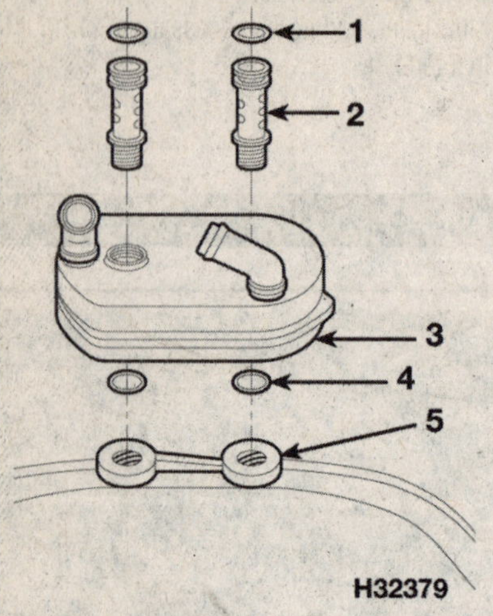

8.3b Transaxle fluid cooler details (typical)

1 *O-ring*
2 *Banjo bolt*
3 *Cooler*
4 *O-ring*
5 *Transaxle*

4 Unscrew the banjo bolts securing the cooler, then lift the cooler from the transaxle. Be careful not to drip coolant or transmission fluid on the vehicle's paint.

5 Installation is the reverse of removal. Be sure to replace the O-rings with new ones, and tighten the banjo bolts to the torque listed in this Chapter's Specifications.

6 After installing the cooler, check the coolant and automatic transaxle fluid, adding as necessary to bring them to the appropriate levels.

9 Automatic transaxle - removal and installation

Refer to illustrations 9.17a, 9.17b, 9.17c, 9.20, 9.23a and 9.23b

WARNING 1:

Wait until the engine is completely cool before beginning this procedure.

WARNING 2:

The manufacturer recommends replacing the transaxle mount bolts and mount support bolts with new ones whenever they are removed.

CAUTION 1:

These models are equipped with an anti-theft radio. Before performing a procedure that requires disconnecting the battery, make sure you have the activation code.

CAUTION 2:

Disconnecting the battery can cause driveability problems that require a scan tool to remedy. If the vehicle exhibits driveability problems after the battery is reconnected, it may be necessary to take it to a dealer service department or other qualified repair shop to reprogram the ECM. See Chapter 5, Section 1 for information on the use of an auxiliary voltage input device (memory saver) before disconnecting the battery.

REMOVAL

1 Select a solid, level surface to park the vehicle upon. Give yourself enough space to move around it easily. Apply the parking brake and chock the rear wheels.

2 Refer to Chapter 11 and remove the hood from its hinges.

3 Remove the air filter housing and air intake duct (see Chapter 4A).

4 Remove the battery and battery tray (see Chapter 5).

5 Remove the windshield wiper arms (see Chapter 12), the cowl cover and plenum close-out panel (see Chapter 11).

6 Refer to Section 3 and disconnect the shift cable from the transaxle selector shaft. Also remove the clip and detach the cable from its bracket.

7 Unbolt the ground strap from the transaxle.

8 Clamp off the coolant hoses leading to and from the transmission fluid cooler unit, then loosen the hose clamps and pull off the hoses (see Section 8). Absorb any coolant that escapes with old rags.

9 Disconnect the wiring harness from the transaxle at the multi-way connectors, labeling each one to aid installation later.

10 Support the engine from above with an engine support fixture, using the engine lifting points (see Chapter 2C; these are the same points that are used when attaching an engine hoist for engine removal). You can obtain one of these at most equipment rental yards.

11 If the vehicle is equipped with a 2.0L engine, remove the Vehicle Speed Sensor (see Chapter 6). Also, remove the exhaust manifold heat shield (see Chapter 2A).

12 Raise the front of the vehicle and support it securely on jackstands. Apply the parking brake.

13 Remove the under-vehicle splash shield and both lower inner fender panels.

14 Unbolt the protective plate from the underside of the transaxle oil pan.

15 Refer to Chapter 5 and remove the starter motor.

16 If the vehicle is equipped with a diesel engine, remove the duct between the turbocharger and intercooler (see Chapter 4B).

17 Remove the access plug, then unscrew each torque converter-to-driveplate nut in turn (see illustrations). As each nut is removed, rotate the crankshaft using a wrench and socket on the crankshaft pulley bolt to expose the next nut. There are three nuts, spaced 120-degrees apart.

18 Remove the engine support strut (see Chapter 2A, Section 17).

19 Support the transaxle with a jack (preferably a special jack made for this purpose; these can be obtained at most equipment rental yards). Safety chains will help steady the transaxle on the jack.

20 Refer to Chapter 8 and unbolt the driveaxles from the transaxle drive flanges. Remove the CV joint shield above the right driveaxle. Turn the right-side drive flange so the flat portion of the flange is vertical (see illustration).

9.17a Pry out this access plug (arrow) . . .

9.17b Turn the crankshaft to bring one of the torque converter nuts in line with the hole . . .

9.17c . . . then unscrew the nut and mark the relationship of the stud to the driveplate

21 Refer to Chapter 10 and separate the balljoint from the left control arm. Detach the stabilizer bar link from the left control arm. Pass the left hand driveaxle over the suspension arm and swing it towards the rear of the vehicle, out of the way.

22 Suspend the right hand driveaxle as high as possible using cable-ties or wire. Turn the steering to full left lock.

23 Remove the transaxle mount bolts, then unbolt the mount support from the transaxle (see illustrations).

24 Remove the transaxle mounting bolts that are accessible from underneath.

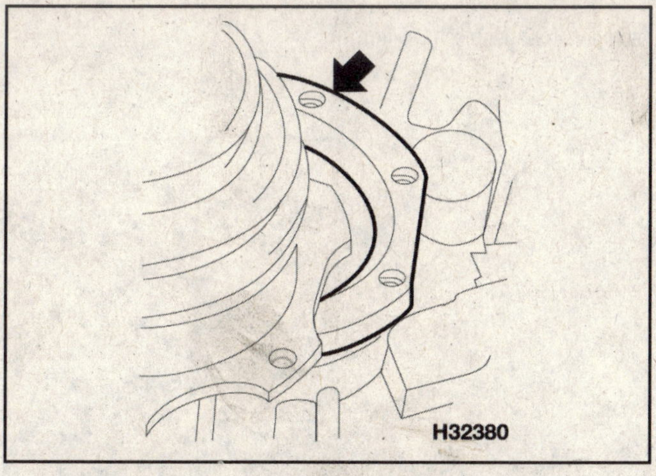

9.20 Turn the right-side drive flange so its flat is vertical - this will allow it to clear the engine block as the transaxle is removed

25 Check that nothing remains connected to the transaxle before attempting to separate it from the engine.

26 Remove the bolts from the top of the bellhousing and pull the transaxle away from the engine.

27 When all the locating dowels are clear of their mounting holes, slowly lower the transaxle down, turning it as necessary to maneuver it out from between the engine and the left side of the chassis. Strap a restraining bar across the front of the bellhousing to keep the torque converter in position.

CAUTION:

Take care to prevent the torque converter from falling out as the transaxle is removed.

INSTALLATION

28 Installation of the transaxle is essentially a reversal of the removal procedure, but note the following points:

a) *As the torque converter is installed, ensure that the drive pins at the center of the torque converter hub engage with the recesses in the automatic transmission fluid pump inner wheel.*
b) *When installing the transaxle, be sure to align the marked torque converter stud with its marked hole.*
c) *Tighten the bellhousing bolts and torque converter-to-driveplate bolts to the specified torque.*
d) *Refer to Chapter 2A and tighten the engine support strut bolts to the specified torque.*

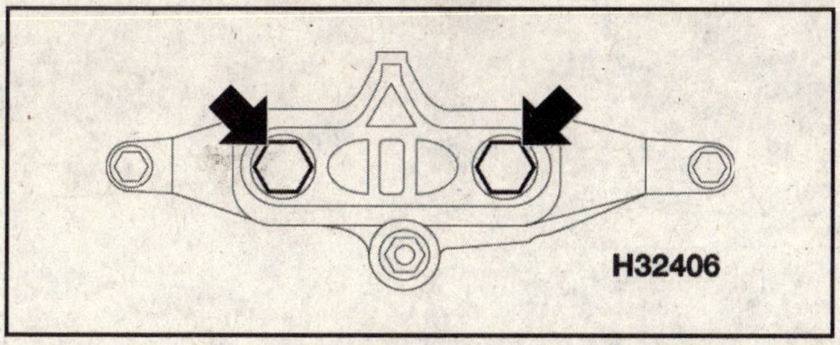

9.23a Unscrew the two transaxle mount bolts (arrows) . . .

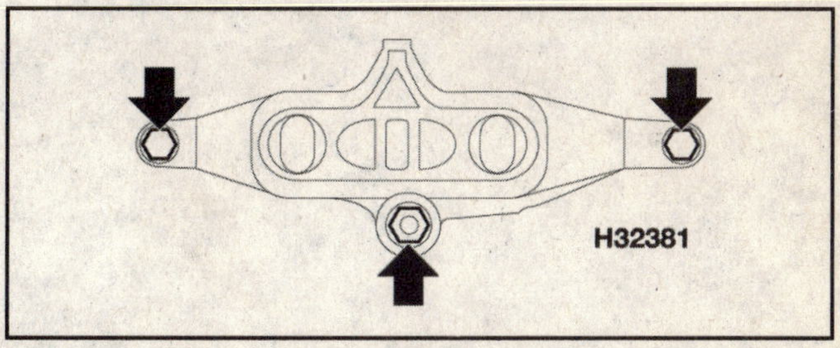

9.23b . . . then unbolt the mount support from the transaxle

e) Tighten the transmission mount bolts and the mount support bolts to the torque listed in this Chapter's Specifications.

f) Tighten the driveaxle-to-drive flange bolts to the torque listed in the Chapter 8 Specifications.

g) Refer to Section 3 and check the shift cable adjustment.

h) Add the recommended type and amount of automatic transmission fluid (see Chapter 1).

i) Check the coolant level, adding as necessary (see Chapter 1).

Specifications

Transaxle fluid type and capacity	See Chapter 1

Torque specifications

Torque specifications	Nm	Ft-lbs (unless otherwise indicated)
Shift cable adjusting bolt	5	44 in-lbs
Transaxle fluid cooler banjo bolts	35	26
Torque converter-to-driveplate nuts	60	44
Transaxle-to-engine bolts		
M10	45	33
M12	80	59
Transaxle mount support-to-transaxle		
Step 1	60	44
Step 2	Tighten an additional 90-degrees (1/4-turn)	
Transaxle mount-to-transaxle support		
Step 1	60	44
Step 2	Tighten an additional 90-degrees (1/4-turn)	

Section

Reference to other Chapters

Clutch/brake fluid level check - See Chapter 1
Driveaxle boot check - See Chapter 1
Flywheel - removal and installation - See Chapter 2A

8

CLUTCH AND DRIVEAXLES

1 General information

The information in this Chapter deals with the components from the rear of the engine to the front wheels, except for the transaxle, which is dealt with in Chapter 7A and 7B. For the purposes of this Chapter, these components are grouped into two categories: Clutch and driveaxles. Separate Sections within this Chapter offer general descriptions and checking procedures for both groups.

Since nearly all the procedures covered in this Chapter involve working under the vehicle, make sure it's securely supported on sturdy jackstands or a hoist where the vehicle can be easily raised and lowered.

2 Clutch - description and check

➧ Refer to illustration 2.1

1 All models with a manual transaxle use a single dry plate, diaphragm spring type clutch (see illustration). The clutch disc has a splined hub which allows it to slide along the splines of the transaxle input shaft. The clutch and pressure plate are held in contact by spring pressure exerted by the diaphragm in the pressure plate.

2 The clutch release system is hydraulically operated. The release system consists of the clutch pedal, the clutch master cylinder, the clutch release cylinder, the hydraulic line between the master cylinder and release cylinder, the clutch release bearing and the clutch release lever.

3 When pressure is applied to the clutch pedal to release the clutch, the clutch master cylinder transmits this movement to the clutch release cylinder, which moves the clutch release lever. As the lever pivots, the release bearing pushes against the fingers of the diaphragm spring of the pressure plate assembly, which in turn releases the clutch plate.

4 Terminology can be a problem regarding the clutch components because common names have in some cases changed from that used by the manufacturer. For example, the clutch release cylinder is sometimes referred to as a slave cylinder, the driven plate is also called the clutch plate or disc, the pressure plate assembly is also known as the clutch cover, and the clutch release bearing is sometimes called a throw-out bearing.

5 Other than replacing components that have obvious damage, some preliminary checks should be performed to diagnose a clutch system failure.

a) *To check clutch "spin down" time, run the engine at normal idle speed with the transaxle in Neutral (clutch pedal up, engaged). Disengage the clutch (pedal down), wait several seconds and shift the transaxle into Reverse. No grinding noise should be heard. A grinding noise would most likely indicate a problem in the pressure plate or the clutch disc.*

b) *To check for complete clutch release, run the engine (with the parking brake applied to prevent movement) and hold the clutch pedal approximately 1/2-inch from the floor. Shift the transaxle between 1st gear and Reverse several times. If the shift is not smooth, component failure is indicated.*

c) *Visually inspect the clutch pedal bushing at the top of the clutch pedal to make sure there is no sticking or excessive wear.*

d) *Make sure that the hydraulic lines aren't leaking at either the master cylinder or the release cylinder (see Sections 3 and 4). Bleed the system if necessary (see Section 5).*

2.1 Clutch components

1 *Bolt*
2 *Pressure plate*
3 *Clutch disc*
4 *Flywheel*

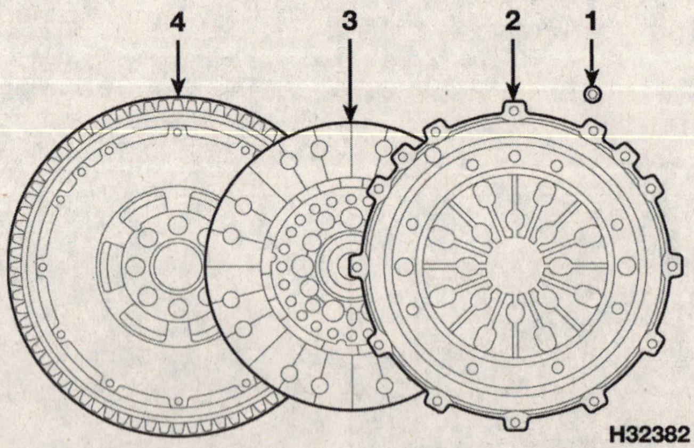

3 Clutch master cylinder - removal and installation

⁂ WARNING:

The manufacturer recommends replacing the clutch pedal mounting bracket nuts and the reinforcement plate nuts with new ones whenever they are removed.

⁂ CAUTION 1:

These models are equipped with an anti-theft radio. Before performing a procedure that requires disconnecting the battery, make sure you have the activation code.

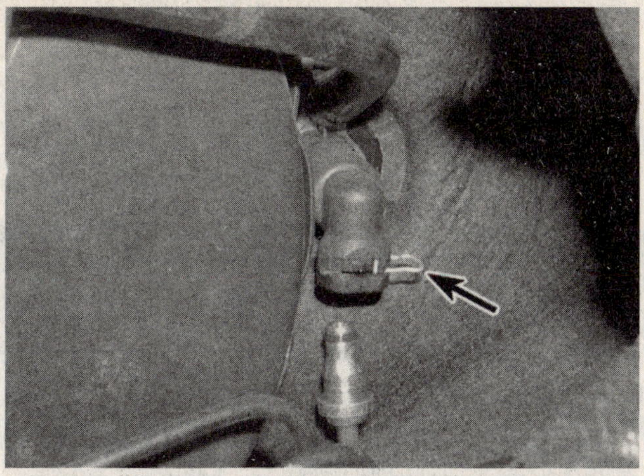

3.4 Pull out the clip retaining the clutch fluid pressure line to the clutch master cylinder, then pull the line out of the fitting

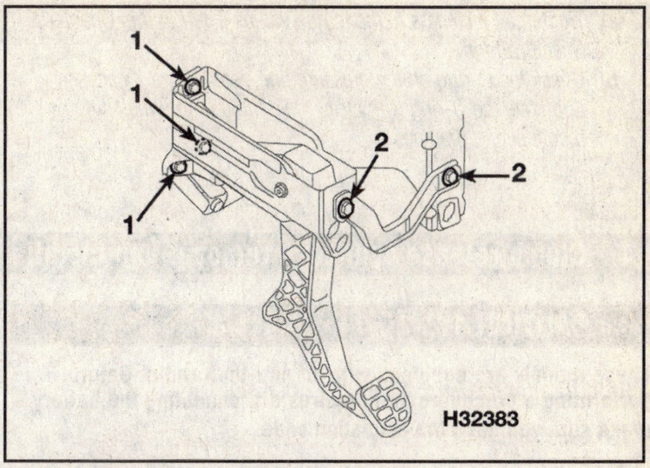

3.7 Clutch pedal mounting bracket details

1 *Clutch pedal mounting bracket nuts*
2 *Reinforcement plate nuts*

CAUTION 2:

Disconnecting the battery can cause driveability problems that require a scan tool to remedy. If the vehicle exhibits driveability problems after the battery is reconnected, it may be necessary to take it to a dealer service department or other qualified repair shop to reprogram the ECM. See Chapter 5, Section 1 for information on the use of an auxiliary voltage input device (memory saver) before disconnecting the battery.

REMOVAL

Refer to illustrations 3.4, 3.7, 3.8, 3.9 and 3.10

1 Disconnect the cable from the negative terminal of the battery.

2 Remove the air filter housing and the air intake duct (see Chapter 4A).

3 Detach the fluid supply hose from the brake master cylinder reservoir. Have a plug ready and immediately plug the port on the reservoir.

4 Remove the clip securing the pressure line to the clutch master cylinder, then separate the line from the cylinder (see illustration).

5 Working inside the vehicle, remove the left-side under-dash panel.

6 Remove the clutch start switch (see Section 8).

7 Remove the reinforcement plate connecting the clutch pedal bracket and the brake pedal bracket, then unbolt the clutch pedal bracket from the firewall (see illustration). Remove the clutch pedal, bracket and master cylinder as an assembly.

8 Disconnect the pushrod from the clutch pedal. To do this you'll have to pry the pushrod retaining tabs inward, then pull the pushrod and retainer out of the clutch pedal (see illustration).

9 Twist the clutch pedal stop counterclockwise and remove it from the pedal bracket (see illustration).

10 Push the clutch master cylinder down and away from the pedal over-center spring, then separate it from the bracket (see illustration).

INSTALLATION

11 Installation is the reverse of removal, with the following points:

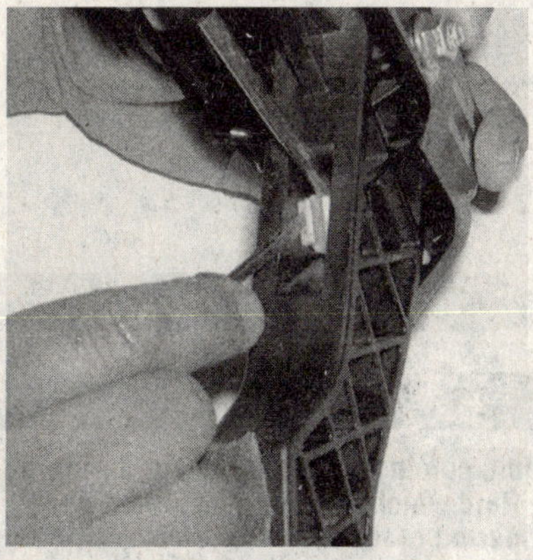

3.8 Pry the pushrod retainer tabs from the holes on either side of the clutch pedal, then pull the pushrod and retainer out of the pedal slot

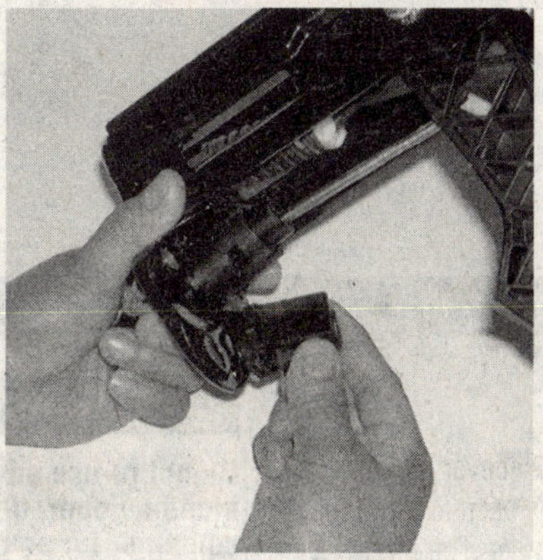

3.9 Turn the clutch pedal stop counterclockwise to detach it from the bracket

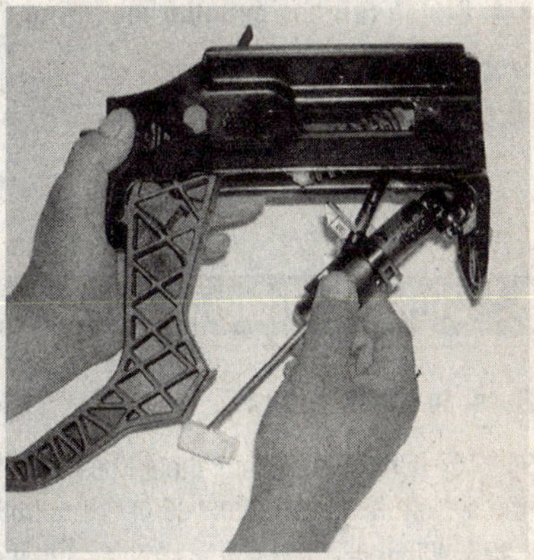

3.10 Push the master cylinder away from the over-center spring, angle it down and detach it from the pedal bracket

a) Tighten all fasteners to the torque values listed in this Chapter's Specifications.
b) Check the O-ring on the pressure line, replacing it if necessary. Lubricate the O-ring with clean brake fluid before installing or reconnecting the line.
c) Bleed the clutch hydraulic system (see Section 5).
d) Check the brake fluid level in the brake fluid reservoir, adding as necessary to bring it to the appropriate level (see Chapter 1).
e) Wash off any spilled brake fluid with water.

4 Clutch release cylinder - removal and installation

CAUTION 1:

These models are equipped with an anti-theft radio. Before performing a procedure that requires disconnecting the battery, make sure you have the activation code.

CAUTION 2:

Disconnecting the battery can cause driveability problems that require a scan tool to remedy. If the vehicle exhibits driveability problems after the battery is reconnected, it may be necessary to take it to a dealer service department or other qualified repair shop to reprogram the ECM. See Chapter 5, Section 1 for information on the use of an auxiliary voltage input device (memory saver) before disconnecting the battery.

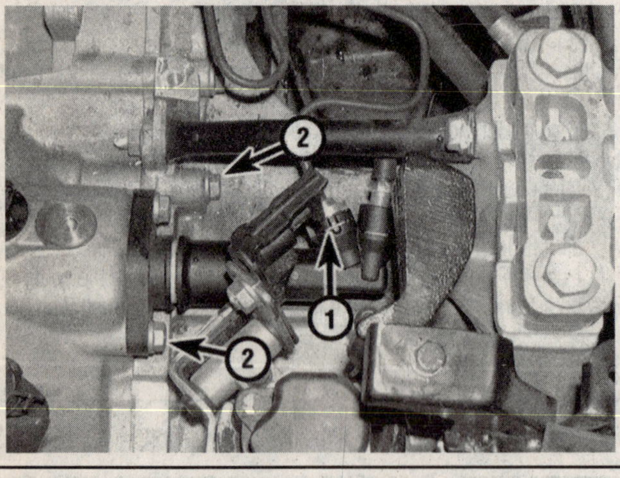

4.4 Clutch release cylinder mounting details

1 Pressure line retaining clip
2 Mounting bolts

REMOVAL

Refer to illustration 4.4

1 Disconnect the cable from the negative terminal of the battery.
2 Remove the air filter housing (see Chapter 4A).
3 Detach the shift cables from the levers at the transaxle (see Chapter 7A).
4 Remove the clip retaining the pressure line to the release cylinder, then detach the line (see illustration). Have some rags handy, as some fluid will be spilled when the line is disconnected.
5 Remove the release cylinder mounting bolts and detach the cylinder from the transaxle.

Note: On some models there is a brace from the clutch pedal bracket to the brake pedal bracket - remove the brace.

INSTALLATION

6 Lubricate the end of the pushrod with a little moly-base grease, then install the release cylinder on the transaxle. Make sure the pushrod is seated in the release lever pocket. Install the mounting bolts and tighten them to the torque listed in this Chapter's Specifications.
7 Check the O-ring on the pressure line, replacing it if necessary. Lubricate the O-ring with clean brake fluid before installing or reconnecting the line. Connect the line to the release cylinder and install the retaining clip.
8 Bleed the system (see Section 5).
9 Check the brake fluid level in the brake fluid reservoir, adding as necessary to bring it to the appropriate level (see Chapter 1).
10 Wash off any spilled brake fluid with water.
11 Install the air filter housing and reconnect the battery.

5 Clutch hydraulic system - bleeding

Refer to illustration 5.3

1 The hydraulic system should be bled of all air whenever any part of the system has been removed or if the fluid level has been allowed to fall so low that air has been drawn into the master cylinder. The procedure is similar to bleeding a brake system.
2 Fill the brake master cylinder with new brake fluid conforming to DOT 4 specifications.

WARNING:

Do not re-use any of the fluid coming from the system during the bleeding operation or use fluid which has been inside an open container for an extended period of time.

3 Remove the air filter housing (see Chapter 4A). Locate the bleeder screw on the clutch release cylinder. Remove the dust cap from the

bleeder screw and push a length of snug-fitting (preferably clear) hose over the valve. Place the other end of the hose into a clear container with about two inches of brake fluid in it. The hose end must be submerged in the fluid (see illustration).

4 Have an assistant depress the clutch pedal and hold it. Open the bleeder screw on the release cylinder, allowing fluid to flow through the hose. Close the bleeder screw when fluid stops flowing from the hose. Once closed, have your assistant release the pedal.

5 Continue this process until all air is evacuated from the system, indicated by a full, solid stream of fluid being ejected from the bleeder screw each time and no air bubbles in the hose or container. Keep a close watch on the fluid level inside the brake master cylinder reservoir; if the level drops too low, air will be sucked back into the system and the process will have to be started over again.

6 Install the dust cap on the bleeder screw. Check carefully for proper operation before placing the vehicle in normal service.

7 Recheck the brake fluid level.

8 Install the air filter housing.

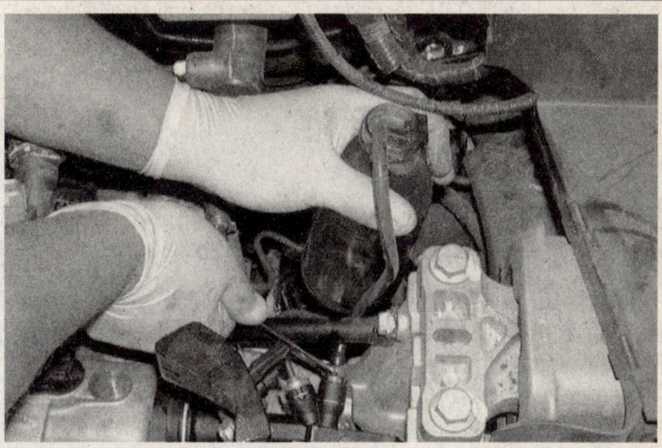

5.3 Push one end of a clear hose over the bleeder screw on the release cylinder and submerge the other end of the hose in a container of clear brake fluid - open the bleeder screw when the pedal is depressed, then close it before the pedal is released

6 Clutch components - removal, inspection and installation

WARNING:

Dust produced by clutch wear and deposited on clutch components is hazardous to your health. DO NOT blow it out with compressed air and DO NOT inhale it. DO NOT use gasoline or petroleum-based solvents to remove the dust. Brake system cleaner should be used to flush the dust into a drain pan. After the clutch components are wiped clean with a rag, dispose of the contaminated rags and cleaner in a labeled, covered container.

REMOVAL

Refer to illustrations 6.5 and 6.7

1 Access to the clutch components is normally accomplished by removing the transaxle, leaving the engine in the vehicle. If, of course, the engine is being removed for major overhaul, then the opportunity should always be taken to check the clutch for wear and replace worn components as necessary. However, the relatively low cost of the clutch components compared to the time and labor involved in gaining access to them warrants their replacement any time the engine or transaxle is removed, unless they are new or in near-perfect condition. The following procedures assume that the engine will stay in place.

2 Remove the transaxle from the vehicle (see Chapter 7A). Support the engine while the transaxle is out. Preferably, an engine hoist or support fixture should be used to support it from above. However, if a jack is used underneath the engine, make sure a piece of wood is used between the jack and oil pan to spread the load.

3 The release lever and release bearing can remain attached to the transaxle; however, you should inspect them (see Section 7) while the transaxle is removed.

4 To support the clutch disc during removal, install a clutch alignment tool through the clutch disc hub.

5 Carefully inspect the flywheel and pressure plate for indexing marks. The marks are usually an X, an O or a white letter. If they cannot be found, scribe marks yourself so the pressure plate and the flywheel will be in the same alignment during installation (see illustration). Of course, this won't be necessary if you're planning on replacing the pressure plate with a new one.

6 Slowly loosen the pressure plate-to-flywheel bolts. Work in a criss-cross pattern and loosen each bolt a little at a time until all spring pressure is relieved.

7 Hold the pressure plate securely and completely remove the bolts, followed by the pressure plate and clutch disc (see illustration).

6.5 Mark the relationship of the pressure plate to the flywheel (in case you're going to re-use the same pressure plate)

6.7 When removing the pressure plate, be careful not to let the clutch disc fall out

On later models with O2M transaxles, there are several small square plates behind the clutch cover perimeter that lock to the flywheel. The plates must be held with an open-end wrench in order to remove the pressure plate bolts.

INSPECTION

➧ Refer to illustrations 6.10, 6.12a and 6.12b

8 Ordinarily, when a problem occurs in the clutch, it can be attributed to wear of the clutch driven plate assembly (clutch disc). However, all components should be inspected at this time.

9 Inspect the flywheel for cracks, heat checking, score marks and other damage. If the imperfections are slight, a machine shop can resurface it to make it flat and smooth. Refer to Chapter 2A for the flywheel removal procedure.

10 Inspect the lining on the clutch disc. There should be at least 1/16-inch of lining above the rivet heads. Check for loose rivets, distortion, cracks, broken springs and other obvious damage (see illustration). As mentioned above, ordinarily the clutch disc is replaced as a matter of course, so if in doubt about the condition, replace it with a new one.

11 The release bearing should be replaced along with the clutch disc (see Section 7).

12 Check the machined surface and the diaphragm spring fingers of the pressure plate (see illustrations). If the surface is grooved or otherwise damaged, replace the pressure plate assembly. Also check for obvious damage, distortion, cracking, etc. Light glazing can be removed with emery cloth or sandpaper. If a new pressure plate is indicated, new or factory rebuilt units are available.

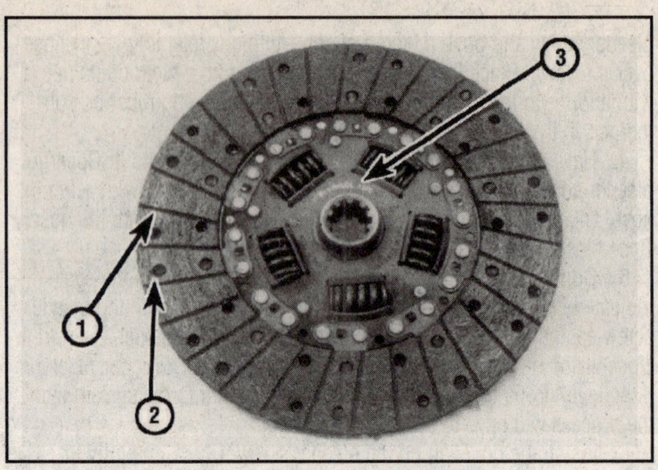

6.10 The clutch disc

1 ***Lining*** *- this will wear down in use*
2 ***Rivets*** *- these secure the lining and will damage the flywheel or pressure plate if allowed to contact the surfaces*
3 ***Hub*** *- Be sure this is installed facing the proper direction (see the text)*

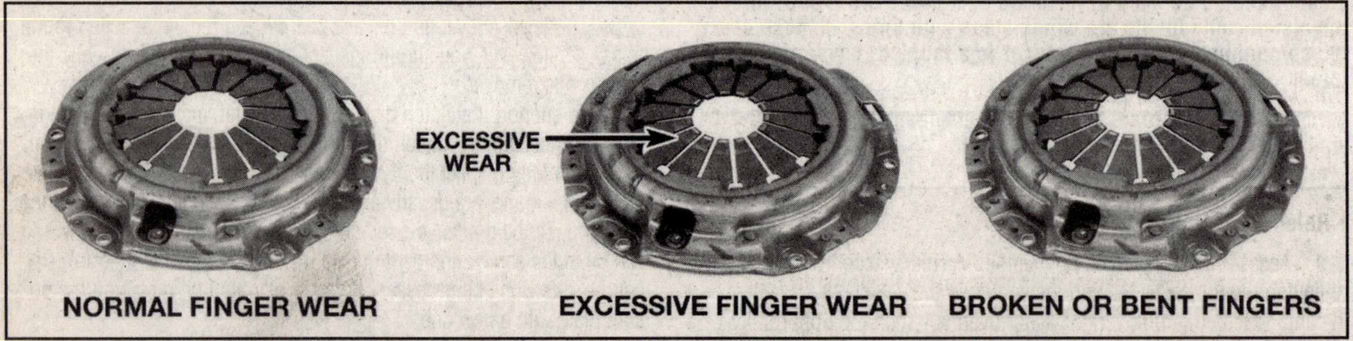

6.12a Replace the pressure plate if excessive wear is noted

6.12b Examine the pressure plate friction surface for score marks, cracks and evidence of overheating

6.15 Center the clutch disc with a clutch alignment tool, then tighten the pressure plate bolts a little at a time, in a criss-cross pattern, to the torque listed in this Chapter's Specifications

INSTALLATION

Refer to illustration 6.15

13 Position the clutch disc and pressure plate with the clutch held in place with an alignment tool. Make sure the disc is installed properly. If your vehicle is equipped with a one-piece flywheel, the spring cage must face the pressure plate. If the vehicle is equipped with a two-piece flywheel, the shorter hub end must face the pressure plate.

14 Tighten the pressure plate-to-flywheel bolts only finger tight, working around the pressure plate.

15 Center the clutch disc by ensuring the alignment tool is through the splined hub and into the recess in the crankshaft (see illustration). Wiggle the tool up, down or side-to-side as needed to bottom the tool. Tighten the pressure plate-to-flywheel bolts a little at a time, working in a criss-cross pattern to prevent distortion of the cover. After all of the bolts are snug, tighten them to the torque listed in this Chapter's Specifications. Remove the alignment tool.

16 Using moly-base grease, lubricate the inner surface of the release bearing (see Section 7), and to the face of the bearing where it contacts the fingers of the pressure plate diaphragm. Also place grease on the release lever contact areas and the transaxle input shaft.

CAUTION:

Don't use too much grease.

17 Install the clutch release bearing (see Section 7).

18 Install the transaxle and all components removed previously, tightening all fasteners to the proper torque specifications.

7 Clutch release bearing and lever - removal, inspection and installation

WARNING:

Dust produced by clutch wear and deposited on clutch components is hazardous to your health. DO NOT blow it out with compressed air and DO NOT inhale it. DO NOT use gasoline or petroleum-based solvents to remove the dust. Brake system cleaner should be used to flush it into a drain pan. After the clutch components are wiped clean with a rag, dispose of the contaminated rags and cleaner in a labeled, covered container.

CAUTION 1:

These models are equipped with an anti-theft radio. Before performing a procedure that requires disconnecting the battery, make sure you have the activation code.

CAUTION 2:

Disconnecting the battery can cause driveability problems that require a scan tool to remedy. If the vehicle exhibits driveability problems after the battery is reconnected, it may be necessary to take it to a dealer service department or other qualified repair shop to reprogram the ECM. See Chapter 5, Section 1 for information on the use of an auxiliary voltage input device (memory saver) before disconnecting the battery.

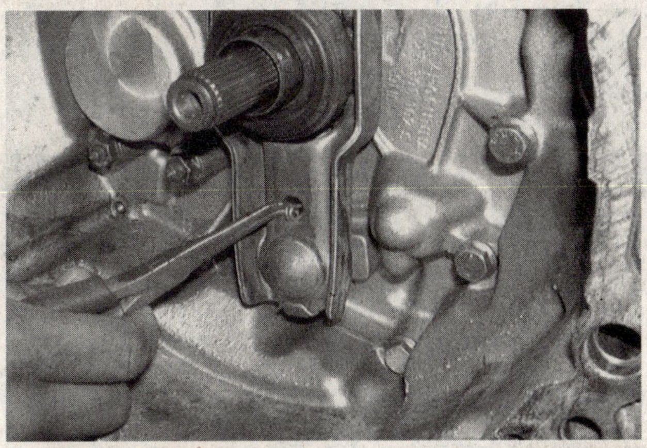

7.3 Pull the release lever off the ball stud

REMOVAL

Refer to illustration 7.3

1 Disconnect the cable from the negative terminal of the battery.

2 Remove the transaxle (see Chapter 7A).

3 Disengage the clutch release lever from the ball stud, then remove the bearing and lever (see illustration).

INSPECTION

Refer to illustration 7.4

4 Detach the bearing from the lever. Hold the bearing by the outer race and rotate the inner race while applying pressure (see illustration). If the bearing doesn't turn smoothly or if it's noisy, replace the bearing/hub assembly with a new one. Wipe the bearing with a clean rag and inspect it for damage, wear and cracks. Don't immerse the bearing in solvent; it's sealed for life and to do so would ruin it. Also check the release lever for cracks and bends.

INSTALLATION

Refer to illustration 7.6

5 Lubricate the surface of the guide sleeve with a light film of moly-

7.4 To check the bearing, hold it by the outer race and rotate the inner race while applying pressure; if the bearing doesn't turn smoothly or if it's noisy, replace it

base grease.
6 Lubricate the release lever ball socket and release cylinder pushrod socket with moly-base grease (see illustration).
7 Attach the retainer spring and the release bearing to the release lever.
8 Slide the release bearing and lever onto the guide sleeve, then push the release lever onto the ball stud until it's firmly seated.
9 Apply a light coat of high-temperature grease to the face of the release bearing where it contacts the pressure plate diaphragm fingers.
10 The remainder of installation is the reverse of removal.

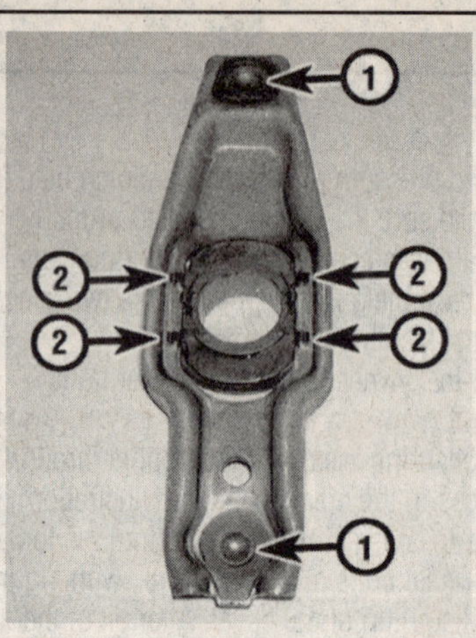

7.6 Apply moly-base grease to the release lever ball socket and release cylinder ball socket (1). When assembling the release bearing to the lever, make sure the retaining clips engage correctly (2)

8 Clutch start switch - check and replacement

CHECK

Refer to illustration 8.1

1 The clutch start switch is located on the clutch pedal bracket (see illustration).
2 Verify that the engine will not start when the clutch pedal is released.
3 Verify that the engine will start when the clutch pedal is depressed all the way.
4 If the clutch start switch doesn't perform as described above, check switch continuity.
5 Remove the left-side under-dash panel. Unplug the electrical connector from the switch and verify that there is continuity between the two clutch start switch terminals when the pedal is depressed.
6 Verify that no continuity exists between the switch terminals when the pedal is released.
7 If the switch fails either of these continuity tests, replace it.

REPLACEMENT

8 Remove the left-side under-dash panel.
9 Unplug the electrical connector from the switch.
10 Rotate the switch 90-degrees counterclockwise and pull it out of the bracket.

8.1 The clutch start switch (1) is mounted on the clutch pedal bracket; the other switch on the pedal bracket (2) is the cruise control cut-off switch

11 To install the new switch, depress the clutch pedal, insert the switch into the bracket (position the switch 1/4-turn from its installed position), rotate it 90-degrees clockwise and release the pedal.
12 Reconnect the electrical connector and check the operation of the clutch start switch.
13 Install the under-dash panel.

9 Driveaxles - general information and inspection

1 Power is transmitted from the transaxle to the wheels through a pair of driveaxles. The inner end of each driveaxle is bolted to a drive flange protruding from the differential; the outer end of each driveaxle has a stub shaft that is splined to the front hub and bearing assembly and locked in place with a large nut.
2 The inner ends of the driveaxles are equipped with sliding constant velocity (CV) joints, which are capable of both angular and axial motion. Each inner CV joint assembly consists of a either a triple rotor-type bearing or a ball-and-cage type bearing and a housing in which the joint is free to slide in-and-out as the driveaxle moves up-and-down with the wheel.
3 The outer ends of the driveaxles are equipped with "ball-and-cage" type CV joints, which are capable of angular but not axial movement. Each outer CV joint consists of six ball bearings running between an inner race and an outer cage.
4 The boots should be inspected periodically for damage and leaking lubricant. Torn CV joint boots must be replaced immediately or the joints will be damaged. If either boot of a driveaxle is damaged, that

driveaxle must be removed in order to replace the boot (see Section 10).

5 Should a boot be damaged, the CV joint can be disassembled and cleaned, but if any parts are damaged, the entire driveaxle assembly may have to be replaced as a unit - check with your local auto parts store regarding the availability of replacement parts and CV joints (see Section 11).

6 The most common symptom of worn or damaged CV joints, besides lubricant leaks, is a clicking noise in turns, a clunk when accelerating after coasting and vibration at highway speeds. To check for wear in the CV joints and driveaxle shafts, grasp each axle (one at a time) and rotate it in both directions while holding the CV joint housings, feeling for play indicating worn splines or sloppy CV joints. Also check the driveaxle shafts for cracks, dents and distortion.

10 Driveaxle - removal and installation

WARNING:

The manufacturer recommends replacing the driveaxle/hub nut with a new one whenever it is removed.

REMOVAL

Refer to illustrations 10.1, 10.3 and 10.6

1 Remove the wheel trim/hub cap (as applicable) and loosen the driveaxle/hub nut with the vehicle resting on its wheels (see illustration). Also loosen the wheel lug bolts.

2 Block the rear wheels of the car, firmly apply the parking brake, then raise the front of the vehicle and support it securely on jackstands. Remove the front wheel. On models with a diesel engine, remove the duct between the turbocharger and the intercooler (see Chapter 4B).

Note: On models with an automatic transmission, it may be necessary to unbolt the transmission support from the subframe and the transmission and pry the engine/transmission forward to provide clearance for driveaxle removal.

3 Unscrew the bolts securing the inner CV joint to the transaxle drive flange and remove the retaining plates from underneath the bolts (see illustration). Support the driveaxle by suspending it with wire or string - do not allow it to hang under its weight, or the joint may be damaged.

4 Using a marking pen, draw around the end of the suspension control arm, marking the correct installed position of the balljoint.

10.1 Loosen the driveaxle/hub nut while the wheel is still on the ground

Unscrew the balljoint retaining bolts and remove the retaining plate from the top of the control arm (see Chapter 10).

5 Remove the driveaxle/hub nut. Also remove the inner fender panel (see Chapter 11) and the under-vehicle splash shield.

6 Carefully pull the hub assembly outwards and withdraw the driveaxle outer constant velocity joint from the hub assembly (see illustration). If the splines of the outer joint are stuck in the hub, tap the joint out of the hub using a soft-faced mallet. If this fails to free it from the

10.3 Remove the bolts retaining the inner CV joint to the drive flange

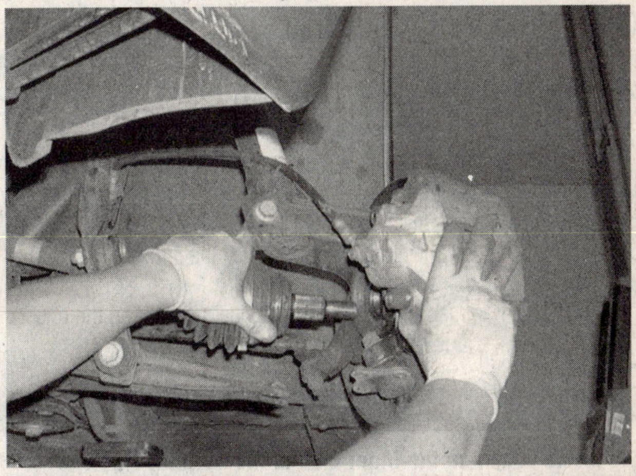

10.6 Pull out on the steering knuckle and detach the driveaxle from the hub

hub, the joint will have to be pressed out using a puller which is bolted to the hub.

7 Maneuver the driveaxle out from underneath the vehicle and recover the gasket from the end of the inner constant velocity joint, if present. Discard the gasket - a new one should be used on installation.

8 Don't allow the vehicle to rest on its wheels with one or both driveaxle(s) removed, as damage to the wheel bearing(s) may result. If moving the vehicle is unavoidable, temporarily insert the outer end of the driveaxle(s) in the hub(s) and tighten the driveaxle/hub nut(s); in this case, the inner end(s) of the driveaxle(s) must be supported, for example by suspending with string from the vehicle underbody. Do not allow the driveaxle to hang down under its weight, or the joint may be damaged.

INSTALLATION

➧ Refer to illustration 10.15

9 Ensure that the transaxle drive flange and inner joint mating surfaces are clean and dry. Install a new gasket to the joint by peeling off its backing foil and sticking it in position.

10 Ensure that the outer joint and hub splines and threads are clean, then lubricate the splines with a light coat of multi-purpose grease.

11 Maneuver the driveaxle into position and engage the outer joint with the hub. Ensure that the threads are clean and use the nut to draw the joint fully into position.

12 Connect the balljoint to the control arm and install the retaining bolts, tightening them to the torque listed in the Chapter 10 Specifications, using the marks made on removal to ensure that the balljoint is correctly positioned.

13 Align the driveaxle inner joint with the transaxle flange and install the retaining bolts and plates. Tighten the bolts to the torque listed in this Chapter's Specifications.

14 Ensure that the outer joint is drawn fully into position, then install the wheel and lower the vehicle to the ground.

15 The remainder of installation is the reverse of removal, with the following points:

a) *Tighten the driveaxle/hub nut to torque and angle of rotation listed in this Chapter's Specifications. If you don't have a torque angle meter, you can mark one point of the twelve-point nut and make another mark on the wheel hub, directly above the next point to the right (see illustration); when tightening the nut in the final Step of the torque sequence, turn the nut until both marks align (the distance between each point of a twelve-point nut is 30-degrees).*
b) *Once the driveaxle/hub nut is correctly tightened, tighten the wheel lug bolts to the torque listed in the Chapter 1 Specifications and install the wheel trim/hub cap.*

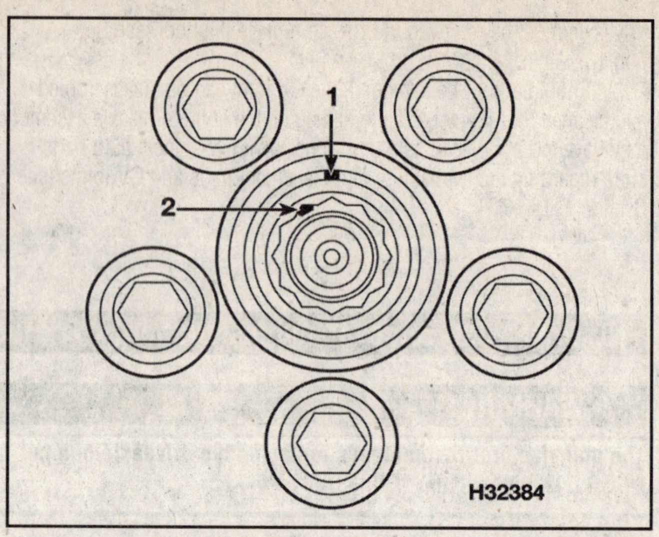

10.15 The distance between each point of a twelve-point nut is 30-degrees - make a mark on one of the points on the nut, then another mark above the next point of the nut to the right; during the final tightening Step, aligning these marks will give the proper angle of rotation

11 Driveaxle boot replacement

1 Remove the driveaxle from the vehicle as described in Section 10.

OUTER CV JOINT (ALL MODELS)

➧ Refer to illustrations 11.4, 11.5, 11.7, 11.8, 11.9, 11.10, 11.17, 11.18 and 11.22

2 Secure the driveaxle in a vise equipped with soft jaws, then loosen the two outer joint boot retaining clamps. If necessary, the clamps can be cut off.

3 Slide the boot down the shaft to expose the constant velocity (CV) joint and wipe off as much grease as possible.

4 Using a hammer and a brass punch, tap the joint off the end of the driveaxle (see illustration).

⁂ CAUTION:

Place the punch on the inner race of the CV joint only.

5 Remove the circlip from the driveaxle groove, then slide off the thrust washer and dished washer, noting which way they are installed

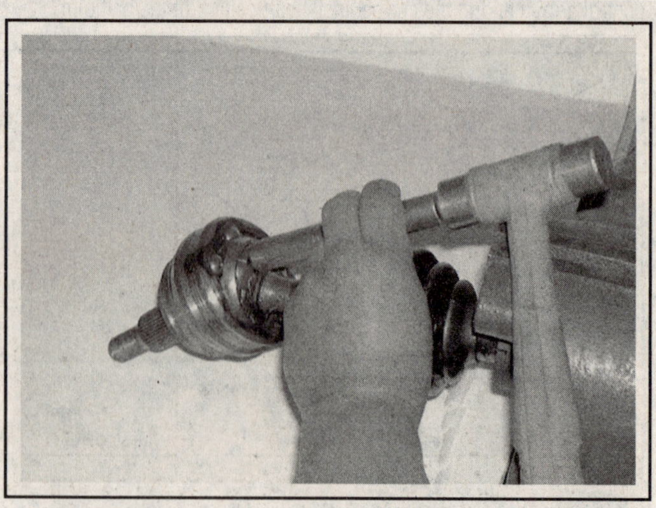

11.4 Drive the outer CV joint off the shaft with a hammer and a brass punch

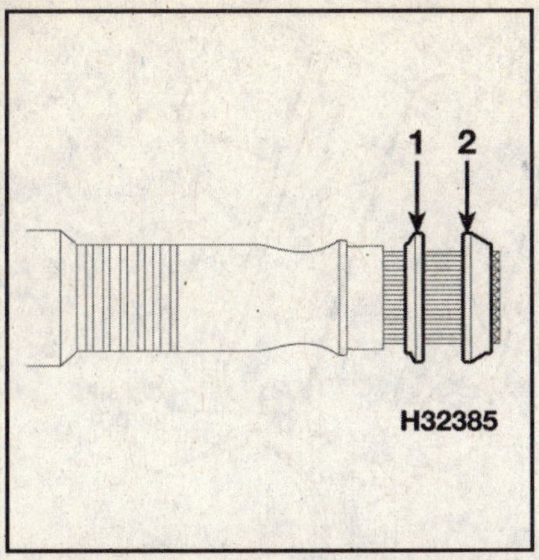

11.5 Proper positioning of the dished washer (1) and thrust washer (2) for the outer CV joint

11.7 Mark the relationship of the bearing cage, inner race and housing

11.8 If necessary, pry the balls out with a screwdriver

(see illustration).

6 Slide the boot off the driveaxle and discard it.

7 Clean the outer CV joint assembly to remove as much grease as possible. Mark the relative position of the bearing cage, inner race and housing (see illustration).

8 Mount the outer CV joint in a vise equipped with soft jaws. Push down on one side of the cage and remove the ball bearing from the opposite side. Repeat this procedure until all of the balls are removed (see illustration). If the joint is tight, tap on the inner race (not the cage) with a hammer and brass punch.

9 Remove the cage and inner race assembly from the housing by tilting it vertically and aligning two opposing cage windows in the area between the ball grooves (see illustration).

10 Turn the inner race 90-degrees to the cage and align one of the spherical lands with a cage window. Raise the land into the window and swivel the inner race out of the cage (see illustration).

11 Clean all of the parts with solvent and dry them off.

12 Inspect the housing, splines, balls and races for damage, corrosion, wear and cracks. Check the inner race, for wear and scoring in the races. If any of the components are not serviceable, the entire CV joint assembly must be replaced with a new one. If the joint is in satisfactory condition, obtain a boot replacement kit; kits usually contain a new boot and retaining clamps, a constant velocity joint snap-ring and the correct type of grease. If grease isn't included in the kit, be sure to obtain some CV joint grease.

13 Coat all of the CV joint components with CV joint grease before beginning reassembly.

14 Install the inner race in the cage and align the marks made in Step 7.

15 Install the inner race and cage assembly into the CV joint housing, aligning the marks on the inner race and cage assembly with the mark on the housing.

16 Install the balls into the holes, one at a time, until they are all in place.

17 Apply CV joint grease through the hole in the inner race, then force a wooden dowel down through the hole (see illustration). This will force the grease into the joint. Continue this procedure until the joint is completely packed. Joints with an outer diameter of 81 mm (3.2-inches) will require 80 grams (2.8 ounces) of grease; joints with an outer diameter of 90 mm (3.5-inches) will require 120 grams (4.2 ounces) of grease. Pack the joint with as much grease as you can, then place the remainder of the grease in the boot.

18 Place the axleshaft in the vise. Clean the end of the axleshaft, then slide the new clamp and boot into place.

11.9 Tilt the inner race and cage 90-degrees, then align the windows in the cage with the lands and rotate the inner race up and out of the outer race

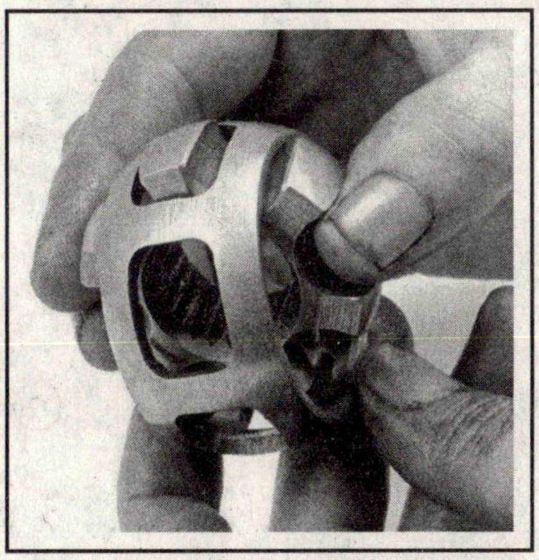

11.10 Align the inner race with the cage windows and rotate the inner race out of the cage

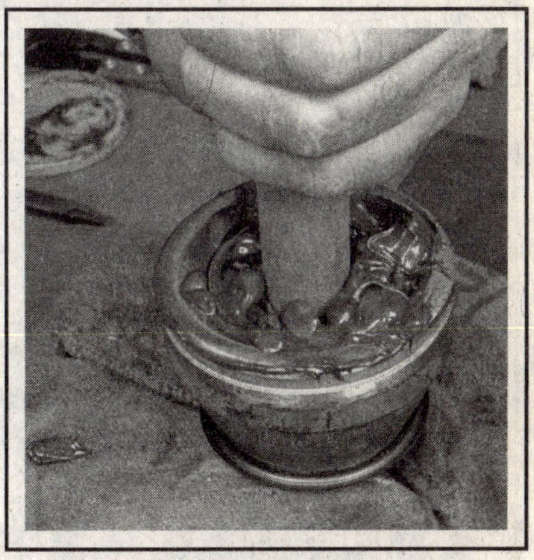

11.17 Apply grease through the splined hole, then insert a wooden dowel into the hole and push down - the dowel will force the grease into the joint

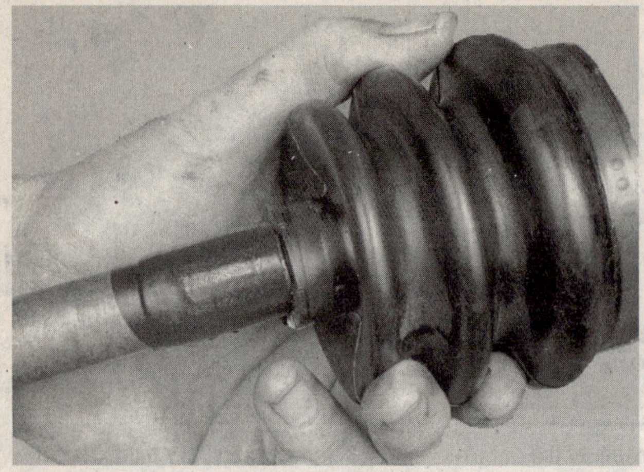

11.18 Before sliding the boot onto the shaft, it's a good idea to wrap the splines of the shaft with electrical tape to prevent damaging the boot

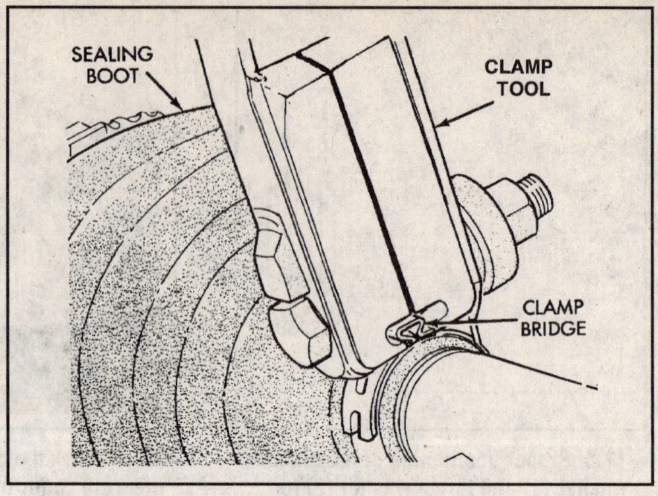

11.22 Secure the boot clamps with a clamp crimping tool like this, available at most auto parts stores

➡Note: It's a good idea to wrap the axleshaft splines with electrical tape to prevent damage to the boot (see illustration).

Apply the remainder of the grease from the kit into the CV joint boot.

19 Remove the protective tape from the driveaxle splines. Slide on the dished washer, convex side first, followed by the thrust washer (see illustration 11.5).

20 Install a new circlip in the groove on the driveaxle, then tap the joint onto the driveaxle until the circlip engages with the groove in the inner race. Make sure the joint is securely retained by the circlip.

21 Ease the boot over the joint, making sure that the boot lips are correctly located on both the driveaxle and CV joint. Lift the outer sealing lip of the boot to equalize air pressure within the boot.

22 Install the large retaining clamp on the boot. Pull the clamp as tight as possible and locate the hooks on the clamp in their slots. Tighten the clamp by crimping the raised area with a special boot clamp tool (see illustration). Due to the relatively hard composition of the boots, this type of tool is required to apply adequate crimping force on the clamps. Secure the small retaining clamp using the same procedure.

23 Make sure the constant velocity joint moves freely in all directions, then install the driveaxle as described in Section 10.

11.25a Exploded view of a driveaxle with a triple-rotor inner joint

1 *Driveaxle/hub nut*
2 *Outer CV joint boot*
3 *Clamp*
4 *Axleshaft*
5 *Clamp*
6 *Inner CV joint housing*
7 *Bolt*
8 *Rollers*
9 *Triple rotor spider*
10 *Snap-ring*
11 *O-ring (discard)*
12 *Square-section O-ring (replacement)*
13 *Cover (discard)*
14 *Clamp*
15 *Inner CV joint boot*
16 *Clamp*
17 *Dished washer*
18 *Thrust washer*
19 *Circlip*
20 *Outer CV joint*

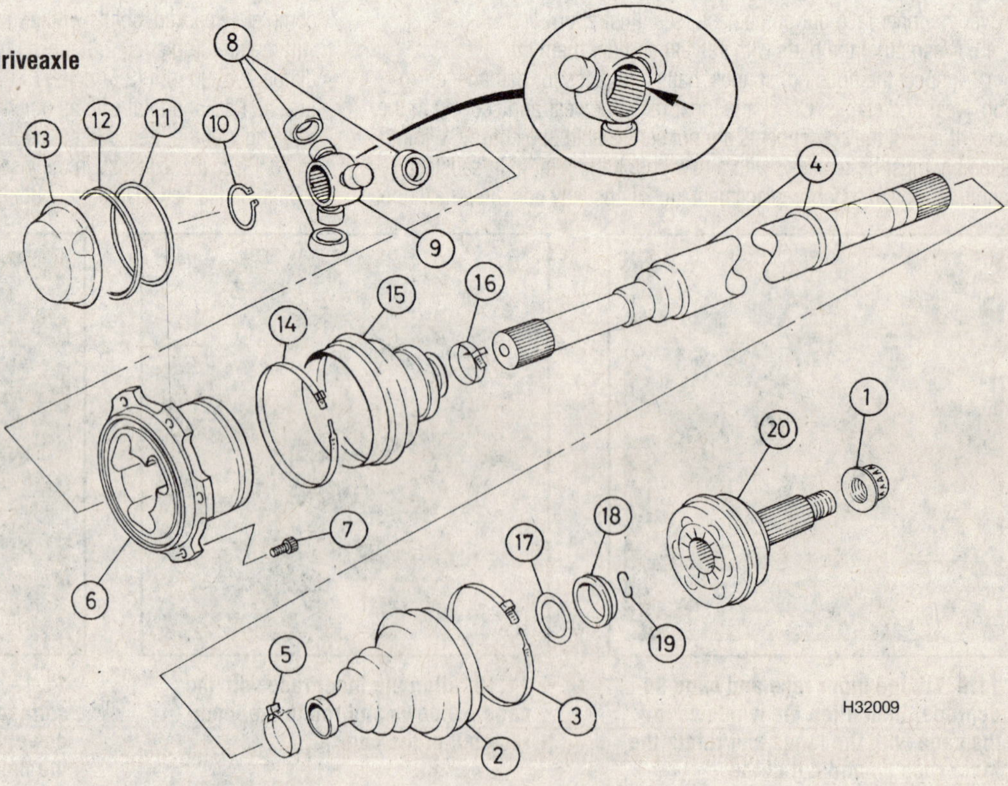

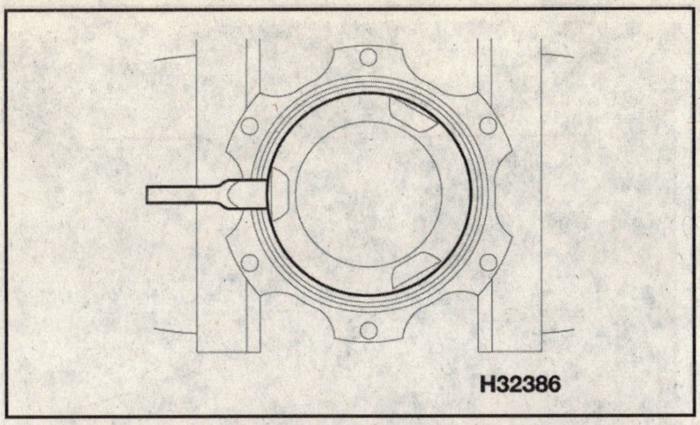

11.25b If this is the first time the inner boot is being replaced, you'll have to pry off this cover and remove the O-ring from the groove underneath

INNER CV JOINT

Triple-rotor type joint

Refer to illustrations 11.25a, 11.25b, 11.26, 11.27, 11.28, 11.36 and 11.38

24 Remove the boot clamps and discard them, then pull the boot back on the shaft (or it can be cut off).

25 Mount the driveaxle in a vise equipped with soft jaws, then pry the cover off the inner end of the joint and remove the O-ring from the groove (see illustrations). Discard the cover and O-ring - it isn't necessary to install a new cover or O-ring, as a square-section O-ring will take its place.

26 Mark the relationship of the housing, triple-rotor spider and the end of the axleshaft (see illustration), then remove the axleshaft from the vise and slide it down the shaft.

27 Mark the relationship of the rollers to the spider (see illustration).

28 Remove the snap-ring from the end of the axleshaft with a pair of snap-ring pliers, then slide the spider off the shaft (see illustration).

Note: If the spider won't slide off or tap off easily, it will be necessary to push it off with a hydraulic press.

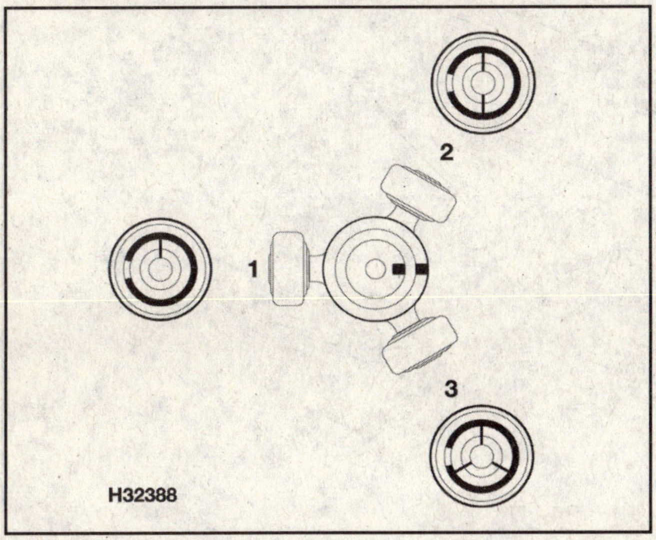

11.27 Use a felt-tip pen to mark the position of each roller to its post on the spider

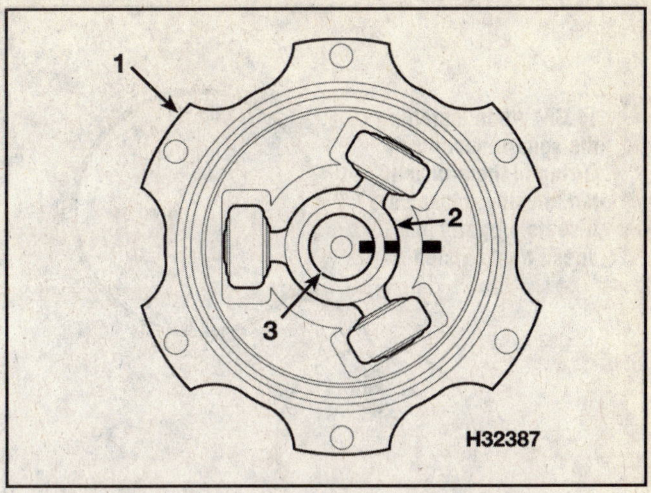

11.26 Mark the relationship of the inner joint housing, spider and axleshaft

1 Inner joint housing
2 Triple-rotor spider
3 Axleshaft

29 Remove the housing from the shaft, then clean all of the components with solvent. Inspect all components for pitting and other signs of wear (shiny, polished spots are normal and won't affect operation). If any signs of wear are found, replace the entire joint.

30 Slide the joint housing onto the shaft, then place the shaft back in the vise.

31 Install the small clamp and the boot onto the shaft. It's a good idea to wrap the splines of the shaft with electrical tape to prevent damage to the boot (see illustration 11.18).

32 Install the spider on the shaft, aligning the marks made in Step 26.

Note: The chamfered side of the spider splines face the opposite side of the shaft.

If necessary, use a deep socket or a piece of pipe to drive the spider onto the shaft until it contacts its stop. Install a new snap-ring, making sure it seats completely in its groove.

33 Lubricate the posts of the spider with CV joint grease, then install the rollers onto their respective posts. Now coat the outside of the rollers with CV joint grease.

34 Release the shaft from the vise, then slide the housing up onto the triple-rotor spider, aligning the marks made in Step 26.

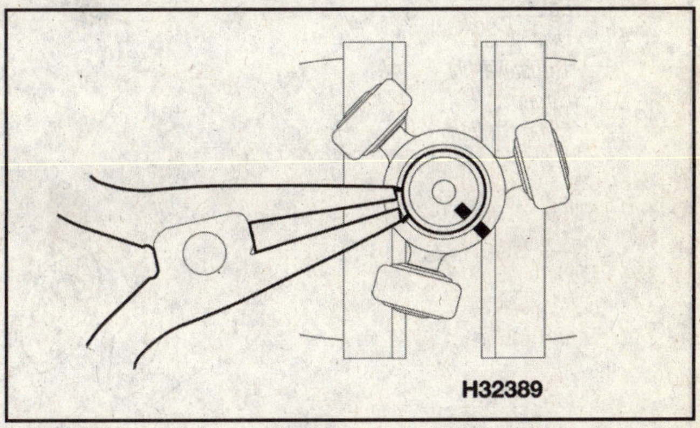

11.28 Remove the snap-ring from the end of the axleshaft, then remove the spider

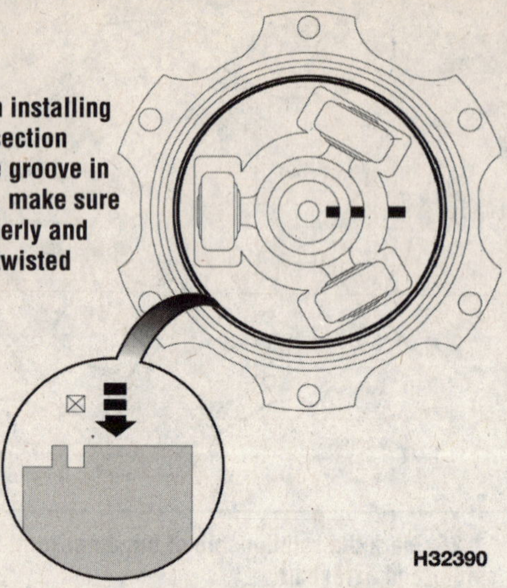

11.36 When installing the square-section O-ring in the groove in the housing, make sure it seats properly and doesn't get twisted

11.38 Clamp crimping pliers like these are available at most auto parts stores

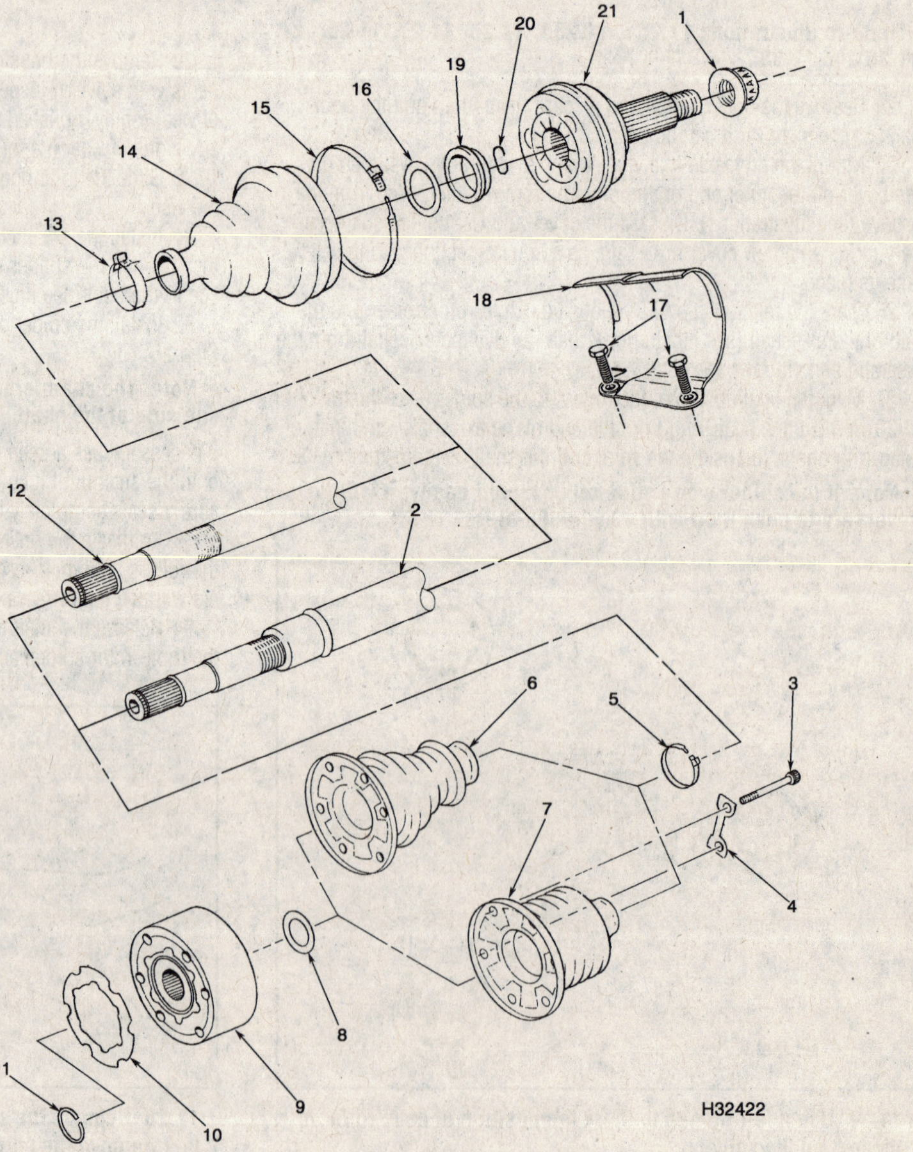

11.42 Exploded view of a driveaxle with a ball-and-cage inner joint

1 *Driveaxle/hub nut*
2 *Axleshaft (right side)*
3 *Bolt*
4 *Retaining plate*
5 *Clamp*
6 *Inner CV joint boot (left side)*
7 *Inner CV joint boot (right side)*
8 *Dished washer*
9 *Inner CV joint*
10 *Gasket*
11 *Snap-ring*
12 *Axleshaft (left side)*
13 *Clamp*
14 *Outer CV joint boot*
15 *Clamp*
16 *Dished washer*
17 *Bolt*
18 *Heat shield*
19 *Thrust washer*
20 *Circlip*
21 *Outer CV joint*

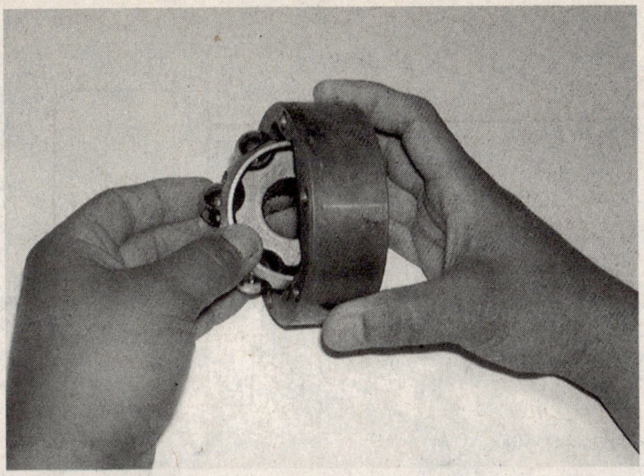

11.45 Turn the cage and inner race 90-degrees and rotate it out of the housing

35 Clamp the housing in the vise, allowing the shaft to hang straight down.

36 Install the square-section O-ring (included in the kit) into the groove in the housing (see illustration).

37 Fill the joint with approximately 3.2 ounces (90 grams) of CV joint grease from the top, then remove the housing from the vise and place the same amount of grease into the back side of the joint.

38 Install the boot onto the housing, making sure it isn't stretched or twisted, then place the clamps in position and tighten them with a pair of clamp crimping pliers (see illustration).

39 Make sure the constant velocity joint moves freely in all directions, then install the driveaxle as described in Section 10.

CAUTION:

When handling the driveaxle, be careful not to allow the housing to become pushed back onto the shaft - if this happens, the triple rotor spider and rollers may protrude from the housing and fall apart. Also, some of the grease will be forced from the housing.

Ball-and-cage type joint

Refer to illustrations 11.42, 11.45, 11.52, 11.53a and 11.53b

40 Remove the boot clamp and discard it.

41 Mount the driveaxle in a vise equipped with soft jaws. Using a hammer and a punch, knock the boot cap off the inner CV joint.

42 Using a pair of snap-ring pliers, remove the snap-ring from its groove in the end of the driveaxle (see illustration).

43 Pull the inner joint off the end of the axleshaft. If it is stuck, use a hammer and a brass punch to drive it off the shaft; apply force to the inner race of the joint only. If it still won't come off, it'll be necessary to push it off with a hydraulic press. Remove the dished washer from the shaft, then pull off the boot.

44 Wipe the grease off the joint and mark the relationship of the inner race, cage and housing.

45 Rotate the cage and inner race 90-degrees and remove it from the housing (see illustration).

46 Remove each ball bearing from the cage, keeping track of their positions so they can be reinstalled in the same spot.

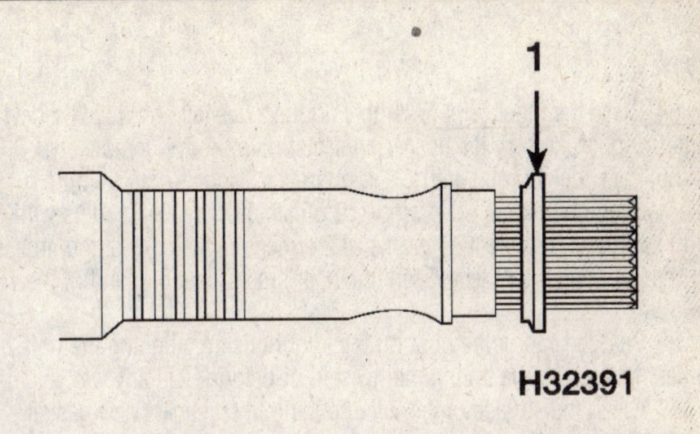

11.52 The concave side of the dished washer (1) must face the end of the axleshaft (ball-and-cage inner CV joint)

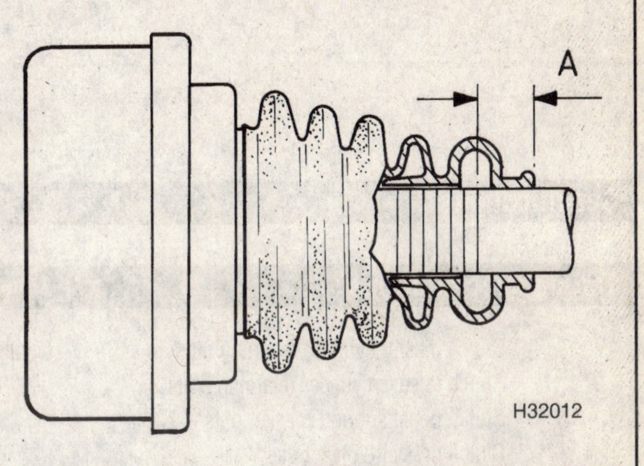

11.53a The inner end of the left inner CV joint boot must be positioned 11/16-inch (17 mm) past the inner groove on the axleshaft (dimension "A")

47 Turn the inner race 90-degrees in the cage, align one of the grooves with the edge of the cage and rotate the inner race out.

48 Clean all of the components and inspect for worn or damaged splines, race grooves, ball bearings and cage. Shiny spots are normal and won't affect operation. Replace the joint with a new one if any of the components show signs of wear.

49 Coat the components of the joint with CV joint grease, then assemble the inner race and cage, aligning the marks made in Step 44.

50 Press the ball bearings into their openings, then insert the inner race, cage and balls into the housing. The chamfered side of the splines must face the larger diameter side of the housing. When the components are rotated into place, the wide-spaced grooves of the inner race must be lined up with the wide-spaced grooves in the housing.

51 Install the new boot and clamp on the axleshaft. It's a good idea to wrap the splines of the axleshaft with electrical tape to prevent damage to the boot (see illustration 11.18). Remove the tape.

52 Install the dished washer on the axleshaft with the concave side facing the end of the shaft (see illustration).

53 If you're replacing the boot on the left-side driveaxle, adjust the inner end of the boot so that it is 11/16-inch past the innermost groove on the axleshaft (see illustration). If you're replacing the boot on the right-side driveaxle, seat the inner end of the boot on the larger diameter portion of the axleshaft (see illustration).

54 Place the inner joint assembly on the axleshaft and install a new snap-ring. Make sure the snap-ring seats in its groove completely.

55 Pack the CV joint with CV joint grease. Models with 94 mm diameter joints require 90-grams (3.2 ounces) of grease; models with 100 mm joints require 120-grams (4.2 ounces). Place 1/3 of the grease in the joint and the other 2/3 on the inner side of the joint and in the boot.

56 Seat the cap of the boot on the joint housing, aligning the bolt holes. Make sure the boot is not twisted or deformed in any way.

57 Clean the surface of the joint housing, then stick a new gasket onto the housing.

58 Make sure the constant velocity joint moves freely in all directions, then install the driveaxle as described in Section 10

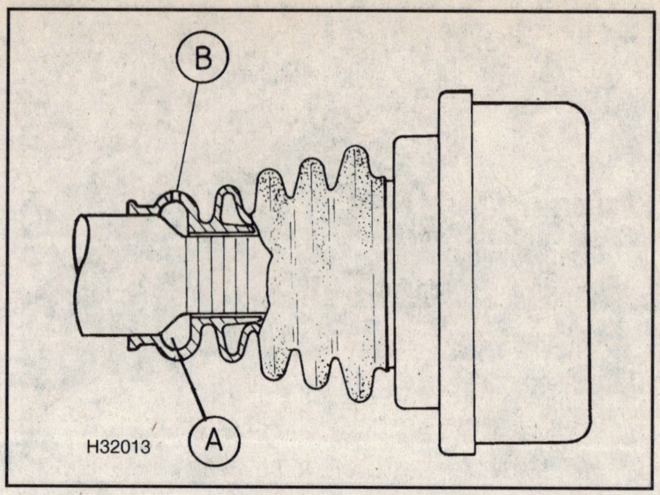

11.53b The inner end of the right inner CV joint boot must be positioned on the axleshaft like this, so the edge of the vent chamber (A) seats on the tube; the vent hole (B) must not be obstructed

Specifications

Torque specifications	Nm	Ft-lbs (unless otherwise indicated)
Clutch master cylinder retaining nuts	25	18
Clutch release cylinder retaining bolts	25	18
Clutch pedal support bracket nuts*	25	18
Clutch pressure plate bolts		
One-piece flywheel	20	15
Two-piece flywheel	13	120 in-lbs
Flywheel bolts	See Chapter 2A or 2B	
Inner CV joint-to-drive flange bolts	40	30
Driveaxle/hub nut*		
Step 1	300	221
Step 2	Loosen one turn	
Step 3	50	37
Step 4	Tighten an additional 30-degrees	

**Replace with new nut(s)*

9

BRAKES

Section

Reference to other Chapters

Brake fluid level check - See Chapter 1
Brake system check - See Chapter 1

1 General information

The vehicles covered by this manual are equipped with hydraulically operated front and rear brake systems. The front brakes are disc type and the rear brakes are disc or drum type. Both the front and rear brakes are self adjusting. The disc brakes automatically compensate for pad wear, while the drum brakes incorporate an adjustment mechanism which is activated as the parking brake is applied.

HYDRAULIC SYSTEM

The hydraulic system consists of two separate circuits. The master cylinder has separate reservoirs for the two circuits, and, in the event of a leak or failure in one hydraulic circuit, the other circuit will remain operative. A pressure regulator valve on non-ABS models provides brake balance between the front and rear brakes.

POWER BRAKE BOOSTER

The power brake booster, utilizing engine manifold vacuum and atmospheric pressure to provide assistance to the hydraulically operated brakes, is mounted on the firewall in the engine compartment.

PARKING BRAKE

The parking brake operates the rear brakes only, through cable actuation. It's activated by a lever mounted in the center console.

SERVICE

After completing any operation involving disassembly of any part of the brake system, always test drive the vehicle to check for proper braking performance before resuming normal driving. When testing the brakes, perform the tests on a clean, dry, flat surface. Conditions other than these can lead to inaccurate test results.

Test the brakes at various speeds with both light and heavy pedal pressure. The vehicle should stop evenly without pulling to one side or the other. Avoid locking the brakes, because this slides the tires and diminishes braking efficiency and control of the vehicle.

Tires, vehicle load and wheel alignment are factors which also affect braking performance.

2 Anti-lock Brake System (ABS) - general information and component removal and installation

GENERAL INFORMATION

Refer to illustration 2.1

Note: On models equipped with traction control, the ABS unit is a dual function unit, controlling both the anti-lock braking system (ABS) and the electronic differential locking (EDL) system functions.

ABS is available as an option on the models covered in this manual. The system comprises a hydraulic unit (which contains the hydraulic solenoid valves and accumulators) (see illustration), and four wheel sensors (one installed on each wheel), the ABS control module and the brake light switch. The purpose of the system is to prevent the wheel(s) locking during heavy braking. This is achieved by automatic release of the brake on the relevant wheel, followed by re-application of the brake.

2.1 The ABS hydraulic unit/control module is mounted to the left of the brake master cylinder, or, on later models, below the master cylinder

The solenoids are controlled by the control unit, which itself receives signals from the four wheel sensors (one installed on each hub), which monitor the speed of rotation of each wheel. By comparing these signals, the ECU can determine the speed at which the vehicle is traveling. It can then use this speed to determine when a wheel is decelerating at an abnormal rate, compared to the speed of the vehicle, and therefore predicts when a wheel is about to lock. During normal operation, the system functions in the same way as a non-ABS braking system.

If the control unit senses that a wheel is about to lock, it operates the relevant solenoid valve in the modulator block, which then isolates the brake caliper on the wheel which is about to lock from the master cylinder, effectively sealing-in the hydraulic pressure.

If the speed of rotation of the wheel continues to decrease at an abnormal rate, the control unit switches on the electrically driven return pump, which pumps the brake fluid back into the master cylinder, releasing pressure on the brake caliper so that the brake is released. Once the speed of rotation of the wheel returns to an acceptable rate, the pump stops; the solenoid valve opens, allowing the master cylinder hydraulic pressure to return to the caliper, which then re-applies the brake. This cycle can be carried out at up to 10 times a second.

The action of the solenoid valves and return pump creates pulses in the hydraulic circuit. When the ABS system is functioning, these pulses can be felt through the brake pedal.

The operation of the ABS system is entirely dependent on electrical signals. To prevent the system responding to any inaccurate signals, a built-in safety circuit monitors all signals received by the control unit. If an inaccurate signal or low battery voltage is detected, the ABS system is automatically shut down, and the warning light on the instrument panel is illuminated, to inform the driver that the ABS system is not operational. Normal braking should still be available, however.

If a fault does develop in the ABS system, the vehicle must be taken to a VW dealer service department or other qualified repair shop for fault diagnosis and repair.

COMPONENT REMOVAL AND INSTALLATION

Hydraulic unit

1 Removal and installation of the hydraulic unit should be entrusted to a VW dealer. Great care has to be taken not to allow any fluid to escape from the unit as the lines are disconnected. If the fluid is allowed to escape, air can enter the unit, causing air locks which cause the hydraulic unit to malfunction.

Control module

2 The ABS control module is mounted to the underside of the hydraulic unit and can only be removed after the hydraulic unit has been removed. New control modules are not coded, but must be coded prior to vehicle operation. This requires a VW scan tool, and even with the proper scan tool the control unit can only be coded after a "dealership code" has been entered into the tool. For this reason, any work involving the ABS control module must be left to a dealer service department or other properly equipped repair facility.

WHEEL SPEED SENSOR

Removal

Refer to illustration 2.4

3 Loosen the wheel bolts. Chock the rear wheels, then firmly apply the parking brake, raise the front of the vehicle and support it securely on jackstands. Remove the appropriate front wheel.

4 Disconnect the electrical connector from the wheel speed sensor (see illustration).

5 Remove the bolt securing the sensor to the steering knuckle, and remove the sensor from the knuckle.

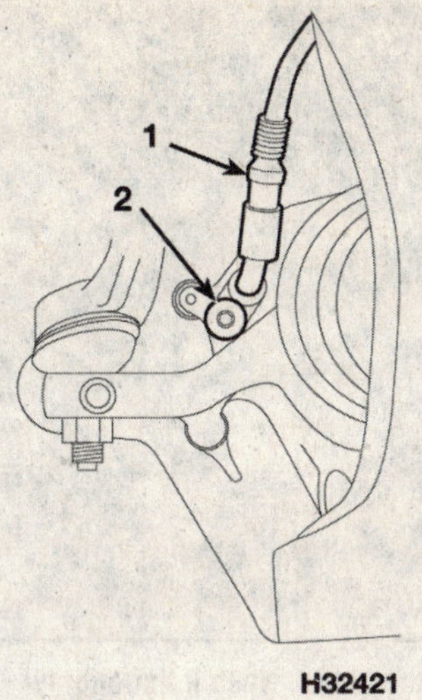

2.4 Wheel speed sensor mounting details (front shown, rear similar)

1 Electrical connector
2 Mounting bolt

Installation

6 Prior to installation, apply a thin coat of multi-purpose grease to the sensor tip.

7 Ensure that the sensor and steering knuckle sealing faces are clean, then install the sensor on the knuckle. Install the retaining bolt and tighten it to the torque listed in this Chapter's Specifications.

8 Plug in the electrical connector.

9 Install the wheel, then lower the vehicle to the ground and tighten the wheel bolts to the torque listed in the Chapter 1 Specifications.

RELUCTOR RINGS

10 The reluctor rings are an integral part of the wheel hubs. Examine the rings for damage such as chipped or missing teeth. If replacement is necessary, the complete hub assembly must be disassembled and the bearings replaced as described in Chapter 10.

3 Disc brake pads - replacement

Refer to illustrations 3.5a through 3.5n or 3.6a through 3.6i

WARNING:

Disc brake pads must be replaced on both front or both rear wheels at the same time - never replace the pads on only one wheel. Also, the dust created by the brake system is harmful to your health. Never blow it out with compressed air and don't inhale any of it. An approved filtering mask should be worn when working on the brakes. Do not, under any circumstances, use petroleum-based solvents to clean brake parts. Use brake system cleaner only!

CAUTION:

Don't depress the brake pedal with the caliper removed.

1 Remove the cap from the brake fluid reservoir. Remove about two-thirds of the fluid from the reservoir, then reinstall the cap.

WARNING:

Brake fluid is poisonous - never siphon it by mouth. Use a suction gun or old poultry baster. If a baster is used, never again use it for the preparation of food.

CAUTION:

Brake fluid will damage paint. If any fluid is spilled, wash it off immediately with plenty of clean, cold water.

2 Loosen the front or rear wheel bolts, raise the front or rear of the vehicle and support it securely on jackstands. Block the wheels at the opposite end.

3.5a Before disassembling the brake, wash it thoroughly with brake system cleaner and allow it to dry - position a drain pan under the brake to catch the residue - DO NOT use compressed air to blow off the brake dust!

3.5b To make room for the new pads, use a C-clamp to depress the piston(s) into the caliper before removing the caliper and pads - do this a little at a time, keeping an eye on the fluid level in the master cylinder to make sure it doesn't overflow

3.5c If you're working on a model with an FN 3 caliper, pry the ends of the retaining spring out of the holes in the caliper, then remove the spring

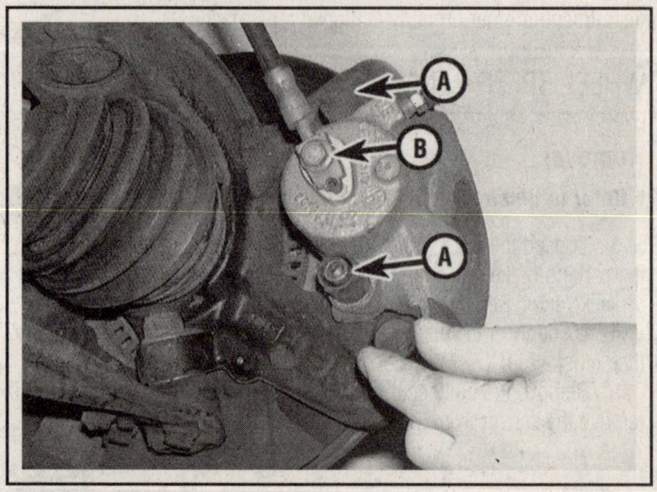

3.5d Remove the caps from the caliper guide pins (A), then unscrew the guide pins. Don't remove the brake hose inlet fitting bolt (B) unless the caliper or hose is being replaced

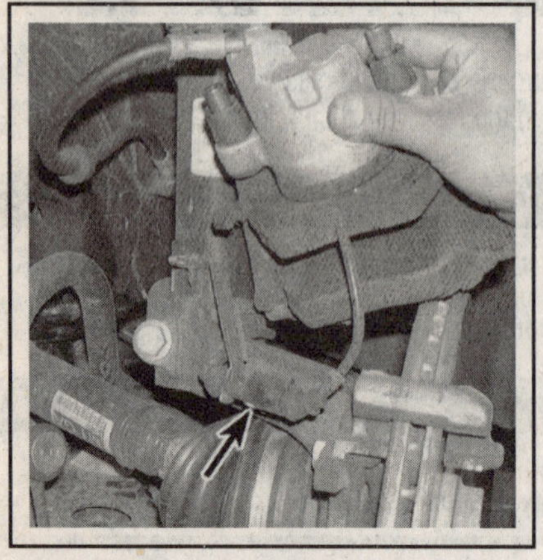

3.5e If the inner pad is equipped with a wear sensor, unplug this electrical connector (arrow)

3.5f Remove the inner brake pad (FS III caliper shown) . . .

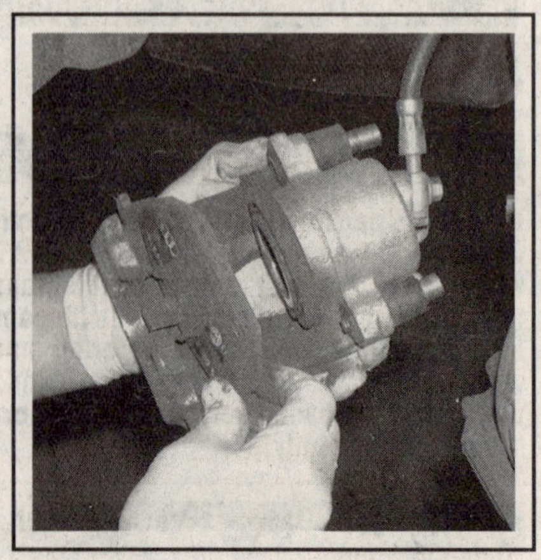

3.5g . . . then remove the outer pad from the caliper frame (FS III caliper shown)

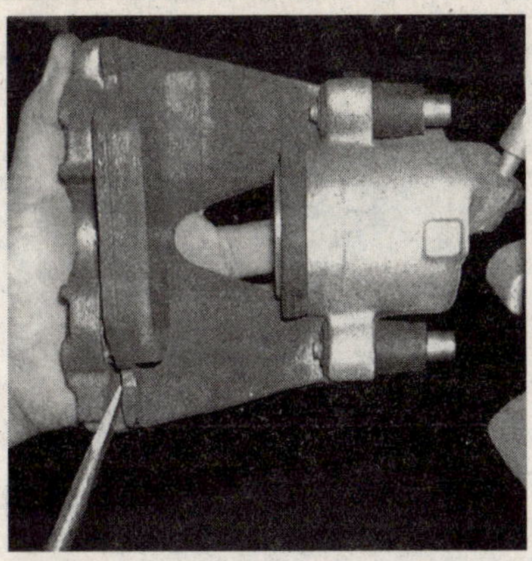

3.5h On an FN 3 caliper the brake pad is stuck to the caliper frame with an adhesive backing - it will probably be necessary to pry it off. On later models, first pry off the large spring clip securing the pad

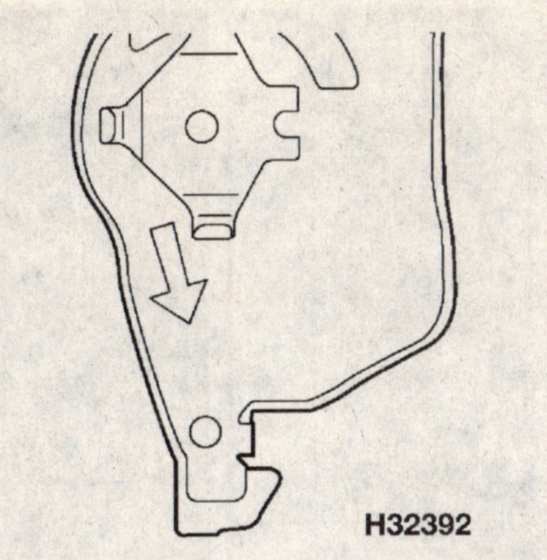

3.5i The inner pad in an FN 3 caliper is directional - when installed, the arrow must point downward

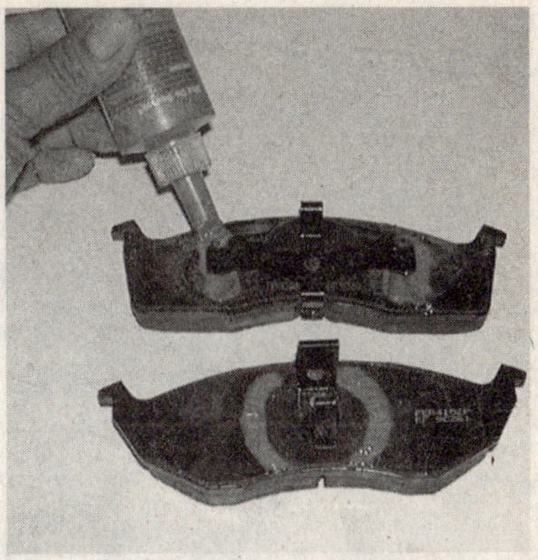

3.5j If the new pads are adhesive-backed, peel the foil from the backing plates before installing them. If they are not adhesive-backed, apply a film of anti-squeal compound to the backing plates (follow the instructions on the product)

3.5k Install the inner pad . . .

3.5l . . . and the outer pad (FS III caliper shown)

3.5m On models with FN 3 calipers, remove the foil from the backing plate and install the outer brake pad in the caliper mounting bracket

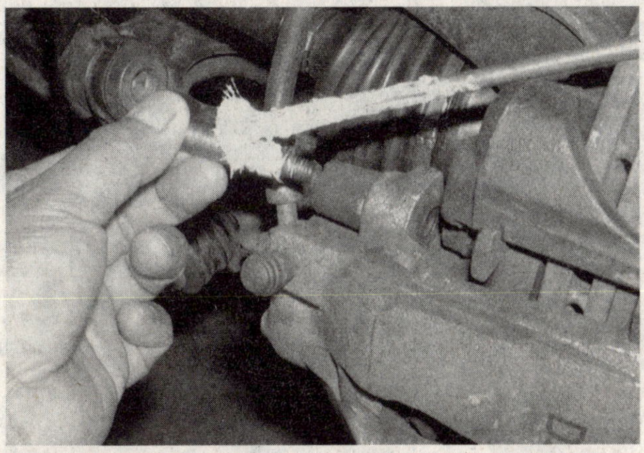

3.5n Install the caliper over the disc, then clean, lubricate (use high-temperature grease) and install the guide pins, tightening them to the specified torque. On FN 3 calipers, install the retaining spring. If equipped with wear sensors, plug in the electrical connector

3 Remove the wheels. Work on one brake assembly at a time, using the assembled brake for reference if necessary.

4 Inspect the brake disc carefully as outlined in Section 5. If machining is necessary, follow the information in that Section to remove the disc, at which time the pads can be removed as well.

5 If you are replacing the front brake pads, follow the first photo sequence (see illustrations 3.5a through 3.5n). Be sure to stay in order and read the caption under each illustration.

➡Note: There are two different types of front brake calipers which may be encountered: FS III and FN 3. The FS III caliper mounts directly to the steering knuckle by two guide pins, and the brake pads are retained to the caliper by clips. The FN 3 caliper has two guide pins that attach the caliper to a mounting bracket which is bolted to the steering knuckle. It also has a retaining spring on the front side of the caliper.

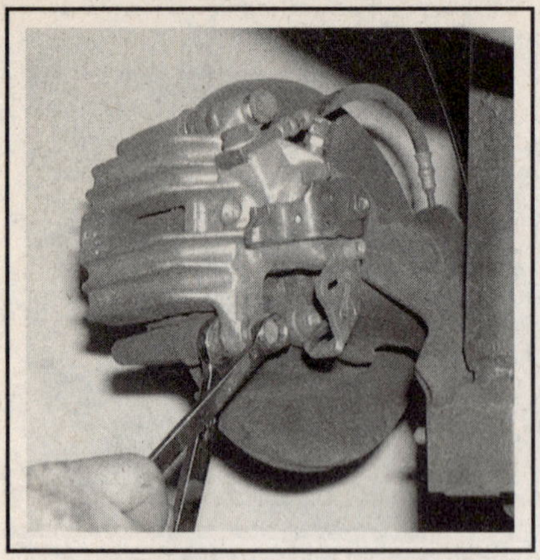

3.6a Clean the brake with brake system cleaner (see illustration 3.5a), then hold the caliper slide pins with an open-end wrench and unscrew the mounting bolts with another wrench . . .

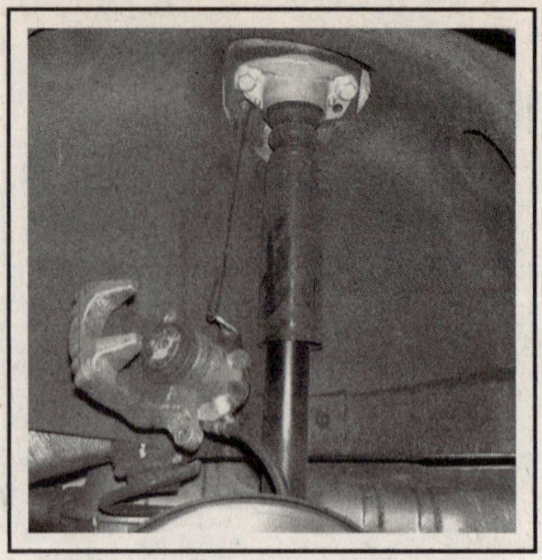

3.6b . . . then hang the caliper with a length of wire (don't let it hang by the hose!)

3.6c Remove the inner brake pad from the caliper mounting bracket . . .

3.6d . . . then remove the outer pad

3.6e Remove the pad retaining springs from the caliper mounting bracket and replace them with new ones

3.6f If the new pads are adhesive-backed, peel the foil from the backing plates before installation. If not, apply anti-squeal compound to the backing plates (follow the instructions on the product), then install the inner pad . . .

3.6g . . . and the outer pad

3.6h To provide room for the new pads, the piston must be retracted - to do this, rotate the piston clockwise while pushing in on it. Piston rotating tools, like the one shown here, are available at most auto parts stores

6 If you're replacing the rear brake pads, follow the second photo sequence (see illustrations 3.6a through 3.6i). Be sure to stay in order and read the caption under each illustration. On later model rear disc brakes, the parking brake cable operates a lever on the caliper. Release the cable end from the bracket before removing the caliper (see illustration 14.8).

7 When reinstalling the caliper, be sure to tighten the mounting bolts to the torque listed in this Chapter's Specifications.

8 After the job has been completed, firmly depress the brake pedal a few times to bring the pads into contact with the disc. Check the level of the brake fluid, adding some if necessary. Check the operation of the brakes carefully before placing the vehicle into normal service.

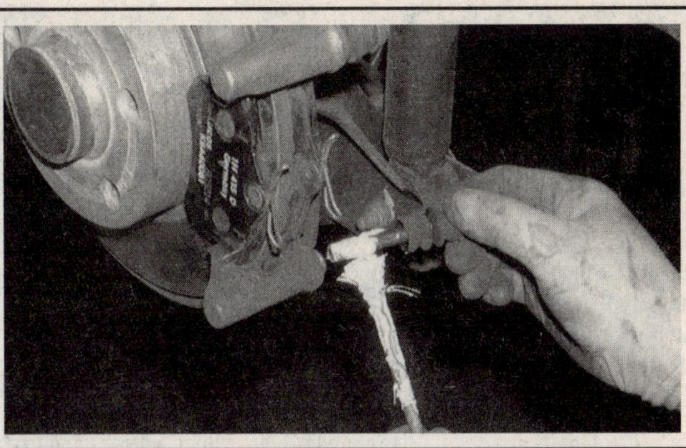

3.6i Clean the caliper slide pins and lubricate them with high-temperature grease, then install them in the caliper mounting bracket, making sure the boots seat properly. Install the caliper and tighten the mounting bolts to the torque listed in this Chapter's Specifications

4 Brake caliper - removal and installation

Refer to illustrations 4.3 and 4.5

WARNING:

The dust created by the brake system is harmful to your health. Never blow it out with compressed air and don't inhale any of it. An approved filtering mask should be worn when working on the brakes. Do not, under any circumstances, use petroleum-based solvents to clean brake parts. Use brake system cleaner only!

REMOVAL

1 Loosen the front or rear wheel bolts, raise the front or rear of the vehicle and place it securely on jackstands. Block the wheels at the opposite end. Remove the front or rear wheel.

2 To disconnect the parking brake cable from the rear caliper, unbolt the cable bracket from the caliper and disengage the cable from the toggle lever.

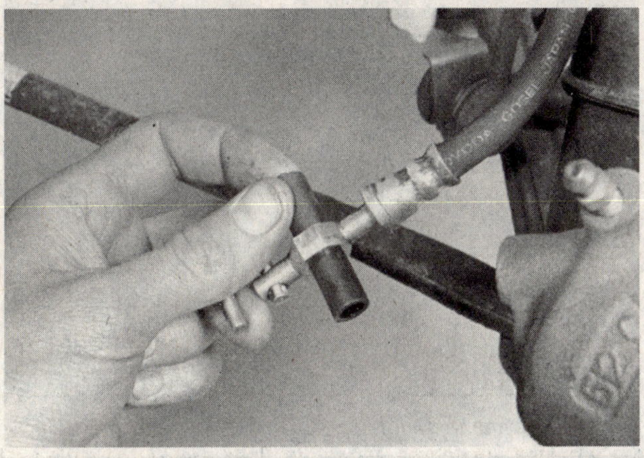

4.3 The brake hose can be plugged using a snug-fitting piece of tubing

3 Remove the inlet fitting bolt and discard the old sealing washers (see illustration 3.5d). Disconnect the brake hose from the caliper. Plug the brake hose to keep contaminants out of the brake system and to prevent losing any more brake fluid than is necessary (see illustration).

Note: If you're removing the caliper for access to other components, don't disconnect the hose. Suspend the caliper with a piece of wire to prevent damaging the brake hose.

4 Remove the caliper guide pins (or on the rear, the mounting bolts) and detach the caliper (see illustration 3.5d [front] or 3.6a [rear]). If the brake pads are equipped with wear sensors, disconnect the electrical connector (see illustration 3.5e).

INSTALLATION

5 Installation is the reverse of removal. Don't forget to use new sealing washers on each side of the brake hose inlet fitting (see illustration).

6 Bleed the brake system (see Section 11). Make sure there are no leaks from the hose connections. Pump the brake pedal several times before driving the vehicle, and test the brakes carefully before returning the vehicle to normal service.

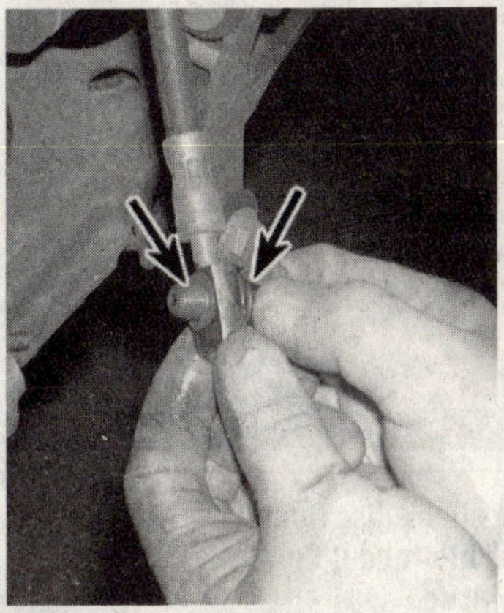

4.5 There is a sealing washer on either side of the brake hose inlet fitting; be sure to replace these with new ones when reconnecting the hose

5 Brake disc - inspection, removal and installation

INSPECTION

➧ Refer to illustrations 5.2, 5.3, 5.4a, 5.4b, 5.5a, 5.5b and 5.5c

1 Loosen the wheel lug bolts, raise the vehicle and support it securely on jackstands. Remove the wheel and install a couple of lug bolts to hold the disc in place.

➡Note: If the lug bolts don't contact the disc when screwed on all the way, install washers under them.

2 Remove the brake caliper (see Section 4). It isn't necessary to disconnect the brake hose. After removing the caliper bolts, suspend the caliper out of the way with a piece of wire (see illustration).

3 Visually inspect the disc surface for score marks and other damage. Light scratches and shallow grooves are normal after use and may not always be detrimental to brake operation, but deep scoring requires disc removal and refinishing by an automotive machine shop. Be sure to check both sides of the disc (see illustration). If pulsating has been noticed during application of the brakes, suspect disc runout.

4 To check disc runout, place a dial indicator at a point about 1/2-inch from the outer edge of the disc (see illustration). Set the indicator to zero and turn the disc. The indicator reading should not exceed the specified allowable runout limit. If it does, the disc should be refinished by an automotive machine shop.

➡Note: When replacing the brake pads, it's a good idea to resurface the discs regardless of the dial indicator reading, as this will impart a smooth finish and ensure a perfectly flat surface, eliminating any brake pedal pulsation or other undesirable symptoms related to questionable discs. At the very least, if you elect not to have the discs resurfaced, remove the glaze from the surface with emery cloth or sandpaper, using a swirling motion (see illustration).

5 It's absolutely critical that the disc not be machined to a thickness under the specified minimum allowable thickness. The minimum (or discard) thickness is cast into the disc (see illustrations). The disc thickness can be checked with a micrometer (see illustration).

5.2 Hang the caliper with a piece of wire - don't let it hang by the brake hose

5.3 The brake pads on this vehicle were obviously neglected, as they wore down to the rivets and cut deep grooves into the disc - wear this severe means the disc must be replaced

5.4a Use a dial indicator to check disc runout - if the reading exceeds the specified allowable runout limit, the disc will have to be machined or replaced

5.4b Using a swirling motion, remove the glaze from the disc surface with sandpaper or emery cloth

REMOVAL

Refer to illustrations 5.6a, 5.6b and 5.7

6 If you're removing a front disc on a model with an FN 3 caliper, or a rear disc on any model, remove the two caliper mounting bracket bolts and detach the mounting bracket (see illustrations).

7 Remove the bolts which you installed to hold the disc in place. If equipped, remove the disc retaining screw (see illustration), then slide the disc off the hub.

INSTALLATION

8 Place the disc in position on the hub flange, aligning the bolt holes. Install the retaining screw and tighten it securely.

9 If equipped, install the caliper mounting bracket and tighten the bolts to the torque listed in this Chapter's Specifications.

10 Install the caliper, tightening the guide pins (front) or mounting bolts (rear) to the torque listed in this Chapter's Specifications.

11 Install the wheel, lower the vehicle and tighten the wheel bolts to the torque listed in the Chapter 1 Specifications. Depress the brake pedal a few times to bring the brake pads into contact with the disc. Bleeding won't be necessary unless the brake hose was disconnected from the caliper. Check the operation of the brakes carefully before driving the vehicle.

5.5a On the front disc, the minimum thickness is cast into the outside of the disc, between the friction surface and the hub area

5.5b On the rear disc, the minimum thickness is cast into the side of the hub area

5.5c Use a micrometer to measure disc thickness at several points

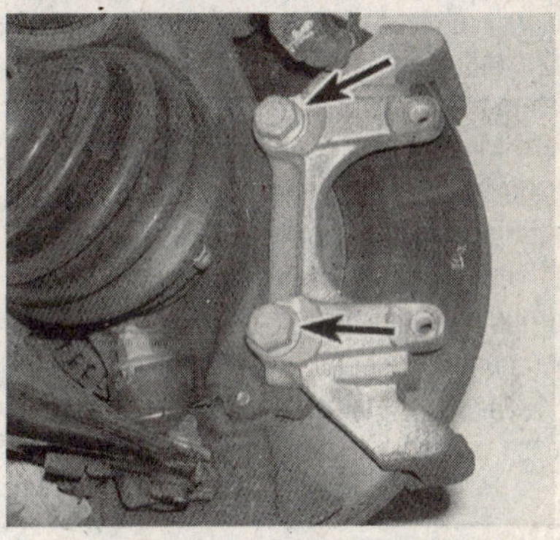

5.6a Remove the caliper mounting bracket bolts and detach the mounting bracket - this is a front caliper mounting bracket for an FN 3 caliper . . .

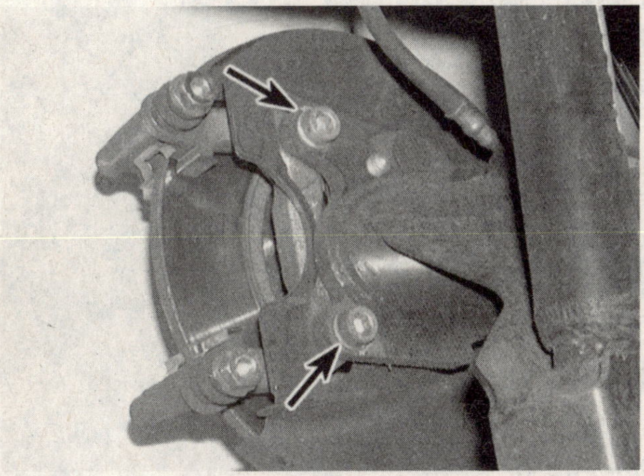

5.6b . . . and this is a rear caliper mounting bracket

5.7 Sometimes disc retaining screws become "frozen" and are too tight to remove by hand; if so, use an impact screwdriver

6 Drum brake shoes - replacement

➧ Refer to illustrations 6.1a, 6.1b, 6.7a, 6.7b, 6.7c, 6.8, 6.9a, 6.9b, 6.10, 6.15a, 6.15b, 6.15c, 6.16, 6.17a, 6.17b, 6.18 and 6.26

⁂ WARNING:

Brake shoes must be replaced on both rear wheels at the same time - never replace the shoes on only one wheel, as uneven braking may result. Also, the dust created by the brake system is harmful to your health. Never blow it out with compressed air and don't inhale any of it. An approved filtering mask should be worn when working on the brakes. Do not, under any circumstances, use petroleum-based solvents to clean brake parts. Use brake system cleaner only!

1 Raise the rear of the vehicle and support it securely on jackstands, then chock the front wheels to prevent the vehicle from rolling. Release the parking brake, remove the brake drum retaining screw (similar to the disc retaining screw shown in illustration 5.7), then slide the drum off the hub. If the drum won't come off, insert a screwdriver through one of the wheel bolt holes and pry up on the adjuster wedge; this will retract the shoes away from the drum (see illustrations).

2 Before beginning work, wash off the brake assembly with brake system cleaner and allow the residue to drain into a drip pan.

3 Measure the thickness of the friction material of each brake shoe at several points; if either shoe is worn at any point to the specified minimum thickness or less, all four shoes must be replaced as a set. The shoes should also be replaced if any are fouled with oil or grease; there is no satisfactory way of degreasing friction material once it has been contaminated.

4 If any of the brake shoes are worn unevenly, or fouled with oil or grease, trace and repair the cause before reassembly.

5 To replace the brake shoes, continue as follows. If all is well, install the brake drum.

6 Note the position of the brake shoes and springs, and mark the webs of the shoes, if necessary, to aid installation.

7 Remove the hold-down cups and springs by depressing and turning them 90-degrees. This can be accomplished with a special hold-down spring tool or, if you're careful, a pair of pliers. With the cups removed, lift off the springs and withdraw the hold-down pins (see illustrations).

8 Ease the shoes out one at a time from the lower pivot point to release the tension of the return spring, then disconnect the lower return spring from both shoes (see illustration).

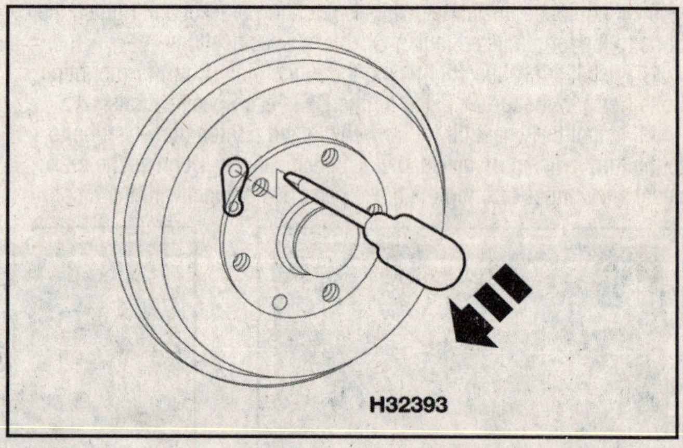

6.1a If the brake drum is tight, release the brake shoes by inserting a flat-bladed screwdriver through a hole in the drum and pry up the adjuster wedge

6.1b Drum brake components - exploded view

1 Plug
2 Hold-down pins
3 Bolt
4 Backing plate
5 Wheel cylinder
6 Upper return spring
7 Trailing shoe
8 Locating spring
9 Adjuster wedge
10 Leading shoe
11 Pushrod
12 Adjuster spring
13 Lower return spring
14 Hold-down spring
15 Spring cup

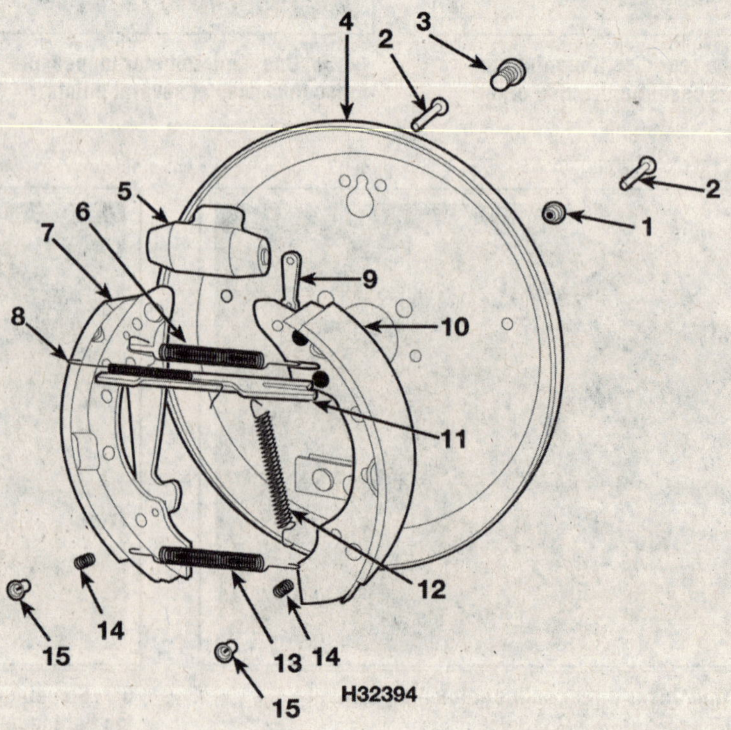

6.7a Remove the spring cup . . .

6.7b . . . then lift off the hold-down spring . . .

6.7c . . . and withdraw the hold-down pin from the rear of the backing plate

6.8 Unhook the shoes from the lower pivot point and remove the lower return spring

6.9a Free the shoes from the wheel cylinder (note the rubber band (arrow) used to retain pistons) . . .

9 Ease the upper end of both shoes out from their wheel cylinder locations, taking care not to damage the wheel cylinder seals, and disconnect the parking brake cable from the lever on the trailing shoe. The brake shoe assembly can then be maneuvered out of position and away from the backing plate. Do not depress the brake pedal until the brakes are reassembled; wrap a strong rubber band around the wheel cylinder pistons to retain them (see illustrations).

10 Make a note of the correct installed positions of all components (see illustration), then unhook the upper return spring, and disengage the adjuster wedge spring.

6.9b . . . then detach the parking brake cable and remove the shoe assembly from the vehicle

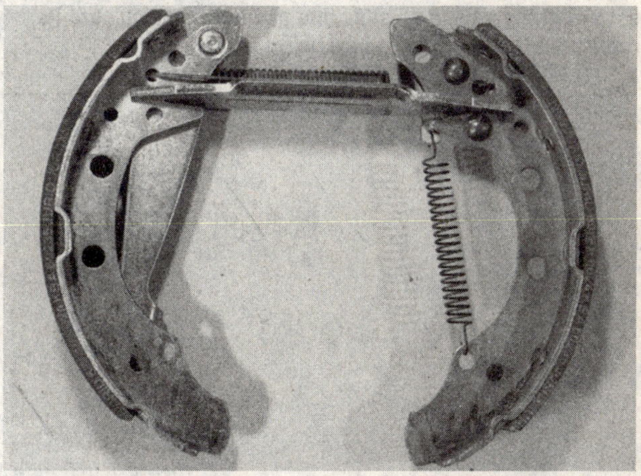

6.10 Prior to dismantling, note the correct installed location of the shoe assembly components

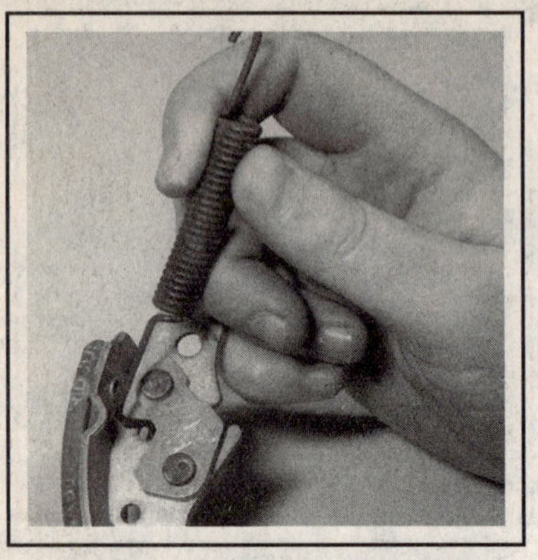

6.15a Hook the locating spring into the trailing shoe . . .

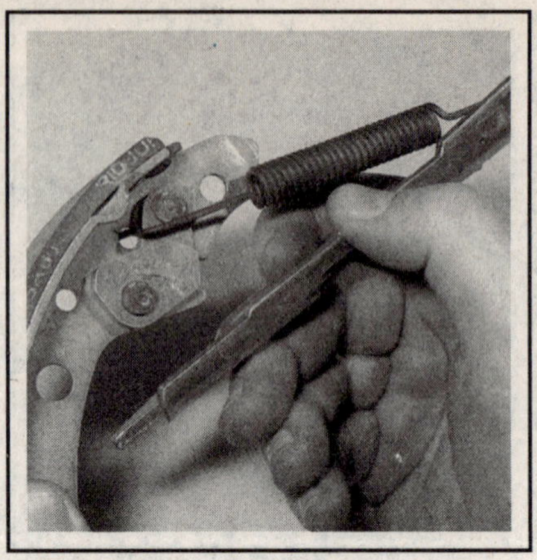

6.15b . . . then engage the pushrod with the opposite end of the spring . . .

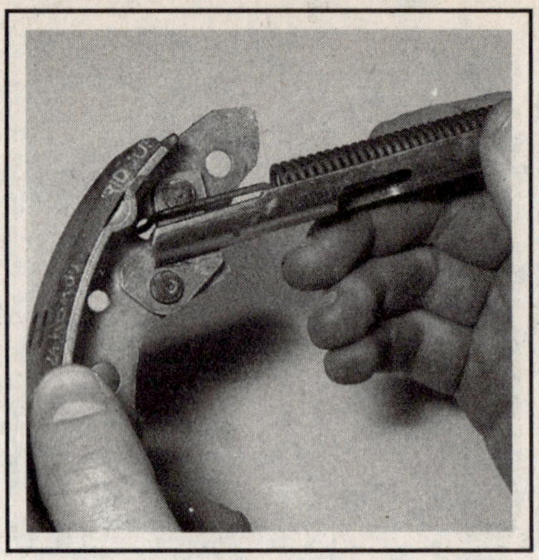

6.15c . . . and pivot the pushrod into position on the shoe

11 Unhook the locating spring, and remove the pushrod from the trailing shoe, together with the wedge key.

12 Examine all components for signs of wear or damage and replace as necessary.

➡Note: All return springs should be replaced, regardless of their apparent condition.

13 Peel back the protective caps, and check the wheel cylinder for fluid leaks or other damage; check that both cylinder pistons are free to move easily. Refer to Section 7, if necessary, for the wheel cylinder replacement procedure.

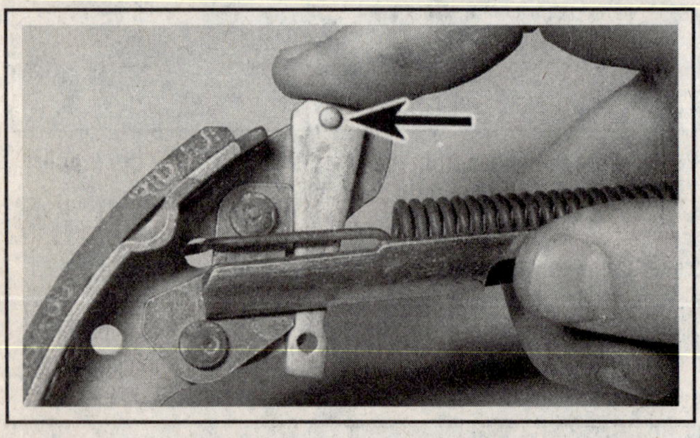

6.16 Slot the adjuster wedge into position, making sure its raised dot (arrow) is facing away from the shoe

14 Apply a little high-temperature brake grease to the contact areas of the pushrod and parking brake lever.

15 Hook the locating spring into the trailing shoe. Engage the pushrod with the opposite end of the spring, and pivot the pushrod into position on the trailing shoe (see illustrations).

16 Install the adjuster wedge between the trailing shoe and pushrod, making sure it is installed correctly (see illustration).

17 Locate the parking brake lever on the leading shoe in the pushrod, and install the upper return spring using a pair of pliers (see illustrations).

18 Install the spring on the adjuster wedge and hook it onto the trailing shoe (see illustration).

19 Prior to installation, clean the backing plate and apply a thin film of high-temperature brake grease or anti-seize compound to the shoe contact areas on the backing plate and to the wheel cylinder pistons and lower pivot point. Do not allow the lubricant to contact the friction material.

20 Remove the rubber band installed on the wheel cylinder, then install the shoe assembly.

21 Connect the parking brake cable to the parking brake lever, then locate the top of the shoes in the wheel cylinder piston slots.

22 Install the lower return spring between the shoes, then lever the bottom of the shoes onto the bottom anchor.

23 Tap the shoes to centralize them with the backing plate, then install the shoe hold-down pins and springs, securing them in position

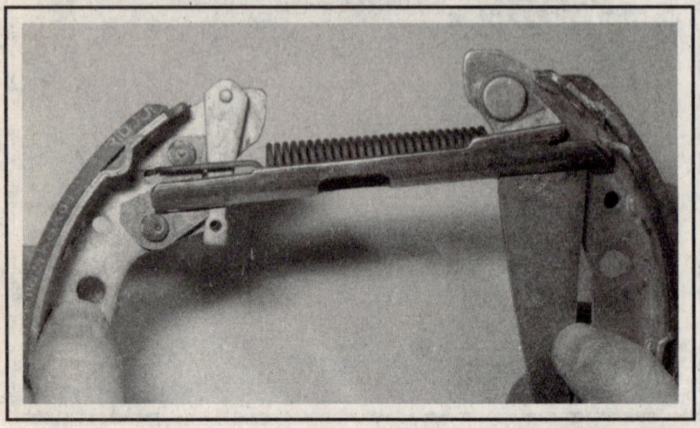

6.17a Locate the leading shoe in the pushrod . . .

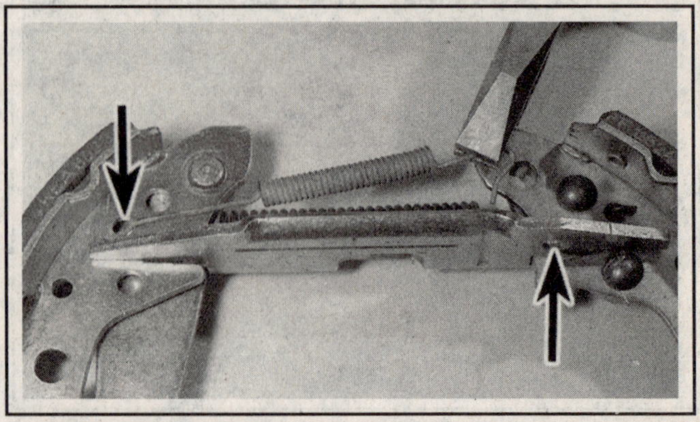

6.17b . . . and hook the upper return spring into position in the leading shoe and pushrod (arrows)

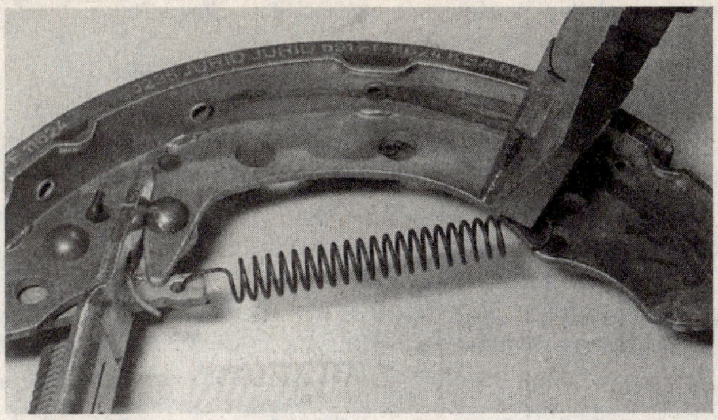

6.18 Fit the spring to the wedge key, and hook it onto the trailing shoe

with the spring cups.

24 Install the brake drum and retaining screw.

25 Repeat the above procedure on the other rear brake.

26 Before installing the drum it should be checked for cracks, score marks, deep scratches and hard spots, which will appear as small discolored areas. If the hard spots cannot be removed with sandpaper or emery cloth, or if any of the other conditions listed above exist, the drum must be taken to an automotive machine shop to have it resurfaced.

➡Note: Professionals recommend resurfacing the drums whenever a brake job is done. Resurfacing will eliminate the possibility of out-of-round or tapered drums. If the drums are worn so much that they can't be resurfaced without exceeding the maximum allowable diameter (stamped into the drum), then new ones will be required. At the very least, if you elect not to have the drums machined, remove the glazing from the surface with sandpaper or emery cloth using a swirling motion (see illustration).

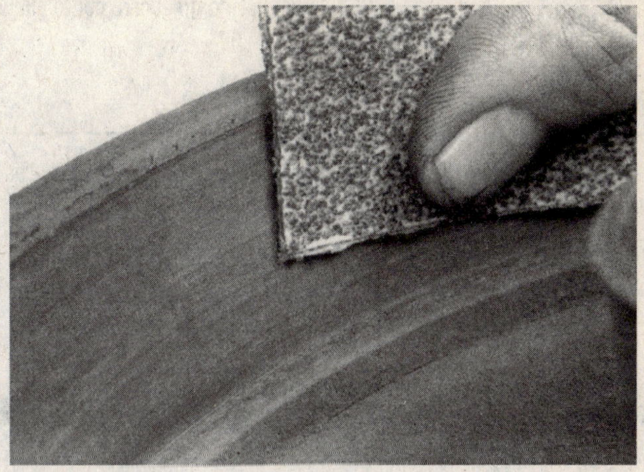

6.26 Remove the glaze from the drum surface with sandpaper or emery cloth

27 Once both sets of rear shoes have been replaced, install the brake drums and adjust the lining-to-drum clearance by repeatedly depressing the brake pedal until normal (non-assisted) pedal pressure returns.

28 Check and, if necessary, adjust the parking brake as described in Section 14.

29 On completion, check the brake fluid level as described in Chapter 1.

➡Note: New shoes will not give full braking efficiency until they have bedded-in. Be prepared for this, and avoid hard braking as far as possible for the first hundred miles or so after shoe replacement.

7 Wheel cylinder - removal and installation

➧ Refer to illustration 7.4

⁂ WARNING:

The dust created by the brake system is harmful to your health. Never blow it out with compressed air and don't inhale any of it. An approved filtering mask should be worn when working on the brakes. Do not, under any circumstances, use petroleum-based solvents to clean brake parts. Use brake system cleaner only!

REMOVAL

1 Remove the brake drum as described in Section 6. Clean the brake assembly with brake system cleaner and allow the residue to drain into a drip pan.

2 Using pliers, carefully unhook the upper brake shoe return spring and remove it from both brake shoes. Pull the upper ends of the shoes away from the wheel cylinder to disengage them from the pistons.

3 Minimize fluid loss by first removing the master cylinder reservoir cap, then tightening it down onto a piece of cellophane to obtain an airtight seal.

4 Using a flare-nut wrench, if available, unscrew the brake line fitting nut at the wheel cylinder (see illustration). Carefully ease the line out of the wheel cylinder and plug its end to prevent dirt entry. Wipe off any spilled fluid immediately.

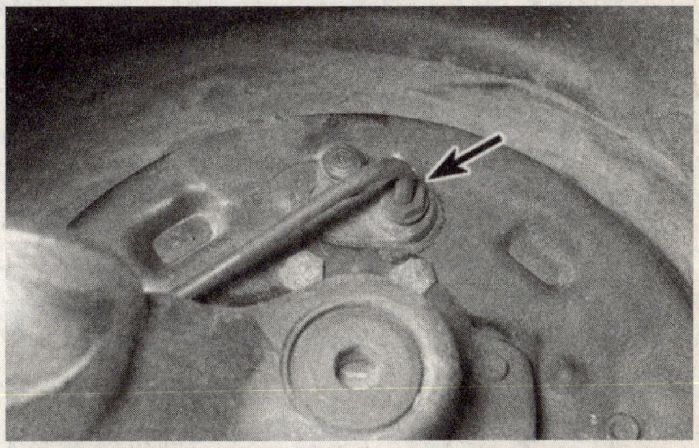

7.4 Unscrew the brake line fitting (arrow) from the wheel cylinder with a flare-nut wrench to prevent rounding off the corners of the fitting

5 Unscrew the two wheel cylinder retaining bolts from the rear of

the backing plate and remove the cylinder, taking great care not to allow brake fluid to contaminate the brake shoe linings.

INSTALLATION

6 Ensure that the backing plate and wheel cylinder mating surfaces are clean, then spread the brake shoes and maneuver the wheel cylinder into position.

7 Engage the brake line and screw in the fitting nut two or three turns to ensure that the thread has started.

8 Insert the two wheel cylinder retaining bolts and tighten them to the torque listed in this Chapter's Specifications. Now securely tighten the brake line fitting nut.

9 Remove the cellophane from the master cylinder reservoir and reinstall the cap.

10 Ensure that the brake shoes are correctly located in the cylinder pistons, then carefully install the brake shoe upper return spring, using a screwdriver to stretch the spring into position.

11 Install the brake drum.

12 Bleed the brake hydraulic system as described in Section 11. Providing precautions were taken to minimize loss of fluid, it should only be necessary to bleed the relevant rear brake.

8 Master cylinder - removal and installation

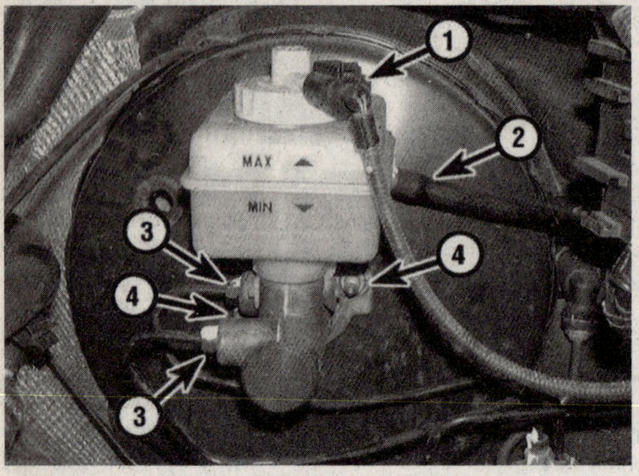

8.2a Master cylinder mounting details (non-ABS)

1 *Fluid level sensor electrical connector*
2 *Supply hose for clutch master cylinder*
3 *Fluid line fittings*
4 *Mounting nuts*

WARNING:

The manufacturer recommends replacing the master cylinder mounting nuts with new ones whenever they are removed.

REMOVAL

Refer to illustrations 8.2a and 8.2b

1 On models through 2000, remove the windshield wiper arms (see Chapter 12), the cowl cover and plenum close-out panel (see Chapter 11). On all models, remove the air filter housing (see Chapter 4).

2 Unplug the electrical connector for the fluid level warning switch (see illustrations). On ABS-equipped models, disconnect the brake pressure switch electrical connector at the underside of the master cylinder. On some models there are two sensors and two connectors.

3 Remove as much fluid as possible from the reservoir with a syringe, suction gun or poultry baster.

WARNING:

Brake fluid is poisonous - never siphon it by mouth. If a baster is used, never again use it for the preparation of food.

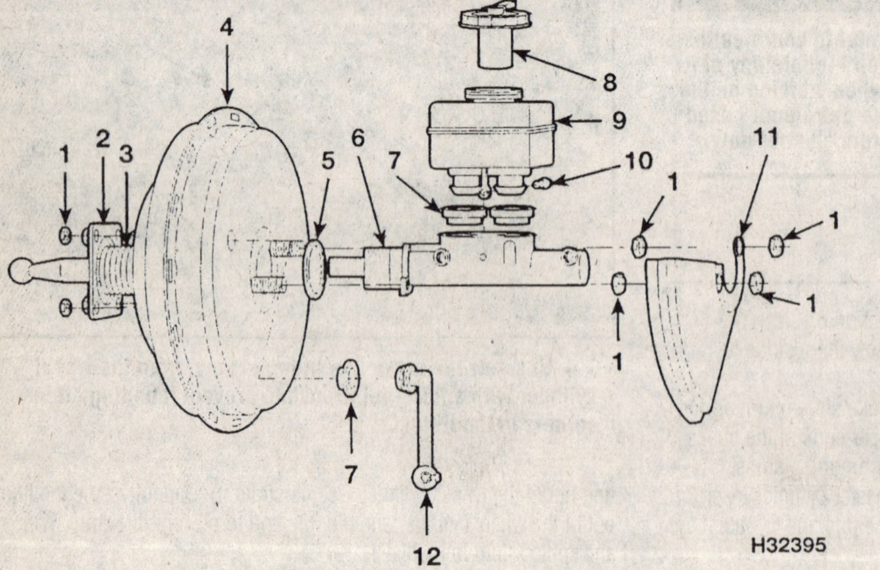

8.2b Exploded view of the master cylinder, reservoir, and power brake booster

1 *Nut*
2 *Gasket*
3 *Boot*
4 *Power brake booster*
5 *O-ring*
6 *Master cylinder*
7 *Grommet*
8 *Cap/fluid level sensor*
9 *Reservoir*
10 *Reservoir retaining pin*
11 *Heat shield (1.8L turbo models only)*
12 *Vacuum hose*

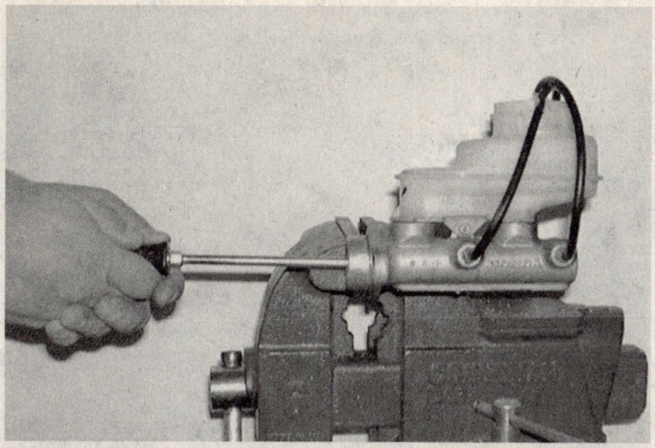

8.9 The best way to bleed air from the master cylinder before installing it on the vehicle is with a pair of bleeder tubes that direct brake fluid into the reservoir during bleeding

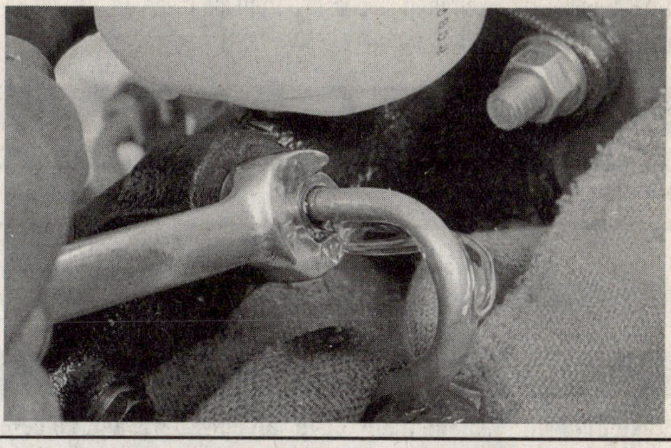

8.17 Have an assistant depress the brake pedal and hold it down, then loosen the fitting nut, allowing air and fluid to escape; repeat this procedure on both fittings until the fluid is clear of air bubbles

CAUTION:

Brake fluid will damage paint. If any fluid is spilled, wash it off immediately with plenty of clean, cold water.

4 Slide the relay box that's adjacent to the master cylinder forward, out of its bracket. Detach the hose leading to the clutch master cylinder from the brake fluid reservoir.

5 Place rags under the fittings and prepare caps or plastic bags to cover the ends of the lines once they're disconnected.

CAUTION:

Brake fluid will damage paint. Cover all body parts and be careful not to spill fluid during this procedure. Loosen the fittings at the ends of the brake lines where they enter the master cylinder. To prevent rounding off the flats, use a flare-nut wrench, which wraps around the fitting hex.

6 Pull the brake lines away from the master cylinder and plug the ends to prevent contamination.

7 Remove the nuts attaching the master cylinder to the power booster (see illustration 8.2a). Pull the master cylinder off the studs to remove it. Again, be careful not to spill the fluid as this is done. Remove and discard the old O-ring on the master cylinder.

INSTALLATION

➧ Refer to illustrations 8.9 and 8.17

8 Bench bleed the new master cylinder before installing it. Mount the master cylinder in a vise, with the jaws of the vise clamping on the mounting flange.

9 Attach a pair of master cylinder bleeder tubes to the outlet ports of the master cylinder (see illustration).

10 Fill the reservoir with brake fluid of the recommended type (see Chapter 1).

11 Slowly push the pistons into the master cylinder (a large Phillips screwdriver can be used for this) - air will be expelled from the pressure chambers and into the reservoir. Because the tubes are submerged in fluid, air can't be drawn back into the master cylinder when you release the pistons.

12 Repeat the procedure until no more air bubbles are present.

13 Remove the bleed tubes, one at a time, and install plugs in the open ports to prevent fluid leakage and air from entering. Install the reservoir cap.

14 Install the master cylinder over the studs on the power brake booster and tighten the attaching nuts only finger tight at this time. Don't forget to install a new O-ring.

15 Thread the brake line fittings into the master cylinder. Since the master cylinder is still a bit loose, it can be moved slightly so the fittings thread in easily. Don't strip the threads as the fittings are tightened.

16 Tighten the mounting nuts to the torque listed in this Chapter's Specifications. Tighten the brake line fittings securely.

17 Fill the master cylinder reservoir with fluid, then bleed the lines at the master cylinder, followed by bleeding the remainder of the brake system (see Section 11). To bleed the lines at the master cylinder, have an assistant depress the brake pedal and hold it down. Loosen the fitting to allow air and fluid to escape (see illustration). Tighten the fitting, then allow your assistant to return the pedal to its rest position. Repeat this procedure on both fittings until the fluid is free of air bubbles, then bleed the rest of the system. Check the operation of the brake system carefully before driving the vehicle.

WARNING:

If you do not have a firm brake pedal at the end of the bleeding procedure, or have any doubts as to the effectiveness of the brake system, DO NOT drive the vehicle. Have it towed to a dealer service department or other qualified repair shop for diagnosis.

RESERVOIR/GROMMET REPLACEMENT

➧ Refer to illustration 8.20

➡Note: The brake fluid reservoir can be replaced separately from the master cylinder body if it becomes damaged. If there is leakage between the reservoir and the master cylinder body, the grommets in the master cylinder body can be replaced.

18 Remove as much fluid as possible from the reservoir with a suction gun, large syringe or a poultry baster.

⁂ WARNING:

If a poultry baster is used, never again use it for the preparation of food.

19 Place rags under the master cylinder to absorb any fluid that may spill out once the reservoir is detached from the master cylinder.

⁂ CAUTION:

Brake fluid will damage paint. Cover all body parts and be careful not to spill fluid during this procedure.

20 Using a pair of pliers, pull out the retaining pin that retains the reservoir to the master cylinder (see illustration).

21 Pull the reservoir out of the master cylinder body.

22 If you are simply replacing the grommets, carefully pry the old grommets out of the master cylinder body and install new ones.

23 Lubricate the grommets with clean brake fluid, then press the reservoir into place on the master cylinder body and secure it with the retaining pin.

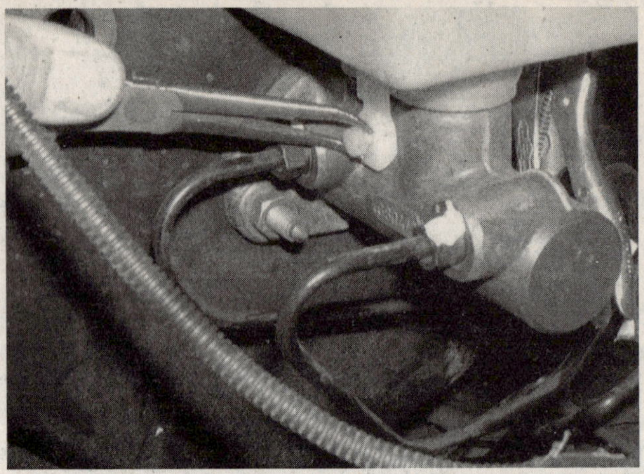

8.20 Pull the retaining pin out with a pair of pliers, then rock the reservoir back-and-forth while pulling up to remove it

24 Refill the reservoir with the recommended brake fluid (see Chapter 1) and check for leaks.

25 Bleed the master cylinder (see illustration 8.17).

9 Brake hoses and lines - inspection and replacement

INSPECTION

1 About every six months, with the vehicle raised and supported securely on jackstands, the rubber hoses which connect the steel brake lines with the front and rear brake assemblies should be inspected for cracks, chafing of the outer cover, leaks, blisters and other damage. These are important and vulnerable parts of the brake system and inspection should be complete. A light and mirror will be helpful for a thorough check. If a hose exhibits any of the above conditions, replace it with a new one.

REPLACEMENT

Front brake hose

➧ Refer to illustrations 9.3, 9.4 and 9.6

2 Loosen the wheel bolts, raise the vehicle and support it securely on jackstands. Remove the wheel.

3 At the bracket, unscrew the brake line fitting from the hose (see illustration). Use a flare-nut wrench to prevent rounding off the corners,

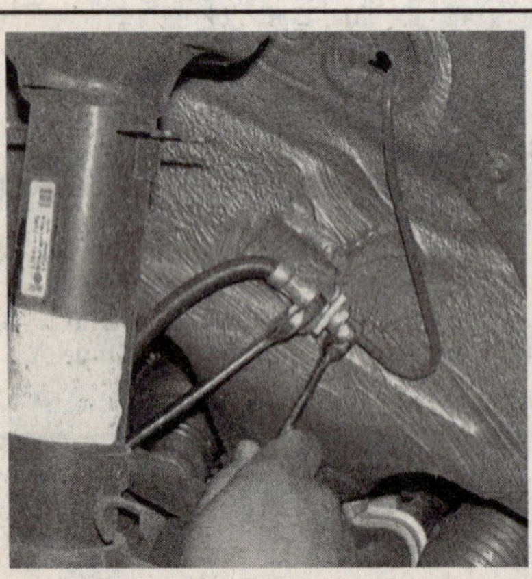

9.3 Using a flare-nut wrench, unscrew the threaded fitting on the brake line, while holding the hose end stationary with an open-end wrench . . .

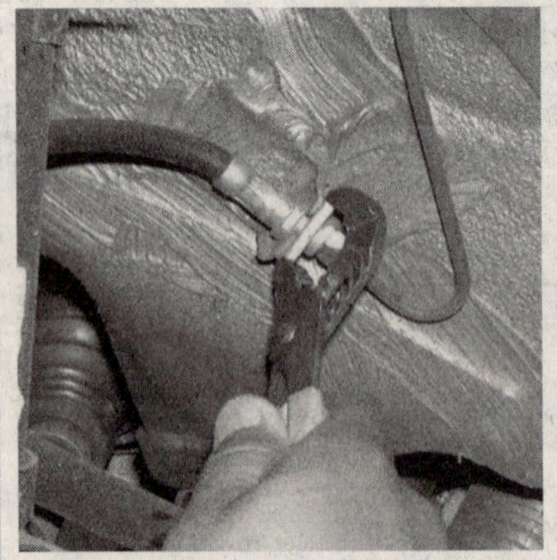

9.4 . . . then remove the U-clip and detach the hose from the bracket

9.6 Unclip the hose from the bracket on the suspension strut

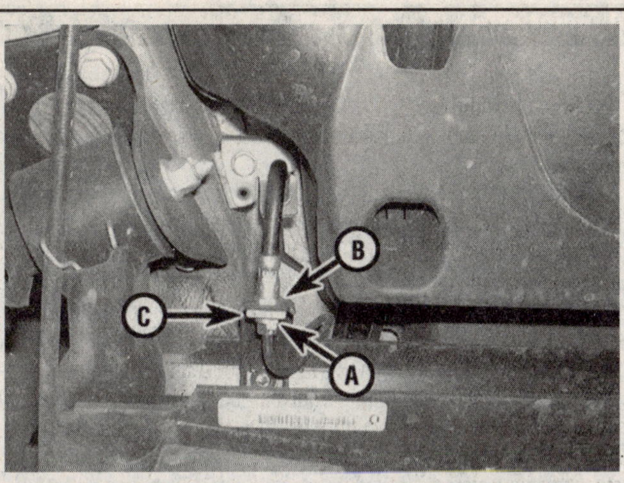

9.13 At each end of the chassis-to-rear axle brake hose, unscrew the line fitting (A) with a flare-nut wrench while holding the end of the brake hose (B) with an open-end wrench, then remove the U-clip (C)

and hold the hose fitting with another wrench to prevent the line or bracket from twisting.

4 Remove the U-clip from the female fitting at the bracket with a pair of pliers (see illustration), then pass the hose through the bracket.

5 At the caliper end of the hose, remove the inlet fitting bolt, then separate the hose from the caliper. Note that there are two sealing washers on either side of the inlet fitting - they should be replaced with new ones during installation (see illustration 4.5).

6 Detach the hose from the bracket on the strut (see illustration).

7 To install the hose, connect the fitting to the caliper with the inlet fitting bolt and new sealing washers.

8 Route the hose into the frame bracket, making sure it isn't twisted, then connect the brake line fitting, starting the threads by hand. Install the U-clip and tighten the fitting securely.

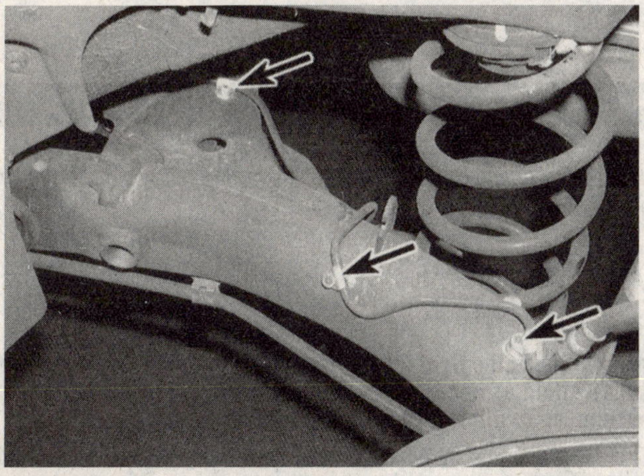

9.18 The rear brake line is secured to the rear axle trailing arm with plastic clips (arrows)

9 Push the hose into the bracket on the strut.

10 Bleed the caliper (see Section 11).

11 Install the wheel, lower the vehicle and tighten the wheel bolts to the torque listed in the Chapter 1 Specifications.

Rear brake hose

12 Loosen the rear wheel lug bolts, raise the rear of the vehicle and support it securely on jackstands. Remove the wheel.

Chassis-to-rear axle hose

➧ Refer to illustration 9.13

13 There's a flexible brake hose on each side of the rear axle beam, connecting the rigid lines on the chassis to the rigid lines leading to the rear brakes (see illustration). These hoses are replaced using the same technique as described in Steps 3 and 4.

14 After replacement, bleed the caliper or wheel cylinder served by the hose that was replaced (see Section 11).

Axle-to-caliper or wheel cylinder line

➧ Refer to illustration 9.18

15 Disconnect the forward end of the line using the technique described in Step 3. There's no need to remove the U-clip or detach the hose from its bracket.

16 On models with rear disc brakes, remove the inlet fitting bolt and detach the hose from the caliper. Discard the sealing washers - new ones should be used on installation.

17 On models with rear drum brakes, unscrew the brake line fitting from the wheel cylinder (see illustration 7.4).

18 Detach the line from the plastic clips along the rear axle (see illustration) and remove the line from the vehicle.

19 Installation is the reverse of removal. If you're working on a model with rear disc brakes, be sure to use new sealing washers on either side of the inlet fitting (see illustration 4.5), and tighten the inlet fitting bolt to the torque listed in this Chapter's Specifications.

20 If you're working on a model with rear drum brakes, tighten the brake line fitting securely.

21 Bleed the caliper or wheel cylinder served by the line that was replaced (see Section 11). Install the wheel, lower the vehicle and tighten the wheel bolts to the torque listed in the Chapter 1 Specifications.

Metal brake lines

22 When replacing brake lines, be sure to use the correct parts. Don't use copper tubing for any brake system components. Purchase steel brake lines from a dealer or auto parts store.

23 Prefabricated brake line, with the tube ends already flared and fittings installed, is available at auto parts stores and dealer parts departments. These lines must be bent to the proper shapes using a tubing bender.

24 When installing the new line, make sure it's securely supported in the brackets and has plenty of clearance between moving or hot components.

25 After installation, check the master cylinder fluid level and add fluid as necessary. Bleed the brake system (see Section 11) and test the brakes carefully before driving the vehicle in traffic.

10 Brake pressure regulator valve - check, removal and installation

➡Note: Models equipped with ABS are not equipped with a rear brake pressure regulator valve; the function is automatically controlled by the ABS unit.

REMOVAL

➧ Refer to illustration 10.1

1 The valve is mounted next to the rear axle, attached to the axle by a spring (see illustration). As the load being carried by the vehicle is altered, the suspension moves in relation to the vehicle body, altering the tension in the spring. The spring then adjusts the pressure regulator valve lever so that the correct pressure is applied to the rear brakes to suit the load being carried.

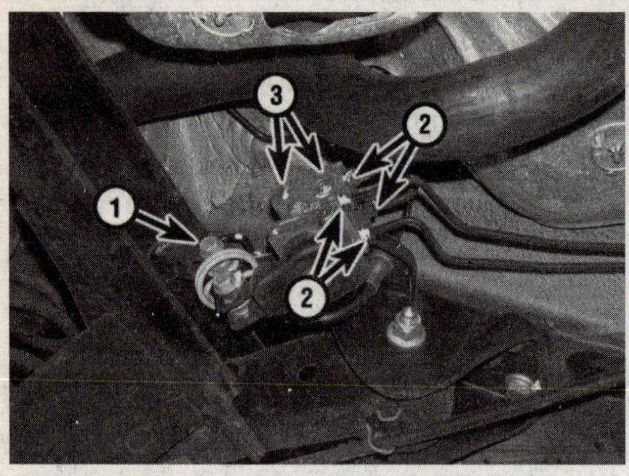

10.1 Pressure regulator valve mounting details

1 Spring-to-axle bolt
2 Brake line fittings
3 Valve mounting bolts

2 Minimize fluid loss by first removing the master cylinder reservoir cap, then tightening it down onto a piece of cellophane to obtain an airtight seal.

3 Unscrew the nut and bolt securing the valve spring to the axle.

4 Wipe clean the area around the brake line fittings on the valve, and place absorbent rags beneath the line fittings to catch any fluid that spills out. Make identification marks on the brake lines; these marks can then be used on installation to ensure each line is correctly reconnected.

5 Using a flare-nut wrench, if available, loosen the fitting nuts and disconnect the brake lines from the valve. Plug or tape over the line ends and valve orifices, to minimize the loss of brake fluid and to prevent the entry of dirt into the system. Wash off any spilled fluid immediately with cold water.

6 Unscrew the bolts and remove the pressure regulator valve and spring.

INSTALLATION

7 Installation is the reverse of the removal procedure, noting the following points:

a) If a new valve is being installed, set the spring adjustment bolt to the same position as the one on the old valve, and tighten it securely.
b) Ensure that the brake pipes are correctly connected to the valve, and that their fitting nuts are securely tightened.
c) Coat the ends of the spring with grease prior to installation.
d) Remove the cellophane from under the reservoir cap, then reinstall the cap.
e) Bleed the complete brake system as described in Section 11.
f) On completion, take the vehicle to a dealer service department or other qualified repair shop to have the valve operation (brake balance) checked and, if necessary, adjusted.

11 Brake hydraulic system - bleeding

➧ Refer to illustration 11.8

⁂ WARNING:

Wear eye protection when bleeding the brake system. If the fluid comes in contact with your eyes, immediately rinse them with water and seek medical attention.

➡Note: Bleeding the hydraulic system is necessary to remove any air that manages to find its way into the system when it's been opened during removal and installation of a hose, line, caliper or master cylinder.

1 You'll probably have to bleed the system at all four brakes if air has entered it due to low fluid level, or if the brake lines have been disconnected at the master cylinder.

2 If a brake line was disconnected only at a wheel, then only that caliper or wheel cylinder must be bled.

3 If a brake line is disconnected at a fitting located between the master cylinder and any of the brakes, that part of the system served by the disconnected line must be bled.

4 Remove any residual vacuum from the brake power booster by applying the brake several times with the engine off.

5 Remove the master cylinder reservoir cap and fill the reservoir with brake fluid. Reinstall the cap.

➡Note: Check the fluid level often during the bleeding operation and add fluid as necessary to prevent the fluid level from falling low enough to allow air bubbles into the master cylinder.

6 Have an assistant on hand, as well as a supply of new brake fluid, a clear plastic container partially filled with clean brake fluid, a length of clear tubing to fit over the bleeder valve and a wrench to open and close the bleeder valve.

7 Beginning at the right rear wheel, loosen the bleeder valve slightly, then tighten it to a point where it's snug but can still be loosened quickly and easily.

8 Place one end of the tubing over the bleeder valve and submerge the other end in brake fluid in the container (see illustration).

9 Have the assistant slowly depress the brake pedal, then hold the pedal down firmly.

10 While the pedal is held down, open the bleeder valve just enough to allow a flow of fluid to leave the valve. Watch for air bubbles to exit the submerged end of the tube. When the fluid flow slows, close the valve and have your assistant release the pedal.

11 Repeat Steps 9 and 10 until no more air is seen leaving the tube, then tighten the bleeder valve and proceed to the left rear wheel, the right front wheel and the left front wheel, in that order, and perform the same procedure. Be sure to check the fluid in the master cylinder reservoir frequently.

12 Never use old brake fluid. It contains moisture which can cause the fluid to boil, rendering the brake system inoperative.

13 Refill the master cylinder with fluid at the end of the operation.

14 Check the operation of the brakes. The pedal should feel solid when depressed, with no sponginess. If necessary, repeat the entire process.

⁂ WARNING:

Do not operate the vehicle if you're in doubt about the effectiveness of the brake system, or if the ABS light on the instrument panel does not go out.

11.8 When bleeding the brakes, a hose is connected to the bleeder valve at the caliper or wheel cylinder and then submerged in brake fluid - air will be seen as bubbles in the tube and container (all air must be expelled before moving to the next wheel)

12 Power brake booster - check, removal and installation

➧ Refer to illustrations 12.9, 12.11a, 12.11b, 12.11c and 12.11d

⁂ WARNING:

The manufacturer recommends replacing the booster mounting nuts with new ones whenever they are removed.

CHECK

Operating check

1 Depress the brake pedal several times with the engine off and make sure there's no change in the pedal reserve distance.

2 Depress the pedal and start the engine. If the pedal goes down slightly, operation is normal. If the pedal does not go down, check the vacuum hose for a leak. If the hose is good, check the intake manifold vacuum with a vacuum gauge (gasoline engines, see Chapter 2C) or the vacuum pump (diesel engines, see Section 13).

Airtightness check

3 Start the engine and turn it off after one or two minutes. Depress the brake pedal slowly several times. If the pedal depresses less each time, the booster is airtight.

4 Depress the brake pedal while the engine is running, then stop the engine with the pedal depressed. If there's no change in the pedal reserve travel after holding the pedal for 30 seconds, the booster is airtight.

Check valve and hose check

5 Detach the vacuum hose from the booster and blow into the hose; air should flow through.

6 Attach a hand-held vacuum pump to the hose and apply vacuum; the valve and hose should hold vacuum.

⁂ WARNING:

It isn't a good idea to perform this check by sucking on the valve with your mouth. If the valve is faulty, you could inhale fuel fumes.

7 If the valve (or hose) fails either test, replace the valve and hose as an assembly.

➡Note: When installed, the arrow on the check valve must point towards the engine.

REMOVAL

➡Note 1: A special tool (VW T 10006, or equivalent) is required to disconnect the power brake booster pushrod from the brake pedal. Check on the availability of this tool before proceeding.

➡Note 2: On early models equipped with ABS, it is not possible to remove the power brake booster without first removing the ABS hydraulic unit (see Section 2). Therefore, booster unit removal and installation on models with ABS should be left to a VW dealer service department or other qualified repair shop.

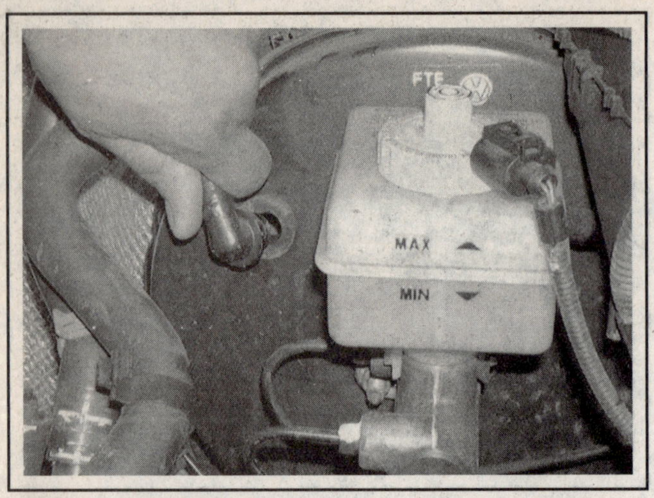

12.9 Pull the vacuum hose out of the booster grommet

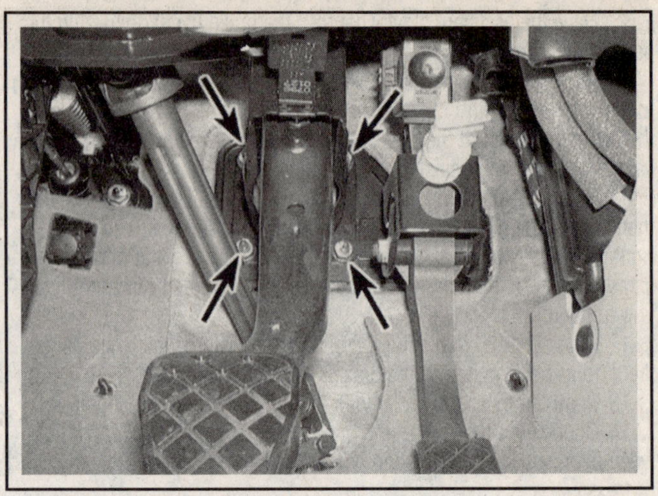

12.11a The power brake booster is retained by these four nuts (arrows)

8 Remove the master cylinder as described in Section 8.

9 Remove the heat shield (if equipped) from the front of the booster, then carefully ease the vacuum hose out from the grommet in the booster (see illustration).

10 From inside the vehicle, remove the brake light switch as described in Section 16. Where equipped, remove the plastic cover on the inside of the firewall, for access to the booster-retaining nuts.

11 Unscrew the four nuts securing the booster unit to the pedal mounting bracket (see illustration). Reach up behind the brake pedal, carefully expand the lugs of the pushrod retaining clips and detach the pushrod ball from the pedal. A special tool will be required to do this (see illustrations).

12 Return to the engine compartment and maneuver the booster unit out of position, noting the gasket which is installed on the rear of the unit.

INSTALLATION

13 Check the booster unit vacuum hose sealing grommet for signs of damage or deterioration and replace it if necessary.

14 Install a new gasket on the rear of the booster unit, then reposition the unit in the engine compartment.

15 From inside the vehicle, ensure that the booster pushrod is correctly engaged with the brake pedal, then clip the pedal onto the pushrod ball. Check that the pedal is securely retained, then install the booster mounting nuts and tighten them to the torque listed in this Chapter's Specifications.

16 Carefully ease the vacuum hose back into position in the booster, taking great care not to displace the sealing grommet. On models so equipped, install the heat shield to the booster.

17 Install the master cylinder as described in Section 8 of this Chapter.

18 Install the brake light switch as described in Section 16.

19 On completion, start the engine and check for a vacuum leak at the vacuum hose-to-booster grommet. Check the operation of the brake system in an isolated area before returning the vehicle to normal service.

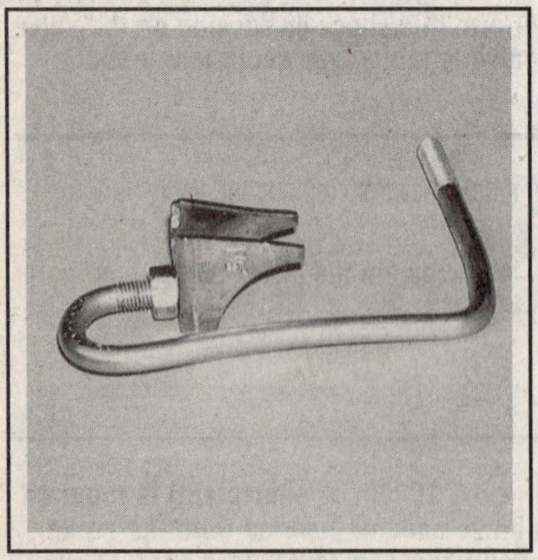

12.11b Improvised special tool constructed from a modified exhaust clamp, threaded rod and nut, used to release the brake pedal from the booster pushrod

12.11c Using the tool to release the brake pedal from the booster pushrod

12.11d Rear view of the brake pedal (pedal removed) showing plastic lugs (arrowed) securing pedal to booster pushrod

13 Vacuum pump - check, removal and installation

Refer to illustration 13.5

Note: This Section applies to models with diesel engines only.

CHECK

1 The most common symptom of a failed vacuum pump is a hard brake pedal. Check the operation of the power brake booster as described in Section 12.

2 If the booster doesn't operate properly, remove the vacuum hose from the power brake booster and connect a vacuum gauge to it.

3 Start the engine and watch the gauge. Vacuum should register on the gauge almost immediately.

4 If vacuum does not register on the gauge, replace the vacuum pump.

REMOVAL AND INSTALLATION

5 Loosen the clamp and disconnect the vacuum hose from the pump (see illustration).

6 Detach the wiring harness retainer from the pump.

7 Unscrew the mounting fasteners and remove the pump from the engine.

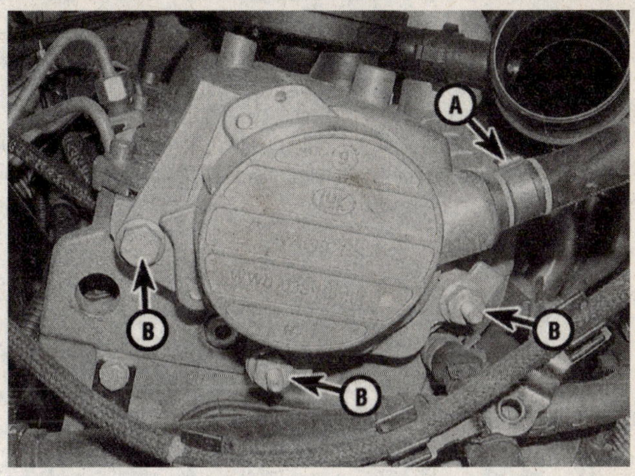

13.5 Vacuum pump mounting details

A Hose clamp *B Mounting bolts*

8 Installation is the reverse of removal. Be sure the pump is correctly aligned with the camshaft when mating it to the cylinder head. Tighten the fasteners securely.

14 Parking brake - adjustment

Refer to illustrations 14.3 and 14.8

1 To check the parking brake adjustment, first apply the brake pedal firmly several times to establish correct shoe-to-drum/pad-to-disc clearance, then apply and release the parking brake several times.

2 Applying normal, moderate pressure, pull the parking brake lever to the fully applied position, counting the number of clicks emitted from the parking brake ratchet mechanism. If adjustment is correct, there should be approximately 4 to 7 clicks before the parking brake is fully applied. If this is not the case, adjust as follows.

3 Remove the rear section of the center console as described in Chapter 11 to gain access to the parking brake adjusting nut (see illustration).

4 Block the front wheels, then raise the rear of the vehicle and support it securely on jackstands. Continue as described under the relevant sub-heading.

REAR DRUM BRAKE MODELS

5 With the parking brake set on the fourth notch of the ratchet mechanism, tighten the adjusting nut until it is difficult to turn both rear wheels. Once this is so, fully release the parking brake lever and check that the wheels/hubs rotate freely. Check the adjustment by applying the parking brake fully, counting the clicks emitted from the parking brake ratchet and, if necessary, re-adjust.

6 Once adjustment is correct hold the adjusting nuts and securely tighten the locknuts. Install the center console section/ashtray (as applicable).

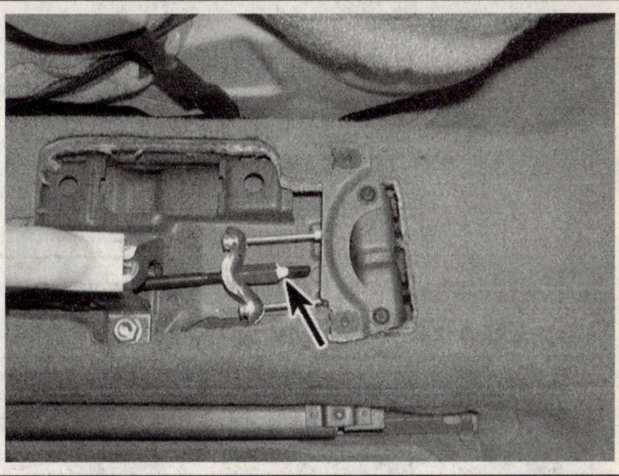

14.3 Parking brake adjusting nut (arrow)

REAR DISC BRAKE MODELS

7 With the parking brake fully released, equally loosen the parking brake adjusting nut until the parking brake levers on both rear calipers are back against their stops.

8 From this point, tighten the adjusting nut until both parking brake levers just move off the caliper stops. Ensure that the gap between each caliper parking brake lever and its stop is less than 1.5 mm (1/16-inch),

and ensure both the right- and left-hand gaps are equal (see illustration). Check that both wheels/discs rotate freely, then check the adjustment by applying the parking brake fully, counting the clicks emitted from the parking brake ratchet. If necessary, re-adjust.

9 Once adjustment is correct, hold the adjusting nuts and securely tighten the locknuts. Install the center console section/ashtray (as applicable).

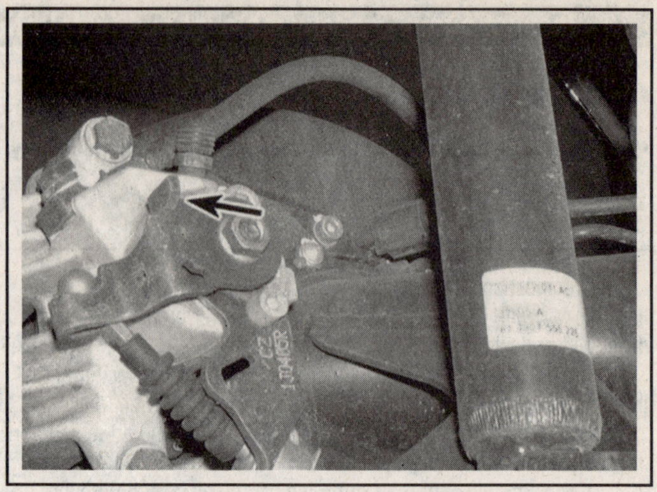

14.8 On models with rear disc brakes, adjust the parking brake so the clearance between the parking brake lever and its stop (arrow) is as specified

15 Parking brake cables - removal and installation

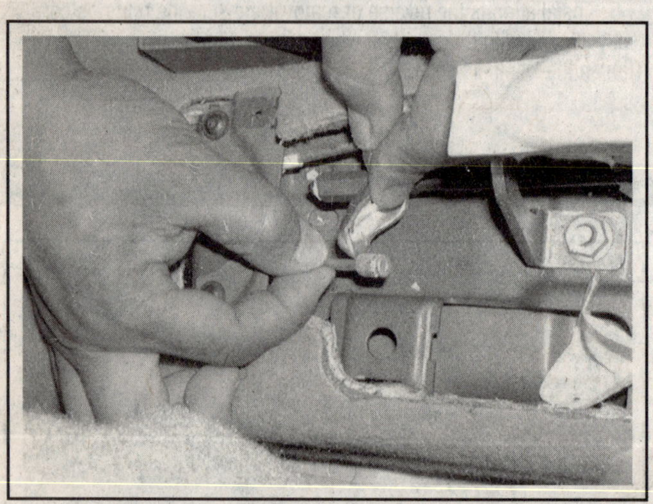

15.2 Detaching the parking brake cable from the equalizer

➧ Refer to illustrations 15.2, 15.5, 15.6 and 15.7

REMOVAL

1 Remove the rear section of the center console as described in Chapter 11 to gain access to the parking brake lever. Each rear brake has its own cable, which is connected to the lever by an equalizer plate.

2 Loosen the parking brake adjuster nut enough to detach the parking brake cable(s) from the equalizer (see illustration).

3 Block the front wheels, then raise the rear of the vehicle and support it securely on jackstands.

4 From the vehicle underbody, free the front end of the outer cable from the body and withdraw the cable from its guide tube.

5 Work back along the length of the cable, noting its correct routing, and free it from all of the retaining clips (see illustration).

6 On models with rear drum brakes, remove the rear brake shoes

15.5 Release the parking brake cable from all of the retaining clips

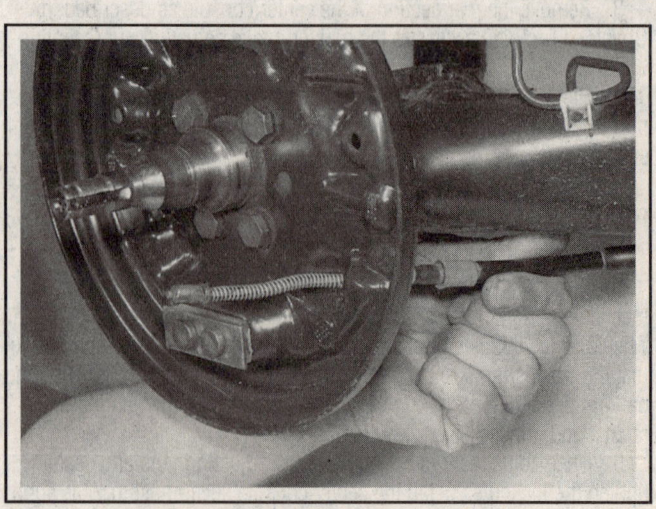

15.6 On models with rear drum brakes, remove the brake shoes and detach the cable from the backing plate

as described in Section 6. Using a hammer and pin punch, carefully tap the outer cable out from the brake backing plate, and remove it from underneath the vehicle (see illustration).

7 On models with rear disc brakes, disengage the inner cable from the caliper parking brake lever (see illustration), then remove the outer cable retaining clip and detach the cable from the caliper.

INSTALLATION

8 Installation is a reversal of the removal procedure. Prior to installing the center console, adjust the parking brake as described in Section 14.

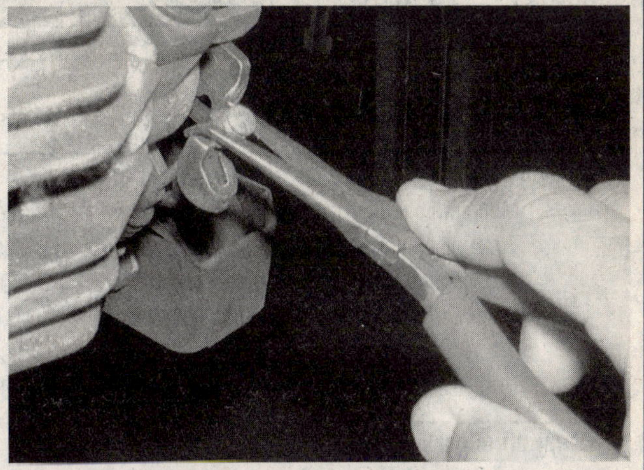

15.7 On models with rear disc brakes, detach the cable from the lever at the caliper, then remove the clip and disconnect the cable casing from the bracket on the caliper

16 Brake light switch - check, adjustment and replacement

➧ Refer to illustration 16.1

CHECK

1 The brake light switch is located on the brake pedal mounting bracket (see illustration). The switch activates the brake lights at the rear of the vehicle when the pedal is depressed. To gain access to the switch, remove the left-side under-dash panel and the heater/air conditioning duct.

2 If the brake lights are inoperative, check the fuse first (see Chapter 12).

3 If the fuse is good, check for voltage to the switch on the feed wire (refer to the wiring diagrams at the end of this manual for the proper color wire to check). If no voltage is present, repair the wire between the switch and the fuse box.

4 If voltage is present, depress the brake pedal and check for voltage at the output wire terminal (again, refer to the wiring diagrams). If no voltage is present, replace the switch.

5 If voltage is present, check for power on the brake light wires at the tail light housings (with the brake pedal depressed). If voltage is not present, repair the circuit between the switch and the brake lights.

6 If voltage is present, check for a bad ground; using a jumper wire connected to a good ground, probe the ground wire terminal at the tail light connector. If the brake lights go on, repair the ground circuit (follow the ground wire from the tail light housing).

7 Keep in mind that the brake light bulbs could be burned out, but the likelihood of all the bulbs being burned out is very slim.

ADJUSTMENT

8 The brake light switch on these vehicles is not adjustable. If it doesn't work as described above, replace it.

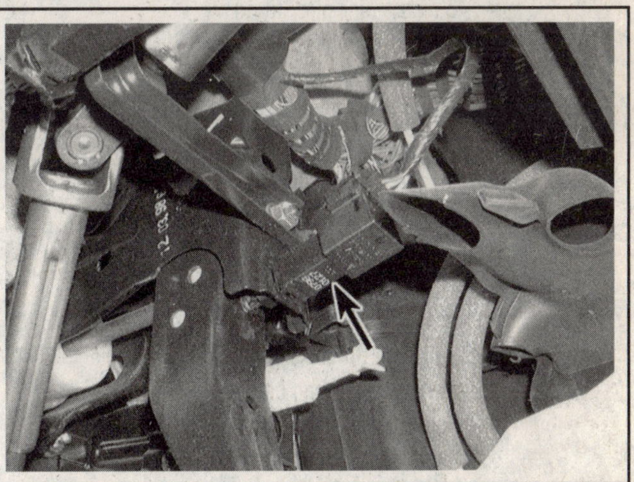

16.1 The brake light switch is mounted just behind the brake pedal on the pedal bracket - twist it 1/4-turn to remove it

REPLACEMENT

9 Remove left-side under-dash panel.

10 Unplug the electrical connector from the switch.

11 Rotate the switch 90-degrees in either direction and pull it out of the bracket.

12 To install the new switch, depress the brake pedal, insert the switch into the bracket (position the switch 1/4-turn from its installed position), rotate it 90-degrees and release the brake pedal.

13 Reconnect the electrical connector and check the operation of the brake lights.

14 Install the under-dash panel.

Specifications

General	
Brake fluid type	See Chapter 1
Disc brakes	
Minimum pad thickness	See Chapter 1
Brake disc minimum thickness	Cast into disc
Maximum disc runout (front and rear)	0.004 inch (0.1 mm)
Maximum disc thickness variation (parallelism)	
Front	0.0004 inch (0.01 mm)
Rear	0.0008 inch (0.02 mm)
Rear drum brakes	
Shoe friction material minimum thickness	See Chapter 1
Maximum inside diameter	Cast into drum
Maximum out-of-round	0.0012 (0.03 mm)

Torque specifications

Torque specifications	Nm	Ft-lbs (unless otherwise indicated)
Brake booster mounting nuts*	20	15
Brake caliper		
Caliper guide pins (front)	28	21
Caliper mounting bolts (rear)	35	26
Caliper mounting bracket bolts		
Front (FN 3 caliper only)	125	92
Rear	65	48
Brake hose-to-caliper inlet fitting bolt		
Front		
FS III caliper	35	26
FN 3	15	132 in-lbs
Rear	35	26
Master cylinder-to-brake booster retaining nuts*	20	15
Wheel cylinder retaining bolts	8	71 in-lbs
Wheel speed sensor bolt		
Front		
FS III caliper	8	71 in-lbs
FN 3 caliper	10	84 in-lbs
Rear	8	71 in-lbs

Replace with new nuts

10

SUSPENSION AND STEERING SYSTEMS

Section

Reference to other Chapters

Power steering fluid level check - See Chapter 1
Steering and suspension check - See Chapter 1
Tire and tire pressure checks - See Chapter 1
Tire rotation - See Chapter 1

1 General information

Refer to illustrations 1.1 and 1.2

FRONT SUSPENSION

The independent front suspension is of the MacPherson strut type, incorporating coil springs and integral telescopic shock absorbers (see illustration). The MacPherson struts are located by transverse control arms, which use rubber inner mounting bushings, and incorporate a balljoint at the outer ends. The front steering knuckles, which carry the wheel bearings, brake calipers and the hub/disc assemblies, are bolted to the MacPherson struts, and connected to the control arms through the balljoints. A stabilizer bar, connected to the chassis and both control arms, reduces body roll when cornering.

REAR SUSPENSION

The rear suspension consists of a torsion axle with shock absorbers and coil springs (see illustration). A stabilizer bar is incorporated into the rear axle assembly (it's integral with the axle beam and isn't removable).

STEERING

The steering column incorporates a universal joint, and is connected to the steering gear by a second individual universal joint.

The steering gear is mounted onto the front subframe, and is connected by two tie-rods, with balljoints at their outer ends, to the steering arms projecting rearwards from the steering knuckles.

Power-assisted steering is standard equipment. The hydraulic steering system is powered by a belt-driven pump, which is driven off the crankshaft pulley.

PRECAUTIONS

Frequently, when working on the suspension or steering system components, you may come across fasteners which seem impossible to loosen. These fasteners on the underside of the vehicle are continually subjected to water, road grime, mud, etc., and can become rusted or "frozen," making them extremely difficult to remove. In order to unscrew these stubborn fasteners without damaging them (or other components), be sure to use lots of penetrating oil and allow it to soak in for a while.

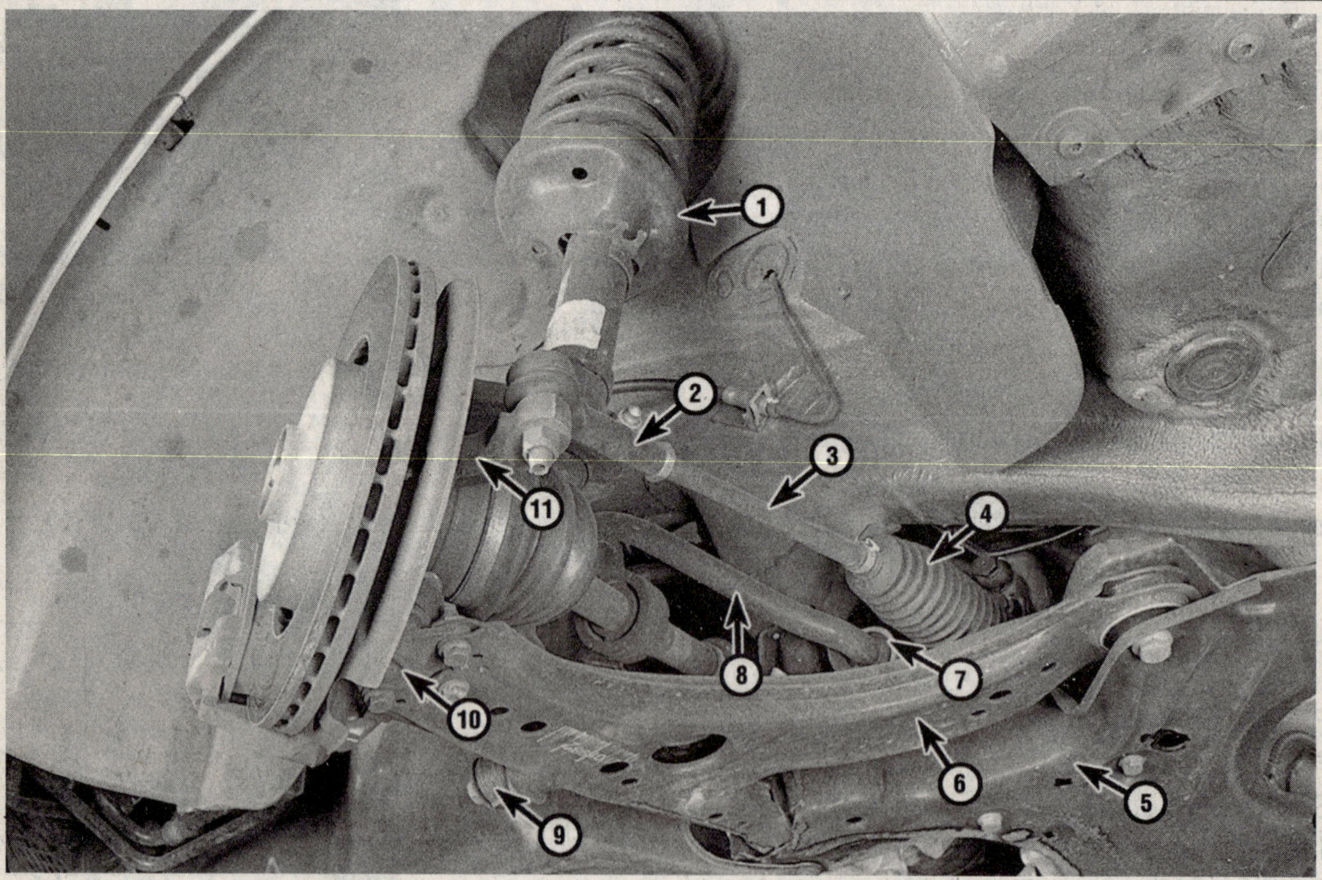

1.1 Front suspension and steering components

1 *Strut/coil spring assembly*
2 *Tie-rod end*
3 *Tie-rod*
4 *Steering gear boot*
5 *Subframe*
6 *Control arm*
7 *Stabilizer bar bushing/clamp*
8 *Stabilizer bar*
9 *Stabilizer bar link*
10 *Balljoint*
11 *Steering knuckle*

Using a wire brush to clean exposed threads will also ease removal of the nut or bolt and prevent damage to the threads. Sometimes a sharp blow with a hammer and punch will break the bond between a nut and bolt threads, but care must be taken to prevent the punch from slipping off the fastener and ruining the threads. Heating the stuck fastener and surrounding area with a torch sometimes helps too, but isn't recommended because of the obvious dangers associated with fire. Long breaker bars and extension, or "cheater," pipes will increase leverage, but never use an extension pipe on a ratchet - the ratcheting mechanism could be damaged. Sometimes tightening the nut or bolt first will help to break it loose. Fasteners that require drastic measures to remove should always be replaced with new ones.

Since most of the procedures dealt with in this Chapter involve jacking up the vehicle and working underneath it, a good pair of jackstands will be needed. A hydraulic floor jack is the preferred type of jack to lift the vehicle, and it can also be used to support certain components during various operations.

WARNING:

Never, under any circumstances, rely on a jack to support the vehicle while working on it. Whenever any of the suspension or steering fasteners are loosened or removed they must be inspected and, if necessary, replaced with new ones of the same part number or of original equipment quality and design. Torque specifications must be followed for proper reassembly and component retention. Never attempt to heat or straighten any suspension or steering components. Instead, replace any bent or damaged part with a new one.

1.2 Rear suspension components

1 *Shock absorber*
2 *Coil spring*
3 *Rear axle*

2 Strut/coil spring assembly (front) - removal, inspection and installation

Refer to illustrations 2.8, 2.9 and 2.11

WARNING:

The manufacturer recommends replacing the strut upper mounting nut and the strut-to-steering knuckle pinch bolt with new ones whenever they are removed.

REMOVAL

1 Remove the windshield wiper arms (see Chapter 12), the cowl cover and plenum close-out panel (see Chapter 11).

2 Loosen the wheel bolts, raise the front of the vehicle and support it securely on jackstands. Remove the wheel.

3 Without detaching the hose from the caliper, remove the brake caliper (see Chapter 9) and suspend it with a piece of wire (don't let it

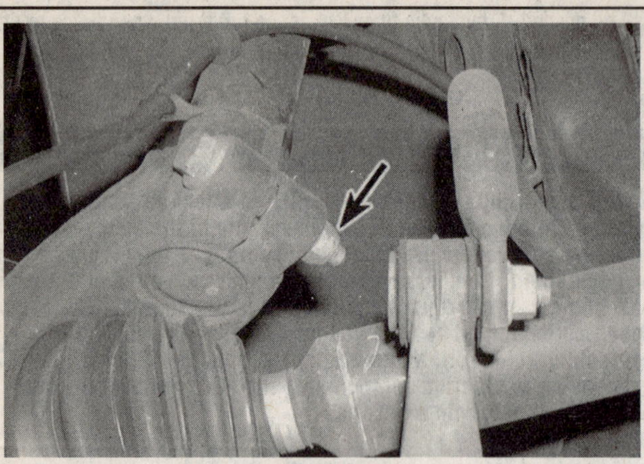

2.8 Remove the strut-to-steering knuckle pinch bolt/nut . . .

2.11 Remove the strut upper mounting nut with a spark plug socket and a wrench, and a long hex bit inserted through the socket and into the damper shaft (to prevent the shaft from turning)

hang by the brake hose). Unclip the hose from the bracket on the strut. If the inner brake pad is equipped with a wear sensor, detach the wiring harness bracket from the strut.

4 If the vehicle is equipped with an Anti-lock Braking System (ABS), detach the wheel speed sensor harness from the bracket on the strut.

5 On diesel engine models, remove the splash shield and the air intake duct connecting the turbocharger and the intercooler (see Chapter 4B).

6 If you're removing the right-side strut, unbolt the inner CV joint

2.9 . . . then spread the steering knuckle apart slightly by driving a chisel into the gap

from the drive flange (see Chapter 8).

7 Unbolt the stabilizer bar link from the control arm (see Section 6).

8 Remove the strut-to-steering knuckle pinch bolt (see illustration).

9 Insert a chisel into the slot in the steering knuckle and drive it in slightly to spread the joint apart (this will make separating the strut from the knuckle easier) (see illustration).

10 Pull the steering knuckle down, off the strut. If it sticks, apply some penetrating oil to the area where the strut meets the knuckle.

11 Remove the strut upper mounting nut and stop plate. This will require holding the strut rod from turning by using a hex key and turning the nut with a deep socket that has a hex for a wrench, such as a spark plug socket (see illustration). An assistant would be helpful at this point to hold the strut while the nut is loosened.

12 Remove the strut from the fenderwell.

INSTALLATION

13 Installation is the reverse of removal, with the following points:

a) *Install a new upper mounting nut.*
b) *Install a new pinch bolt and nut.*
c) *Tighten the strut upper mounting nut and the pinch bolt nut to the torque values listed in this Chapter's Specifications.*
d) *Tighten the brake caliper guide pins to the torque listed in the Chapter 9 Specifications.*
e) *If the right-side strut was removed, connect the inner CV joint to the drive flange and tighten the bolts to the torque listed in the Chapter 8 Specifications.*
f) *Tighten the wheel bolts to the torque listed in the Chapter 1 Specifications.*

14 Have the front wheel alignment checked and, if necessary, adjusted.

3 Strut or coil spring (front) - replacement

➧ Refer to illustrations 3.3, 3.4a, 3.4b, 3.4c, 3.4d, 3.4e, 3.5a, 3.5b, 3.5c and 3.11

WARNING:

Always replace the struts or coil springs in pairs - never replace just one of them.

1 If the struts or coil springs exhibit the telltale signs of wear (leaking fluid, loss of damping capability, chipped, sagging or cracked coil springs), explore all options before beginning any work. The strut/shock absorber assemblies are not serviceable and must be replaced if a problem develops. However, strut assemblies complete with springs may be available on an exchange basis, which eliminates much time and work.

3.3 Install the spring compressor following the tool manufacturer's instructions and compress the spring until all pressure is removed from the upper spring seat

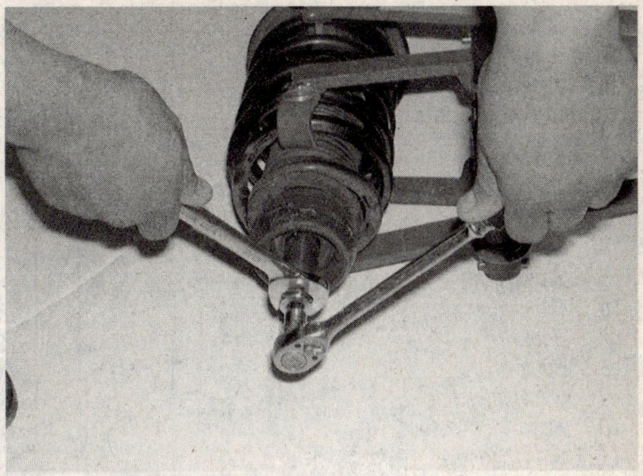

3.4a Remove the damper shaft nut . . .

3.4b . . . lift off the upper mount . . .

3.4c . . . followed by the bearing . . .

Whichever route you choose to take, check on the cost and availability of parts before disassembling your vehicle.

⁂ WARNING:

Disassembling a strut is potentially dangerous and utmost attention must be directed to the job, or serious injury may result. Use only a high-quality spring compressor and carefully follow the manufacturer's instructions furnished with the tool. After removing the coil spring from the strut assembly, set it aside in a safe, isolated area.

2 Remove the strut and spring assembly following the procedure described in the previous Section. Mount the strut assembly in a vise. Line the vise jaws with wood or rags to prevent damage to the unit and don't tighten the vise excessively.

3 Following the tool manufacturer's instructions, install the spring compressor (which can be obtained at most auto parts stores or equipment yards on a daily rental basis) on the spring and compress it sufficiently to relieve all pressure from the upper spring seat (see illustration). This can be verified by wiggling the spring.

4 Loosen and remove the damper shaft nut, while retaining the damper shaft with a hex bit, then remove the upper mount, bearing and spring seat (see illustrations). Also remove the bushing, if equipped (does not apply to U.S. or Canada models).

3.4d . . . and spring seat

3.4e Exploded view of the suspension strut and related components

1 Strut body
2 Nut
3 Pinch bolt
4 Steering knuckle
5 Spring seat
6 Coil spring
7 Boot
8 Bump stop
9 Upper mounting nut
10 Stop
11 Strut tower
12 Damper shaft nut
13 Upper mount
14 Bearing
15 Bushing (not installed on US or Canadian models)

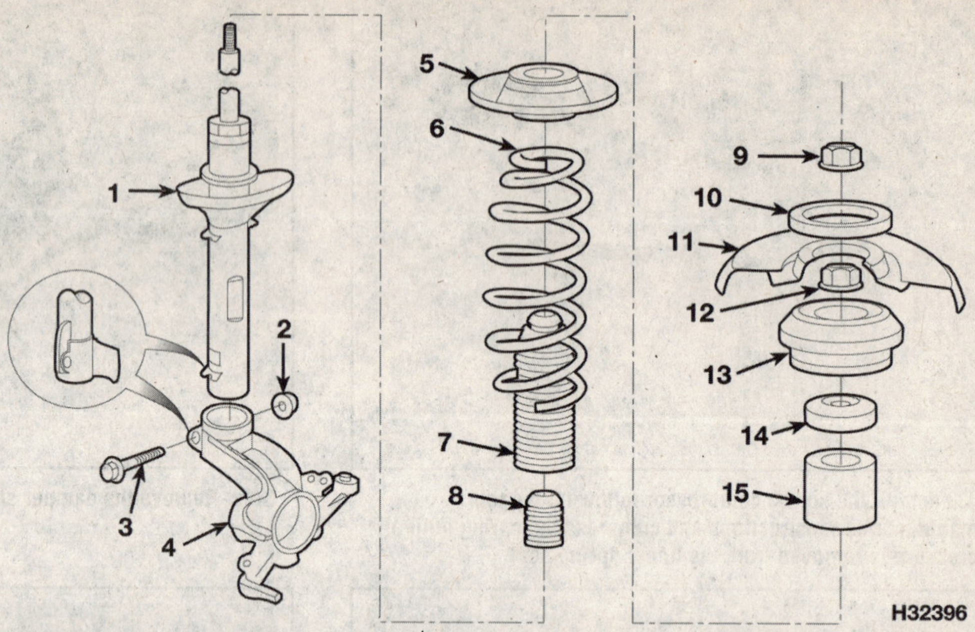

3.5a Remove the compressed spring, keeping the ends of the spring pointed away from your body

5 Remove the coil spring, then slide off the boot and rubber bump stop (see illustrations).

⁂ WARNING:

Keep the ends of the spring away from your body.

6 With the strut now completely disassembled, examine all the components for wear, damage or deformation, and check the bearing for smoothness of operation. Replace any of the components as necessary.

7 Examine the strut for signs of fluid leakage (a slight amount of seepage is normal). Check the strut piston for signs of pitting along its entire length, and check the strut body for signs of damage. While holding it in an upright position, test the operation of the strut by moving the damper shaft through a full stroke, and then through several short strokes. In both cases, the resistance felt should be smooth and continuous. If the resistance is jerky, or uneven, or if there is any visible sign of wear or damage to the strut, replacement is necessary.

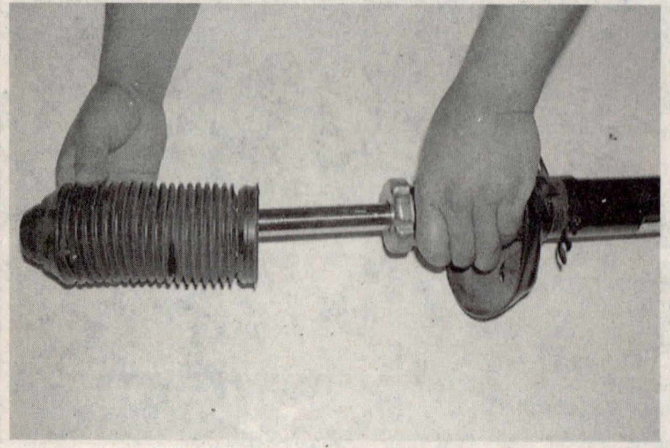

3.5b Remove the boot . . .

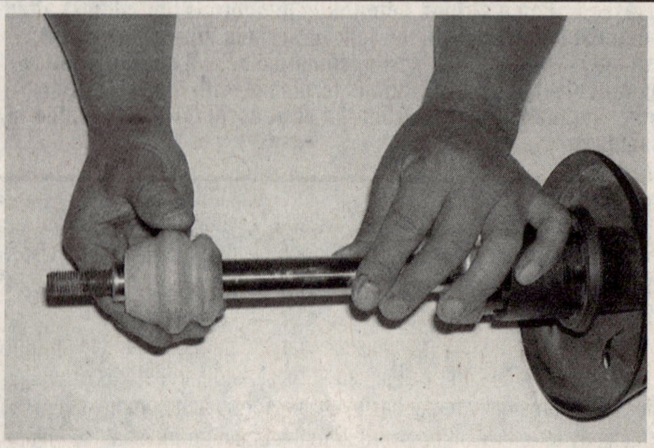

3.5c . . . and the bump stop

8 If any doubt exists about the condition of the coil spring, carefully remove the spring compressor and check the spring for distortion and signs of cracking. Replace the spring if it is damaged or distorted, or if there is any doubt as to its condition.

9 Inspect all other components for signs of damage or deterioration, and replace any that are suspect.

10 Slide the bump stop and boot onto the strut piston rod.

11 Install the coil spring onto the strut, making sure its end is correctly located against the spring seat stop (see illustration).

12 Install the upper spring seat, bushing (if equipped), bearing and upper mount, and screw on the damper shaft nut. Tighten the retaining nut to the torque listed in this Chapter's Specifications while preventing the piston rod from turning.

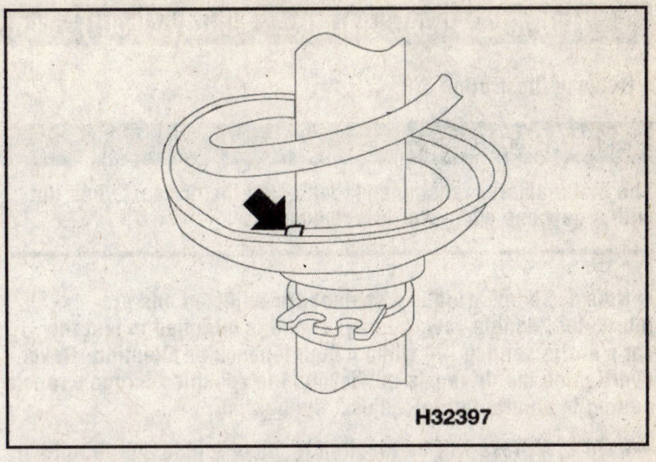

3.11 Make sure the end of the coil spring is positioned against the stop on the spring seat

4 Steering knuckle and hub - removal and installation

⁂ WARNING:

The manufacturer recommends replacing the driveaxle/hub nut, strut-to-steering knuckle pinch bolt and nut, and the tie-rod end nut with new ones whenever they are removed.

REMOVAL

1 Remove the wheel cover and loosen the driveaxle retaining nut with the vehicle resting on its wheels. Also loosen the wheel bolts.

2 Chock the rear wheels of the car, firmly apply the parking brake, then raise the front of the car and support it securely on jackstands. Remove the front wheel.

3 Remove the driveaxle retaining nut.

4 On models with ABS, remove the wheel speed sensor as described in Chapter 9.

5 Remove the brake caliper (don't disconnect the hose) and brake disc (see Chapter 9). Using a piece of wire or string, tie the caliper to the coil spring - don't let the caliper hang by the hose.

6 Detach the tie-rod end from the steering knuckle (see Section 17).

7 Mark the positions of the balljoint bolt heads to the control arm. Unscrew the balljoint retaining bolts and remove the retaining plate from the top of the control arm (see Section 8).

8 Detach the steering knuckle from the strut (see Section 2).

9 Carefully pull the hub assembly outwards while pushing the driveaxle from the hub. If necessary, tap the joint out of the hub using a soft-faced hammer. If this fails to free it from the hub, the joint will have to be pressed out using a puller.

⁂ CAUTION:

Be careful not to overextend the inner CV joint.

10 If necessary, loosen the balljoint nut a few turns, then break loose the balljoint stud from the steering knuckle with a balljoint removal tool. Remove the nut and separate the balljoint from the knuckle.

INSTALLATION

11 Lubricate the splines of the driveaxle with multi-purpose grease.

12 If removed, attach the balljoint to the steering knuckle, install a new nut, then tighten the nut to the torque listed in this Chapter's Specifications.

13 Maneuver the hub assembly into position and engage it with the driveaxle stub shaft. Install a new driveaxle/hub nut, but don't attempt to tighten it yet.

14 Engage the steering knuckle with the strut. Install a new pinch bolt and nut, then tighten the nut to the torque listed in this Chapter's Specifications.

15 Connect the balljoint to the control arm. Install the nut plate and the bolts, tightening the bolts to the torque listed in this Chapter's Specifications.

16 Engage the tie-rod end with the steering knuckle, then install a new retaining nut and tighten it to the torque listed in this Chapter's Specifications.

17 Install the brake disc and caliper, tightening the caliper mounting bracket bolts (if equipped) and caliper guide pins to the torque listed in the Chapter 9 Specifications.

18 If equipped, install the ABS wheel speed sensor as described in Chapter 9.

19 Install the wheel and lower the vehicle to the ground.

20 Tighten the driveaxle retaining nut to the torque listed in the Chapter 8 Specifications, then tighten the wheel bolts to the torque listed in the Chapter 1 Specifications.

5 Hub and wheel bearing assembly (front) - removal, bearing replacement and installation

➧ Refer to illustration 5.3

⁂ WARNING:

The manufacturer recommends replacing the driveaxle/hub nut with a new one whenever it is removed.

➡Note 1: The bearing is a sealed, pre-adjusted and pre-lubricated, double-row roller type, and is intended to last the car's entire service life without maintenance or attention. Never overtighten the driveaxle nut beyond the specified torque wrench setting in an attempt to "adjust" the bearing.

➡Note 2: A press will be required to disassemble and rebuild the assembly; if such a tool is not available, take the steering knuckle and hub assembly to an automotive machine shop to have the old bearing pressed out and the new one pressed in. The bearing's inner races are an interference fit on the hub; if the inner race remains on the hub when it is pressed out of the hub carrier, a knife-edged bearing puller will be required to remove it.

H32398

5.3 The hub bearing is retained by a snap-ring

1 Remove the steering knuckle assembly as described in Section 4.

2 Support the steering knuckle securely on blocks or in a vise. Using a tubular spacer which bears only on the inner end of the hub flange, press the hub flange out of the bearing. If the bearing's outboard inner race remains on the hub, remove it using a bearing puller (see **Note** above).

3 Extract the bearing retaining snap-ring from the steering knuckle assembly (see illustration).

4 Securely support the outer face of the steering knuckle. Using a tubular spacer, press the complete bearing assembly out of the steering knuckle.

5 Thoroughly clean the hub and steering knuckle, removing all traces of dirt and grease, and polish away any burrs or raised edges which might hinder reassembly. Check both for cracks or any other signs of wear or damage, and replace them if necessary. Replace the snap-ring, regardless of its apparent condition.

6 On reassembly, apply a light coating of moly-based grease to the bearing outer race and bearing surface of the steering knuckle.

7 Securely support the steering knuckle, and locate the bearing in the hub. Press the bearing fully into position, ensuring that it enters the hub squarely, using a tubular spacer which bears only on the bearing outer race.

8 Install the snap-ring, making sure it seats properly. The opening between the ends of the snap-ring must be pointing down (towards the balljoint) when the knuckle is in its installed position.

9 Securely support the outer face of the hub flange, and locate the steering knuckle bearing inner race over the end of the hub flange. Press the bearing onto the hub, using a tubular spacer which bears only on the inner race of the hub bearing, until it seats against the hub shoulder. Check that the hub flange rotates freely, and wipe off any excess grease.

10 Install the steering knuckle assembly as described in Section 4.

6 Stabilizer bar and bushings (front) - removal and installation

➧ Refer to illustrations 6.8, 6.9 and 6.11

⁂ WARNING:

The manufacturer recommends replacing the balljoint mounting bolts, steering gear bolts and the subframe mounting bolts with new ones whenever they are removed.

REMOVAL

1 Remove the wheel covers and loosen the driveaxle retaining nuts with the vehicle resting on its wheels. Loosen the wheel bolts, raise the front of the vehicle and support it securely on jackstands placed under the unibody frame rails or rocker panel flanges. Remove the wheels.

2 Remove the under-vehicle splash shield.

3 On diesel engine models, remove the splash shield and the air intake duct connecting the turbocharger and the intercooler (see Chapter 4B).

4 Remove the driveaxles (see Chapter 8).

5 Unbolt the transmission support from the subframe and the transmission (see Chapter 7A).

6 Unbolt the steering gear from the subframe (see Section 19).

7 If the vehicle is equipped with a diesel engine, unbolt the exhaust system from the subframe.

8 Unbolt the stabilizer bar links from the stabilizer bar (see illustration).

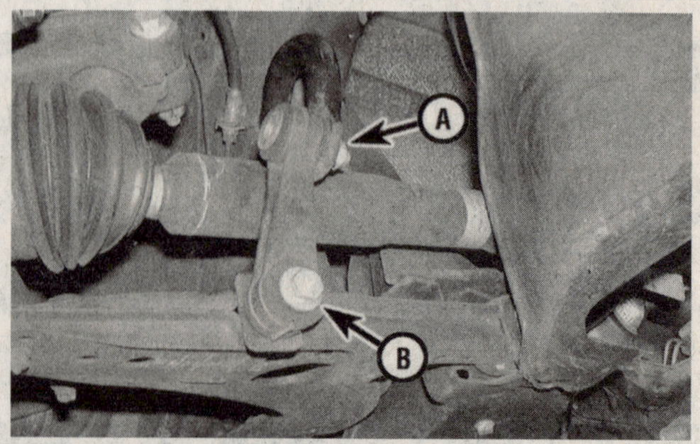

6.8 Remove the nut and bolt (A) securing the stabilizer bar link to the bar - remove bolt B if you're removing the control arm

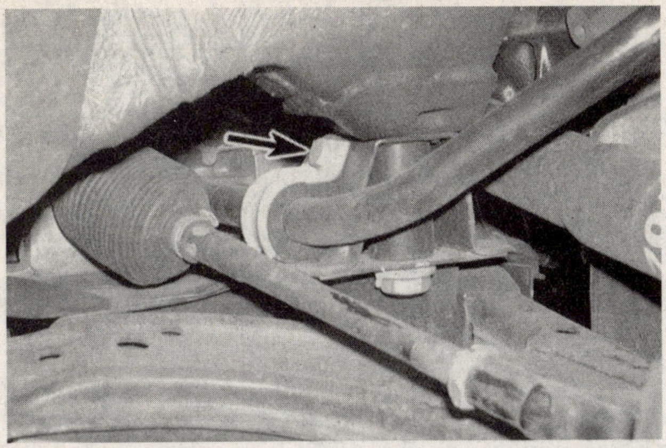

6.9 The stabilizer bar bushing clamps are each retained by one bolt

6.11 Subframe mounting bolts (arrows)

9 Unbolt the stabilizer bar bushing clamps from the subframe (see illustration).

10 Support the subframe with a jack - preferably a floor jack equipped with a transmission jack head adapter.

11 Remove the subframe mounting bolts (see illustration). Carefully lower the subframe enough to maneuver the stabilizer bar out.

12 Remove the stabilizer bar. If necessary, unbolt the stabilizer bar links from the control arms.

13 Inspect the clamp bushings and the link bushings. If they're cracked, hardened or deteriorated in any way, replace them.

INSTALLATION

14 Begin by positioning the stabilizer bar on the subframe, making sure it is centered. Install the clamps and bolts, but don't tighten the bolts yet.

15 Raise the subframe up and connect the steering gear to it, making sure it is positioned properly. Install new steering gear mounting bolts, but don't tighten them completely yet.

16 Attach the subframe to the vehicle, using new bolts. Tighten the bolts to the torque listed in this Chapter's Specifications.

17 Tighten the steering gear mounting bolts to the torque listed in this Chapter's Specifications.

18 Tighten the stabilizer bar bushing clamp bolts to the torque listed in this Chapter's Specifications.

19 Install the stabilizer bar links, tightening the fasteners to the torque listed in this Chapter's Specifications.

20 The remainder of installation is the reverse of removal, following the appropriate Sections in this Chapter, Chapter 7A and Chapter 8. Tighten all fasteners to the proper torque Specifications.

21 Install the wheels and lower the vehicle, then tighten the wheel bolts to the torque listed in the Chapter 1 Specifications, and the drive-axle/hub nuts to the torque listed in the Chapter 8 Specifications.

22 Have the front end alignment checked and, if necessary, adjusted.

7 Control arm (front) - removal, bushing replacement and installation

⁂ WARNING:

The manufacturer recommends replacing the control arm pivot bolt, the rear mounting bolt and the balljoint securing bolts with new ones whenever they are removed.

REMOVAL

➧ Refer to illustrations 7.4a and 7.4b

1 Loosen the wheel bolts. Chock the rear wheels, firmly apply the parking brake, then raise the front of the vehicle and support it securely on jackstands. Remove the wheel.

2 Unbolt the stabilizer bar link from the control arm (see illustration 6.8).

3 Mark the position of the balljoint bolt heads to the control arm. Unscrew the balljoint retaining bolts and remove the retaining plate from the top of the control arm (see Section 8).

4 Loosen and remove the control arm pivot bolt and rear mounting bolt (see illustrations).

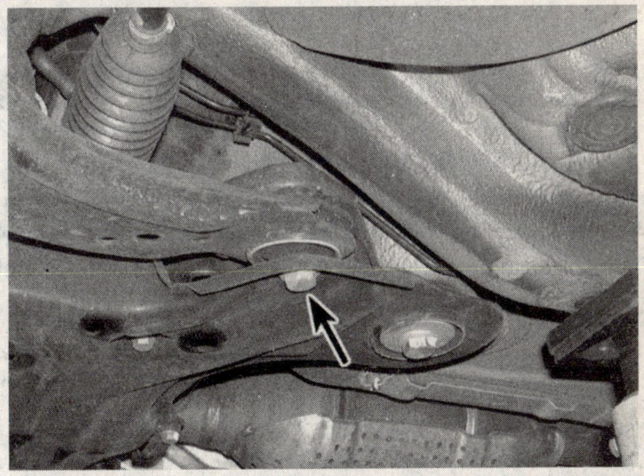

7.4a Control arm rear mounting bolt (arrow)

5 Lower the arm out of position, and remove it from underneath the vehicle.

BUSHING REPLACEMENT

Refer to illustration 7.7

6 Thoroughly clean the control arm and the area around the bushings, removing all traces of dirt and underseal if necessary, then check carefully for cracks, distortion or any other signs of wear or damage, paying particular attention to the pivot and rear mounting bushings. If either bushing requires replacement, the control arm should be taken to an automotive machine shop. A hydraulic press and suitable spacers are required to press the rear bushing out of the arm and install the new one, and a drawbolt-type bushing removal tool is required to replace the front pivot bushing.

7 If you do have access to the necessary equipment, be sure to orient the rear bushing properly (see illustration).

INSTALLATION

8 Maneuver the control arm into position, engaging it with the balljoint.

9 Install the new pivot bolt and rear mounting bolt.

10 Position the retaining plate on the top of the arm, then install the control arm balljoint retaining bolts. Align the balljoint with the marks made prior to removal, then tighten the retaining bolts to the torque listed in this Chapter's Specifications.

11 Tighten the control arm rear mounting bolt to the torque listed in this Chapter's Specifications. Tighten the pivot bolt lightly only at this stage.

12 Using a floor jack, raise the outer end of the control arm to simulate normal ride height, then tighten the control arm front pivot bolt to the torque listed in this Chapter's Specifications.

13 Connect the stabilizer bar link to the control arm, tightening the bolt to the torque listed in this Chapter's Specifications.

14 Install the wheel, then lower the vehicle and tighten the wheel bolts to the torque listed in the Chapter 1 Specifications.

15 Have the front end alignment checked and, if necessary, adjusted.

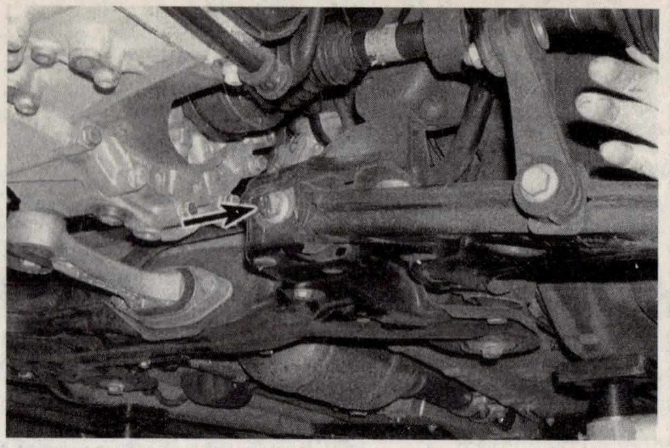

7.4b Control arm pivot bolt (arrow)

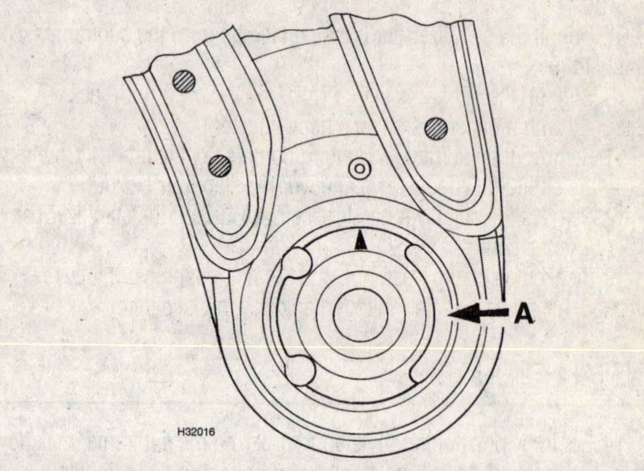

7.7 The control arm rear bushing must be installed with one of the arrows pointing towards the front pivot bushing, and the kidney-shaped cutout (A) facing the center of the vehicle

8 Balljoints - check and replacement

Refer to illustrations 8.5, 8.6a and 8.6b

WARNING:

The manufacturer recommends replacing the balljoint-to-control arm bolts and the balljoint stud nut with new ones whenever they are removed.

1 Loosen the wheel bolts, raise the front of the vehicle and support it securely on jackstands. Remove the wheel.

CHECK

2 Grasp the control arm and attempt to move it up-and-down - you shouldn't be able to feel any movement. Now grasp the bottom of the tire and try to move it in-and-out. If you can feel movement during either of these checks, replace the balljoint.

Note: Don't confuse play in the wheel bearings with balljoint wear.

3 Inspect the balljoint boot. If it is torn, replace the balljoint.

REPLACEMENT

4 Loosen the balljoint-to-steering knuckle nut a few turns, but don't remove it.

5 Break loose the balljoint stud from the steering knuckle with a balljoint removal tool (see illustration). Remove the nut and separate the balljoint from the knuckle.

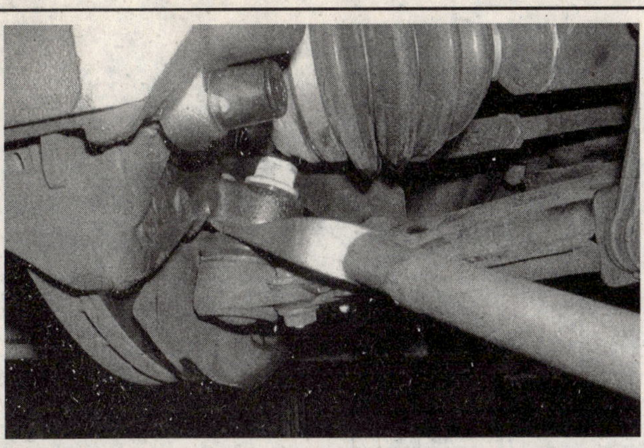

8.5 Separating the balljoint from the steering knuckle (the use of a "picklefork" type balljoint separator, shown here, will most likely damage the balljoint boot)

6 Mark the position of the balljoint bolt heads to the control arm. Unscrew the balljoint retaining bolts and remove the retaining plate from the top of the control arm (see illustrations). Detach the balljoint from the control arm.

7 Installation is the reverse of removal. Align the balljoint-to-control arm bolt heads with the marks made in Step 6, then tighten the balljoint-to-control arm bolts, and the balljoint stud nut, to the torque values listed in this Chapter's Specifications.

8 Install the wheel, lower the vehicle and tighten the wheel bolts to the torque listed in the Chapter 1 Specifications.

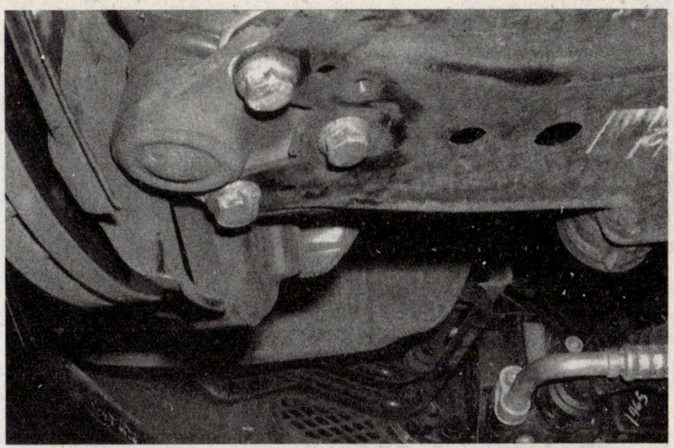

8.6a Mark the positions of the balljoint bolt heads on the lower control arm . . .

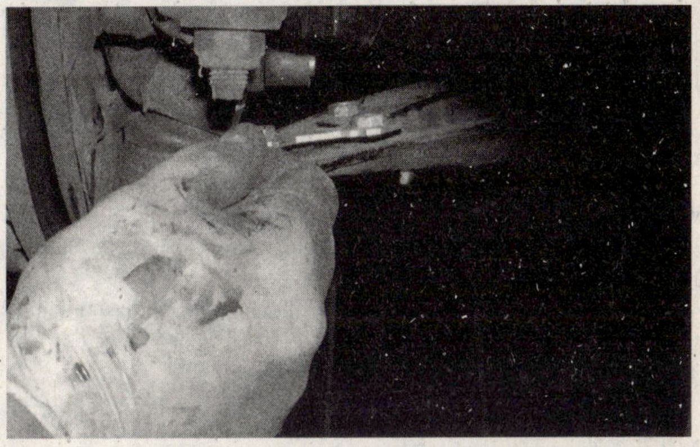

8.6b . . . then remove the bolts and the nut retaining plate

9 Shock absorber (rear) - removal and installation

➧ Refer to illustrations 9.3, 9.4 and 9.5

✻✻ WARNING 1:

Always replace the shock absorbers in pairs - never replace just one of them.

9.3 Shock absorber lower mounting bolt/nut

✻✻ WARNING 2:

The manufacturer recommends replacing the shock absorber upper mounting bolts, damper rod nut, and lower mounting bolt/nut with new ones whenever they are removed.

1 Loosen the rear wheel bolts. Chock the front wheels to keep the vehicle from rolling, then raise the rear of the vehicle and support it securely on jackstands placed underneath the rocker panel flanges. Remove the rear wheels.

2 Support the rear axle with a floor jack placed under the coil spring pocket.

✻✻ WARNING:

The jack must remain in this position until the shock absorber is reinstalled.

3 Remove the shock absorber lower mounting bolt/nut (see illustration).

4 Remove the shock absorber upper mounting bolts and remove

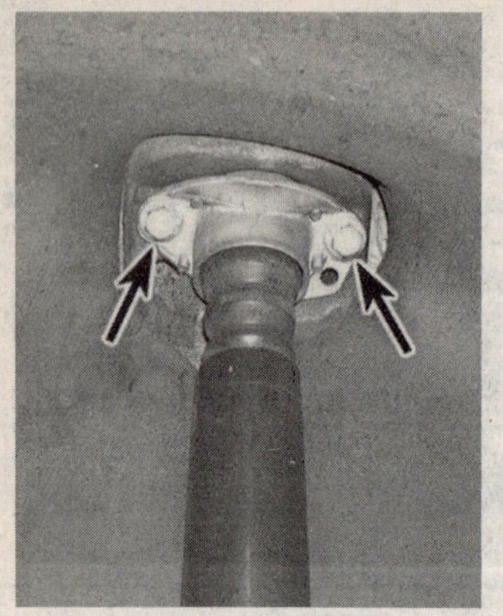

9.4 Shock absorber upper mounting bolts

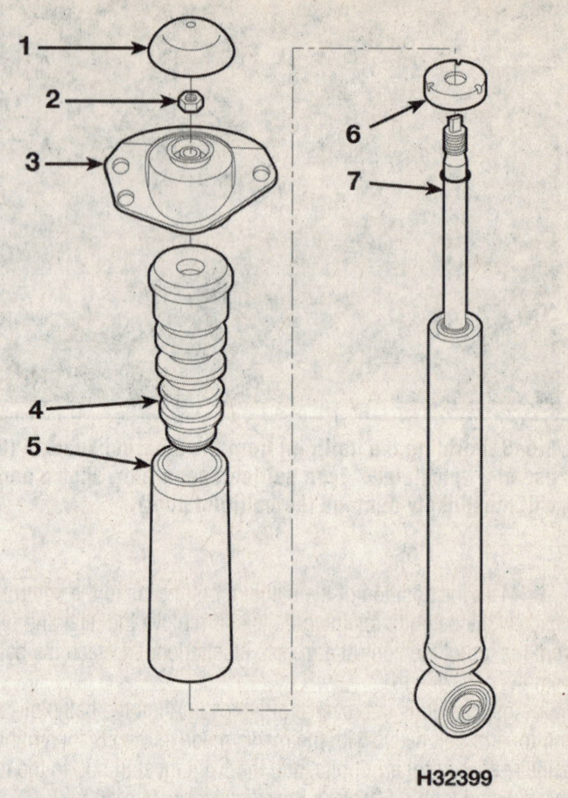

9.5 Shock absorber and upper mount - exploded view

1	*Cover*	5	*Boot*
2	*Damper rod nut*	6	*Cap*
3	*Upper mount*	7	*Shock absorber*
4	*Bump stop*		

the shock absorber (see illustration).

5 Remove the cover from the top of the shock absorber upper mount, then unscrew the damper rod nut (see illustration). It'll probably be necessary to hold the damper rod with another wrench to prevent it from turning when loosening the nut.

6 Remove the upper mount, bump stop, boot and cap from the damper rod.

7 Install the cap, boot, bump stop and upper mount on the new shock absorber. Install a new nut on the damper shaft and tighten it to the torque listed in this Chapter's Specifications. Install the cover on the upper mount.

8 Guide the shock absorber into position and install new upper mounting bolts and a new lower mounting bolt and nut. Don't tighten the lower mounting bolt/nut yet.

9 Tighten the upper mounting bolts to the torque listed in this Chapter's Specifications.

10 Raise the rear axle to simulate normal ride height, then tighten the lower mounting bolt/nut to the torque listed in this Chapter's Specifications.

11 Repeat the procedure to replace the other rear shock absorber.

12 Install the wheels and lower the vehicle. Tighten the wheel bolts to the torque listed in the Chapter 1 Specifications.

10 Coil spring (rear) - removal and installation

Refer to illustration 10.7

WARNING 1:

Always replace the coil springs in pairs - never replace just one of them.

WARNING 2:

The manufacturer recommends replacing the shock absorber upper mounting bolts with new ones whenever they are removed.

1 Loosen the wheel bolts. Chock the front wheels to prevent the vehicle from rolling, then raise the rear of the vehicle and support it securely on jackstands placed under the rocker panel flanges. Remove the wheel.

Note: It is not absolutely necessary to remove the wheels, but doing so greatly improves access to the springs.

2 Support the rear axle with a floor jack placed under one of the spring pockets. Raise the jack slightly to take the spring pressure off the shock absorber upper mount.

3 Remove the shock absorber upper mounting bolts (see illustration 9.4).

4 Slowly lower the floor jack; at this point the spring will not be fully extended, as the shock absorber on the other side is limiting the downward travel of the axle.

5 Place the floor jack under the other spring seat, remove the shock absorber upper mounting bolts, then slowly lower the jack until the coil

springs are fully extended.

6 Remove the spring, upper insulator and lower bushing. Check the spring for cracks and chips, replacing the springs as a set if any defects are found. Also check the upper insulator and the lower zinc bushing for damage and deterioration, replacing them as necessary.

7 Installation is the reverse of the removal procedure, but make sure the coil springs are positioned properly against their upper mounts (see illustration).

8 Tighten the shock absorber upper mounting bolts to the torque listed in this Chapter's Specifications. Tighten the wheel bolts to the torque listed in the Chapter 1 Specifications.

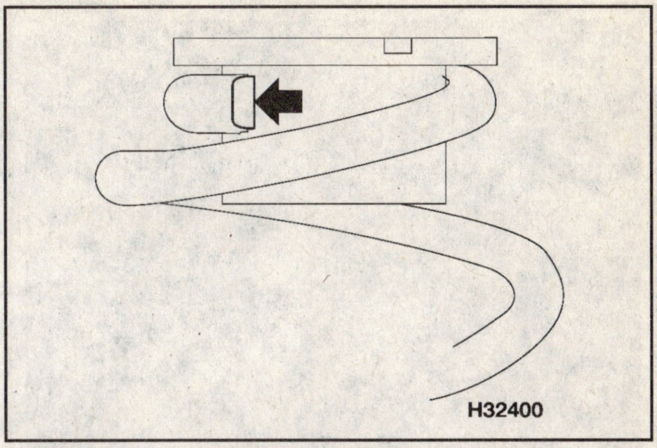

10.7 Make sure the upper ends of the coil springs are positioned like this, against the stop in the upper seat

11 Hub and wheel bearing assembly (rear) - removal and installation

➧ Refer to illustrations 11.3, 11.4, 11.5, 11.6a, 11.6b and 11.9

⁂ WARNING:

The manufacturer recommends replacing the hub nut whenever it is removed. The dust cap should also be replaced with a new one upon reassembly.

REMOVAL

1 Loosen the rear wheel bolts, raise the rear of the vehicle, support it securely on jackstands and remove the wheels.

2 Remove the brake caliper (don't detach the hose), mounting bracket and disc, or the brake drum (see Chapter 9). On models with rear disc brakes, hang the caliper with a piece of wire - don't let it hang by the brake hose.

3 Remove the dust cap (see illustration).

4 Remove the hub nut (see illustration).

5 Using a puller, remove the hub flange and outer bearing from the stub axle (see illustration). The inner bearing will most likely remain on the stub axle.

11.3 Remove the dust cap from the hub flange

11.4 Unscrew the hub nut

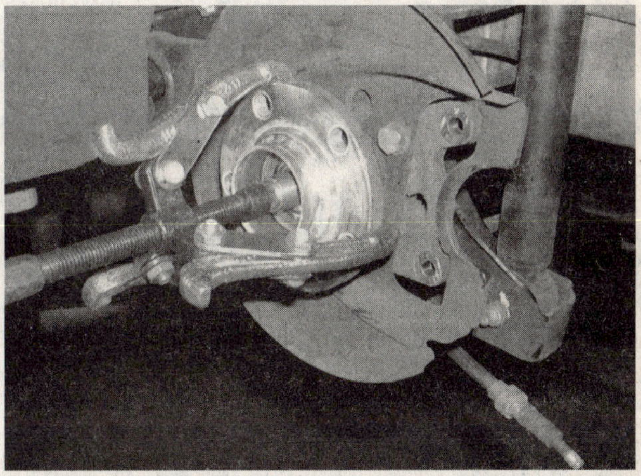

11.5 Removing the hub flange and outer bearing with a puller

11.6a Knock the inner race of the inner bearing away from the shoulder of the stub axle . . .

11.6b . . . then remove it the rest of the way with a puller

6 Using a hammer and chisel, drive the inner race of the inner bearing away from the shoulder of the stub axle, far enough to allow the jaws of a puller to grasp the inner race (see illustration). Now, using a puller, remove the inner bearing assembly from the stub axle (see illustration).

11.9 Drive the hub and bearing assembly onto the stub shaft with a large socket that contacts the inner race of the bearing

INSTALLATION

7 Before installing the hub and bearing assembly, clean the stub axle and apply a film of wheel bearing grease to the area on the stub axle where the bearings ride.

8 Push the hub and bearing assembly onto the stub axle as far as possible by hand. Make sure it goes on straight - not cocked to the side.

9 Using a hammer and a large socket that fits inside the hub opening and mates with the inner race of the outer bearing, drive the hub and bearing assembly onto the stub axle far enough to allow you to install the hub nut (see illustration).

10 Install the old hub nut onto the stub axle and tighten it to the torque listed in this Chapter's Specifications. This will push the hub and bearing assembly into place.

11 Remove the old hub nut and install the new one, tightening it to the torque listed in this Chapter's Specifications.

12 Install a new dust cap.

13 The remainder of installation is the reverse of removal. On models with rear disc brakes, tighten the caliper mounting bracket bolts and the caliper bolts to the torque listed in the Chapter 9 Specifications.

14 Install the wheel, lower the vehicle and tighten the wheel bolts to the torque listed in the Chapter 1 Specifications.

12 Stub axle (rear) - removal and installation

Refer to illustration 12.6

WARNING:

The manufacturer recommends replacing the hub nut and the stub axle mounting bolts with new ones whenever they are removed.

1 Loosen the wheel bolts. Chock the front wheels, then raise the rear of the vehicle and support it securely on jackstands. Remove the rear wheel.

2 Remove the hub and bearing assembly (see Section 11). On models with disc brakes, be sure to hang the caliper with a piece of wire - don't let it hang by the brake hose.

3 Inspect the stub axle surface for signs of damage such as scoring, and replace if necessary. Do not attempt to straighten the stub axle. Also check the threads; if they are damaged in any way, replace the stub axle. Don't attempt to "clean up" the threads with a thread file or a die.

4 If the vehicle is equipped with an Anti-lock Braking System (ABS), remove the rear wheel speed sensor (see Chapter 9).

5 If the vehicle is equipped with rear drum brakes, unscrew the brake line fitting from the wheel cylinder. Use a flare-nut wrench, if available, to prevent rounding-off the fitting.

6 Remove the stub axle mounting bolts (see illustration). Remove

the stub axle along with the disc splash shield or, on models with rear drum brakes, the brake backing plate/brake shoe assembly.

7 Installation is the reverse of removal, with the following points:

a) *Make sure the stub axle mating surfaces and the mounting bolt holes on the axle beam are clean.*
b) *Use new stub axle mounting bolts and tighten them to the torque listed in this Chapter's Specifications.*
c) *On models with rear drum brakes, tighten the brake line fitting securely.*
d) *Install a new hub nut and tighten it to the torque listed in this Chapter's Specifications.*
e) *On models with rear disc brakes, tighten the caliper mounting bracket bolts and the caliper bolts to the torque listed in the Chapter 9 Specifications.*
f) *On models with rear drum brakes, bleed the brake system (see Chapter 9).*

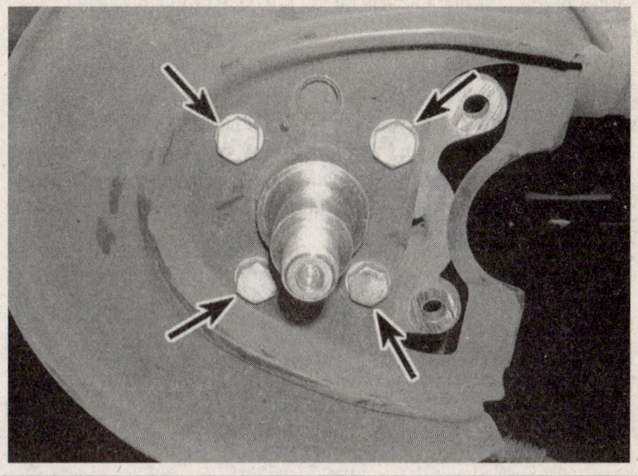

12.6 Stub axle mounting bolts

13 Stabilizer bar (rear) - removal and installation

The rear suspension stabilizer bar runs along the length of the axle beam. It is an integral part of the axle assembly, and cannot be removed. If the stabilizer bar is damaged, which is very unlikely, the complete axle assembly must be replaced.

14 Rear axle assembly - removal and installation

WARNING:

The manufacturer recommends replacing all suspension fasteners with new ones whenever they are removed.

REMOVAL

Refer to illustrations 14.13 and 14.15

1 Loosen the rear wheel bolts. Chock the front wheels, then raise the rear of the vehicle and support it securely on jackstands. Remove both rear wheels.

2 Referring to Chapter 9, fully loosen the parking brake cable adjuster nut.

3 On models with rear drum brakes, remove the brake drums.

4 Loosen the parking brake cable adjuster nut, then detach the parking brake cables from the calipers or, on models with rear drum brakes, from the parking brake levers (see Chapter 9).

5 On models with rear disc brakes, remove the brake calipers, mounting brackets and discs (see Chapter 9).

6 On models equipped with an Anti-lock Braking System (ABS), remove the rear wheel speed sensors.

7 If the axle is to be replaced with a new one, remove the hub and bearing assemblies (see Section 11) and the stub axles (see Section 12).

8 Detach the parking brake cables from the retaining clips along the axle trailing arms.

9 Unscrew the brake line fittings from the brake hoses on each side of the rear axle (near the pivot bushings). Use a flare-nut wrench, if available, to prevent rounding-off the fittings (see Chapter 9). Plug the hoses to minimize fluid loss and prevent the entry of dirt into the hydraulic system. If the rear axle is to be replaced with a new one, detach the brake lines from the clips on the axle.

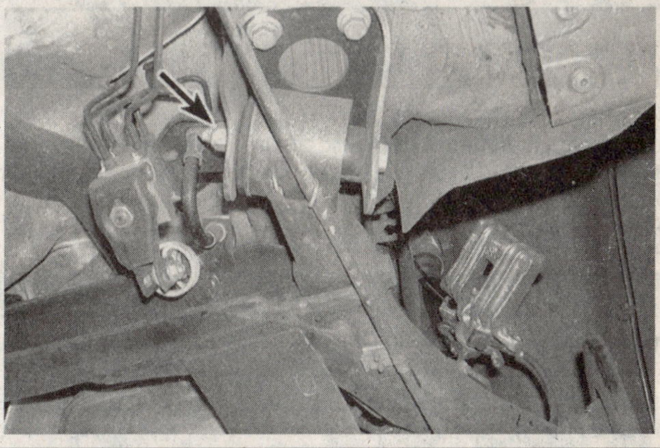

14.13 Rear axle pivot bolt/nut (the nut should only be tightened with the suspension at normal ride height)

10 On models without ABS, unbolt the brake pressure regulator spring from the left side of the axle beam (see Chapter 9). Also unbolt the valve from its bracket (it isn't necessary to detach the brake lines from it, but unbolting it from the bracket will help prevent damage to the valve as the axle is lowered).

11 Remove the coil springs (see Section 10).

12 Make a final check that all necessary components have been disconnected and positioned so that they will not hinder the removal procedure, then position a floor jack beneath the center of the rear axle assembly. Raise the jack until it is just supporting the weight of the axle.

13 Remove the nuts from the axle pivot bolts on each side of the vehicle, then remove the bolts (see illustration). Slowly lower the jack,

being careful not to let the axle hit the pressure regulator valve (models without ABS only).

14 Inspect the axle pivot bushings for signs of damage or deterioration. If replacement is necessary, they can be removed with a drawbolt-type bushing replacement tool (and suitable adapters) or a slide hammer, and installed with the drawbolt-type bushing replacement tool (and suitable adapters). If you don't have access to the necessary tools, take the axle to an automotive machine shop to have the old bushings removed and the new ones installed.

15 If the bushings are in need of replacement, make sure the new ones are installed properly (see illustration).

16 Don't unbolt the rear axle mounting brackets from the floor pan unless they are damaged (bent, cracked, or if the pivot bolt holes are worn). If replacement is necessary, mark the position of the bracket to the floor pan and install the new bracket in the same position (this will preserve rear wheel alignment).

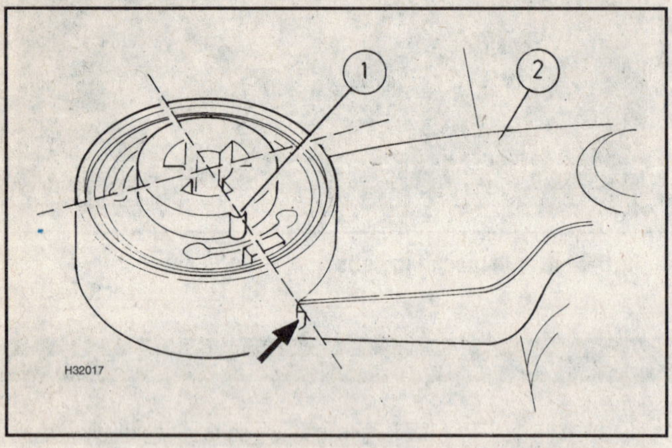

14.15 The rear axle pivot bushing must be installed with the mark (1) pointing toward the area where the trailing arm portion of the axle (2) connects to the bushing boss (arrow)

INSTALLATION

17 Installation of the rear axle is the reverse of the removal procedure, with the following points:

a) *Replace all suspension fasteners with new ones.*
b) *Ensure that the brake lines, parking brake cables and wiring (as applicable) are correctly routed, and retained by all the necessary retaining clips.*
c) *Don't tighten the rear axle pivot bolts or the shock absorber lower mounting bolts until the weight of the vehicle is on its wheels (or until the rear suspension has been raised to simulate normal ride height). This will prevent the bushings from "winding-up," which could eventually damage them.*
d) *Tighten all fasteners to the proper torque Specifications.*
e) *Bleed the brake system (see Chapter 9).*
f) *If either mounting bracket was removed from the floor pan, have the rear wheel alignment checked and, if necessary, adjusted.*

15 Steering wheel - removal and installation

⁂ WARNING 1:

These models are equipped with airbags. Always disable the airbag system before working in the vicinity of any airbag system component to avoid the possibility of accidental deployment of the airbag(s), which could cause personal injury (see Chapter 12).

⁂ WARNING 2:

Do not use a memory saving device to preserve the ECM's memory when working on or near airbag system components.

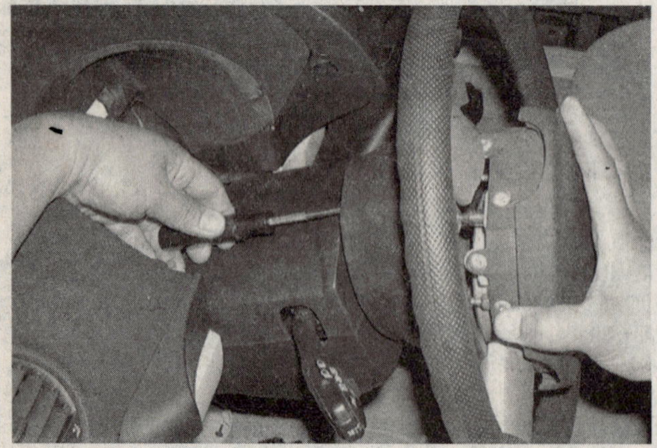

15.3a Insert a screwdriver into the hole in the steering wheel and pry up, which will release the clip securing one side of the airbag module (repeat this on the other side)

⁂ CAUTION 1:

These models are equipped with an anti-theft radio. Before performing a procedure that requires disconnecting the battery, make sure you have the activation code.

⁂ CAUTION 2:

Disconnecting the battery can cause driveability problems that require a scan tool to remedy. If the vehicle exhibits driveability problems after the battery is reconnected, it may be necessary to take it to a dealer service department or other qualified repair shop to reprogram the ECM.

REMOVAL

➧ Refer to illustrations 15.3a, 15.3b, 15.4, 15.6, 15.7a and 15.7b

1 Park the vehicle with the wheels pointing straight ahead. Disconnect the cable from the negative terminal of the battery.

2 Refer to Chapter 12 and disable the airbag system, then remove the instrument cluster (also in Chapter 12).

3 Turn the steering wheel 90-degrees to gain access to the hole in the backside (the side facing the dash) of the steering wheel. Insert a screwdriver into the hole for the spring clip that retains the airbag module and push the spring aside to release the pin (see illustrations). Now turn the steering wheel 180-degrees in the other direction and do the same thing to release the other pin.

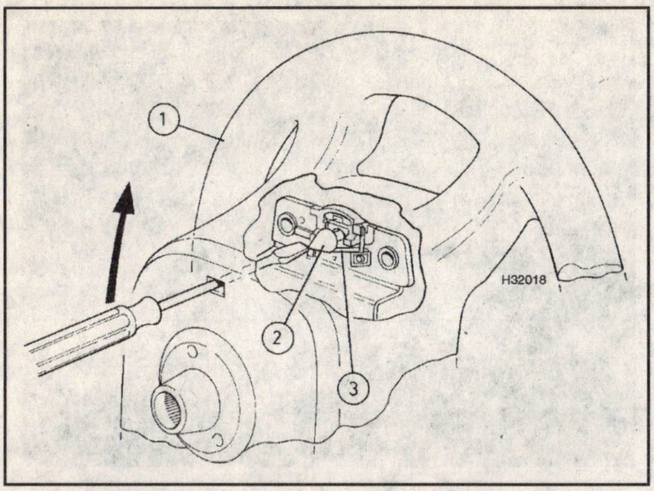

15.3b Here's a ghost view showing the screwdriver prying the spring clip away from the airbag module retaining post

1 Steering wheel 2 Post 3 Spring clip

➡Note: This step can be very difficult and frustrating. Take your time and be careful not to tear the rubber steering wheel trim when prying.

4 Disconnect the electrical connector from the airbag module (see illustration). Set the module aside in a safe, isolated area, with the airbag side of the module facing UP.

⁂ WARNING:

When carrying the airbag module, keep the driver's (trim) side of it away from your body and, when you set it down, make sure the driver's side is facing up.

5 Center the steering wheel.

6 Unplug the airbag/horn wiring harness from the clockspring and airbag securing plate (see illustration).

7 Using a 12 mm 12-point spline-drive bit, remove the steering wheel retaining bolt and mark the position of the steering wheel to the shaft, if marks don't already exist or don't line up (see illustrations).

15.7a Unscrew the steering wheel bolt with a 12 mm, 12-point spline-drive bit

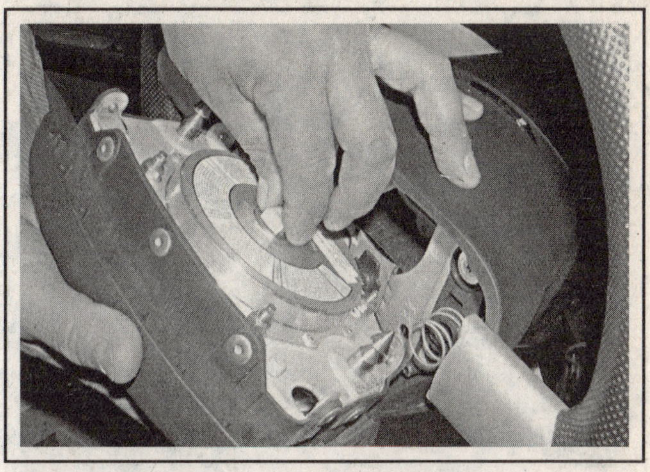

15.4 Pull the electrical connector straight out of the airbag module

15.6 Unplug the wiring harness from the airbag clockspring and securing plate

8 Remove the steering wheel from the shaft. If it is tight, tap it up near the center using the palm of your hand, or twist it from side-to-side while pulling upwards.

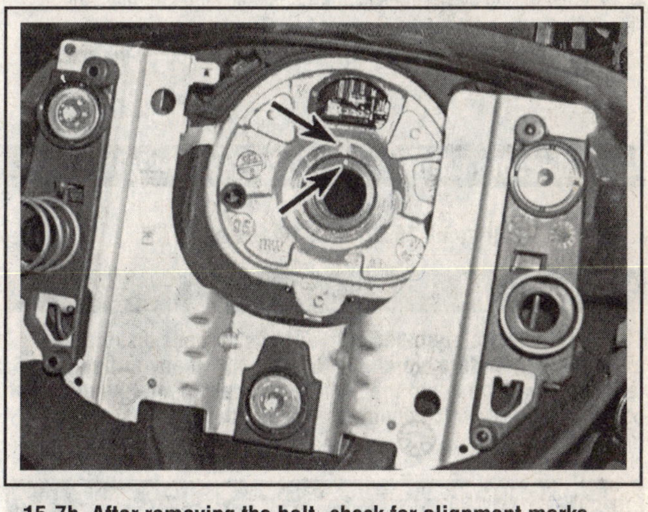

15.7b After removing the bolt, check for alignment marks on the steering wheel and steering shaft - if there aren't any, make your own

15.10 After centering the clockspring, make sure the yellow flag appears in the window

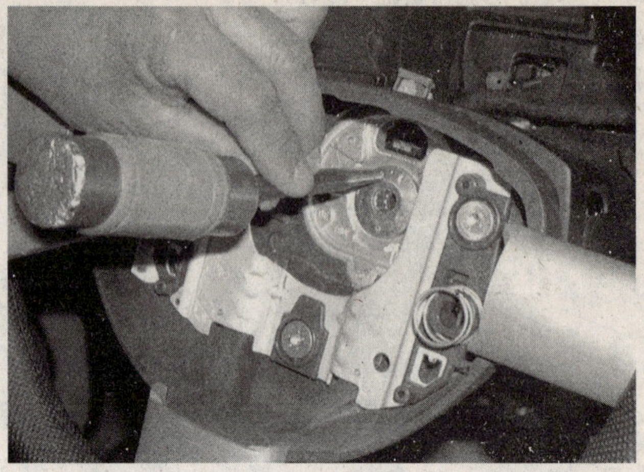

15.14 Make a punch mark on the steering wheel bolt after installing it (it can be used up to five times)

⁂ CAUTION:

Don't hammer on the shaft to remove the wheel.

9 Lift the steering wheel from the shaft.

⁂ WARNING:

Don't allow the steering shaft to turn with the steering wheel removed. If the shaft turns, the airbag clockspring will become uncentered, which may cause the wire inside to break when the vehicle is returned to service.

INSTALLATION

➧ Refer to illustrations 15.10 and 15.14

10 Before installing the steering wheel, make sure the airbag clockspring is centered (see illustration).

11 If the airbag system clockspring is not centered, remove the steering column covers (see Chapter 11). Release the locking tabs, unplug the electrical connector and lift the clockspring off the steering column. Unplug the electrical connector.

12 To center the clockspring, depress the spring-loaded plunger and turn the hub in either direction until it stops (don't apply too much force). Now, rotate the hub in the other direction, counting the number of turns it takes to reach the opposite stop. Divide that number by two, then turn the hub back that many turns, approximately, until the yellow flag appears in the window (see illustration 15.10). Install the clockspring, making sure the locking tabs engage securely. Install the steering column covers.

13 Install the wheel on the steering shaft, aligning the marks.

14 Install the steering wheel bolt and tighten it to the torque listed in this Chapter's Specifications.

➡Note: This bolt can be re-used up to five times. Using a center punch, place a mark on the head of the bolt after you tighten it, to keep track of how many times the bolt has been tightened (see illustration).

15 Plug in the electrical connector for the airbag and horn into the clockspring. Attach the ground leads to the airbag securing plate.

16 Connect the airbag connector to the back of the airbag module.

17 Position the airbag module on the steering wheel and push it in until the pins on the module engage with the spring clips.

18 Refer to Chapter 12 for the procedure to enable the airbag system.

16 Steering column - removal and installation

➧ Refer to illustrations 16.3, 16.9, 16.10 and 16.11

⁂ WARNING 1:

These models are equipped with airbags. Always disable the airbag system before working in the vicinity of any airbag system component to avoid the possibility of accidental deployment of the airbag(s), which could cause personal injury (see Chapter 12).

⁂ WARNING 2:

Do not use a memory saving device to preserve the ECM's memory when working on or near airbag system components.

⁂ CAUTION 1:

These models are equipped with an anti-theft radio. Before performing a procedure that requires disconnecting the battery, make sure you have the activation code.

⁂ CAUTION 2:

Disconnecting the battery can cause driveability problems that require a scan tool to remedy. If the vehicle exhibits driveability problems after the battery is reconnected, it may be necessary to take it to a dealer service department or other qualified repair shop to reprogram the ECM.

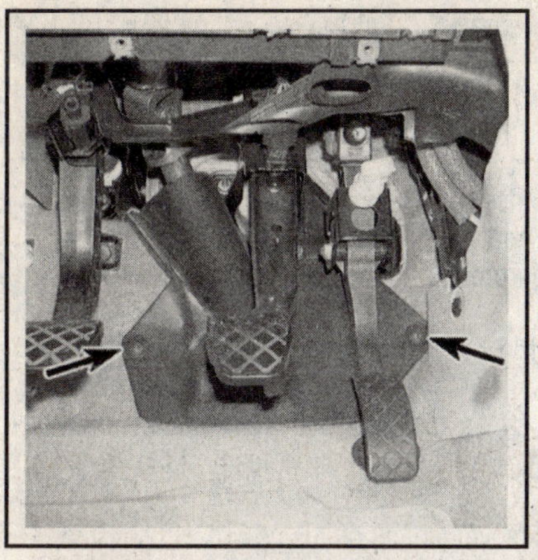

16.3 Unscrew these fasteners and remove the cover for access to the steering shaft universal joint

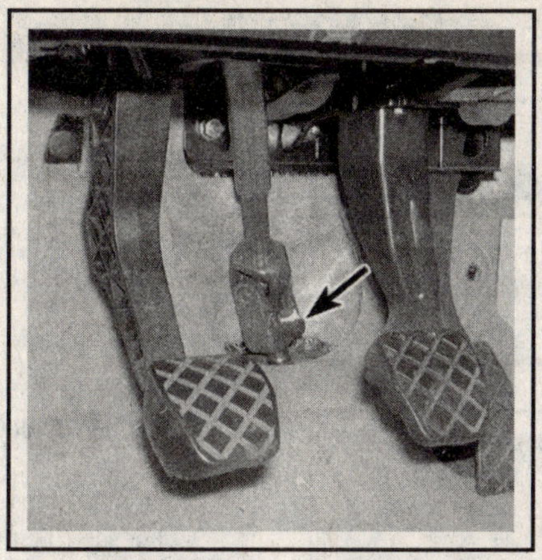

16.9 Remove the pinch bolt from the steering shaft U-joint

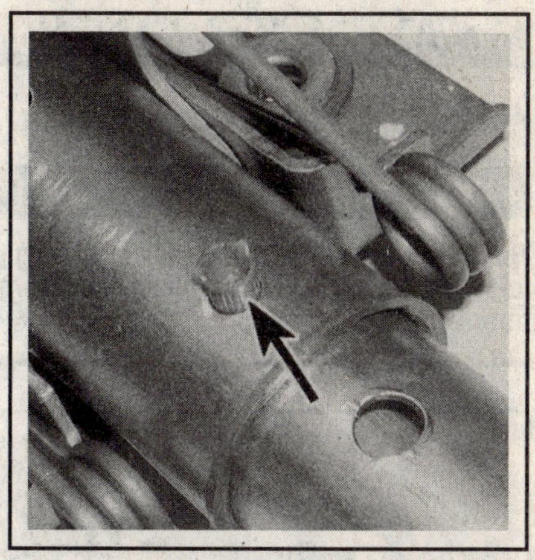

16.10 Align the holes in the steering shafts, then insert a clip or pin to hold them together

REMOVAL

1 Park the vehicle with the wheels in the straight-ahead position. Disconnect the cable from the negative terminal of the battery. Disable the airbag system (see Chapter 12).

2 Remove the steering wheel (see Section 15).

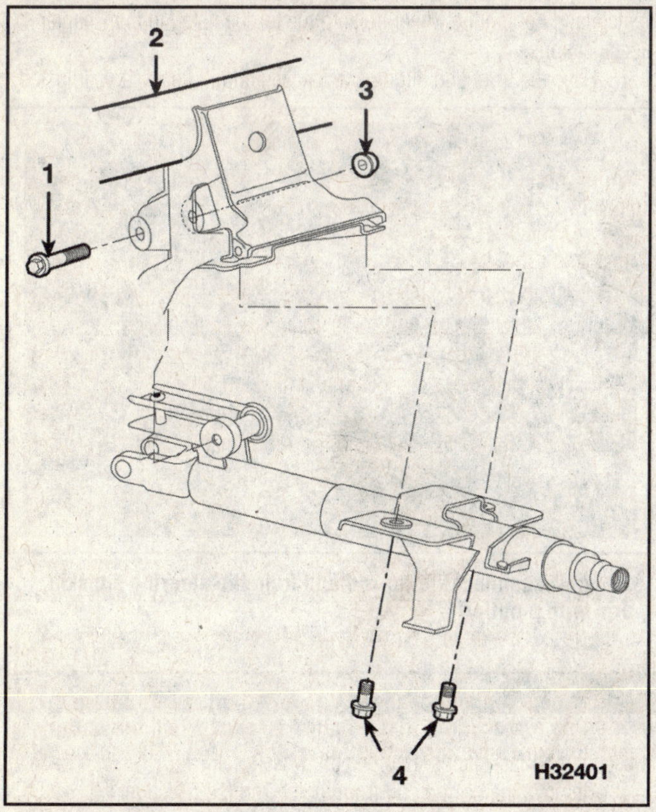

16.11 Steering column mounting details

1 Lower mounting bolt
2 Support brace
3 Nut
4 Upper mounting bolts

3 Remove the lower instrument panel trim (under the steering column) (see Chapter 11). Also remove the cover from the bottom of the steering column, below the brake pedal (see illustration).

4 Remove the steering column covers (see Chapter 11).

5 Remove the airbag system clockspring (see Section 15).

6 Remove the multi-function switch (see Chapter 12).

7 Remove the plastic cover from over the ignition switch shear-head bolts (see Chapter 12).

8 On models equipped with an automatic transmission, detach the shift interlock cable from the ignition switch (see Chapter 7B).

9 Mark the relationship of the steering shaft lower universal joint to the steering gear input shaft, then remove the pinch bolt from the universal joint (see illustration). Separate the U-joint from the steering gear input shaft.

10 Push or pull the lower steering shaft in-or-out of the upper portion of the shaft until the holes in the shaft are aligned. Insert a pin or clip into the hole to hold the shafts in this relationship (see illustration).

11 Remove the steering column mounting bolts (see illustration), then guide the column out from the instrument panel.

INSTALLATION

12 Guide the column into position, connecting the U-joint with the steering gear input shaft.

13 Install the mounting bolts. Tighten the lower mounting bolt/nut to the torque listed in this Chapter's Specifications.

14 Tighten the upper mounting bolts to the torque listed in this Chapter's Specifications.

15 Remove the clip installed in Step 10. If a new steering column was installed, remove the plastic clip holding the shafts in proper relationship.

16 Tighten the U-joint pinch bolt to the torque listed in this Chapter's Specifications.

17 The remainder of installation is the reverse of the removal procedure. Be sure the airbag clockspring is centered, and tighten the steering wheel bolt to the torque listed in this Chapter's Specifications.

17 Tie-rod ends - removal and installation

➧ Refer to illustrations 17.2, 17.3a and 17.3b

WARNING:

The manufacturer recommends replacing the tie-rod end-to-steering knuckle nuts with new ones whenever they are removed.

17.2 Hold the tie-rod still with a wrench, then break loose the jam nut

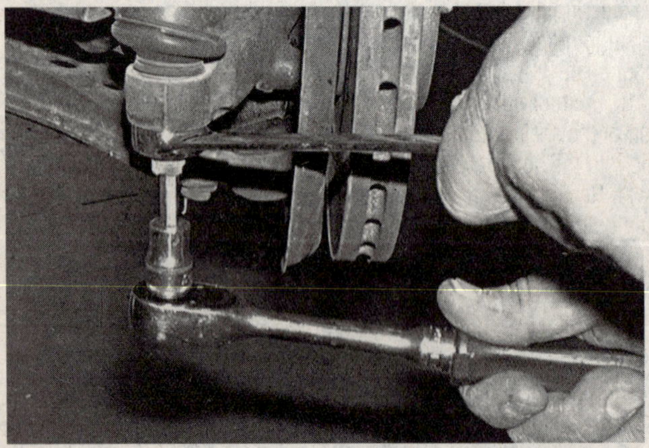

17.3a If the ballstud turns while attempting to loosen the nut, hold it with a hex bit

REMOVAL

1 Loosen the wheel bolts, raise the front of the vehicle and support it securely on jackstands. Apply the parking brake and block the rear wheels to keep the vehicle from rolling off the jackstands. Remove the wheel.

2 Break loose the tie-rod end jam nut (see illustration). Don't back the nut off; once it has just been loosened, it will serve as the point to which the tie-rod end will be threaded. If you are removing the tie-rod end to replace the steering gear boot, mark the threads of the tie-rod on the inner side of the nut.

3 Loosen (but don't remove) the nut on the tie-rod end ballstud, then disconnect the tie-rod end from the steering knuckle arm with a puller (see illustrations).

4 Unscrew the tie-rod end from the tie-rod.

INSTALLATION

5 If the jam nut was removed, thread it onto the tie-rod until it meets the mark applied in Step 2. Thread the tie-rod end onto the tie-rod until it contacts the jam nut, then connect the tie-rod end to the steering arm. Install the ballstud nut and tighten it to the torque listed in this Chapter's Specifications.

6 Tighten the jam nut securely and install the wheel. Lower the vehicle and tighten the wheel bolts to the torque listed in the Chapter 1 Specifications.

7 Have the front end alignment checked and, if necessary, adjusted.

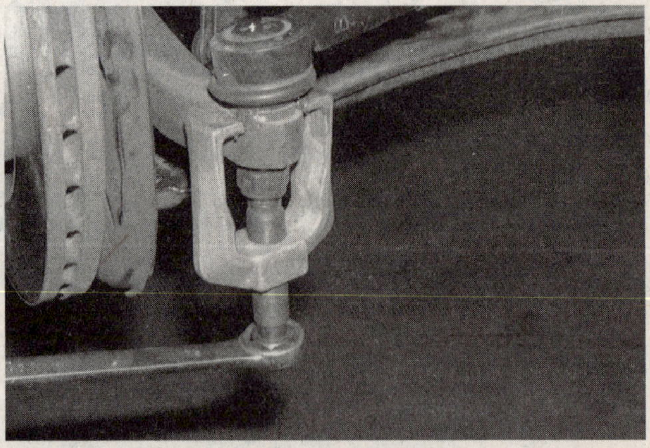

17.3b Disconnect the tie-rod end from the steering knuckle arm with a puller

18 Steering gear boots - replacement

➧ Refer to illustrations 18.4 and 18.8

WARNING:

The manufacturer recommends replacing the tie-rod end-to-steering knuckle nuts with new ones whenever they are removed.

1 Loosen the wheel bolts, raise the front of the vehicle and support it securely on jackstands. Remove the wheels.

2 Remove the tie-rod end from the tie-rod (see Section 17).

3 Remove the tie-rod end jam nut.

4 Remove the inner and outer boot clamps and discard them (see illustration). The clamps can be pried apart at the crimped area, or cut off with a pair of cutting pliers.

5 Remove the boot.

6 Install a new clamp on the inner end of the boot.

7 Slide the boot into place.

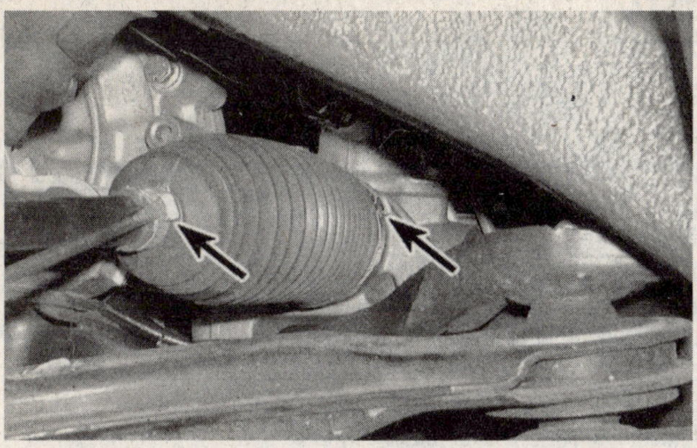

18.4 Pry open or cut off the steering gear boot clamps

8 Make sure the boot isn't twisted, then tighten the inner clamp with a pair of clamp crimping pliers (see illustration).

9 Install and tighten the outer clamp.

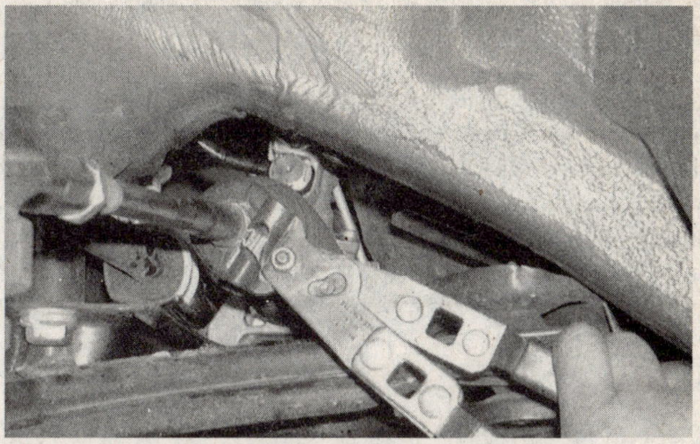

18.8 Clamp crimping pliers, available at most auto parts stores, are needed to tighten the boot clamps properly

10 Install the tie-rod ends (see Section 17).

11 Have the front end alignment checked and, if necessary, adjusted.

19 Steering gear - removal and installation

⁂ WARNING 1:

These models are equipped with airbags. Always disable the airbag system before working in the vicinity of the any airbag system component to avoid the possibility of accidental deployment of the airbag, which could cause personal injury (see Chapter 12). Also, don't allow the steering wheel to turn after the steering gear has been removed. To prevent this, pass the seat belt through the steering wheel and plug it into its latch.

⁂ WARNING 2:

The manufacturer recommends replacing the subframe mounting bolts, tie-rod end-to-steering knuckle nuts, steering gear retaining nuts, and steering shaft U-joint pinch bolt with new ones whenever they are removed.

REMOVAL

➧ Refer to illustrations 19.7 and 19.8

1 Loosen the wheel bolts. Chock the rear wheels, firmly apply the parking brake, then raise the front of the vehicle and support it securely on jackstands. Remove both front wheels.

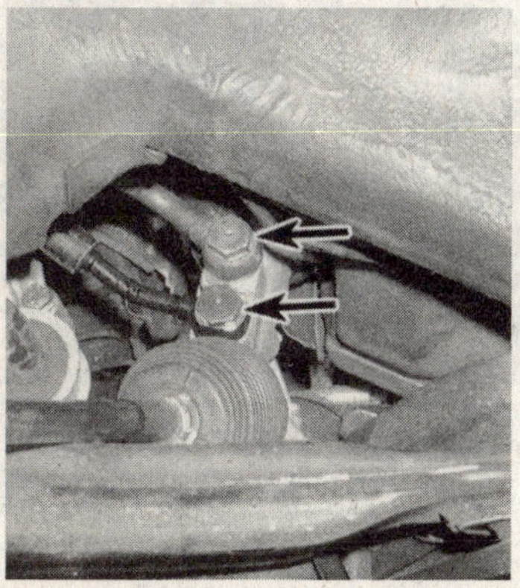

19.7 Unscrew the banjo bolts from the pressure and return line fittings at the steering gear, then plug the fittings to prevent leakage and the entry of contaminants

2 Detach the tie-rod ends from the steering knuckles (see Section 17).

3 Remove the cover from the bottom of the steering column, below the brake pedal. Mark the relationship of the steering shaft lower universal joint to the steering gear input shaft, then remove the pinch bolt from the universal joint. Separate the U-joint from the steering gear input shaft (see Section 16).

4 Remove the under-vehicle splash shield.

5 Unbolt the transmission support from the subframe and the transmission (see Chapter 7A).

6 If the vehicle is equipped with a diesel engine, unbolt the exhaust system from the subframe.

7 Remove the power steering pressure and return line banjo bolts (see illustration). Detach the lines from the steering gear and plug them or wrap them tightly with plastic bags and tape to prevent fluid leakage and contamination. Discard the sealing washers - new ones should be used during installation.

8 Loosen the steering gear mounting bolts (see illustration).

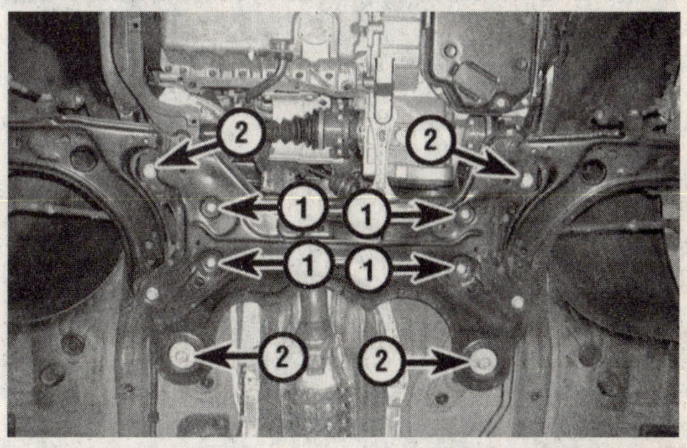

19.8 The power steering gear is bolted to the subframe

1 Steering gear mounting bolts
2 Subframe mounting bolts

9 Support the subframe with a jack - preferably a floor jack equipped with a transmission jack head adapter.

10 Remove the subframe mounting bolts (see illustration 19.8). Carefully lower the subframe enough to maneuver the steering gear out. If it won't lower far enough to get the steering gear out, unbolt the stabilizer bar link from the control arms (see Section 6).

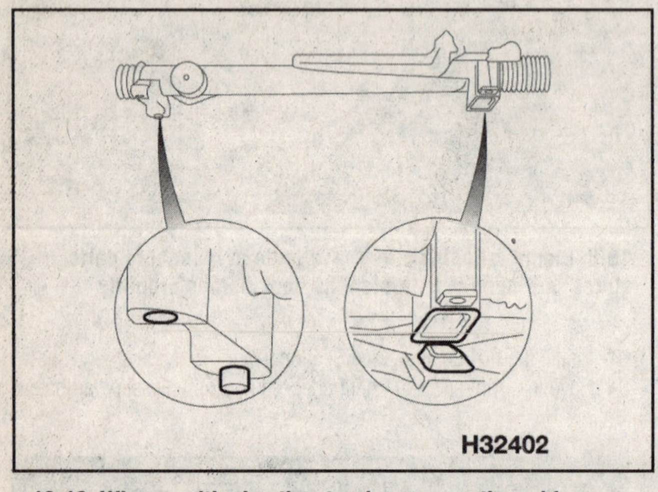

19.12 When positioning the steering gear on the subframe, the threaded post on the left side mount must engage with the hole in the subframe, and the right side of the steering gear must engage with the raised rectangle

INSTALLATION

Refer to illustration 19.12

11 Make sure the mounting areas on the subframe are clean. Also make sure the steering gear rack is in the centered position. Inspect the gasket around the steering gear input shaft, replacing it if necessary.

12 Guide the rack into position on the subframe, then install the bolts finger tight. The threaded sleeve at the left front mounting bolt hole of the steering gear must be engaged with the hole in the subframe, and the right side steering gear mount must be engaged with the rectangular protrusion on the subframe (see illustration).

13 Raise the subframe into position and install the mounting bolts, tightening them to the torque listed in this Chapter's Specifications.

14 Tighten the steering gear mounting bolts to the torque listed in this Chapter's Specifications.

15 Using new sealing washers, connect the power steering pressure and return lines to the steering gear.

Note: The manufacturer recommends lubricating the sealing washers with liquid soap before installation. Tighten the banjo fitting bolts to the torque listed in this Chapter's Specifications.

16 The remainder of installation is the reverse of removal, with the following points:

a) *Tighten all fasteners to the recommended torque specifications.*
b) *Bleed the power steering system (see Section 21).*
c) *Have the front end alignment checked and, if necessary, adjusted.*

20 Power steering pump - removal and installation

Refer to illustrations 20.4, 20.5, 20.6a, 20.6b and 20.6c

Note: On some models the power steering pump is mounted to the lower portion of the mounting bracket on the front of the engine. On other models it's mounted to the upper portion of the bracket.

1 Loosen the right front wheel bolts. Raise the front of the vehicle and support it securely on jackstands. Remove the wheel.

2 Remove inner fender panel (see Chapter 11).

3 Remove the serpentine drivebelt (see Chapter 1).

4 Remove the bolts and detach the pulley from the pump (see illustration). To prevent the pulley from rotating, insert a screwdriver through one of the slots in the pulley and brace it against the pump, or hold the pulley by inserting the proper-sized hex key in the center bolt.

5 Clamp the pressure and feed lines shut, then detach them from the power steering pump (see illustration).

6 Remove the pump mounting bolts (see accompanying illustrations and illustration 20.5) and detach the pump from the mounting bracket.

20.4 The pulley is held to the power steering pump with three bolts

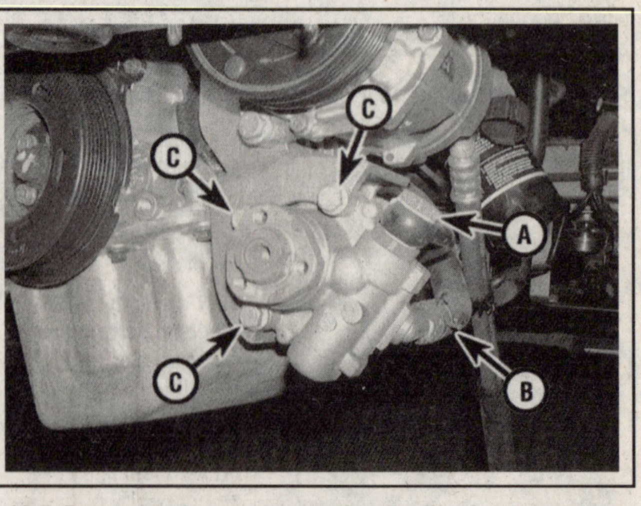

20.5 Power steering pump mounting details

A *Pressure line banjo fitting*
B *Return hose clamp*
C *Mounting bolts (3 of 4)*

7 Installation is the reverse of removal, with the following points:

a) *Use new sealing washers on either side of the pressure line banjo fitting. The manufacturer recommends lubricating these washers with liquid soap before installation.*

b) *Tighten mounting fasteners and the pressure line banjo fitting bolt to the torque values listed in this Chapter's Specifications.*

c) *Tighten the wheel bolts to the torque listed in the Chapter 1 Specifications.*

d) *Bleed the power steering system (see Section 21).*

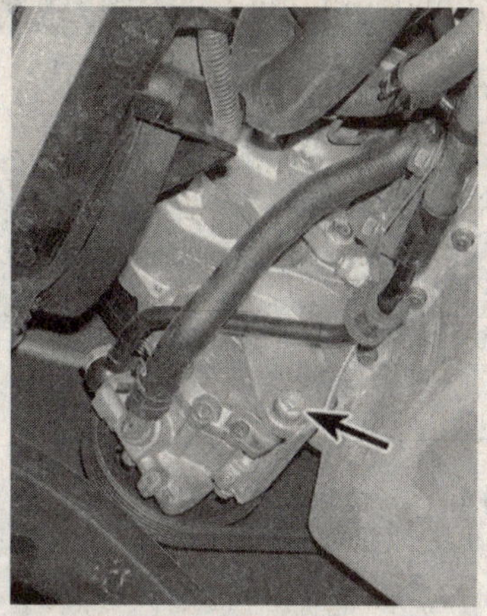

20.6a There's one more mounting bolt on the back of the pump

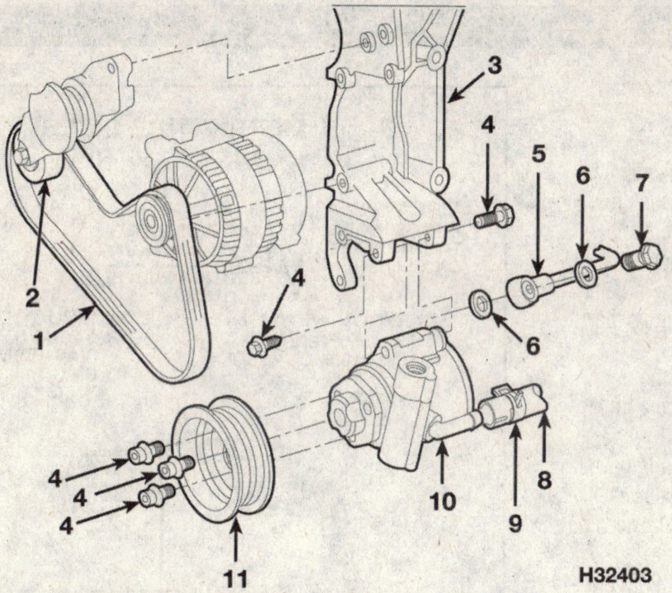

20.6b Power steering pump mounting details - low-mounted pump

1 *Drivebelt*
2 *Tensioner*
3 *Mounting bracket*
4 *Bolt*
5 *Pressure line*
6 *Sealing washers*
7 *Banjo fitting bolt*
8 *Return hose*
9 *Hose clamp*
10 *Power steering pump*
11 *Pulley*

20.6c Power steering pump mounting details - high-mounted pump

1 *Bolt*
2 *Drivebelt*
3 *Tensioner*
4 *Mounting bracket*
5 *Power steering pump*
6 *Return hose fitting*
7 *Banjo fitting bolt*
8 *Sealing washers*
9 *Pressure line*
10 *Pulley*

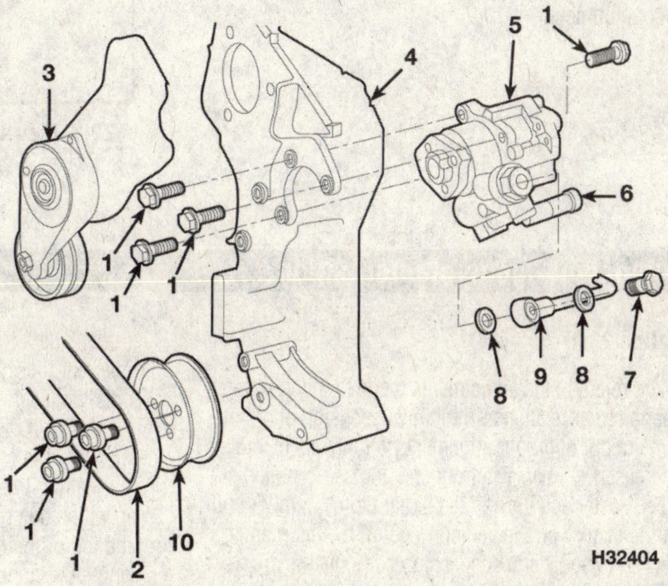

21 Power steering system - bleeding

1 Following any operation in which the power steering fluid lines have been disconnected, the power steering system must be bled to remove all air and obtain proper steering performance.

2 With the front wheels in the straight ahead position, check the power steering fluid level and, if low, add fluid until it is between the Cold marks on the reservoir.

3 Raise the front of the vehicle and support it securely on jackstands.

4 Turn the steering wheel back-and-forth repeatedly, lightly hitting the stops.

5 Start the engine and allow it to run at fast idle. Recheck the fluid level and add more if necessary to reach the Cold marks.

6 Bleed the system by turning the wheels from side to side, just barely contacting the stops. This will work the air out of the system. Keep the reservoir full of fluid as this is done.

7 When the air is worked out of the system and the fluid level stabilizes, return the wheels to the straight ahead position and leave the vehicle running for several more minutes before shutting it off. Lower the vehicle.

8 Road test the vehicle to be sure the steering system is functioning normally and noise-free.

9 Recheck the fluid level to be sure it is between the Hot marks on the reservoir while the engine is at normal operating temperature. Add fluid if necessary (see Chapter 1).

22 Wheels and tires - general information

Refer to illustration 22.1

1 All vehicles covered by this manual are equipped with metric-sized steel belted radial tires (see illustration). Use of other size or type of tires may affect the ride and handling of the vehicle. Don't mix different types of tires, such as radials and bias belted, on the same vehicle as handling may be seriously affected. It's recommended that tires be replaced in pairs on the same axle, but if only one tire is being replaced, be sure it's the same size, structure and tread design as the other.

2 Because tire pressure has a substantial effect on handling and wear, the pressure on all tires should be checked at least once a month or before any extended trips (see Chapter 1).

3 Wheels must be replaced if they are bent, dented, leak air, have elongated bolt holes, are heavily rusted, out of vertical symmetry or if the lug bolts won't stay tight. Wheel repairs that use welding or peening are not recommended.

4 Tire and wheel balance is important in the overall handling, braking and performance of the vehicle. Unbalanced wheels can adversely affect handling and ride characteristics as well as tire life. Whenever a tire is installed on a wheel, the tire and wheel should be balanced by a shop with the proper equipment.

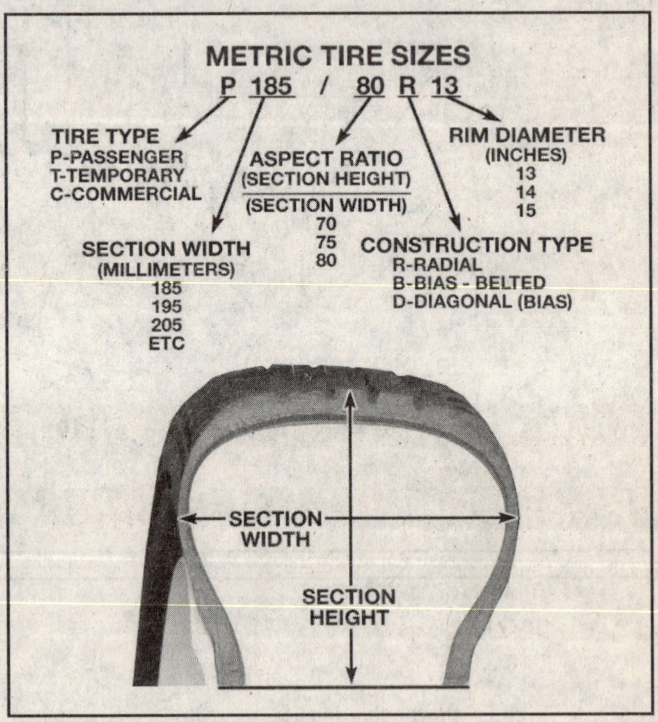

22.1 Metric tire size code

23 Wheel alignment - general information

Refer to illustration 23.1

A wheel alignment refers to the adjustments made to the wheels so they are in proper angular relationship to the suspension and the ground. Wheels that are out of proper alignment not only affect vehicle control, but also increase tire wear. The front end angles normally measured are camber, caster and toe-in (see illustration). Camber and caster are preset at the factory on the vehicle covered by this manual, but are usually checked to see if any suspension components are worn or damaged; front toe-in is the only adjustable angle on these vehicles. The rear toe-in should also be checked to make sure it is equal on each side, but each wheel is not adjustable individually (the entire axle can be shifted if the toe angles are not equal. Camber and caster are usually measured to check for a bent axle.

Getting the proper wheel alignment is a very exacting process, one in which complicated and expensive machines are necessary to perform the job properly. Because of this, you should have a technician with the proper equipment perform these tasks. We will, however, use this space to give you a basic idea of what is involved with a wheel alignment so you can better understand the process and deal intelligently with the shop that does the work.

Toe-in is the turning in of the wheels. The purpose of a toe specification is to ensure parallel rolling of the wheels. In a vehicle with zero toe-in, the distance between the front edges of the wheels will be the same as the distance between the rear edges of the wheels. The actual amount of toe-in is normally only a fraction of an inch. On the front end, toe-in is controlled by the tie-rod end position on the tie-rod. Incorrect toe-in will cause the tires to wear improperly by making them scrub against the road surface.

Camber is the tilting of the wheels from vertical when viewed from one end of the vehicle. When the wheels tilt out at the top, the camber is said to be positive (+). When the wheels tilt in at the top the camber is negative (-). The amount of tilt is measured in degrees from vertical and this measurement is called the camber angle. This angle affects the amount of tire tread which contacts the road and compensates for changes in the suspension geometry when the vehicle is cornering or traveling over an undulating surface.

Caster is the tilting of the front steering axis from the vertical. A tilt toward the rear is positive caster and a tilt toward the front is negative caster.

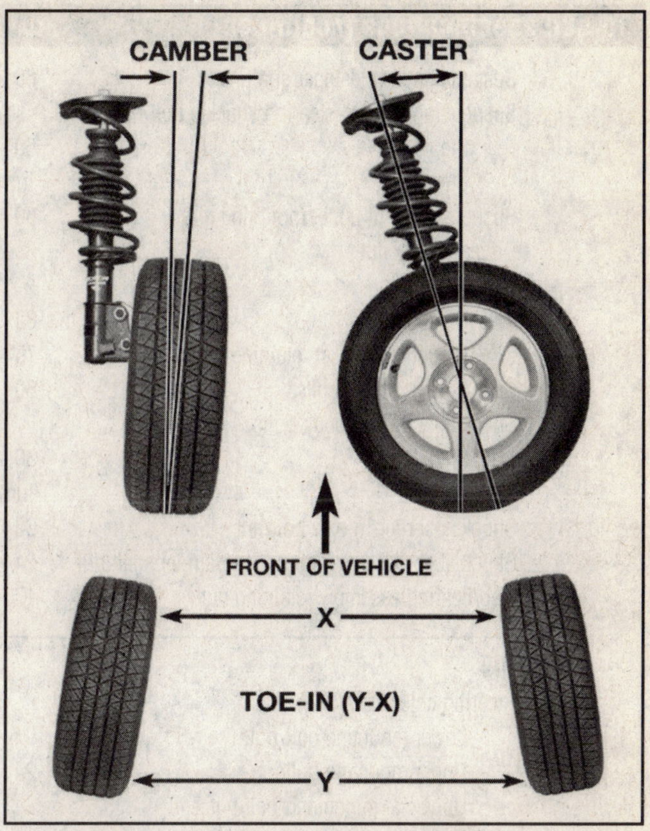

23.1 Camber, caster and toe-in angles

Specifications

Torque specifications	Nm	Ft-lbs (unless otherwise indicated)
Front suspension		
Stabilizer bar		
Link-to stabilizer bar bolt/nut	30	22
Link-to-control arm bolt	45	33
Clamp bolts	24	18
Balljoint-to-control arm bolts		
Step 1	20	15
Step 2	Turn an additional 90-degrees	
Balljoint-to-steering knuckle nut	45	33
Control arm pivot bolt		
Step 1	71	52
Step 2	Turn an additional 90-degrees	
Control arm rear mounting bolt		
Step 1	71	52
Step 2	Turn an additional 90-degrees	
Subframe mounting bolts		
Step 1	100	74
Step 2	Turn an additional 90-degrees	

Torque specifications (continued)	Nm	Ft-lbs (unless otherwise indicated)
Suspension strut damper shaft nut	60	44
Suspension strut-to-steering knuckle bolt/nut		
Step 1	50	37
Step 2	Turn an additional 90-degrees	
Suspension strut upper mounting nut	60	44
Driveaxle/hub nut	See Chapter 8	
Rear axle		
Pivot bolt/nut	80	59
Mounting bracket retaining bolts	75	55
Stub axle mounting bolts	60	44
Shock absorber-to-body bolts		
Step 1	30	22
Step 2	Turn an additional 90-degrees	
Shock absorber-to-axle bolt/nut	60	44
Shock absorber upper mount-to-damper shaft nut	24	18
Hub/bearing assembly retaining nut	175	129
Steering		
Steering column		
Lower mounting bolt/nut	9.5	84 in-lbs
Upper mounting bolts	24	18
Universal joint pinch bolt/nut		
Step 1	20	15
Step 2	Turn an additional 90-degrees	
Steering gear		
Mounting bolts		
Step 1	20	15
Step 2	Turn an additional 90-degrees	
Power steering hose banjo fitting bolts		
Pressure line	45	33
Return line	41	30
Tie-rod end-to-steering knuckle nut		
Through 2004	45	33
2005	50	37
Steering wheel bolt		
Through 2004	50	37
2005	44	33
Power steering pump		
Mounting bolts	24	18
Feed line banjo fitting bolt	38	28
Pulley retaining bolts	24	18
Wheels		
Wheel bolts	See Chapter 1	

11

BODY

Section

1 General information

The Volkswagen New Beetle features a "unibody" layout, using a floor pan with front and rear frame side rails which support the body components, front and rear suspension systems and other mechanical components. Certain components are particularly vulnerable to accident damage and can be unbolted and repaired or replaced. Among these parts are the bumpers, front and rear fenders, doors, the hood and hatch. Only general body maintenance practices and body panel repair procedures within the scope of the do-it-yourselfer are included in this chapter.

Although all covered models are very similar, some procedures may differ somewhat from one body to another.

⁂ CAUTION 1:

These models are equipped with an anti-theft radio. Before performing a procedure that requires disconnecting the battery, make sure you have the proper activation code.

⁂ CAUTION 2:

Disconnecting the battery can cause driveability problems that require a scan tool to remedy. See Chapter 5 for the use of an auxiliary voltage input device before disconnecting the battery.

2 Body - maintenance

1 The condition of your vehicle's body is very important, because the resale value depends a great deal on it. It's much more difficult to repair a neglected or damaged body than it is to repair mechanical components. The hidden areas of the body, such as the wheel wells, the frame and the engine compartment, are equally important, although they don't require as frequent attention as the rest of the body.

2 Once a year, or every 12,000 miles, it's a good idea to have the underside of the body steam cleaned. All traces of dirt and oil will be removed and the area can then be inspected carefully for rust, damaged brake lines, frayed electrical wires, damaged cables and other problems. The front suspension components should be greased after completion of this job.

3 At the same time, clean the engine and the engine compartment with a steam cleaner or water-soluble degreaser.

4 The wheel wells should be given close attention, since undercoating can peel away and stones and dirt thrown up by the tires can cause the paint to chip and flake, allowing rust to set in. If rust is found, clean down to the bare metal and apply an anti-rust paint.

5 The body should be washed about once a week. Wet the vehicle thoroughly to soften the dirt, then wash it down with a soft sponge and plenty of clean soapy water. If the surplus dirt is not washed off very carefully, it can wear down the paint.

6 Spots of tar or asphalt thrown up from the road should be removed with a cloth soaked in solvent.

7 Once every six months, wax the body and chrome trim. If a chrome cleaner is used to remove rust from any of the vehicle's plated parts, remember that the cleaner also removes part of the chrome, so use it sparingly. After cleaning chrome trim, apply paste wax to preserve it.

3 Vinyl trim - maintenance

Don't clean vinyl trim with detergents, caustic soap or petroleum-based cleaners. Plain soap and water works just fine, with a soft brush to clean dirt that may be ingrained. Wash the vinyl as frequently as the rest of the vehicle. After cleaning, application of a high-quality rubber and vinyl protectant will help prevent oxidation and cracks. The protectant can also be applied to weatherstripping, vacuum lines and rubber hoses, which often fail as a result of chemical degradation, and to the tires.

4 Upholstery and carpets - maintenance

1 Every three months remove the floormats and clean the interior of the vehicle (more frequently if necessary). Use a stiff whiskbroom to brush the carpeting and loosen dirt and dust, then vacuum the upholstery and carpets thoroughly, especially along seams and crevices.

2 Dirt and stains can be removed from carpeting with basic household or automotive carpet shampoos available in spray cans. Follow the directions and vacuum again, then use a stiff brush to bring back the "nap" of the carpet.

3 Most interiors have cloth or vinyl upholstery, either of which can be cleaned and maintained with a number of material-specific cleaners or shampoos available in auto supply stores. Follow the directions on the product for usage, and always spot-test any upholstery cleaner on an inconspicuous area (bottom edge of a backseat cushion) to ensure that it doesn't cause a color shift in the material.

4 After cleaning, vinyl upholstery should be treated with a protectant.

➡Note: Make sure the protectant container indicates the product can be used on seats - some products may make a seat too slippery.

⁂ CAUTION:

Do not use protectant on vinyl-covered steering wheels.

5 Leather upholstery requires special care. It should be cleaned regularly with saddlesoap or leather cleaner. Never use alcohol, gasoline, water, nail polish remover or thinner to clean leather upholstery.

6 After cleaning, regularly treat leather upholstery with a leather conditioner, rubbed in with a soft cotton cloth. Never use car wax on leather upholstery.

7 In areas where the interior of the vehicle is subject to bright sunlight, cover leather seating areas of the seats with a sheet if the vehicle is to be left out for any length of time.

5 Body repair - minor damage

PLASTIC BODY PANELS

The following repair procedures are for minor scratches and gouges. Repair of more serious damage should be left to a dealer service department or qualified auto body shop. Below is a list of the equipment and materials necessary to perform the following repair procedures on plastic body panels. Although a specific brand of material may be mentioned, it should be noted that equivalent products from other manufacturers may be used instead.

Wax, grease and silicone removing solvent
Cloth-backed body tape
Sanding discs
Drill motor with three-inch disc holder
Hand sanding block
Rubber squeegees
Sandpaper
Non-porous mixing palette
Wood paddle or putty knife
Curved-tooth body file
Flexible parts repair material

1 Remove the damaged panel, if necessary or desirable. In most cases, repairs can be carried out with the panel installed.

2 Clean the area(s) to be repaired with a wax, grease and silicone removing solvent applied with a water-dampened cloth.

3 If the damage is structural, that is, if it extends through the panel, clean the backside of the panel area to be repaired as well. Wipe dry.

4 Sand the rear surface about 1-1/2 inches beyond the break.

5 Cut two pieces of fiberglass cloth large enough to overlap the break by about 1-1/2 inches. Cut only to the required length.

6 Mix the adhesive from the repair kit according to the instructions included with the kit, and apply a layer of the mixture approximately 1/8-inch thick on the backside of the panel. Overlap the break by at least 1-1/2 inches.

7 Apply one piece of fiberglass cloth to the adhesive and cover the cloth with additional adhesive. Apply a second piece of fiberglass cloth to the adhesive and immediately cover the cloth with additional adhesive in sufficient quantity to fill the weave.

8 Allow the repair to cure for 20 to 30 minutes at 60-degrees to 80-degrees F.

9 If necessary, trim the excess repair material at the edge.

10 Remove all of the paint film over and around the area(s) to be repaired. The repair material should not overlap the painted surface.

11 With a drill motor and a sanding disc (or a rotary file), cut a "V" along the break line approximately 1/2-inch wide. Remove all dust and loose particles from the repair area.

12 Mix and apply the repair material. Apply a light coat first over the damaged area; then continue applying material until it reaches a level slightly higher than the surrounding finish.

13 Cure the mixture for 20 to 30 minutes at 60-degrees to 80-degrees F.

14 Roughly establish the contour of the area being repaired with a body file. If low areas or pits remain, mix and apply additional adhesive.

15 Block sand the damaged area with sandpaper to establish the actual contour of the surrounding surface.

16 If desired, the repaired area can be temporarily protected with several light coats of primer. Because of the special paints and techniques required for flexible body panels, it is recommended that the vehicle be taken to a paint shop for completion of the body repair.

STEEL BODY PANELS

➧ See photo sequence

Repair of minor scratches

17 If the scratch is superficial and does not penetrate to the metal of the body, repair is very simple. Lightly rub the scratched area with a fine rubbing compound to remove loose paint and built up wax. Rinse the area with clean water.

18 Apply touch-up paint to the scratch, using a small brush. Continue to apply thin layers of paint until the surface of the paint in the scratch is level with the surrounding paint. Allow the new paint at least two weeks to harden, then blend it into the surrounding paint by rubbing with a very fine rubbing compound. Finally, apply a coat of wax to the scratch area.

19 If the scratch has penetrated the paint and exposed the metal of the body, causing the metal to rust, a different repair technique is required. Remove all loose rust from the bottom of the scratch with a pocketknife, then apply rust inhibiting paint to prevent the formation of rust in the future. Using a rubber or nylon applicator, coat the scratched area with glaze-type filler. If required, the filler can be mixed with thinner to provide a very thin paste, which is ideal for filling narrow scratches. Before the glaze filler in the scratch hardens, wrap a piece of smooth cotton cloth around the tip of a finger. Dip the cloth in thinner and then quickly wipe it along the surface of the scratch. This will ensure that the surface of the filler is slightly hollow. The scratch can now be painted over as described earlier in this Section.

Repair of dents

20 When repairing dents, the first job is to pull the dent out until the affected area is as close as possible to its original shape. There is no point in trying to restore the original shape completely as the metal in the damaged area will have stretched on impact and cannot be restored to its original contours. It is better to bring the level of the dent up to a point that is about 1/8-inch below the level of the surrounding metal. In cases where the dent is very shallow, it is not worth trying to pull it out at all.

21 If the backside of the dent is accessible, it can be hammered out gently from behind using a soft-face hammer. While doing this, hold a block of wood firmly against the opposite side of the metal to absorb the hammer blows and prevent the metal from being stretched.

22 If the dent is in a section of the body which has double layers, or some other factor makes it inaccessible from behind, a different technique is required. Drill several small holes through the metal inside the damaged area, particularly in the deeper sections. Screw long, self-tapping screws into the holes just enough for them to get a good grip in the metal. Now pulling on the protruding heads of the screws with locking pliers can pull out the dent.

23 The next stage of repair is the removal of paint from the damaged area and from an inch or so of the surrounding metal. This is easily done with a wire brush or sanding disk in a drill motor, although it can be done just as effectively by hand with sandpaper. To complete the preparation for filling, score the surface of the bare metal with a screwdriver or the tang of a file or drill small holes in the affected area. This will provide a good grip for the filler material. To complete the repair, see the Section on filling and painting.

These photos illustrate a method of repairing simple dents. They are intended to supplement Body repair - minor damage in this Chapter and should not be used as the sole instructions for body repair on these vehicles.

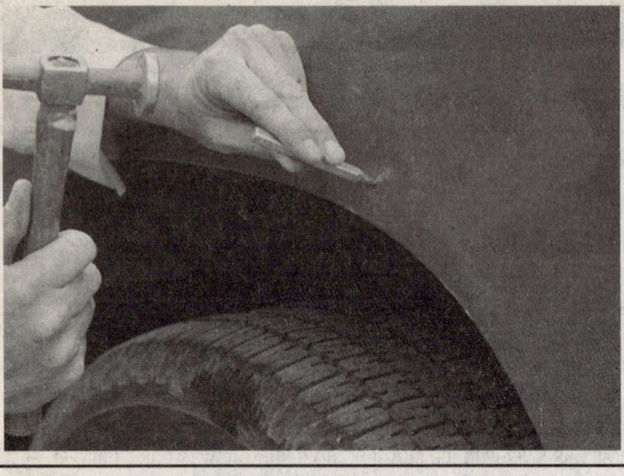

1 If you can't access the backside of the body panel to hammer out the dent, pull it out with a slide-hammer-type dent puller. In the deepest portion of the dent or along the crease line, drill or punch hole(s) at least one inch apart . . .

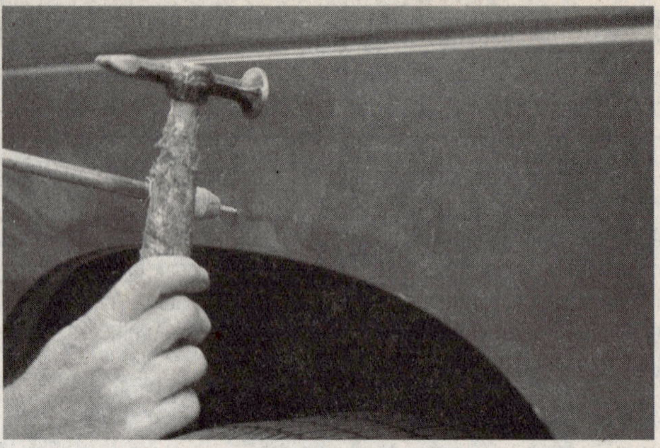

2 . . . then screw the slide-hammer into the hole and operate it. Tap with a hammer near the edge of the dent to help 'pop' the metal back to its original shape. When you're finished, the dent area should be close to its original contour and about 1/8-inch below the surface of the surrounding metal

3 Using coarse-grit sandpaper, remove the paint down to the bare metal. Hand sanding works fine, but the disc sander shown here makes the job faster. Use finer (about 320-grit) sandpaper to feather-edge the paint at least one inch around the dent area

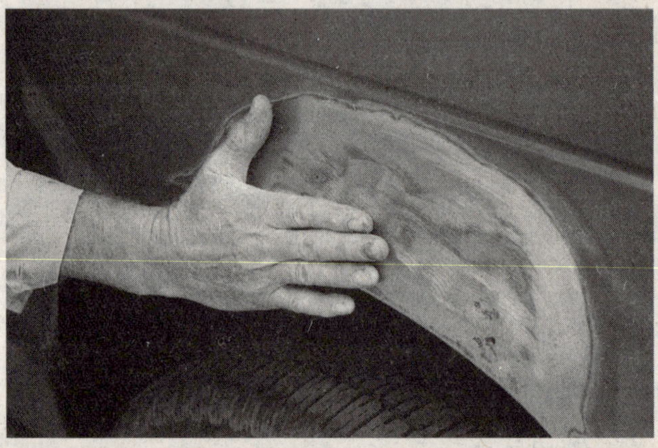

4 When the paint is removed, touch will probably be more helpful than sight for telling if the metal is straight. Hammer down the high spots or raise the low spots as necessary. Clean the repair area with wax/silicone remover

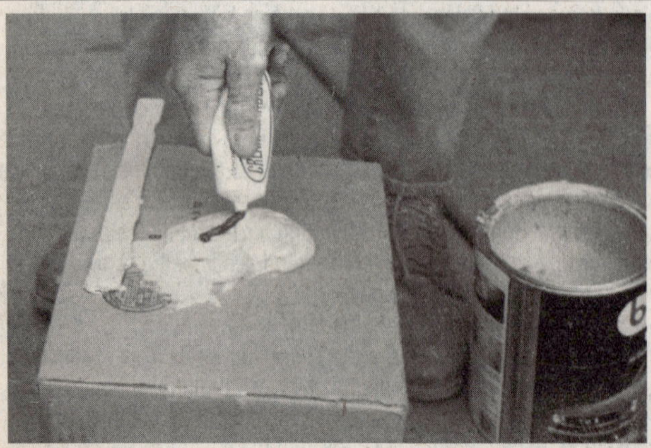

5 Following label instructions, mix up a batch of plastic filler and hardener. The ratio of filler to hardener is critical, and, if you mix it incorrectly, it will either not cure properly or cure too quickly (you won't have time to file and sand it into shape)

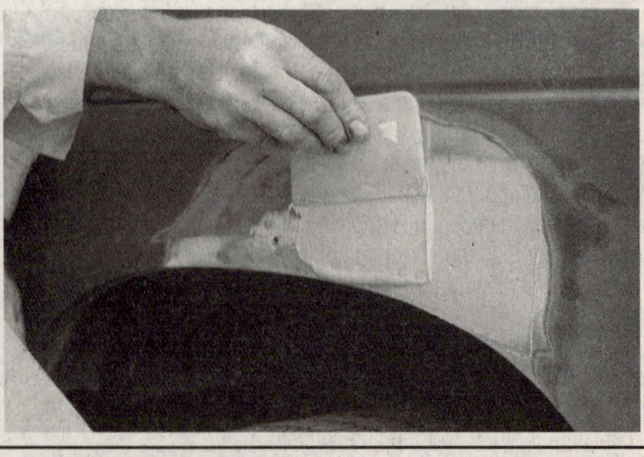

6 Working quickly so the filler doesn't harden, use a plastic applicator to press the body filler firmly into the metal, assuring it bonds completely. Work the filler until it matches the original contour and is slightly above the surrounding metal

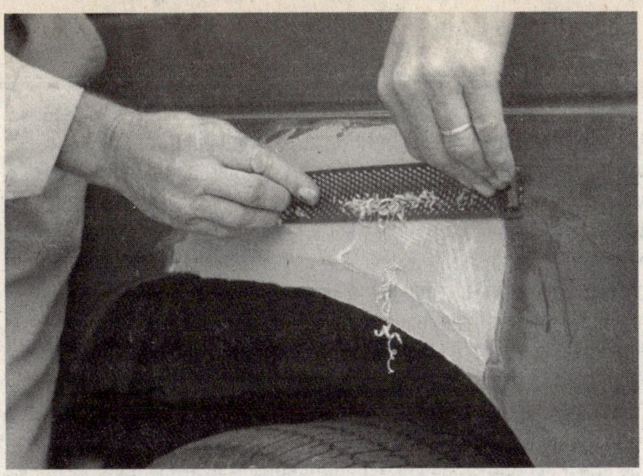

7 Let the filler harden until you can just dent it with your fingernail. Use a body file or Surform tool (shown here) to rough-shape the filler

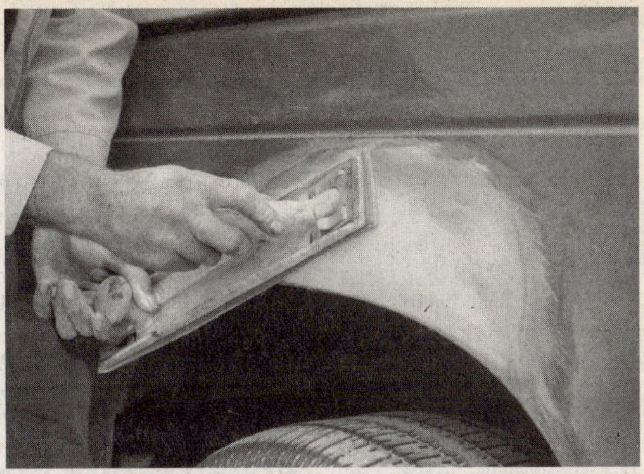

8 Use coarse-grit sandpaper and a sanding board or block to work the filler down until it's smooth and even. Work down to finer grits of sandpaper - always using a board or block - ending up with 360 or 400 grit

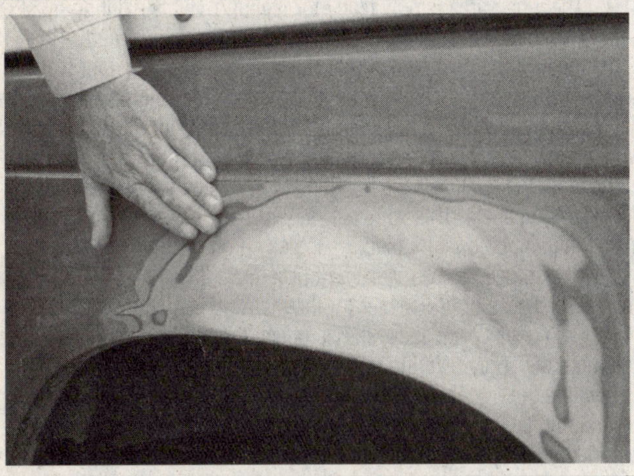

9 You shouldn't be able to feel any ridge at the transition from the filler to the bare metal or from the bare metal to the old paint. As soon as the repair is flat and uniform, remove the dust and mask off the adjacent panels or trim pieces

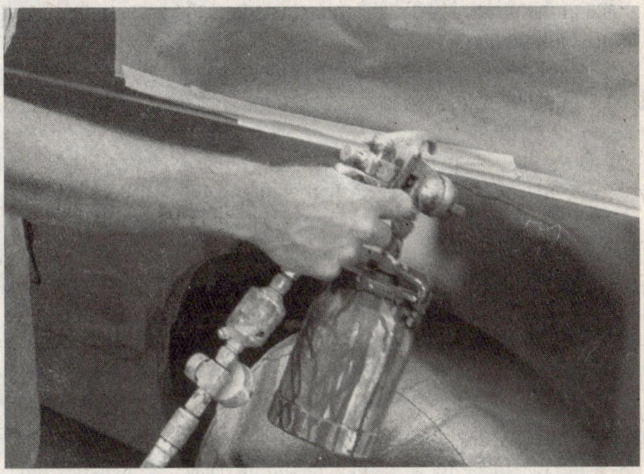

10 Apply several layers of primer to the area. Don't spray the primer on too heavy, so it sags or runs, and make sure each coat is dry before you spray on the next one. A professional-type spray gun is being used here, but aerosol spray primer is available inexpensively from auto parts stores

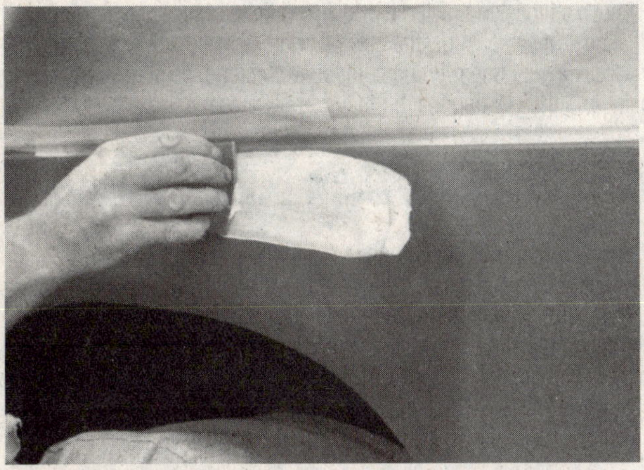

11 The primer will help reveal imperfections or scratches. Fill these with glazing compound. Follow the label instructions and sand it with 360 or 400-grit sandpaper until it's smooth. Repeat the glazing, sanding and respraying until the primer reveals a perfectly smooth surface

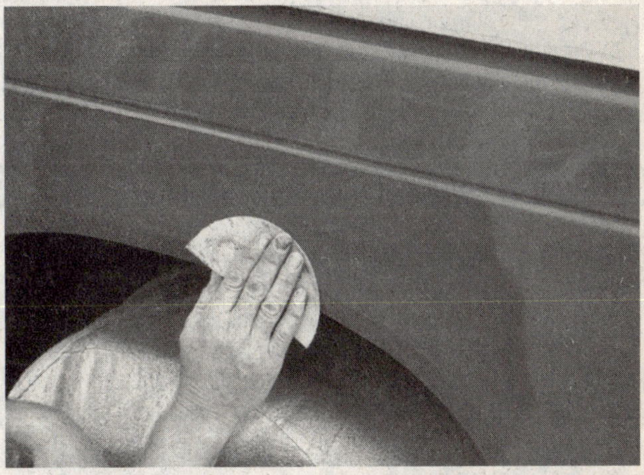

12 Finish sand the primer with very fine sandpaper (400 or 600-grit) to remove the primer overspray. Clean the area with water and allow it to dry. Use a tack rag to remove any dust, then apply the finish coat. Don't attempt to rub out or wax the repair area until the paint has dried completely (at least two weeks)

Repair of rust holes or gashes

24 Remove all paint from the affected area and from an inch or so of the surrounding metal using a sanding disk or wire brush mounted in a drill motor. If these are not available, a few sheets of sandpaper will do the job just as effectively.

25 With the paint removed, you will be able to determine the severity of the corrosion and decide whether to replace the whole panel, if possible, or repair the affected area. New body panels are not as expensive as most people think and it is often quicker to install a new panel than to repair large areas of rust.

26 Remove all trim pieces from the affected area except those which will act as a guide to the original shape of the damaged body, such as headlight shells, etc. Using metal snips or a hacksaw blade, remove all loose metal and any other metal that is badly affected by rust. Hammer the edges of the hole in to create a slight depression for the filler material.

27 Wire-brush the affected area to remove the powdery rust from the surface of the metal. If the back of the rusted area is accessible, treat it with rust inhibiting paint.

28 Before filling is done, block the hole in some way. This can be done with sheet metal riveted or screwed into place, or by stuffing the hole with wire mesh.

29 Once the hole is blocked off, the affected area can be filled and painted. See the following subsection on filling and painting.

Filling and painting

30 Many types of body fillers are available, but generally speaking, body repair kits which contain filler paste and a tube of resin hardener are best for this type of repair work. A wide, flexible plastic or nylon applicator will be necessary for imparting a smooth and contoured finish to the surface of the filler material. Mix up a small amount of filler on a clean piece of wood or cardboard (use the hardener sparingly). Follow the manufacturer's instructions on the package, otherwise the filler will set incorrectly.

31 Using the applicator, apply the filler paste to the prepared area. Draw the applicator across the surface of the filler to achieve the desired contour and to level the filler surface. As soon as a contour that approximates the original one is achieved, stop working the paste. If you continue, the paste will begin to stick to the applicator. Continue to add thin layers of paste at 20-minute intervals until the level of the filler is just above the surrounding metal.

32 Once the filler has hardened, the excess can be removed with a body file. From then on, progressively finer grades of sandpaper should be used, starting with a 180-grit paper and finishing with 600-grit wet-or-dry paper. Always wrap the sandpaper around a flat rubber or wooden block, otherwise the surface of the filler will not be completely flat. During the sanding of the filler surface, the wet-or-dry paper should be periodically rinsed in water. This will ensure that a very smooth finish is produced in the final stage.

33 At this point, the repair area should be surrounded by a ring of bare metal, which in turn should be encircled by the finely feathered edge of good paint. Rinse the repair area with clean water until all of the dust produced by the sanding operation is gone.

34 Spray the entire area with a light coat of primer. This will reveal any imperfections in the surface of the filler. Repair the imperfections with fresh filler paste or glaze filler and once more smooth the surface with sandpaper. Repeat this spray-and-repair procedure until you are satisfied that the surface of the filler and the feathered edge of the paint are perfect. Rinse the area with clean water and allow it to dry completely.

35 The repair area is now ready for painting. Spray painting must be carried out in a warm, dry, windless and dust free atmosphere. These conditions can be created if you have access to a large indoor work area, but if you are forced to work in the open, you will have to pick the day very carefully. If you are working indoors, dousing the floor in the work area with water will help settle the dust that would otherwise be in the air. If the repair area is confined to one body panel, mask off the surrounding panels. This will help minimize the effects of a slight mismatch in paint color. Trim pieces such as chrome strips, door handles, etc., will also need to be masked off or removed. Use masking tape and several thickness of newspaper for the masking operations.

36 Before spraying, shake the paint can thoroughly, then spray a test area until the spray painting technique is mastered. Cover the repair area with a thick coat of primer. The thickness should be built up using several thin layers of primer rather than one thick one. Using 600-grit wet-or-dry sandpaper, rub down the surface of the primer until it is very smooth. While doing this, the work area should be thoroughly rinsed with water and the wet-or-dry sandpaper periodically rinsed as well. Allow the primer to dry before spraying additional coats.

37 Spray on the top coat, again building up the thickness by using several thin layers of paint. Begin spraying in the center of the repair area and then, using a circular motion, work out until the whole repair area and about two inches of the surrounding original paint is covered. Remove all masking material 10 to 15 minutes after spraying on the final coat of paint. Allow the new paint at least two weeks to harden, then use a very fine rubbing compound to blend the edges of the new paint into the existing paint. Finally, apply a coat of wax.

6 Body repair - major damage

1 Major damage must be repaired by an auto body shop specifically equipped to perform unibody repairs. These shops have the specialized equipment required to do the job properly.

2 If the damage is extensive, the body must be checked for proper alignment or the vehicle's handling characteristics may be adversely affected and other components may wear at an accelerated rate.

3 Due to the fact that all of the major body components (hood, fenders, etc.) are separate and replaceable units, any seriously damaged components should be replaced rather than repaired. Sometimes the components can be found in a wrecking yard that specializes in used vehicle components, often at considerable savings over the cost of new parts.

7 Hinges and locks - maintenance

Once every 3000 miles, or every three months, the hinges and latch assemblies on the doors, hood and trunk should be given a few drops of light oil or lock lubricant. The door latch strikers should also be lubricated with a thin coat of grease to reduce wear and ensure free movement. Lubricate the door and trunk locks with spray-on graphite lubricant.

8 Windshield and fixed glass - replacement

Replacement of the windshield and fixed glass requires the use of special fast-setting adhesive/caulk materials and some specialized tools and techniques. These operations should be left to a dealer service department or a shop specializing in glass work.

9 Radiator grille - removal and installation

Refer to illustration 9.1

1 From behind the bumper cover, release the plastic clips and detach the grille (see illustration).

2 On models with fog lights, disconnect the electrical connectors from the fog lights, and remove the fog lamp mounting screws before detaching the grille (see Chapter 12).

3 To install, place the grille in position and seat the clips into the holes in the bumper cover.

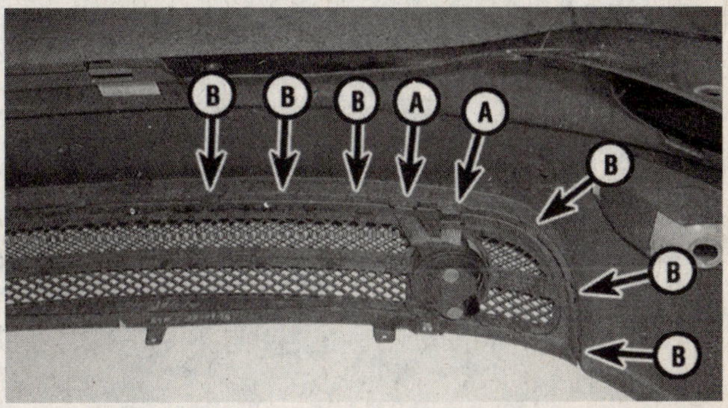

9.1 Remove the fog light mounting screws (A), then release the tabs (B) from behind the front bumper cover to release the grille (right half shown, left side similar)

10 Hood - removal, installation and adjustment

Note: The hood is heavy and somewhat awkward to remove and install - at least two people should perform this procedure.

REMOVAL AND INSTALLATION

Refer to illustration 10.2

1 Use blankets or pads to cover the cowl area of the body and the fenders. This will protect the body and paint as the hood is lifted off.

2 Scribe alignment marks around the hinge plate to insure proper alignment during installation (a permanent-type felt-tip marker also will work for this) (see illustration).

3 Disconnect the windshield washer hoses from the jets at the rear of the hood, and the underhood light, if applicable.

4 Have an assistant support the weight of the hood, and remove the hinge-to-hood bolts (see illustration 10.2).

10.2 Use a marking pen to outline the hinge plate

5 Lift off the hood.

6 Installation is the reverse of removal.

ADJUSTMENT

Refer to illustrations 10.10 and 10.11

7 Fore-and-aft and side-to-side adjustment of the hood is done by moving the hood in relation to the hinge plates after loosening the bolts.

8 Scribe or trace a line around the entire hinge plate so you can judge the amount of movement (see illustration 10.2).

9 Loosen the bolts and move the hood into correct alignment. Move it only a little at a time. Tighten the hinge bolts or nuts and carefully lower the hood to check the alignment.

10 Adjust the hood bumpers on the front edge of the hood, so the hood is flush with the fenders when closed (see illustration).

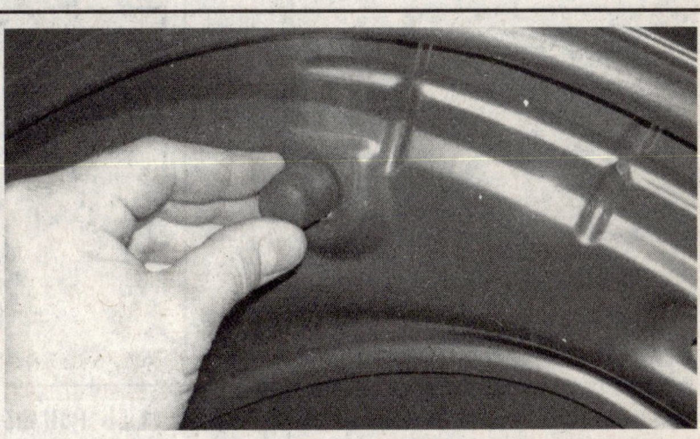

10.10 Adjust the hood height by screwing the hood bumpers in or out

11 The safety latch assembly can also be adjusted up-and-down and side-to-side after loosening the nuts (see illustration).

12 The hood latch assembly, as well as the hinges, should be periodically lubricated with white lithium-base grease to prevent sticking and wear.

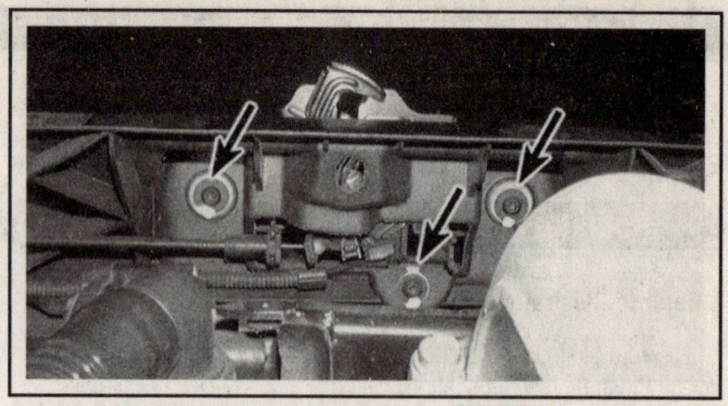

10.11 Loosen the bolts (arrows) and move the latch to adjust the hood fit in the closed position

11 Hood latch and release cable - removal and installation

➧ Refer to illustrations 11.3, 11.6a, 11.6b and 11.7

LATCH

1 Open the hood. On 2001 and later models, there is an exterior pull handle. To separate the handle from the latch, pry open the tabs on the handle with a screwdriver until the handle separates from the pin on the latch assembly.

2 Remove the nuts and detach the latch assembly (see illustration 10.11).

3 Detach the cable end from the back of the latch assembly, then remove the latch (see illustration).

4 Installation is the reverse of removal.

CABLE

5 Use a screwdriver to release the cable end, then detach the cable case from the bracket on the back of the latch (see illustration 11.3).

6 In the passenger compartment, remove the driver's side kick panel, pry out the clip and remove the hood release handle (see illustrations).

7 Remove the hood release handle mounting screws, pull the handle mount out and release the cable (see illustration).

8 Under the dash, remove the cable grommet from the firewall.

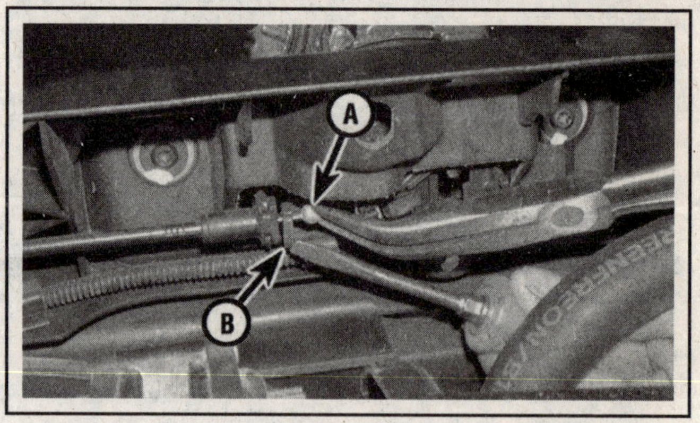

11.3 Detach the cable end from the latch (A) and the cable body from the bracket (B)

9 Release the clips at the inner edge of the left front fenderwell until the cable is free of the body.

10 Attach a wire or string to the hood end of the old cable and pull the cable through into the vehicle's interior.

11 Connect the string or wire to the new cable and pull it through the firewall into the engine compartment.

12 The remainder of installation is the reverse of removal.

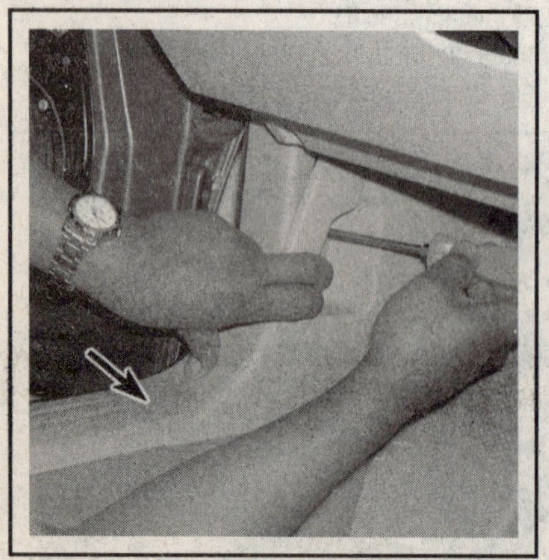

11.6a Pull back the hood release handle to remove the kick panel screw, and pry up at the door sill end (arrow)

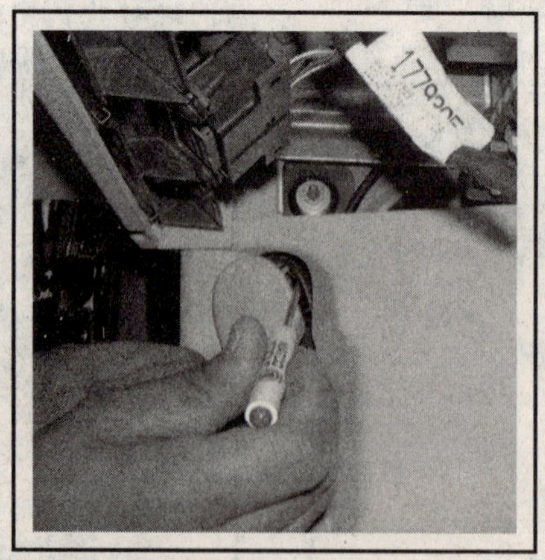

11.6b Pull the hood release lever back enough to allow using a small screwdriver to pull the clip out, then remove the handle and the kick panel

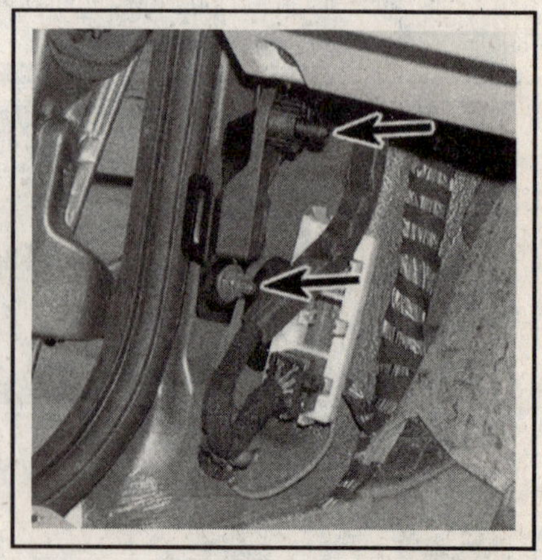

11.7 Hood release handle mounting screws (arrows)

12 Bumpers - removal and installation

1 Front and rear bumpers on all models are composed of a fascia, or exterior cover, and a foam-covered structural beam.

FRONT BUMPER

➧ Refer to illustration 12.4

2 Apply the parking brake, raise the vehicle and support it securely on jackstands. Disconnect the fog lights (if equipped), and remove the radiator grille (see Section 9). Remove the fenderwell liners (see Section 13).

3 The front bumper cover is part of the front clip, which consists of the two front fenders and the bumper cover. The clip should be removed as a unit (see Section 13).

4 With the clip removed, remove the bolts and nuts securing the front bumper beam to the impact dampers at the chassis (see illustration).

5 Installation is the reverse of the removal procedure.

12.4 Remove the bolts/nuts (arrows) to remove the front bumper beam

REAR BUMPER

➧ Refer to illustration 12.7

6 The rear bumper cover is part of the rear clip, which includes the bumper cover and the rear fenders. Like the front clip, it's very difficult to remove the rear bumper cover while the rear fenders remain on the vehicle. See Section 13 for removal of the rear clip (fenders and bumper cover together).

7 If the rear bumper reinforcement beam is to be removed, remove the bolt and nuts (see illustration).

8 Installation is the reverse of removal.

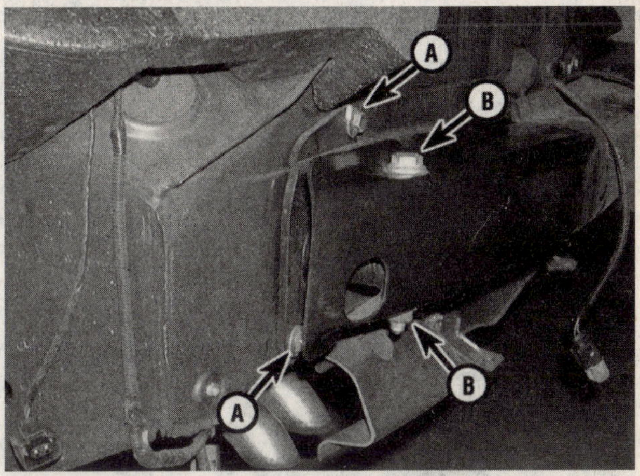

12.7 Remove the bolts/nuts (arrows) to remove the rear bumper reinforcement beam

A Impact damper mounting bolts
B Bumper beam mounting bolt/nut

13 Fenders - removal and installation

FRONT

➧ Refer to illustrations 13.3a, 13.3b, 13.4a, 13.4b, 13.4c, 13.5, 13.6a, 13.6b and 13.7

1 Loosen the front wheel lug bolts, raise the vehicle, support it securely on jackstands and remove the front wheels. The front bumper cover is part of the front body clip, which includes the bumper cover and both fenders. It's best to remove the clip as a unit, then separate the fender(s).

➡Note: This is a job for two people.

2 Apply wide masking tape along the front edge of the cowl and doors to prevent scratching the paint during fender removal.

3 Remove the screws and take out the plastic fenderwell liner (see illustrations).

4 Remove the fender-to-cowl mounting bolts (see illustrations).

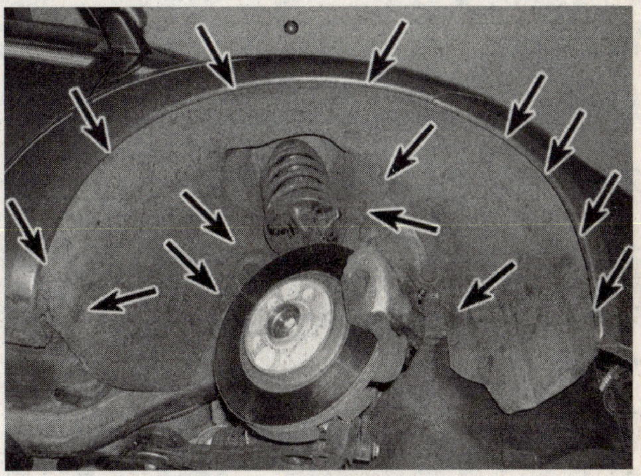

13.3a Remove these screws (arrows) and the fenderwell liner

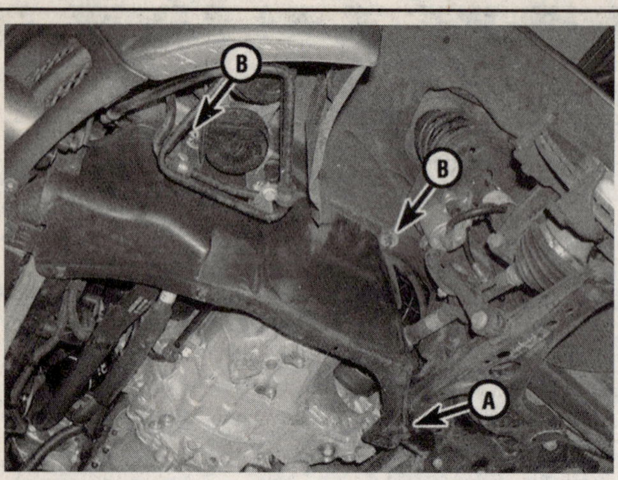

13.3b To remove the lower splash shield (not required for fender removal), remove the screw (A) and two star fasteners (B)

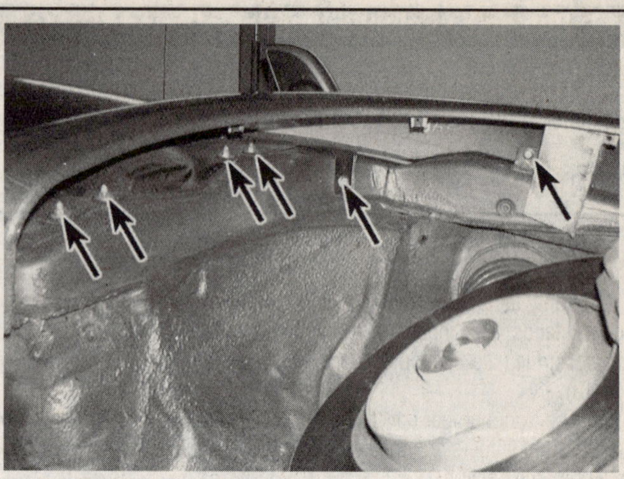

13.4a Remove the bolts (arrows) inside the fenderwell

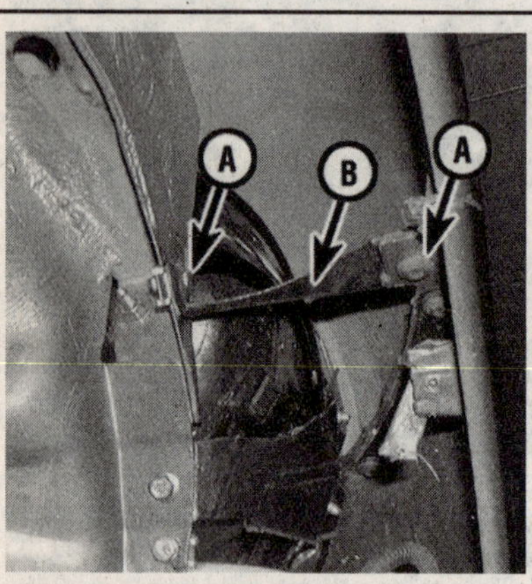

13.4b At the front of the fenderwell, remove the fender brace bolts (A) and the brace (B)

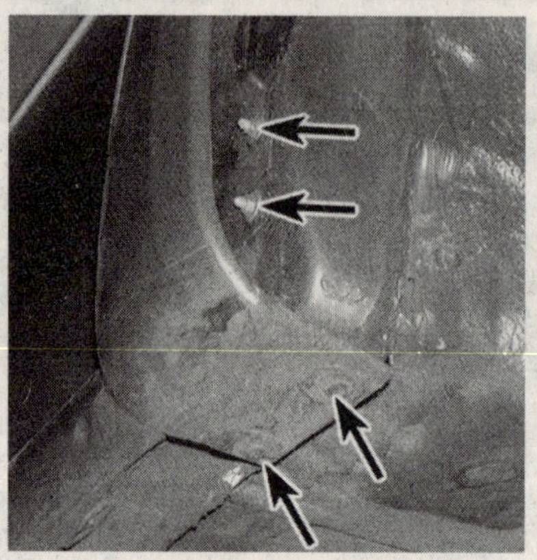

13.4c At the rear of the fenderwell, remove the brace bolts and the two lower fender-to-cowl bolts (arrows)

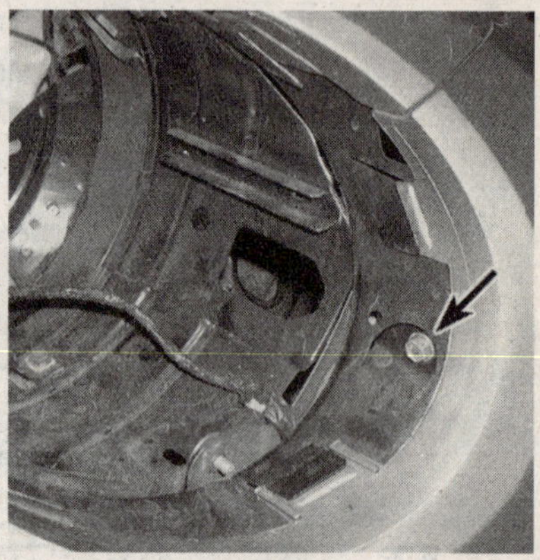

13.5 Remove this one bolt (arrow) securing the headlight housing cavity to the front bumper cover

5 Remove the headlight housings, disconnect the turn signal lights, and if equipped, disconnect the electrical connectors at the back of the fog lights (see Chapter 12). Remove the one body bolt inside the headlight housing opening (see illustration).

6 Remove the screws along the bottom edge of the front bumper cover and the top of the hood opening (see illustrations).

13.6a Remove the screws (arrows) along the bottom front of the bumper cover . . .

13.6b . . . and the two in the front of the hood opening (arrows)

13.7 Where each front fender joins the body, remove the screws (arrows)

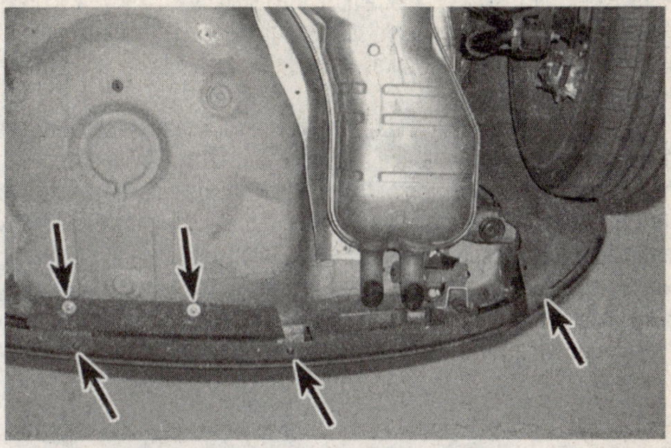

13.13 Remove the screws (arrows) securing the rear bumper cover to the bottom of the body

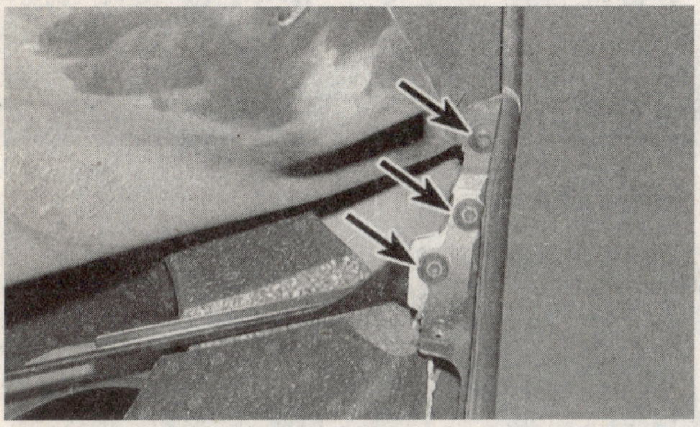

13.14a Pull back the plastic fenderwell liner to access these screws (arrows)

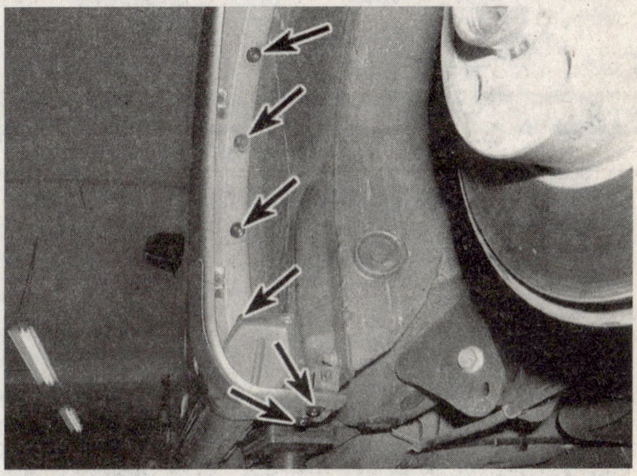

13.14b At the front of the rear fenderwell, remove these bolts (arrows) holding the fender to the body . . .

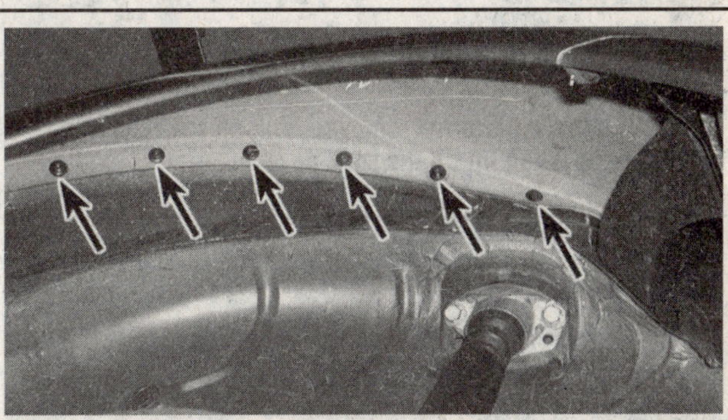

13.14c . . . and these (arrows) along the top edge

7 Open the hood, and remove the screws securing the front clip to the body (see illustration).

8 Have an assistant help you remove the front clip from the vehicle, pulling the back of the fenders out slightly while pulling the whole clip forward. To replace the fender, remove the bolts holding the front of the fender to the bumper cover.

9 Installation is the reverse of removal. Make sure that the foam cushion for the top rear of the fender is in place and that the anti-squeak strip along the top of the body is aligned with the holes in the body.

➡Note: Most body bolts on the New Beetle pass through two body part flanges; on one side they must be fitted with reinforcement strips or zinc washers, and they thread into a metal strip with nuts welded on. These should all be in place finger-tight until the final fender-to-bumper cover fit is exact before final tightening.

10 Tighten all nuts, bolts and screws securely.

REAR

➧ Refer to illustrations 13.13, 13.14a, 13.14b, 13.14c and 13.17

11 Loosen the rear wheel lug bolts, raise the vehicle, support it securely on jackstands and remove the rear wheels. As with the front clip, the rear fenders and bumper cover are best removed as a unit, then separated for individual part replacement.

12 Apply wide masking tape along the rear quarter panel to prevent scratching the paint during fender removal.

13 From below, remove the fasteners retaining the bottom edge of the rear bumper cover to the body (see illustration). Disconnect the electrical connectors at the license plate light and backup lights.

14 In each rear fenderwell, remove the screws securing the fenderwell liner to the bumper cover ends, and the bolts holding the fender to the body (see illustrations).

15 Refer to Chapter 12 and remove the taillights.

16 At the right-side fender, disconnect the fuel door release cable from the fuel door opening by prying out the small retaining clip inside the fuel door opening.

17 Open the rear hatch and remove the remaining bolts securing the fender and bumper cover to the body (see illustration). With an assistant, spread the front portion of the plastic rear fenders apart enough to clear the body and lift off the rear clip. With the clip off, there is access to the bolts and reinforcement strips that hold the fenders to the rear bumper cover.

18 Installation is the reverse of the removal procedure. Make sure the anti-squeak material is in place at the fender/body joint.

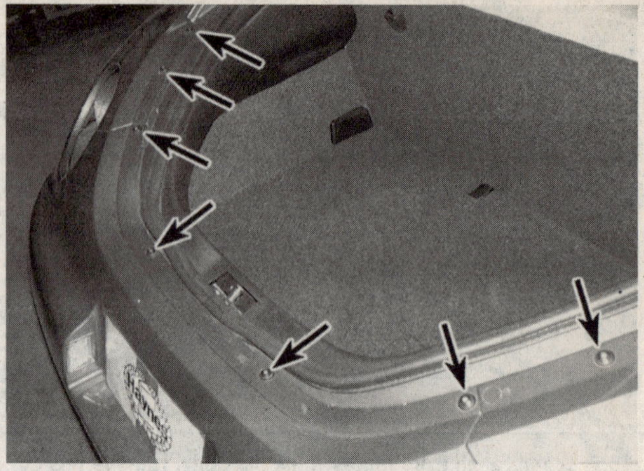

13.17 Remove the bolts (arrows) around the rear hatch opening (not all are shown in this view)

14 Cowl cover and plenum closeout panel - removal and installation

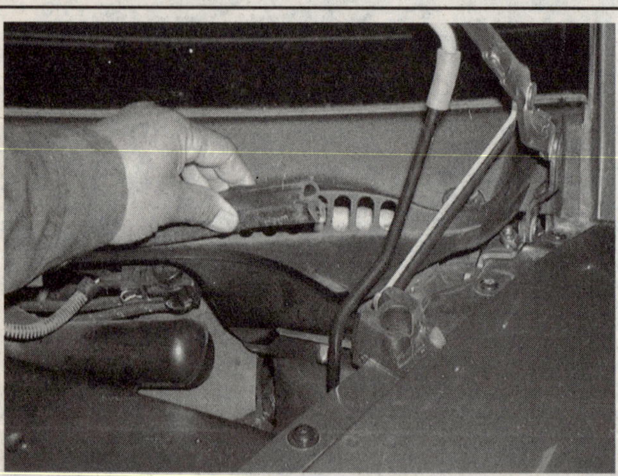

14.2 With the wiper arms removed, pull off the rubber weatherstrip, then lift off the cowl cover

➧ Refer to illustrations 14.2, 14.4 and 14.5

1 Open the hood, then refer to Chapter 12 and remove the windshield wiper arms.

2 Remove the rubber weatherstrip at the front of the cowl cover (see illustration).

3 Lift off the plastic cowl cover.

4 To remove the plenum closeout panel, first separate the Secondary Air Injection solenoid from its bracket on the plenum closeout panel (see illustration).

5 Remove the fasteners and lift out the plenum closeout panel (see illustration).

6 Installation is the reverse of removal.

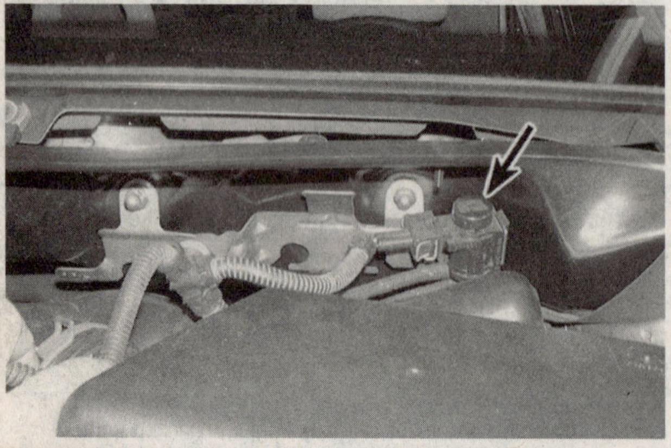

14.4 Remove the SAI solenoid (arrow) from the front of the plenum closeout panel

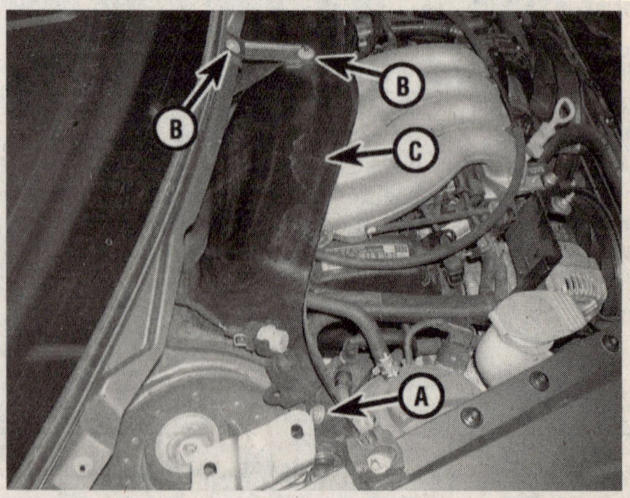

14.5 Remove the bolt (A) at each side of the panel, the two bolts (B) at the center bracket, then remove the closeout panel (C)

15 Door trim panels - removal and installation

➧ Refer to illustrations 15.2a, 15.2b, 15.3, 15.4, 15.6a and 15.6b

1 Disconnect the negative battery cable, but read the **Cautions** in Section 1.

2 Remove the door pull cover and the two door panel screws behind it (see illustrations).

3 If equipped with manual windows, slide the spacer (between the handle and the door trim panel) away from the knob end of the handle to release the handle clip (see illustration). Remove the handle.

4 Remove the screws retaining the bottom of the door panel to the door (see illustration).

5 Using a trim tool, pry along the sides of the door panel to disengage the clips.

6 Grasp the trim panel and pull up sharply to detach it from the door. Disconnect the inner door handle lock rod from the handle and unplug any electrical connectors (see illustrations).

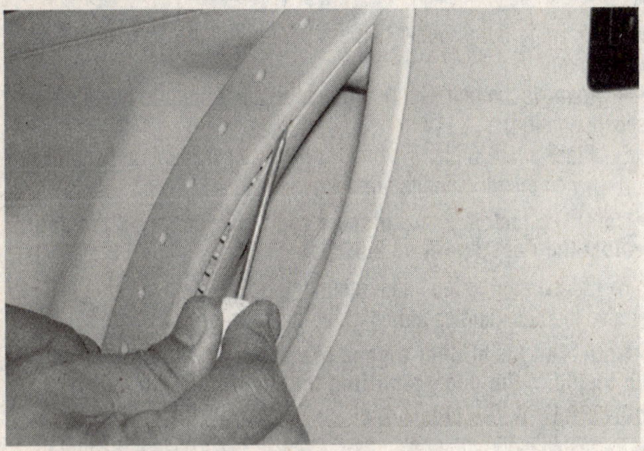

15.2a Pry off the door pull cover . . .

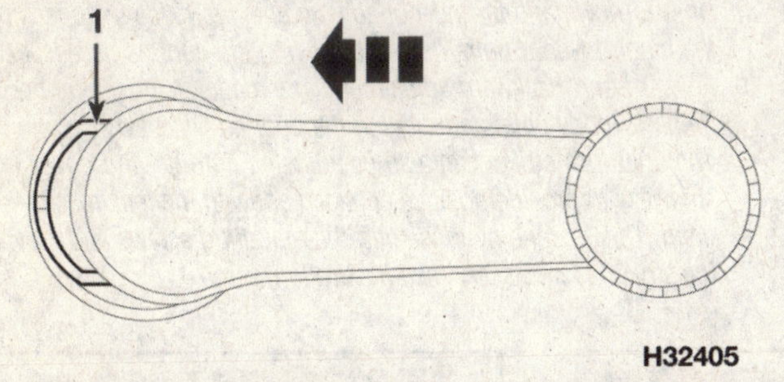

15.3 Push the spacer (1) in the direction of the arrow to disengage the clip behind the manual window handle

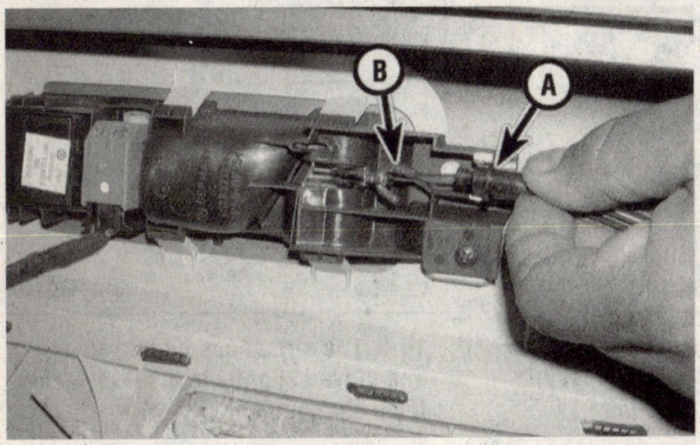

15.6a Pry the door handle cable end (A) from the handle assembly at the back of the door panel, then twist to disengage the hook (B)

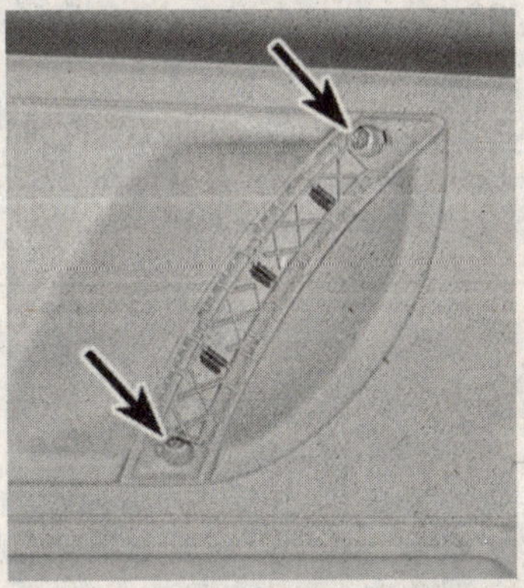

15.2b . . . then remove the two screws (arrows)

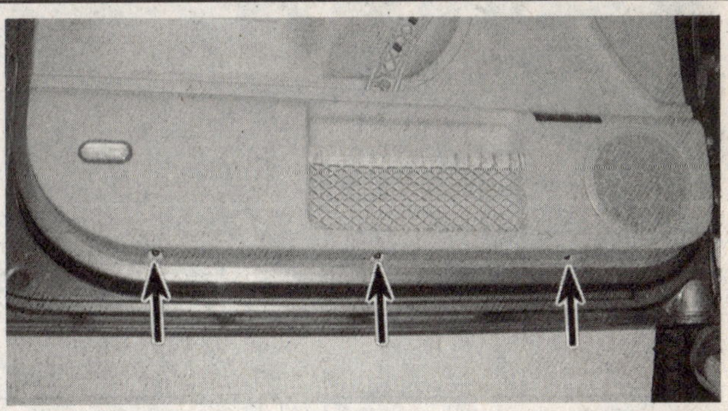

15.4 Remove the screws (arrows) securing the door panel - then use a trim tool along the front and rear edges of the panel to release the clips in back - later models also have a bolt at the lower rear of the door panel

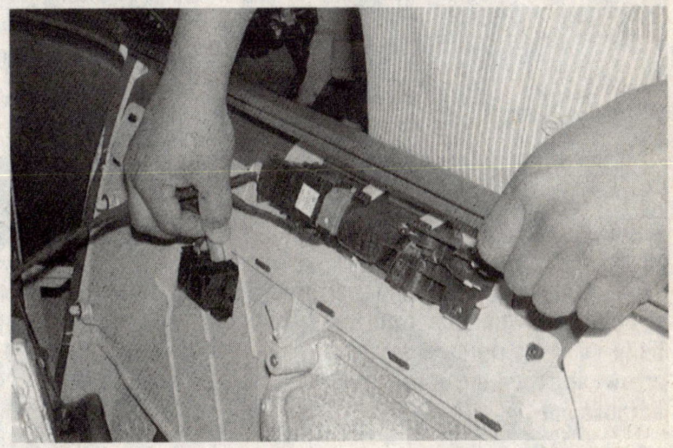

15.6b Disconnect the electrical connectors at the door panel

7 To install the panel, connect the wire harness connectors and door handle rod, then place the panel in position in the door. Press the trim panel down into place until the clips are seated.

8 Install the screws and screw covers, where used. Reconnect the negative battery cable.

16 Door - removal, installation and adjustment

➧ Refer to illustrations 16.3, 16.4, 16.6 and 16.8

1 Remove the door trim panel (see Section 15).

2 If removing the driver's door, refer to Section 11 and remove the hood release handle assembly from the kick panel. On either side, pull first at the top, then the bottom to release the kick panel from its clips.

3 At the front door pillar, disconnect the electrical connectors (see illustration). Pull the rubber harness boot out of the body and feed the harness out.

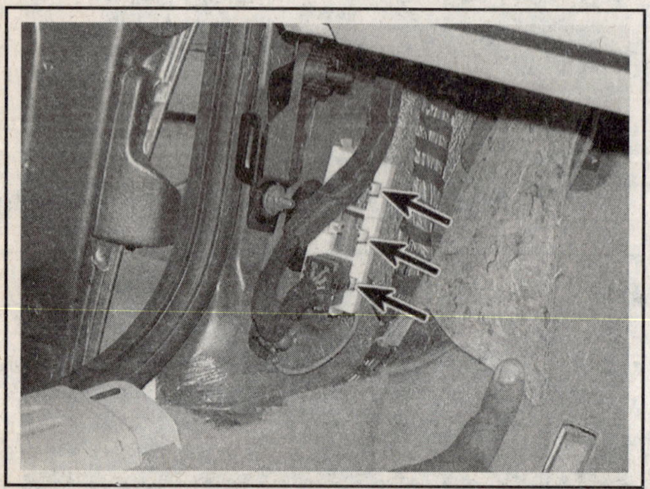

16.3 Disconnect the electrical connectors (arrows) in the door jamb and feed the wires out through the body pillar

4 Remove the bolt securing the door check strap to the doorjamb (see illustration).

5 Place a jack under the door or have an assistant on hand to support it when the hinge bolts are removed.

➡Note: If a jack is used, place a rag between it and the door to protect the door's painted surfaces.

6 Scribe around the hinge mounting bolts with a marking pen, remove the bolts and carefully lift off the door (see illustration).

➡Note: The lower door hinges are attached with bolts from the inside of the door, requiring that the inner door carrier be removed (see Section 17).

7 Installation is the reverse of removal, making sure to align the hinge with the marks made during removal before tightening the bolts.

8 Following installation of the door, check the alignment and adjust it if necessary as follows:

a) *Up-and-down and in-and-out adjustments are made by loosening the hinge-to-door bolts/nuts and moving the door as necessary.*
b) *Forward-and-backward adjustments are made by loosening the hinge-to-body bolts and moving the door as necessary.*
c) *The door lock striker can also be adjusted both up-and-down and sideways to provide positive engagement with the lock mechanism. This is done by loosening the mounting screws and moving the striker as necessary (see illustration).*

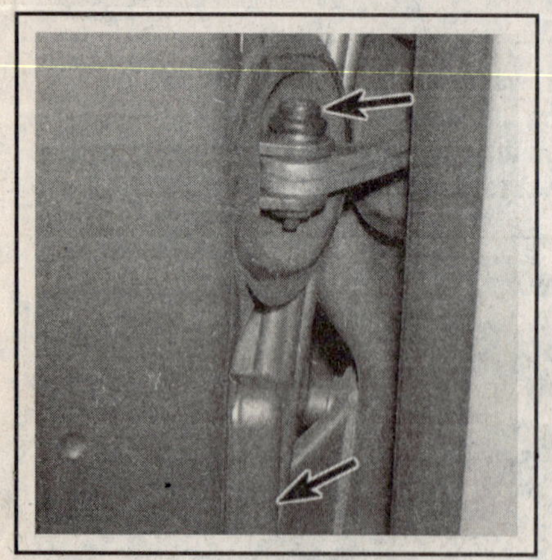

16.4 Remove the bolt (upper arrow) to detach the door stay from the body - pull the rubber boot out (lower arrow) and feed the door wiring harness out

16.6 Remove the hinge bolts (arrows) and lift the door off (upper hinge shown, lower hinge similar)

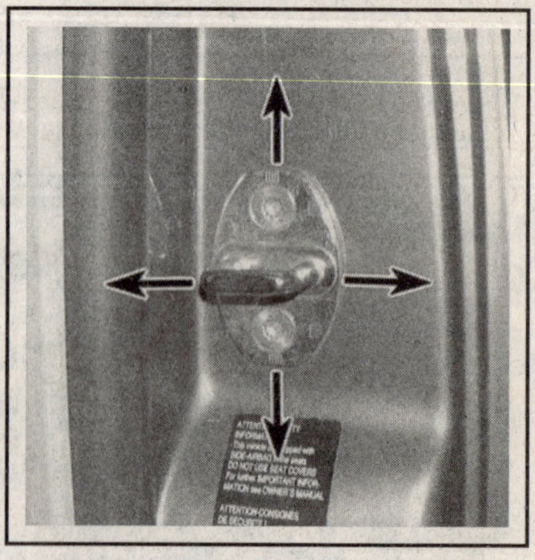

16.8 The striker can be loosened and moved slightly to achieve secure latch engagement

17 Door latch, lock cylinder and outside handle - removal and installation

LATCH

Refer to illustrations 17.5, 17.6 and 17.7

1 Remove the door trim panel (see Section 15). The door latch, radio speaker and window regulator are all attached to a large metal carrier, which must be removed from the door as a unit.

2 Refer to Steps 9 and 10 and remove the lock cylinder.

3 Lower the window glass until the glass track bolts can be loosened (see Section 18). With the glass separated from the regulator, pull the glass up by hand and tape it in the UP position.

4 Refer to Steps 2 and 3 in Section 16 and disconnect the electrical harness for the door, feeding the wires into the door cavity.

5 Remove the two Torx-head mounting screws from the end of the door (it may be necessary to use an impact-type screwdriver to loosen them) and detach the latch from the door (see illustration).

6 Remove the bolts securing the carrier to the door, then lift it up and out, working towards the hinged end of the door (see illustration). At the rear of the carrier, disconnect the electrical connector at the door lock solenoid.

7 To remove the latch from the carrier, use a punch to drive out the two pins, then flip the latch over and disconnect the cable and the linkage rod (see illustration).

8 Installation is the reverse of the removal procedure.

17.5 Remove the two Torx bolts (arrows) securing the latch to the door frame

LOCK CYLINDER

Refer to illustration 17.9

9 Remove the plastic plug in the latch end of the door and loosen the setscrew at the lock cylinder (see illustration).

CAUTION:

Do not turn the setscrew all the way out, or it will fall into the door.

10 Insert the key in the lock cylinder and pull the cylinder out of the door.

11 Installation is the reverse of removal.

OUTSIDE HANDLE

Refer to illustration 17.13

12 Refer to Steps 9 and 10 and remove the lock cylinder.

13 Remove the clip, swing the key-end of the handle outward from the door, then rearward (see illustration).

14 Installation is the reverse of removal.

17.6 Plastic carrier mounting bolts (arrows)

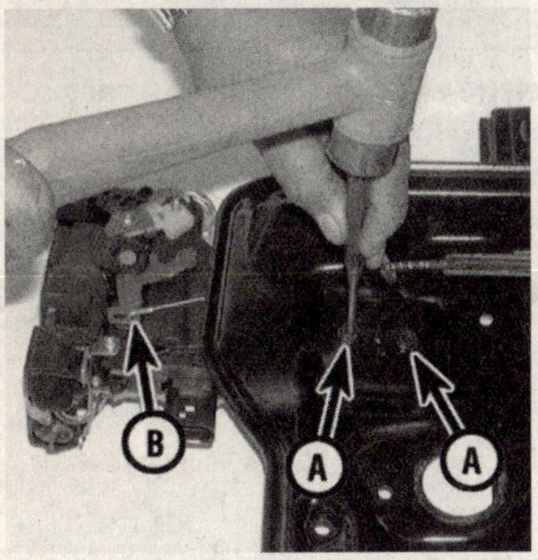

17.7 Tap out these pins (A) with a punch and hammer, then disconnect the cable (B)

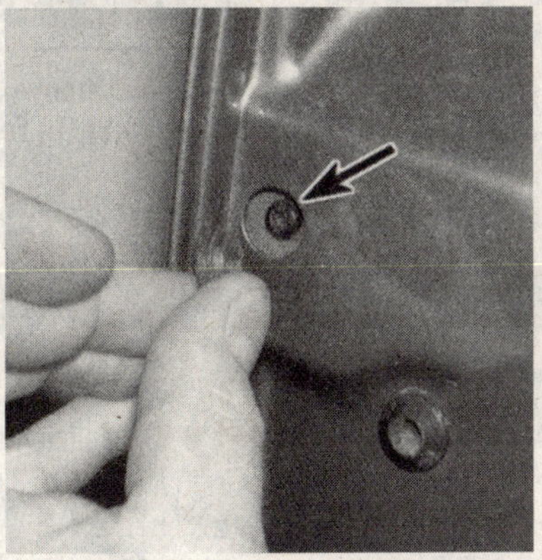

17.9 Loosen the screw through this hole (arrow) after removing the plug

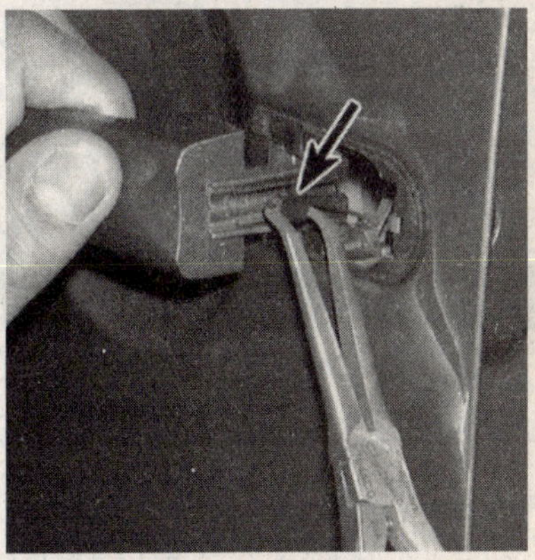

17.13 Pull this clip (arrow), then remove the door handle

18 Door window glass - removal and installation

REMOVAL

➧ Refer to illustration 18.2

1 Remove the door trim panel (see Section 15).
2 Working through the two holes in the carrier, use a socket and extension to loosen the bolts retaining the regulator glass channel to the glass (see illustration). Separate the clamps to free the glass.
3 Secure the glass in this position with tape, then lower the window regulator all the way down.
4 Tilt the rear end of the glass up and out of the door until the glass clears the door.

18.2 Remove the two rubber plugs from the carrier, then loosen the glass track bolts (arrows) - you may have to spread the clips with a screwdriver to replace the glass

INSTALLATION

5 Raise the regulator until the glass channel bolts align with the holes in the carrier (see illustration 18.2).
6 Lower the glass into the door front end first, then tilt it until the clamps fit around the glass and the bolts can be tightened.
7 The remainder of the installation is the reverse of the removal procedure.

19 Door window glass regulator/motor - removal and installation

➧ Refer to illustration 19.5

1 Remove the door trim panel (see Section 15).
2 The window regulator and motor is part of the plastic inner door carrier.
3 The regulator is part of the carrier assembly and can't be replaced separately, although the window winder (manual windows) or electric motor (power windows) can be replaced.
4 On manual-window models, remove the carrier from the door (see Section 17), access the fasteners on the back of the window winder and remove the winder from the front.
5 On models with power windows, remove the fastener(s) at the front of the carrier and remove the motor (see illustration). The carrier does not have to be removed from the door.
6 Installation is the reverse of removal.

19.5 Remove the bolts (arrows) at the front of the carrier assembly to remove the power window motor

20 Mirrors - removal and installation

INTERIOR

➧ Refer to illustrations 20.1 and 20.2

1 Pull down the panel at the front of the overhead console (see illustration).
2 Depress the spring clip where the base of the mirror fits through the overhead panel, and slide the mirror out (see illustration).
3 Installation is the reverse of removal.

EXTERIOR

➧ Refer to illustrations 20.6 and 20.7

4 Refer to Section 15 and remove the door trim panel.

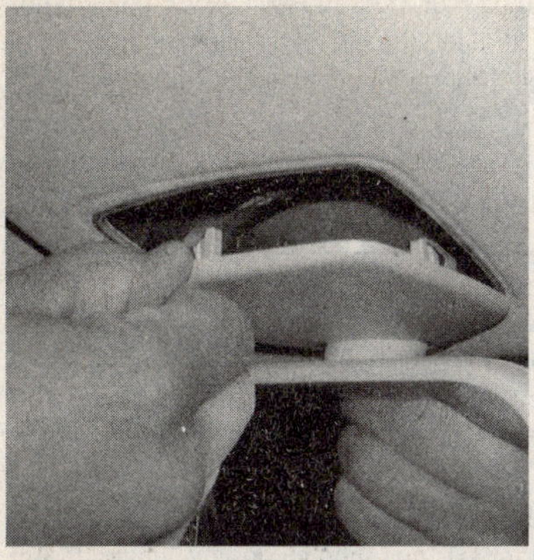

20.1 Pull down the panel in the overhead console and disconnect the electrical connector to the overhead light

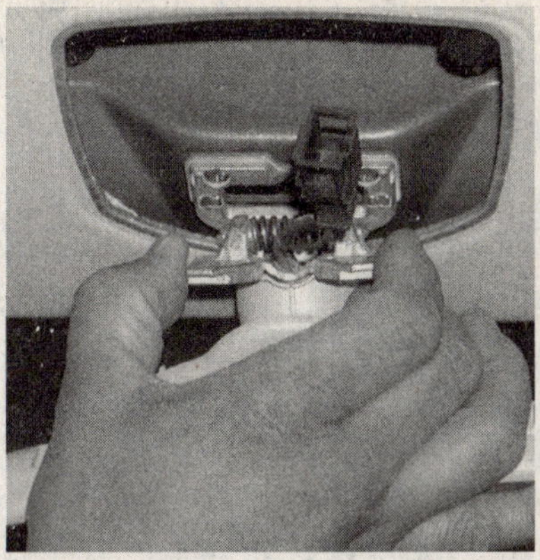

20.2 Push inward on the two clips, then slide the mirror base rearward out of the panel

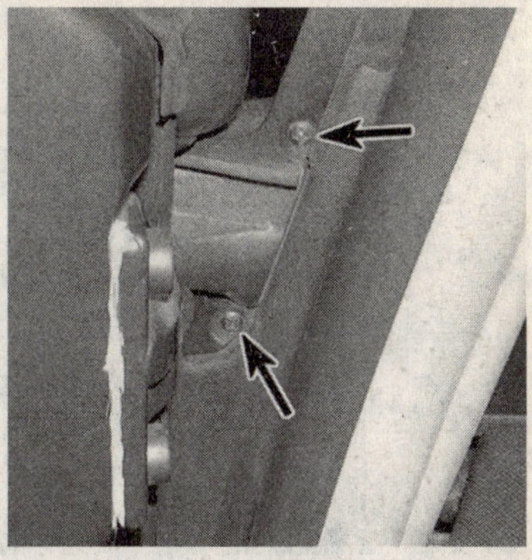

20.6 Remove the mirror hinge cover mounting screws (arrows)

5 Disconnect the electrical connector for the mirror, inside the door, then apply tape to the connector so that it can be withdrawn through the mirror harness hole in the outside of the door without snagging.

6 With the door open, remove the two screws securing the mirror's hinge cover, then remove the hinge cover (see illustration).

7 Remove the screws and detach the mirror from the door (see illustration). Work the mirror harness through the hole in the door.

8 Installation is the reverse of removal.

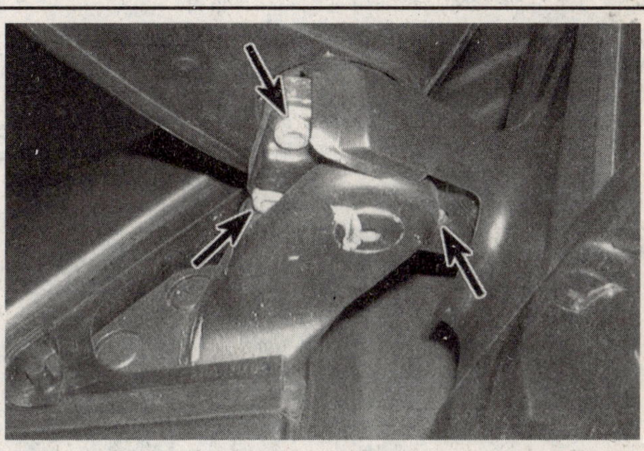

20.7 Swivel the mirror so that all three of the screws (arrows) can be removed, then take the mirror off its base

21 Rear hatch and hatch support struts - removal, installation and adjustment

➡Note 1: The hatch is heavy and somewhat awkward to remove and install - at least two people should perform this procedure.

➡Note 2: The procedure is virtually the same for 2003 and later convertible rear lid.

HATCH

▸ Refer to illustrations 21.3, 21.8a, 21.8b and 21.8c

1 Open the hatch and cover the edges of the trunk compartment with pads or cloths to protect the painted surfaces when the lid is removed.

2 Disconnect the electrical connector at the rear window defogger, if equipped.

3 Use a marking pen to make alignment marks around the hinge bolt heads (see illustration).

4 Prop the hatch open with a long prop like a broom handle. Refer to Steps 9, 10 and 11 and remove the two support struts.

5 While an assistant supports the weight of the hatch, remove the hinge bolts from both sides and lift the hatch off.

6 Installation is the reverse of removal.

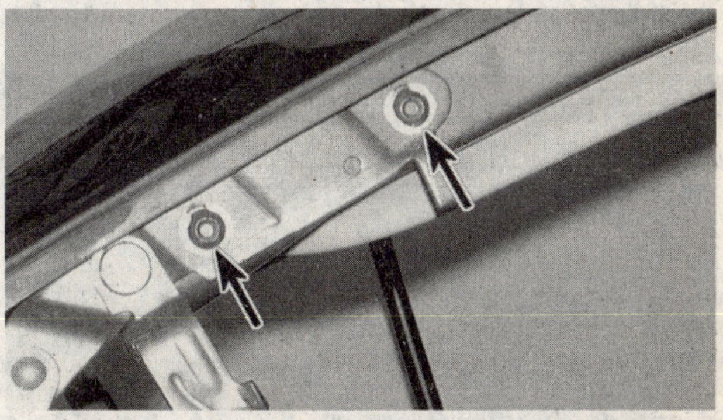

21.3 Draw around the bolt heads (arrows) with a marking pen so the hatch can be installed in the same position

➡Note: When reinstalling the hatch, align the hinge bolt heads with the marks made during removal.

7 After installation, close the hatch and see if it's in proper alignment with the surrounding panels. Fore-and-aft and side-to-side adjust-

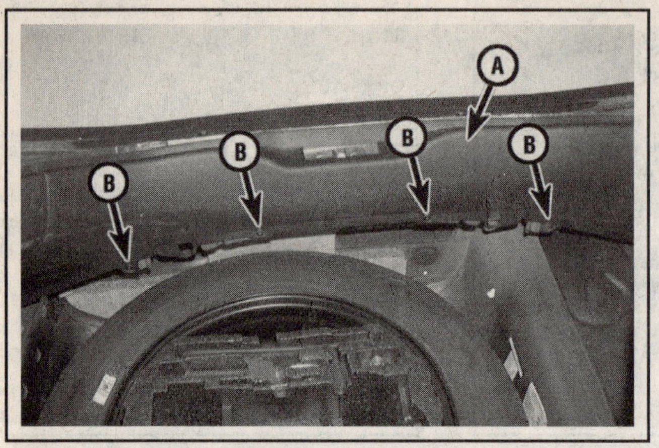

21.8a To access the striker, remove the screws (B) and this lower trim panel (A) in the trunk

ments of the hatch are controlled by the position of the hinge bolts in the holes. To adjust it, loosen the hinge bolts, reposition the hatch and retighten the bolts.

8 The height of the hatch in relation to the surrounding body panels when closed can be adjusted by loosening the lock striker bolts, repositioning the striker and retightening the bolts (see illustrations). Some adjustment can also be made through positioning the adjustment "buffer" cushion (see illustration).

HATCH SUPPORT STRUTS

➧ Refer to illustration 21.11

9 The hatch supports are attached to upper and lower studs on the body and the hatch.

10 To replace hatch supports, open the hatch and support it with a prop.

11 Pry out the locking caps at the top and bottom of the support struts and pull the struts from their mounting studs (see illustration).

➡Note 1: Always replace support struts as a pair, but install the new one on one side of the vehicle before removing the old one from the other side. Always have one strut in place.

➡Note 2: Only pry out the clips just enough to release the strut from the ball-stud. If the clip comes completely off, the strut will have to be replaced.

12 Installation is the reverse of removal.

21.8b To adjust the striker, mark around the edges with a pen, then loosen the bolts (arrows) and move the striker slightly and retighten

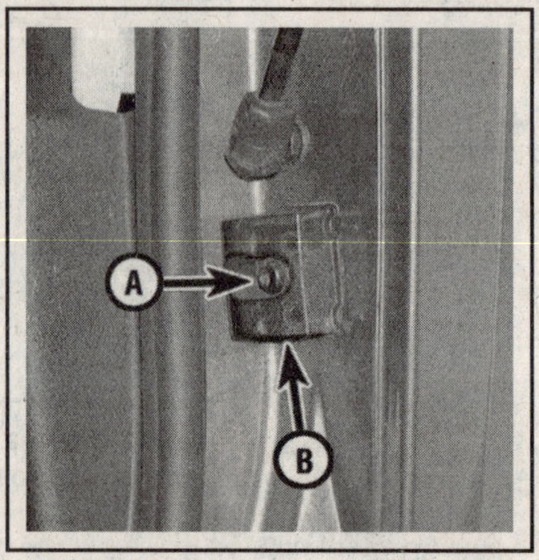

21.8c Loosen the bolt (A) and raise the buffer (B) - close and open the hatch, then check the height of the buffer; raise it 1 or 2 mm and tighten the bolt

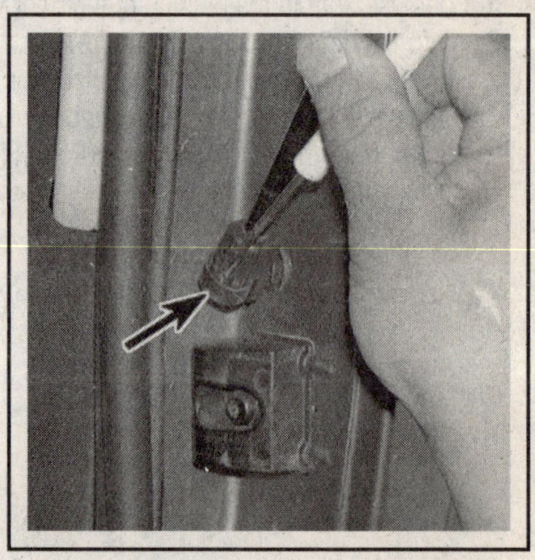

21.11 Pry the hatch support strut locking clips out (arrow), then pry the ends from their mounting studs (lower end shown, upper end similar)

22 Rear hatch latch and lock cylinder - removal and installation

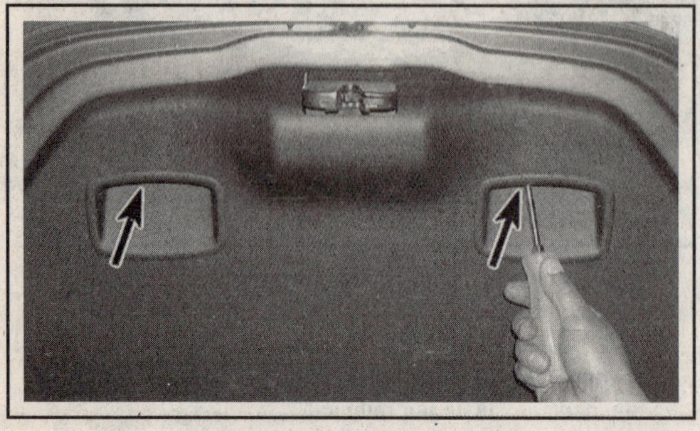

22.1 Remove the screws (arrows), then use a trim tool to pry around the edges of the hatch trim panel

LATCH

➧ Refer to illustrations 22.1 and 22.3

1 Open the hatch and remove the two screws, then use a trim tool to pry off the hatch trim panel (see illustration).

2 Scribe a line around the latch assembly for a reference point to aid the installation procedure.

3 Disconnect the electrical connector (on power models), pop the link from the rod connecting the latch and the lock cylinder, and remove the latch mounting screws (see illustration).

4 Installation is the reverse of removal.

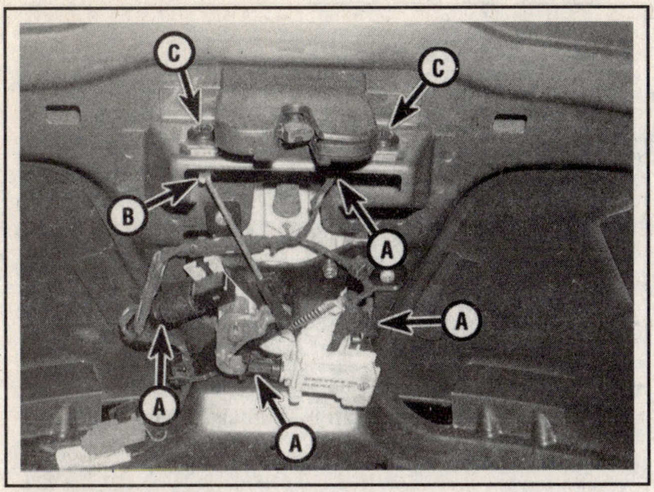

22.3 Disconnect the electrical connectors (A, only one is at the latch itself), pry off the control rod (B), then remove the mounting screws (C)

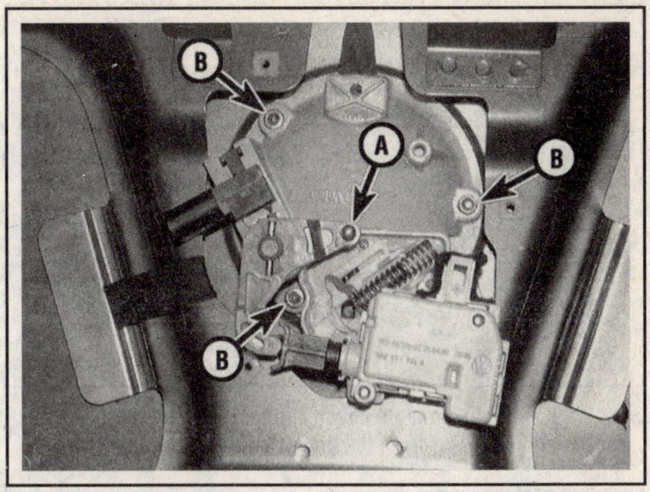

22.6 Disconnect the lock cylinder rod from the stud (A), then remove the screws (B) - when you pull down the lock cylinder and carrier, the VW emblem on the outside will also come off

LOCK CYLINDER

Refer to illustration 22.6

5 Open the hatch and remove the two screws, then use a trim tool to pry off the hatch lower trim panel (see illustration 22.1).

6 Disconnect the lock cylinder rod and remove the three screws (see illustration).

7 Lower the lock cylinder and its carrier down from the hatch.

Note: The exterior VW emblem will comes off as you take down the carrier, since the carrier screws go through the hatch and into the emblem.

8 Installation is the reverse of removal.

23 Center console - removal and installation

WARNING:

The models covered by this manual are equipped with Supplemental Restraint Systems (SRS), more commonly known as airbags. Always disable the airbag system before working in the vicinity of any airbag system components to avoid the possibility of accidental deployment of the airbags, which could cause personal injury (see Chapter 12).

CONSOLE SHIFT BEZEL

Refer to illustrations 23.1a, 23.1b, 23.2a, 23.2b, 23.3a, 23.3b and 23.3c

1 Remove the front and rear cupholders to access the screws holding the console cover (see illustrations).

2 Remove the one screw near the parking brake handle and pry up the console cover to release it from the clips (see illustration). Discon-

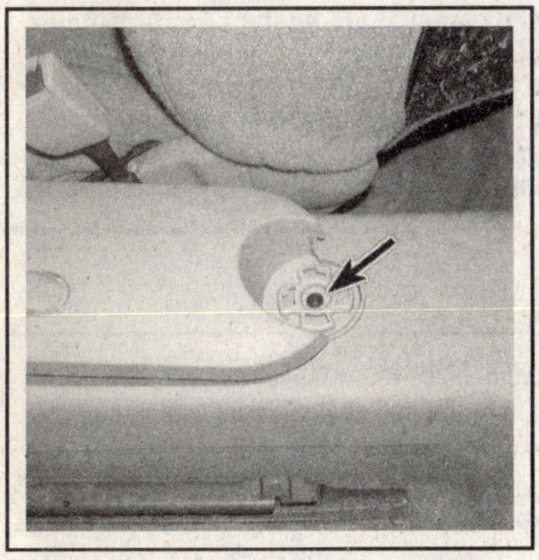

23.1a Remove the rear cupholder to access this screw (arrow) . . .

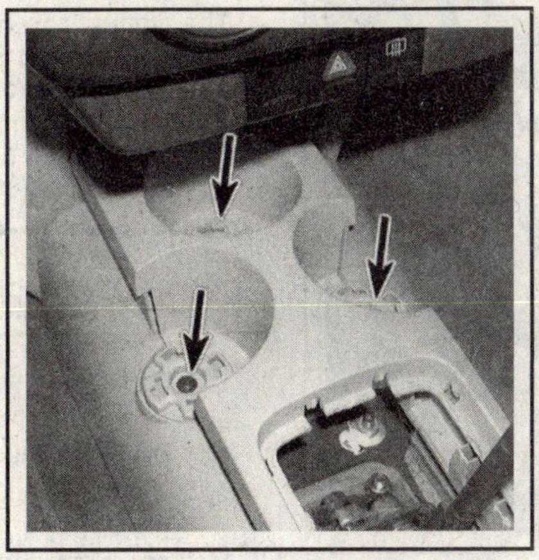

23.1b . . . and the cupholders at the front of the console to access these screws (arrows)

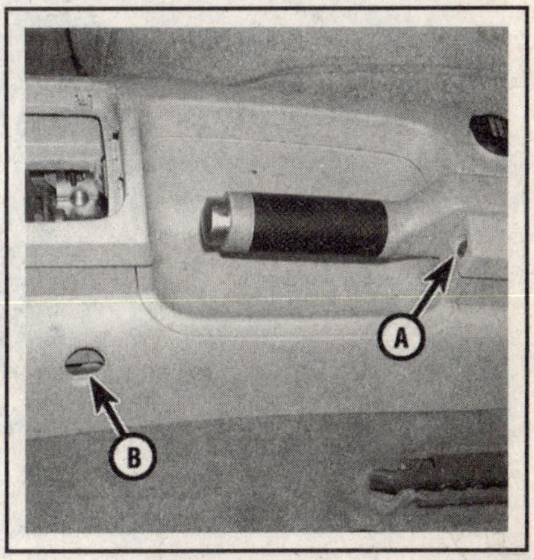

23.2a Remove this screw (A) at the rear of the parking brake cavity, and remove the plastic caps to access a screw on each side of the console (B)

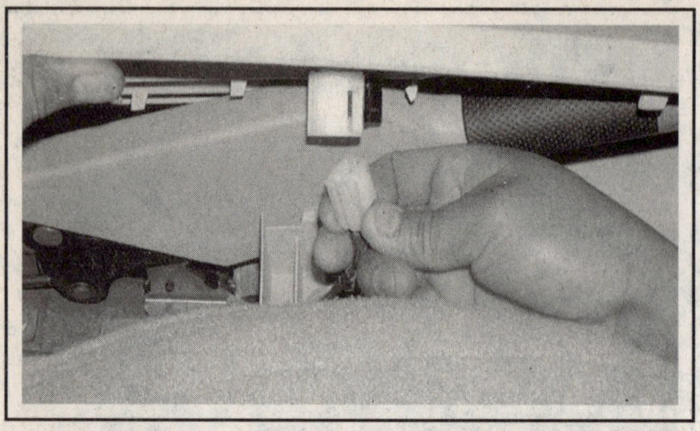

23.2b Pry up the console cover, then disconnect the electrical connector from the power outlet in the cover

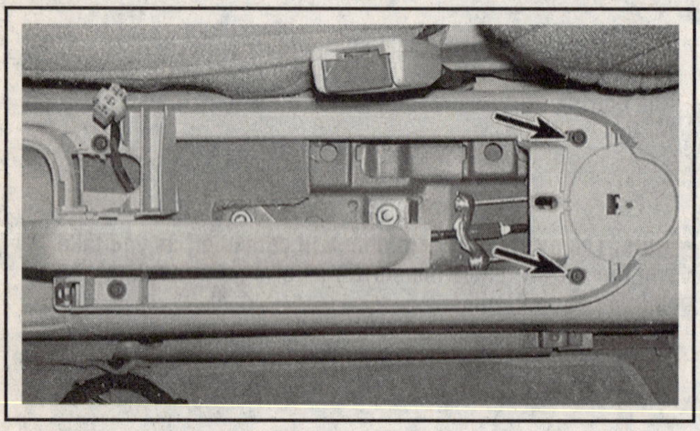

23.3b . . . then these two (arrows) at the rear of the console . . .

nect the electrical connector to the power outlet (see illustration).

3 Remove the screws and lift the console out (see illustrations).

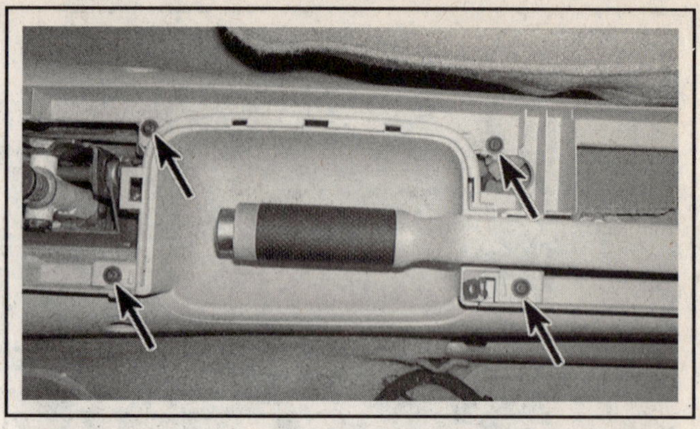

23.3a Remove the screws (arrows) around the parking brake handle . . .

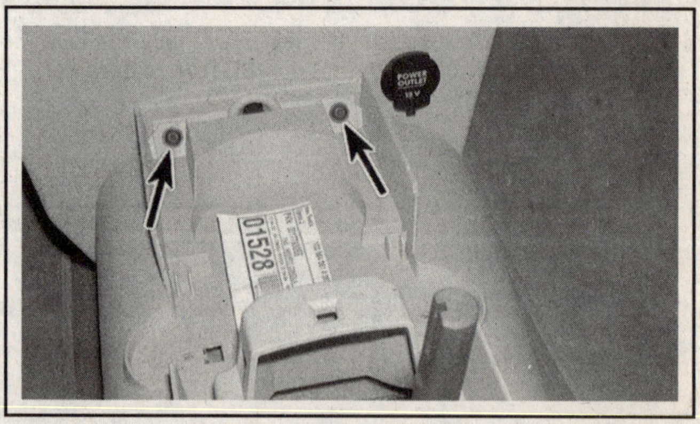

23.3c . . . and these two (arrows) at the front of the console

➡Note: The screw locations vary with the year and model, but are visible on all models when the upper panel is removed.

4 Installation is the reverse of the removal procedure.

24 Dashboard trim panels - removal and installation

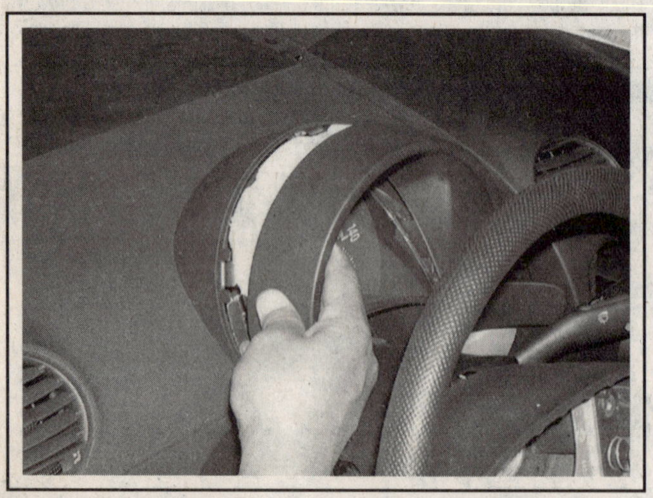

24.5 Pull straight back on the instrument cluster bezel to remove it

WARNING:

The models covered by this manual are equipped with Supplemental Restraint Systems (SRS), more commonly known as airbags. Always disable the airbag system before working in the vicinity of any airbag system components to avoid the possibility of accidental deployment of the airbags, which could cause personal injury (see Chapter 12).

DASHBOARD END CAPS

1 The end caps are held in place by clips.

2 Grasp the cover securely and pull sharply to remove it. If necessary, gently pry with a trim tool.

3 Installation is the reverse of removal.

INSTRUMENT CLUSTER BEZEL

➧ Refer to illustration 24.5

4 Lower the tilt steering wheel to its lowest position.

24.7 The bolster is held by screws (arrows) at the side, bottom and clips at the top

24.9 Remove the screws (arrows) to remove the knee bolster reinforcement panel

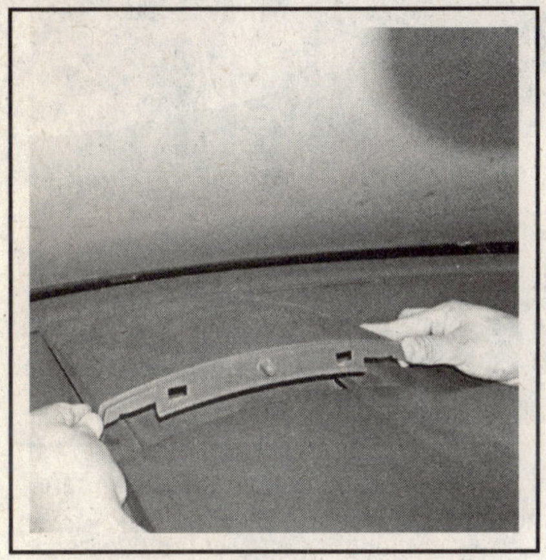

24.11 Push forward on the center trim panel forward extension, then lift it up to clear the clips

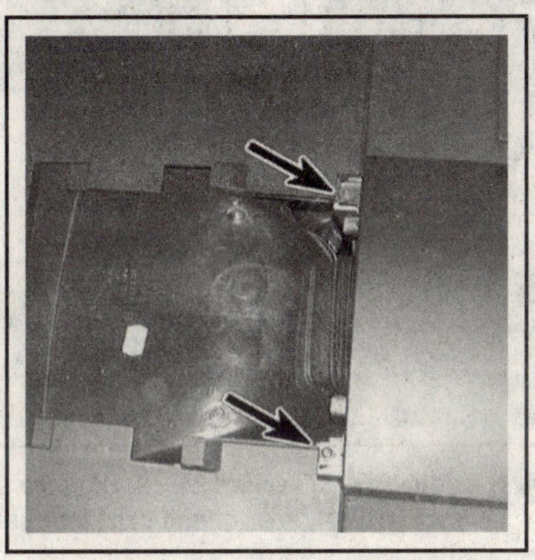

24.13a At the top of the center panel, remove the two screws (arrows) . . .

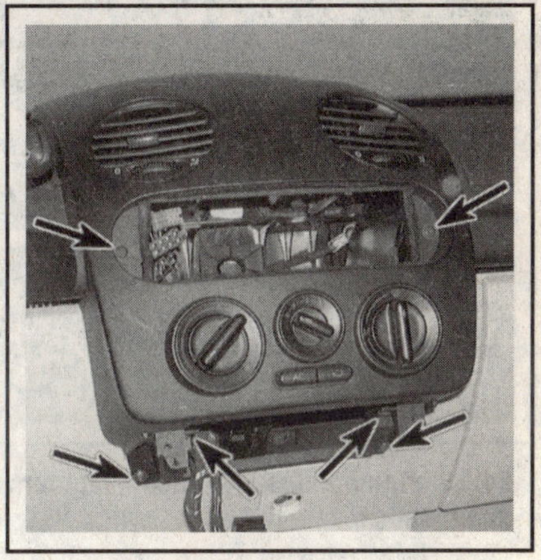

24.13b . . . then remove the remaining screws (arrows) and lift off the center trim panel

5 Pull the instrument cluster bezel straight back to release the three clips (see illustration).

6 Installation is the reverse of removal.

DRIVER'S KNEE BOLSTER

➧ Refer to illustrations 24.7 and 24.9

7 Remove the screws retaining the bottom of the bolster to the instrument panel (see illustration). Remove the left dashboard end cap to access two side screws.

8 Pull backward on the top of the bolster to release it from the clips, and disconnect the electrical connectors at the headlight and dimmer switches.

9 If the bolster reinforcement panel needs to be removed to perform a repair procedure, remove the reinforcement panel screws (see illustration).

10 Installation is the reverse of removal.

CENTER TRIM PANELS

➧ Refer to illustrations 24.11, 24.13a and 24.13b

11 To access the upper screws for the center trim panel, push forward on the center panel's front extension until it comes free of its clips, then lift it up and off (see illustration).

12 Refer to Chapter 12 and remove the radio.

13 Remove the screws at the top of the center trim panel, then the two that had been hidden by the radio (see illustrations). Lift out the trim panel.

14 Installation is the reverse of removal.

UPPER DASH COVERS AND PLENUM CLOSEOUT PANELS

➧ Refer to illustrations 24.16a, 24.16b and 24.18

15 Remove the center trim panel forward extension (see illustration 24.11). On models so equipped, remove the flower vase and pry the vase bracket from its clips.

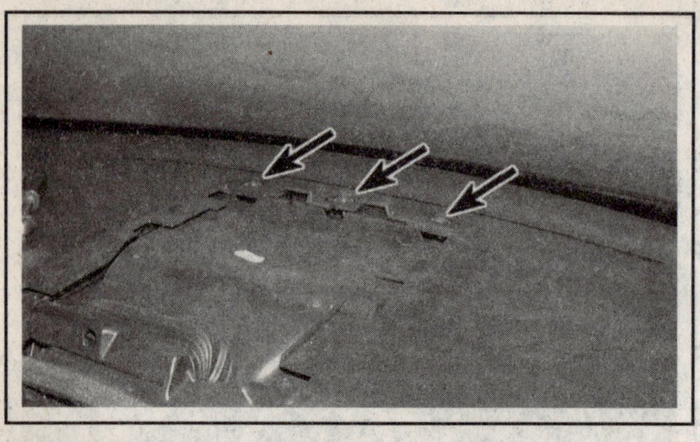

24.16a Remove these three screws (arrows) . . .

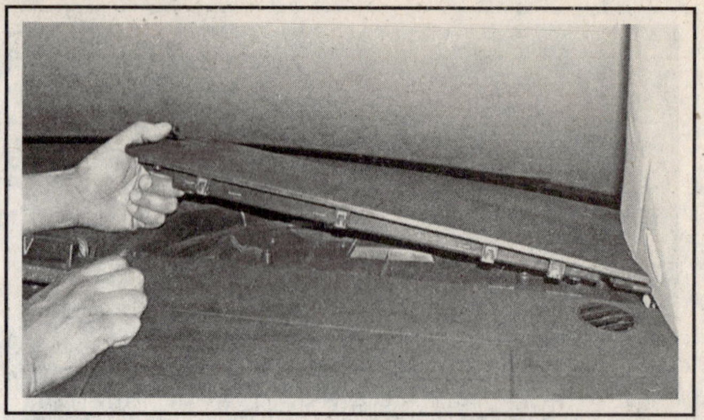

24.16b . . . then remove the left and right dash covers (right shown)

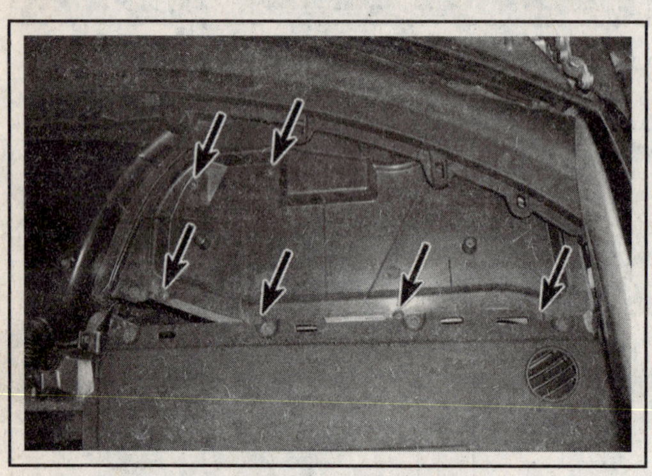

24.18 Plenum closeout panel fasteners (arrows) - right panel shown

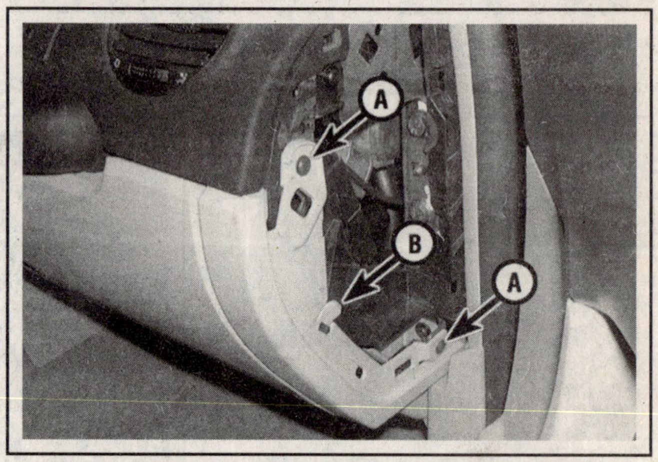

24.20 Remove screws (A) after removing the right dashboard end cap, then release the clip (B)

16 Remove the screws at the center of the instrument panel securing the two upper dash covers, then remove the covers (see illustrations). On later models, there is a panel of switches below the HVAC controls. Remove the screws (pry off the caps for access), then lower the panel and disconnect the electrical connectors.

17 Below the plastic dash covers are left and right plenum closeout panels (metal). These need to be removed for access to other components (the Engine Control Module is under the left cover, the air conditioning filter under the right).

18 To remove the plenum closeout panels, remove the bolts and angle the panels out (see illustration).

19 Installation is the reverse of the removal procedure.

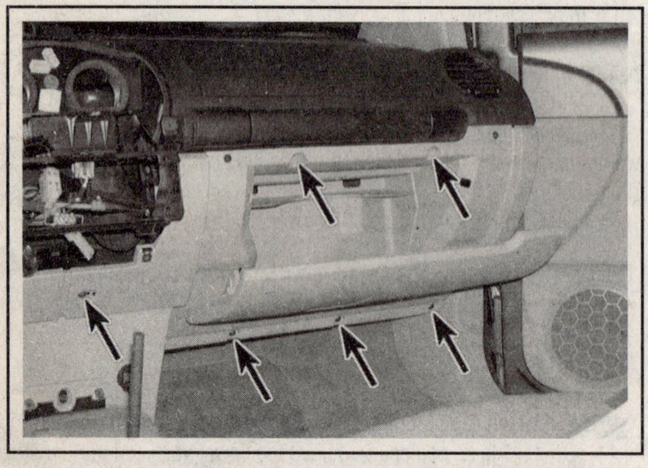

24.21a Remove the screws (arrows) to take out the glove box and panel

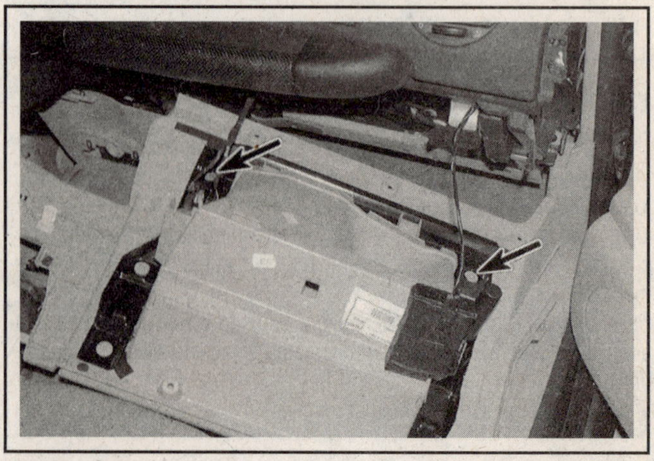

24.21b Disconnect the electrical connectors (arrows) behind the glove box

GLOVE BOX

➧ Refer to illustrations 24.20, 24.21a, 24.21b and 24.22

20 Snap out the right side dashboard end cap and remove the glovebox screws (see illustration).

21 Remove the mounting screw (under a plastic cap) at the left end of the glove box assembly and the screws below the glove box (see illustrations).

22 For some repair procedures, the frame behind the glovebox may need to be removed. Remove the screws and the frame (see illustration).

23 Installation is the reverse of removal.

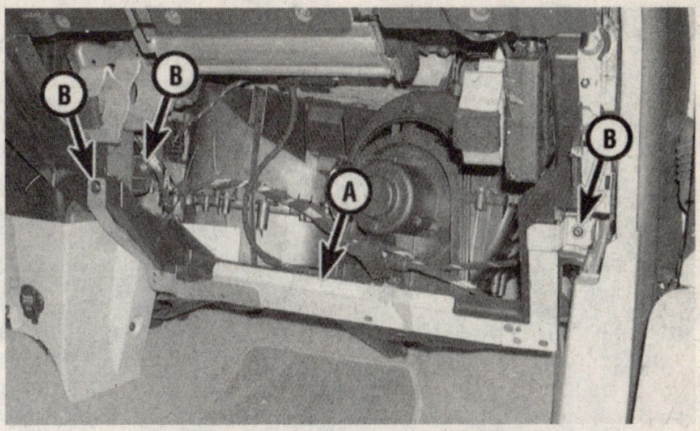

24.22 To remove the glove box frame (A), remove these fasteners (B)

25 Steering column covers - removal and installation

➧ Refer to illustration 25.1

1 Remove the screws from the lower steering column cover, then remove the upper cover (see illustration). On later models, remove the two bolts and the upper cover first, then the three bolts and the lower cover.

2 To remove the lower cover, the steering wheel must be removed (see Chapter 10). Pull down on the steering column tilt lever, then remove the lower cover.

3 Installation is the reverse of removal.

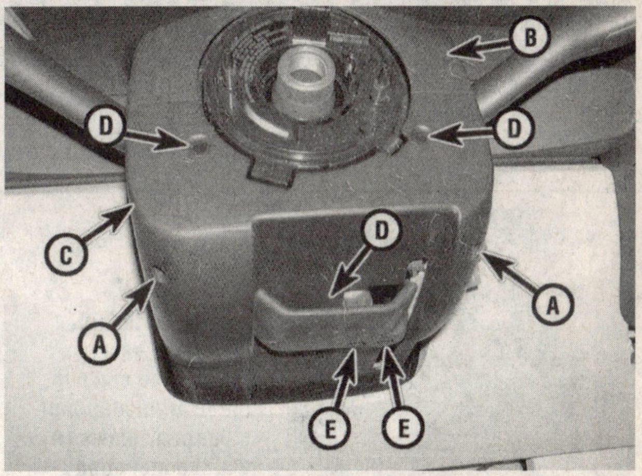

25.1 Steering column cover mounting details

A Upper cover mounting screws
B Upper column cover
C Lower column cover
D Lower cover mounting screws
E Tilt lever handle mounting screws

26 Instrument panel - removal and installation

➧ Refer to illustrations 26.7, 26.10, 26.11a, 26.11b, 26.12a, 26.12b, 26.12c, 26.12d, 26.12e and 26.12f

⁂ WARNING:

The models covered by this manual are equipped with Supplemental Restraint Systems (SRS), more commonly known as airbags. Always disable the airbag system before working in the vicinity of any airbag system components to avoid the possibility of accidental deployment of the airbags, which could cause personal injury (see Chapter 12).

⁂ CAUTION:

This is a difficult procedure for the home mechanic, involving tedious disassembly and the disconnection/reconnection of numerous electrical connectors. If you do attempt this procedure, make sure you take good notes and mark all matching connectors (and their mounting points) to aid reassembly.

1 Disconnect the negative battery cable (see Chapter 5). Read the **Cautions** in Section 1.

2 Remove the center console (see Section 23).

3 Remove the steering column covers (see Section 25).

4 Remove the dashboard trim panels (see Section 24).

5 Remove the audio components, the multi-function switch, the

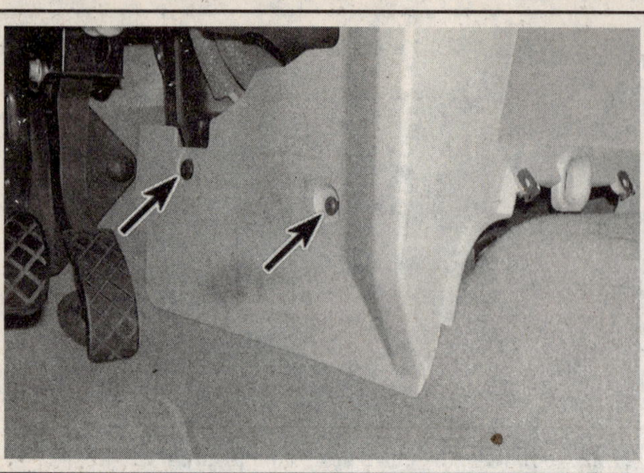

26.7 Remove the screws (arrows, two are on the left side, one on the right) and the panel below the center of the dash

instrument cluster, the center dash switch panel and the passenger air-bag (see Chapter 12).

6 Refer to Chapter 10 and remove the steering wheel.

7 Remove the two screws and the trim panel below the driver's side of the instrument panel. Remove the screws and the panel below the center of the dash (see illustration).

➡**Note: Disconnect the electrical connector at the back of the**

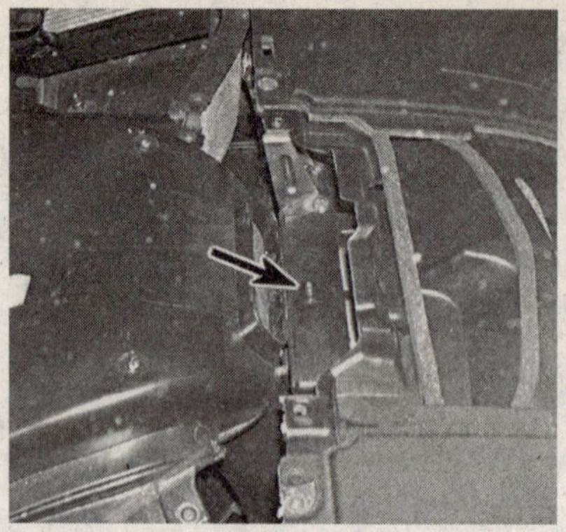

26.10 Center instrument panel mounting screw (arrow)

power outlet.

8 At the upper center of the instrument panel, remove the screws and the left and right upper dash panels, and the plenum closeout panels below them (see Section 24).

9 Refer to Chapter 3 and remove the heating/air conditioning controls.

10 At the top center of the dash, remove the mounting screw (see illustration).

11 Remove the instrument panel screws at the left and right end of the dash (see illustrations).

12 Remove all of the fasteners securing the instrument panel, and any electrical connectors still attached to the panel (see illustrations). Have an assistant help you pull the panel back and out of the vehicle.

13 Installation is the reverse of removal.

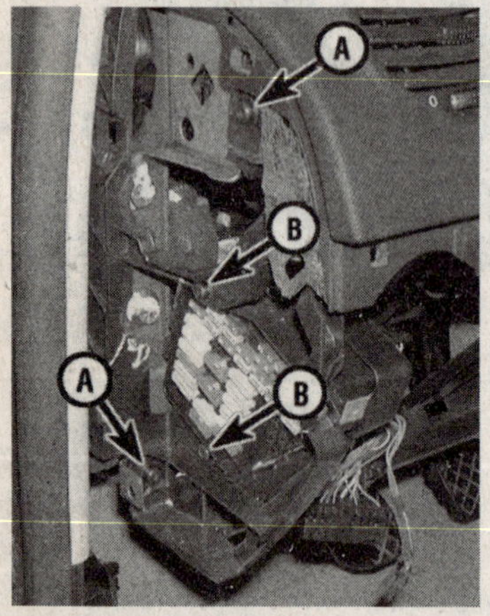

26.11a At the left end of the dash, remove the end cap, the instrument panel mounting screws (A) and the mounting screws for the fuse/relay box (B)

COWL REINFORCEMENT BEAM

➧ **Refer to illustrations 26.16, 26.17a, 26.17b, 26.18a and 26.18b**

14 With the instrument panel removed, the cowl reinforcement beam can be removed if necessary for other repair procedures, such as heating/air conditioning work.

⁂ WARNING:

This is a difficult job for the home mechanic and involves safety-related critical fasteners.

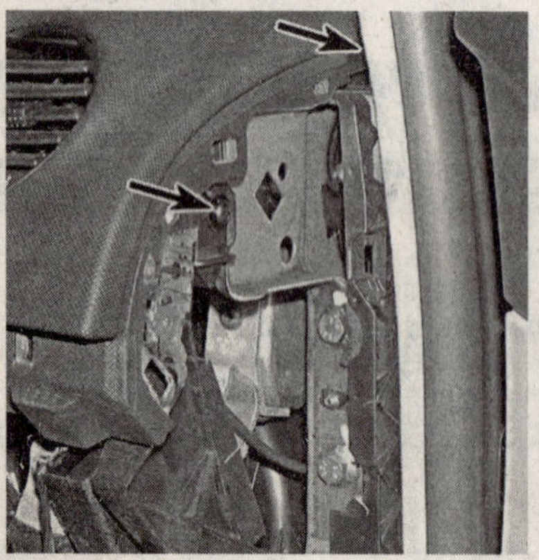

26.11b Remove the mounting screws (arrows) at the right end of the instrument panel

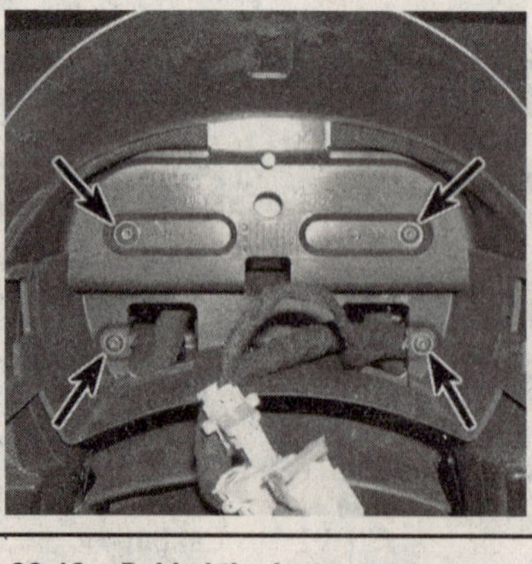

26.12a Behind the instrument cluster, remove these mounting screws (arrows) and remove the cluster housing

26.12b With the cluster housing removed, one more dash screw (arrow) is revealed

26.12c Pry off the grab bar covering, then remove these two screws (arrows)

26.12d Arrows here indicate electrical connectors to be disconnected at the left end of the instrument panel - there are others at the lower center and right end

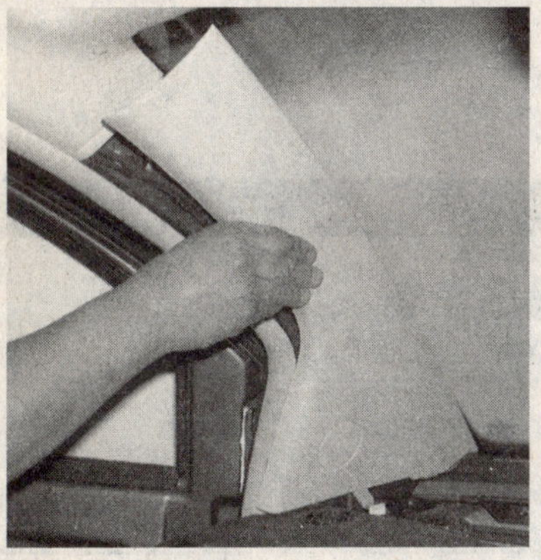

26.12e Use a trim tool to remove the inner windshield post trim covers

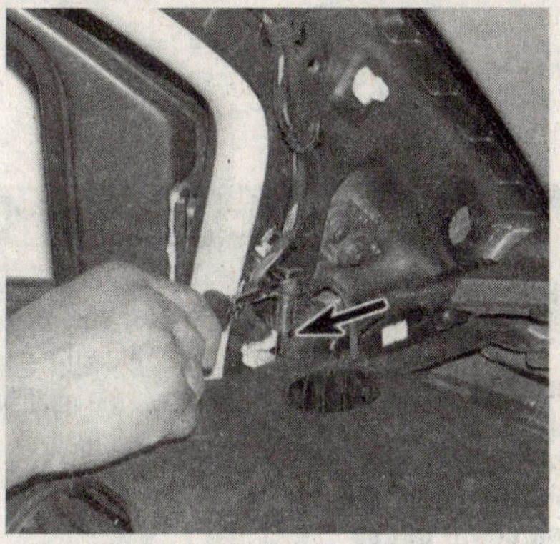

26.12f At each upper corner of the instrument panel, remove the upper mounting bolt and remove the instrument panel

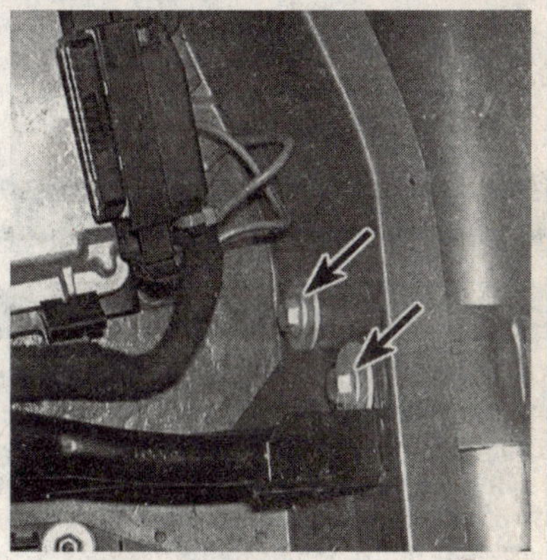

26.16 Remove the beam-to-firewall bracket bolts (arrows)

15 Many electrical harnesses and connectors are attached with clips to the beam. Detach them all, taking notes and marking items with masking tape and marking pen to aid in reassembly.

16 Under the center of the dash, remove the bolts securing the beam to the brace from the firewall (see illustration).

17 Also at the center of the dash, remove the beam-to-floor-bracket bolts (see illustrations).

26.17a Remove the left beam-to-bracket bolts (arrows) . . .

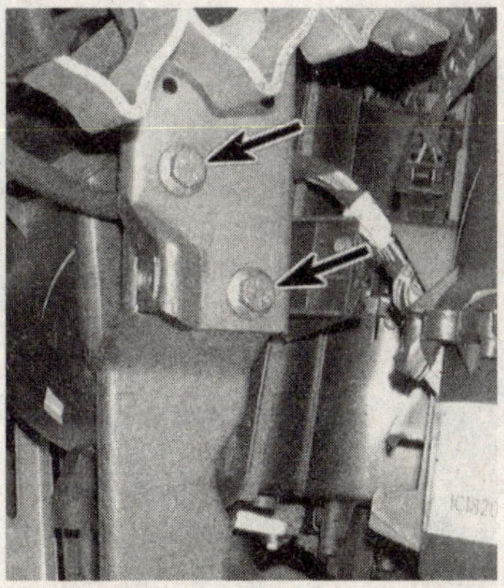

26.17b . . . and the right bolts (arrows)

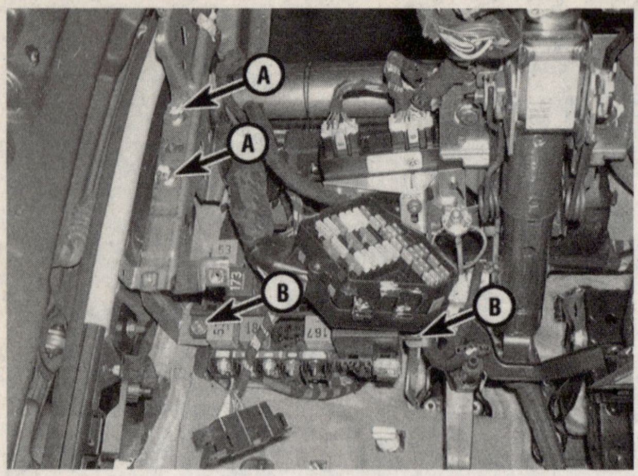

26.18a Remove the mounting bolts (A) at the left end of the beam, and the fuse/relay box mounting bolts (B) - there are two bolts at the far right end of the beam also

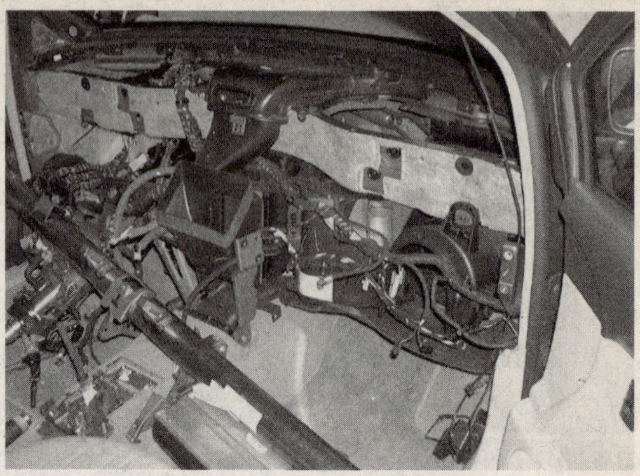

26.18b With all fasteners and electrical harness removed, the beam can be pulled back for access to heating/air conditioning components

18 At each end of the beam, remove the fasteners, then pull the beam away from the cowl (see illustrations).

19 Installation is the reverse of removal.

27 Seats - removal and installation

⁂ WARNING:

The models covered by this manual are equipped with Supplemental Restraint Systems (SRS), more commonly known as airbags. Always disable the airbag system before working in the vicinity of any airbag system components to avoid the possibility of accidental deployment of the airbags, which could cause personal injury (see Chapter 12).

FRONT

Refer to illustrations 27.2a and 27.2b

1 The front seats on some models are equipped with side-impact airbags at the upper outside of the seatback. Refer to Chapter 12 to disable the airbag system before working on front seats.

2 Slide the seat forward and remove the seat track covers, move the seat rearward and remove the front seat track bolts and unplug any electrical connectors attached to the seat (see illustrations).

3 Slide the seat all the way to the back of the tracks and off. Remove the seat from the vehicle.

4 Installation is the reverse of removal.

REAR

Refer to illustrations 27.5 and 27.6

5 Remove the seat bottom cushion by pushing in on the front edge of the cushion in the vicinity of the clips, to disengage part of the seat's wire frame from the clips (see illustration). With the clips unhooked, lift up and remove the seat bottom from the vehicle.

6 Fold the seat back down, then remove the two center bolts (see illustration). Push the spring-loaded hook back to release the seat pivot at each side, then pull out the seat back.

7 Installation is the reverse of removal. Torque the two seat back bolts to 15 ft-lbs.

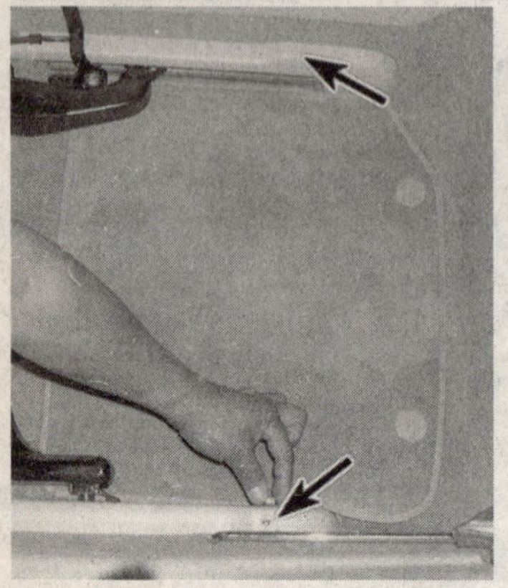

27.2a Remove the plastic buttons in the seat track covers, then remove the screws (arrows) and the covers

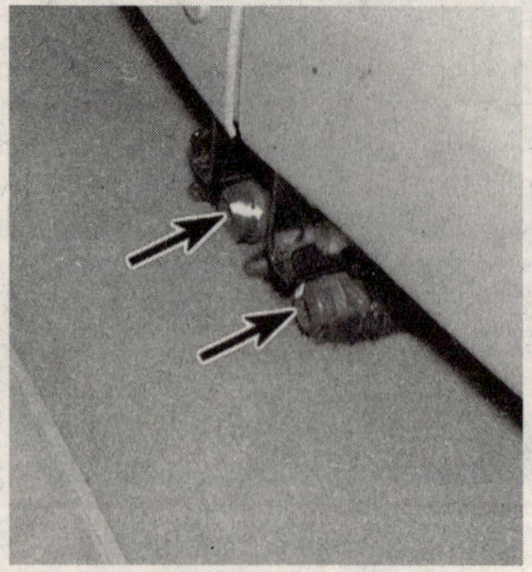

27.2b Remove the front seat front mounting bolts (arrows), then slide the seat all the way back and off the tracks

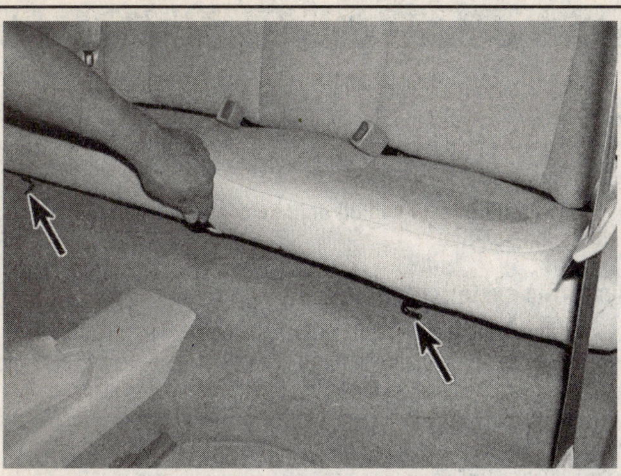

27.5 Pull up the center tab of the rear seat cushion to disengage the cushion from two clips (arrows)

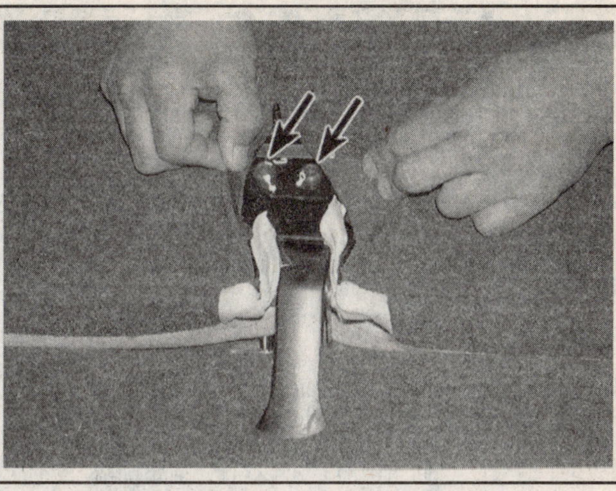

27.6 Unzip the upholstery, then remove the two bolts (arrows) at the bottom-center of the seat back

28 Sunroof - adjustment

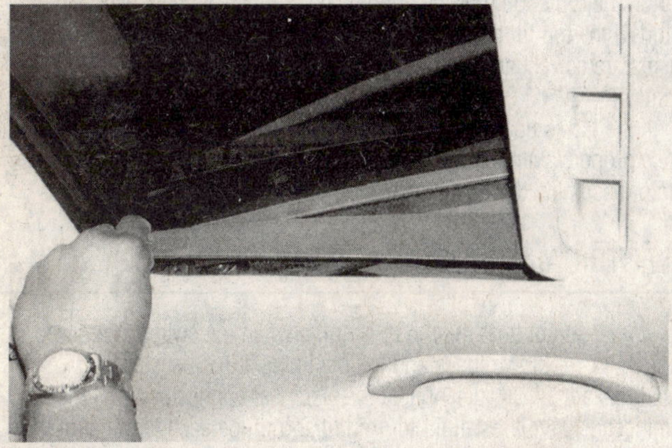

28.3a Unsnap and pull out the first sunroof track plastic cover . . .

Refer to illustrations 28.3a, 28.3b and 28.4

1 The position of the glass panel can be adjusted in the following manner.

2 Open the interior sunshade all the way back and operate the sunroof until it is tilted open.

3 There are glass mounting screws on each side of the sunroof opening, inside the vehicle. Remove the plastic covers (see illustrations).

4 Loosen the front screws and lower the sunroof (see illustration). Set the front edge of the glass approximately 0.039-inch (1 mm) below the surface of the roof when closed. Tighten the front screws.

5 Loosen the rear glass adjustment screws and adjust the glass approximately 0.039-inch (1 mm) above the roof at the rear. Tighten the rear screws.

6 When the adjustment is correct, tilt the sunroof open and reinstall the plastic covers over the glass rails.

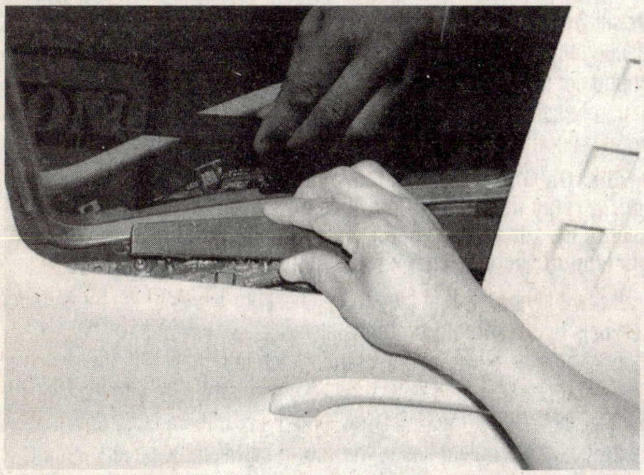

28.3b . . . then the second cover to reveal the tracks

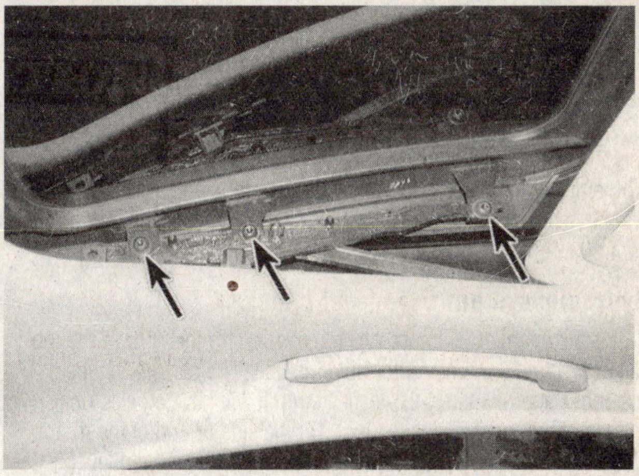

28.4 The sunroof glass has three screws on each side (arrows) used to adjust the fit of the glass to the roof

29 Convertible top - general information

➡Note: This information is general in nature, as it is intended to apply to all vehicle types, years and models.

ADJUSTMENTS

➡Note: The following are typical adjustments. Some of these adjustments may not be applicable to your vehicle.

Latches

The latches secure the convertible top frame to the upper edge of the windshield frame. If the latches are too loose, the top will rattle and move side to side. If the latches are too tight, they will be difficult or impossible to secure. The hooks on some latches are threaded so they can be screwed in or out to tighten or loosen the latch; pliers are often necessary, so make sure you protect the finish of the hook with a rag. Other latches have set-screws that lock the hooks in place - loosen the set-screws (usually with an Allen or Torx wrench), position the hooks as desired, then tighten the set-screws.

Also keep in mind that weatherstrip is attached to the front edge of the top. As this weatherstrip deteriorates over time, it will cause the latches and the front edge of the convertible top to become loose. The proper fix for this problem is to replace the deteriorated weatherstrip, which will also reduce wind noise.

POWER-TOP TROUBLESHOOTING

Power tops are operated by hydraulic cylinders that are mounted between the vehicle body (at the left-rear) and the convertible-top framework. The cylinder receives hydraulic pressure from an electric motor/pump assembly that is located in the luggage compartment. When the switch is pressed to raise the top, the motor drives the pump in a direction that sends hydraulic pressure to the bottom end of each hydraulic cylinder, driving the pistons out of the cylinders, which raises the top. When the switch is pressed to lower the top, the motor turns in the opposite direction, sending hydraulic pressure to the top of each cylinder, driving the pistons down and lowering the top.

As a first step in troubleshooting, listen for the whirring sound of the motor as the switch is pressed. If there is no sound from the motor, proceed to *Electrical Troubleshooting*. If you can hear the motor running but the top does not raise, proceed to *Hydraulic/mechanical troubleshooting*.

Hydraulic/mechanical troubleshooting

Mechanical and hydraulic problems will cause the top to not open (or close) or to get stuck part-way through the process. If the motor is operating normally, there are generally two possible causes for these problems: 1) A hydraulic system malfunction (low fluid level, air in the system, inadequate pump pressure) or 2) A mechanical binding in the top framework.

Hydraulic system fluid level check

Locate the hydraulic motor/pump attached to the floor in the luggage compartment. The motor/pump is usually a canister-shaped assembly, approximately 12-inches long. It will have an electrical connector and hoses attached to it. On the side or end of the motor/pump assembly will be a screw-type check plug. The fluid level is seen through the side of the reservoir. The level should be between Min and Max marks. If necessary, add fluid through the check-plug hole to bring the fluid level to normal, but do not exceed the Max mark.

Check your owner's manual for specific fluid recommendations, but generally, DEXRON III Automatic Transmission Fluid (ATF) is used. If you find a low fluid level, check for leaks at the pump, hoses and hydraulic cylinders.

➡Note: The fluid level should only be checked when the top is in the open position.

Hydraulic system bleeding

Air can get into the hydraulic system through leaks, and if the fluid level is too low. Air in the system will generally cause excess noise and, in extreme cases, will cause the top to raise or lower only part-way.

To bleed the hydraulic system, first check the fluid level, then start the engine and raise and lower the top. If the fluid has excessive air, the top might not raise or lower, so you will need an assistant to hold the switch while you raise and lower the top slowly by hand. Raise and lower the top several times to work out all the air. Check the fluid level again, since it will likely drop as the air is expelled.

Checking for binding

If the hydraulic system seems to be operating properly, disconnect the hydraulic cylinders from the top framework and operate the top by hand. The top should go up and down smoothly, without excessive effort. If the top binds during manual operation, make sure all adjustments are correct and spray penetrating lubricant on all framework joints.

If the top is operating smoothly during manual operation but will not raise properly with the hydraulic cylinders connected, suspect a pump/motor assembly that is not providing adequate pressure.

ELECTRICAL TROUBLESHOOTING

If you cannot hear the motor running when the switch is pressed, first check for blown fuses and fusible links. If the fuses and fusible links are all OK, disconnect the electrical connector at the motor. Connect a 12-volt test-light to ground and probe each terminal of the disconnected wiring-harness connector while an assistant operates the switch. There should be power (light will illuminate) at one terminal with the switch in the TOP UP position and power at the other terminal with the switch in the TOP DOWN position.

If there is power, but the motor does not operate, check the ground circuit. On most models, there is a third wire from the motor that is attached to ground somewhere near the motor. It is common for this ground connection to become loose or corroded, causing the motor to stop functioning. If the motor is receiving power and has a good ground, but it is still not functioning, the motor itself is the problem.

➡Note: On some vehicles there is no separate ground connection. A relay in the system alternately grounds one wire and powers the other, depending on switch position. On models with this type of system, check for a solid ground at the relay.

If there is no power at the motor, check for power at the three switch terminals by backprobing with the electrical connector still attached. There should be power at one of the switch terminals with the switch in the neutral position, and power at two of the terminals with the switch in either the TOP UP or TOP DOWN position. If there is no power at the switch, the problem lies in the wiring between the battery and the switch. If there is power at only one terminal in the TOP UP or TOP DOWN position, the switch is bad. If the switch tests are OK, the problem lies in the relay(s) (if equipped) or the wiring.

12 CHASSIS ELECTRICAL SYSTEM

Section

1 General information

The electrical system is a 12-volt, negative ground type. Power for the lights and all electrical accessories is supplied by a lead/acid-type battery that is charged by the alternator.

This Chapter covers repair and service procedures for the various electrical components not associated with the engine. Information on the battery, alternator, ignition system and starter motor can be found in Chapter 5.

It should be noted that when portions of the electrical system are serviced, the negative cable should be disconnected from the battery to prevent electrical shorts and/or fires.

CAUTION 1:

These models are equipped with an anti-theft radio. Before performing a procedure that requires disconnecting the battery, make sure you have the proper activation code.

CAUTION 2:

Disconnecting the battery can cause driveability problems that require a scan tool to remedy. See Chapter 5 for the use of an auxiliary voltage input device before disconnecting the battery.

2 Electrical troubleshooting - general information

Refer to illustrations 2.5a, 2.5b, 2.6, 2.9 and 2.15

A typical electrical circuit consists of an electrical component, any switches, relays, motors, fuses, fusible links or circuit breakers related to that component and the wiring and connectors that link the component to both the battery and the chassis. To help you pinpoint an electrical circuit problem, wiring diagrams are included at the end of this Chapter.

Before tackling any troublesome electrical circuit, first study the appropriate wiring diagrams to get a complete understanding of what makes up that individual circuit. Trouble spots, for instance, can often be narrowed down by noting if other components related to the circuit are operating properly. If several components or circuits fail at one time, chances are the problem is in a fuse or ground connection, because several circuits are often routed through the same fuse and ground connections.

Electrical problems usually stem from simple causes, such as loose or corroded connections, a blown fuse, a melted fusible link or a failed relay. Visually inspect the condition of all fuses, wires and connections in a problem circuit before troubleshooting the circuit.

If test equipment and instruments are going to be utilized, use the diagrams to plan ahead of time where you will make the necessary connections in order to accurately pinpoint the trouble spot.

The basic tools needed for electrical troubleshooting include a circuit tester or voltmeter (a 12-volt bulb with a set of test leads can also be used), a continuity tester, which includes a bulb, battery and set of test leads, and a jumper wire, preferably with a circuit breaker incorporated, which can be used to bypass electrical components (see illustrations). Before attempting to locate a problem with test instruments, use the wiring diagram(s) to decide where to make the connections.

VOLTAGE CHECKS

Voltage checks should be performed if a circuit is not functioning properly. Connect one lead of a circuit tester to either the negative battery terminal or a known good ground. Connect the other lead to a connector in the circuit being tested, preferably nearest to the battery or fuse (see illustration). If the bulb of the tester lights, voltage is present, which means that the part of the circuit between the connector and the battery is problem free. Continue checking the rest of the circuit in the same fashion. When you reach a point at which no voltage is present, the problem lies between that point and the last test point with voltage. Most of the time the problem can be traced to a loose connection.

Note: Keep in mind that some circuits receive voltage only when the ignition key is in the Accessory or Run position.

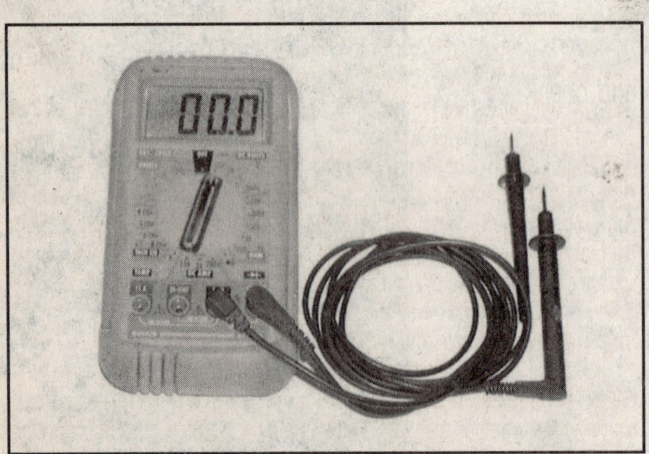

2.5a The most useful tool for electrical troubleshooting is a digital multimeter that can check volts, amps, and test continuity

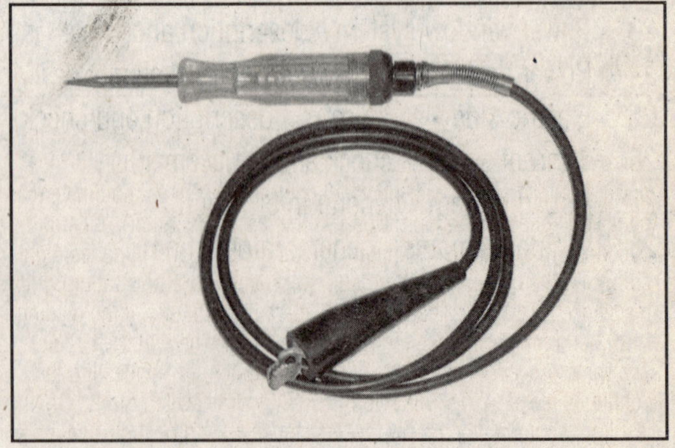

2.5b A simple test light is very handy, especially when testing for voltage

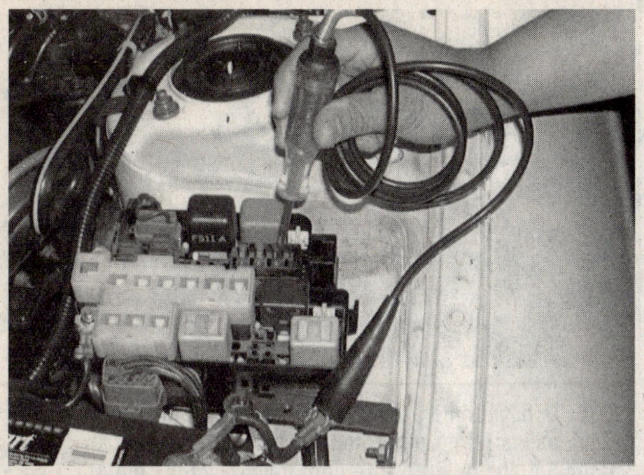

2.6 In use, a basic test light's lead is clipped to a known good ground, then the pointed probe can test connectors, wires or electrical sockets - if the bulb lights, the circuit being tested has battery voltage

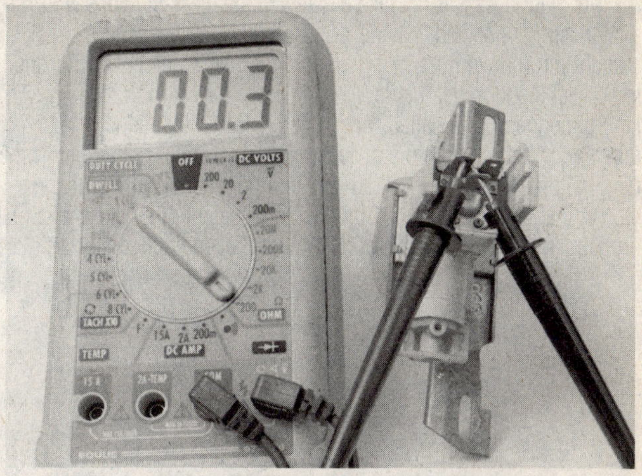

2.9 With a multimeter set to the ohms scale, resistance can be checked across two terminals - when checking for continuity, a low reading indicates continuity, a high reading or infinity indicates lack of continuity

FINDING A SHORT

One method of finding shorts in a live circuit is to remove the fuse and connect a test light in place of the fuse terminals (fabricate two jumper wires with small spade terminals, plug the jumper wires into the fuse box and connect the test light). There should be voltage present in the circuit. Move the suspected wiring harness from side-to-side while watching the test light. If the bulb goes off, there is a short to ground somewhere in that area, probably where the insulation has rubbed through.

GROUND CHECK

Perform a ground test to check whether a component is properly grounded. Disconnect the battery and connect one lead of a continuity tester or multimeter (set to the ohms scale), to a known good ground. Connect the other lead to the wire or ground connection being tested. If the resistance is low (less than 5 ohms), the ground is good. If the bulb on a self-powered test light does not go on, the ground is not good.

CONTINUITY CHECK

A continuity check is done to determine if there are any breaks in a circuit - if it is passing electricity properly. With the circuit off (no power in the circuit), a self-powered continuity tester or multimeter can be used to check the circuit. Connect the test leads to both ends of the circuit (or to the "power" end and a good ground), and if the test light comes on the circuit is passing current properly (see illustration). If the resistance is low (less than 5 ohms), there is continuity; if the reading is 10,000 ohms or higher, there is a break somewhere in the circuit. The same procedure can be used to test a switch, by connecting the continuity tester to the switch terminals. With the switch turned On, the test light should come on (or low resistance should be indicated on a meter).

FINDING AN OPEN CIRCUIT

When diagnosing for possible open circuits, it is often difficult to locate them by sight because the connectors hide oxidation or terminal misalignment. Merely wiggling a connector on a sensor or in the wiring harness may correct the open circuit condition. Remember this when an open circuit is indicated when troubleshooting a circuit. Intermittent problems may also be caused by oxidized or loose connections.

Electrical troubleshooting is simple if you keep in mind that all electrical circuits are basically electricity running from the battery, through the wires, switches, relays, fuses and fusible links to each electrical component (light bulb, motor, etc.) and to ground, from which it is passed back to the battery. Any electrical problem is an interruption in the flow of electricity to and from the battery.

CONNECTORS

Most electrical connections on these vehicles are made with multiwire plastic connectors. The mating halves of many connectors are secured with locking clips molded into the plastic connector shells. The mating halves of large connectors, such as some of those under the instrument panel, are held together by a bolt through the center of the connector.

To separate a connector with locking clips, use a small screwdriver to pry the clips apart carefully, then separate the connector halves. Pull only on the shell, never pull on the wiring harness as you may damage the individual wires and terminals inside the connectors. Look at the connector closely before trying to separate the halves. Often the locking clips are engaged in a way that is not immediately clear. Additionally, many connectors have more than one set of clips.

Each pair of connector terminals has a male half and a female half. When you look at the end view of a connector in a diagram, be sure to understand whether the view shows the harness side or the component side of the connector. Connector halves are mirror images of each other, and a terminal shown on the right side end-view of one half will be on the left side end view of the other half.

It is often necessary to take circuit voltage measurements with a connector connected. Whenever possible, carefully insert a small straight pin (not your meter probe) into the rear of the connector shell to contact the terminal inside, then clip your meter lead to the pin. This kind of connection is called "backprobing" (see illustration). When inserting a test probe into a male terminal, be careful not to distort the terminal opening. Doing so can lead to a poor connection and corrosion at that terminal later. Using the small straight pin instead of a meter probe results in less chance of deforming the terminal connector.

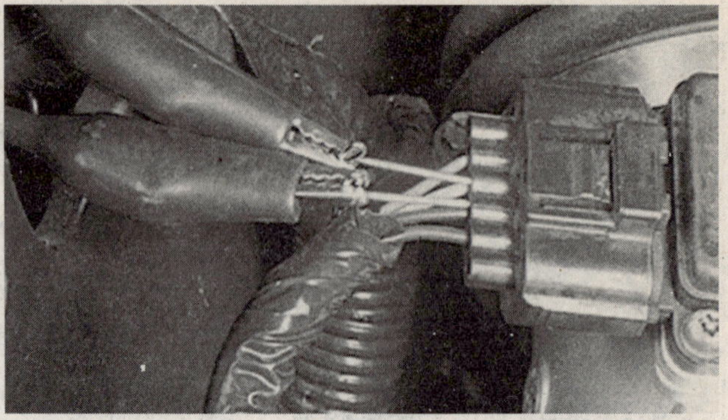

2.15 To backprobe a connector, insert a small, sharp probe (such as a straight-pin) into the back of the connector alongside the desired wire until it contacts the metal terminal inside; connect your meter leads to the probes - this allows you to test a functioning circuit

3 Fuses - general information

3.1a The main fuse/relay box is at the left end of the instrument panel, behind the end cover - the inside of the cover has a legend to identify the fuses and relays

➧ Refer to illustrations 3.1a, 3.1b, 3.1c, 3.3a and 3.3b

The electrical circuits of the vehicle are protected by a combination of fuses, circuit breakers and fusible links. The interior fuse/relay panel is located at the left end of the instrument panel, while the main fuse-relay panel is in the engine compartment (see illustrations).

Each of the fuses is designed to protect a specific circuit, and the various circuits are identified on the fuse panel itself.

Several sizes of fuses are employed in the fuse blocks. There are small, medium and large sizes of the same design, all with the same blade terminal design, as well as five "metal" fuses located in the engine compartment fuse panel. The medium and large fuses can be removed with your fingers, but the small fuses require the use of pliers or the small plastic fuse-puller tool found in most fuse boxes. The metal fuses are for heavy loads, and if the metal strip melts due to an overload, it is easily seen, although the battery should be disconnected while replacing this type fuse (see the **Cautions** in Section 1). If an electrical component fails, always check the fuse first. The best way to

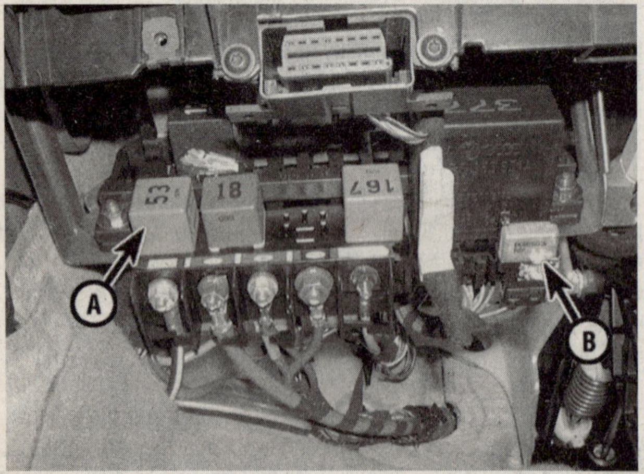

3.1b The under-dash fuse/relay panel is located under the sound insulator at the left end of the instrument panel - this panel contains relays (A), fuses (B) and circuit breakers if equipped

3.1c The fuse panel on top of the battery contains large capacity fuses - three are standard type 30-amp (A), and there are five metal fuses (B)

3.3a All three of these fuses are of 30-amp rating, yet are different sizes, at left is a small fuse, the center is a medium, and at right is a large - make sure you get the right amperage and size when purchasing replacement fuses

check the fuses is with a test light. Check for power at the exposed terminal tips of each fuse (see illustration). If power is present at one side of the fuse but not the other, the fuse is blown. A blown fuse can also be identified by visually inspecting it (see illustration).

Be sure to replace blown fuses with the correct type. Fuses (of the same physical size) of different ratings may be physically interchangeable, but only fuses of the proper rating should be used. Replacing a fuse with one of a higher or lower value than specified is not recommended. Each electrical circuit needs a specific amount of protection.

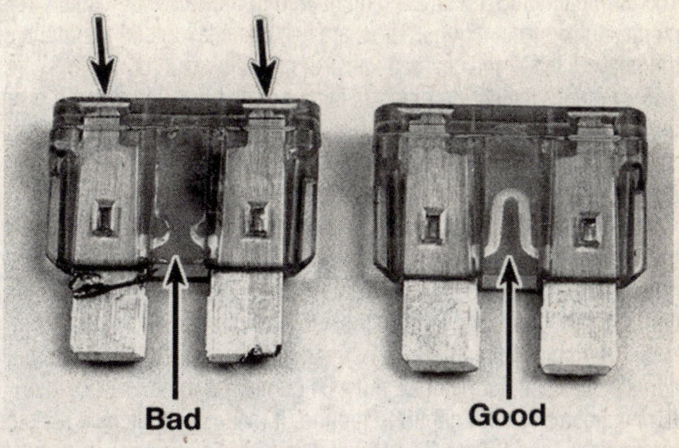

3.3b When a fuse blows, the element between the terminals melts - the fuse on the left is blown, the one on the right is good - these fuses can be tested on top (arrows) with a test light without removing the fuse

The amperage value of each fuse is molded into the top of the fuse body.

If the replacement fuse immediately fails, don't replace it again until the cause of the problem is isolated and corrected. In most cases, this will be a short circuit in the wiring caused by a broken or deteriorated wire.

4 Circuit breakers - general information and check

Circuit breakers protect certain circuits, such as the power windows or heated seats. Depending on the vehicle's accessories, there may be one or two 25-amp circuit breakers, located in the under-dash fuse/relay box under the left end of the dashboard (see illustration 3.1b).

Because the circuit breakers reset automatically, an electrical overload in a circuit-breaker-protected system will cause the circuit to fail momentarily, then come back on. If the circuit does not come back on, check it immediately.

For a basic check, pull the circuit breaker up out of its socket on the fuse panel, but just far enough to probe with a voltmeter. The breaker should still contact the sockets.

With the voltmeter negative lead on a good chassis ground, touch each end prong of the circuit breaker with the positive meter probe. There should be battery voltage at each end. If there is battery voltage only at one end, the circuit breaker must be replaced.

5 Relays - general information and testing

GENERAL INFORMATION

1 Several electrical accessories in the vehicle, such as the fuel injection system, horns, starter, and fog lamps use relays to transmit the electrical signal to the component. Relays use a low-current circuit (the control circuit) to open and close a high-current circuit (the power circuit). If the relay is defective, that component will not operate properly. Most relays are mounted in the under-dash fuse/relay box (see illustration 3.1b). If a faulty relay is suspected, it can be removed and tested using the procedure below or by a dealer service department or a repair shop. Defective relays must be replaced as a unit.

TESTING

➧ Refer to illustrations 5.2a and 5.2b

2 Most of the relays used in these vehicles are of a type often called "ISO" relays, which refers to the International Standards Organization.

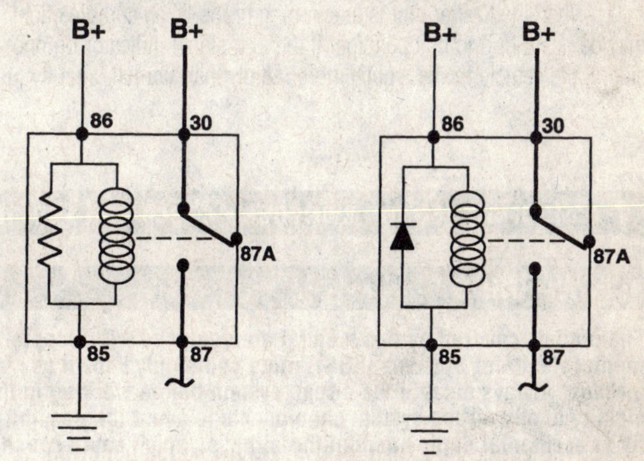

5.2a Typical ISO relay designs, terminal numbering and circuit connections

The terminals of ISO relays are numbered to indicate their usual circuit connections and functions. There are two basic layouts of terminals on the relays used in the covered vehicles (see illustrations).

3 Refer to the wiring diagram for the circuit to determine the proper connections for the relay you're testing. If you can't determine the correct connection from the wiring diagrams, however, you may be able to determine the test connections from the information that follows.

4 Two of the terminals are the relay control circuit and connect to the relay coil. The other relay terminals are the power circuit. When the relay is energized, the coil creates a magnetic field that closes the larger contacts of the power circuit to provide power to the circuit loads.

5 Terminals 85 and 86 are normally the control circuit. If the relay contains a diode, terminal 86 must be connected to battery positive (B+) voltage and terminal 85 to ground. If the relay contains a resistor, terminals 85 and 86 can be connected in either direction with respect to B+ and ground.

6 Terminal 30 is normally connected to the battery voltage (B+) source for the circuit loads. Terminal 87 is connected to the ground side of the circuit, either directly or through a load. If the relay has several alternate terminals for load or ground connections, they usually are numbered 87A, 87B, 87C, and so on.

7 Use an ohmmeter to check continuity through the relay control coil.

- *a) Connect the meter according to the polarity shown in illustration 5.2a for one check; then reverse the ohmmeter leads and check continuity in the other direction.*
- *b) If the relay contains a resistor, resistance will be indicated on the meter, and should be the same value with the ohmmeter in either direction.*
- *c) If the relay contains a diode, resistance should be higher with the ohmmeter in the forward polarity direction than with the meter leads reversed.*
- *d) If the ohmmeter shows infinite resistance in both directions, replace the relay.*

5.2b Most relays are marked on the outside to easily identify the control circuit and power circuits - this one is of the four-terminal type

8 Remove the relay from the vehicle and use the ohmmeter to check for continuity between the relay power circuit terminals. There should be no continuity between terminal 30 and 87 with the relay de-energized.

9 Connect a fused jumper wire to terminal 86 and the positive battery terminal. Connect another jumper wire between terminal 85 and ground. When the connections are made, the relay should click.

10 With the jumper wires connected, check for continuity between the power circuit terminals. Now, there should be continuity between terminals 30 and 87.

11 If the relay fails any of the above tests, replace it.

6 Turn signal and hazard flashers - check and replacement

1 The flasher function on the covered models is performed by the hazard-warning switch, located on the switch panel at the lower center of the dash. For testing and replacement of the switch, refer to Section 9 and the wiring diagrams at the end of this Chapter.

2 When the flasher unit is functioning properly, an audible click may be heard during its operation. If the turn signal indicator on one side of the vehicle flashes much more rapidly than normal, a faulty turn signal bulb is indicated.

3 If both turn signals fail to blink, the problem may be due to a blown fuse, a faulty flasher unit switch or a loose or open connection. If a quick check of the fuse box indicates that the turn signal fuse has blown, check the wiring for a short before installing a new fuse.

4 If the flasher switch is not the problem, refer to Section 7 and test the turn signal portion of the multi-function switch.

7 Steering column switches - check and replacement

WARNING:

The models covered by this manual are equipped with Supplemental Restraint Systems (SRS), more commonly known as airbags. Always disable the airbag system before working in the vicinity of any airbag system components to avoid the possibility of accidental deployment of the airbags, which could cause personal injury (see Section 27).

MULTI-FUNCTION SWITCH

1 The multi-function switch is located on the top of the steering column. It incorporates into one switch the turn signal, headlight dimmer, windshield wiper/washer and, if equipped, cruise control functions.

➡Note: The following checks can be made with simple equipment, but thorough testing of the systems controlled by the multi-function switch can only be made at a dealer or other repair shop with a scan diagnosis tool.

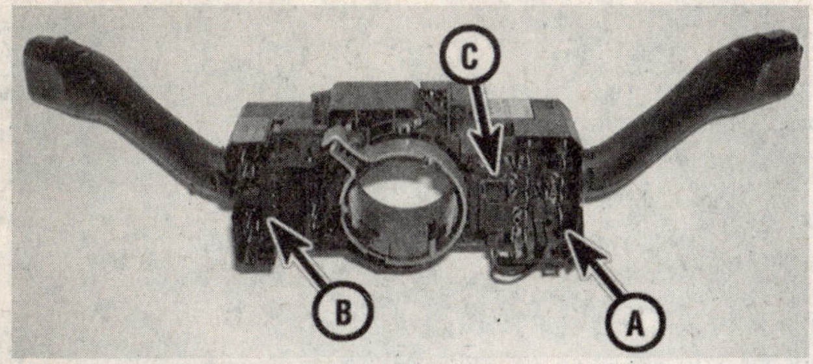

7.3 Multi-function switch harness connectors: (A) is used for the turn signals and headlight dimming, (B) for the wipe-wash functions, and (C) for the cruise control

Check

➧ Refer to illustration 7.3

2 Remove the steering column covers and the knee bolster for access (see Chapter 11). Follow the wiring diagrams at the end of this Chapter to determine which terminals supply the left and right turn signal circuits, the cruise control and the headlight dimming systems.

3 Use a test light to check for power at the designated terminals with the switch in each position. Look at the wiring diagrams at the end of this Chapter to see how power and ground are routed. If any portion of the multi-function switch fails the tests, the switch must be replaced as a unit. The 12-pin connector handles the turn signals, headlight dimmer and parking lights, the 6-pin and 8-pin connectors control the wipe/wash functions, and the cruise control connector is the 10-pin connector (see illustration).

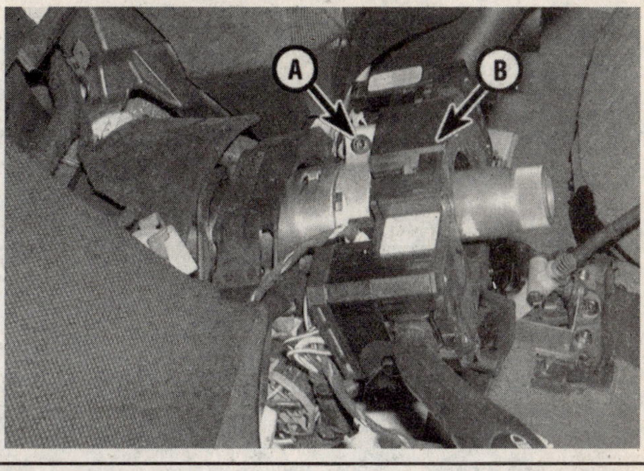

7.6 Remove the Allen bolt (A) at the back of the multi-function switch (B)

Replacement

➧ Refer to illustration 7.6

4 Remove the steering wheel (see Chapter 10).

5 Remove the steering column covers (see Chapter 11).

6 Remove the Allen bolt at the rear of the multi-function switch (see illustration).

7 Pull the multi-function switch out from the steering column enough to disconnect the electrical connectors.

8 Installation is the reverse of removal.

8 Ignition switch and key lock cylinder - check and replacement

➧ Refer to illustrations 8.7a, 8.7b and 8.8

⁂ WARNING:

The models covered by this manual are equipped with Supplemental Restraint Systems (SRS), more commonly known as airbags. Always disable the airbag system before working in the vicinity of any airbag system components to avoid the possibility of accidental deployment of the airbags, which could cause personal injury (see Section 27).

1 The ignition switch, located under the steering column (switch on the left, key-lock cylinder on the right), is comprised of a cast-metal housing, an ignition lock cylinder and an electrical component, the switch device.

CHECK

2 Access the switch and disconnect the electrical connector from it (see Steps 6 and 7).

3 Backprobe the wires from the switch with an ohmmeter. Power is supplied to the switch through the two red wires (attached to the terminals marked "30" on the back of the switch). Check for continuity between the red and other wires in different switch positions. With the switch in Off, there should be no continuity. With the switch in Run, there should be continuity at the black/red and the black wires, and in the Start position, only at the red/black and brown/red wires.

4 If the switch does not have the correct continuity, replace it.

8.7a Remove the electrical connector (arrow) from the ignition switch

REPLACEMENT

5 Disconnect the negative battery cable (see the **Cautions** in Section 1).

6 Remove the steering column covers (see Chapter 11). On models with automatic transmissions, the shift linkage must be in Park before removing/installing the ignition switch.

7 Disconnect the large electrical connector on the left side of the switch, then remove the two switch mounting screws and pull the switch out of the housing (see illustrations).

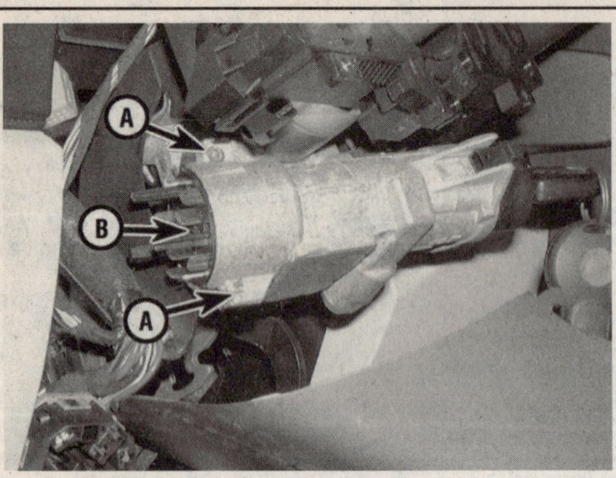

8.7b With the connector off, remove the two screws (A) securing the switch (B) to the housing

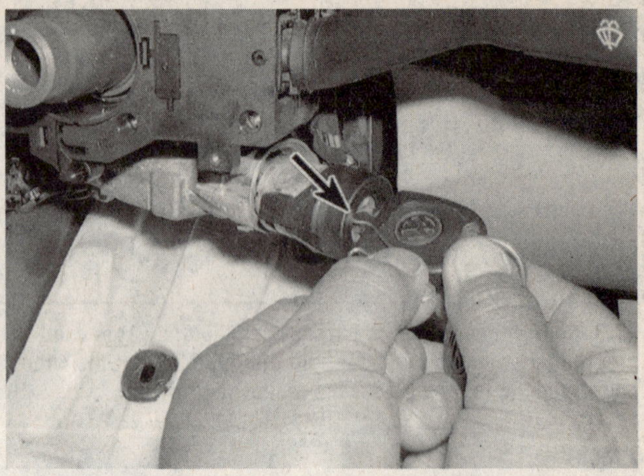

8.8 Depress the retainer with a straightened paper clip (arrow) and withdraw the cylinder by holding the key

➡Note: The screw heads may be coated with a special "locking" coating. You'll have to chip this off to remove the screws.

8 To remove the key lock cylinder, place the key in the lock and rotate clockwise to the Run position. Insert the straightened end of a large paper clip into the hole in the cylinder and pull the key and lock cylinder from the housing (see illustration).

9 When installing the cylinder, align the cylinder as it was (still in Run position with the key in), then push the cylinder in until it snaps in position. Turn the key to the Lock position and remove the key.

10 The remainder of installation is the reverse of removal.

9 Instrument panel switches - check and replacement

⁂ WARNING:

The models covered by this manual are equipped with Supplemental Restraint Systems (SRS), more commonly known as airbags. Always disable the airbag system before working in the vicinity of any airbag system components to avoid the possibility of accidental deployment of the airbags, which could cause personal injury (see Section 27).

HEADLIGHT SWITCH

Check

1 The headlight switch must be removed from the instrument panel for testing. See Steps 4 and 5 for removal.

2 Using an ohmmeter or self-powered continuity tester, check the switch for proper continuity between the terminals. Refer to the wiring diagrams at the end of this Chapter for the wire colors and circuits. There should be continuity between the power side and the lighting side when the switch is in the indicated position. If the switch fails any of the tests, replace the switch.

Replacement

3 Disconnect the negative cable at the battery (see Chapter 5). See the **Cautions** in Section 1.

4 Turn the headlight switch knob counterclockwise until it stops. Push in on the switch and twist it to the right (clockwise) then withdraw it from the instrument panel.

5 Pull the switch out far enough to disconnect the electrical connector.

6 Installation is the reverse of removal.

DASH LIGHT DIMMER SWITCH

➧ Refer to illustration 9.8

7 The dimmer is located just to the right of the headlight switch.

8 Use a screwdriver or a trim tool to pry the switch housing out of the instrument panel (see illustration).

9 Apply battery voltage to the input side of the switch (see the wiring diagrams at the end of this Chapter), and use a digital ohmmeter to check for resistance on the output side. If the resistance doesn't vary as the knob is turned, replace the switch.

10 Installation is the reverse of removal.

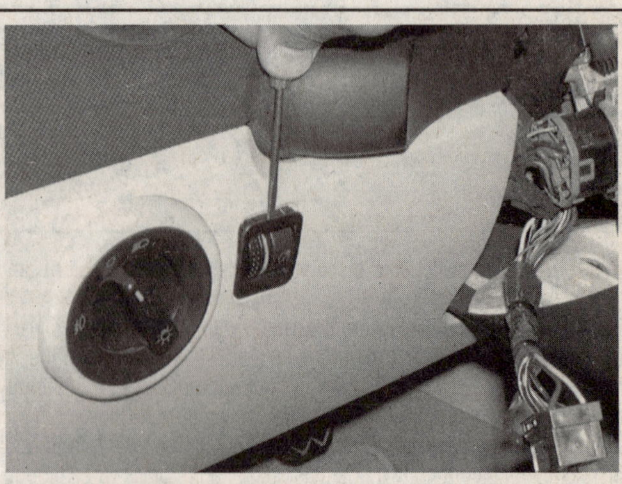

9.8 Pry the dashboard light dimmer out with a trim tool

ON/OFF SWITCHES

Refer to illustration 9.13

11 Depending on the options of the vehicle, there may be one or more switches on the instrument panel, including seat heaters, hazard flasher and rear window defogger.

12 All of the above-mentioned switches are located in the switch panel at the center of the dashboard, just below the radio and heater controls.

13 All of the switches need to be removed for testing, and all are removed the same way. Use a screwdriver or trim tool to pry behind the lip of the switches to release them from the switch panel, then pull them out far enough to disconnect the electrical connector (see illustration).

14 With simple on/off switches, use an ohmmeter or self-powered continuity tester to check the switch for proper continuity between the terminals. Refer to the wiring diagrams at the end of this Chapter for input and output terminals and wire colors. There should be continuity between input and output terminals only when the switch is engaged. If any switch fails the test, replace the switch.

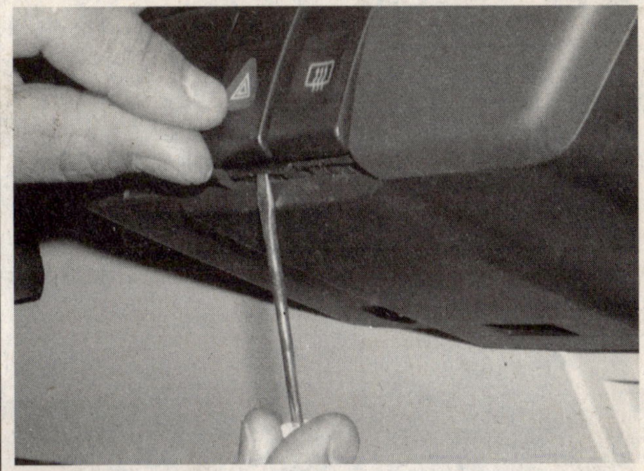

9.13 Remove on/off type switches by prying behind the switch lip and pulling the switch out of the instrument panel

10 Fuel gauge - check

WARNING:

The models covered by this manual are equipped with Supplemental Restraint Systems (SRS), more commonly known as airbags. Always disable the airbag system before working in the vicinity of any airbag system components to avoid the possibility of accidental deployment of the airbags, which could cause personal injury (see Section 27).

Note: This procedure applies to conventional analog type gauges (NON-digital) only.

1 All tests below require the ignition switch to be turned to Off position before testing.

2 If the gauge pointer does not move from the empty position, check the fuse. If the fuse is OK, locate the sending unit for the circuit (see Chapter 4A for fuel sending unit location). Connect the sending unit connector to ground with a jumper wire.

3 Turn the ignition key to On momentarily. If the pointer goes to the full position, replace the sending unit.

Note: Turn the key Off right away; grounding the sending unit for too long could damage the gauge.

If the pointer stays in the same position, use a jumper wire to ground the sending unit terminal on the back of the instrument cluster (violet/black wire). Refer to the wiring diagrams at the end of this Chapter and refer to Section 11 for access to the connector at the back of the instrument cluster. If the pointer moves, the problem lies in the wiring between the gauge and the sending unit. If the pointer does not move with the sending unit terminal on the back of the gauge grounded, check for voltage at the other terminal of the gauge (the brown/white wire). There should not be voltage.

4 If the problem is in the gauge panel itself, the instrument cluster must be replaced (it is not user serviceable). See your dealer for a new or rebuilt instrument cluster.

Note: Problems within the instrument cluster, including the fuel gauge, will set DTC's (Diagnostic Trouble Codes) that can be retrieved with a scan tool.

11 Instrument cluster - removal and installation

Refer to illustrations 11.4 and 11.5

WARNING:

The models covered by this manual are equipped with Supplemental Restraint Systems (SRS), more commonly known as airbags. Always disable the airbag system before working in the vicinity of any airbag system components to avoid the possibility of accidental deployment of the airbags, which could cause personal injury (see Section 27).

1 Disconnect the negative cable from the battery (see Chapter 5). See **Cautions** in Section 1.

2 Lower the steering column to the bottom of its range, and pull it out as far as it will go.

3 Remove the instrument cluster bezel (see Chapter 11).

4 Remove the screws securing the cluster to the instrument panel

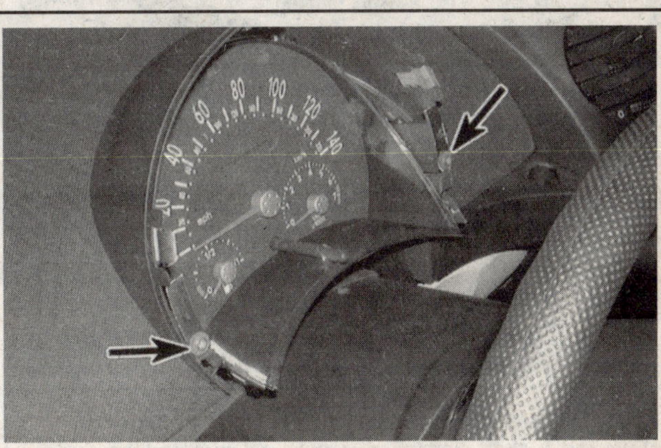

11.4 Remove the instrument cluster mounting screws (arrows)

(see illustration).

5 Pull the cluster forward enough to disengage the two large electrical connectors at the back, then pull the cluster out, tilting the bottom out first (see illustration).

6 Installation is the reverse of removal.

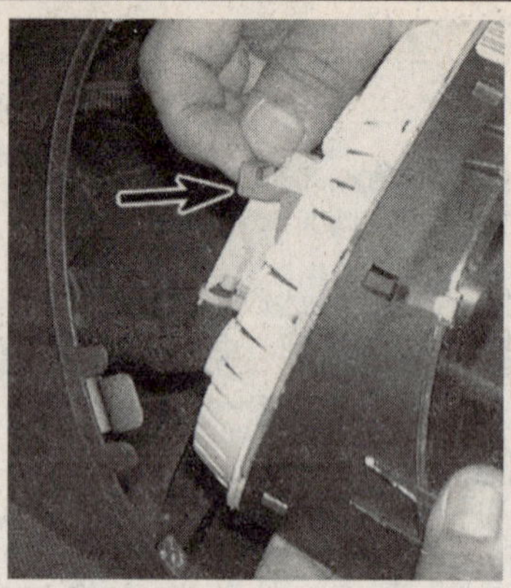

11.5 Pull the cluster out enough to disconnect the electrical connectors (arrow indicates one)

12 Radio and speakers - removal and installation

✻✻ WARNING:

The models covered by this manual are equipped with Supplemental Restraint Systems (SRS), more commonly known as airbags. Always disable the airbag system before working in the vicinity of any airbag system components to avoid the possibility of accidental deployment of the airbags, which could cause personal injury (see Section 27).

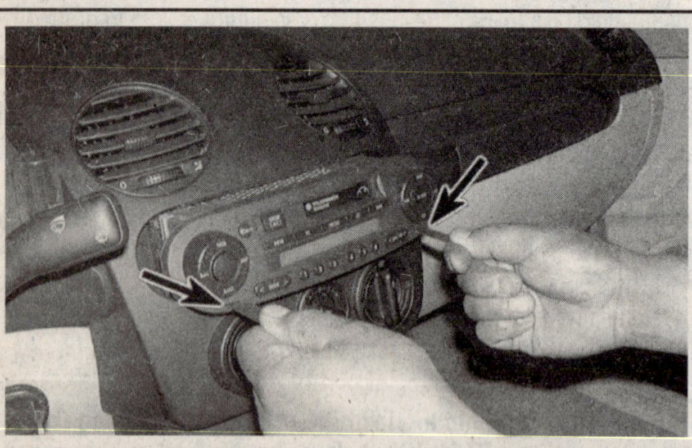

12.3a Push the tools (arrows) into the slots and pull the radio out

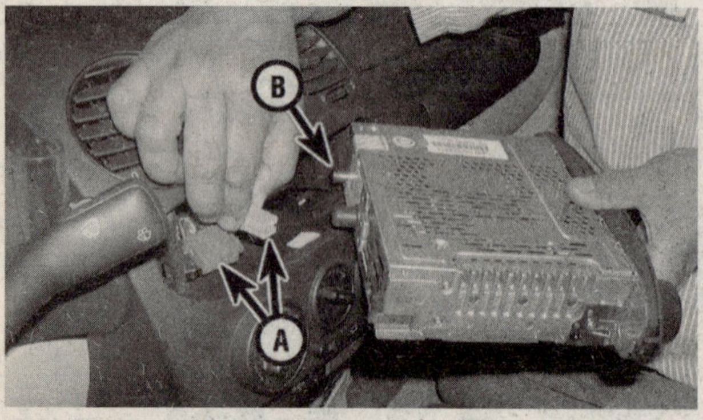

12.3b Disconnect the two connectors (A) from the back of the radio, then disconnect the ground wire and the antenna cable (B)

➡Note: The audio system is part of the diagnostic network of the vehicle. Any problems with the radio, antenna or speakers may set a DTC that can be retrieved with a scan tool.

RADIO

➧ Refer to illustrations 12.3a and 12.3b

1 Refer to Chapter 5 and disconnect the negative battery cable. Refer to the **Cautions** in Section 1.

2 Two special tools, available at auto parts stores must be used to remove the VW radio.

3 Push the two tools into the slots on either side of the radio, pull the radio out of the instrument panel, disconnect the connectors, then remove it from the vehicle (see illustrations).

4 Installation is the reverse of removal. To disengage the plastic tools from the radio before installation, depress the lugs on the side of the radio.

SPEAKERS

➧ Refer to illustration 12.7

5 The are small tweeter speakers in each of the front "A" pillars at

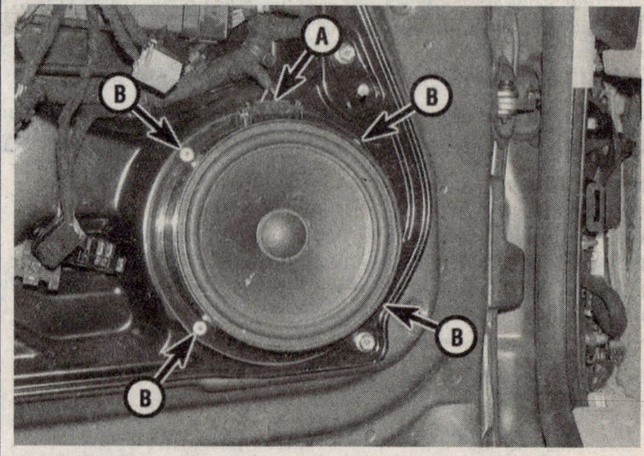

12.7 To remove a door or rear speaker, disconnect the connector (A), then drill out the rivets (B)

either side of the windshield. These speakers are part of the trim panel for the "A" pillar, and if found defective, the pillar trim/speaker must be replaced as a unit. The remaining speakers include one in each door and one in each quarter panel.

6 To remove the door or quarter-panel speakers, remove the interior trim panel first (see Chapter 11).

7 Disconnect the electrical connector at the speaker, then drill out the rivets securing it to the door carrier (see illustration).

8 Installation is the reverse of the removal procedure. Self-tapping screws may be used to replace the speaker mounting rivets.

13 Antenna - removal and installation

➧ Refer to illustration 13.1

➡Note: On convertible models, the radio antenna is a grid type, embedded in the top of the windshield glass. It cannot be repaired or replaced without replacing the windshield.

1 Grip the base of the antenna mast firmly and twist to unscrew it from the mounting base (see illustration).

2 If the base or the cable itself must be replaced, the headliner must be removed to access the nut on the bottom of the antenna base, on the interior side. This is a complex job for the home mechanic, and should be left to a dealer or professional auto stereo shop.

3 Installation of the antenna mast is the reverse of the removal procedure.

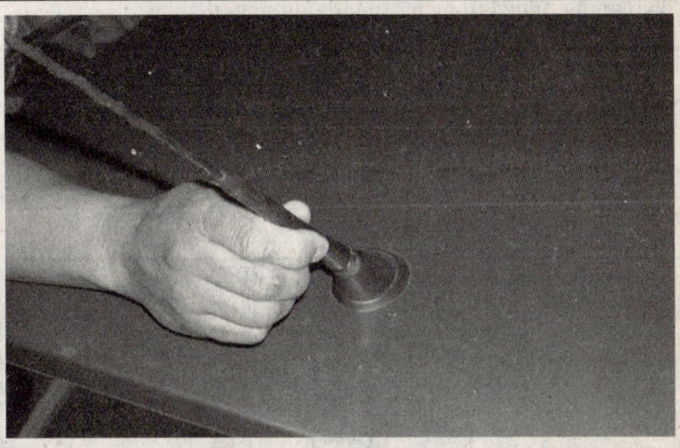

13.1 The antenna mast can be unscrewed from its base by twisting it by hand

14 Headlight bulb - replacement

➧ Refer to illustrations 14.2a and 14.2b

⁂ WARNING:

Halogen bulbs are gas-filled and under pressure and may shatter if the surface is scratched or the bulb is dropped. Wear eye protection and handle the bulbs carefully, grasping only the base whenever possible. Don't touch the surface of the bulb with your fingers because the oil from your skin could cause it to overheat and fail prematurely. If you do touch the bulb surface, clean it with rubbing alcohol.

1 Refer to Section 16 and remove the headlight housing.

2 Remove the retaining clip to remove the upper (low beam) or lower (high beam) bulbholder from the back of the headlight housing (see illustrations). Remove the old bulb from the holder.

3 Handling the new bulb only with gloves or a clean rag, insert the new bulb in the holder.

4 Installation of the housing is the reverse of the removal procedure.

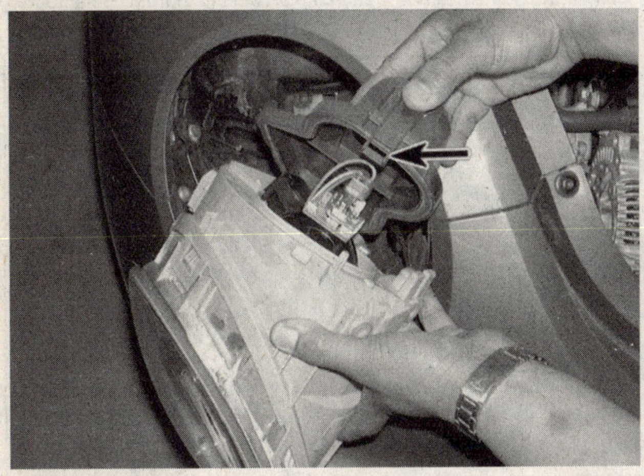

14.2a Release the clip (arrow) at the top and pull back the bulbholder cover

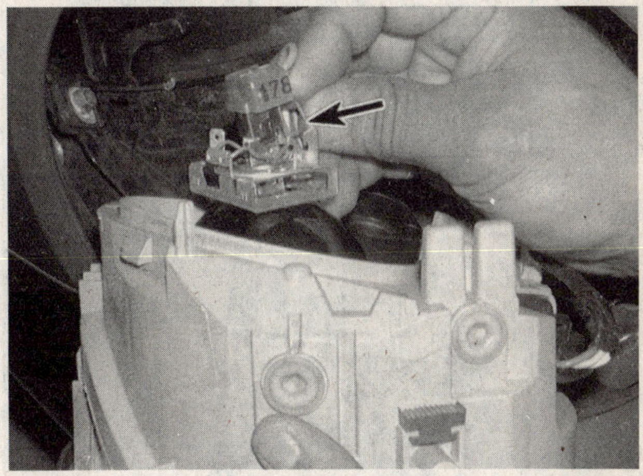

14.2b Remove the retaining clip (arrow), then pull out the headlight bulb holder (low beam shown, high beam bulb is below this on the housing)

15 Headlights and fog lights - adjustment

WARNING:

The headlights must be aimed correctly. If adjusted incorrectly, they could temporarily blind the driver of an oncoming vehicle and cause an accident or seriously reduce your ability to see the road. The headlights should be checked for proper aim every 12 months and any time a new headlight is installed or front-end bodywork is performed. The following procedure is only an interim step to provide temporary adjustment until the headlights can be adjusted by a properly equipped shop.

HEADLIGHTS

Refer to illustrations 15.1 and 15.2

1 These models are equipped with headlight housings with two adjustment screws, one controlling left-and-right movement and one for up-and-down movement (see illustration). A 6mm Allen wrench is used on the screws. The type with a ball-end works best.

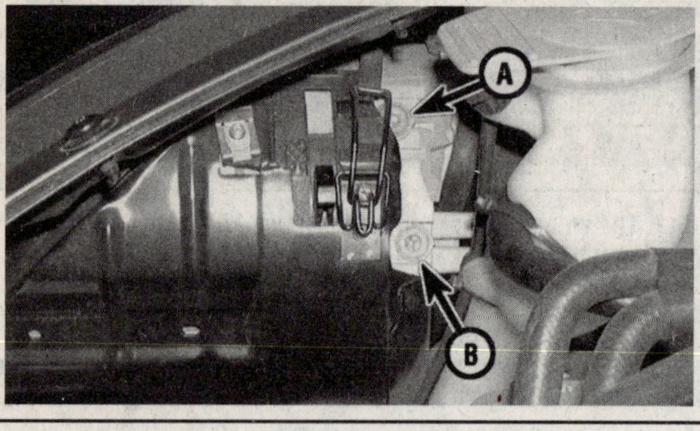

15.1 Open the hood and adjust the headlight alignment with these screws - (A) is the vertical adjuster, (B) is the horizontal adjuster

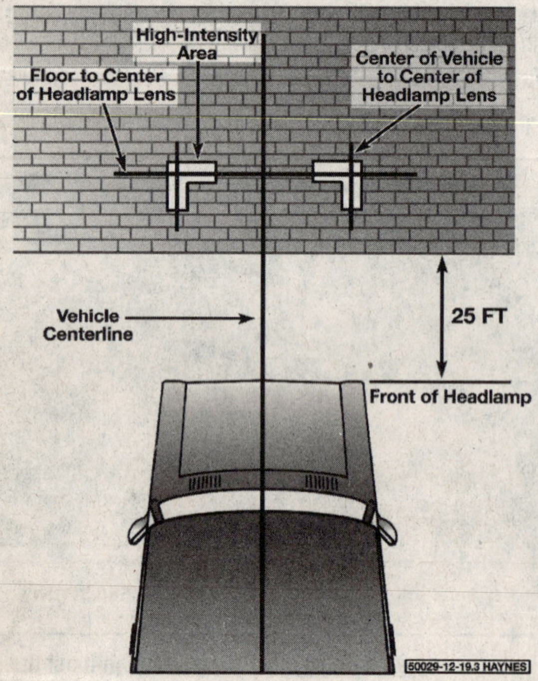

15.2 Headlight adjustment details

2 There are several methods of adjusting the headlights. The simplest method requires an open area with a blank wall and a level floor (see illustration).

3 Position masking tape vertically on the wall in reference to the vehicle centerline and the centerlines of both headlights.

4 Position a horizontal tape line in reference to the centerline of all the headlights.

Note: It may be easier to position the tape on the wall with the vehicle parked only a few inches away.

5 Adjustment should be made with the vehicle parked 25 feet from the wall, sitting level, the gas tank half-full and no unusually heavy load in the vehicle.

6 Starting with the low beam adjustment, position the high intensity zone so it is two inches below the horizontal line and two inches to the side of the vertical headlight line away from oncoming traffic. Twist the adjustment screws until the desired level has been achieved.

7 With the high beams on, the high intensity zone should be vertically centered with the exact center just below the horizontal line.

Note: It may not be possible to position the headlight aim exactly for both high and low beams. If a compromise must be made, keep in mind that the low beams are the most used and have the greatest effect on driver safety.

8 Have the headlights adjusted by a dealer service department at the earliest opportunity.

FOG LIGHTS

Refer to illustration 15.11

9 Some models have optional fog lights that can be aimed just like headlights (see illustration 15.2).

10 Tape a horizontal line on the wall that represents the height of the fog lamps, and another tape line four inches below that line. Park the vehicle 25 feet from the wall.

11 Using the adjusting screws on the fog lamps, adjust the pattern on the wall so that the top of the fog lamp beam meets the lower line on the wall, and that the beam is centered horizontally in front of the fog lamps (see illustration).

15.11 Turn this screw (arrow) in the grille area to adjust the aim of optional fog lights

16 Headlight housing - replacement

➧ **Refer to illustrations 16.1 and 16.3**

1 Open the hood and pull up the release lever at the back of the headlight housing (see illustration). The headlight housing slides in and out of the body in tracks.

2 Pull, the headlight housing far enough forward to disconnect the electrical connectors, then slide the headlight housing all the way out.

3 To install the housing, slide it back in, aligning the ridges on each side with the two tracks in the body (see illustration).

4 When the housing is in far enough, reach through from inside the engine compartment to reattach the electrical connectors. The remainder of installation is the reverse of the removal procedure. Refer to Section 15 for aiming procedures after the headlight housing is installed.

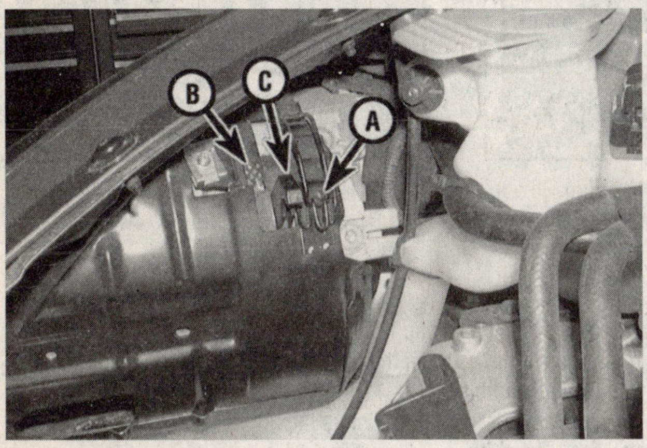

16.1 Release the spring clamp (A), depress the tab (B), then rotate the headlight housing release lever (C) upward and pull the housing forward enough to disconnect the electrical connectors (some models may not be equipped with a spring clamp)

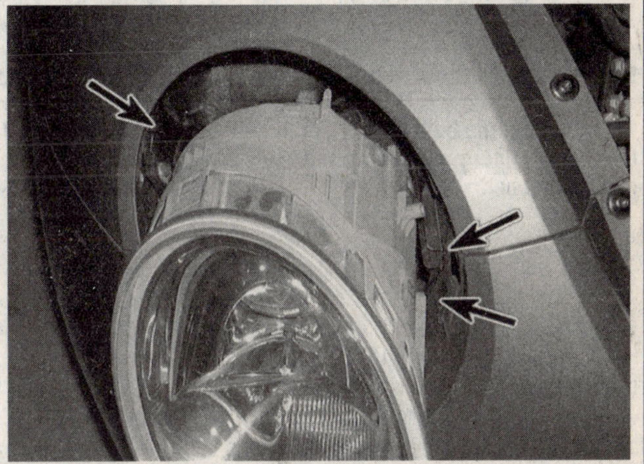

16.3 Align the headlight housing ribs with the tracks (arrows) in the body

17 Bulb replacement

⁂ WARNING:

Bulbs can remain hot for up to twenty minutes after they're turned off. Be sure bulbs are off and cool before you touch them.

TURN SIGNAL/SIDE-MARKER LIGHTS

➧ **Refer to illustrations 17.1 and 17.3**

➥ **Note: Turn signal lights on 1998 models are only on the front fenders; 1999 and later models have additional small turn signal lights on the sides of the cowl.**

1 Remove the screw securing one end of the turn-signal housing, then swing the housing out away from the vehicle (see illustration).

2 Disconnect the electrical connectors.

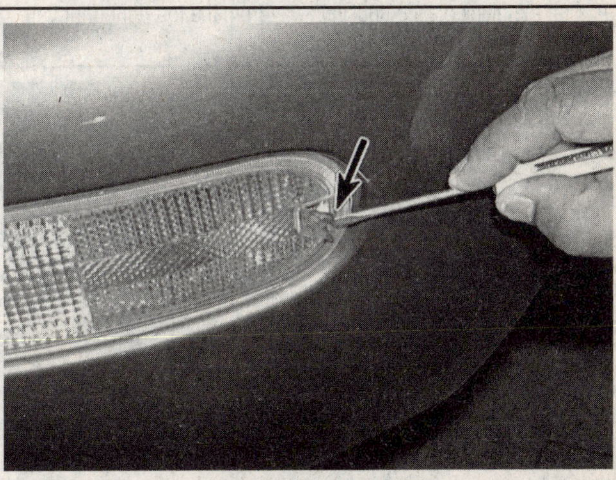

17.1 Pry back the plastic cover, then remove the screw (arrow) and swing the turn signal housing out away from the vehicle

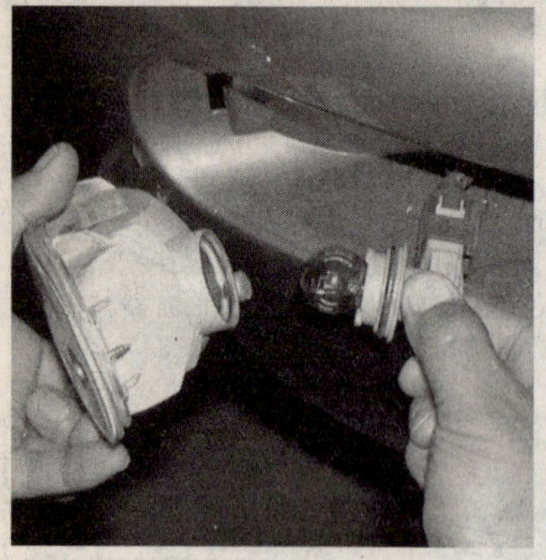

17.3 Pull out the bulb holder for the turn signal/side-marker lamp and replace the bulb

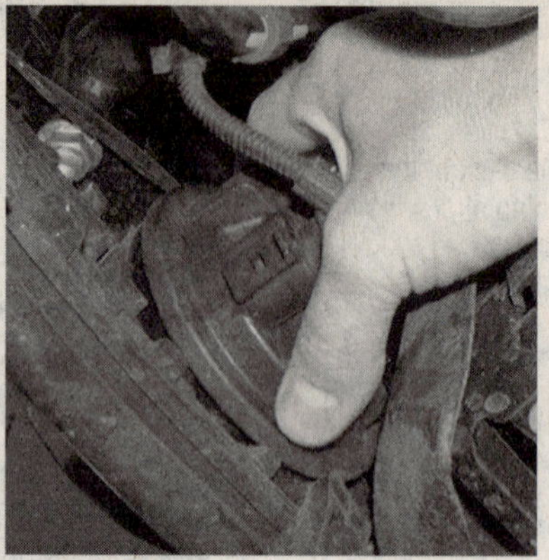

17.8 Twist the fog light housing counterclockwise to disengage it from the light

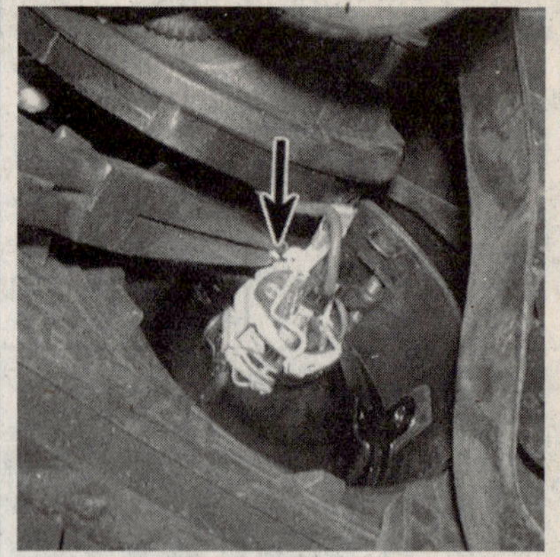

17.9 With the housing pulled to the side, release the clip (arrow) and pull out the bulbholder

3 Each bulb is replaced by twisting the bulbholder out (see illustration).

4 Installation is the reverse of removal.

5 To replace the bulbs in the 1999 and later cowl-mounted lights, first apply masking tape around the light to prevent scratching the paint.

6 These lights have a spring-clip at one end and a notch at the other. Carefully push the light toward the spring-clip side (towards the door) until the notch is disengaged, then remove the light from the body. Remove the bulbholder to replace the bulb on the backside of the light.

FOG LIGHTS

Refer to illustrations 17.8 and 17.9

7 Raise the front of the vehicle and support it securely on jackstands.

8 From behind the grille area, grasp the back of the fog light housing and twist it counterclockwise (see illustration).

9 Pull the black plastic housing away from the fog light until you have access to the metal clip securing the bulbholder (see illustration).

10 Install the new bulb in the holder and secure it in the fog light housing with the clip.

11 The remainder of installation is the reverse of removal.

TAIL/STOP/TURN/BACKUP LIGHT

Refer to illustrations 17.13 and 17.14

12 On all models, these rear lights are all in one housing, mounted on the rear fenders.

13 Refer to Chapter 11 and remove the trim panel at the rear of the luggage area. Remove the plastic nut securing the taillight housing (see illustration).

14 Pull the taillight housing out and disconnect the electrical connectors. Twist the bulbholders counterclockwise to remove them for bulb replacement (see illustration).

15 Installation is the reverse of removal. Make sure the washer and spacer are in place before installing the plastic nut on the housing.

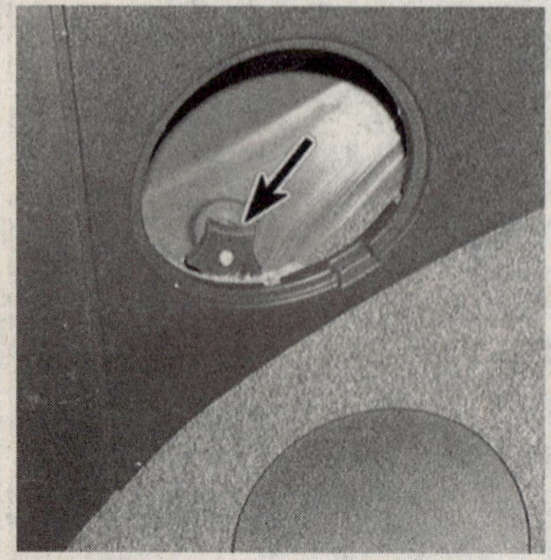

17.13 Remove the plastic trim cover in the trunk to access the taillight housing mounting nut (arrow)

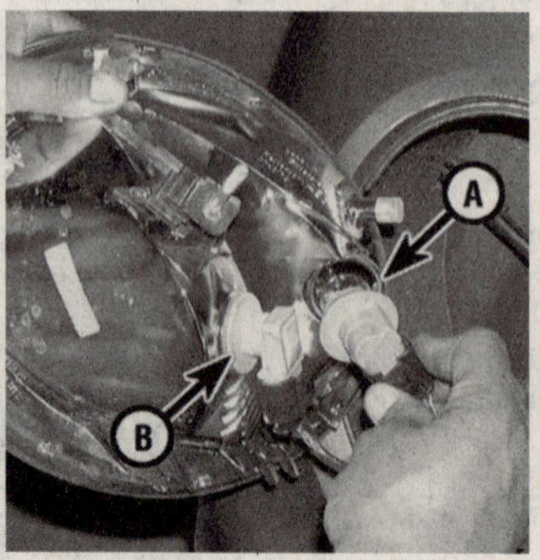

17.14 Rotate the bulb holder and pull it out to remove it - upper bulb (A) is for the turn signals, (B) is the brake/taillight bulb

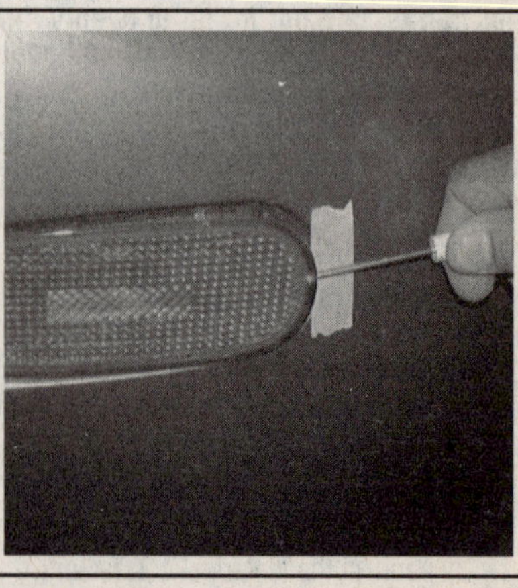

17.17 Mask the body to protect the paint, then use a small screwdriver to push in on the tab to release the rear side marker light

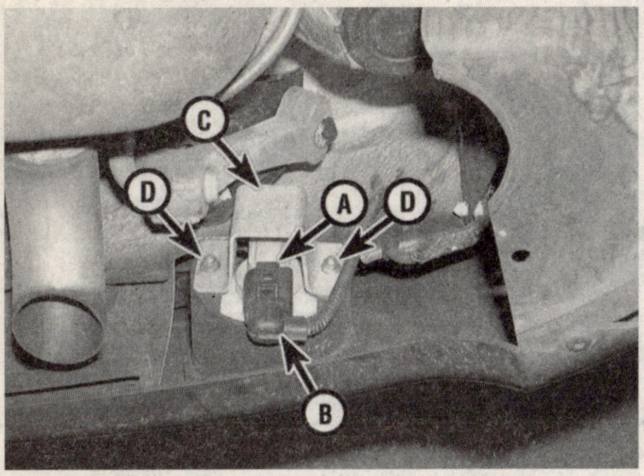

17.21a Backup light mounting details

A Bulb housing
B Electrical connector
C Guard bracket
D Lens mounting nuts

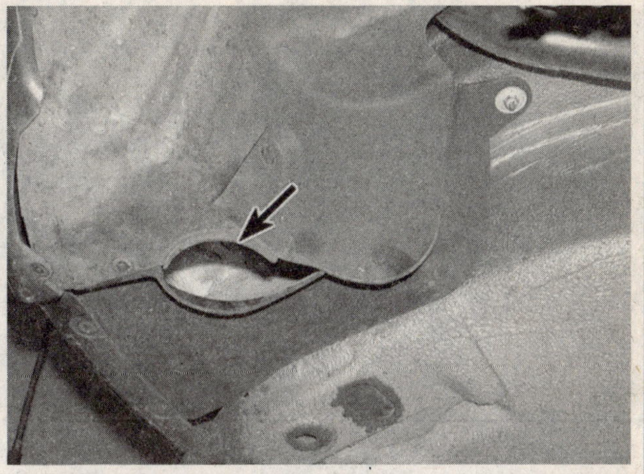

17.21b To access the right-side backup light, work through this opening (arrow) in the plastic panel at the back of the fenderwell

REAR SIDE-MARKER LIGHTS

➧ Refer to illustration 17.17

16 The side marker lights are in the lower portion of each rear fender. Refer to Chapter 11 and remove the plastic rear fenderwell liner.

17 Use a small screwdriver to depress the tab on the back of the marker light housing and push the light housing to the outside (see illustration).

18 Twist the bulbholder counterclockwise to remove it and replace the bulb.

19 Installation is the reverse of the removal procedure. Make sure the hook end of the light housing goes in first, then snap the housing in place until the tab is engaged.

BACKUP LIGHTS

➧ Refer to illustrations 17.21a and 17.21b

20 The backup lights are located at the bottom of the rear bumper cover.

21 Remove the nuts on the backside of the bumper cover to pull out the light housing (see illustration). On the right side, you'll have to work through an opening in the plastic cover at the rear of the fenderwell for access (see illustration).

22 Twist the bulbholder counterclockwise to remove for bulb replacement.

23 Installation is the reverse of the removal procedure.

CENTER HIGH-MOUNTED STOP LIGHT (CHMSL)

➧ Refer to illustrations 17.25 and 17.26

24 Refer to Chapter 11 and remove the inner trim panel from the rear hatch.

25 Remove the three screws for the CHMSL (see illustration).

26 Pull the housing from the hatch and disconnect the electrical connector (see illustration). Twist the bulbholders counterclockwise, then replace the bulbs.

27 Installation is the reverse of the removal procedure. Make sure the gasket on the back of the housing is intact, or a water leak could develop.

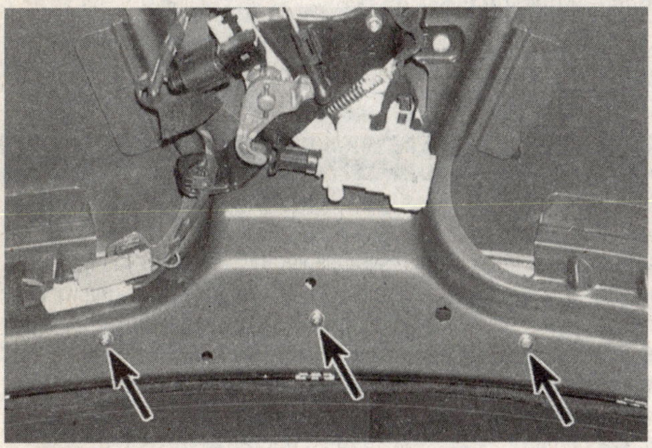

17.25 Remove the screws (arrows) and pull the CHMSL housing out

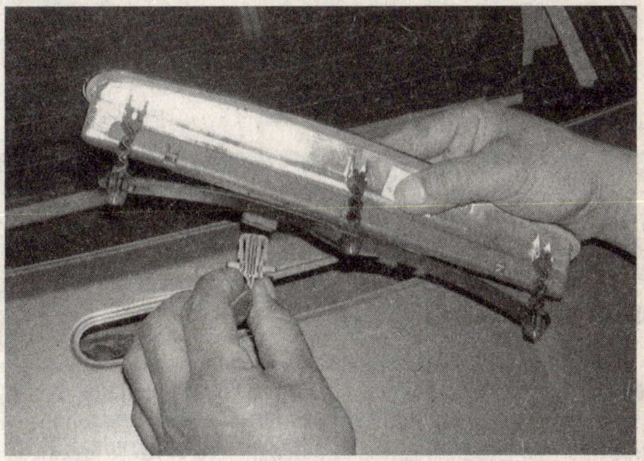

17.26 Disconnect the electrical connector from the back of the CHMSL

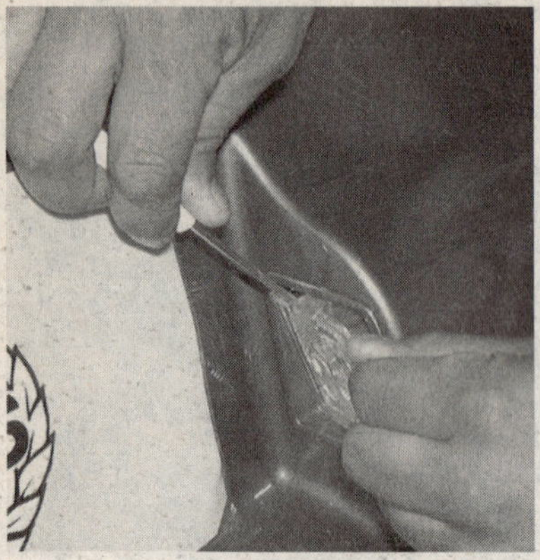

17.28 Use a small screwdriver to depress the clip securing the license plate light to the rear bumper cover

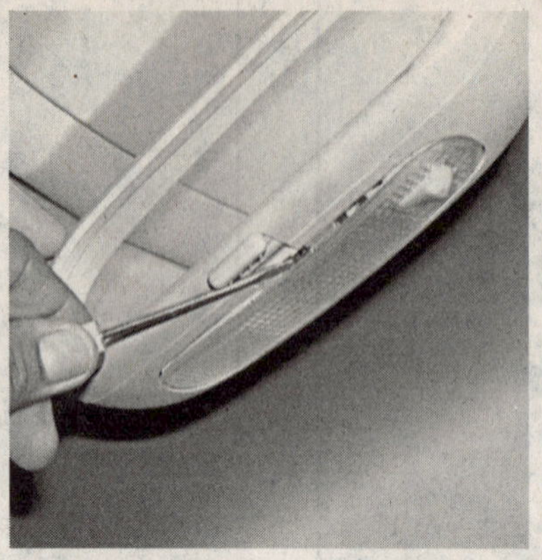

17.32 Remove the overhead light from the interior rearview mirror

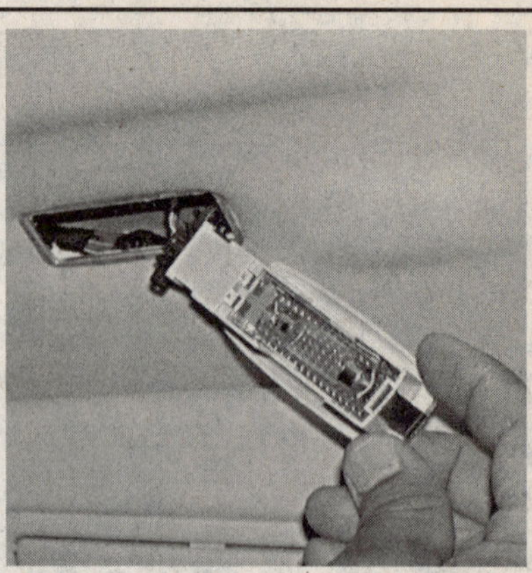

17.34 Pry the vanity light from the headliner, then replace the bulb from the back side

LICENSE PLATE BULB

Refer to illustration 17.28

28 Use a small screwdriver to depress the clip at the top of the license plate light, then tilt and pull out the light housing (see illustration).

29 Remove the bulbholder from the backside to replace the bulb.

30 Installation is the reverse of removal.

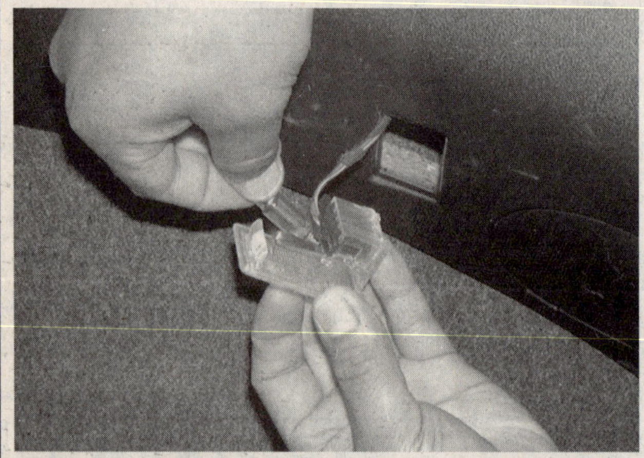

17.36 Remove the bulb from the back of the trunk light

INSTRUMENT CLUSTER LIGHTS

31 On the covered models, the instrument cluster is illuminated by LED's that are part of the printed circuit board. Thus, there are no user-replaceable bulbs behind the instrument cluster.

OVERHEAD LIGHT

Refer to illustration 17.32

32 On all models, use a small screwdriver to pry the reading light from the bottom of the interior rearview mirror, then remove the bulb (see illustration).

33 Installation is the reverse of removal.

Vanity lights

Refer to illustration 17.34

34 Pry the light out of the headliner carefully with a small screwdriver, then replace the bulb by pulling it straight out (see illustration).

35 Installation is the reverse of removal.

TRUNK LIGHT

Refer to illustration 17.36

36 Use a small screwdriver to pry the trunk light out of the trunk liner, then remove the bulb from the holder (see illustration).

18 Wiper motor - check and replacement

WIPER MOTOR CIRCUIT CHECK

Refer to illustration 18.2

Note: Refer to the wiring diagrams for wire colors and locations in the following checks. When checking for voltage, probe a grounded 12-volt test light to each terminal at a connector until it lights; this verifies voltage (power) at the terminal. If the following checks fail to locate the problem, have the system diagnosed by a dealer service department or other properly equipped repair facility.

1 If the wipers work slowly, make sure the battery is in good condition and has a strong charge (see Chapter 5). If the battery is in good condition, remove the wiper motor (see below) and operate the wiper arms by hand. Check for binding linkage and pivots. Lubricate or repair the linkage or pivots as necessary. Reinstall the wiper motor. If the wipers still operate slowly, check for loose or corroded connections, especially the ground connection. If all connections look OK, replace the motor.

2 If the wipers fail to operate when activated, check the fuse. If the fuse is OK, connect a jumper wire between the wiper motor's ground

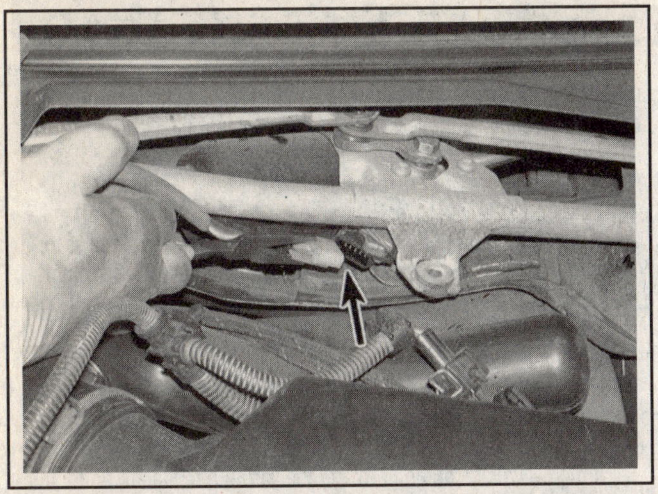

18.2 Refer to the wiring diagrams and check for proper ground and voltage at the wiper motor connector (arrow)

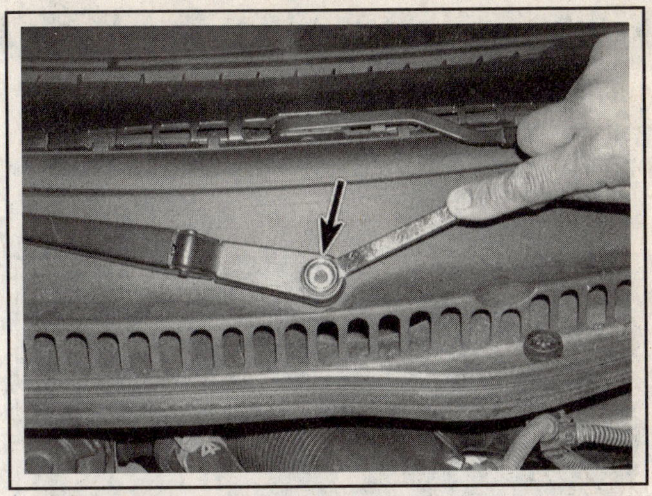

18.7 Lift the cover, use a wrench to remove the nut (arrow) on the wiper shaft, then grasp the wiper arm and use a rocking motion to detach it from the shaft

terminal and ground, then retest. If the motor works now, repair the ground connection. If the motor still doesn't work, turn the wiper switch to the HI position and check for voltage at the motor (see illustration).

➡**Note: The cowl cover and plenum panel will have to be removed (see Chapter 11) to access the electrical connector.**

3 If there's voltage at the connector, remove the motor and check it off the vehicle with fused jumper wires from the battery. If the motor now works, check for binding linkage (see Step 1 above). If the motor still doesn't work, replace it. If there's no voltage to the motor, check for voltage at the wiper control relays. If there's voltage at the wiper control relays and no voltage at the wiper motor, check the switch for continuity (see Section 7). If the switch is OK, the wiper control relay is probably bad. See Section 5 for relay testing.

4 If the interval (delay) function is inoperative, check the continuity of all the wiring between the switch and wiper control module.

5 If the wipers stop at the position they're in when the switch is turned off (fail to park), check for voltage at the park feed wire of the wiper motor connector when the wiper switch is OFF but the ignition is ON. If no voltage is present, check for an open circuit between the wiper motor and the fuse panel.

REPLACEMENT

➧ **Refer to illustrations 18.7, 18.9 and 18.10**

6 Disconnect the negative cable from the battery (see Chapter 5). Refer to the **Cautions** in Section 1.

7 Mark the positions of the wiper arms on the windshield, then remove the wiper arms (see illustration).

➡**Note: Disconnect the washer hose from the wiper arm.**

8 Remove the cowl cover (see Chapter 11).

9 Remove the wiper motor/linkage mounting bolts (see illustration).

10 Remove the assembly, disconnect the electrical connector and unbolt the motor from the linkage (see illustration).

11 Installation is the reverse of removal.

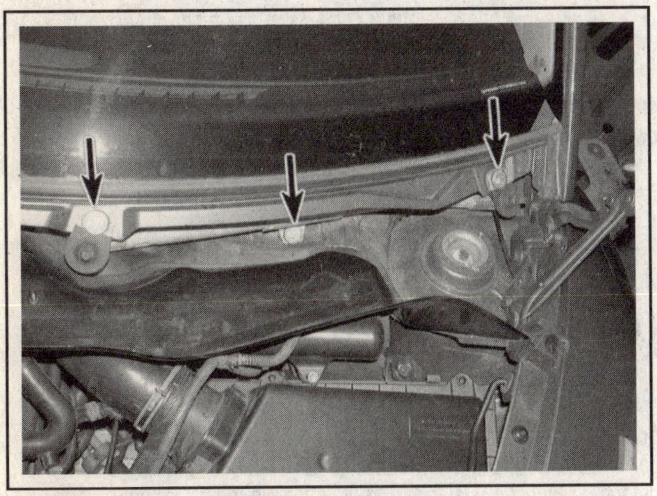

18.9 The wiper linkage mounting bolts (arrows) can be removed even with the plenum panel in place, but the plenum panel must be removed to take out the motor/linkage

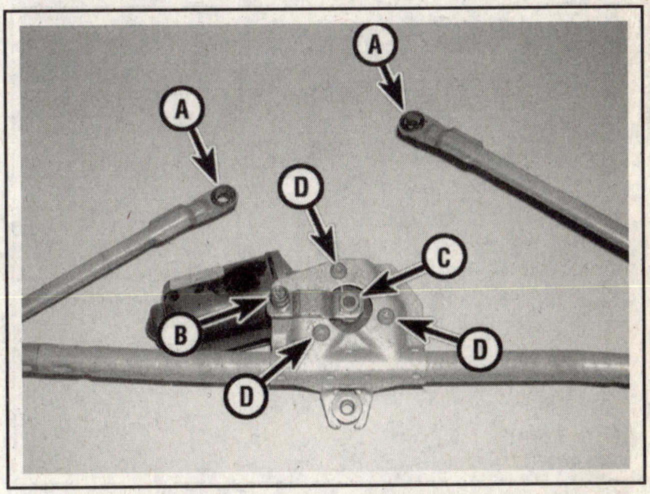

18.10 Remove the links (A) from the motor arm stud (B), then remove the motor arm nut (C) and remove the motor-to-linkage mounting bolts (D)

19 Horn - check and replacement

CHECK

Refer to illustration 19.2

1 Remove the cover from the under-dash fuse/relay box (see illustration 3.1b) and check the relay (see Section 5). Check the horn fuse in the fuse/relay panel at the left end of the instrument panel.

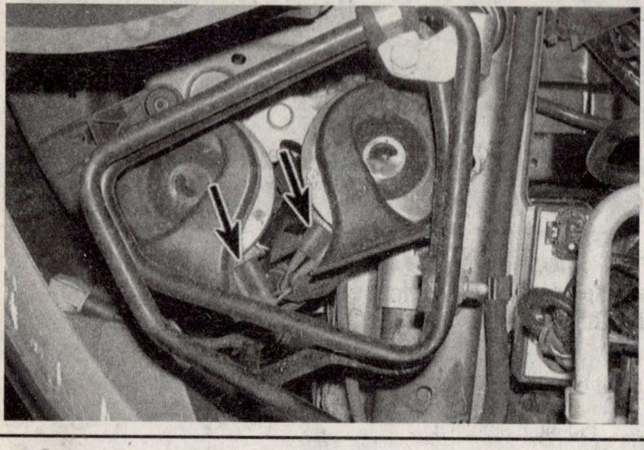

19.2 Disconnect the horn connectors (arrows) - shown here from below and behind the front bumper

2 Disconnect the electrical connectors from the horns (see illustration). There are two horns, located in the left front fenderwell, just behind the front bumper cover.

3 Have an assistant press the horn button and use a voltmeter to make sure there is battery voltage at the voltage supply wire terminal to the horns (see the wiring diagrams at the end of this Chapter). If the relay is good and there's no voltage at the horn, the wire (which leads to the relay) has a fault.

4 Use an ohmmeter to measure the resistance between the wiring connector brown wire and a good ground. There should be less than 5.0 ohms. If not, repair the fault in the ground circuit.

5 If there's voltage at the horn and the wiring circuits are good, the horn is faulty and must be replaced.

REPLACEMENT

6 Disconnect the electrical connectors, remove the mounting bolts and detach the horns (see illustration 19.2).

7 Installation is the reverse of removal.

20 Daytime Running Lights (DRL) - general information

The Daytime Running Lights (DRL) system used on all models illuminates the low-beam headlight bulbs (at reduced power), whenever the ignition is On. The only exception is with the engine running and the parking brake engaged (standard-shift models) or with the shift lever in Park (automatic-transmission models). Once the parking brake is released or the shift lever is moved, the lights will remain on as long as the ignition switch is on.

21 Rear window defogger - check and repair

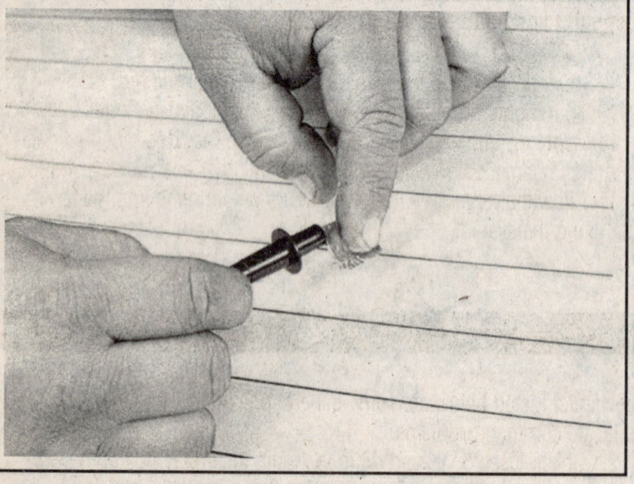

21.5 When measuring the voltage at the rear window defogger grid, wrap a piece of aluminum foil around the positive probe of the voltmeter and press the foil against the wire with your finger

1 The rear window defogger consists of a number of horizontal heating elements baked onto the inside surface of the glass. Power is supplied through a large fuse from the underhood fuse/relay box in the engine compartment. The heater is controlled by the instrument panel switch.

2 Small breaks in the element can be repaired without removing the rear window.

CHECK

Refer to illustrations 21.5, 21.6 and 21.8

3 Turn the ignition switch and defogger switch to the ON position.

4 Using a voltmeter, place the positive probe against the defogger grid positive terminal and the negative probe against the ground terminal. If battery voltage is not indicated, check the fuse, defogger switch, defogger relay and related wiring. If voltage is indicated, but all or part of the defogger doesn't heat, proceed with the following tests.

5 When measuring voltage during the next two tests, wrap a piece of aluminum foil around the tip of the voltmeter positive probe and

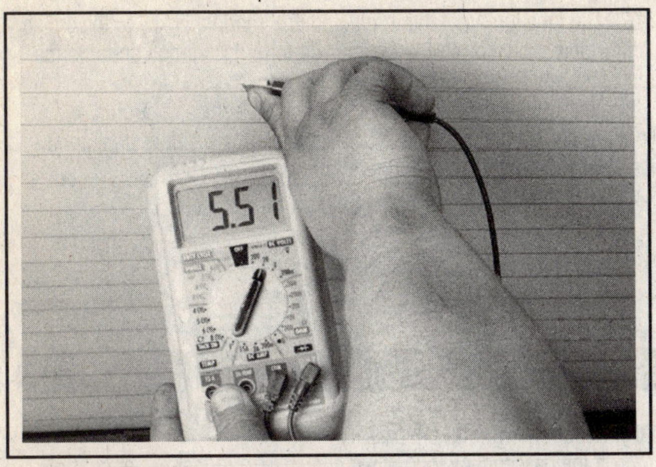

21.6 To determine if a heating element has broken, check the voltage at the center of each element - if the voltage is 6-volts, the element is unbroken

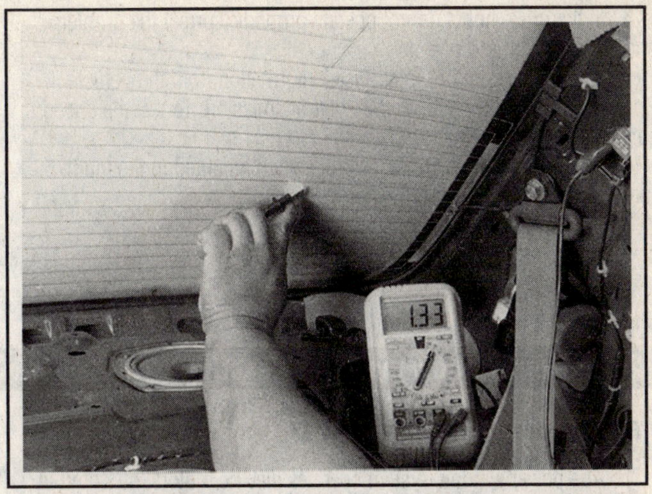

21.8 To find the break, place the voltmeter negative lead against the defogger ground terminal, place the voltmeter positive lead with the foil strip against the heat wire at the positive terminal end and slide it toward the negative terminal end - the point at which the voltmeter deflects from several volts to zero volts is the point at which the wire is broken

press the foil against the heating element with your finger (see illustration). Place the negative probe on the defogger grid ground terminal.

6 Check the voltage at the center of each heating element (see illustration). If the voltage is 5 to 6 volts, the element is okay (there is no break). If the voltage is 0 volts, the element is broken between the center of the element and the positive end. If the voltage is 10 to 12 volts the element is broken between the center of the element and the ground side. Check each heating element.

7 If none of the elements are broken, connect the negative probe to a good chassis ground. The voltage reading should stay the same, if it doesn't the ground connection is bad.

8 To find the break, place the voltmeter negative probe against the defogger ground terminal. Place the voltmeter positive probe with the foil strip against the heating element at the positive side and slide it toward the negative side. The point at which the voltmeter deflects from several volts to zero is the point where the heating element is broken (see illustration).

REPAIR

➧ **Refer to illustration 21.14**

9 Repair the break in the element using a repair kit specifically for this purpose, such as Dupont paste No. 4817 (or equivalent). The kit includes conductive plastic epoxy.

10 Before repairing a break, turn off the system and allow it to cool for a few minutes.

11 Lightly buff the element area with fine steel wool; then clean it thoroughly with rubbing alcohol.

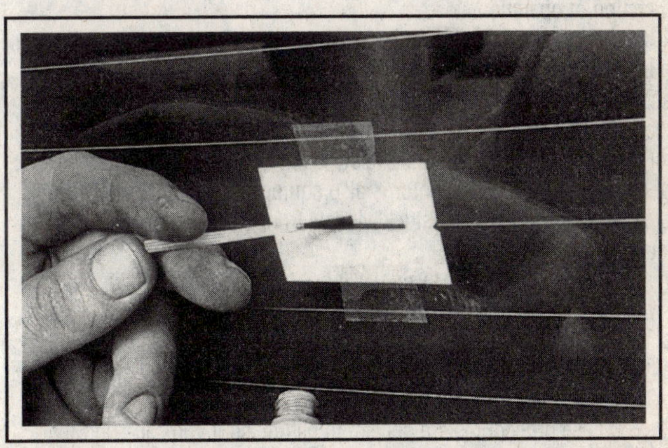

21.14 To use a defogger repair kit, apply masking tape to the inside of the window at the damaged area, then brush on the special conductive coating

12 Use masking tape to mask off the area being repaired.

13 Thoroughly mix the epoxy, following the kit instructions.

14 Apply the epoxy material to the slit in the masking tape, overlapping the undamaged area about 3/4-inch on either end (see illustration).

15 Allow the repair to cure for 24 hours before removing the tape and using the system.

22 Cruise control system - description and check

1 The New Beetle has an unusual cruise control system. Instead of having a separate cruise control module, the input from the vehicle speed sensor and cruise control switches at the steering column is directed to the Engine Control Module (ECM), which controls vehicle speed electronically through the electronic throttle control module on the throttle body. Some features of the system require special testers and diagnostic procedures that are beyond the scope of the home mechanic. Listed below are some general procedures that may be used to locate common problems.

2 Check fuses 5, 13 and 43 in the main fuse/relay box at the left end of the instrument panel (see Section 3).

3 The brake pedal position (BPP) switch (or stop lamp switch) deactivates the cruise control system. Have an assistant press the brake pedal while you check the stop lamp operation.

4 If the brake lights do not operate properly, correct the problem and retest the cruise control.

5 The cruise control system uses information from the Vehicle Speed Sensor, which is located in the transmission. To test the speed sensor, see Chapter 6A.

6 The testing of the cruise control switch mounted in the multi-function lever is covered in Section 7.

7 The functioning of the throttle control module is key to the cruise control operation, so refer to Chapter 6A for a description and information about any tests for this device.

8 Test-drive the vehicle to determine if the cruise control is now working. If it isn't, take it to a dealer service department or an automotive electrical specialist for further diagnosis.

23 Power window system - description and check

Refer to illustrations 23.10a and 23.10b

1 The power window system operates electric motors, mounted in the doors, which lower and raise the windows. The system consists of the control switches, the motors, regulators, glass mechanisms and associated wiring.

2 The power windows can be lowered and raised from the master control switch by the driver or by the switch located at the passenger window. Each window has a separate motor that is reversible. The position of the control switch determines the polarity and therefore the direction of operation.

3 The circuit is protected by fuses and a circuit breaker. Check the number 5 fuse in the fuse panel at the left end of the instrument panel, and the 30-amp fuse at the top of the fuse/relay panel near the base of the steering column. The circuit breaker (the one with the red connector) is at the bottom of the fuse/relay panel. See Section 4 for testing of circuit breakers. Each motor is also equipped with an internal circuit breaker; this prevents one stuck window from disabling the whole system.

4 The power window system will only operate when the ignition switch is ON. In addition, many models have a window lockout switch at the master control switch which, when activated, disables the switch at the passenger's window also. Always check these items before troubleshooting a window problem.

5 These procedures are general in nature, so if you can't find the problem using them, take the vehicle to a dealer service department or other properly equipped repair facility.

6 Check the wiring between the switches and fuse panel for continuity. Repair the wiring, if necessary.

7 If the passenger window is inoperative from the master control switch, try the other control switch at the passenger window.

8 If the same window works from one switch, but not the other, check the switch for continuity.

9 If the switch tests OK, check for a short or open in the circuit between the affected switch and the window motor.

10 If one window is inoperative from both switches, use a flat-bladed trim tool to pry up and remove the switch panel from the affected door. Check for voltage at the switch (see illustrations) and at the motor (refer to Chapter 11 for door panel removal) while the switch is operated.

11 If voltage is reaching the motor, disconnect the glass from the regulator (see Chapter 11). Move the window up and down by hand while checking for binding and damage. Also check for binding and damage to the regulator. If the regulator is not damaged and the window moves up and down smoothly, replace the motor. If there's binding or damage, lubricate, repair or replace parts, as necessary.

12 If voltage isn't reaching the motor, check the wiring in the circuit for continuity between the switches and motors. You'll need to consult the wiring diagram at the end of this Chapter.

13 Test the windows after you are done to confirm proper repairs.

Note: This system is governed by the Central Comfort Control Module, which is located under the left end of the instrument panel, just above the foot pedals. Problems within this module must be diagnosed by a technician with the manufacturer's scan tool. If you have eliminated the obvious causes of a problem, have the vehicle checked at a dealership.

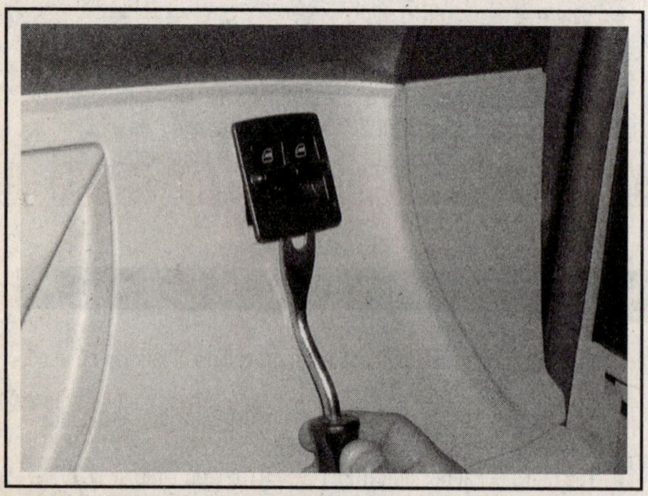

23.10a Use a trim tool to pry the door switches from the door panel

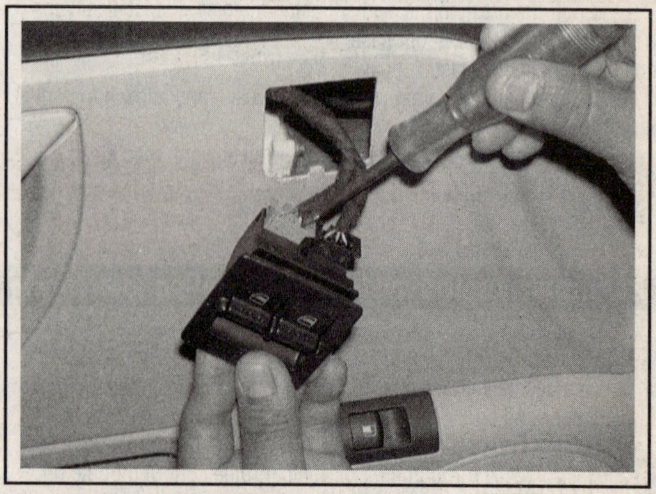

23.10b Pry up the driver's door switch assembly from the door panel and backprobe the connector

24 Power door lock and keyless entry system - description and check

Refer to illustration 24.10

1 The power door lock system operates the door lock actuators mounted in each door. The system consists of the switches, actuators, relays and associated wiring. Diagnosis can usually be limited to simple checks of the wiring connections and actuators for minor faults that can be easily repaired.

2 Power door lock systems are operated by bi-directional solenoids located in the doors. The lock switches have two operating positions: Lock and Unlock. These switches activate a relay, which in turn connects voltage to the door lock solenoids. Depending on which way the relay is activated, it reverses polarity, allowing the two sides of the circuit to be used alternately as the feed (positive) and ground side.

3 Some models may have keyless entry, electronic control modules and anti-theft systems incorporated into the power locks. If you are unable to locate the trouble using the following general steps, consult your dealer service department.

4 Always check the circuit protection first. Refer to the wiring diagrams at the end of this Chapter. Check the two fuses at the upper right of the under-dash fuse/relay panel located near the base of the steering column.

5 Operate the door lock switches in both directions (Lock and Unlock) with the engine off. Listen for the faint click of the relay operating.

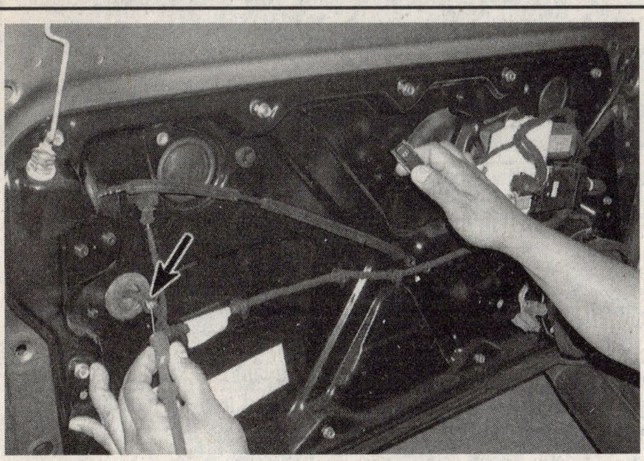

24.10 To test a door lock solenoid, check that power is getting to the solenoid connector (arrow) while the switch is operated

6 If there's no click, check for voltage at the switches. If no voltage is present, check the wiring between the fuse panel and the switches for shorts and opens.

7 If voltage is present but no click is heard, test the switch for continuity. Replace it if there's not continuity in both switch positions. To remove the switch, use a flat-bladed trim tool to pry out the door/window switch assembly (see Chapter 11).

8 If the switch has continuity but the relay doesn't click, check the wiring between the switch and relay for continuity. Repair the wiring if there's no continuity.

9 If the relay is receiving voltage from the switch but is not sending voltage to the solenoids, check for a bad ground at the relay case. If the relay case is grounding properly, replace the relay. Relay tests are described in Section 5.

10 If only one lock solenoid operates, remove the trim panel from the affected door (see Chapter 11) and check for voltage at the solenoid while the lock switch is operated (see illustration). One of the wires should have voltage in the Lock position; the other should have voltage in the Unlock position.

11 If the inoperative solenoid is receiving voltage, replace the solenoid.

12 If the inoperative solenoid isn't receiving voltage, check for an open or short in the wire between the lock solenoid and the relay.

Note: It's common for wires to break in the portion of the harness between the body and door (opening and closing the door fatigues and eventually breaks the wires).

Note: This system is governed by the Central Comfort Control Module, which is located under the left end of the instrument panel, just above the foot pedals. Problems within this module must be diagnosed by a technician with a factory scan tool. If you have eliminated the obvious causes of a problem, have the vehicle checked at a dealership.

KEYLESS ENTRY SYSTEM

13 The keyless entry system consists of a remote control transmitter that sends a coded infrared signal to a receiver, which then operates the door-lock system.

14 Replace the battery when the transmitter doesn't operate the locks at a distance of 10 feet. Normal range should be about 30 feet.

15 Use a small screwdriver to carefully separate the case halves.

16 Replace the two (3-volt) CR2016 lithium batteries.

17 Snap the case halves together.

25 Electric side view mirrors - description and check

1 The electric rear view mirrors use two motors to move the glass; one for up and down adjustments and one for left-right adjustments.

2 The control switch has a selector portion that sends voltage to the left or right side mirror. With the ignition ACC position and the engine OFF, roll down the windows and operate the mirror control switch through all functions (left-right and up-down) for both the left and right side mirrors.

Note: On some models, the mirrors are also heated electrically, by turning the outside mirror "joystick" control to the "heat" position, which is between the left and right positions.

3 Listen carefully for the sound of the electric motors running in the mirrors.

4 If the motors can be heard but the mirror glass doesn't move, there's probably a problem with the drive mechanism inside the mirror. Power mirrors have no user-serviceable parts inside, a defective mirror must be replaced as a unit (see Chapter 11).

5 If the mirrors don't operate and no sound comes from the mirrors, check the Comfort system fuse in the fuse panel at the left end of the instrument panel.

6 If the fuses are OK, remove the switch panel for access to the back of the mirror control switch without disconnecting the wires attached to it. Turn the ignition ON and check for voltage at the switch.

There should be voltage at one terminal. If there's no voltage at the switch, check for an open or short in the wiring between the fuse panel and the switch.

7 If there's voltage at the switch, disconnect it. Check the switch for continuity in all its operating positions. If the switch does not have continuity, replace it.

8 Locate the wire going from the switch to ground. Leaving the switch connected, connect a jumper wire between this wire and ground. If the mirror works normally with this wire in place, repair the faulty ground connection.

9 If the mirror still doesn't work, remove the mirror and check the wires at the mirror for voltage. Check with ignition ON and the mirror selector switch on the appropriate side. Operate the mirror switch in all its positions. There should be voltage at one of the switch-to-mirror wires in each switch position (except the neutral "off" position).

10 If voltage is absent in each switch position, check the wiring between the mirror and control switch for opens and shorts.

11 If there's voltage, remove the mirror and test it off the vehicle with jumper wires. Replace the mirror if it fails this test.

➡Note: This system is governed by the Central Comfort Control Module, which is located under the left end of the instrument panel, just above the foot pedals. Problems within this module must be diagnosed by a technician with the manufacturer's scan tool. If you have eliminated the obvious causes of a problem, have the vehicle checked at a dealership.

26 Sunroof switch - check and replacement

Refer to illustration 26.4

1 The sunroof switch is located in a panel in the headliner, above the rear-view mirror. The switch is accessible, but the sunroof module and drive can only be accessed by removing the vehicle headliner, which is a job for a dealership service department or other properly equipped repair facility.

2 To check the switch, the panel must be pulled down (see Steps 6 and 7).

3 Refer to Sections 3 and 5 and check the fuse and relay first.

4 Disconnect the electrical connector on the back of the sunroof switch and check for continuity (see illustration).

5 With the switch Off, there should be no continuity between terminal 2 and 4 or 5. With the switch turned either to Open or Closed, there should be continuity from terminal 2 with 4 in one direction or 5 in the other direction. If the switch fails the tests, replace the switch.

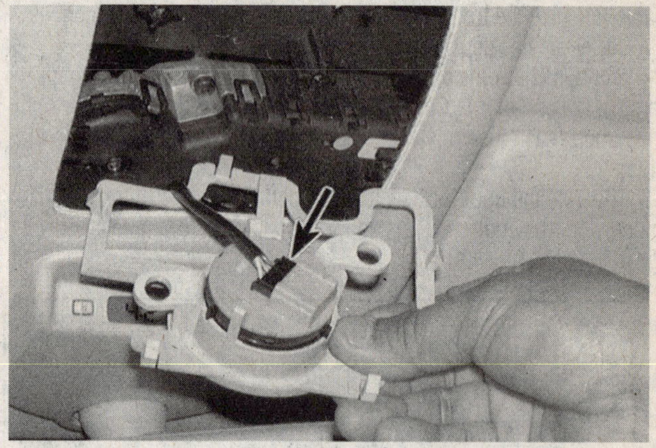

26.4 Remove the sunroof switch connector (arrow) and check the switch for continuity

REPLACEMENT

Refer to illustrations 26.6, 26.7 and 26.9

6 Pull down the sunroof switch panel from the headliner (see illustration).

7 Remove the two screws and pull down the switch and it's frame (see illustration).

8 The switch can be removed from the frame by disconnecting the electrical connector (see illustration 26.4) and releasing the two clips at the back of the switch.

9 In case of sunroof switch or module failure, the sunroof can be operated manually. A crank handle is snapped into clips on the back of the switch panel. Use the crank handle in the hole in the sunroof drive to open or close the sunroof (see illustration).

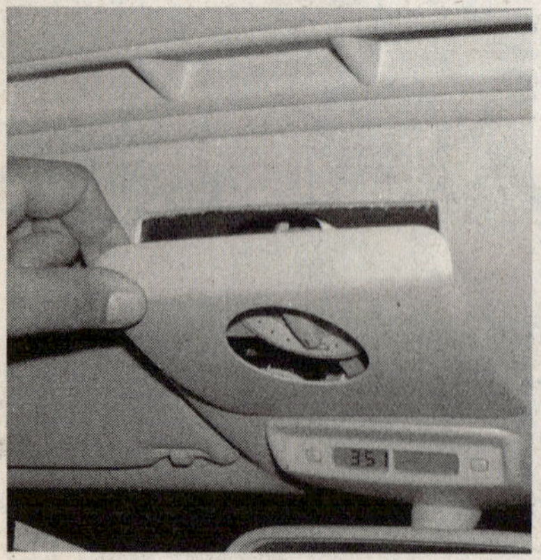

26.6 Use a trim tool to pry down the switch panel from the headliner

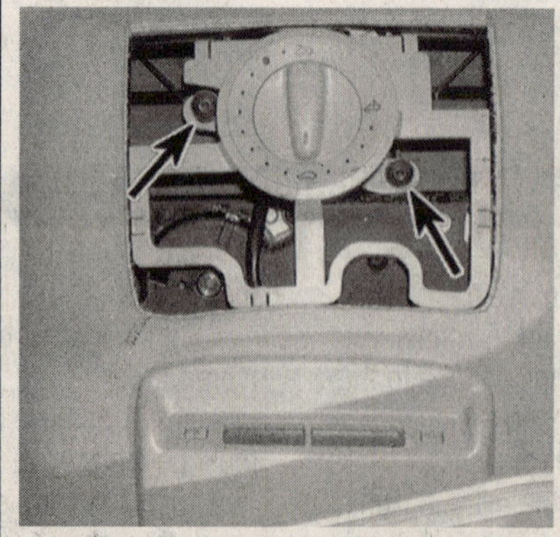

26.7 Remove the screws (arrows), pull down the switch frame, and disconnect the two clips on the back securing the switch

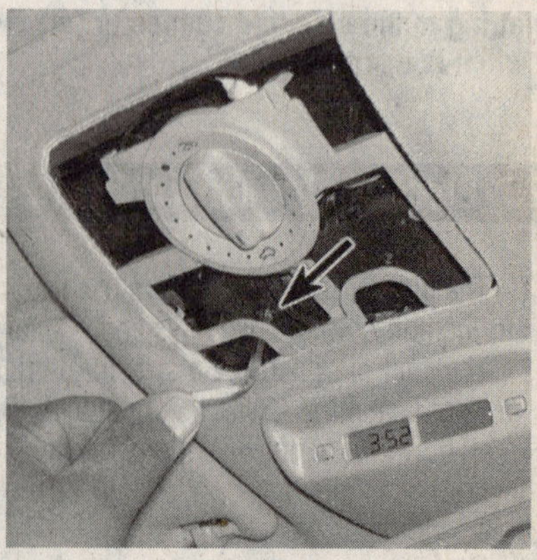

26.9 Use the emergency sunroof crank handle in the hole in the drive (arrow)

27 Airbag system - general information

Refer to illustrations 27.1 and 27.7

1 These models are equipped with a Supplemental Restraint System (SRS), more commonly known as airbags, designed to protect the driver and front seat passenger from serious injury in the event of a head-on or frontal collision. All models have a sensing/diagnostic control unit, located on the floor under the center of the instrument panel, next to the firewall (see illustration).

WARNING:

If your vehicle is ever involved in a flood, or the interior carpeting is soaked for any reason, disconnect the battery and do not start the vehicle until the airbag system can be checked by your dealer. If the SRS system is subjected to flooding, the airbags could go off upon starting the vehicle, even without an accident taking place.

AIRBAG MODULES

2 The airbag modules consist of a housing incorporating the cushion (airbag) and inflator unit. The inflator assembly is mounted on the back of the housing over a hole through which gas is expelled, inflating the bag almost instantaneously when an electrical signal is sent from the system. The specially-wound wire on the driver's side that carries this signal to the driver's module is called a clockspring. The clockspring is a flat, ribbon-like electrically conductive tape that is wound many times so that it can transmit an electrical signal regardless of steering wheel position. Airbag modules are located in the steering wheel, on the passenger side above the glove box, and at the upper side of each front seat (side-impact airbags).

SENSING/DIAGNOSTIC CONTROL UNIT AND SENSORS

3 The sensing/diagnostic control unit contains an on-board microprocessor which monitors the operation of the system, and also contains a crash sensor. It checks this system every time the vehicle is started, causing the "AIRBAG" light to illuminate for five seconds, then go off, if the system is operating properly. If there is a fault in the system, the light may not come on at all, the light will go on and continue, either illuminated steadily or blinking, and the unit will store fault codes indicating the nature of the fault.

27.1 The airbag control module is located on the floor ahead of the console - do not tamper with the yellow connectors attached to it

OPERATION

4 For the airbag(s) to deploy, one or both impact sensors and the safing sensor must be activated. When this condition occurs, the circuit to the airbag inflator is closed and the airbag inflates. If the battery is destroyed by the impact, or is too low to power the inflator, a back-up power unit inside the SRS system provides power.

SELF-DIAGNOSIS SYSTEM

5 A self-diagnosis circuit in the SRS unit displays a light on the instrument panel when the ignition switch is turned to the On position. If the system is operating normally, the light should go out after about five seconds. If the light doesn't come on, or doesn't go out after a short time, or if it comes on while you're driving the vehicle, or if it blinks at any time, there's a malfunction in the SRS system. Have it inspected and repaired as soon as possible. Do not attempt to troubleshoot or service the SRS system yourself. Even a small mistake could cause the SRS system to malfunction when you need it.

SERVICING COMPONENTS NEAR THE SRS SYSTEM

6 Nevertheless, there are times when you need to remove the steering wheel, radio or service other components on or near the dashboard. At these times, you'll be working around components and wire harnesses for the SRS system. The SRS wiring harnesses are easy to identify: They're all covered with a bright yellow conduit. Do not unplug the connectors for these wires. And do not use electrical test equipment on airbag system wires; it could cause the airbag(s) to deploy. ALWAYS DISABLE THE SRS SYSTEM BEFORE WORKING NEAR THE SRS SYSTEM COMPONENTS OR RELATED WIRING.

DISABLING THE SRS SYSTEM

WARNING:

Any time you are working in the vicinity of airbag wiring or components, DISABLE THE SRS SYSTEM.

7 To disable the airbag system, perform the following steps:

a) *Turn the steering wheel to the straight-ahead position and turn the*

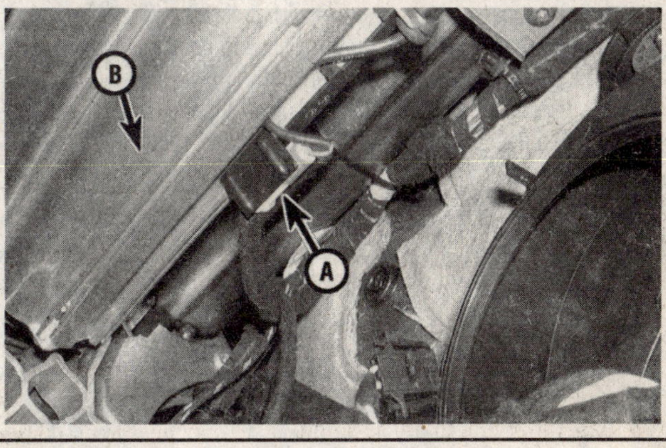

27.7 Disconnect the connector (A) from the passenger airbag (B), below the right side of the dash

ignition switch to the Lock position, then remove the key.

b) *Disconnect the negative battery cable (see Chapter 5). Refer to the* **Cautions** *in Section 1 of this Chapter.*

c) *Wait at least two minutes for the backup power supply to be depleted before beginning work.*

d) *Remove the driver's knee bolster (see Chapter 11) and disconnect the driver's airbag connector at the steering column (see Chapter 10).*

e) *Open and drop the glovebox door (see Chapter 11) and disconnect the connector to the passenger airbag (see illustration).*

ENABLING THE SYSTEM

8 To enable the airbag system, perform the following steps:

a) *Turn the ignition switch to the Lock position and remove the key.*

b) *Reconnect the passenger and driver's airbag connectors, making sure the CPA (Connector Positive Assurance) clips are in place so the connectors can't accidentally disengage.*

c) *Connect the battery cable.*

d) *With your body out of the path of the airbag, turn the ignition switch to the On position. Confirm that the airbag warning light glows for 5 seconds, then goes out, indicating the system is functioning properly.*

REMOVAL AND INSTALLATION

Driver's side airbag

9 Refer to Chapter 10 for removal and installation of the driver's side airbag.

Passenger side airbag

10 Disable the airbag system, see the **Warning** above.

11 Remove the glovebox and its dash panel (see Chapter 11).

12 From under the instrument panel, disconnect the passenger airbag connector (see illustration 27.7).

13 The instrument panel must be removed for access to the passenger airbag mounting screws (see Chapter 11).

➡Note: This is a difficult job for the home mechanic, involving tedious disassembly and many electrical connections to keep track of.

14 Remove the screws and gently pry the airbag module from the brackets on the cowl support beam.

⁂ WARNING:

Whenever handling an airbag module, always carry the airbag module with the trim panel side facing away from your body. Place the airbag module in a safe location with the trim panel side facing up.

⁂ CAUTION:

The airbag assembly is heavier than it looks, use both hands when removing it from the dash.

15 Installation is the reverse of the removal procedure.

IMPACT SEAT BELT RETRACTORS

16 All models are equipped with pyrotechnic (explosive) units in the front seat belt retracting mechanisms for both the lap and shoulder belts. During an impact that would trigger the airbag system, the airbag control unit also triggers the seat belt retractors. When the pyrotechnic charges go off, they accelerate the retractors to instantly take up any slack in the seat belt system to more fully prepare the driver and front seat passenger for impact.

17 The airbag system should be disabled anytime work is done to or around the seats or the body "B" pillars.

⁂ CAUTION:

Never strike the pillars or floorpan with a hammer or impact-driver tool unless the system is disabled.

28 Wiring diagrams - general information

Since it isn't possible to include all wiring diagrams for every year and model covered by this manual, the following diagrams are those that are typical and most commonly needed.

Prior to troubleshooting any circuits, check the fuse and circuit breakers (if equipped) to make sure they're in good condition. Make sure the battery is properly charged and check the cable connections (see Chapter 1).

When checking a circuit, make sure that all connectors are clean, with no broken or loose terminals. When disconnecting a connector, do not pull on the wires. Pull only on the connector housings themselves.

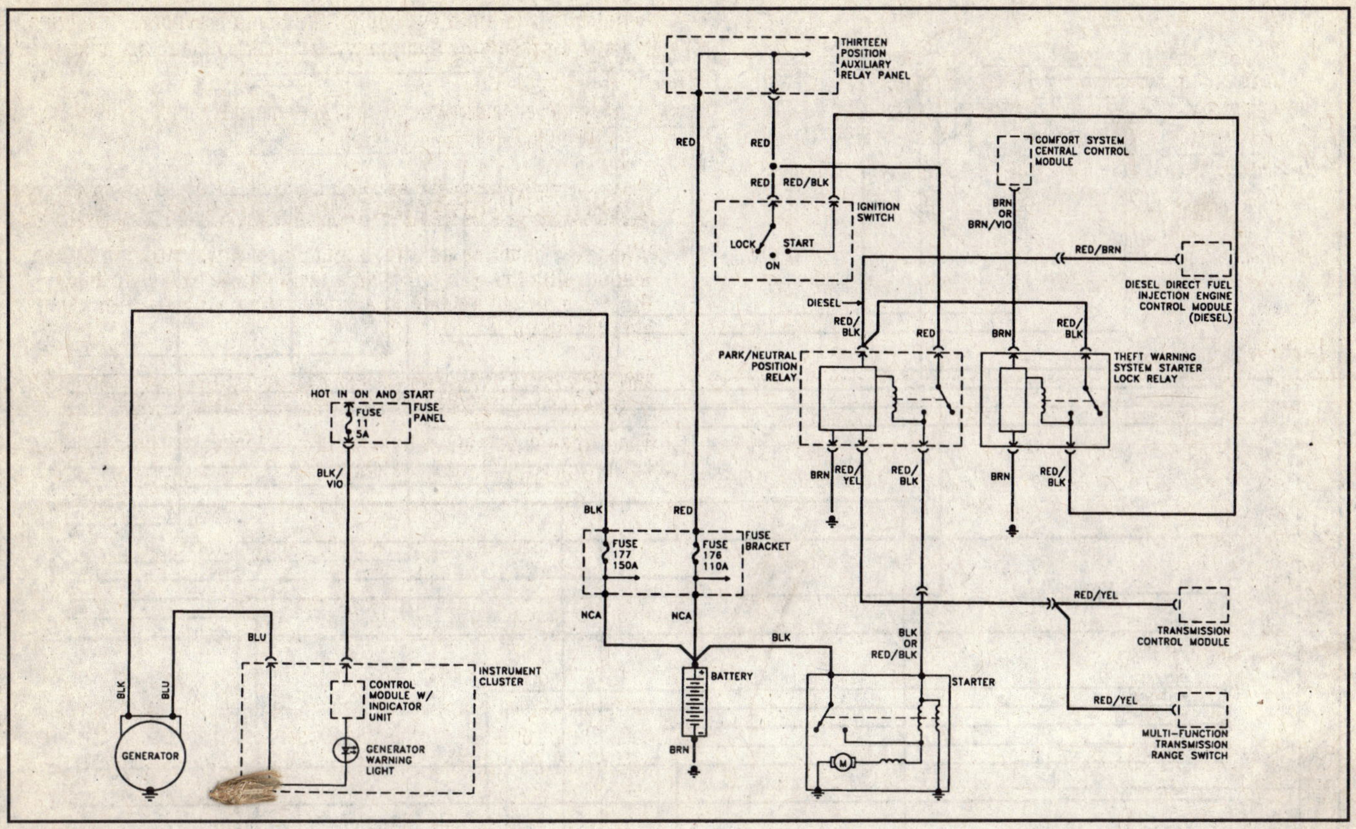

Starting and charging system - automatic transaxle

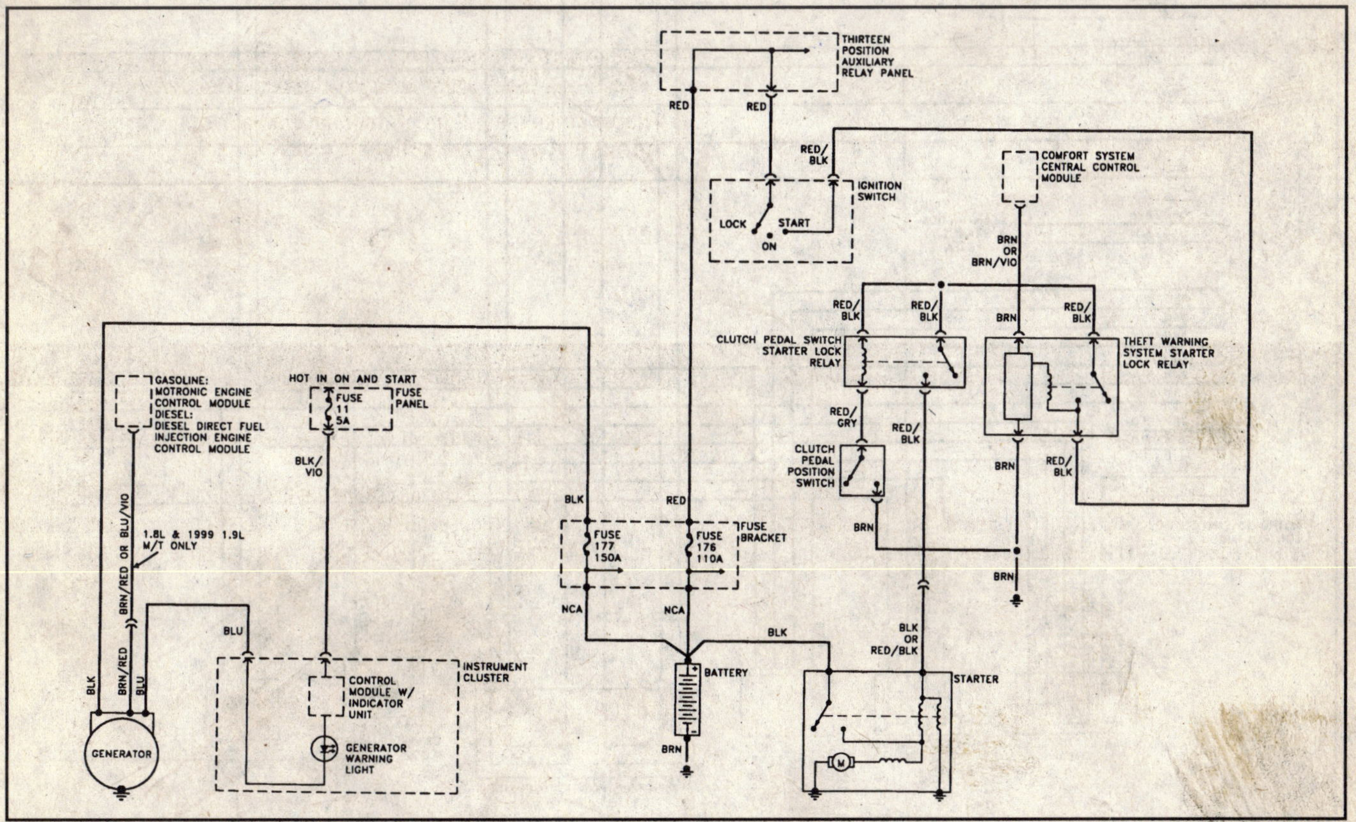

Starting and charging system - manual transaxle

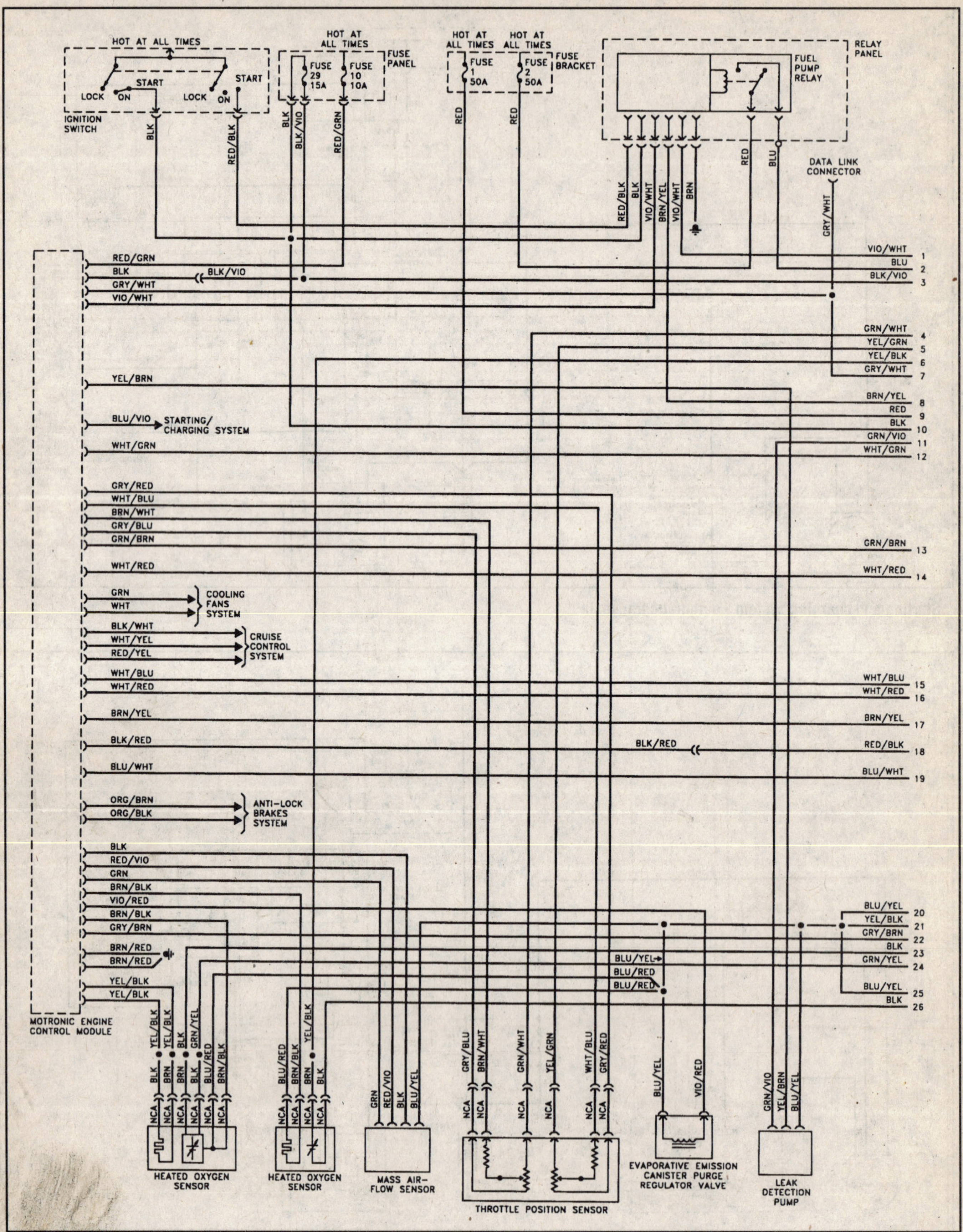

Engine control system - 1.8L gasoline engine (1 of 3)

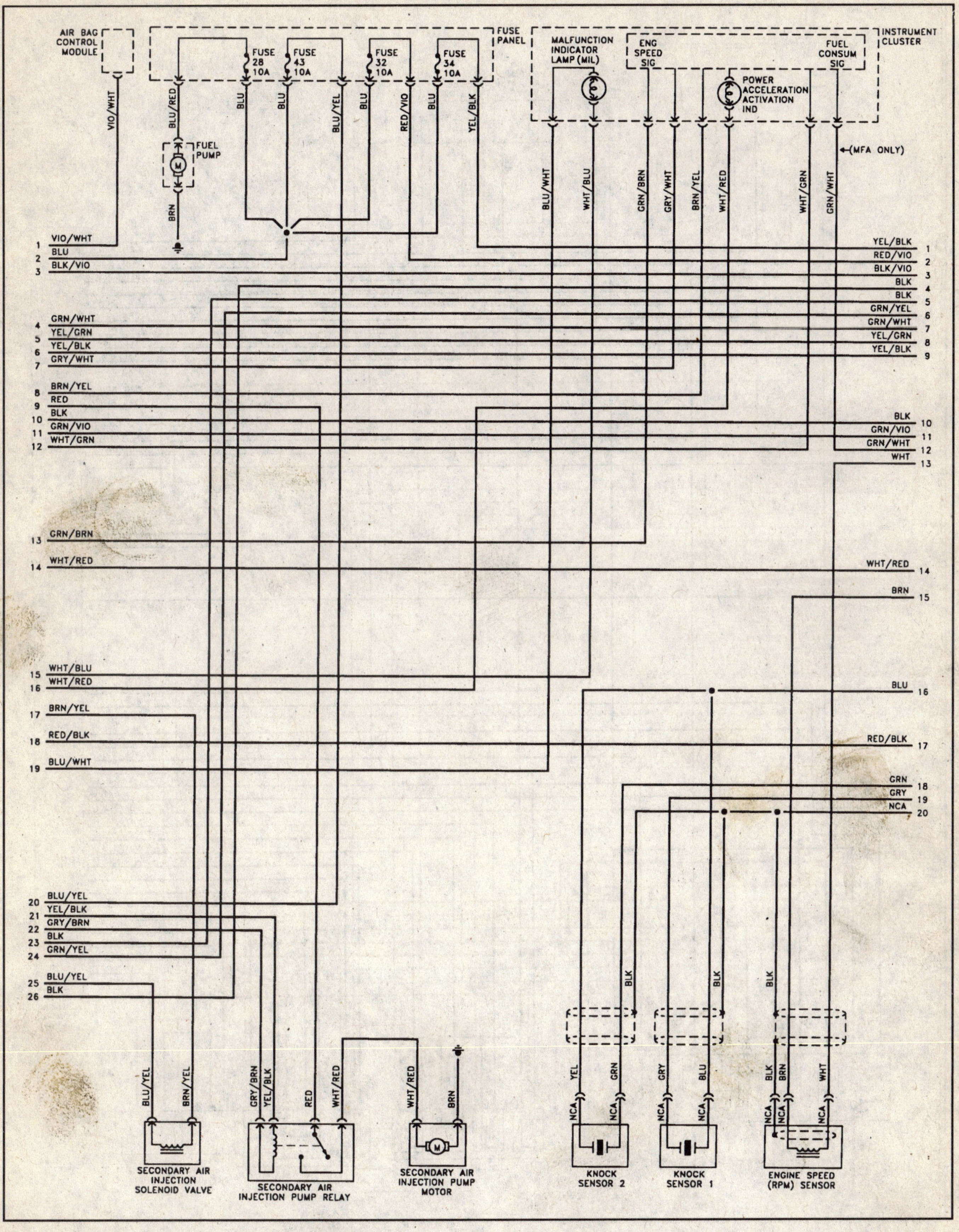

Engine control system - 1.8L gasoline engine (2 of 3)

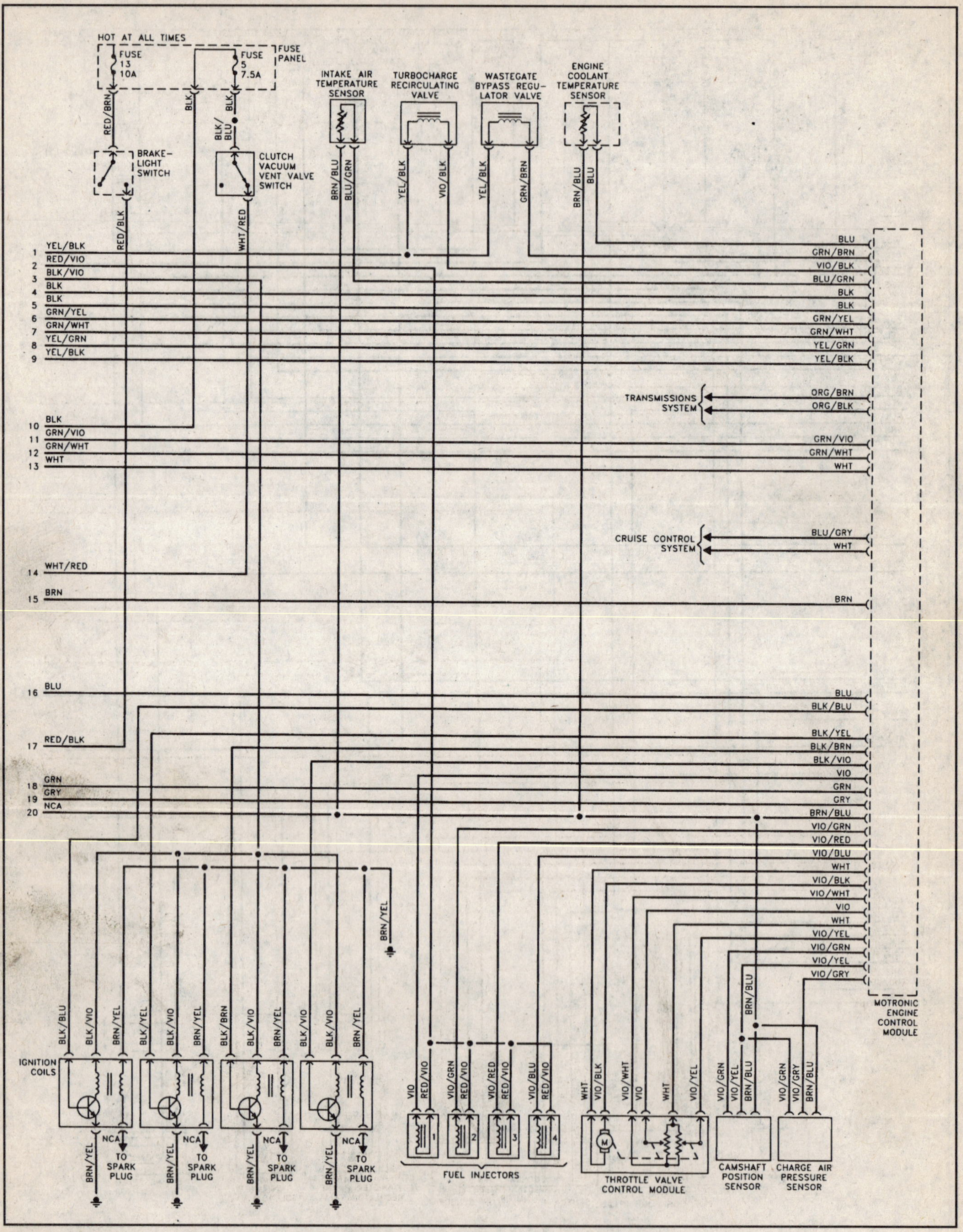

Engine control system - 1.8L gasoline engine (3 of 3)

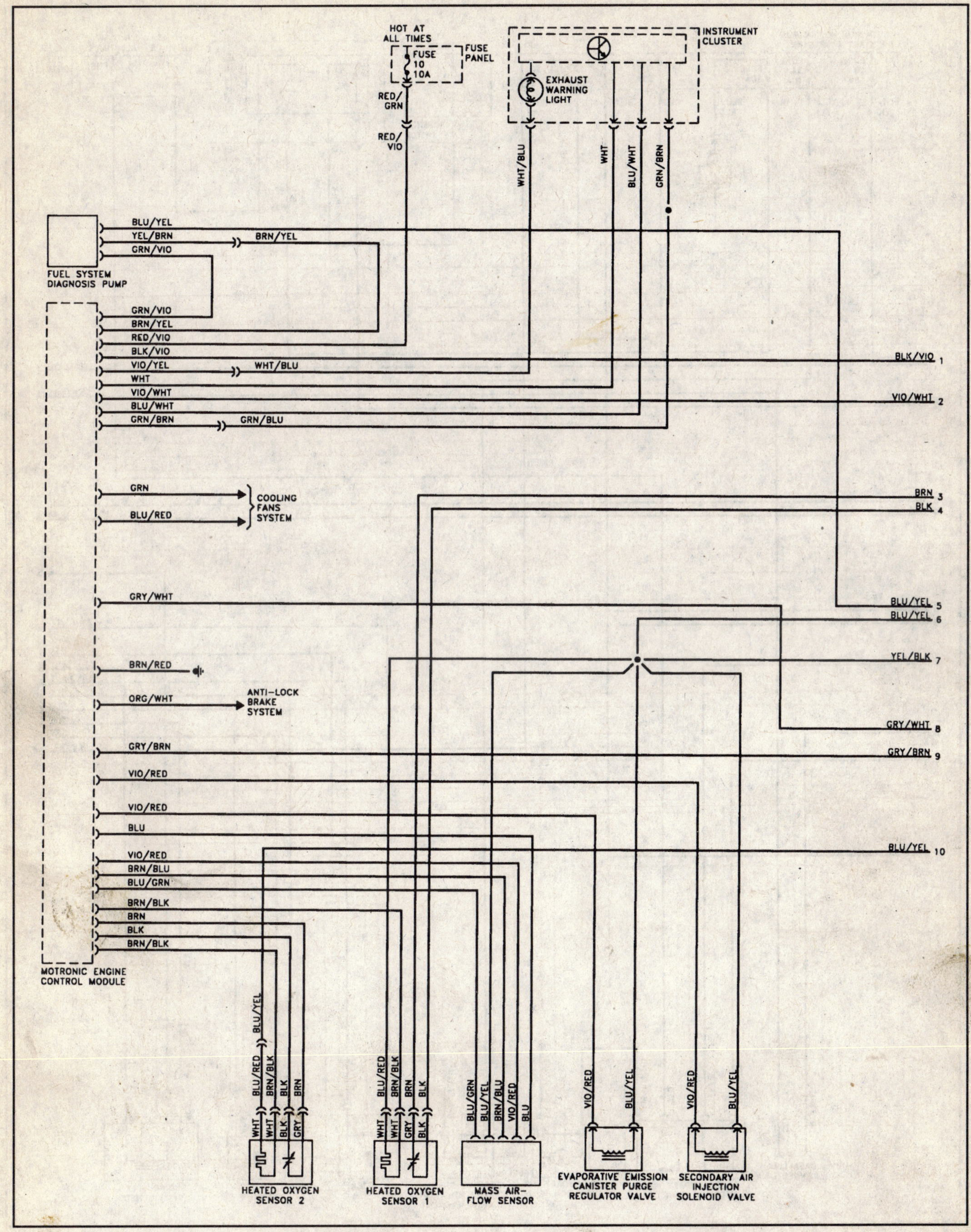

Engine control system - 2.0L gasoline engine (1 of 3)

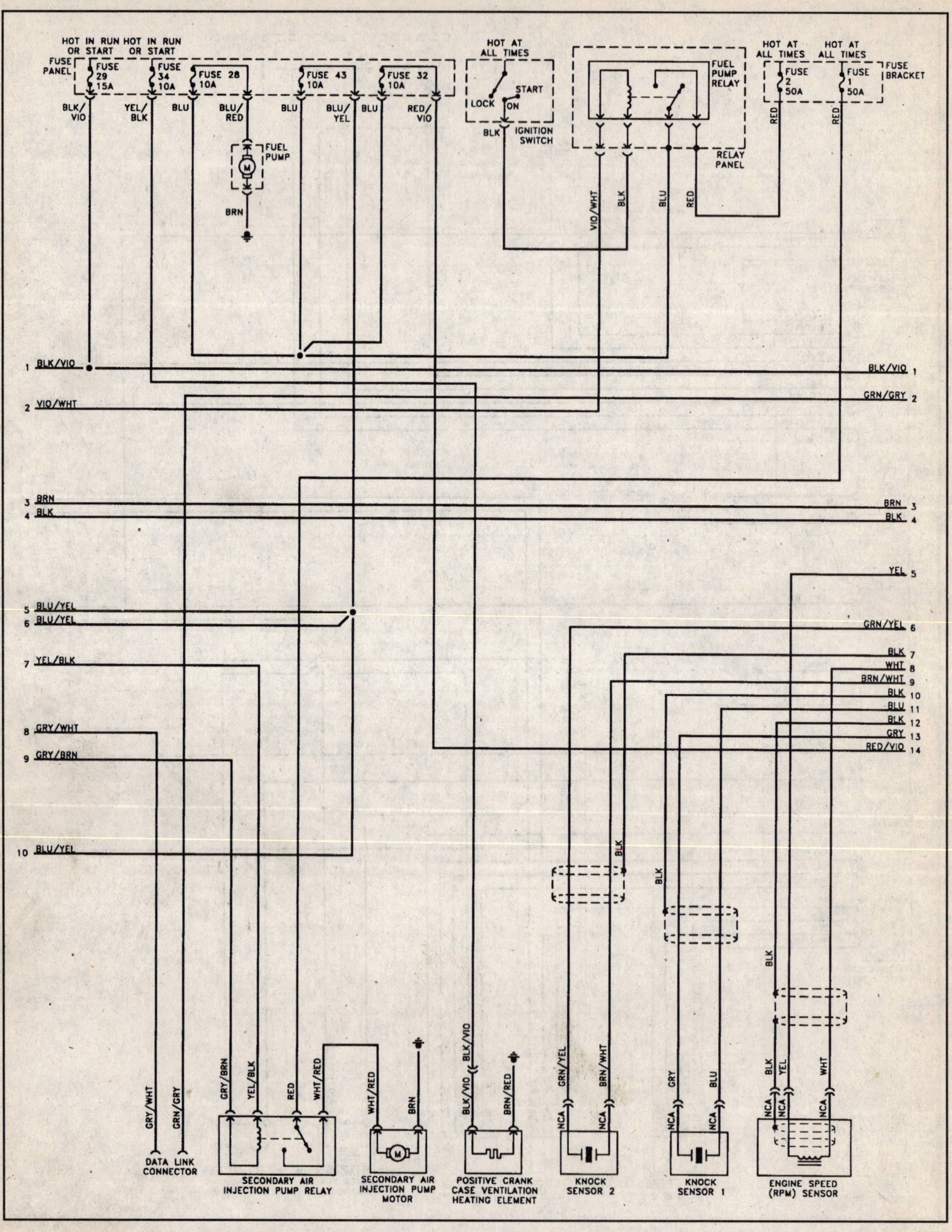

Engine control system - 2.0L gasoline engine (2 of 3)

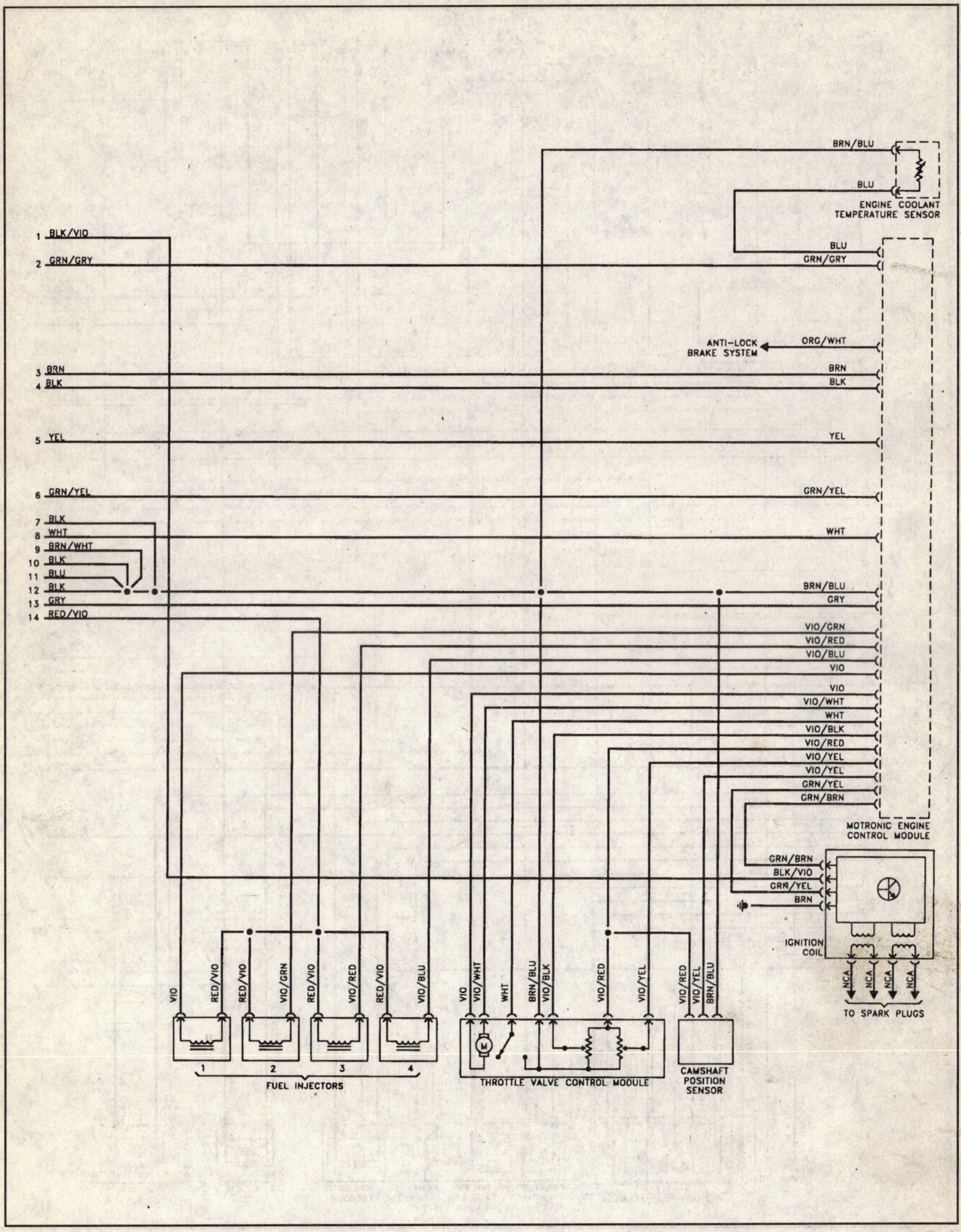

Engine control system - 2.0L gasoline engine (3 of 3)

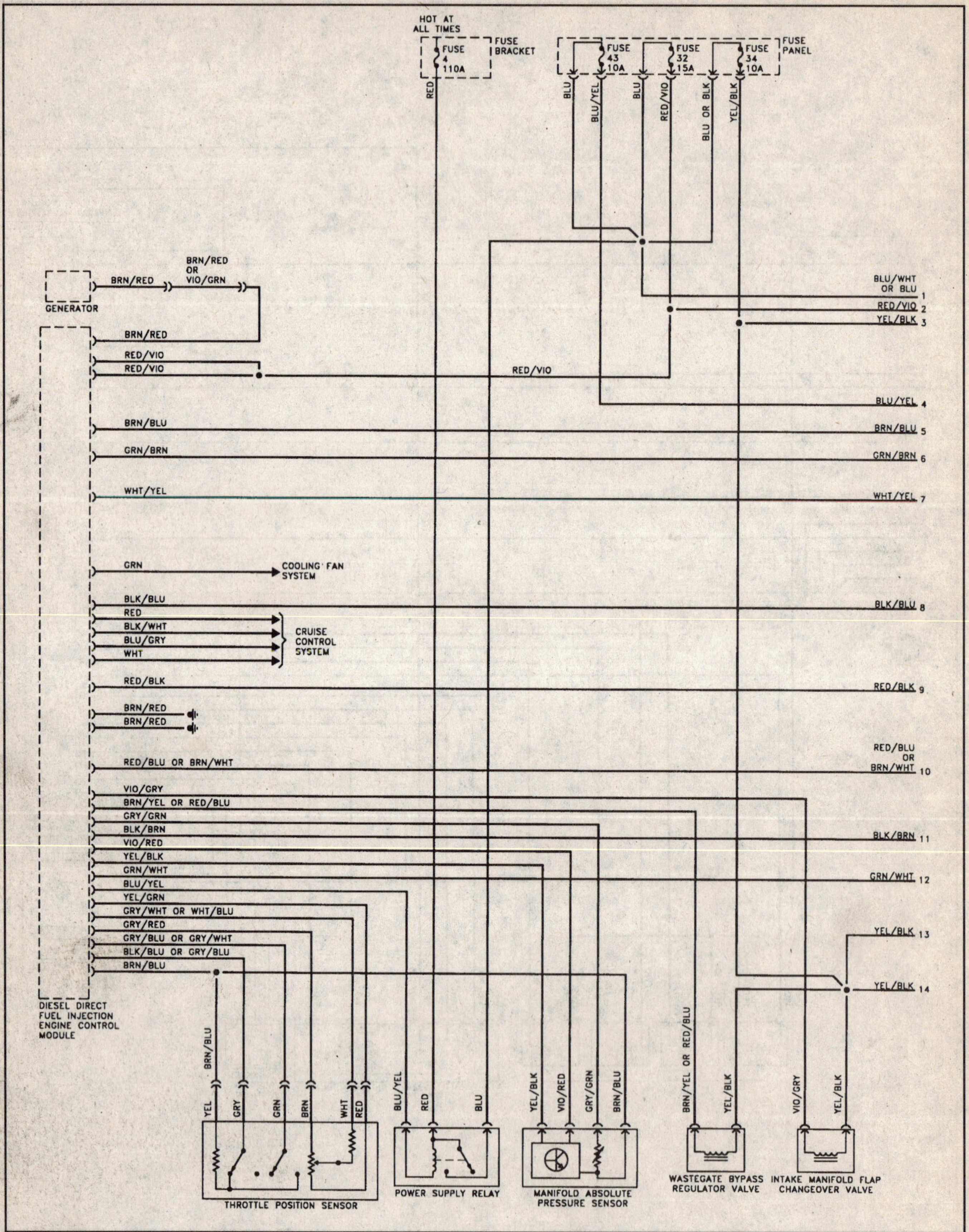

Engine control system - 1.9L diesel engine (1 of 3)

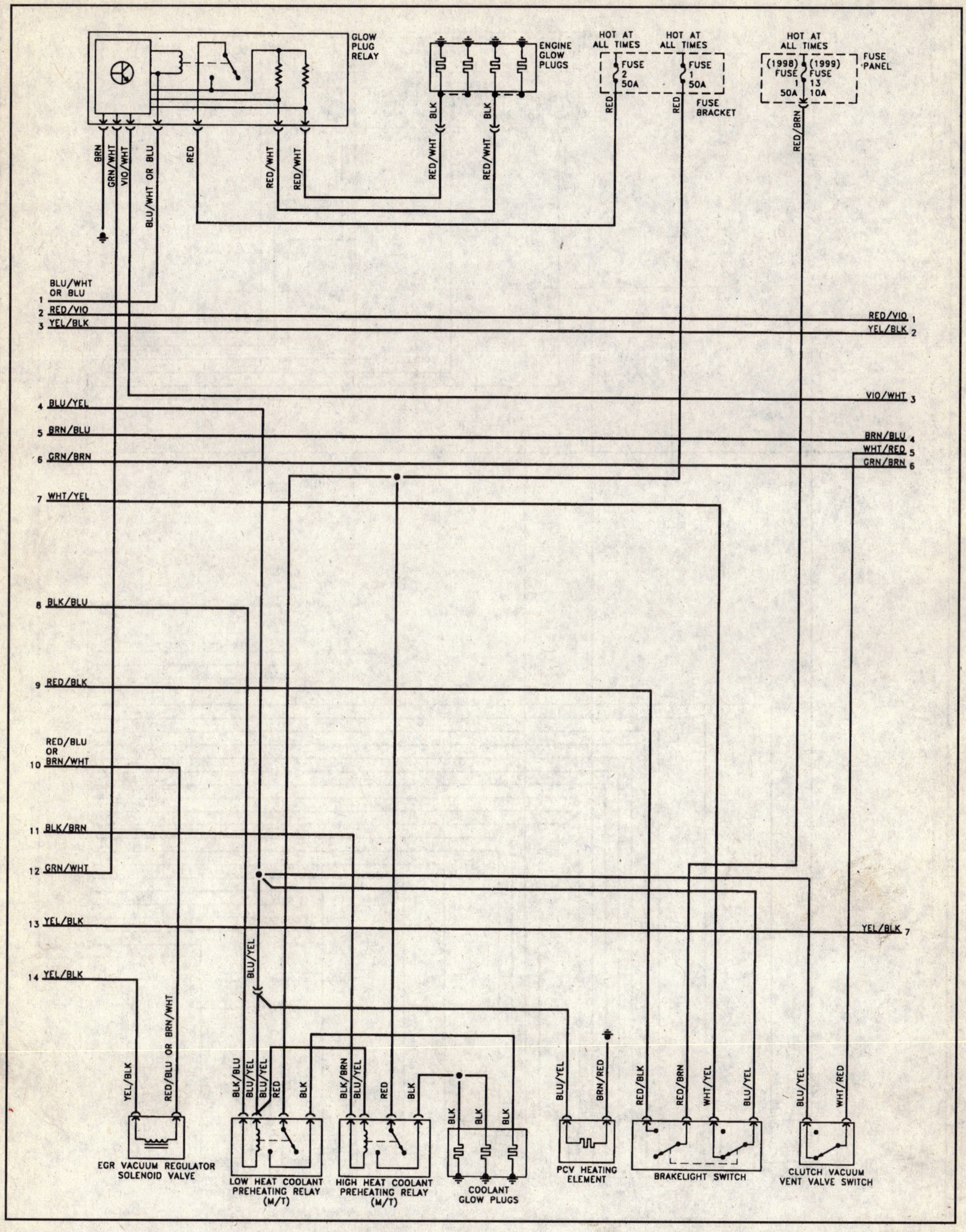

Engine control system - 1.9L diesel engine (2 of 3)

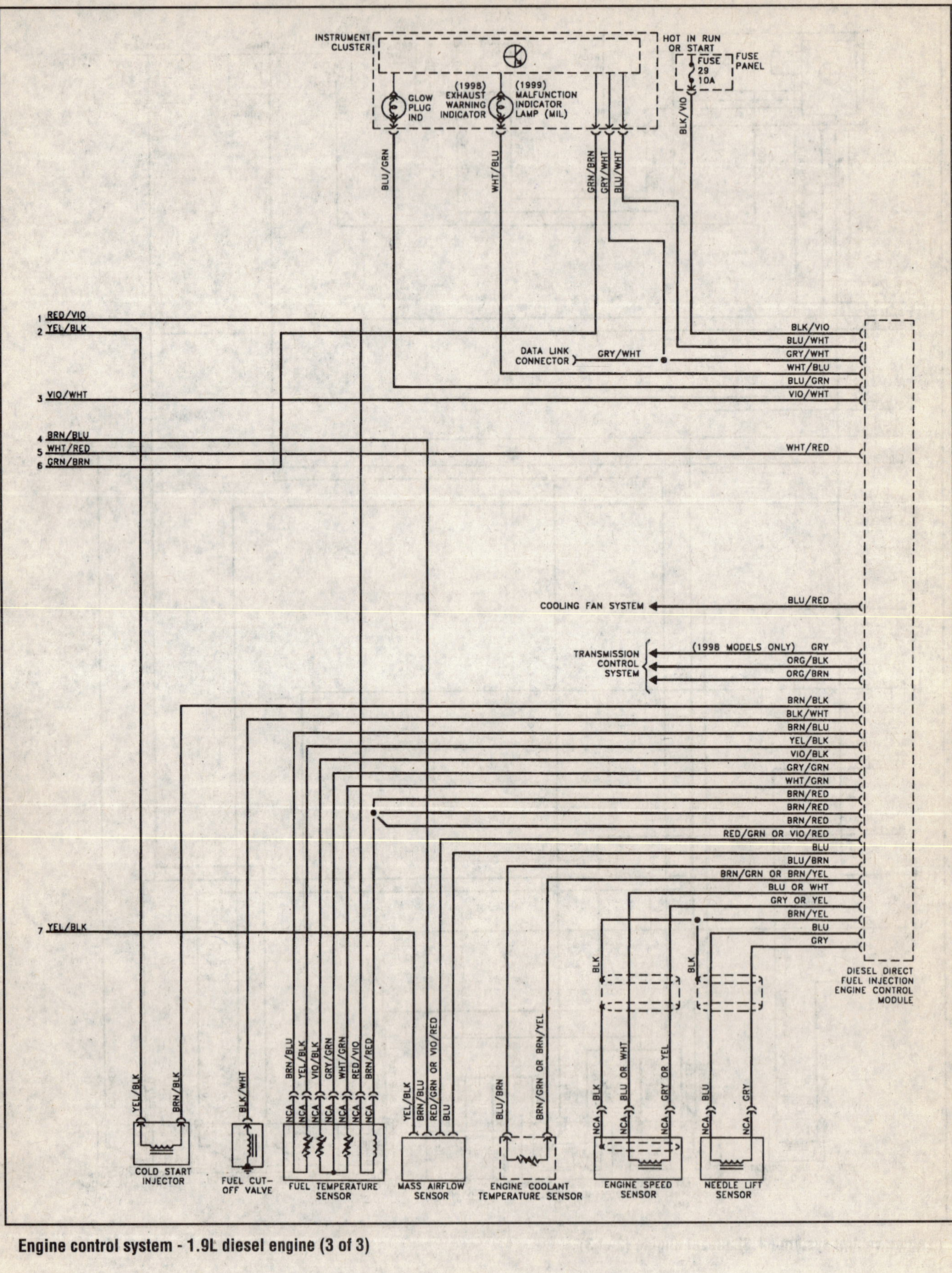

Engine control system - 1.9L diesel engine (3 of 3)

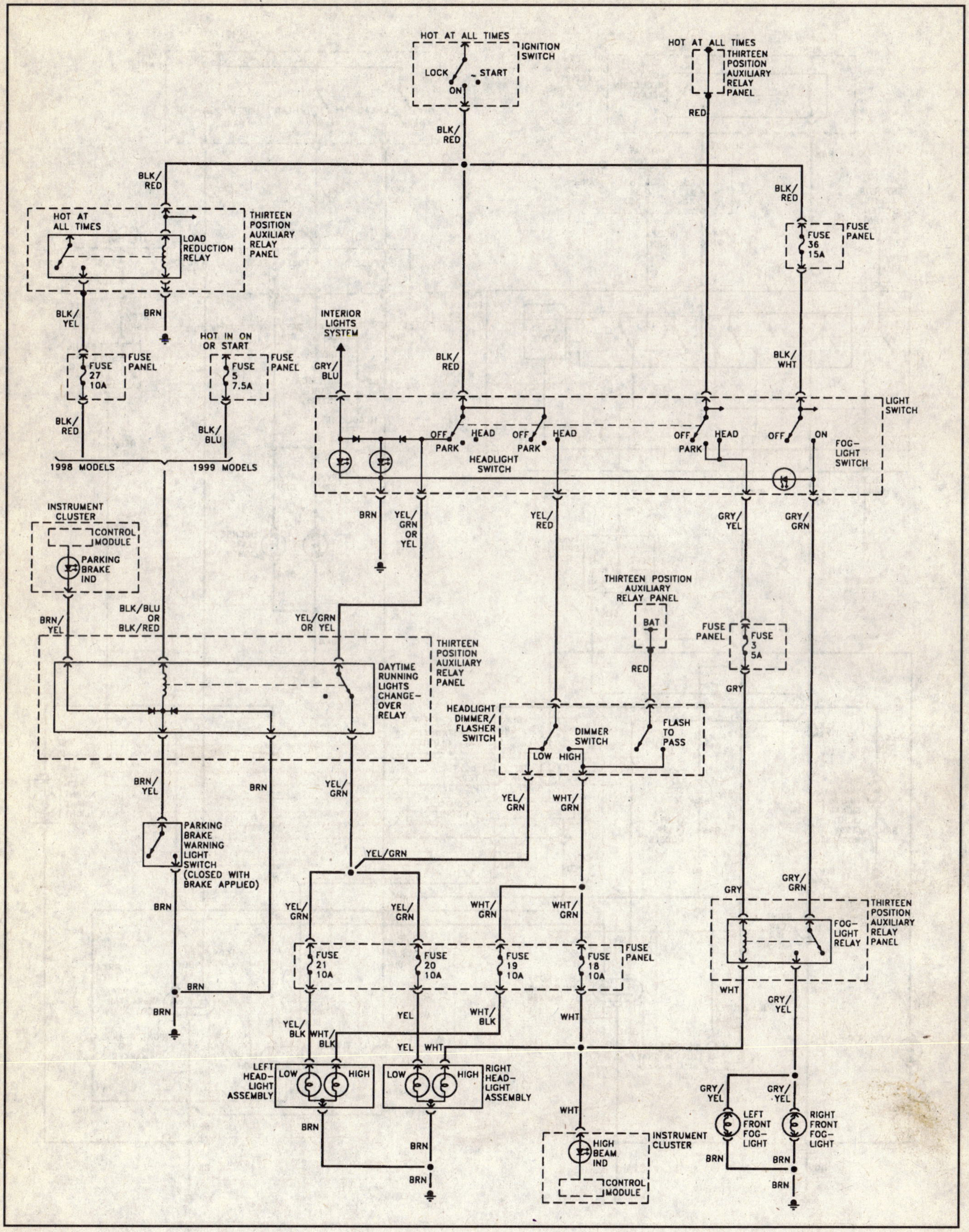

Headlight system

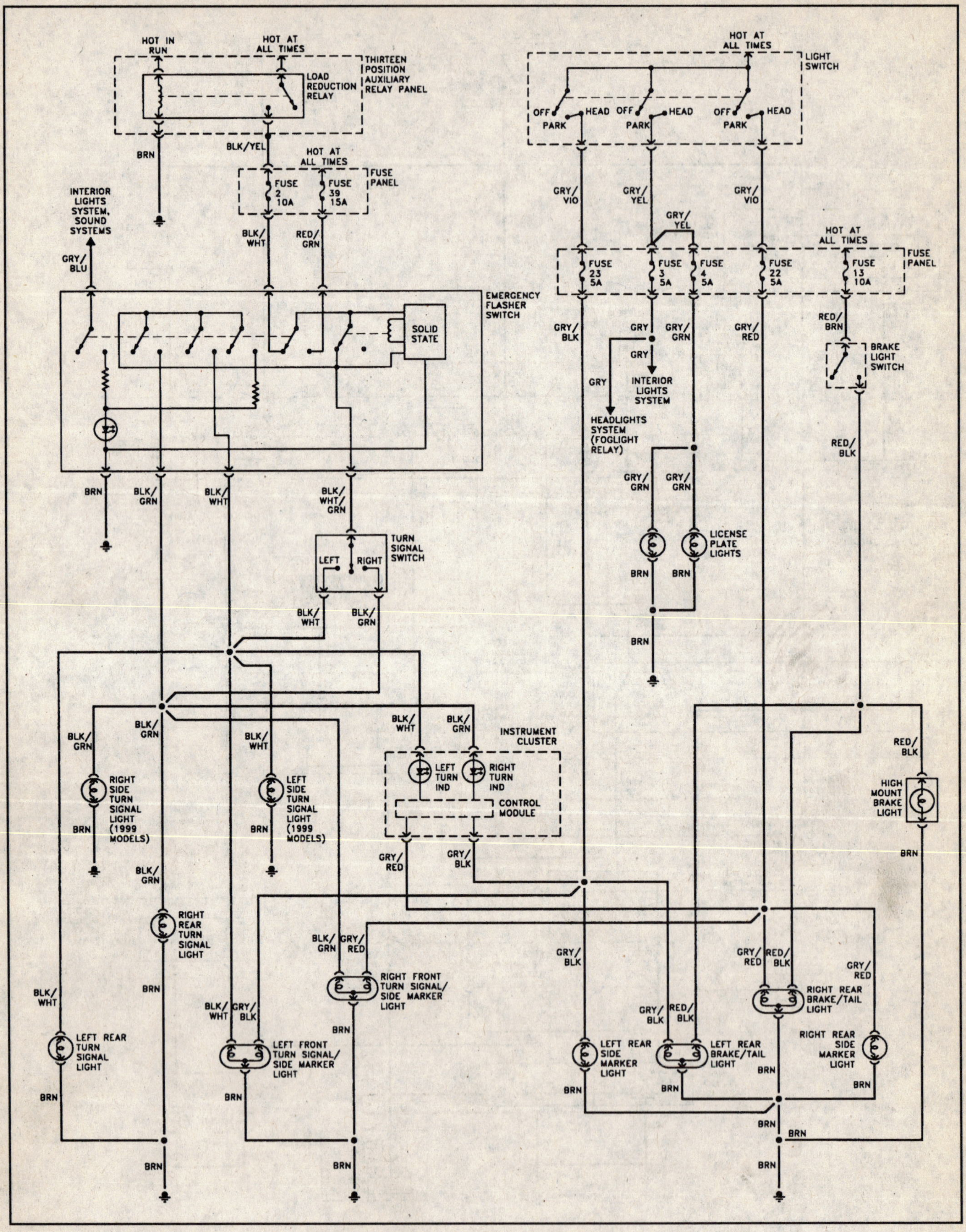

Exterior lighting system

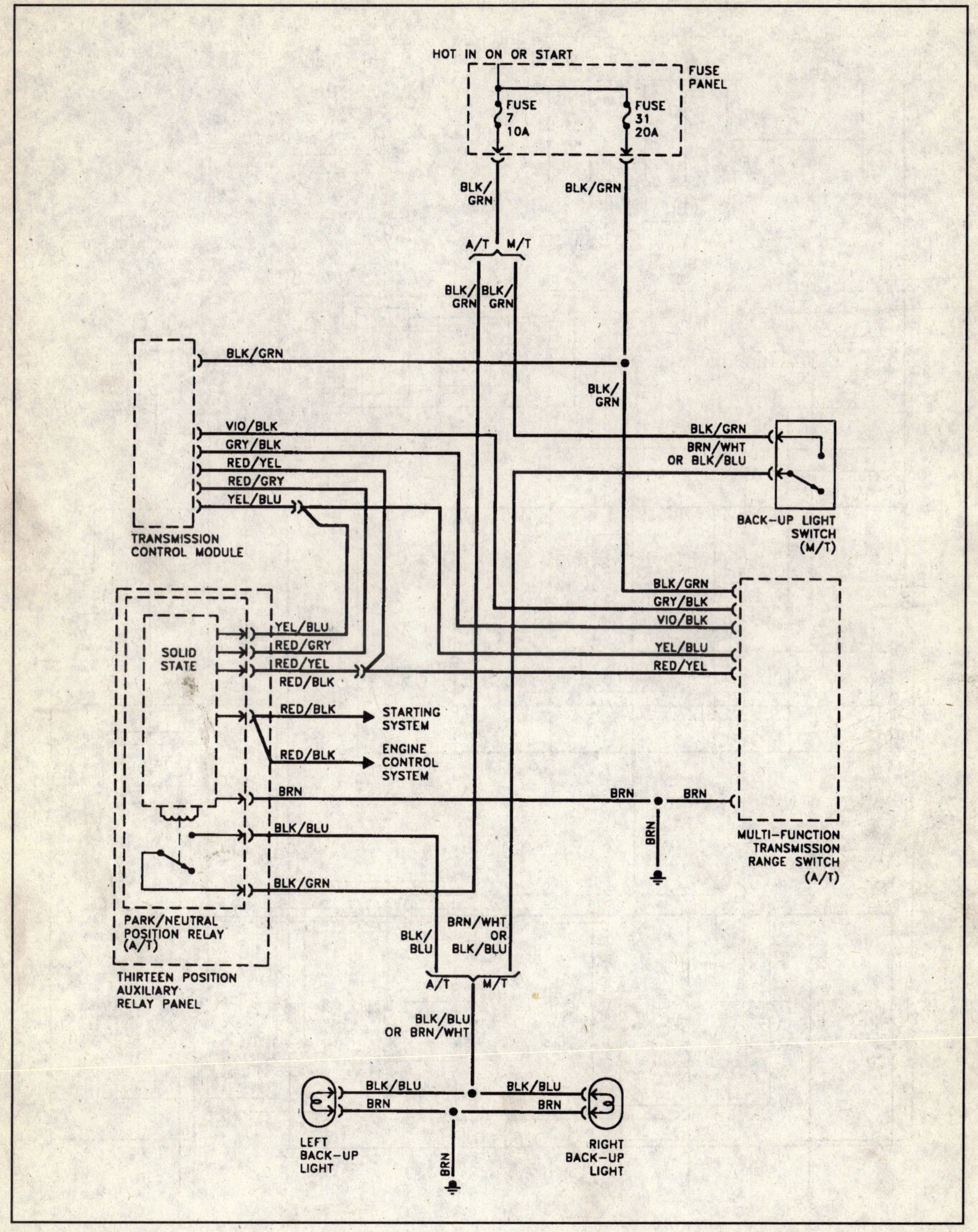

Back-up light system

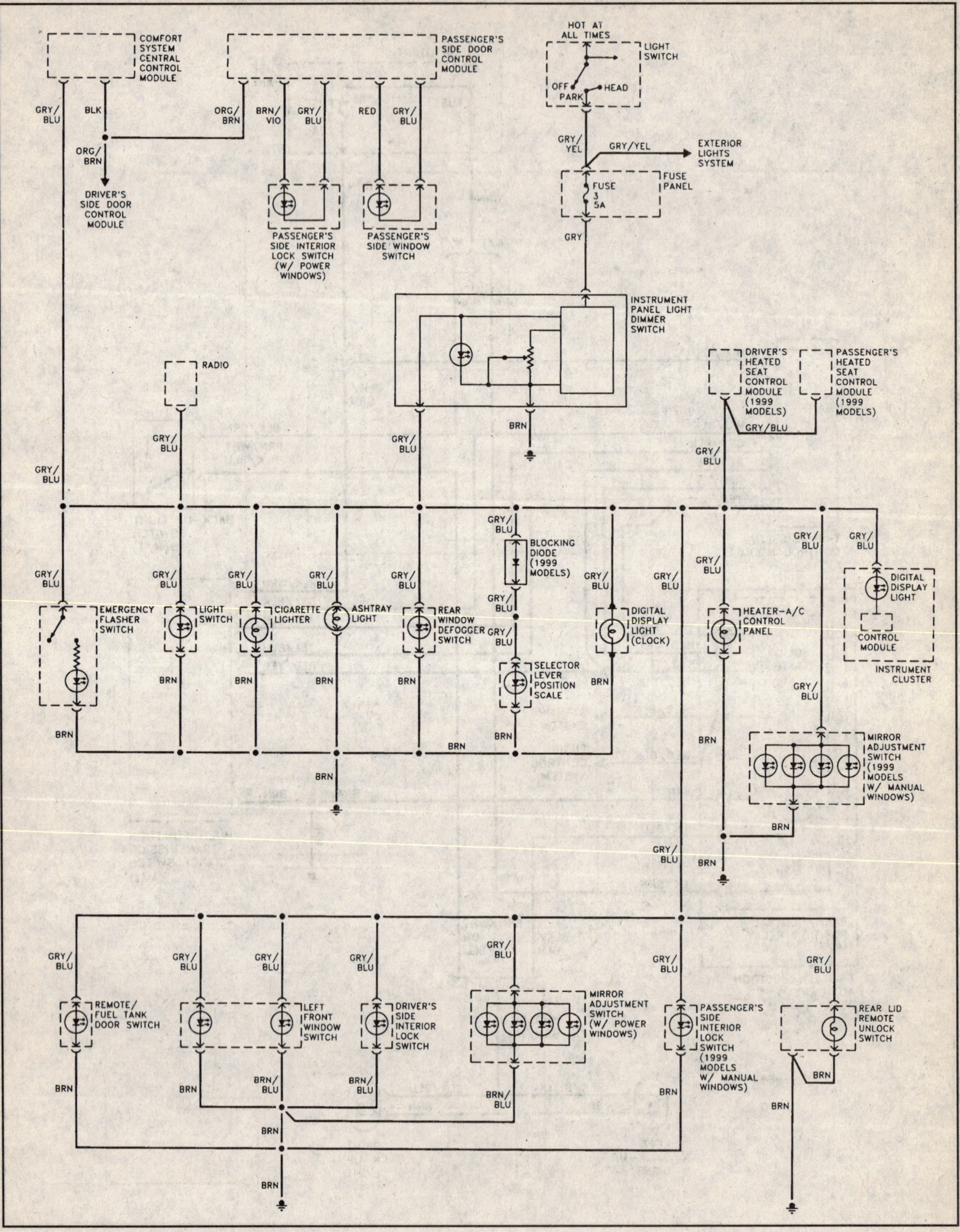

Interior lighting system - illumination lights

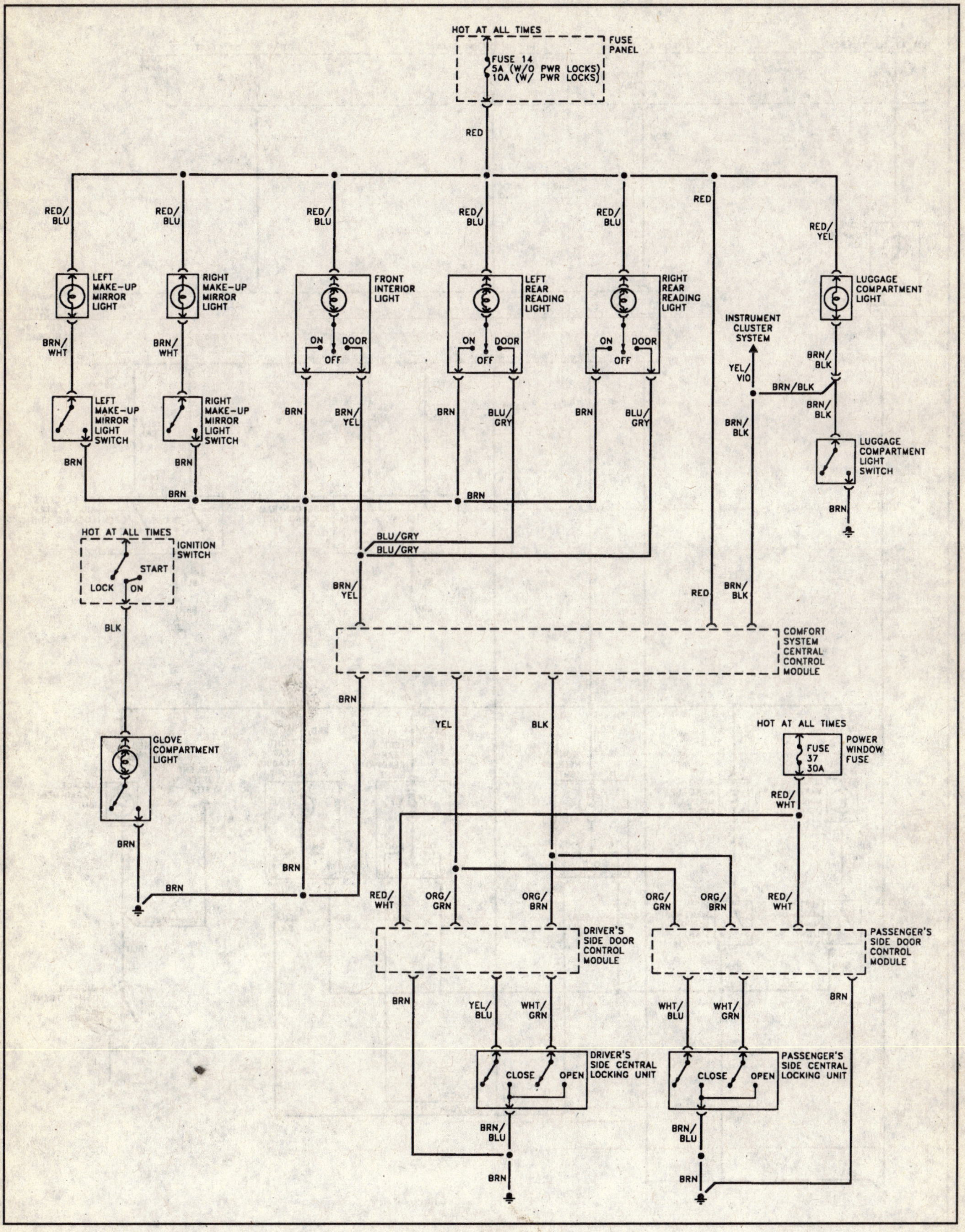

Courtesy light system - 1998 models

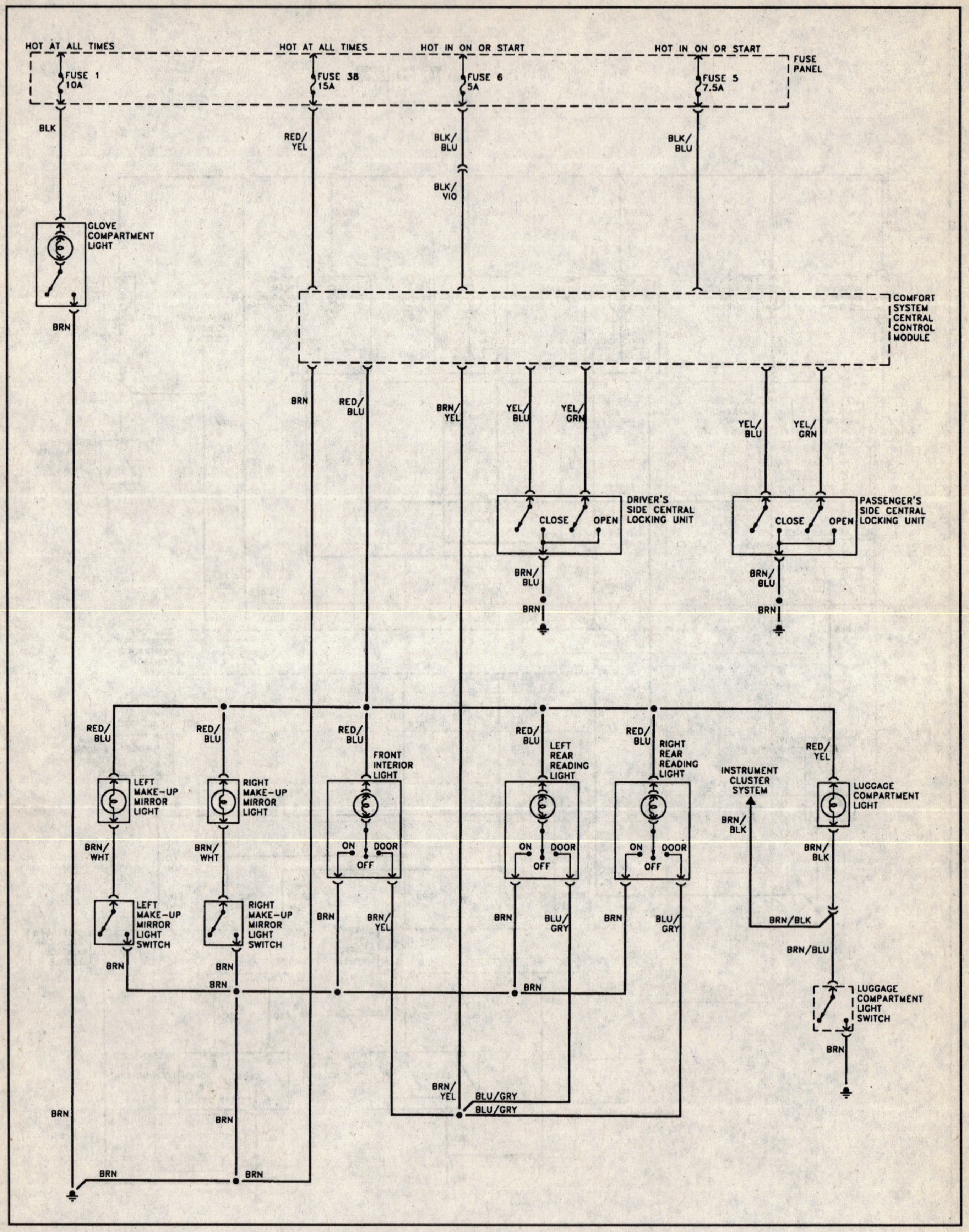

Courtesy light system - 1999 and later models without power windows

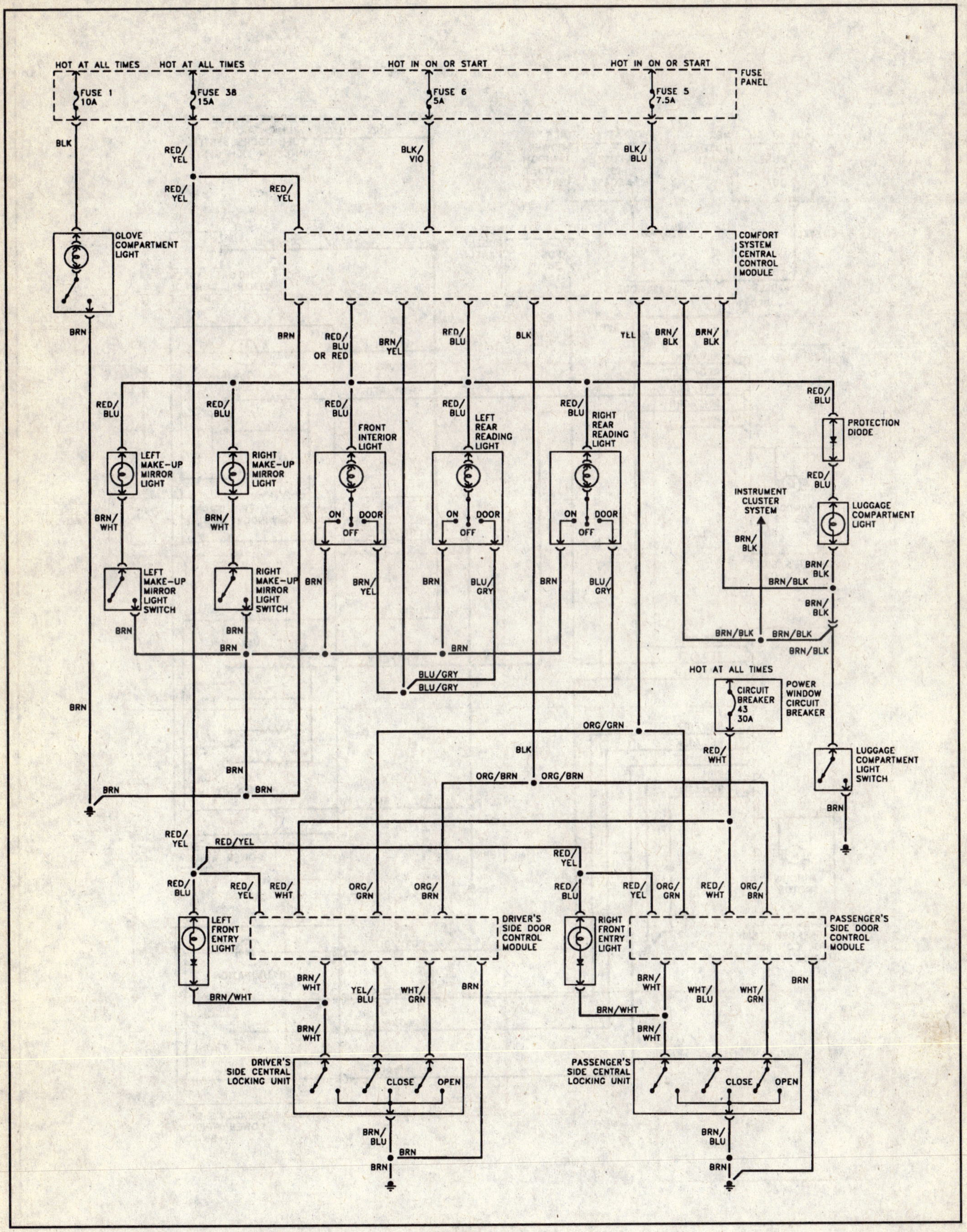

Courtesy light system - 1999 and later models with power windows

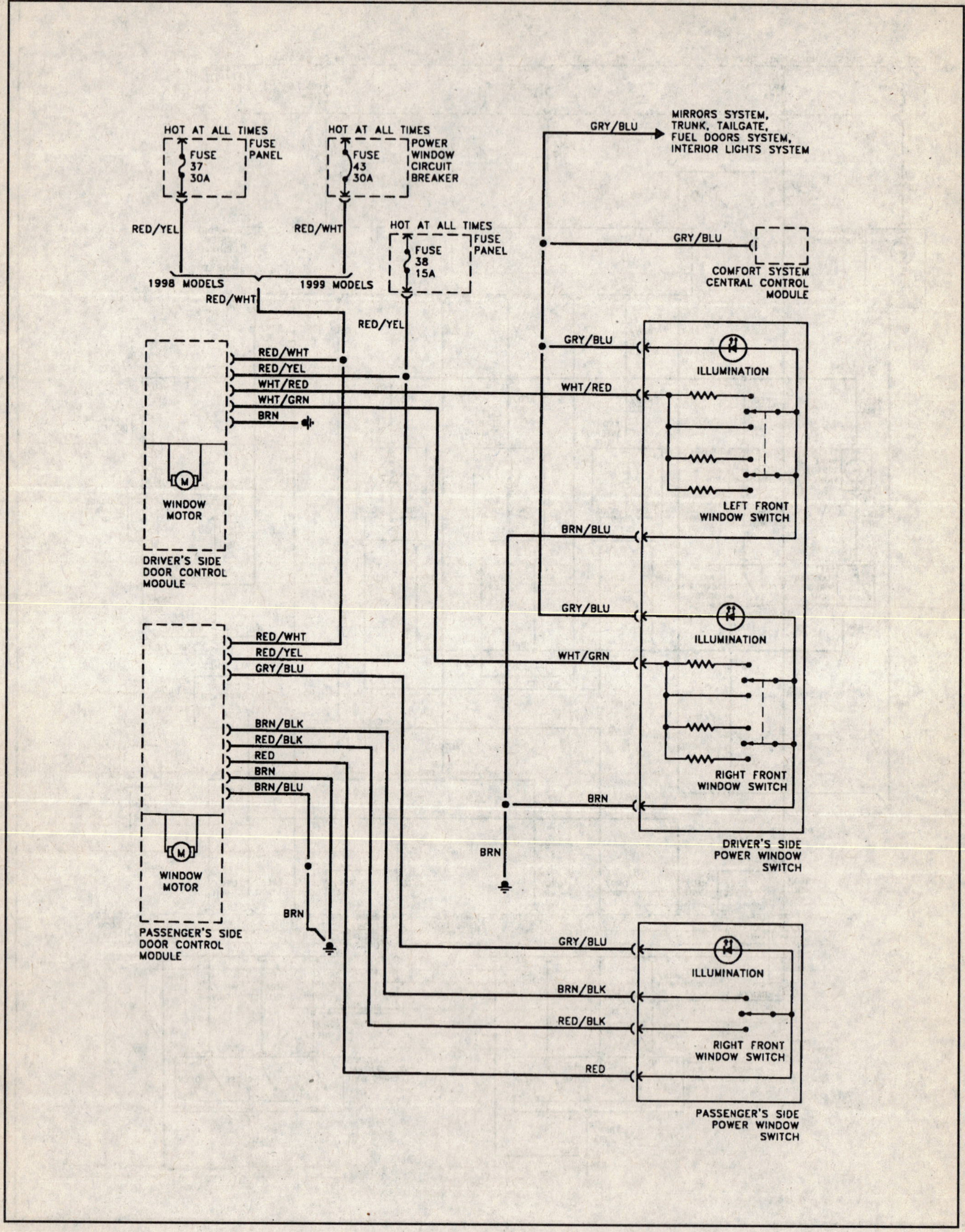

Power window system

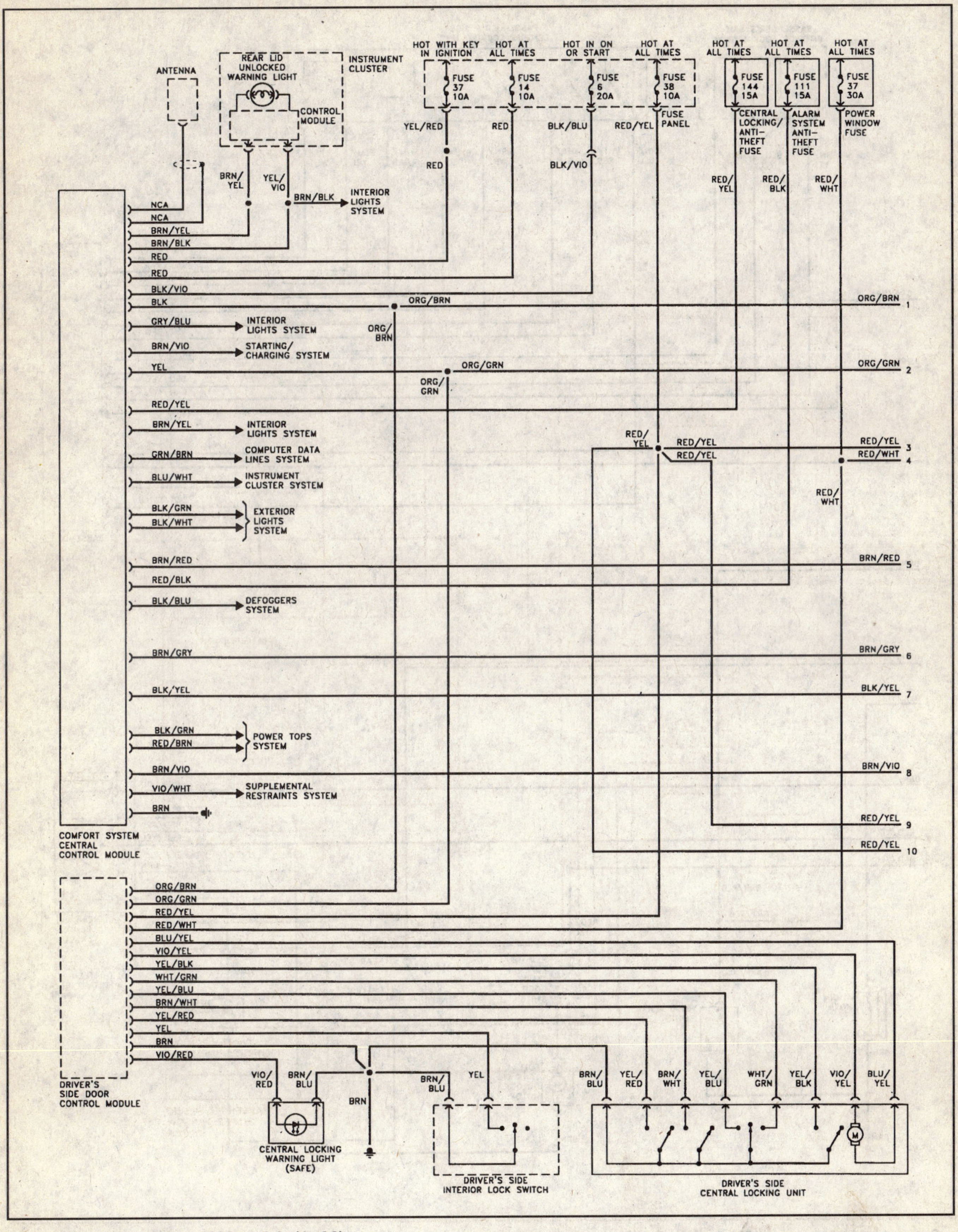

Power door lock system - 1998 models (1 of 2)

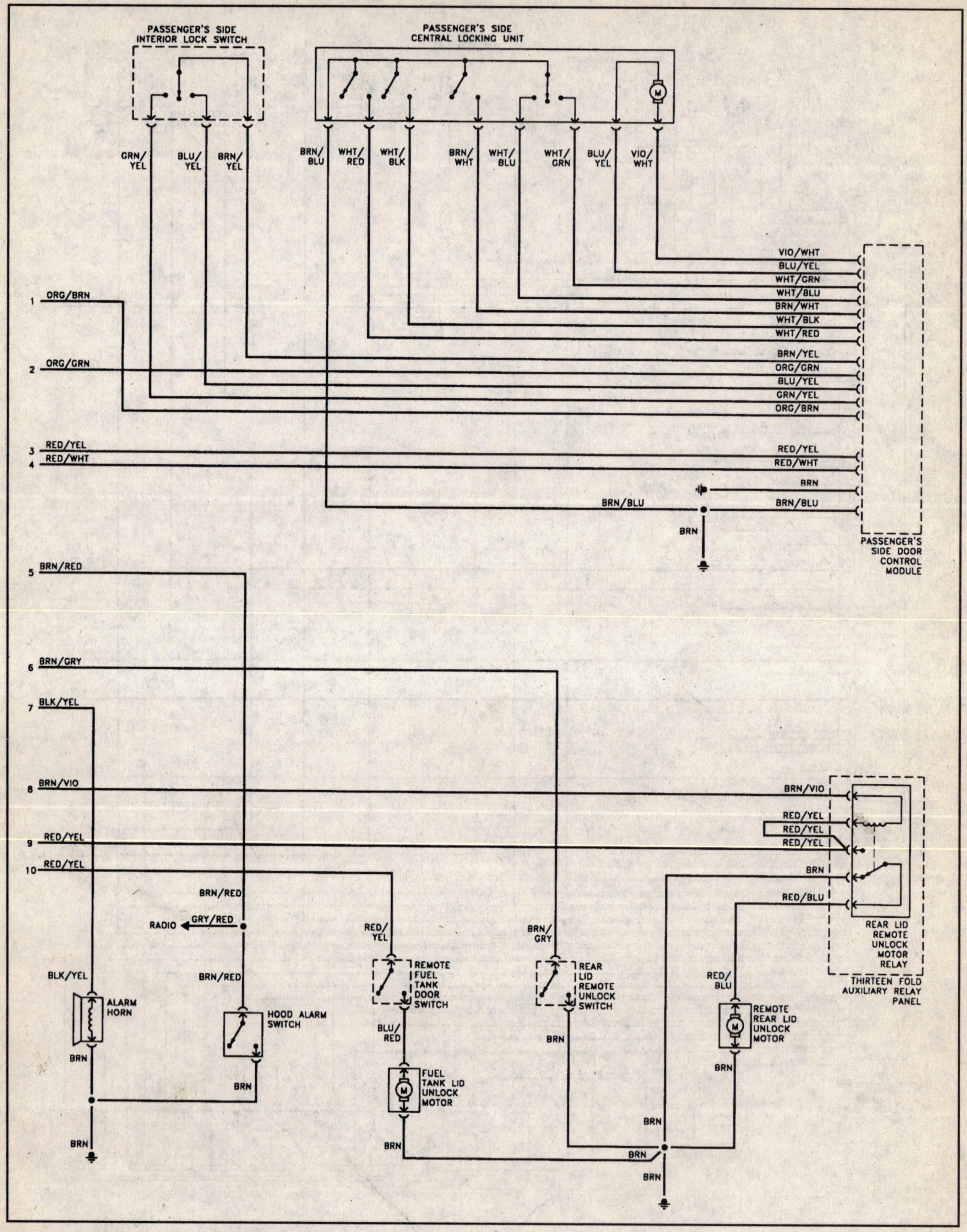

Power door lock system - 1998 models (2 of 2)

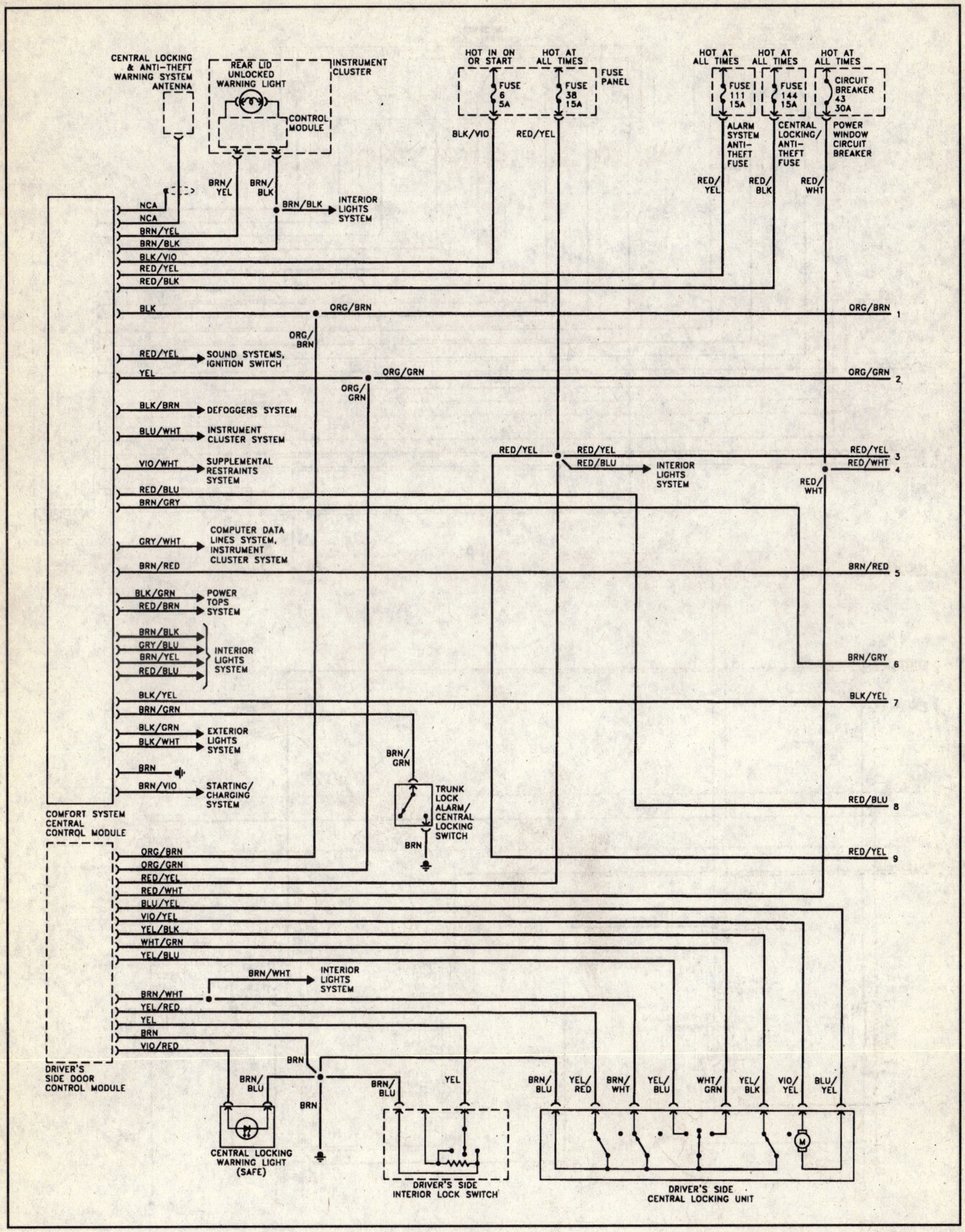

Power door lock system - 1999 and later models with power windows (1 of 2)

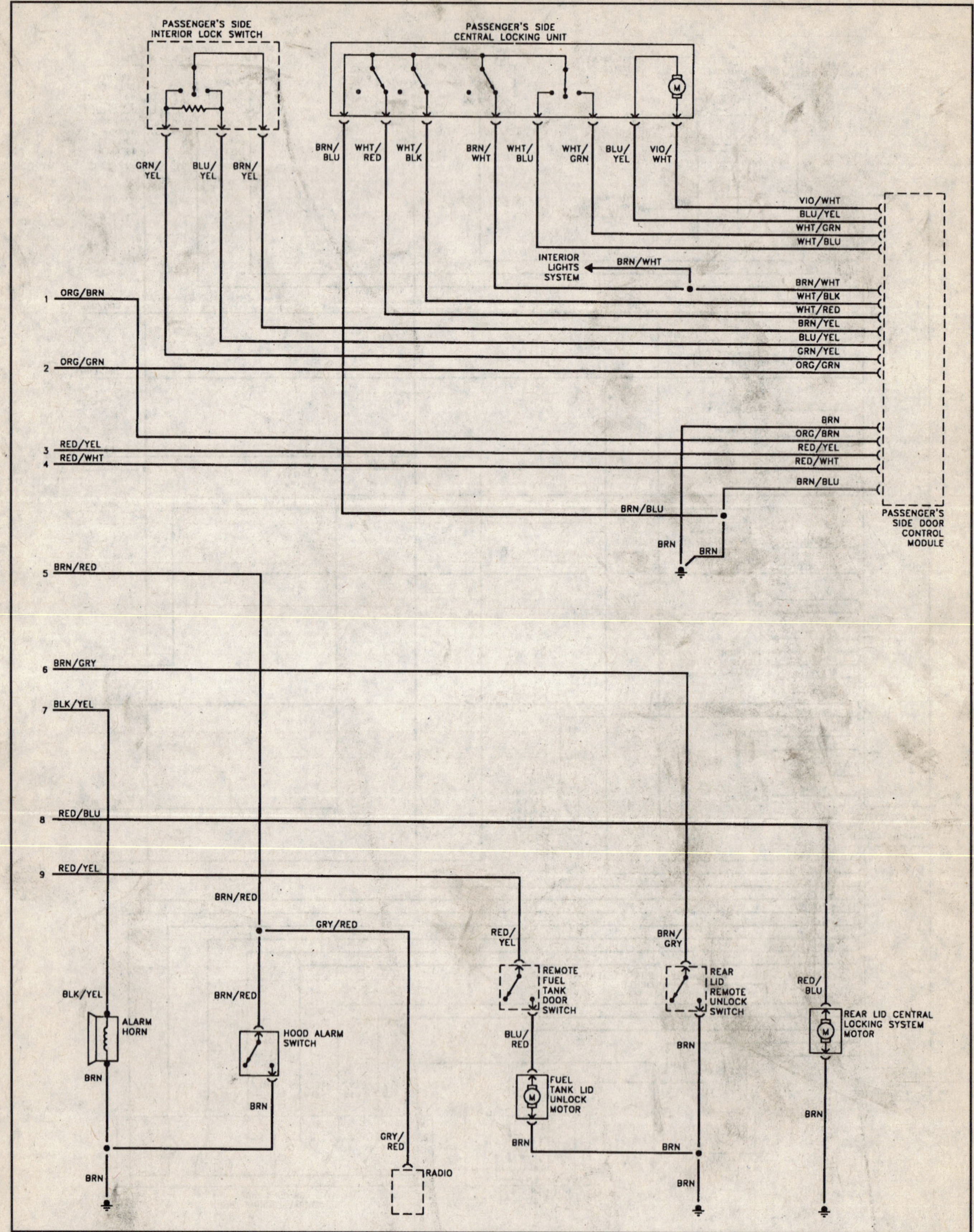

Power door lock system - 1999 and later models with power windows (2 of 2)

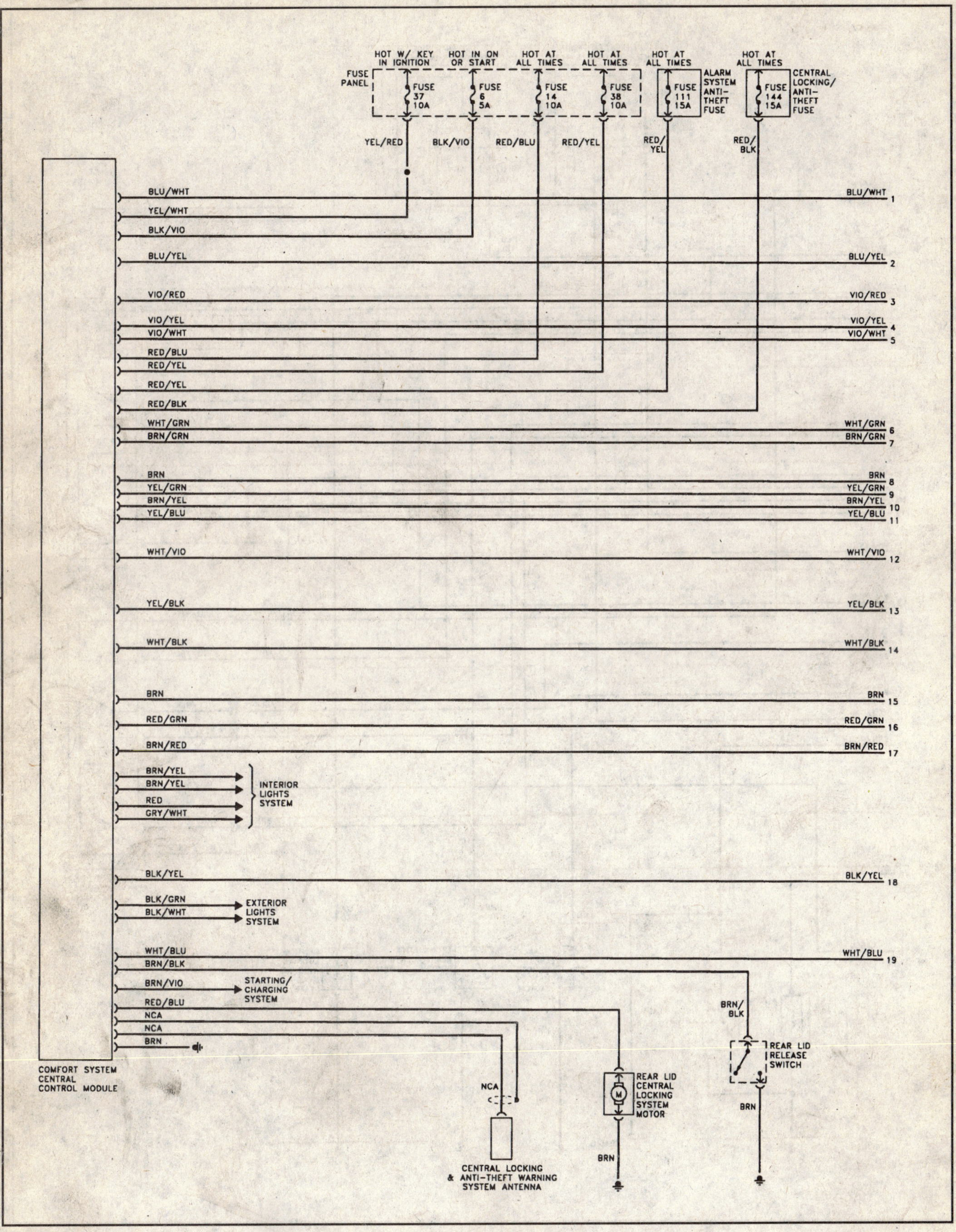

Power door lock system - 1999 and later models without power windows (1 of 2)

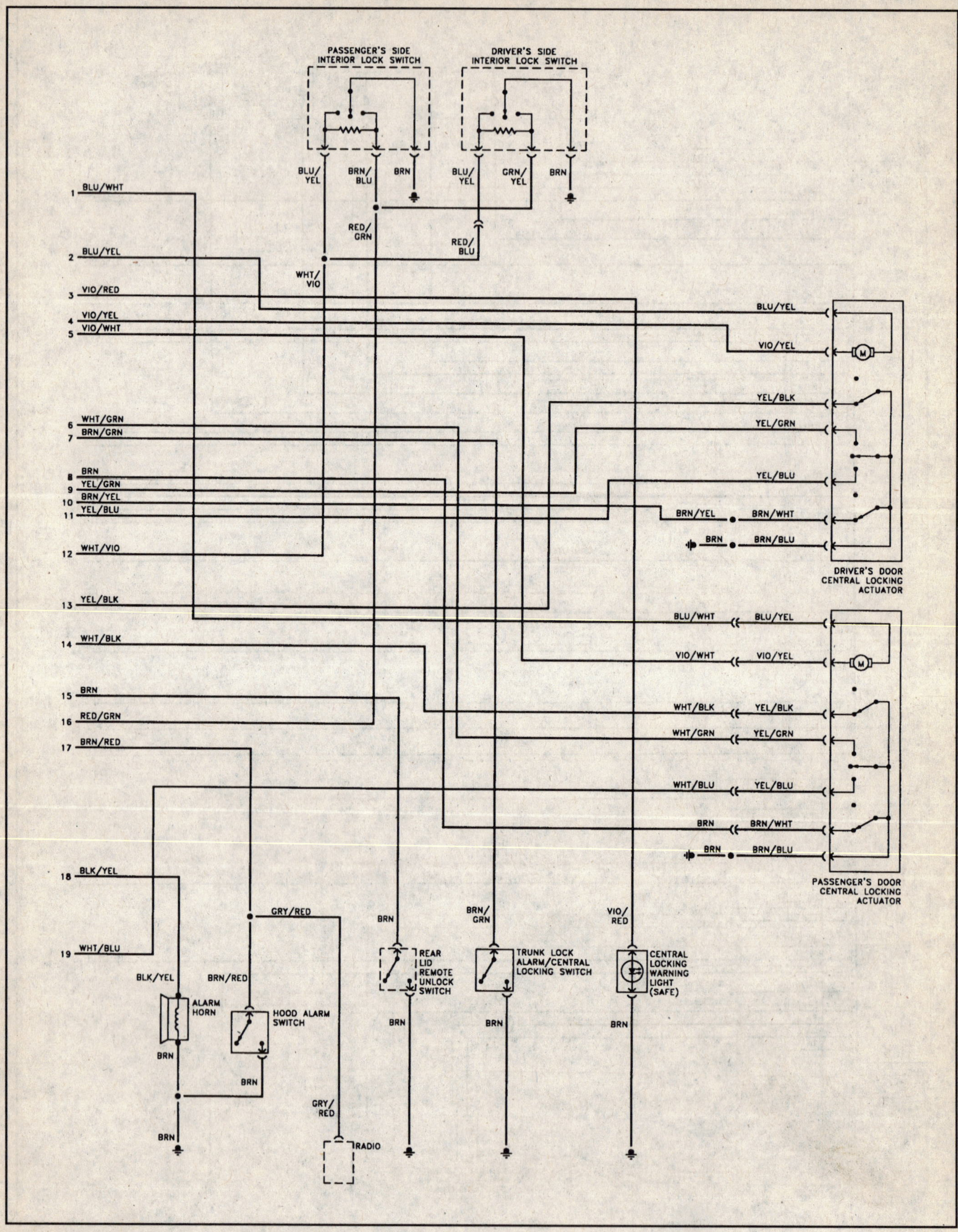

Power door lock system - 1999 and later models without power windows (2 of 2)

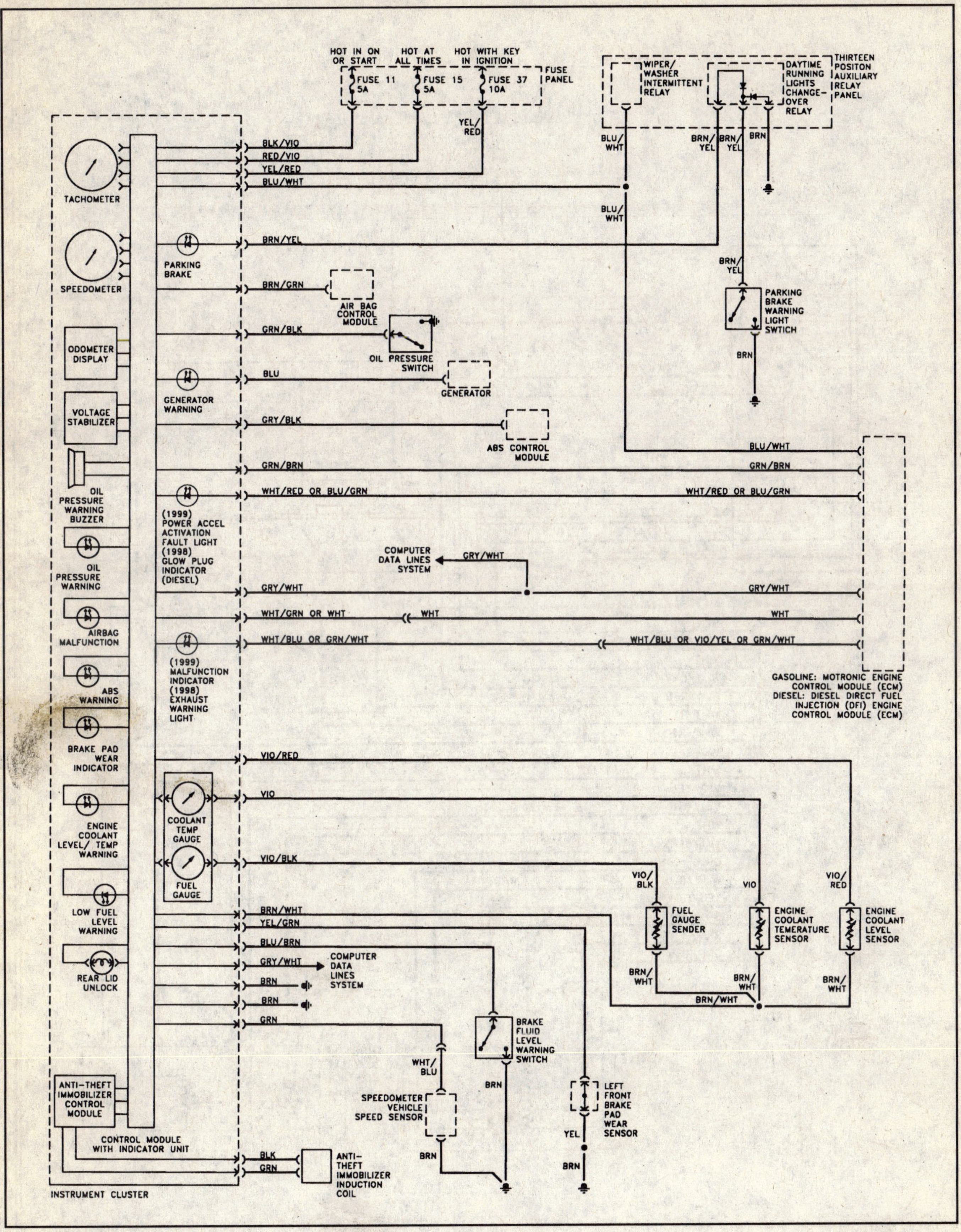

Instrument panel gauges and warning light system

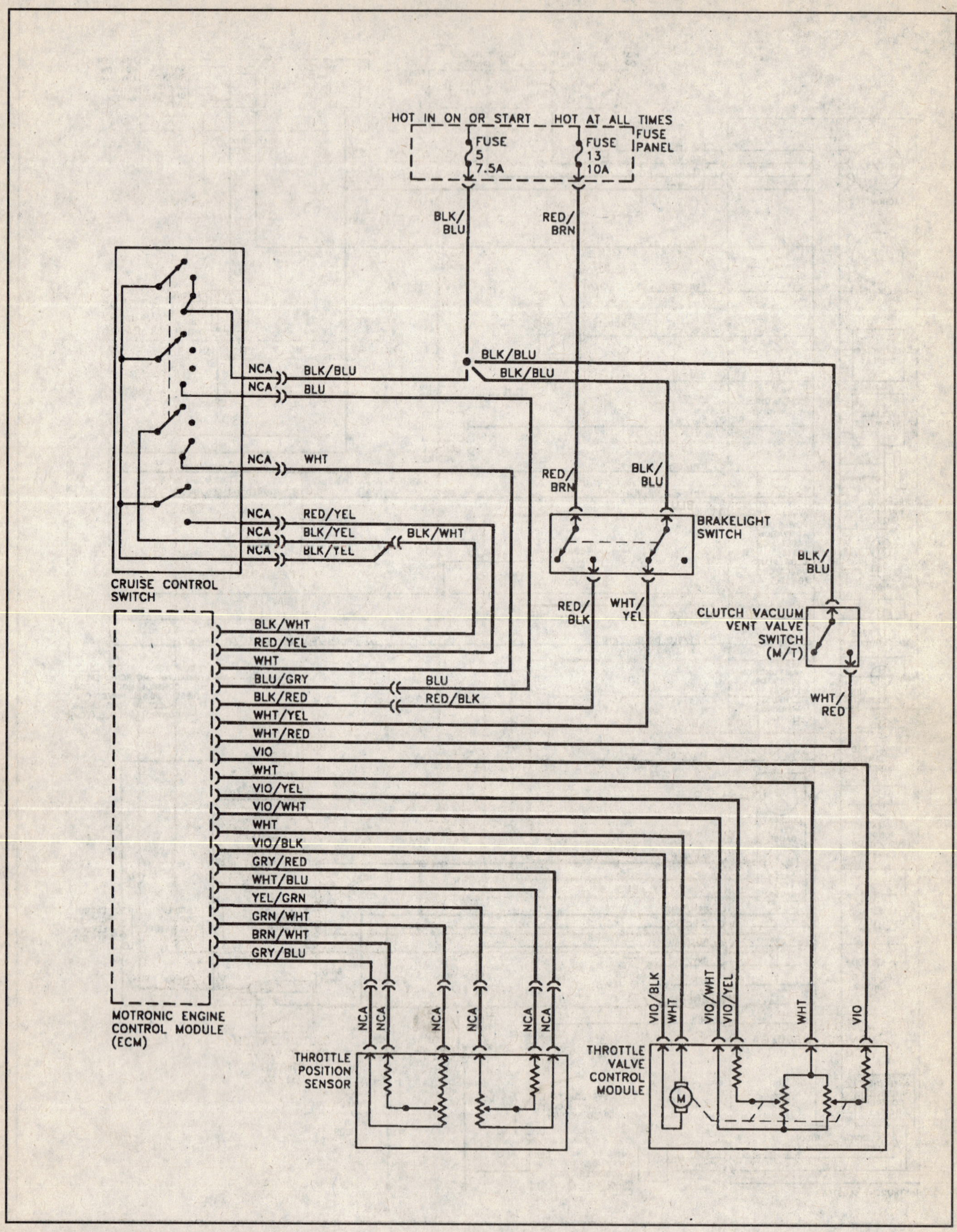

Cruise control system - 1.8L gasoline engine

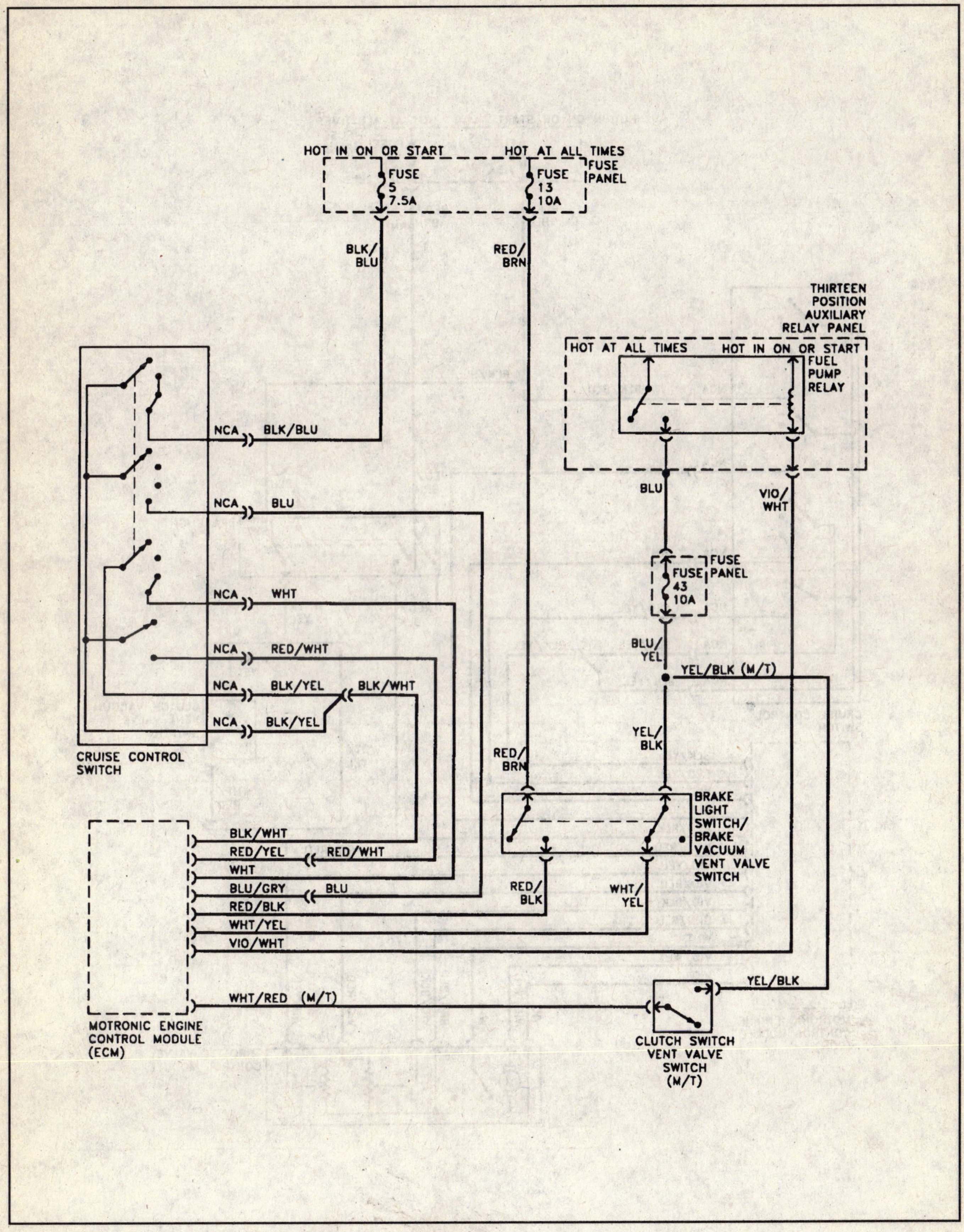

Cruise control system - 1998 2.0L gasoline engine

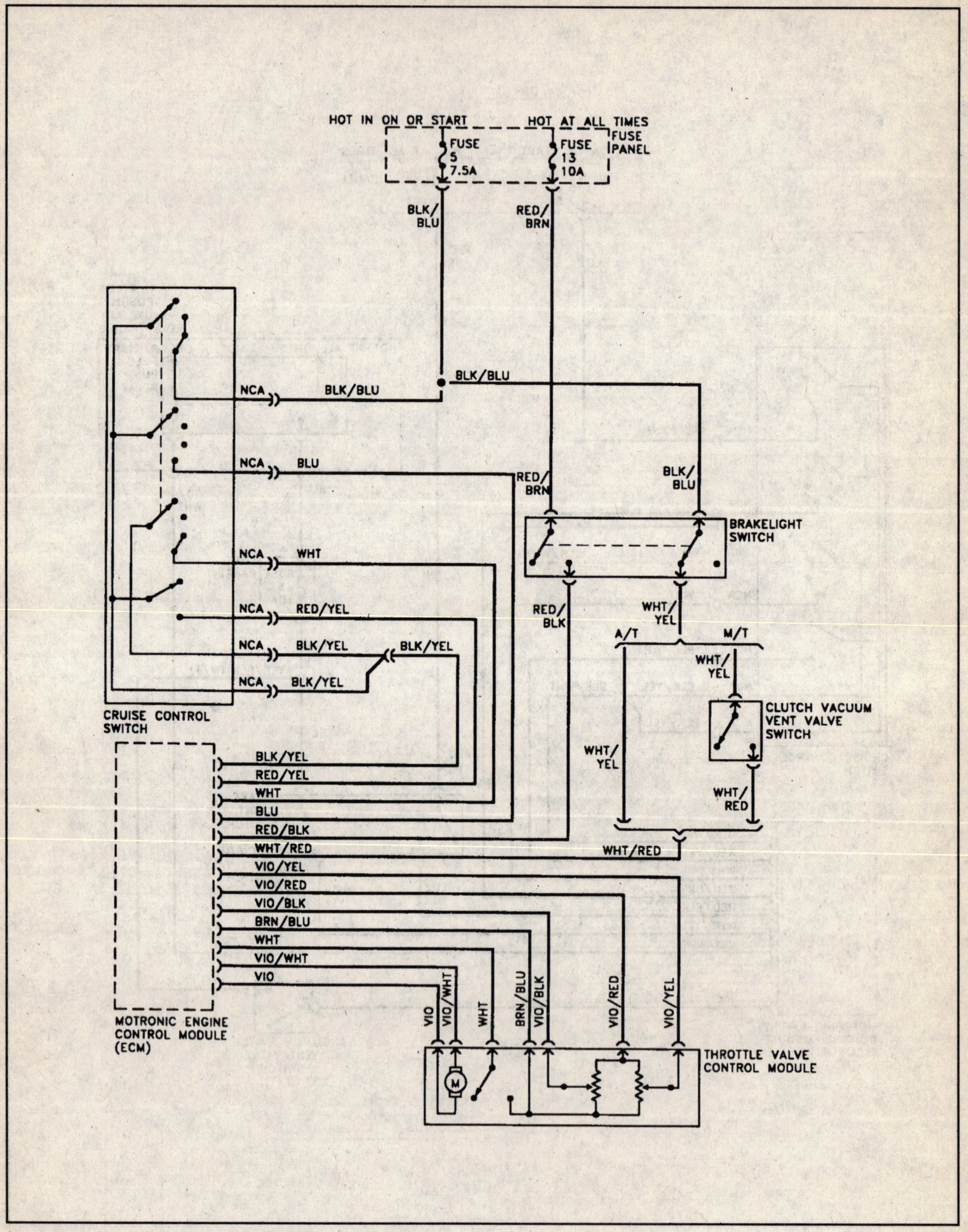

Cruise control system - 1999 and later 2.0L gasoline engine

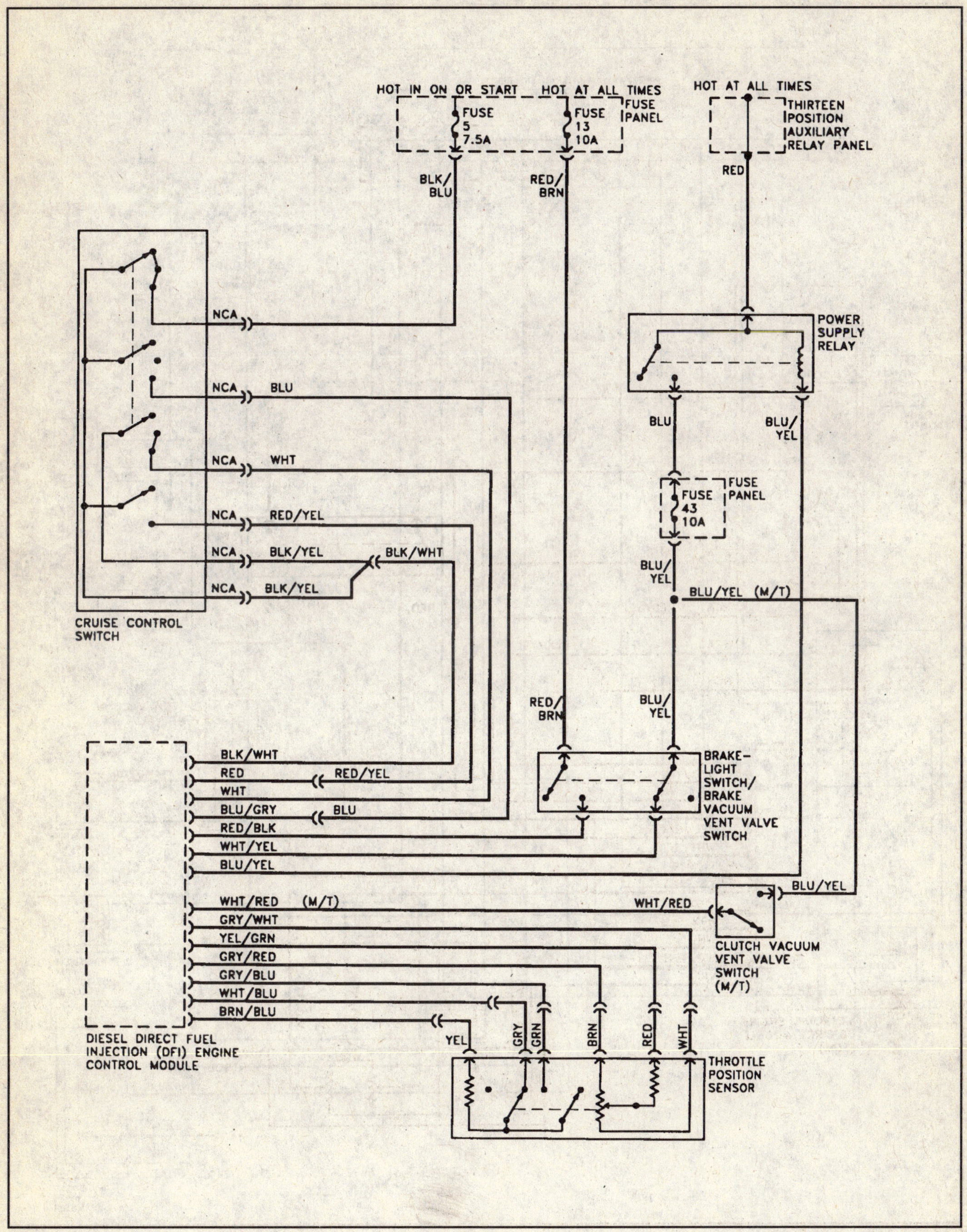

Cruise control system - 1.9L diesel engine

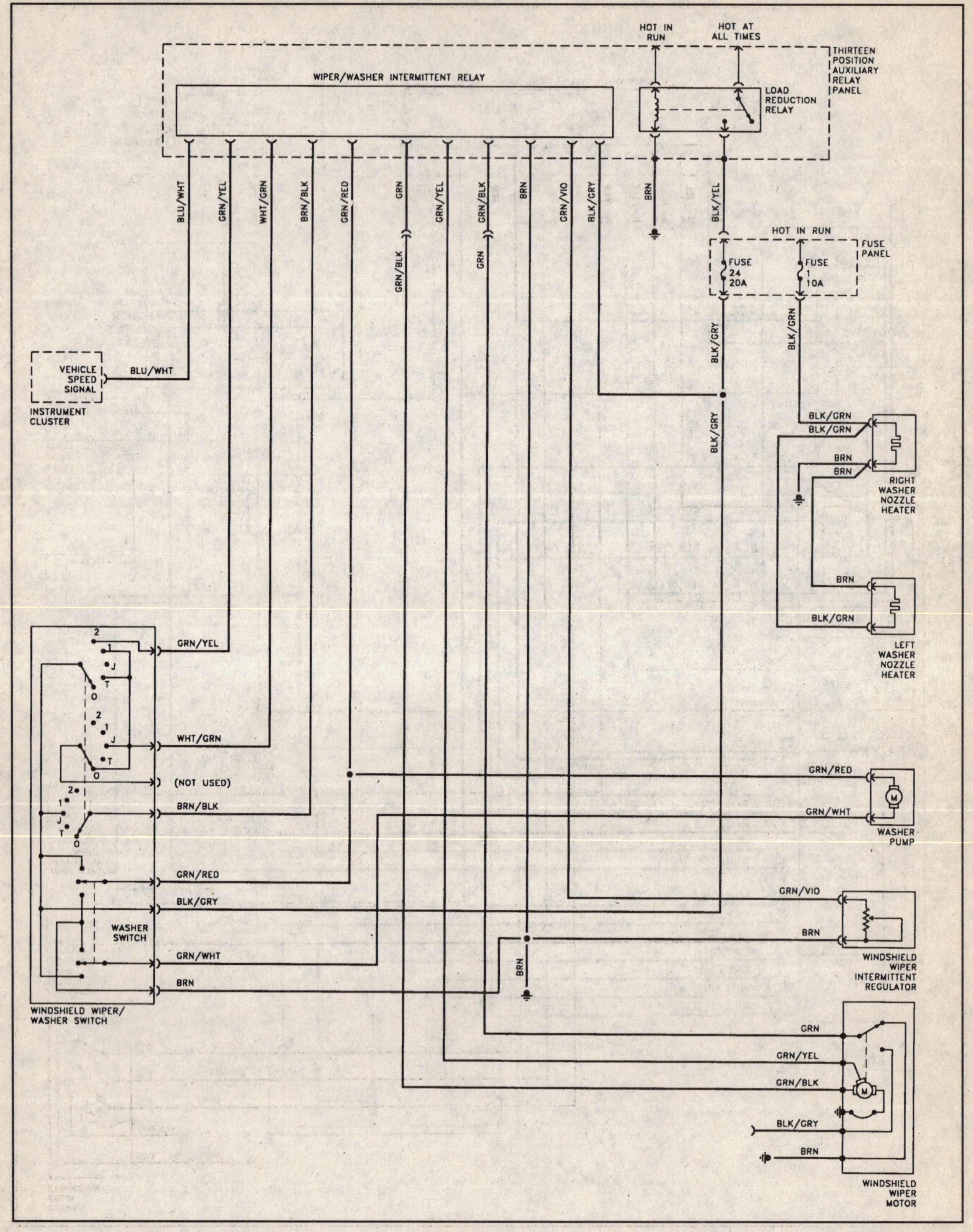

Windshield wiper system - 1998 models

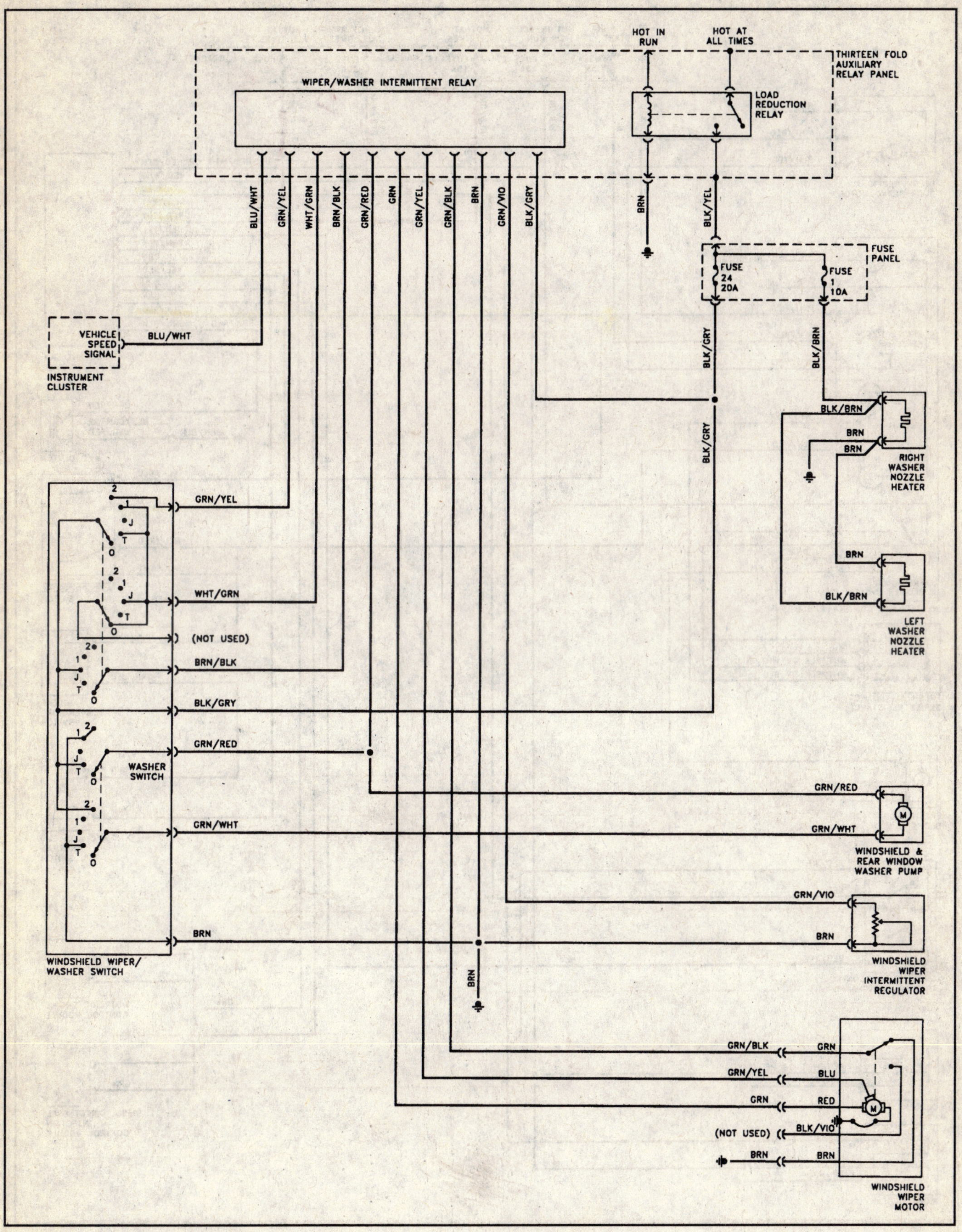

Windshield wiper system - 1999 and later models

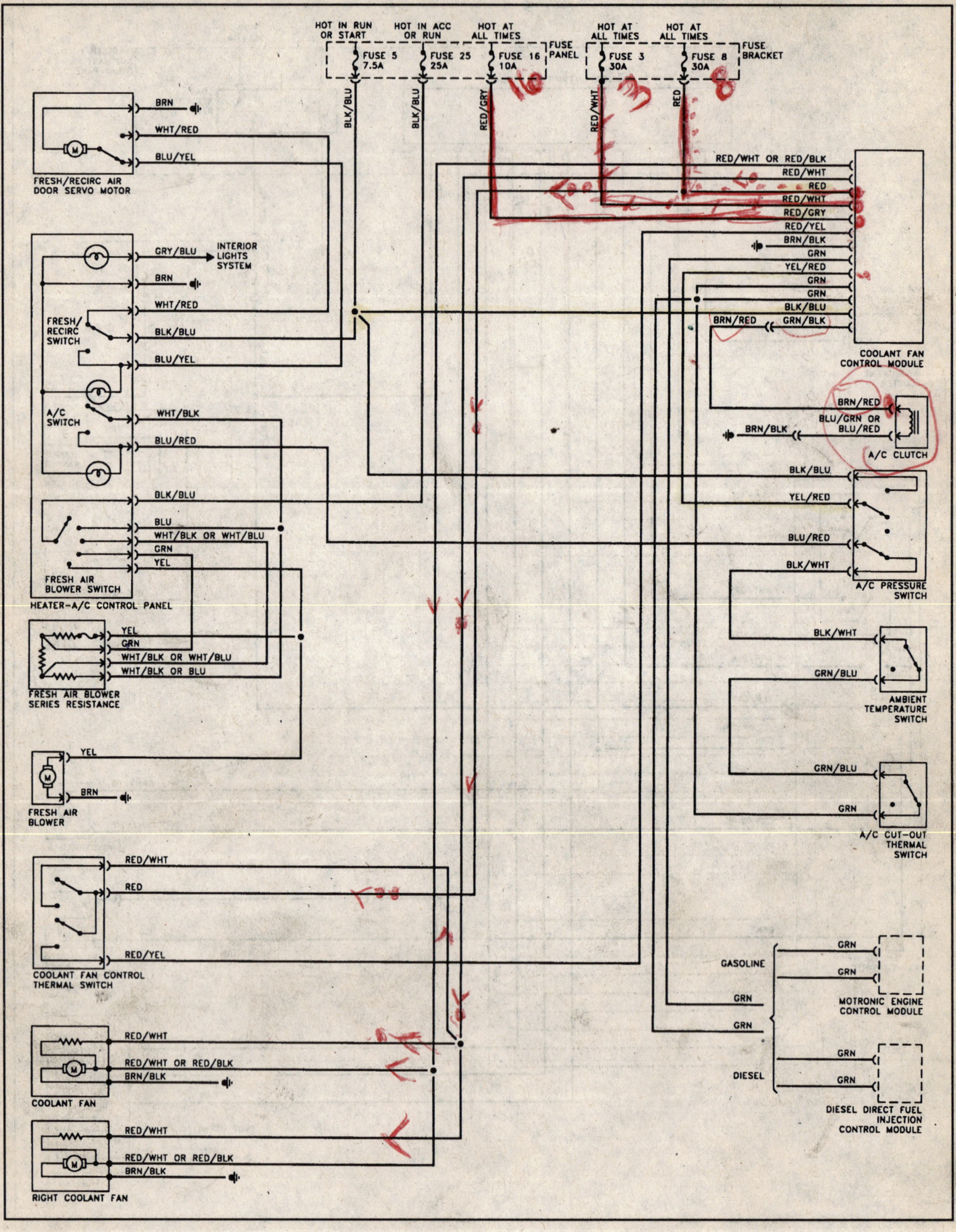

Heating and air conditioning system (including cooling fans)

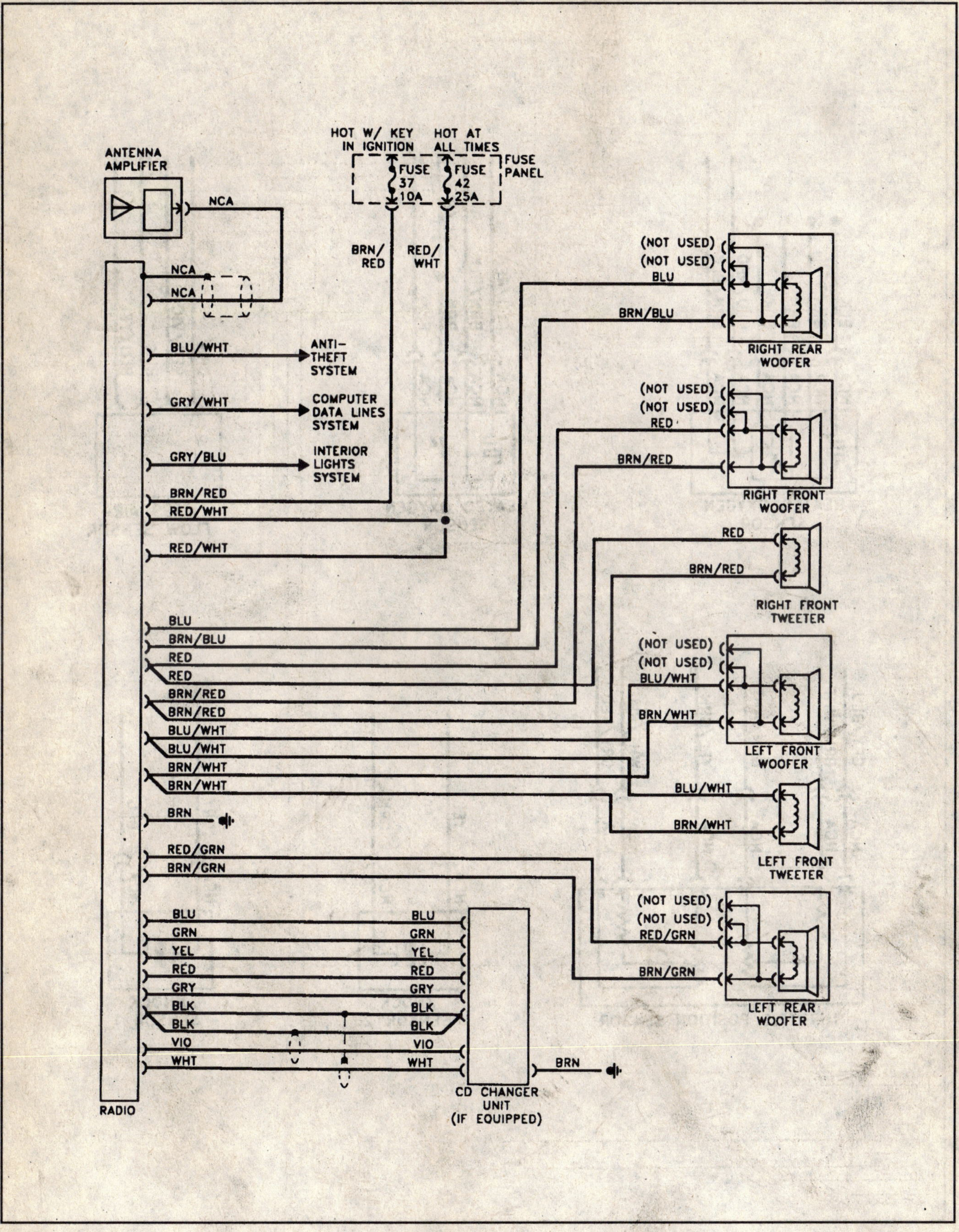

Typical audio system

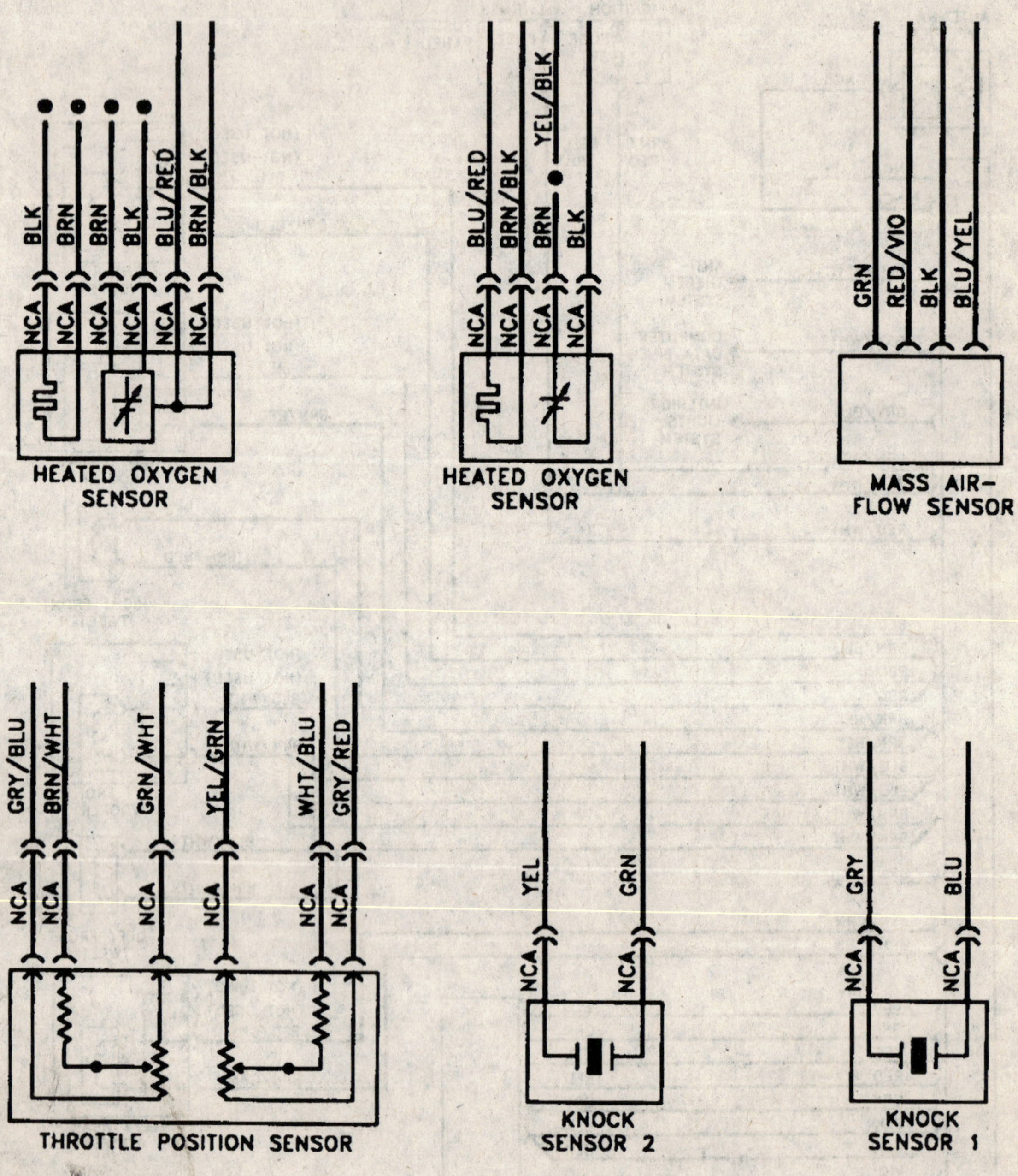

Sensors - 1.8L gasoline engine (1 of 2)

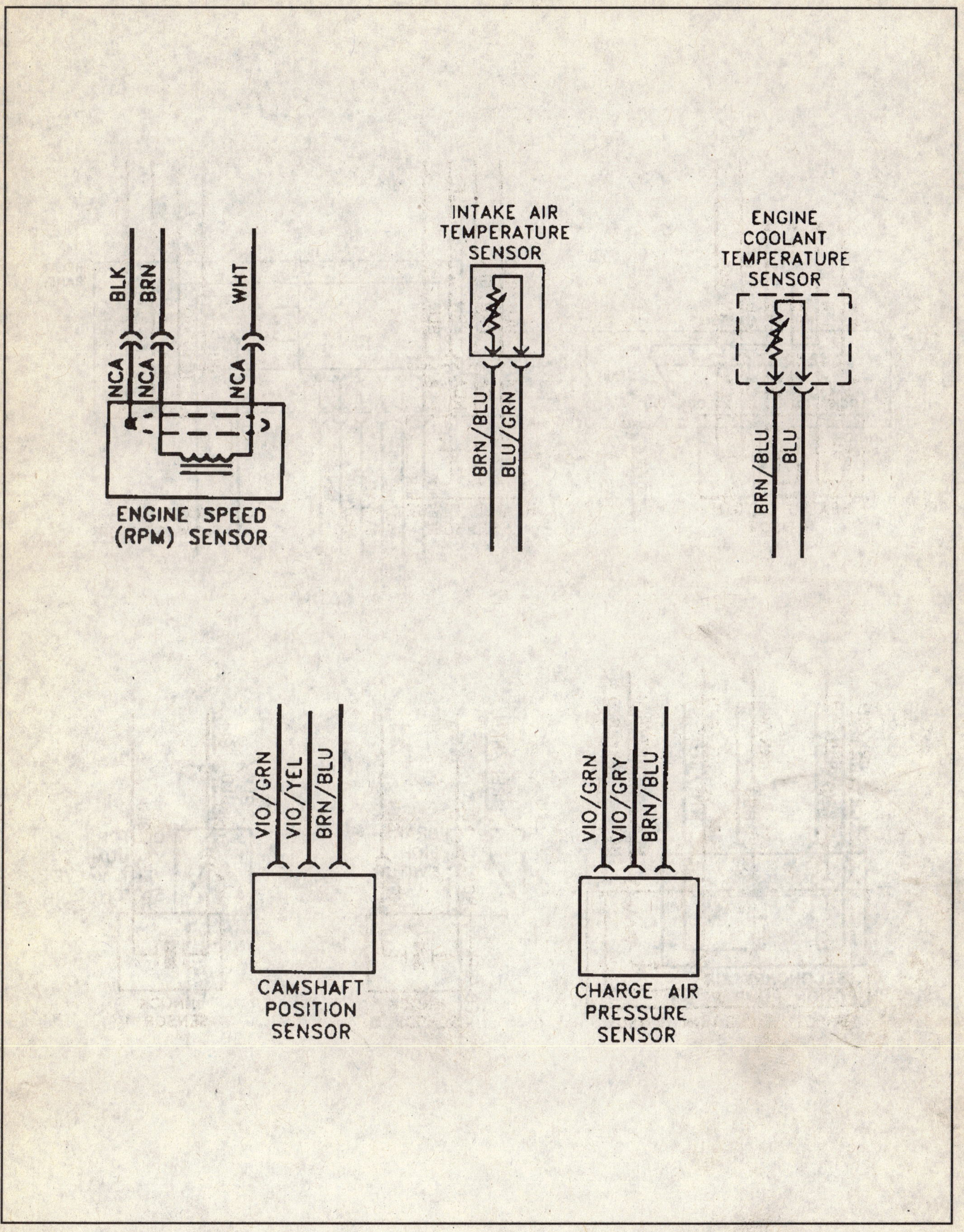

Sensors - 1.8L gasoline engine (2 of 2)

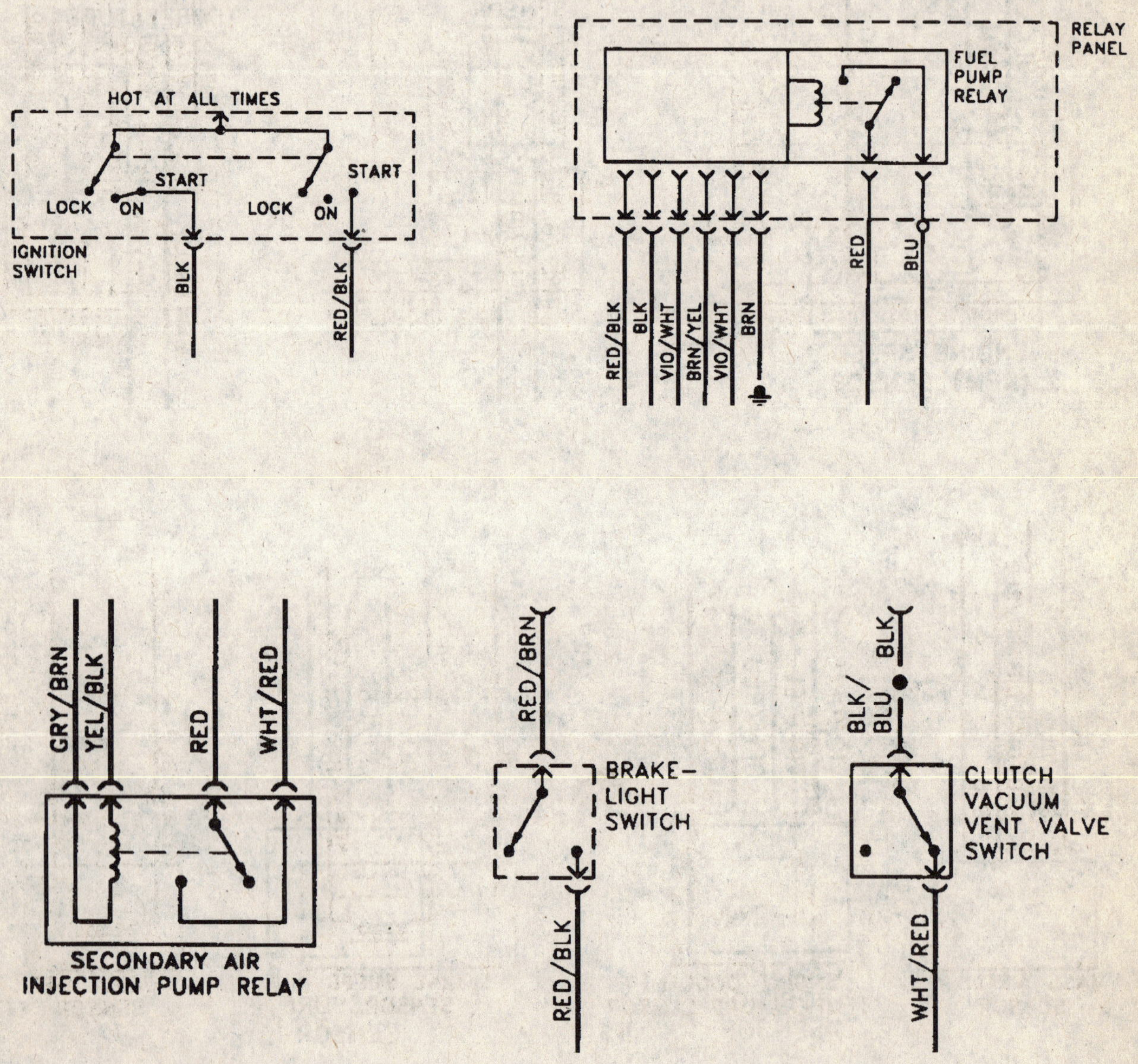

Switches and relays - 1.8L gasoline engine

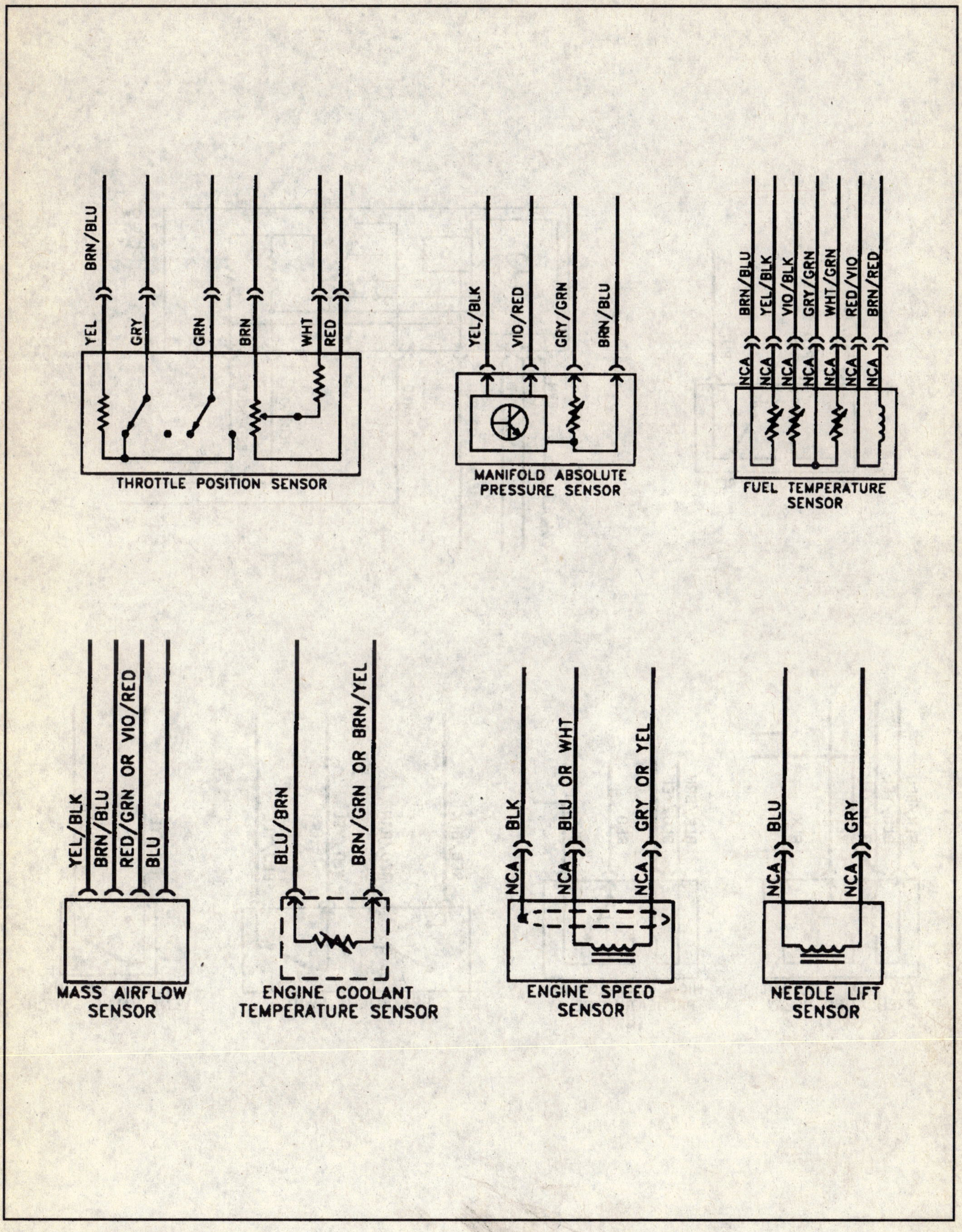

Sensors - 1.9L diesel engine

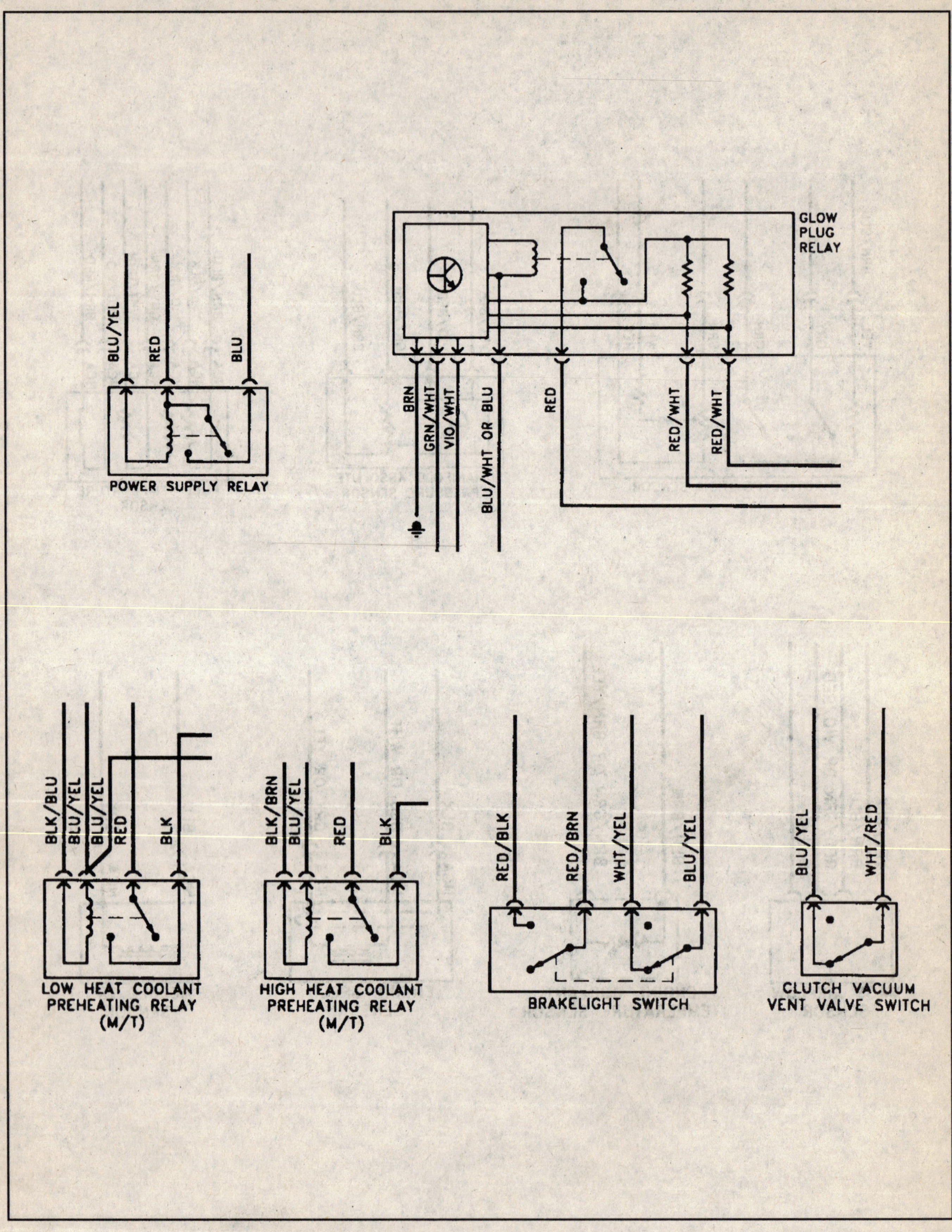

Switches and relays - 1.9L diesel engine

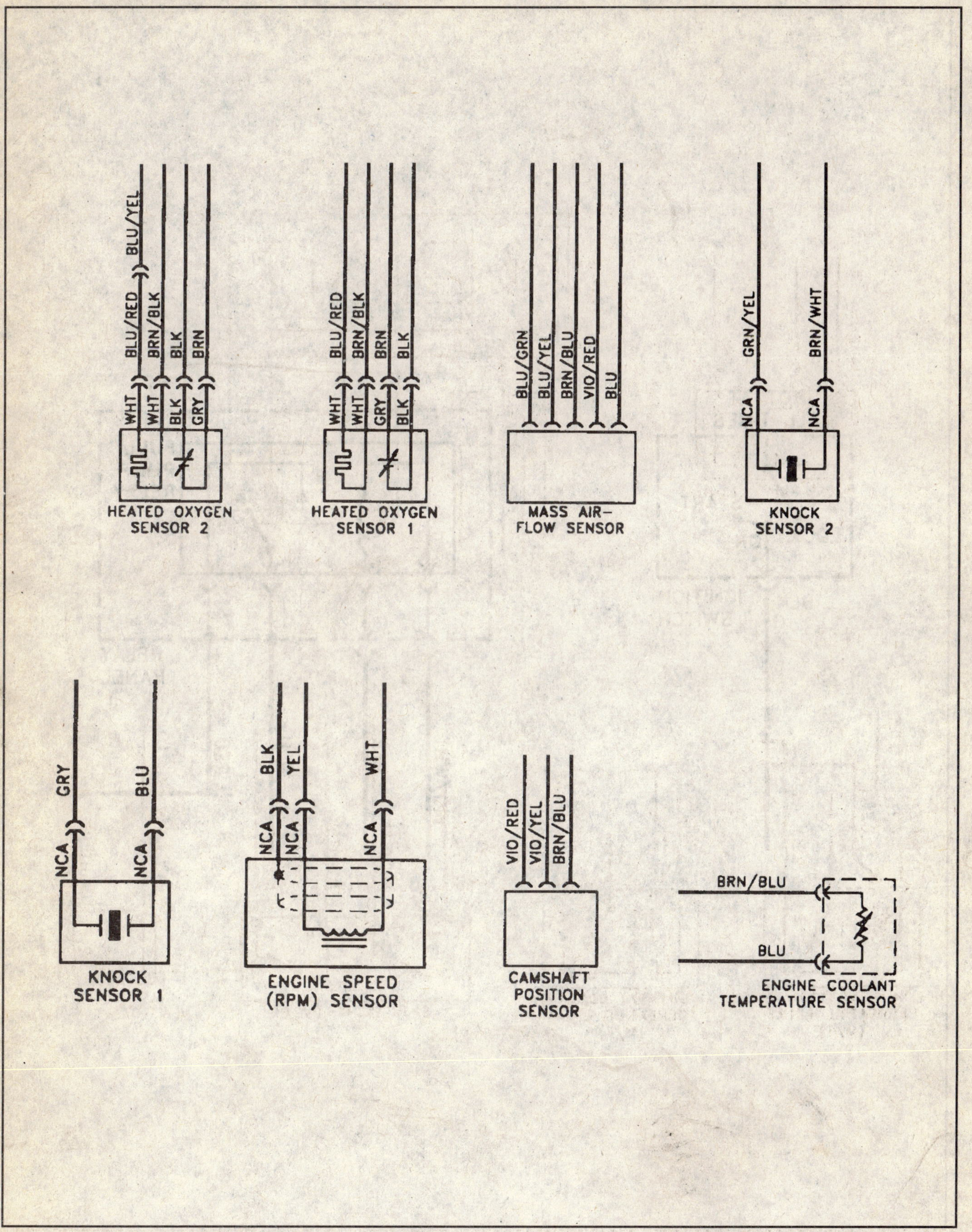

Sensors - 2.0L gasoline engine

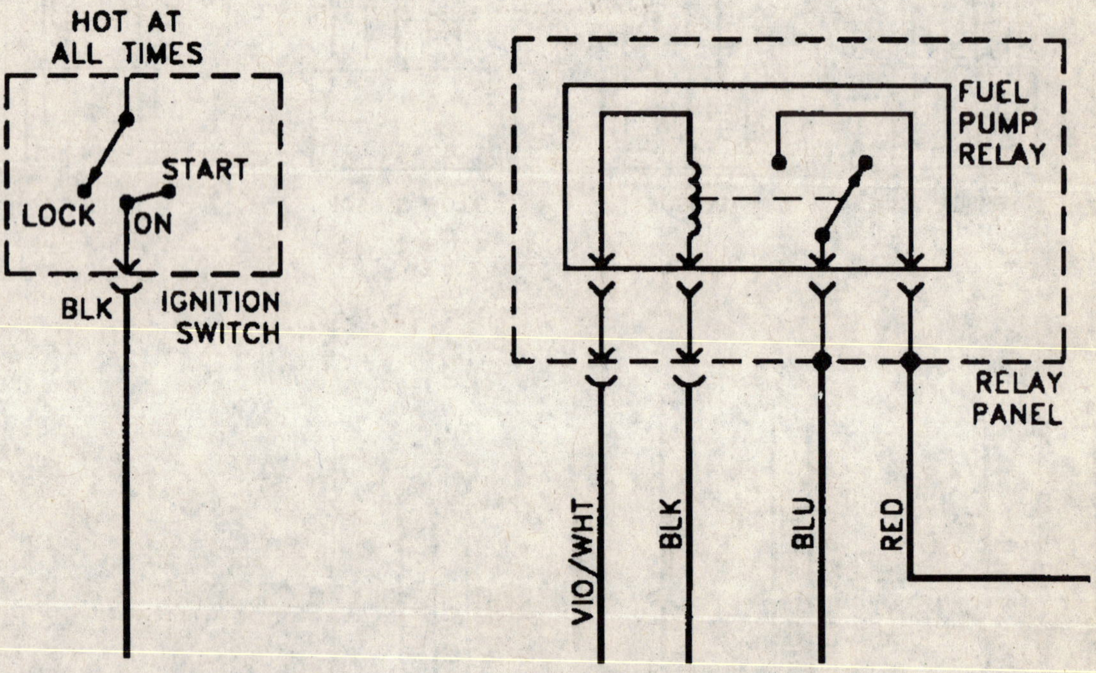

Switches and relays - 2.0L gasoline engine

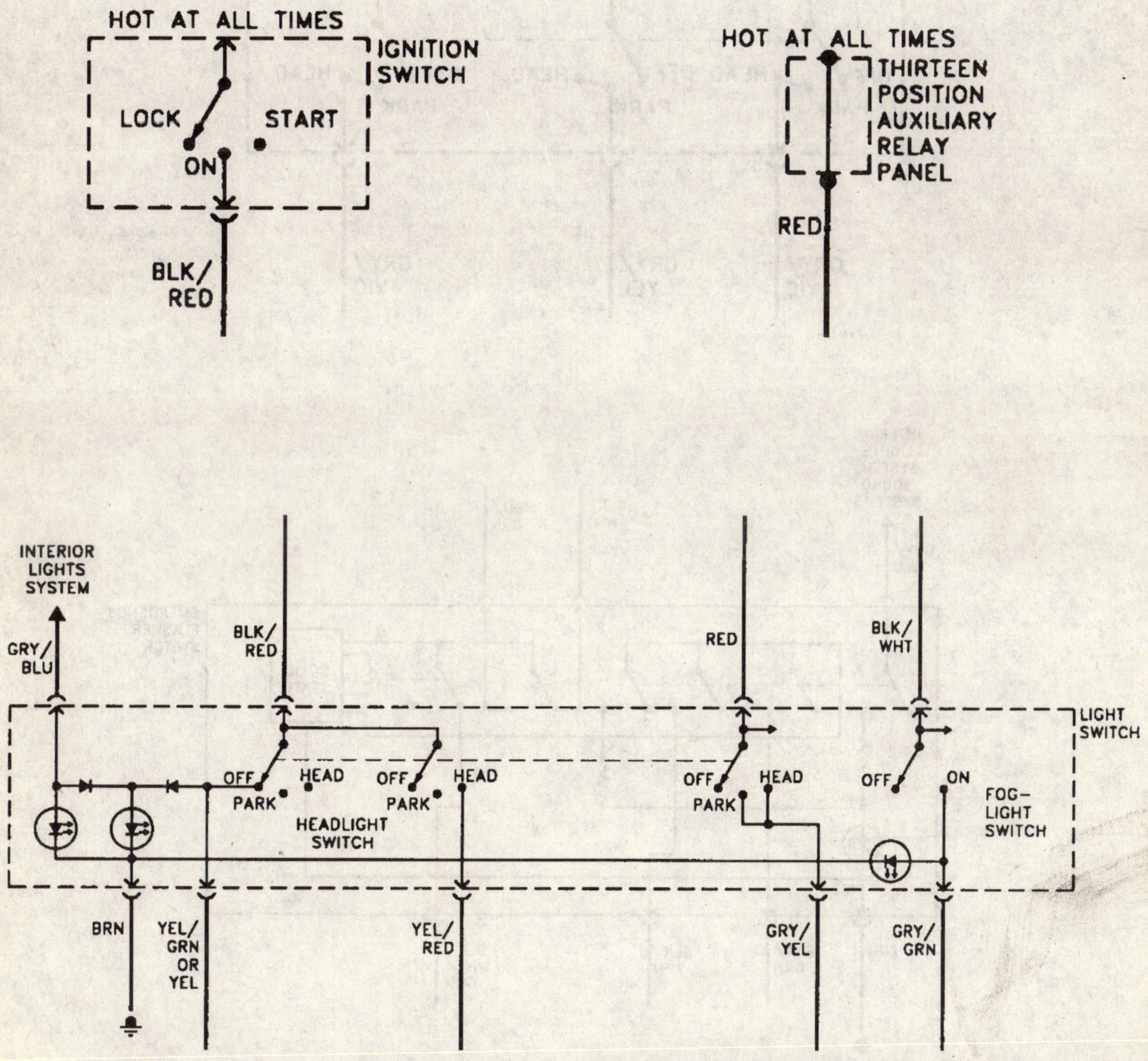

Switches and relays - All engines (1 of 4)

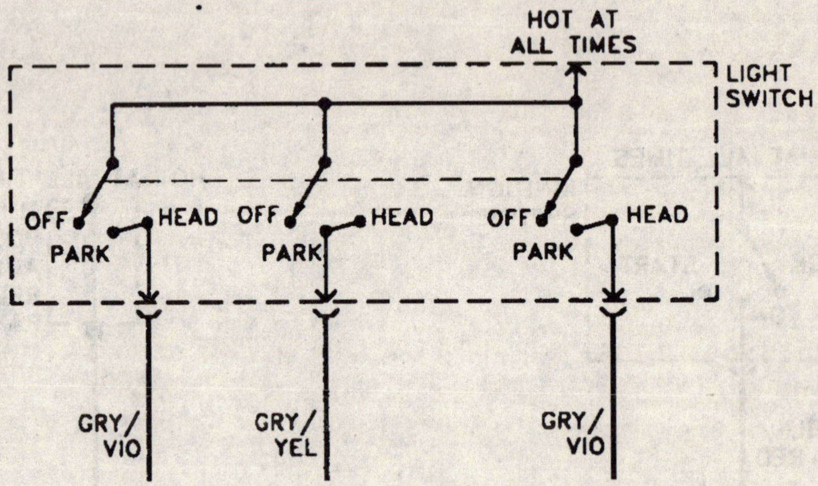

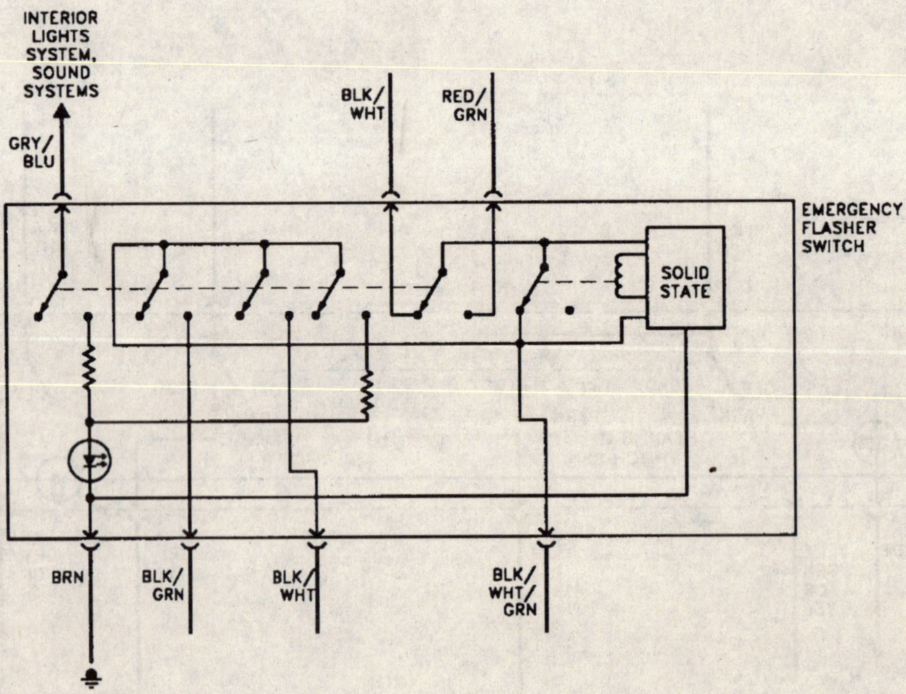

Switches and relays - All engines (2 of 4)

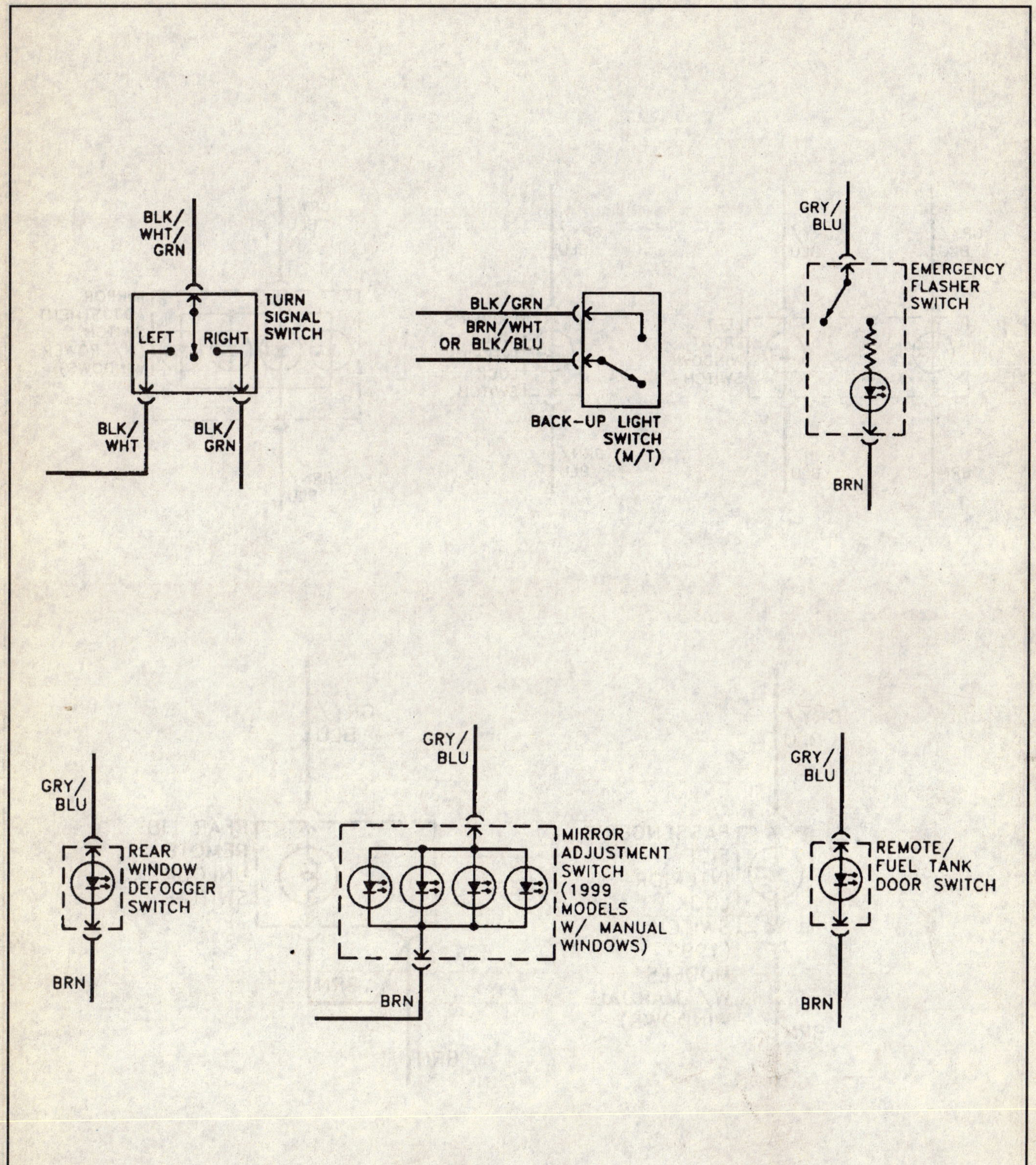

Switches and relays - All engines (3 of 4)

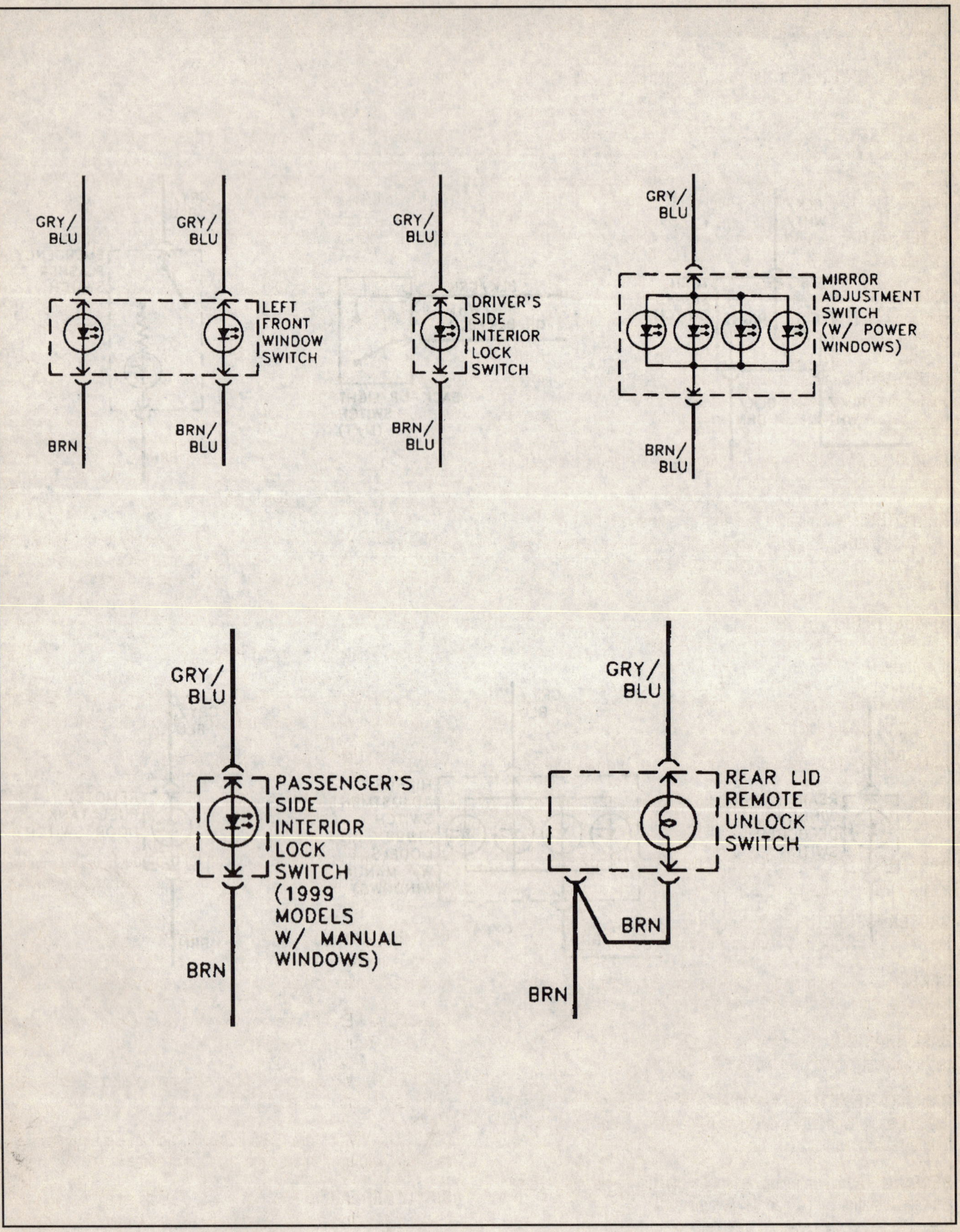

Switches and relays - All engines (4 of 4)

GLOSSARY

AIR/FUEL RATIO: The ratio of air-to-gasoline by weight in the fuel mixture drawn into the engine.

AIR INJECTION: One method of reducing harmful exhaust emissions by injecting air into each of the exhaust ports of an engine. The fresh air entering the hot exhaust manifold causes any remaining fuel to be burned before it can exit the tailpipe.

ALTERNATOR: A device used for converting mechanical energy into electrical energy.

AMMETER: An instrument, calibrated in amperes, used to measure the flow of an electrical current in a circuit. Ammeters are always connected in series with the circuit being tested.

AMPERE: The rate of flow of electrical current present when one volt of electrical pressure is applied against one ohm of electrical resistance.

ANALOG COMPUTER: Any microprocessor that uses similar (analogous) electrical signals to make its calculations.

ARMATURE: A laminated, soft iron core wrapped by a wire that converts electrical energy to mechanical energy as in a motor or relay. When rotated in a magnetic field, it changes mechanical energy into electrical energy as in a generator.

ATMOSPHERIC PRESSURE: The pressure on the Earth's surface caused by the weight of the air in the atmosphere. At sea level, this pressure is 14.7 psi at 32°F (101 kPa at 0°C).

ATOMIZATION: The breaking down of a liquid into a fine mist that can be suspended in air.

AXIAL PLAY: Movement parallel to a shaft or bearing bore.

BACKFIRE: The sudden combustion of gases in the intake or exhaust system that results in a loud explosion.

BACKLASH: The clearance or play between two parts, such as meshed gears.

BACKPRESSURE: Restrictions in the exhaust system that slow the exit of exhaust gases from the combustion chamber.

BAKELITE: A heat resistant, plastic insulator material commonly used in printed circuit boards and transistorized components.

BALL BEARING: A bearing made up of hardened inner and outer races between which hardened steel balls roll.

BALLAST RESISTOR: A resistor in the primary ignition circuit that lowers voltage after the engine is started to reduce wear on ignition components.

BEARING: A friction reducing, supportive device usually located between a stationary part and a moving part.

BIMETAL TEMPERATURE SENSOR: Any sensor or switch made of two dissimilar types of metal that bend when heated or cooled due to the different expansion rates of the alloys. These types of sensors usually function as an on/off switch.

BLOWBY: Combustion gases, composed of water vapor and unburned fuel, that leak past the piston rings into the crankcase during normal engine operation. These gases are removed by the PCV system to prevent the buildup of harmful acids in the crankcase.

BRAKE PAD: A brake shoe and lining assembly used with disc brakes.

BRAKE SHOE: The backing for the brake lining. The term is, however, usually applied to the assembly of the brake backing and lining.

BUSHING: A liner, usually removable, for a bearing; an anti-friction liner used in place of a bearing.

CALIPER: A hydraulically activated device in a disc brake system, which is mounted straddling the brake rotor (disc). The caliper contains at least one piston and two brake pads. Hydraulic pressure on the piston(s) forces the pads against the rotor.

CAMSHAFT: A shaft in the engine on which are the lobes (cams) which operate the valves. The camshaft is driven by the crankshaft, via a belt, chain or gears, at one half the crankshaft speed.

CAPACITOR: A device which stores an electrical charge.

CARBON MONOXIDE (CO): A colorless, odorless gas given off as a normal byproduct of combustion. It is poisonous and extremely dangerous in confined areas, building up slowly to toxic levels without warning if adequate ventilation is not available.

CARBURETOR: A device, usually mounted on the intake manifold of an engine, which mixes the air and fuel in the proper proportion to allow even combustion.

CATALYTIC CONVERTER: A device installed in the exhaust system, like a muffler, that converts harmful byproducts of combustion into carbon dioxide and water vapor by means of a heat-producing chemical reaction.

CENTRIFUGAL ADVANCE: A mechanical method of advancing the spark timing by using flyweights in the distributor that react to centrifugal force generated by the distributor shaft rotation.

CHECK VALVE: Any one-way valve installed to permit the flow of air, fuel or vacuum in one direction only.

CHOKE: A device, usually a moveable valve, placed in the intake path of a carburetor to restrict the flow of air.

CIRCUIT: Any unbroken path through which an electrical current can flow. Also used to describe fuel flow in some instances.

CIRCUIT BREAKER: A switch which protects an electrical circuit from overload by opening the circuit when the current flow exceeds a predetermined level. Some circuit breakers must be reset manually, while most reset automatically.

COIL (IGNITION): A transformer in the ignition circuit which steps up the voltage provided to the spark plugs.

COMBINATION MANIFOLD: An assembly which includes both the intake and exhaust manifolds in one casting.

COMBINATION VALVE: A device used in some fuel systems that routes fuel vapors to a charcoal storage canister instead of venting them into the atmosphere. The valve relieves fuel tank pressure and allows fresh air into the tank as the fuel level drops to prevent a vapor lock situation.

COMPRESSION RATIO: The comparison of the total volume of the cylinder and combustion chamber with the piston at BDC and the piston at TDC.

CONDENSER: 1. An electrical device which acts to store an electrical charge, preventing voltage surges. 2. A radiator-like device in the air conditioning system in which refrigerant gas condenses into a liquid, giving off heat.

CONDUCTOR: Any material through which an electrical current can be transmitted easily.

CONTINUITY: Continuous or complete circuit. Can be checked with an ohmmeter.

COUNTERSHAFT: An intermediate shaft which is rotated by a mainshaft and transmits, in turn, that rotation to a working part.

CRANKCASE: The lower part of an engine in which the crankshaft and related parts operate.

CRANKSHAFT: The main driving shaft of an engine which receives reciprocating motion from the pistons and converts it to rotary motion.

CYLINDER: In an engine, the round hole in the engine block in which the piston(s) ride.

CYLINDER BLOCK: The main structural member of an engine in which is found the cylinders, crankshaft and other principal parts.

CYLINDER HEAD: The detachable portion of the engine, usually fastened to the top of the cylinder block and containing all or most of the combustion chambers. On overhead valve engines, it contains the valves and their operating parts. On overhead cam engines, it contains the camshaft as well.

DEAD CENTER: The extreme top or bottom of the piston stroke.

DETONATION: An unwanted explosion of the air/fuel mixture in the combustion chamber caused by excess heat and compression, advanced timing, or an overly lean mixture. Also referred to as "ping".

DIAPHRAGM: A thin, flexible wall separating two cavities, such as in a vacuum advance unit.

DIESELING: A condition in which hot spots in the combustion chamber cause the engine to run on after the key is turned off.

DIFFERENTIAL: A geared assembly which allows the transmission of motion between drive axles, giving one axle the ability to turn faster than the other.

DIODE: An electrical device that will allow current to flow in one direction only.

DISC BRAKE: A hydraulic braking assembly consisting of a brake disc, or rotor, mounted on an axle, and a caliper assembly containing, usually two brake pads which are activated by hydraulic pressure. The pads are forced against the sides of the disc, creating friction which slows the vehicle.

DISTRIBUTOR: A mechanically driven device on an engine which is responsible for electrically firing the spark plug at a predetermined point of the piston stroke.

DOWEL PIN: A pin, inserted in mating holes in two different parts allowing those parts to maintain a fixed relationship.

DRUM BRAKE: A braking system which consists of two brake shoes and one or two wheel cylinders, mounted on a fixed backing plate, and a brake drum, mounted on an axle, which revolves around the assembly.

DWELL: The rate, measured in degrees of shaft rotation, at which an electrical circuit cycles on and off.

ELECTRONIC CONTROL UNIT (ECU): Ignition module, module, amplifier or igniter. See Module for definition.

ELECTRONIC IGNITION: A system in which the timing and firing of the spark plugs is controlled by an electronic control unit, usually called a module. These systems have no points or condenser.

END-PLAY: The measured amount of axial movement in a shaft.

ENGINE: A device that converts heat into mechanical energy.

EXHAUST MANIFOLD: A set of cast passages or pipes which conduct exhaust gases from the engine.

FEELER GAUGE: A blade, usually metal, or precisely predetermined thickness, used to measure the clearance between two parts.

FIRING ORDER: The order in which combustion occurs in the cylinders of an engine. Also the order in which spark is distributed to the plugs by the distributor.

FLOODING: The presence of too much fuel in the intake manifold and combustion chamber which prevents the air/fuel mixture from firing, thereby causing a no-start situation.

FLYWHEEL: A disc shaped part bolted to the rear end of the crankshaft. Around the outer perimeter is affixed the ring gear. The starter drive engages the ring gear, turning the flywheel, which rotates the crankshaft, imparting the initial starting motion to the engine.

FOOT POUND (ft. lbs. or sometimes, ft.lb.): The amount of energy or work needed to raise an item weighing one pound, a distance of one foot.

FUSE: A protective device in a circuit which prevents circuit overload by breaking the circuit when a specific amperage is present. The device is constructed around a strip or wire of a lower amperage rating than the circuit it is designed to protect. When an amperage higher than that stamped on the fuse is present in the circuit, the strip or wire melts, opening the circuit.

GEAR RATIO: The ratio between the number of teeth on meshing gears.

GENERATOR: A device which converts mechanical energy into electrical energy.

HEAT RANGE: The measure of a spark plug's ability to dissipate heat from its firing end. The higher the heat range, the hotter the plug fires.

HUB: The center part of a wheel or gear.

HYDROCARBON (HC): Any chemical compound made up of hydrogen and carbon. A major pollutant formed by the engine as a byproduct of combustion.

HYDROMETER: An instrument used to measure the specific gravity of a solution.

INCH POUND (inch lbs.; sometimes in.lb. or in. lbs.): One twelfth of a foot pound.

INDUCTION: A means of transferring electrical energy in the form of a magnetic field. Principle used in the ignition coil to increase voltage.

INJECTOR: A device which receives metered fuel under relatively low pressure and is activated to inject the fuel into the engine under relatively high pressure at a predetermined time.

INPUT SHAFT: The shaft to which torque is applied, usually carrying the driving gear or gears.

INTAKE MANIFOLD: A casting of passages or pipes used to conduct air or a fuel/air mixture to the cylinders.

JOURNAL: The bearing surface within which a shaft operates.

KEY: A small block usually fitted in a notch between a shaft and a hub to prevent slippage of the two parts.

MANIFOLD: A casting of passages or set of pipes which connect the cylinders to an inlet or outlet source.

MANIFOLD VACUUM: Low pressure in an engine intake manifold formed just below the throttle plates. Manifold vacuum is highest at idle and drops under acceleration.

MASTER CYLINDER: The primary fluid pressurizing device in a hydraulic system. In automotive use, it is found in brake and hydraulic clutch systems and is pedal activated, either directly or, in a power brake system, through the power booster.

MODULE: Electronic control unit, amplifier or igniter of solid state or integrated design which controls the current flow in the ignition primary circuit based on input from the pick-up coil. When the module opens the primary circuit, high secondary voltage is induced in the coil.

NEEDLE BEARING: A bearing which consists of a number (usually a large number) of long, thin rollers.

OHM: (Ω) The unit used to measure the resistance of conductor-to-electrical flow. One ohm is the amount of resistance that limits current flow to one ampere in a circuit with one volt of pressure.

OHMMETER: An instrument used for measuring the resistance, in ohms, in an electrical circuit.

OUTPUT SHAFT: The shaft which transmits torque from a device, such as a transmission.

OVERDRIVE: A gear assembly which produces more shaft revolutions than that transmitted to it.

OVERHEAD CAMSHAFT (OHC): An engine configuration in which the camshaft is mounted on top of the cylinder head and operates the valve either directly or by means of rocker arms.

OVERHEAD VALVE (OHV): An engine configuration in which all of the valves are located in the cylinder head and the camshaft is located in the cylinder block. The camshaft operates the valves via lifters and pushrods.

OXIDES OF NITROGEN (NOx): Chemical compounds of nitrogen produced as a byproduct of combustion. They combine with hydrocarbons to produce smog.

OXYGEN SENSOR: Use with the feedback system to sense the presence of oxygen in the exhaust gas and signal the computer which can reference the voltage signal to an air/fuel ratio.

PINION: The smaller of two meshing gears.

PISTON RING: An open-ended ring with fits into a groove on the outer diameter of the piston. Its chief function is to form a seal between the piston and cylinder wall. Most automotive pistons have three rings: two for compression sealing; one for oil sealing.

PRELOAD: A predetermined load placed on a bearing during assembly or by adjustment.

PRIMARY CIRCUIT: the low voltage side of the ignition system which consists of the ignition switch, ballast resistor or resistance wire, bypass, coil, electronic control unit and pick-up coil as well as the connecting wires and harnesses.

PRESS FIT: The mating of two parts under pressure, due to the inner diameter of one being smaller than the outer diameter of the other, or vice versa; an interference fit.

RACE: The surface on the inner or outer ring of a bearing on which the balls, needles or rollers move.

REGULATOR: A device which maintains the amperage and/or voltage levels of a circuit at predetermined values.

RELAY: A switch which automatically opens and/or closes a circuit.

RESISTANCE: The opposition to the flow of current through a circuit or electrical device, and is measured in ohms. Resistance is equal to the voltage divided by the amperage.

RESISTOR: A device, usually made of wire, which offers a preset amount of resistance in an electrical circuit.

RING GEAR: The name given to a ring-shaped gear attached to a differential case, or affixed to a flywheel or as part of a planetary gear set.

ROLLER BEARING: A bearing made up of hardened inner and outer races between which hardened steel rollers move.

ROTOR: 1. The disc-shaped part of a disc brake assembly, upon which the brake pads bear; also called, brake disc. 2. The device mounted atop the distributor shaft, which passes current to the distributor cap tower contacts.

SECONDARY CIRCUIT: The high voltage side of the ignition system, usually above 20,000 volts. The secondary includes the ignition coil, coil wire, distributor cap and rotor, spark plug wires and spark plugs.

SENDING UNIT: A mechanical, electrical, hydraulic or electromagnetic device which transmits information to a gauge.

SENSOR: Any device designed to measure engine operating conditions or ambient pressures and temperatures. Usually electronic in nature and designed to send a voltage signal to an on-board computer, some sensors may operate as a simple on/off switch or they may provide a variable voltage signal (like a potentiometer) as conditions or measured parameters change.

SHIM: Spacers of precise, predetermined thickness used between parts to establish a proper working relationship.

SLAVE CYLINDER: In automotive use, a device in the hydraulic clutch system which is activated by hydraulic force, disengaging the clutch.

SOLENOID: A coil used to produce a magnetic field, the effect of which is to produce work.

SPARK PLUG: A device screwed into the combustion chamber of a spark ignition engine. The basic construction is a conductive core inside of a ceramic insulator, mounted in an outer conductive base. An electrical charge from the spark plug wire travels along the conductive core and jumps a preset air gap to a grounding point or points at the end of the conductive base. The resultant spark ignites the fuel/air mixture in the combustion chamber.

SPLINES: Ridges machined or cast onto the outer diameter of a shaft or inner diameter of a bore to enable parts to mate without rotation.

TACHOMETER: A device used to measure the rotary speed of an engine, shaft, gear, etc., usually in rotations per minute.

THERMOSTAT: A valve, located in the cooling system of an engine, which is closed when cold and opens gradually in response to engine heating, controlling the temperature of the coolant and rate of coolant flow.

TOP DEAD CENTER (TDC): The point at which the piston reaches the top of its travel on the compression stroke.

TORQUE: The twisting force applied to an object.

TORQUE CONVERTER: A turbine used to transmit power from a driving member to a driven member via hydraulic action, providing changes in drive ratio and torque. In automotive use, it links the driveplate at the rear of the engine to the automatic transmission.

TRANSDUCER: A device used to change a force into an electrical signal.

TRANSISTOR: A semi-conductor component which can be actuated by a small voltage to perform an electrical switching function.

TUNE-UP: A regular maintenance function, usually associated with the replacement and adjustment of parts and components in the electrical and fuel systems of a vehicle for the purpose of attaining optimum performance.

TURBOCHARGER: An exhaust driven pump which compresses intake air and forces it into the combustion chambers at higher than atmospheric pressures. The increased air pressure allows more fuel to be burned and results in increased horsepower being produced.

VACUUM ADVANCE: A device which advances the ignition timing in response to increased engine vacuum.

VACUUM GAUGE: An instrument used to measure the presence of vacuum in a chamber.

VALVE: A device which control the pressure, direction of flow or rate of flow of a liquid or gas.

VALVE CLEARANCE: The measured gap between the end of the valve stem and the rocker arm, cam lobe or follower that activates the valve.

VISCOSITY: The rating of a liquid's internal resistance to flow.

VOLTMETER: An instrument used for measuring electrical force in units called volts. Voltmeters are always connected parallel with the circuit being tested.

WHEEL CYLINDER: Found in the automotive drum brake assembly, it is a device, actuated by hydraulic pressure, which, through internal pistons, pushes the brake shoes outward against the drums.

MASTER INDEX

A

B

C